The Enzymes

VOLUME IX

GROUP TRANSFER

Part B

PHOSPHORYL TRANSFER
ONE-CARBON GROUP TRANSFER
GLYCOSYL TRANSFER
AMINO GROUP TRANSFER
OTHER TRANSFERASES

Third Edition

CONTRIBUTORS

ELIZABETH P. ANDERSON
ERNEST BOREK
ALEXANDER E. BRAUNSTEIN
SIDNEY P. COLOWICK
GÉZA DÉNES
KURT E. EBNER
H. G. HERS
F. M. HUENNEKENS
GARY R. JACOBSON
W. P. JENCKS
M. E. JONES
SYLVIA J. KERR
JEANNE I. RADER
L. RAIJMAN
W. STALMANS
GEORGE R. STARK
ROBERT T. TAYLOR
JAMES B. WALKER
HERBERT WEISSBACH

ADVISORY BOARD

ARTHUR KORNBERG
HENRY LARDY
FRITZ LIPMANN
EARL STADTMAN
HERBERT TABOR

THE ENZYMES

Edited by PAUL D. BOYER

*Molecular Biology Institute and
Department of Chemistry
University of California
Los Angeles, California*

Volume IX
GROUP TRANSFER
Part B

PHOSPHORYL TRANSFER
ONE-CARBON GROUP TRANSFER
GLYCOSYL TRANSFER
AMINO GROUP TRANSFER
OTHER TRANSFERASES

THIRD EDITION

ACADEMIC PRESS New York and London   1973
A Subsidiary of Harcourt Brace Jovanovich, Publishers

COPYRIGHT © 1973, BY ACADEMIC PRESS, INC.
ALL RIGHTS RESERVED.
NO PART OF THIS PUBLICATION MAY BE REPRODUCED OR
TRANSMITTED IN ANY FORM OR BY ANY MEANS, ELECTRONIC
OR MECHANICAL, INCLUDING PHOTOCOPY, RECORDING, OR ANY
INFORMATION STORAGE AND RETRIEVAL SYSTEM, WITHOUT
PERMISSION IN WRITING FROM THE PUBLISHER.

ACADEMIC PRESS, INC.
111 Fifth Avenue, New York, New York 10003

United Kingdom Edition published by
ACADEMIC PRESS, INC. (LONDON) LTD.
24/28 Oval Road, London NW1

Library of Congress Cataloging in Publication Data

Main entry under title:

The Enzymes.

Includes bibliographical references.
CONTENTS:
v.2.   Kinetics and mechanism–v. 3.   Hydrolysis: peptide bonds.–v. 4.   Hydrolysis: other C–N bonds, phosphate esters.   [etc.]
1.   Enzymes.   I.   Boyer, Paul D., ed.
[DNLM: 1.   Enzymes.   QU135 B791e]
QP601.E523   574.1'925   75–117107
ISBN 0–12–122709–X (v.9)

PRINTED IN THE UNITED STATES OF AMERICA

Contents

List of Contributors	ix
Preface	xi
Contents of Other Volumes	xii

1. The Hexokinases
SIDNEY P. COLOWICK

I. Introduction	1
II. The Yeast Hexokinases	2
III. The Mammalian Hexokinases	31
IV. Comparative Aspects	46

2. Nucleoside and Nucleotide Kinases
ELIZABETH P. ANDERSON

I. Introduction	49
II. Nucleoside Kinases	51
III. Nucleotide Kinases	82

3. Carbamate Kinase
L. RAIJMAN AND M. E. JONES

I. Introduction	97
II. Molecular Properties	102
III. Catalytic Properties	105
IV. Function and Relation to Other Enzymes; Metabolite Control of Activity	115

4. N^5-Methyltetrahydrofolate-Homocysteine Methyltransferases

ROBERT T. TAYLOR AND HERBERT WEISSBACH

I. Introduction	121
II. B_{12} Methyltransferase from *Escherichia coli*	122
III. Non-B_{12} Methyltransferase from *Escherichia coli*	154
IV. Other Sources of Non-B_{12} and B_{12} N^5-Methyltetrahydrofolate-Homocysteine Methyltransferases	160

5. Enzymic Methylation of Natural Polynucleotides

SYLVIA J. KERR AND ERNEST BOREK

I. Introduction	167
II. Transfer RNA Methyltransferases	168
III. Ribosomal RNA Methyltransferases	187
IV. DNA Methyltransferases	190

6. Folate Coenzyme-Mediated Transfer of One-Carbon Groups

JEANNE I. RADER AND F. M. HUENNEKENS

I. Introduction	197
II. Transfer of Formate and Its Congeners	198
III. Transfer of Formaldehyde and Its Congeners	209

7. Aspartate Transcarbamylases

GARY R. JACOBSON AND GEORGE R. STARK

I. Introduction	226
II. Isolation and Characterization of the Native Enzyme from *Escherichia coli*	227
III. Isolation and Characterization of Subunits	231
IV. Reconstitution of Native ATCase from the Isolated Subunits	237
V. Detailed Subunit Structure of Native ATCase	239
VI. Mechanism of Catalysis for the *E. coli* Enzyme	243
VII. Cooperative Properties of the *E. coli* Enzyme	268
VIII. Biosynthesis and Genetics of Bacterial Aspartate Transcarbamylases	292
IX. Aspartate Transcarbamylases from Other Organisms	297

8. Glycogen Synthesis from UDPG

W. STALMANS AND H. G. HERS

I. Historical	310
II. Molecular Properties	311
III. General Catalytic Properties	316
IV. The Two Forms of Glycogen Synthetase and Their Interconversion	322
V. Glycogen Synthetase of Mammalian Muscle	332

CONTENTS

VI. Glycogen Synthetase of Mammalian Heart	340
VII. Glycogen Synthetase of Mammalian Liver	341
VIII. Glycogen Synthetase of Other Mammalian Tissues	353
IX. Glycogen Synthetase of Nonmammalian Organisms	357

9. Lactose Synthetase

KURT E. EBNER

I. Introduction	363
II. Molecular Properties	365

10. Amino Group Transfer

ALEXANDER E. BRAUNSTEIN

I. Introduction	379
II. Basic Chemical Features of the Transamination Reaction	387
III. Aspartate:2-Oxoglutarate Aminotransferases	393
IV. Aminotransferases Acting on Other Substrates	462
V. Concluding Remarks	480

11. Coenzyme A Transferases

W. P. JENCKS

I. Introduction	483
II. Properties	485
III. Catalytic Properties	486

12. Amidinotransferases

JAMES B. WALKER

I. Introduction	497
II. Glycine Amidinotransferase	498
III. Inosamine-P Amidinotransferase	505

13. N-Acetylglutamate-5-Phosphotransferase

GÉZA DÉNES

I. Introduction	511
II. Molecular Properties	513
III. Catalytic Properties	513

Author Index 521

Subject Index 552

List of Contributors

Numbers in parentheses indicate the pages on which the authors' contributions begin.

ELIZABETH P. ANDERSON (49), Laboratory of Biochemistry, National Cancer Institute, National Institutes of Health, Bethesda, Maryland

ERNEST BOREK (167), Department of Microbiology, University of Colorado Medical Center, Denver, Colorado

ALEXANDER E. BRAUNSTEIN (379), Institute of Molecular Biology, Academy of Sciences of the U.S.S.R., Moscow, U.S.S.R.

SIDNEY P. COLOWICK (1), Department of Microbiology, Vanderbilt University School of Medicine, Nashville, Tennessee

GÉZA DÉNES (511), Institute of Biochemistry, Biological Research Center, Hungarian Academy of Sciences, Szeged, Hungary

KURT E. EBNER (363), Department of Biochemistry, Oklahoma State University, Stillwater, Oklahoma

H. G. HERS (309), Laboratoire de Chimie Physiologique, Université de Louvain, Louvain, Belgium

F. M. HUENNEKENS (197), Department of Biochemistry, Scripps Clinic and Research Foundation, La Jolla, California

GARY R. JACOBSON (225), Department of Biochemistry, Stanford University School of Medicine, Stanford, California

W. P. JENCKS (483), Graduate Department of Biochemistry, Brandeis University, Waltham, Massachusetts

M. E. JONES (97), Department of Biochemistry, University of Southern California School of Medicine, Los Angeles, California

SYLVIA J. KERR (167), Department of Microbiology, University of Colorado Medical Center, Denver, Colorado

JEANNE I. RADER (197), Department of Biochemistry, Scripps Clinic and Research Foundation, La Jolla, California

L. RAIJMAN (97), Department of Biochemistry, University of Southern California School of Medicine, Los Angeles, California

W. STALMANS (309), Laboratoire de Chimie Physiologique, Université de Louvain, Louvain, Belgium

GEORGE R. STARK (225), Department of Biochemistry, Stanford University School of Medicine, Stanford, California

ROBERT T. TAYLOR (121), Biomedical Division, Lawrence Livermore Laboratory, University of California, Livermore, California

JAMES B. WALKER (497), Department of Biochemistry, Rice University, Houston, Texas

HERBERT WEISSBACH (121), Roche Institute of Molecular Biochemistry, Nutley, New Jersey

Preface

Volumes VIII and IX of this treatise deal with the important and versatile enzymes that catalyze the transfer of a variety of chemical groups. The most extensively studied are enzymes catalyzing phosphoryl group transfer from ATP to various acceptors, known by their trivial name as kinases; thirteen separate chapters on kinases appear in the two volumes. Volume VIII also includes a general chapter dealing with the chemical basis of phosphoryl transfer. Intramolecular phosphoryl transfer was covered earlier in Volume VI with the isomerases, and phosphoryl transfer by phosphatases was covered in Volume IV with the hydrolases.

Another class of transferases of prime metabolic importance are those catalyzing transfer of one-carbon groups. These are covered in this volume. Prominent consideration is given the catalytic versatility of folic acid. Aspartate transcarbamylase also falls within this group because of the elegant researches on its separate catalytic and regulatory subunits that introduced new principles into enzymology. One-carbon group transfer also includes some molecular information about methylation of DNA; rapid growth is anticipated in this area.

As representative of the widespread occurrence of transfer of glycosyl moieties, some key enzymes of carbohydrate metabolism and biosynthesis appear in these volumes. Other miscellaneous group transfer reactions complete the coverage.

As with previous volumes, the principal criterion for inclusion as a separate chapter is considerable information at the molecular level about either the enzyme or the process catalyzed. Response of the best-qualified authors in the field continues to be gratifying. For nearly all the chapters, the author is the first choice of the Advisory Board and the Editor. It is to these authors that the reader is indebted for the excellent coverage. Also it is a pleasure to extend appreciation to the members of the Advisory Board for their invaluable assistance in planning the volumes.

The continued fine professional work by the staff of Academic Press is clearly evident in the product produced.

PAUL D. BOYER

Contents of Other Volumes

Volume I: Structure and Control

X-Ray Crystallography and Enzyme Structure
David Eisenberg

Chemical Modification by Active-Site-Directed Reagents
Elliott Shaw

Chemical Modification as a Probe of Structure and Function
Louis A. Cohen

Multienzyme Complexes
Lester J. Reed and David J. Cox

Genetic Probes of Enzyme Structure
Milton J. Schlesinger

Evolution of Enzymes
Emil L. Smith

The Molecular Basis for Enzyme Regulation
D. E. Koshland, Jr.

Mechanisms of Enzyme Regulation in Metabolism
E. R. Stadtman

Enzymes as Control Elements in Metabolic Regulation
Daniel E. Atkinson

Author Index—Subject Index

Volume II: Kinetics and Mechanism

Steady State Kinetics
  W. W. Cleland

Rapid Reactions and Transient States
  Gordon B. Hammes and Paul R. Schimmel

Stereospecificity of Enzymic Reactions
  G. Popják

Proximity Effects and Enzyme Catalysis
  Thomas C. Bruice

Enzymology of Proton Abstraction and Transfer Reactions
  Irwin A. Rose

Kinetic Isotope Effects in Enzymic Reactions
  J. H. Richards

Schiff Base Intermediates in Enzyme Catalysis
  Esmond E. Snell and Samuel J. Di Mari

Some Physical Probes of Enzyme Structure in Solution
  Serge N. Timasheff

Metals in Enzyme Catalysis
  Albert S. Mildvan

Author Index—Subject Index

Volume III: Hydrolysis: Peptide Bonds

Carboxypeptidase A
  Jean A. Hartsuck and William N. Lipscomb

Carboxypeptidase B
  J. E. Folk

Leucine Aminopeptidase and Other N-Terminal Exopeptidases
  Robert J. DeLange and Emil L. Smith

Pepsin
  Joseph S. Fruton

Chymotrypsinogen: X-Ray Structure
  J. Kraut

The Structure of Chymotrypsin
  D. M. Blow

Chymotrypsin—Chemical Properties and Catalysis
  George P. Hess

Trypsin
  B. Keil

Thrombin and Prothrombin
  Staffan Magnusson

Pancreatic Elastase
  B. S. Hartley and D. M. Shotton

Protein Proteinase Inhibitors—Molecular Aspects
  Michael Laskowski, Jr., and Robert W. Sealock

Cathepsins and Kinin-Forming and -Destroying Enzymes
  Lowell M. Greenbaum

Papain, X-Ray Structure
  J. Drenth, J. N. Jansonius, R. Koekoek, and B. G. Wolthers

Papain and Other Plant Sulfhydryl Proteolytic Enzymes
  A. N. Glazer and Emil L. Smith

Subtilisin: X-Ray Structure
  J. Kraut

Subtilisins: Primary Structure, Chemical and Physical Properties
  Francis S. Markland, Jr., and Emil L. Smith

Streptococcal Proteinase
  Teh-Yung Liu and S. D. Elliott

The Collagenases
  Sam Seifter and Elvin Harper

Clostripain
  William M. Mitchell and William F. Harrington

Other Bacterial, Mold, and Yeast Proteases
  Hiroshi Matsubara and Joseph Feder

Author Index—Subject Index

Volume IV: Hydrolysis: Other C–N Bonds, Phosphate Esters

Ureases
  F. J. Reithel

Penicillinase and Other β-Lactamases
  Nathan Citri

Purine, Purine Nucleoside, Purine Nucleotide Aminohydrolases
  C. L. Zielke and C. H. Suelter

Glutaminase and γ-Glutamyltransferases
  Standish C. Hartman

L-Asparaginase
  John C. Wriston, Jr.

Enzymology of Pyrrolidone Carboxylic Acid
  Marian Orlowski and Alton Meister

Staphylococcal Nuclease X-Ray Structure
  F. Albert Cotton and Edward E. Hazen, Jr.

Staphylococcal Nuclease, Chemical Properties and Catalysis
  Christian B. Anfinsen, Pedro Cuatrecasas, and Hiroshi Taniuchi

Microbial Ribonucleases with Special Reference to
RNases T_1, T_2, N_1, and U_2
  Tsuneko Uchida and Fujio Egami

Bacterial Deoxyribonucleases
  I. R. Lehman

Spleen Acid Deoxyribonuclease
  Giorgio Bernardi

Deoxyribonuclease I
  M. Laskowski, Sr.

Venom Exonuclease
    M. Laskowski, Sr.

Spleen Acid Exonuclease
    Alberto Bernardi and Giorgio Bernardi

Nucleotide Phosphomonoesterases
    George I. Drummond and Masanobu Yamamoto

Nucleoside Cyclic Phosphate Diesterases
    George I. Drummond and Masanobu Yamamoto

E. coli Alkaline Phosphatase
    Ted W. Reid and Irwin B. Wilson

Mammalian Alkaline Phosphatases
    H. N. Fernley

Acid Phosphatases
    Vincent P. Hollander

Inorganic Pyrophosphatase of *Escherichia coli*
    John Josse and Simon C. K. Wong

Yeast and Other Inorganic Pyrophosphatases
    Larry G. Butler

Glucose-6-Phosphatase, Hydrolytic and Synthetic Activities
    Robert C. Nordlie

Fructose-1,6-Diphosphatases
    S. Pontremoli and B. L. Horecker

Bovine Pancreatic Ribonuclease
    Frederic M. Richards and Harold W. Wyckoff

Author Index—Subject Index

Volume V: Hydrolysis (Sulfate Esters, Carboxyl Esters, Glycosides), Hydration

The Hydrolysis of Sulfate Esters
    A. B. Roy

Arylsulfatases
  R. G. Nicholls and A. B. Roy

Carboxylic Ester Hydrolases
  Klaus Krisch

Phospholipases
  Donald J. Hanahan

Acetylcholinesterase
  Harry C. Froede and Irwin B. Wilson

Plant and Animal Amylases
  John A. Thoma, Joseph E. Spradlin, and Stephen Dygert

Glycogen and Starch Debranching Enzymes
  E. Y. C. Lee and W. J. Whelan

Bacterial and Mold Amylases
  Toshio Takagi, Hiroko Toda, and Toshizo Isemura

Cellulases
  D. R. Whitaker

Yeast and *Neurospora* Invertases
  J. Oliver Lampen

Hyaluronidases
  Karl Meyer

Neuraminidases
  Alfred Gottschalk and A. S. Bhargava

Phage Lysozyme and Other Lytic Enzymes
  Akira Tsugita

Aconitase
  Jenny Pickworth Glusker

β-Hydroxydecanoyl Thioester Dehydrase
  Konrad Bloch

Dehydration in Nucleotide-Linked Deoxysugar Synthesis
  L. Glaser and H. Zarkowsky

Dehydrations Requiring Vitamin B_{12} Coenzyme
  Robert H. Abeles

Enolase
  Finn Wold

Fumarase and Crotonase
  Robert L. Hill and John W. Teipel

6-Phosphogluconic and Related Dehydrases
  W. A. Wood

Carbonic Anhydrase
  S. Lindskog, L. E. Henderson, K. K. Kannan, A. Liljas, P. O. Nyman, and B. Strandberg

Author Index—Subject Index

Volume VI: Carboxylation and Decarboxylation (Nonoxidative), Isomerization

Pyruvate Carboxylase
  Michael C. Scrutton and Murray R. Young

Acyl-CoA Carboxylases
  Alfred W. Alberts and P. Roy Vagelos

Transcarboxylase
  Harland G. Wood

Formation of Oxalacetate by CO_2 Fixation on Phosphoenolpyruvate
  Merton F. Utter and Harold M. Kolenbrander

Ribulose-1,5-Diphosphate Carboxylase
  Marvin I. Siegel, Marcia Wishnick, and M. Daniel Lane

Ferredoxin-Linked Carboxylation Reactions
  Bob B. Buchanan

Amino Acid Decarboxylases
  Elizabeth A. Boeker and Esmond E. Snell

Acetoacetate Decarboxylase
  Irwin Fridovich

Aldose–Ketose Isomerases
  Ernst A. Noltmann

Epimerases
  Luis Glaser

Cis–Trans Isomerization
  Stanley Seltzer

Phosphomutases
  W. J. Ray, Jr., and E. J. Peck, Jr.

Amino Acid Racemases and Epimerases
  Elijah Adams

Coenzyme B_{12}-Dependent Mutases Causing Carbon Chain Rearrangements
  H. A. Barker

B_{12} Coenzyme-Dependent Amino Group Migrations
  Thressa C. Stadtman

Isopentenylpyrophosphate Isomerase
  P. W. Holloway

Isomerization in the Visual Cycle
  Joram Heller

Δ^5-3-Ketosteroid Isomerase
  Paul Talalay and Ann M. Benson

Author Index—Subject Index

Volume VII: Elimination and Addition, Aldol Cleavage and Condensation, Other C–C Cleavage, Phosphorolysis, Hydrolysis (Fats, Glycosides)

Tryptophan Synthetase
  Charles Yanofsky and Irving P. Crawford

Pyridoxal-Linked Elimination and Replacement Reactions
  Leodis Davis and David E. Metzler

The Enzymic Elimination of Ammonia
  Kenneth R. Hanson and Evelyn A. Havir

Argininosuccinases and Adenylosuccinases
  Sarah Ratner

Epoxidases
  William B. Jakoby and Thorsten A. Fjellstedt

Aldolases
  B. L. Horecker, Orestes Tsolas, and C. Y. Lai

Transaldolase
  Orestes Tsolas and B. L. Horecker

2-Keto-3-deoxy-6-phosphogluconic and Related Aldolases
  W. A. Wood

Other Deoxy Sugar Aldolases
  David Sidney Feingold and Patricia Ann Hoffee

δ-Aminolevulinic Acid Dehydratase
  David Shemin

δ-Aminolevulinic Acid Synthetase
  Peter M. Jordan and David Shemin

Citrate Cleavage and Related Enzymes
  Leonard B. Spector

Thiolase
  Ulrich Gehring and Feodor Lynen

Acyl-CoA Ligases
  Malcolm J. P. Higgins, Jack A. Kornblatt, and Harry Rudney

α-Glucan Phosphorylases—Chemical and Physical Basis of Catalysis and Regulation
  Donald J. Graves and Jerry H. Wang

Purine Nucleoside Phosphorylase
  R. E. Parks, Jr., and R. P. Agarwal

Disaccharide Phosphorylases
  John J. Mieyal and Robert H. Abeles

Polynucleotide Phosphorylase
  T. Godefroy-Colburn and M. Grunberg-Manago

The Lipases
 P. Desnuelle

β-Galactosidase
 Kurt Wallenfels and Rudolf Weil

Vertebrate Lysozymes
 Taiji Imoto, L. N. Johnson, A. C. T. North, D. C. Phillips, and J. A. Rupley

Author Index—Subject Index

Volume VIII: Group Transfer, Part A: Nucleotidyl Transfer, Nucleosidyl Transfer, Acyl Transfer, Phosphoryl Transfer

Adenylyl Transfer Reactions
 E. R. Stadtman

Uridine Diphosphoryl Glucose Pyrophosphorylase
 Richard L. Turnquist and R. Gaurth Hansen

Adenosine Diphosphoryl Glucose Pyrophosphorylase
 Jack Preiss

The Adenosyltransferases
 S. Harvey Mudd

Acyl Group Transfer (Acyl Carrier Protein)
 P. Roy Vagelos

Chemical Basis of Biological Phosphoryl Transfer
 S. J. Benkovic and K. J. Schray

Phosphofructokinase
 David P. Bloxham and Henry A. Lardy

Adenylate Kinase
 L. Noda

Nucleoside Diphosphokinases
 R. E. Parks, Jr., and R. P. Agarwal

3-Phosphoglycerate Kinase
  R. K. Scopes

Pyruvate Kinase
  F. J. Kayne

Creatine Kinase (Adenosine 5'-Triphosphate-Creatine Phosphotransferase)
  D. C. Watts

Arginine Kinase and Other Invertebrate Guanidino Kinases
  J. F. Morrison

Glycerol and Glycerate Kinases
  Jeremy W. Thorner and Henry Paulus

Microbial Aspartokinases
  Paolo Truffa-Bachi

Protein Kinases
  Donal A. Wlash and Edwin G. Krebs

Author Index—Subject Index

1

The Hexokinases

SIDNEY P. COLOWICK

I. Introduction	1
II. The Yeast Hexokinases	2
A. The Two Native Isoenzymes and Their Modification by Endogenous Proteases	2
B. Modification of Native Hexokinases by Added Proteases	6
C. Molecular Weight and Subunit Structure	7
D. Chemical Studies on Native and Modified Forms	9
E. Enzyme Mechanism	13
F. Regulation of Hexokinase Activity	29
III. The Mammalian Hexokinases	31
A. Occurrence of Multiple Forms	31
B. Relation of Soluble to Insoluble Forms	33
C. Purification and Molecular Properties	37
D. Enzyme Mechanism	41
E. Regulation of Enzyme Activity	44
IV. Comparative Aspects	46

I. Introduction

Since the previous chapter on hexokinases appeared in "The Enzymes" in 1962 (*1*), there has been a considerable advance in our knowledge of both the yeast and mammalian hexokinases. The main advances with the yeast system have been as follows: (a) identification of two native hexokinases in baker's yeast; (b) resolution of the question of the molecular weight and subunit structure of each; (c) delineation of the chemical and physical changes accompanying proteolytic modification of the enzyme; (d) new knowledge on the mechanism of the reaction, including substrate

1. R. K. Crane, "The Enzymes," 2nd ed., Vol. 6, p. 47, 1962.

binding studies as well as kinetic studies on both the phosphorylating and the hydrolytic functions; (e) studies bearing on the question of a phosphoenzyme intermediate in the reaction; (f) discovery of regulatory effects on one of the isoenzymes by metabolites which serve as activators or inhibitors; (g) the finding of "hysteresis," i.e., detectably slow responses to addition of substrates, activators or inhibitors; and (h) the initiation of X-ray crystallographic studies.

With the mammalian hexokinases, progress has been made along the following lines: (a) identification of several isoenzymes with different distribution in different tissues; (b) discovery of factors which determine the distribution of enzyme between soluble and mitochondria-bound forms, and which exert regulatory effects on activity; (c) purification of solubilized enzymes, in some cases to homogeneity; (d) establishment of molecular weight of peptide chain; (e) studies on enzyme mechanism; and (f) studies on factors regulating activity of the enzymes.

In this chapter, we will describe these advances in our knowledge of the yeast and mammalian enzymes. We will also touch briefly upon other hexokinases and on enzyme systems for sugar phosphorylation in which compounds other than nucleoside triphosphates serve as phosphate donor.

II. The Yeast Hexokinases

A. The Two Native Isoenzymes and Their Modification by Endogenous Proteases

The early history of the discovery of yeast hexokinase (2), the establishment of the nature of the reaction catalyzed (3–5), and the isolation of the enzyme in crystalline form in several laboratories (6–8) has been reviewed previously (9). It was not recognized until recently that there are two different naturally occurring isoenzymes with different catalytic and physical properties. The discovery of these two forms, and of derivatives resulting from their proteolytic modification, came about as follows.

2. O. Meyerhof, *Biochem. Z.* **183**, 176 (1927).
3. H. Euler and E. Adler, *Hoppe-Seyler's Z. Physiol. Chem.* **235**, 122 (1935).
4. C. Lutwak-Mann and T. Mann, *Biochem. Z.* **281**, 140 (1935).
5. S. P. Colowick and H. M. Kalckar, *JBC* **148**, 117 (1943).
6. M. Kunitz and M. McDonald, *J. Gen. Physiol.* **29**, 393 (1946).
7. L. Berger, M. W. Slein, S. P. Colowick, and C. F. Cori, *J. Gen. Physiol.* **29**, 379 (1946).
8. K. Bailey and E. C. Webb, *BJ* **42**, 60 (1948).
9. S. P. Colowick, "The Enzymes," 1st ed., Vol. 2, Part 1, p. 114, 1951.

When the crystalline enzyme prepared from dried yeast according to Darrow and Colowick (10) was subjected to DEAE-cellulose chromatography, a multiplicity of fully active species were found (11). Some of these were eluted by a pH gradient of 5 mM succinate buffer between pH 5.4 and 4.8, while others required a salt gradient (0–50 mM NaCl) for elution. It was then found (12, 13) that chromatographic behavior could be altered by treatment with proteolytic enzymes (trypsin or chymotrypsin), i.e., that the more readily elutable material could be converted to a more difficulty elutable form. However, it was not thought at that point that endogenous proteases were responsible for the multiplicity of isoenzymes found since crude extracts were relatively stable to incubation in terms of chromatographic pattern.

In later studies (14–16), however, it became apparent that all of the forms observed in the Darrow–Colowick preparation were proteolytic degradation products derived from the native material. The endogenous proteases, which are very dilute in the crude extracts, become concentrated during purification of hexokinases, and, in addition, appear to be more effective in modifying the hexokinases after contaminating proteins have been removed (16). Remarkably enough, the modification can occur while the enzyme is being stored in the cold as a crystalline suspension in half-saturated ammonium sulfate (16), presumably by protease action in the liquid phase.

The native material, as observed either in crude extracts or in crystals prepared and stored after removal of protease, was eluted much more readily by pH gradient than any of the forms present in a Darrow–Colowick preparation. The native enzyme was found to consist of two distinct fractions occurring as overlapping peaks. These fractions, called P-I and P-II to denote their order of elution in a pH gradient, could also be separated by gel electrophoresis, in which case form P-I migrates ahead of P-II in tris buffer at pH 9. With the preparations which have undergone proteolytic modification, no enzyme is elutable at the positions of P-I and P-II; some is eluted toward the end of the pH gradient and is

10. R. A. Darrow and S. P. Colowick, "Methods in Enzymology," Vol. 5, p. 226, 1962.
11. K. A. Trayser and S. P. Colowick, *ABB* **94**, 177 (1961).
12. A. Kaji, *ABB* **112**, 54 (1965).
13. A. Kaji, K. A. Trayser, and S. P. Colowick, *Ann. N. Y. Acad. Sci.* **94**, 798 (1961).
14. U. Kenkare, I. T. Schulze, J. Gazith, and S. P. Colowick, *Proc. Int. Congr. Biochem., 6th, 1964* Vol. VI, p. 477 (1964).
15. J. Gazith, I. T. Schulze, R. H. Gooding, F. C. Womack, and S. P. Colowick, *Ann. N. Y. Acad. Sci.* **151**, 307 (1968).
16. I. T. Schulze and S. P. Colowick, *JBC* **244**, 2306 (1969).

designated P-III, while the remainder requires salt for elution and is designated as S forms.

A method for preparation of the native hexokinase isoenzymes by means of batchwise elution from DEAE-cellulose has been described (17). In this procedure, the protease remains adsorbed on the DEAE-cellulose so that the eluates are essentially protease-free and suitable for preparation of stable crystalline enzyme. As an alternative or adjunct procedure for dealing with the protease problem, the same report (17) described the use of phenylmethanesulfonyl fluoride, an inhibitor of serine proteases (18) which is relatively nontoxic and therefore safer for laboratory use than diisopropyl fluorophosphate. More recently an improved method of separation has been developed involving as an added step the batchwise elution of P-II, followed by P-I, from a hydroxylapatite column by means of increasing concentrations of phosphate buffer (19).

Concurrently with the development of methods for the isolation of the two native isoenzymes from dried yeast in our laboratory, methods were developed in Dr. Eric Barnard's laboratory for the isolation of the same isoenzymes from fresh yeast (20, 21). The forms which we have named P-I and P-II were found to correspond to Barnard's forms A and B, respectively, when they were compared in the same column chromatography system (experiments performed by courtesy of Dr. Barnard). In the remainder of this chapter, the symbols P-I and P-II will be used to designate the native isoenzymes regardless of the method used for preparation. The symbols P-I and P-II are preferred because the terms A and B were applied previously to modified forms.

The two native forms are chemically (15, 22) and serologically (19) distinct (see below) and cannot be interconverted. They are distinguishable not only by chromatography and electrophoresis but also by catalytic properties. Under standard assay conditions, with glucose as substrate, the purest preparations of P-I and P-II show specific activities of around 200 and 800 μmoles/min/mg protein, respectively, at pH 8–9 and 30° (19). Essentially the same activities are obtained in the direct colori-

17. I. T. Schulze, J. Gazith, and R. H. Gooding, "Methods in Enzymology," Vol. 9, p. 376, 1966.
18. D. E. Fahrney and A. M. Gold, JACS 85, 997 (1963).
19. F. C. Womack, M. K. Welch, J. Nielsen, and S. P. Colowick, ABB (1973) (in press).
20. N. R. Lazarus, A. H. Ramel, Y. M. Rustum, and E. A. Barnard, Biochemistry 5, 4003 (1966).
21. Y. M. Rustum, A. H. Ramel, and E. A. Barnard, Prep. Biochem. 1, 309 (1971).
22. J. J. Schmidt and S. P. Colowick, ABB (1973) (in press).

metric assay of acid production (*19*) and the coupled enzymic assay in which glucose-6-phosphate dehydrogenase is present in excess and the rate of formation of TPNH is measured (*20*).

An important criterion for distinguishing P-I and P-II was discovered by the Barnard group (*20*). They found that the activity of P-I with fructose was 2.6 times that with glucose, whereas with P-II, the fructose: glucose ratio was only 1.3. The mannose:glucose ratios for P-I and P-II were also different: 0.6 and 0.3, respectively. The measurement of the fructose:glucose ratio has proven to be extremely valuable for determining the relative amounts of P-I and P-II present in crude extracts or in purified preparations. The ratio is not altered by proteolytic modification of P-I or P-II. Our own best preparations of P-I and P-II show fructose: glucose ratios of 2.8–3.0 and 1.0–1.2, respectively (*19*).

Since fructose:glucose ratios were also reported for some of the earlier preparations of crystalline enzyme (*6*, *7*) we can now deduce which of the isoenzymes was represented in those preparations. Kunitz and McDonald (*6*) reported a fructose:glucose ratio of 1, which suggests that their enzyme was derived from P-II. Most likely their enzyme was protease-modified, judging from their molecular weight observations (see next section) as well as the instability observed. From similar considerations, the preparation of Berger *et al.* (*7*) is also believed to have been largely a modified form of P-II (*23*). The Darrow–Colowick preparation also appears to consist of modified forms derived exclusively from P-II, judging from the high specific activity toward glucose of all of the forms observed (*11*, *13*).

Thus, the crystalline preparations isolated in earlier investigations seem to have been largely P-II derived. The reason for this may lie in the fact that form P-II and its derivatives crystallize much more rapidly from ammonium sulfate than P-I (*19*); thus, the latter may have been lost on repeated recrystallization. With the modern methods based on column chromatography, P-I is usually isolated in larger amounts (on a weight basis) than P-II.

It is of some importance to recognize that the different commercial preparations of pure yeast hexokinase now available may represent dif-

23. The specific activity reported for the preparation of Berger *et al.*, 134 μmoles/min/mg at 30°, is much lower than that expected for a P-II preparation. However, this low activity results from the lability of the pure enzyme under the manometric assay conditions, which give rates less than one-third those found in the colorimetric assay (R. A. Darrow, Ph.D. Thesis, Johns Hopkins University, Baltimore, Maryland, 1957). The reported specific activity of the preparation of Kunitz and McDonald, corresponding to 70 μmoles/min/mg at 30° is also very low, probably because of the low Mg:ATP ratio (1:6) used in that assay.

ferent isoenzymes. For example, preparations from Boehringer have in our hands shown consistently the high fructose:glucose ratio, low specific activity toward glucose, and the chromatographic and electrophoretic properties of pure unmodified P-I. Other commercial preparations which show much higher specific activities are presumed to be P-II or a derivative of P-II. In view of the marked difference between P-II and P-I with respect to their responses to regulatory metabolites (as described in a later section), it becomes especially important to select the appropriate isoenzyme for certain metabolic studies.

B. Modification of Native Hexokinases by Added Proteases

In the early studies of Berger *et al.* (*7*) it was found that high concentrations of trypsin (e.g., 1% of the weight of hexokinase) caused rapid inactivation of the enzyme and that glucose could prevent this inactivation. The concentration of glucose required for converting one-half of the enzyme to an apparently trypsin-resistant enzyme–glucose complex was later shown (*24*) to be of the same order of magnitude as the kinetically determined K_m for glucose. This in fact provided the first reported measurement of the dissociation constant of the enzyme–glucose complex.

It was then noted that although glucose protected the enzyme from inactivation by proteases, it did not prevent conversion to a new chromatographic form (*12, 13*). In fact, later work (*16*) showed that this process of modification by proteases is actually promoted by glucose. Minute concentrations of trypsin (0.01% of the weight of hexokinase), which alone have no detectable effect on either the activity or chromatography of P-I or P-II, will, in the presence of glucose or high salt concentration, cause complete conversion of the native enzymes to modified forms which are called S-I and S-II, respectively. The role of glucose or salts in promoting the modification reaction is discussed in the next section. The modified forms are essentially unchanged from the parent enzyme in catalytic activity, but they are strikingly altered in chromatographic and electrophoretic behavior as well as in subunit structure (see below).

The modified forms are more acidic than the parent enzymes. This was indicated by greater mobility toward the anode in gel electrophoresis and greater affinity for DEAE-cellulose. Because of the latter property, the pH gradient which is used for elution of the native enzymes P-I and P-II fails to elute the modified forms S-I and S-II, which require a salt gradient of 0–50 mM for elution. The symbols P and S were chosen to in-

24. K. A. Trayser and S. P. Colowick, *ABB* **94,** 169 (1961).

dicate the requirement of pH and salt gradients, respectively, for elution of the native and modified forms, respectively.

The treatment with low concentrations of trypsin thus causes a modification of the pure native hexokinases which is quite analogous to that seen when the hexokinase crystals are contaminated with yeast protease (*16*). However, the products of modification are not identical since the modified form designated P-III (see above) was observed in addition to the S forms in the case of yeast protease action, but never in the case of modification by trypsin (*16*).

C. MOLECULAR WEIGHT AND SUBUNIT STRUCTURE

1. *Molecular Weights in Nondenaturing Solvents*

It is now recognized that many of the early molecular weight measurements (*24a, 25, 26*) were carried out on enzyme which had probably been protease-modified. When Darrow–Colowick preparations were used, the fraction that was probably least modified, which we would now designate P-III, showed the same behavior in molecular weight measurements (*24a, 25, 26*) as is now found (*16, 22, 27*) with the unmodified forms P-I and P-II; that is, the molecular weight remains unchanged at 100,000 over the pH range 5.5–8.5. On the other hand, the fraction in the protease-modified preparations which we would now designate as S form(s) showed (*25, 26*) the same behavior seen now with preparations of S-I or S-II derived by treatment of P-I or P-II with traces of trypsin; that is, the molecular weight is around 100,000 at the low pH, but complete dissociation to the 50,000 size occurs in the neutral range (*16, 22*). The preparation of Kunitz and McDonald (*6*), the first for which molecular weight measurements were reported, was apparently of the latter type [see Kenkare and Colowick (*26*), footnote 1].

It appeared from these studies that the modification by yeast proteases or by trypsin results in the loss of certain peptides which are essential for maintaining the 100,000 molecular weight structure at neutral pH. Apparently when the yeast protease modification stops with the production of form P-III, the remaining structure is still capable of supporting subunit association.

Schachman (*24a*) reported that the enzyme (presumably form P-III)

24a. H. K. Schachman, *Brookhaven Symp. Biol.* **13**, 49 (1960).
25. A. Ramel, E. A. Barnard, and H. K. Schachman, *Angew. Chem., Int. Ed. Engl.* **2**, 745 (1963).
26. U. Kenkare and S. P. Colowick, *JBC* **240**, 4570 (1965).
27. M. Derechin, A. H. Ramel, N. R. Lazarus, and E. A. Barnard, *Biochemistry* **5**, 4017 (1966).

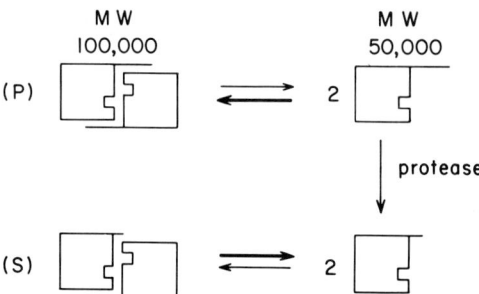

FIG. 1. Schematic drawing of the proposed pathway for the conversion of the native P forms of hexokinase to the protease-modified S forms (16). The small grooves represent the glucose binding sites (63). The peptide which is cleaved by protease is now known to be at the amino end of the chain (29, 38).

which showed a molecular weight of 100,000 at neutral pH could undergo dissociation to a 50,000 species on addition of glucose or phosphate ions, or by raising the pH or ionic strength. Similar effects were later noted by Schulze and Colowick (16) with the native enzymes P-I and P-II. It was found (16) that the dissociating effect of glucose was observed in phosphate but not in tris (hydroxymethyl) aminomethane at the same pH. Kinetic and binding data (described below) suggest that phosphate, by binding at the ATP site on the enzyme, promotes formation of the ternary complex, enzyme–phosphate–glucose.

The finding that glucose and high ionic strength promote dissociation of P-I and P-II provides the explanation for the role of these agents in promoting the modification by trypsin. It appears that the peptide bond(s) which are split in the modification reaction are not accessible to the protease in the 100,000 molecular weight state but become accessible upon dissociation into subunits of 50,000 size. Once these bonds are cleaved, reassociation cannot occur at neutral pH (Fig. 1).

2. *Ultimate Subunit Size in Denaturing Solvents*

The question of the size of the ultimate subunit in the yeast hexokinases has only recently been resolved. There is now general agreement that the smallest unit in denaturing solvents is a peptide chain of molecular weight 50,000 for P-I or P-II. Thus, the 100,000 species is actually a dimer and not a tetramer as was so widely believed.

The major evidence for the minimum peptide chain weight of 50,000, as determined by Schmidt (22, 28, 29) in our laboratory, is as follows:

28. J. J. Schmidt and S. P. Colowick, *Fed. Proc., Fed. Amer. Soc. Exp. Biol.* **29**, 334 (1970).
29. J. J. Schmidt, Ph.D. Thesis, Vanderbilt University, 1971.

(a) carboxypeptidase A digestion showed a single residue of alanine per 50,000 molecular weight for both P-I and P-II; (b) peptide mapping of P-I and P-II showed a total of 40–46 tryptic peptides, approaching that expected for a minimum peptide chain weight of 50,000 and clearly ruling out identical subunits of 25,000 molecular weight; (c) molecular weight measurement by gel electrophoresis in sodium dodecyl sulfate solution showed consistent values near 50,000 for P-I, P-II, S-I, or S-II; and (d) molecular weight measurements by sedimentation equilibrium in 6 M guanidine hydrochloride or 9 M urea, with SH groups of the enzyme blocked by either iodoacetate or methylmercuric iodide, also showed molecular weights close to 50,000. The results in other laboratories (*30, 31*) are now in general agreement with these findings.

The earlier view that the minimum peptide chain weight was of the order of 25,000 (*15, 26, 31–34*) deserves some comment. An important source of error is the occurrence of traces of yeast proteases in some of the preparations used. Pringle (*35*) showed that measurements of molecular weight of commercial hexokinase in sodium dodecyl sulfate by gel electrophoresis gave erroneously low values unless precautions were taken to destroy the protease (e.g., by boiling) prior to the measurement. It is important to note that the proteases did not degrade the hexokinase peptide chain until the preparation was placed in the denaturing solvent; that is, the unfolded hexokinase was much more readily attacked by the protease, which remains active under these conditions. This is undoubtedly the explanation for the low molecular weights of around 25,000 reported from Barnard's laboratory (*33*) with some rather highly purified preparations. Direct evidence for this explanation was obtained (*22, 29*) when one of these samples, supplied to us by Dr. Barnard, was incubated in dodecyl sulfate solution containing carboxypeptidase A. Under these conditions a large number of different C-terminal amino acids were detected, suggesting extensive degradation of the chain by yeast protease(s) of very broad specificity. Samples of hexokinases which are free of this protease do not exhibit this type of degradation and show the expected molecular weight of close to 50,000 in dodecyl sulfate.

The above type of protease contamination could also explain the low

30. Y. M. Rustum, E. J. Massaro, and E. A. Barnard, *Biochemistry* **10**, 3509 (1971).
31. J. S. Easterby and M. A. Rosemeyer, *Eur. J. Biochem.* **28**, 241 (1972).
32. A. Ramel, E. Stellwagen, and H. K. Schachman, *Fed. Proc., Fed. Amer. Soc. Exp. Biol.* **20**, 387 (1961).
33. N. R. Lazarus, M. Derechin, and E. A. Barnard, *Biochemistry* **7**, 2390 (1968).
34. Y. M. Rustum, E. J. Massaro, and E. A. Barnard, *Fed. Proc., Fed. Amer. Soc. Exp. Biol.* **29**, 334 (1970).
35. J. R. Pringle, *BBRC* **39**, 46 (1970).

molecular weights of around 25,000 obtained from sedimentation rates in dodecyl sulfate (*32*) or at a pH of 10–11 (*26*) with preparations which had not been freed of proteases. Alternatively, these low values may result from a change in shape or hydration for which no correction was attempted. In this connection, it may be pointed out that when pure protease-free P-I and P-II were subjected to pH values of 10–11, the molecular weight fell from 100,000 to 50,000, but not lower (*16*). This, in fact, was the earliest clue that the ultimate subunit size was 50,000 rather than 25,000.

It is also necessary to explain earlier chemical data which suggested a minimum molecular weight of 25,000. It was originally reported (*15, 36*) that carboxypeptidase A could give rise to 2 alanines rather than 1 per 50,000. The extra alanine probably resulted from some protease action during the long incubation period. It was also reported (*15, 36*) that only 24–30 tryptic peptides occurred in peptide maps of hexokinase P-II, compared with values of 40–46 reported later (*22, 29*). The latter values, which are now regarded as correct, revealed more peptides because the digestion conditions led to more complete cleavage and the pH used for electrophoretic separation was more effective (*22, 29*). In general, hexokinase samples which have been carboxymethylated in guanidine give the best analytical data because they are the most likely to be protease-free and because they cannot undergo aggregation via disulfide formation.

D. CHEMICAL STUDIES ON NATIVE AND MODIFIED FORMS

1. *Comparison of the Native Isoenzymes*

The studies by Gazith in our laboratory (*15, 36*) showed that there were many similarities, but also several clear differences in amino acid composition between the two native hexokinases P-I and P-II. It was therefore apparent that P-I and P-II were not interconvertible and must be determined by distinct structural genes. The more recent data of Schmidt (*22, 29*), shown in Table I, are in essential agreement with that of Gazith in that clear differences are seen between P-I and P-II with respect to content of histidine, valine, proline, phenylalanine, isoleucine, and leucine. It is also evident that the isoenzymes are identical in their content of many amino acids. The identity in tyrosine and tryptophan content is reflected by the similarity in absorption at 280 nm: $A_{1\,cm}^{1\%}$ = 8.85 for P-I and 9.47 for P-II (*22, 29*). It should be noted that the cys-

36. J. Gazith, Ph.D. Thesis, Vanderbilt University, Nashville, Tennessee, 1967.

TABLE I
AMINO ACID COMPOSITION OF NATIVE YEAST HEXOKINASES P-I AND P-II[a]

Residue	Residues per mole[b]	
	P-I	P-II
Lysine	38	34
Histidine	8	5
Arginine	18	18
Aspartic acid	56	53
Threonine	30	29
Serine	23	25
Glutamic acid	52	54
Proline	25	29
Glycine	42	40
Alanine	32	33
Valine	27	24
Methionine	11	11
Isoleucine	30	36
Leucine	49	35
Tyrosine	15	15
Phenylalanine	18	23
Tryptophan	4	4
Cysteine	4	4
Sum of residue weights	53,017	52,351

[a] Data from Schmidt and Colowick (22) and Schmidt (29).
[b] Lysine was chosen as the basis of calculation. The number of lysine residues was decided on the basis of physical data which indicated molecular weights slightly over 50,000 for both P-I and P-II in denaturing solvents.

teine content is also identical for P-I and P-II, contrary to the earlier report (15, 36).

As indicated in the previous section, P-I and P-II are identical in their C-terminal amino acid (alanine) but differ in the penultimate residue. The amino-terminal is valine for both isoenzymes.

Peptide mapping of tryptic digests confirms the impression gained from amino acid analysis data that there are regions of identity and regions of dissimilarity in P-I and P-II (22, 29). Out of the total of 43–46 tryptic peptides, about 27 appeared to be common to P-I and P-II, while around 16–19 were unique to each form. This suggests that the two isoenzymes arose by gene duplication and divergent evolution.

The extent of the divergence in structure is indicated by serological studies (19). Antisera prepared by immunizing rabbits with pure P-I or P-II are highly specific for the respective isoenzyme, showing little or no cross reaction. Such antisera may be used for detection of P-I and P-II in crude extracts. By such assays, both isoenzymes are found to be present in a haploid strain of Saccharomyces cerevisiae, thus demonstrating that these isoenzymes do not represent alleles of a single gene.

2. Chemical Nature of Proteolytic Modification

Gazith (15, 36) had shown that conversion of P to S forms by trypsin treatment was accompanied by the production of low molecular weight peptide material. The work of Schmidt (29, 37, 38) has established that modification by trypsin involves removal of the following sequence of amino acids from the N-terminus of either P-I or P-II:

Val·His·Leu·Gly·Pro·Lys·Lys·Pro·Gln·Ala·Arg

This sequence ordinarily undergoes cleavage at the Lys·Lys bond as well as at the Arg bond; thus, two peptides are usually found after the modification. In one case the entire 11-amino acid peptide accumulated. That this sequence comes from the N-terminal region of the native enzyme was established by demonstrating the same series of amino acids upon Edman degradation of form P-I.

It is noteworthy that both isoenzymes have retained this same N-terminal sequence during evolution. It suggests that this sequence plays an important role in physiology of the yeast cell. Since this part of the peptide chain is not required for activity, but is needed for maintenance of the dimeric structure of the enzyme (see Fig. 1), it appears that there must be some physiological advantage for the dimeric structure of yeast hexokinases.

It is interesting to note the extremely basic nature of the peptide removed by trypsin modification. This observation is consistent with the fact that S-I and S-II are much more acidic than P-I and P-II, respectively, as measured by mobility in gel electrophoresis (15, 38) or by isoelectric focusing (38). The pertinent data on mobilities and isoelectric points are summarized in Table II (39). One anomaly which remains to be explained is why P-I, which appears to be more basic than P-II as judged by isoelectric focusing or ease of elution from DEAE-cellulose, shows greater mobility as an anion in gel electrophoresis.

The chemical structures of the S forms are in agreement with expectation. Gazith (15, 36) had shown that S-I resembled P-I and S-II resembled P-II very closely. The removal of the 11-amino acid sequence

37. J. J. Schmidt, Fed. Proc., Fed. Amer. Soc. Exp. Biol. 30, 1122 (1971).
38. J. J. Schmidt and S. P. Colowick, ABB (1973) (in press).
39. Kunitz and McDonald (6) reported a value of 4.5–4.8 for the isoelectric point. This supports the view that their preparation corresponded to S-II.

TABLE II

Isoelectric Points and Electrophoretic Mobilities of Native and Modified Yeast Hexokinases[a]

Measurement	P-I	S-I	P-II	S-II
Isoelectric points	5.25	4.99	4.93	4.73
Electrophoretic mobilities (R_f)[b]	0.47–0.50	0.63–0.65	0.40–0.43	0.55–0.57

[a] Data from Schmidt (29) and Schmidt and Colowick (38).
[b] Mobilities are expressed relative to bromthymol blue for polyacrylamide disc gel electrophoresis in tris buffer pH 9.

out of a total of close to 500 residues would not be expected to produce measurable effects on abundant amino acid residues. However, in the case of the less abundant histidine, Gazith (15, 36) had noted a decrease from 8 to 7 in going from P-I to S-I and from 5 to 4 in going from P-II to S-II. These observations are in keeping with the occurrence of histidine in the N-terminal peptide which is removed from the P forms by trypsin.

Similarly, Schmidt (22, 29) found that peptide maps of S-I and S-II were identical with the maps of P-I and P-II, respectively, except for one missing peptide which was common to P-I and P-II maps and gave a positive Pauly test for histidine. Further evidence that this peptide is derived from the N-terminus of the P forms comes from the finding (15, 22) that the S forms have the same C-terminal amino acid (alanine) as the P forms.

E. Enzyme Mechanism

1. Evidence from Kinetic Studies of the Yeast Hexokinase Reaction

There is now general agreement that the reaction involves formation of a ternary complex of enzyme, phosphate donor, and phosphate acceptor prior to the phosphoryl transfer reaction. However, two central questions remain unresolved, namely, the order of addition of substrates and the existence of a phosphoenzyme intermediate.

Concerning the question of the order of addition of substrates and release of products, there is considerable evidence, on the one hand, that these are random processes in the sense that they can occur by alternative routes (40–48) and other evidence that the process is ordered, with glucose

40. F. B. Rudolph and H. J. Fromm, JBC **246**, 6611 (1971).
41. H. J. Fromm, E. Silverstein, and P. D. Boyer, JBC **239**, 3646 (1964).
42. D. P. Kosow and I. A. Rose, JBC **245**, 198 (1970).
43. D. P. Kosow and I. A. Rose, JBC **246**, 2618 (1971).

being the first substrate on and glucose 6-phosphate the last product off (*42, 49–53*). All of the data become reasonably compatible if one grants that the process may be random in the sense that either substrate may be the first to add and either may be the last to leave, but that there are *quantitative* differences that make one pathway predominate over another.

Scheme 1 illustrates this point of view, which we will assume to be

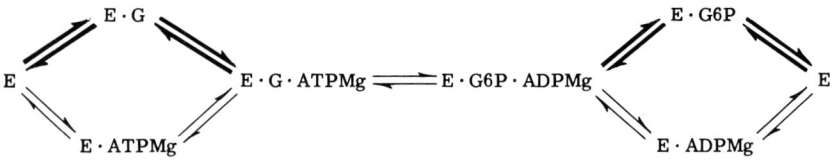

Scheme 1

correct for the purposes of discussion. The heavy arrows indicate the pathway that appears to predominate and that therefore yields certain types of kinetic data that conform to equations for an ordered mechanism. The light arrows represent a slower alternative pathway that is readily detected when other types of kinetic measurements are applied. We will review here some of the more important contributions that have led to the proposal of this scheme, considering first the evidence for the predominant ordered pathway.

Hammes and Kochavi (*52*) were the first to suggest, from classical initial velocity studies at varying substrate concentrations, that yeast hexokinase kinetics were compatible with the ordered mechanism shown as the main pathway in Scheme 1. However, their data, although ruling out other ordered mechanisms, were equally compatible with a random mechanism (*40, 46*). Thus, their results, although clearly establishing that the reaction was sequential rather than "ping-pong," could not dis-

---

44. H. J. Fromm and V. Zewe, *JBC* **237**, 1661 (1962).
45. V. Zewe, H. J. Fromm, and R. Fabiano, *JBC* **239**, 1625 (1964).
46. H. J. Fromm, *Eur. J. Biochem.* **7**, 385 (1969).
47. F. B. Rudolph and H. J. Fromm, *Biochemistry* **9**, 4660 (1970).
48. D. L. Purich and H. J. Fromm, *ABB* **149**, 307 (1972).
49. G. Noat, J. Ricard, M. Borel, and C. Got, *Eur. J. Biochem.* **5**, 55 (1968).
50. J. Ricard, G. Noat, C. Got, and M. Borel, *Eur. J. Biochem.* **31**, 14 (1972).
51. H. G. Britton and J. B. Clark, *BJ* **128**, 104p (1972).
52. G. G. Hammes and D. Kochavi, *JACS* **84**, 2069 and 2073 (1962).
53. D. C. Hohnadel and C. Cooper, *Eur. J. Biochem.* **31**, 180 (1972).

tinguish between random and ordered mechanisms for formation of the ternary complex of the two substrates with the enzyme.

Noat et al. (49) reached the same conclusions as Hammes and Kochavi (52) from a similar set of data. In a later paper (50), in which they studied the effect of alternative substrates (fructose and mannose) on the rate of glucose phosphorylation, they concluded that a random mechanism could be ruled out by this technique, and that the ordered mechanism previously deduced was correct.

In a recent report dealing with the relation between the structure and function of analogs of ATP, that analog in which glucose replaces the ribose moiety was found to be a poor substrate and an effective inhibitor, fully competitive with ATP, but "uncompetitive" with glucose (53). This is the result to be expected for an ordered mechanism in which glucose is the leading substrate. If the mechanism were random, a noncompetitive inhibition versus glucose would have been expected. The "uncompetitive" and "noncompetitive" plots are not always easy to distinguish, especially at low inhibitor concentrations, so that a small contribution from an alternative pathway cannot be ruled out. Nevertheless, the data show a high degree of order, suggesting that glucose binding enhances the binding of this analog of ATP.

Britton and Clark (51) have provided evidence in a preliminary report for a predominant pathway in which glucose is the leading substrate. They modified the flux technique (41) by measuring the ratio of two fluxes (G6P $\rightleftarrows$ ATP/G6P $\rightleftarrows$ G) as a function of increasing concentration of G or ATP at equilibrium. They found no change in ratio as G was increased, but a linear rise from 1 to 5 as the ATP was raised from zero to 1.2 mM. (Equilibrium was maintained by corresponding increases in G6P in both cases.) This is the expected result for an ordered pathway with glucose leading. The alternative pathway, with ATP leading, was not detected and apparently represented less than 10% of the total flux.

Thus, there is now strong evidence from kinetic data in support of the Hammes-Kochavi model for an ordered process with glucose as the leading substrate. We will next summarize some of the evidence that the alternative pathway of substrate addition can also occur and that product release, proceeding mainly by way of E·G6P, can under certain circumstances take the alternative pathway via E·ADP.

In the experiments of Fromm et al. (41), G $\rightleftarrows$ G6P and ADP $\rightleftarrows$ ATP flux rates, measured at equilibrium, showed hyperbolic increases in rate to a plateau value as the concentration of either pair was raised. If a compulsory pathway of binding occurs with glucose as the leading substrate the G ⇌ G6P flux should be suppressed at sufficiently high ATP concentrations. Since no suppression was found, it was concluded that there are alternative pathways of substrate binding. This conclusion is not necessarily in conflict with the finding of Britton and Clark (*51*), with the flux ratio technique, that the process is indeed ordered to a high degree. The latter technique apparently can reveal a preferred pathway even when the alternative pathway is fast enough to prevent suppression of individual fluxes by high substrate concentrations.

In addition to their conclusion that alternative pathways were involved Fromm et al. (*41*) concluded from the fact that the G ⇌ G6P exchange was only about half of the ADP ⇌ ATP exchange, that the dissociation of G and/or G6P from the enzyme was contributing to the rate limitation. This conclusion was greatly strengthened by Kosow and Rose (*42*) who showed a rapid preequilibrium exchange of ADP into ATP under conditions where only the product G6P bound to the enzyme was available for reaction. Preequilibrium exchange of G6P into ATP was not detectable, proving unequivocally that there was a highly preferred order for dissociation of the ternary product complex via E·G6P under the usual assay conditions.

Kosow and Rose (*42*) deduced further that the alternative pathway of decomposition of the ternary product complex via E·ADP, although ordinarily negligible, became the sole pathway at very high ADP concentrations, where the normally preferred route was suppressed. This conclusion was based on the fact that ADP inhibition has a distinct noncompetitive component (intercept effect) at low ADP concentrations, but shows purely competitive inhibitions at ADP concentrations above 18 mM. They assume that the noncompetitive component is due to combination of ADP with E·G6P (with which ATP cannot combine). At very high ADP concentrations, the E·G6P·ADP cannot dissociate appreciably via the "normal" path, so that the residual slow reaction proceeds solely through the alternative path of product release. ADP inhibition at concentrations above 18 mM is purely competitive with ATP because it is then acting solely by combination with the E·G complex.

The "mixed inhibition" phenomenon, which is commonly seen with hexokinases but not other kinases, is presumably related to the fact that the binding of ATP and of G6P to the hexokinases are mutually exclusive (i.e., G6P inhibition is strictly competitive with ATP), so that ADP binding to E·G6P is noncompetitive with ATP. An interesting effect of Mg in promoting the inhibitory effects of ADP on isoenzyme P-II but not on P-I (*42*) is noteworthy with respect to regulatory effects on P-II (see Section II,F).

The work of Kosow and Rose (*42*) described above shows that there is normally a highly ordered path for decomposition of the ternary product complex via E·G6P, but suggests that a second pathway is brought into play at high ADP concentrations. This pathway is considerable in magnitude at saturating ATP levels and high ADP levels. This may help to account for the lack of suppression of the G ⇌ G6P flux at high ADP concentrations in the equilibrium exchange experiments of Fromm *et al.* (*41*). However, it should be noted that the preparation used by Fromm *et al.* (*41*), which was presumably protease-modified, failed to show the intercept effect with ADP (*44*), so it may have had a faster E·G6P dissociation rate than the unmodified enzyme. A change in kinetics during storage of a P-II preparation has been observed (*42*).

Kosow and Rose (*43*) have also noted that at pH 6, where the system is extremely inhibited compared with the rates at pH 8, the normally minor route of decomposition of the ternary complex via E·ADP becomes the preferred route, as determined by equilibrium exchange, where the order of exchange rates was ATP ⇌ G6P > ATP ⇌ ADP = G ⇌ G6P.

A further conclusion that might have been drawn is that the ternary substrate complex dissociated preferentially via E·G, the dissociation of which seemed to be a rate-determining factor for the exchange in this case. This result is compatible with the data of Britton and Clark (*51*) and favors the view that glucose precedes ATP in forming the ternary complex, even under these conditions where the product complex decomposes via E·ADP.

A number of the arguments for a random mechanism come from initial rate studies in Fromm's laboratory (*40, 44–48*). An important example is their report (*40*) that AMP, which inhibits competitively with respect to ATP, apparently inhibits noncompetitively with respect to glucose, suggesting that glucose need not precede nucleotide binding. This result with AMP is in contrast to the finding with the ATP analog used by Hohnadel and Cooper (*53*), where the inhibition was uncompetitive with glucose, and pointed clearly to a promoting effect of glucose on nucleotide binding.

It is important to point out that there is no real discrepancy between the above results with AMP and those with the ATP analog. As we shall see from the information on the effect of sugars on nucleotide binding, discussed in the following sections, the data with the ATP analog probably provide a measure of the predominant pathway, with glucose leading, while the data with AMP reflect the existence of the minor alternative pathway.

2. Evidence from Kinetic Studies of the ATPase Reaction of Yeast Hexokinase

Hexokinase has an intrinsic hydrolytic activity (13, 54, 55) which is extremely low compared with the glucose phosphorylating activity. The rate upon saturation with ATP is about 0.02 μmole/min/mg of protein compared with a value of 800 for sugar phosphorylation. Since the K_m is about 40 times as high as that observed for ATP in the hexokinase reaction (4.0 mM vs. 0.1 mM), it has been proposed that the combination of glucose with the enzyme alters the structure of the protein so that the interaction with nucleotides is altered accordingly (55).

When N-acyl glucosamines and sorbose-1-phosphate, which are known inhibitors of hexokinase activity (56, 57), were tested, they were found to inhibit the ATPase activity with about the same relative order of effectiveness (55). This could mean that the glucose site can be occupied by these analogs in such a manner as to exclude water from the site.

The interesting observation was made by Dela Fuente and Sols (58, 59) that certain pentoses, which are nonphosphorylatable inhibitors of glucose phosphorylation by hexokinase, can promote strikingly the hydrolytic activity of the enzyme. Thus, 0.1 M lyxose causes an approximately 20-fold increase in V_{max} and a 40-fold decrease in K_m for ATP hydrolysis (60). The concentration of lyxose for half-maximum activation is the same as the K_i value for inhibition of hexokinase, suggesting that the lyxose effect on ATP hydrolysis occurs at the same site where glucose is normally attached. Dela Fuente et al. (59) therefore proposed that the binding of sugar at the active site involves an "induced fit" which modifies the structure of the protein with respect to both the ability to bind and the ability to hydrolyze nucleoside triphosphates. Such an induced fit is compatible with an ordered mechanism of reaction. Similar evidence for a role of sugars in the interaction of nucleotides with the enzyme has been obtained from physical studies described below.

Dela Fuente and Sols (58, 61) made another important observation in

54. K. A. Trayser and S. P. Colowick, ABB **94**, 161 (1961).
55. A. Kaji and S. P. Colowick, JBC **240**, 4454 (1965).
56. F. Maley and H. A. Lardy, JBC **214**, 65 (1955).
57. H. A. Lardy, V. D. Wiebelhaus, and K. M. Mann, JBC **187**, 325 (1950).
58. G. Dela Fuente and A. Sols, BJ **89**, 36p (1963).
59. G. Dela Fuente, R. Lagunas, and A. Sols, Eur. J. Biochem. **16**, 226 (1970).
60. The V_{max} values for ATPase and hexokinase reported by Dela Fuente et al. (59) are about one-fourth of those found in our laboratory (55). This is presumably because they were working with a P-I preparation (Boehringer) while ours was probably derived from P-II.
61. G. Dela Fuente, Eur. J. Biochem. **16**, 240 (1970).

the course of their studies on pentose activation of ATP hydrolysis. They noted that the activating effect of D-xylose was short-lived, and was followed by inactivation of the enzyme, as measured by either ATPase or glucose phosphorylating activity. They have attributed this to lability of the structure resulting from the induced fit in this case. Evidence that the inactivation involves phosphorylation of the protein (*62*) is discussed below.

3. *Evidence from Equilibrium Measurements of Enzyme–Substrate Interaction*

The interaction of glucose with hexokinase depends on the state of dissociation of the enzyme. With the native forms P-I and P-II, which are dimers under the conditions of measurement, the binding is poor at low glucose concentrations (K_{diss} ca. 10^{-3} M) but shows positive cooperativity; thus, at high glucose concentrations the final slope of the Scatchard plot corresponds to a K_{diss} of ca. 10^{-4} (*63, 64*). With the modified forms, S-I and S-II, which are monomers under the conditions of measurement, tight binding is observed (K_{diss} = 3×10^{-5} and 3×10^{-4}, respectively) with no evidence of cooperativity, i.e., the Scatchard plots are linear (*63, 64*). This difference between P and S forms has led us to propose (see Fig. 1) that the active site for glucose is readily accessible when the enzyme is in the monomeric state but becomes buried when dimers are formed (*63, 64*).

The requirement of a relatively high concentration of glucose for saturating the native enzyme is not seen in the ordinary catalytic assays where very low concentrations (0.1–2 µg/ml) are used at fairly high pH (8–9) and temperature (30°). It seems likely that both native and modified enzymes occur as monomers under these conditions (*65*). However, when catalytic assays are run at high enzyme concentrations (1 mg/ml) with GTP as phosphate donor under the conditions of the above glucose binding experiments (pH 7 phosphate buffer, 4°), the native enzyme requires much higher glucose concentration for saturation than the modified

---

62. L. Cheng, T. Inagami, and S. P. Colowick, *Fed. Proc., Fed. Amer. Soc. Exp. Biol.* **32**, (1973).

63. S. P. Colowick, F. C. Womack, and J. Nielsen, in "The Role of Nucleotides for the Function and Conformation of Enzymes" (H. M. Kalckar, ed.), p. 15. Munksgaard, Copenhagen, 1969.

64. F. C. Womack and S. P. Colowick, *ABB* (1973) (in press).

65. M. Derechin, A. K. Bhargava, and E. A. Barnard, *Fed. Proc., Fed. Amer. Soc. Exp. Biol.* **31**, 474 (1972).

enzyme (*66, 67*). This is especially striking when P-II and S-II are compared. These results suggest that native hexokinases at the high concentration and low pH in the yeast cell are most likely in the dimer form and that the active sites are relatively inaccessible to glucose. Factors which altered this accessibility could control the rate of glucose uptake.

The stoichiometry of glucose binding seems to be the same with native and modified enzyme (*63, 64, 68*). With modified enzyme (S-I or S-II), which occurs as monomer, there is 1 mole of glucose bound per mole (*63, 64*). With native enzyme (P-I or P-II), under conditions where the dimer form would predominate, about 2 moles of glucose are bound per mole of dimer (*63, 64*). Molecular weight measurements by sedimentation equilibrium (carried out by Dr. William Mitchell) indicate that saturating glucose causes only partial dissociation into monomers since the lowest molecular weights observed were around 70,000. This suggests that the dimer can bind two glucose molecules without dissociation. Thus it appears possible that the dimer itself can function catalytically without the need for dissociation.

Glucose binding is strongly promoted in the presence of 0.05 M phosphate (*15, 63, 64*). This effect is observed with S forms as well as P forms; thus, the increased binding is not attributable to an effect on the state of dissociation of the protein. Since phosphate at this concentration is an inhibitor competitive with ATP (F. C. Womack, unpublished data), this result suggests that normally the binding of glucose may be promoted by ATP.

The converse of this, that is, that nucleotide binding is promoted by sugars, has been suggested by several lines of evidence. The earliest evidence for this was provided by Cohn (*69*), who showed that glucose was required to demonstrate complexing of hexokinase with Mn·ADP, as measured by enhancement of the Mn^{2+} effect on the relaxation rate of the nuclear magnetic resonance of water protons. Later measurements with different enzyme preparations showed smaller qualitatively similar effects (*70*).

There is no doubt now from other lines of evidence that sugars do promote nucleotide binding. This was implicit in the findings from kinetic studies described above in which the K_m for ATP was 40 times higher with water than with glucose as phosphate acceptor (*55*),

66. F. C. Womack and S. P. Colowick, *Abstr. Southeast. Reg. Meet., Amer. Chem. Soc., Nashville, Tenn.* (1971).
67. F. C. Womack and S. P. Colowick, *ABB* (1973) (in press).
68. G. Noat, J. Ricard, M. Borel, and C. Got, *Eur. J. Biochem.* **11**, 106 (1969).
69. M. Cohn, *Biochemistry* **2**, 623 (1963).
70. A. S. Mildvan and M. Cohn, *Advan. Enzymol.* **33**, 1 (1970).

and this was made even more likely by the demonstration that D-lyxose lowered strikingly the K_m for ATP hydrolysis (58). Direct studies of ATP binding have shown clearly that the binding is very poor in the absence of sugars (63, 68, 71), approximating the K_m value of $4 \times 10^{-3} M$ for ATP hydrolysis. This poor binding is seen with both P and S forms (63, 71); thus, it cannot be attributed to the undissociated state of the enzyme as Rudolph and Fromm had proposed (72). The report (45) of tight binding of ATP to the enzyme as measured by fluorescence quenching cannot be reconciled with any of the other binding data.

The occurrence of ATPase activity makes it essential to carry out ATP binding measurements fairly rapidly, e.g., by gel filtration (68) or by the flow dialysis method (63, 71, 73). Even with the latter method, which requires only about 1 min for a binding measurement, it was found that D-lyxose addition caused such rapid hydrolysis of ATP, even at 0°, that the binding measurement could not be made. However, by use of deoxy-ATP, which is hydrolyzed more slowly than ATP, it was possible to show that D-lyxose addition causes greatly increased binding of nucleoside triphosphate substrates (63, 71). It was found convenient to use ADP for the binding studies since it is not hydrolyzed by the enzyme. Addition of 0.1 M D-lyxose causes a dramatic increase in the binding of ADP to the enzyme (63, 71), suggesting the formation of the ternary complex enzyme–lyxose–Mg·ADP. The K_{diss} value for dissociation of Mg·ADP from this complex appears to depend on the protein concentration, the lower K_{diss} values (around $2 \times 10^{-5} M$) being found with form S-I when the protein concentration is kept below 5 mg/ml ($1 \times 10^{-4} M$). When protein concentrations around 15 to 20 mg/ml are used, K_{diss} values as high as 1.2×10^{-4} or 2.2×10^{-3} are found, suggesting that even form S-I may associate extensively at these very high concentrations, thereby interfering with binding of D-lyxose and/or ADP (71).

Addition of glucose in low concentrations has been found to reverse the effect of lyxose on ADP binding (63, 71). This finding suggests that the interactions around carbon 6 of glucose must have subtle effects on the structure of the protein, such that binding of ADP is prevented, while binding of ATP is promoted.

We can conclude from the direct binding data as well as the studies on the ATPase activity of hexokinase that there is some interaction of nucleotides with yeast hexokinases in the absence of sugars. We know furthermore that with some nucleotides there is strong enhancement of

71. F. C. Womack and S. P. Colowick, *ABB* (1973) (in press).
72. F. B. Rudolph and H. J. Fromm, *JBC* **245**, 4047 (1970).
73. S. P. Colowick and F. C. Womack, *JBC* **244**, 774 (1969).

binding by sugars (e.g., D-lyxose effect on ATP, deoxy ATP or ADP), while with other nucleotides there is actually decreased binding in the presence of sugars (e.g., D-lyxose effect on ITP (59), glucose effect on ADP). These differences between different nucleotides, with respect to the effect of sugars on their binding to the enzyme, are probably of importance for explaining the apparent discrepancy between effects of AMP (40) and the glucose analog of ATP (53) in the kinetic experiments, where the former showed noncompetitive inhibition and the latter uncompetitive inhibition versus glucose. It seems quite likely that direct binding studies with AMP and with the glucose-containing ATP analog would show no effect of sugars on the former, but an enhancement by sugars in the latter case.

It would seem preferable in selecting structural analogs for a kinetic study to use those that best mimic the natural substrate in its interaction with the enzyme. In this case, it seems that the glucose-containing ATP analog, which in fact functions weakly as a substrate, would be a better candidate than AMP. Therefore, the conclusion seems justified that the pathway when ATP is the phosphate donor is highly ordered, with glucose as leading substrate. The result with AMP suggests that binding of nucleotides can occur at the ATP site without requirement for sugar, so that it also seems justified to conclude that there is normally a minor alternative pathway.

However, Dela Fuente and Sols (73a) have speculated that, even though nucleotides can bind to the enzymes in the absence of sugars, the positioning of the nucleotide on the surface of the enzyme may be quite different in this case from that occurring on binding to an enzyme–sugar complex. They suggest that subsequent binding of sugar to an enzyme nucleotide complex may lead to abortive ternary complexes. In this case it is possible that in effect a compulsory ordered pathway of substrate addition, with glucose leading, would result.

From the data considered thus far, the following conclusions concerning the enzyme mechanism for yeast hexokinases seem reasonably clear.

1. *Substrate addition* to the enzyme is highly ordered, with glucose adding before ATP. The alternative path of substrate addition accounts for less than 10% of the overall rate.

2. *Product dissociation* from the ternary complex is usually highly ordered with G6P coming off last. The latter process is slow enough to contribute to rate limitation and to account for noncompetitive inhibition by ADP versus ATP. At high ADP concentration or low pH, the alternative pathway of product dissociation via E·ADP becomes prominent.

73a. G. Dela Fuente and A. Sols, *Eur. J. Biochem.* **16**, 234 (1970).

4. On the Question of a Phosphoenzyme Intermediate

In recent years, largely through the work of Spector and his colleagues, direct evidence has been obtained for the occurrence of phosphoenzyme intermediates in several kinase reactions, including acetate kinase (74) and phosphoglycerate kinase (75). In the case of acetate kinase, subsequent kinetic studies confirmed the existence of such an intermediate by the demonstration of ping-pong kinetics, while all kinetic studies on yeast hexokinase have been negative in this respect (48). This tends to rule out a mechanism involving "half-reactions" of the following type:

$$\text{ATP} + \text{enzyme} \rightleftarrows \text{phosphoenzyme} + \text{ADP}$$
$$\text{phosphoenzyme} + \text{glucose} \rightleftarrows \text{enzyme} + \text{G-6-P}$$

However, it does not provide any information on the nature of the intermediates in the phosphate transfer when ternary complexes are involved:

$$\text{Enzyme-ATP-glucose} \rightleftarrows \text{enzyme-ADP-G-6-P}$$

Thus, the possible existence of an intermediate such as phosphoenzyme–ADP–glucose cannot be ruled out by the usual kinetic analysis. This type of intermediate would be difficult to identify or isolate. It is therefore customary in seeking a phosphoenzyme intermediate to study one of the above half-reactions, i.e., involving either the reaction of phosphate donor with enzyme to form the presumed phosphoenzyme, or the reaction of the presumed phosphoenzyme with phosphate acceptor. This is done even though it is recognized that the half-reactions may be "unphysiological" and proceed at low rates compared with those of the "normal" process involving ternary complexes.

The original claim (76) that an acid-insoluble phosphoenzyme could be isolated after incubation of ^{32}P-labeled ATP with the enzyme was not confirmed in later work (54, 55, 77). No labeled phosphoenzyme could be detected in the acid-insoluble fraction, even by quick fixation during the course of glucose phosphorylation (54). Since the possibility of an acid-labile phosphoenzyme was not thereby ruled out, additional experiments were carried out in which the excess ^{32}P-ATP was removed by charcoal adsorption, but no phosphoenzyme was detectable (55). A third approach involved measurement of ADP production after incubation of ATP with stoichiometric amounts of enzyme (54, 55). These experiments were negative, in the sense that no burst of ADP formation was found,

---

74. R. S. Anthony and L. B. Spector, *JBC* **245**, 6379 (1970).
75. C. T. Walsh, Jr. and L. B. Spector, *JBC* **246**, 1255 (1971).
76. G. Agren and L. Engstrom, *Acta Chem. Scand.* **10**, 489 (1956).
77. L. F. Hass, P. D. Boyer, and A. M. Reynard, *JBC* **236**, 2284 (1961).

but led to the discovery of a small hydrolytic activity, i.e., the ATPase function which is intrinsic to the hexokinase molecule (*54, 55*).

Evidence for phosphoenzyme formation has also been sought from isotope exchange experiments involving either of the above-mentioned half-reactions. The finding of Kaufman (*78*) of enzymic ADP exchange into ATP has been confirmed with pure preparations of hexokinase which were free of any trace of glucose or glucose 6-phosphate (*55*). It is therefore well-established that an exchange reaction occurs which cannot be explained trivially in terms of reversible equilibrium of the overall reaction. However, the rate of this exchange can be calculated to be only about 0.08 μmole/min/mg of protein, i.e., only 0.01% of the rate of glucose phosphorylation, but four times the maximum rate of ATP hydrolysis. It should be mentioned that the exchange rate has erroneously been stated to be 5% of the rate of glucose phosphorylation (*79*). The value of 0.01% given here is close to that reported by Solomon and Rose (*80*).

One might conclude from these data that a phosphoenzyme is involved in the reaction and that its formation is slow only because of the absence of the co-substrate, glucose, which might be expected to improve the interaction of nucleotides with the enzyme. However, N-acetylglucosamine, which is known to bind at the glucose site and to inhibit the ATPase, has little or no effect on the exchange rate of ADP into ATP (*55, 80*). Furthermore, D-lyxose, which has a remarkable stimulatory effect on the ATPase (see above), was found not to influence the rate of the ADP–ATP exchange (*80*). Finally, the remarkable inactivation of hexokinase by incubation with ATP and D-xylose (*61*) was found not to be accompanied by any alteration in the rate of the ADP–ATP exchange reaction (*80*). These data have been interpreted to mean that the ADP–ATP exchange cannot involve the catalytic site of the hexokinase reaction.

The second postulated "half-reaction," that of phosphoenzyme with glucose to form G-6-P and enzyme, was originally studied by Gamble and Najjar (*81*). They found no measurable exchange between glucose and G-6-P unless nucleotides were added. The problem has recently been reinvestigated by Walsh and Spector (*79*). They reported that a significant exchange between labeled G-6-P and glucose can be demonstrated in the absence of nucleotides provided that they use a high ratio of G-6-P to glucose (400:1), and a relatively low pH (6.5). The optimal

78. S. Kaufman, *JBC* **216**, 153 (1955).
79. C. T. Walsh, Jr. and L. B. Spector, *ABB* **145**, 1 (1971).
80. F. Solomon and I. A. Rose, *ABB* **147**, 349 (1971).
81. J. L. Gamble, Jr. and V. A. Najjar, *JBC* **217**, 595 (1955).

rate of exchange found is about the same as that for ADP–ATP exchange; they reported an exchange rate of 1% of the overall reaction rate, but our calculation from their data gives a value closer to 0.01%. Walsh and Spector concluded that the exchange reactions are suggestive of the existence of a phosphoenzyme intermediate in the reaction, and they proposed that the failure thus far to isolate the phosphoenzyme intermediate for certain kinase systems is because these kinases exhibit hydrolytic activity (ATPase).

We have recently obtained some support for this concept (*62*). It has been found that when hexokinase is incubated with ATP and D-xylose, the loss of the ATPase and sugar phosphorylating activities (*61*) is accompanied by the accumulation of an inactive phosphoenzyme (*62*). This inactive phosphoenzyme can be converted back to active enzyme by incubation with ADP and D-xylose. It can be calculated that during the hydrolysis of ATP in the presence of D-xylose the enzyme performs about 300 catalytic cycles before undergoing the "lethal" phosphorylation event. It seems possible that catalysis involves an active phosphoenzyme derivative which in the presence of D-xylose undergoes occasional rearrangement to form the inactive derivative. It is also possible, however, that the phosphorylation of the protein occurs only as a side reaction and not as part of the normal catalytic pathway. In any case, the finding of a phosphorylated derivative adds new interest to the phosphoenzyme question and in addition provides a tool for probing the active site of the enzyme.

5. *Role of SH Groups in Enzymic Action*

The initial work implicating SH groups in the activity of yeast hexokinase came from studies during World War II on the mode of action of lachrymators (*82, 83*). Some of the more recent studies are summarized here.

In titrations with silver ion at pH 8, it was reported (*55*) that the hexokinase and ATPase activities show a parallel loss of activity, the inactivation being proportional to the fraction of the total of SH groups titrated (four per monomer of 50,000 molecular weight). However, in titrations with *p*-mercuribenzoate at pH 7 in the presence of glucose, Fasella and Hammes (*84*) reported that all of the SH groups could be titrated without loss of activity. Kaji (*85*) then carried out titrations

82. M. Dixon and D. M. Needham, *Nature (London)* **158**, 432 (1946).
83. M. Dixon, *Biochem. Soc. Symp.* **2**, 39 (1950).
84. P. Fasella and G. G. Hammes, *ABB* **100**, 295 (1963).
85. A. Kaji, *BBA* **122**, 43 (1966).

with p-mercuribenzoate under similar conditions and found that about half of the total SH groups could be titrated without appreciable loss of activity but that further reaction resulted in extensive (though not complete) loss of activity. Later studies by Lazarus et al. (*33*) with better defined enzyme preparations and with methylmercury as titrating agent, led to results similar to those of Kaji (*85*); i.e., the first half of the SH groups could be titrated without loss of activity, but the titration of the second half led to complete loss of catalytic activity. It is not entirely clear, however, whether the inactivation which accompanies the titration of the last 2 residues is the result of reaction with residues at or near the active site or the result of the instability which has been observed (*84, 85*) with enzyme undergoing titration.

There is, in fact, little basis from titrations with thiol reagents for assuming that there are "active-site thiol groups" (*86, 87*) in hexokinase. Although glucose is very effective in preventing inactivation by p-mercuribenzoate (*85*), N-ethylmaleimide (*85*), or iodoacetate (*86–88*), its effect is not necessarily the result of steric interference with these agents at the active site. It seems more likely that glucose acts by causing a change in conformation which renders the entire molecule less susceptible to attack, thereby accounting also for its protective effect against inactivation by proteases (*24*).

Some indirect evidence for the presence of an SH group in the region of the active site can be deduced from the properties of the inactive phosphoenzyme resulting from incubation with ATP and D-xylose (*62*). This phosphoenzyme appears to undergo air oxidation with loss of an SH group unless protected by an added thiol or D-xylose. If oxidation occurs, reactivation by means of ADP will not take place. It will be of interest to determine whether a phosphopeptide can be isolated which contains the cysteine residue under consideration.

A rapid inactivation of hexokinases P-I and P-II by various thiols at pH 8–9 was reported several years ago from this laboratory (*15, 89*). It was also reported that mercaptoethanol at pH 9 caused a marked decrease in s value (*16*). The inactivation was reported to be reversed by simply lowering the pH to around 7 (*15, 89*). Extensive efforts (by J. J. Schmidt and by R. Gooding, unpublished) to repeat the observations on inactivation by thiols have been unsuccessful. No explanation for the earlier observation has been found. Since the inactivation, when

86. E. A. Barnard and A. H. Ramel, *BJ* **84**, 72 (1962).
87. J. G. Jones, *BJ* **115**, 41 (1969).
88. J. G. Jones, *BJ* **119**, 20 (1970).
89. R. H. Gooding and S. P. Colowick, *Fed. Proc., Fed. Amer. Soc. Exp. Biol.* **25**, 407 (1966).

observed, could be prevented by cyanide, an oxidative inactivation by mixed disulfide formation was suspected but not demonstrated.

6. *Conformational Changes in Relation to Enzymic Activity*

The enzyme can be converted by means of acid (pH 2) or alkali (pH 11), especially at high salt concentrations, to inactive forms which remain inactive under conditions of assay at pH above 8 (*26*). The alkali-inactivated enzyme undergoes reactivation in good yield after neutralization to pH 7 in the presence of thiols. Similarly, the acid-inactivated material undergoes reactivation in good yield provided that a metal chelator, ethylenediaminetetraacetate, or a thiol is present during neutralization to pH 7. The inactivation was originally thought to involve dissociation to peptide chains of 25,000 molecular weight (*26*). The present interpretation, in the light of new information on the structure (*22, 28*), is that acid or alkali inactivation and reactivation involve only an unfolding and refolding of the 50,000 molecular weight monomer. This view is compatible with the first-order time course of the reactivation process (*26*). The major evidence for unfolding and refolding during inactivation and reactivation is the reversible loss of helical structure and the reversible appearance of readily titratable SH groups. The hexokinase and ATPase activities undergo parallel inactivation and reactivation.

There is good reason to believe that extensive conformational changes occur when glucose combines with the protein. This is suggested by the marked changes in nucleotide binding which occur in the presence of sugars (*59, 63*). Striking visual evidence for a conformational change was noted when glucose or mannose, added to large crystals of hexokinase P-II, caused shattering of the crystals (Fig. 2). Since smaller nearby crystals were not affected, the shattering effect did not seem to result from an increase in solubility.

Definitive evidence concerning effects of ligands on the conformation of the enzyme has not yet been reported. A preliminary report by Steitz (*90*) on X-ray crystallography of hexokinase P-II indicates a most unusual structural feature in the hexokinase crystal. The two 50,000 molecular weight subunits appear to be identical, but one is translated relative to the other by 3.6 Å along a pseudo-diad axis. This means that interactions between subunits may be heterologous rather than isologous; in other words, at the interface between two subunits, a particular residue on one subunit may find itself in a different local environment from

90. T. A. Steitz, *JMB* **61**, 695 (1971).

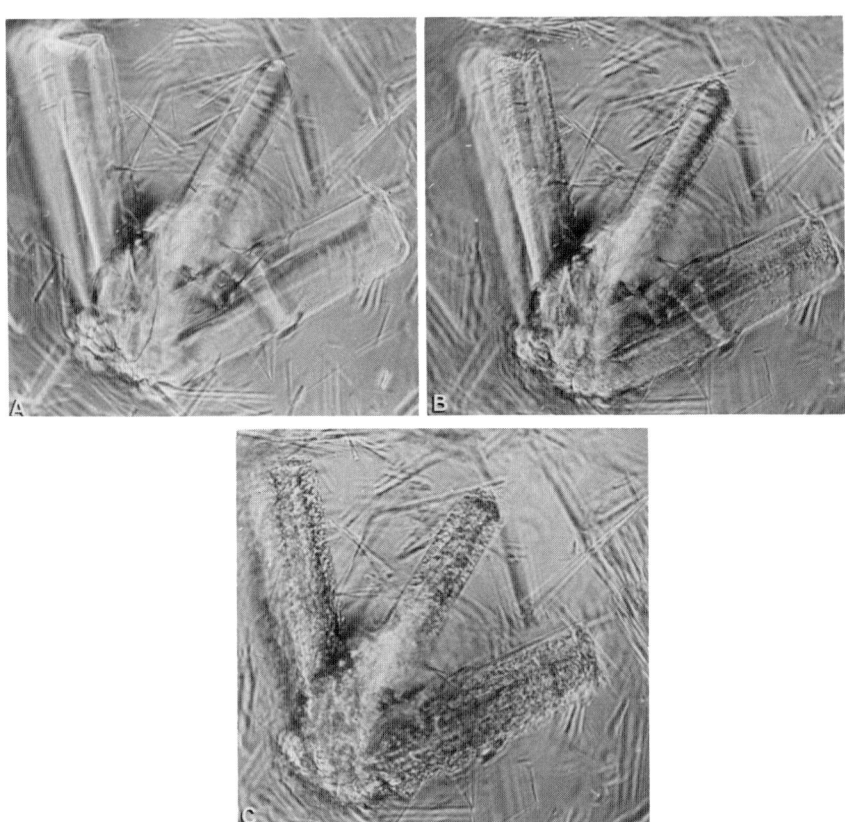

FIG. 2. Shattering of hexokinase crystals by addition of substrate. Photographs were taken at 3 min (A), 39 min (B), and 170 min (C) after addition of D-mannose to a suspension of hexokinase P-II crystals in $2 M$ $(NH_4)_2SO_4$. The final mannose concentration was $0.02 M$. Similar effects were seen with D-glucose but not with D-galactose. The dimensions of the large crystals were about $8 \times 2 \times 2\ \mu$ (J. Nielsen and S. P. Colowick, unpublished).

that of the same residue on the other subunit. Thus, as Steitz pointed out, it would be possible to have two nonidentical binding sites for a given ligand in a dimer consisting of identical subunits. The substrate-binding studies now in progress with another crystal form (*90a*) show that substrates or phosphate can bind to the crystals to produce measurable intensity changes in the X-ray diffraction patterns; thus, further analysis should lead to valuable insights into the conformational changes induced by substrates.

90a. T. A. Steitz, R. J. Fletterick, and K. J. Hwang, *JMB* (1973) (in press).

F. REGULATION OF HEXOKINASE ACTIVITY

An important new dimension was added recently to studies on yeast hexokinase when it was discovered by Kosow and Rose (43) that form P-II of the enzyme is subject to activation by low concentrations of various metabolites when it is at a suboptimal pH (6.5–7) corresponding to that in the yeast cell. They showed that in this pH range, with the sulfonated buffers (e.g., TES buffer) which they used, the saturation curves for ATP were anomolous, reaching a low activity plateau around 0.5 mM, followed by a sharp rise at higher ATP levels ("substrate activation"). By working at the lower ATP concentrations, they could demonstrate pronounced activation by citrate, phosphate, 3-phosphoglycerate, and other metabolites.

In attempting to confirm these results, we found that the inhibition at low pH was not very pronounced when imidazole or "tris" buffer was used (91, 92). Strong inhibition was seen by addition of TES buffer, 50–100 mM, and the activating effects of the various metabolites could then be confirmed. It was then found that the inhibitory effects of TES buffer could be mimicked at low pH by relatively low concentrations (0.2–0.5 mM) of ADP (91) or GDP (92). The inhibition, about 70%, did not increase appreciably with higher concentrations of the nucleotides. These inhibitory effects were much greater with form P-II or S-II than with P-I or S-I. Thus, it appears that one of the two native hexokinase isoenzymes is subject to a control mechanism in which the physiological inhibitors are purine nucleoside diphosphates and the activators are the various metabolites mentioned above.

In addition to the activators found by Kosow and Rose (43), catecholamines also activate yeast hexokinase (92a). Whether this effect is more pronounced with P-I or P-II has not been reported.

There are various unusual kinetic aspects of this regulated system. As already mentioned, the inhibited system (whether inhibited by TES or by nucleoside diphosphates) shows an ATP saturation curve with an intermediary plateau, i.e., what Koshland (93) has called a "bumpy curve." In reciprocal plots, there is superficial resemblance to a negative cooperativity phenomenon, but Koshland (94) has pointed out that

91. F. C. Womack and S. P. Colowick, *Fed. Proc., Fed. Amer. Soc. Exp. Biol.* **30**, 1059 (1971).
92. F. C. Womack and S. P. Colowick, *ABB* (1973) (in press).
92a. W. H. Harrison and R. M. Gray, *ABB* **151**, 357 (1972).
93. A. Levitski and D. E. Koshland, Jr., *Proc. Nat. Acad. Sci. U. S.* **62**, 1121 (1969).
94. J. Teipel and D. E. Koshland, Jr., *Biochemistry* **8**, 4656 (1969).

systems exhibiting bumpy curves can be explained by subunit interactions only if there is a minimum of three subunits and if negative cooperativity occurs between sites 1 and 2 followed by positive cooperativity between sites 2 and 3. Since the hexokinases have a maximum of two subunits and since these regulatory effects can be seen even with form S-II, which is likely to exist as a monomer, it would appear that any explanation in terms of subunit interactions is untenable.

It is probably of considerable significance for explaining the kinetic behavior that the enzyme undergoes slow changes in activity over a period of many seconds or minutes in response to addition of activators and inhibitors. Kosow and Rose (*43*) noted that an inhibited system responded with a measurable lag period after addition of an activator. Shill and Neet (*95*) found a burst of activity, followed by a slower steady-state rate, the time required to reach the steady state being longer in the region of low pH. They confirmed that the steady state kinetics appeared to show negative cooperativity and showed that the burst phenomenon was much more pronounced with form P-II than P-I. We have recently found (*92*) that the inhibition by ADP or GDP, after a fairly rapid onset, undergoes a gradual decrease in intensity over a period of several minutes when EDTA is present.

Ainslie et al. (*96*) have recently attempted to explain the kinetics of this and other "hysteretic enzymes" (*97*) in terms of a mechanism which does not involve subunit interactions but rather involves a monomer which can exist in different conformations. These different conformations are assumed to be slowly interconvertible and to react differently with various ligands. By assuming various rate constants for the ligand-adding and the conformation-changing steps, they can simulate the bursts and lags, the negative and the positive cooperativity characteristic of various hysteretic enzymes. These models assume only a single substrate site per molecule and thus differ from the model of Kosow and Rose (*43*) which assumes substrate binding at both a catalytic and a regulatory site. Similar proposals of nonequilibrium models for systems showing nonclassic kinetics have been made previously [for review, see Whitehead (*98*)]. It has not been shown that any of the systems yet proposed (*43, 96*) is suitable for fitting the bumpy curve.

Physical studies on the binding of activators and inhibitors of hexokinase may be useful in clarifying the mechanism of regulation. Measurements of lyxose-induced ADP binding show only 1 mole of ADP bound

95. J. P. Shill and K. E. Neet, *BJ* **123**, 283 (1971).
96. G. R. Ainslie, Jr., J. P. Shill, and K. E. Neet, *JBC* **247**, 7088 (1972).
97. C. Frieden, *JBC* **245**, 5788 (1970).
98. E. Whitehead, *Progr. Biophys. Mol. Biol.* **21**, 321 (1970).

per monomer at pH 7 in phosphate buffer (see above). Under similar conditions no binding of GDP can be detected (*92*). These results are interpreted to mean that the ADP binding occurs only at the catalytic site and that there is no binding of ADP or GDP at a regulatory site under these conditions. The failure to observe binding of the inhibitory nucleosides may be explained by the evidence of Kosow and Rose (*43*) that the enzyme can be converted to the inhibited state only when both substrates are present. Thus, when enzyme alone (or with one substrate) is exposed to low pH, activators added at the onset of reaction can exert their full effect without lag. Similarly, we find that GDP or ADP addition to enzyme alone does not lead to an inhibited state (*92*). Measurement of binding of inhibitors in the presence of substrates would be of interest.

III. The Mammalian Hexokinases

A. Occurrence of Multiple Forms

It has been known for many years that mammalian tissues contained a hexokinase resembling in specificity that of yeast, but with much lower activity (*1, 9, 99*). The most thorough early study was that of Crane and Sols (*100*) on the brain enzyme. They showed that the enzyme was largely particulate and confirmed its extreme sensitivity to inhibition by glucose 6-phosphate (*101*). Like the yeast enzyme, it could act on glucose, mannose, or fructose, and required relatively high concentrations of the latter for saturation.

The first evidence that more than one type of hexokinase existed in mammalian cells was the discovery of Dipietro et al. (*102*) that liver contained a "glucokinase" that required a very high concentration of glucose for saturation, and thus appeared to account for the requirement of high blood glucose concentrations for glycogen synthesis in the liver. It is now recognized (*103, 104*) that glucokinase is a true hexokinase in the sense that it acts on the same set of hexoses as the

99. R. K. Crane and A. Sols, "Methods in Enzymology," Vol. 1, p. 277, 1955.
100. R. K. Crane and A. Sols, *JBC* **210**, 597 (1954).
101. H. Weil-Malherbe and A. D. Bone, *BJ* **49**, 339 (1951).
102. D. Dipietro, C. Sharma, and S. Weinhouse, *Biochemistry* **1**, 455 (1962).
103. D. G. Walker, *in* "Essays in Biochemistry" (P. N. Campbell and G. D. Greville, eds.), Vol. 2, p. 33. Academic Press, New York, 1966.
104. E. Vinuela, M. Salas, and A. Sols, *JBC* **238**, PC1175 (1963).

yeast or brain enzymes. However, because of its low affinity (high K_M) for all substrates, its activity toward fructose becomes negligible at physiological sugar concentrations. It was noted in several laboratories that liver contained both the high K_M and low K_M activities (104–106), and that the high K_M enzyme disappeared preferentially during starvation or diabetes and reappeared during refeeding or insulin administration (104, 106–108). Later studies showed that the high K_M enzyme is restricted to the hepatocytes (parenchymal cells) whereas the bulk of the low K_M activity is associated with the nonparenchymal tissue of liver (109). The glucokinase was distinguished by its insensitivity to glucose 6-phosphate (104). This strengthened the view that glucokinase was important in glycogen synthesis, since glycogen synthetase is activated by glucose 6-phosphate (110).

A great step forward came with the finding that four distinct enzymes could be separated in extracts of liver by means of column chromatography (111) or starch gel electrophoresis (112). They are numbered I, II, III, and IV on the basis of either the order of elution from a DEAE-cellulose column, or the order of the bands of activity in starch gel electrophoresis, band I being closest to the origin. These isoenzymes were also readily distinguished by means of the striking differences in their K_M values for glucose (112). These are, in terms of order of magnitude, 10^{-5}, 10^{-4}, 10^{-6}, and 10^{-2} M, respectively.

When the same chromatographic and electrophoretic techniques were applied to other tissues, an important unifying observation was made, namely that the same isoenzymes were present, but in varying proportions, in all of the tissues (113). Thus, brain and kidney were shown to contain mainly type I, skeletal muscle mainly type II, and fat pad, heart, and intestine about equal amounts of types I and II. Kidney and intestine contain in addition detectable amounts of type III, while liver contains all four types. Since type III has an extremely low K_M and is inhibited by excess glucose, it is best detected at low glucose concentrations, while type IV requires high glucose concentrations for detection. Although most studies have been done with rat tissues, a

105. D. G. Walker, *BJ* **84**, 118P (1962).
106. C. Sharma, R. Manjeshwar, and S. Weinhouse, *JBC* **238**, 3840 (1963).
107. H. Niemeyer, L. Clark-Turri, and E. Rabajille, *Nature* (*London*) **198**, 1096 (1963).
108. M. Salas, E. Vinuela, and A. Sols, *JBC* **238**, 3535 (1963).
109. M. Sapag-Hagar, R. Marco, and A. Sols, *FEBS Lett.* **3**, 68 (1969).
110. L. F. Leloir, J. M. Olavarria, S. H. Goldenberg, and H. Carminatti, *ABB* **81**, 508 (1959).
111. C. Gonzalez, T. Ureta, R. Sanchez, and H. Niemeyer, *BBRC* **16**, 347 (1964).
112. H. M. Katzen, D. D. Soderman, and H. M. Nitowsky, *BBRC* **19**, 377 (1965).
113. H. M. Katzen and R. T. Schimke, *Proc. Nat. Acad. Sci. U. S.* **54**, 1218 (1965).

similar distribution of the various types of hexokinase has been observed in various species (*114*).

As stated above, the type IV enzyme of liver undergoes marked changes in amount with the hormonal state of the animal. In the fat pad (*115, 116*) and in muscle (*115*), the type II enzyme also appears to decrease with starvation or diabetes and to reappear with refeeding or insulin administration. The effects are of interest because the type II enzyme is characteristically found in the insulin-sensitive tissues. The type II enzyme activity is elevated in cell cultures at high glucose concentrations (*112*). Alterations in type II activity of certain tissues with age have also been noted (*113*).

The partial purification of types I, II, and III from various tissues established unequivocally that a given type had the same characteristic properties regardless of the tissue of origin, as measured by chromatography, gel electrophoresis, K_M for glucose, and sensitivity to trypsin treatment or heat (*117*). The results were compatible with earlier data in which different K_M values for glucose were reported for purified preparations from different tissues, e.g., calf brain (*118*) (type I) and rat skeletal muscle (*119*) (type II). Furthermore, it was demonstrated (*117*) that purified types I, II, and III have many features in common, including extreme sensitivity to glucose 6-phosphate inhibition, which is competitive with ATP but not with glucose, and moderate sensitivity to ADP inhibition, which is not competitive with either ATP or glucose. There are perhaps slight differences in the degree of these inhibitory effects with the different types. The nucleotide specificities are the same for all three types, with ATP giving by far the highest activity, and ITP showing appreciable activity in all cases. All three showed the same broad pH optimum, around pH 8.4, and the same fructose/glucose ratio of about 1.2 (V_{max} values). The three types were all similar in molecular weight as determined by activity measurements after sucrose gradient sedimentation (MW 96,000–100,000).

B. RELATION OF SOLUBLE TO INSOLUBLE FORMS

It has been recognized since the work of Crane and Sols (*99, 100*) that a considerable fraction of the total hexokinase activity in tissue

114. H. M. Katzen, D. D. Soderman, and V. J. Cirillo, *Ann. N. Y. Acad. Sci.* **151**, 351 (1968).
115. H. M. Katzen, *Advan. Enzyme Regul.* **5**, 335 (1967).
116. R. Hansen, S. J. Pilkis, and M. E. Krahl, *Endocrinology* **81**, 1397 (1967).
117. L. Grossbard and R. T. Schimke, *JBC* **241**, 3546 (1966).
118. H. J. Fromm and V. Zewe, *JBC* **237**, 1661 (1962).
119. T. L. Hanson and H. J. Fromm, *JBC* **240**, 4133 (1965).

homogenates is associated with particles. Katzen et al. (*120*) have shown that the hexokinases found bound to the mitochondrial or microsomal fractions are indistinguishable from types I and II found in the soluble fraction, and that type III does not occur in particulate form. A portion of the bound enzyme is present in a "latent" form which becomes available to added substrates only upon disruption of the membrane structure (*120–123*).

It is now generally agreed that the hexokinases in tissue particles are largely associated with mitochondria, as some early studies suggested (*124, 125*). Occurrence in the microsomal fraction could be an artifact of homogenization, which may allow binding of hexokinases to new membrane surfaces derived from the plasma membrane or endoplasmic reticulum. Although Green et al. (*126*) have reported an association of the entire glycolytic system with the plasma membrane of erythrocytes and yeast cells, Emmelot and Bos (*127*) have taken exception to this for the case of plasma membranes from liver cells, which contain no hexokinase. On the other hand they find a tightly bound hexokinase activity in the plasma membrane of hepatocellular rat hepatoma 484, and propose a role for the enzyme in increasing sugar transport and glycolysis in these cells. It would be of interest to determine which type of hexokinase was bound.

The elegant studies of Rose and Warms (*128*) provided an unequivocal demonstration of the association of hexokinase with mitochondria. Using the particulate hexokinase of an ascites tumor, they showed that the specific loading of the mitochondria with calcium phosphate during oxidative phosphorylation resulted in the predicted change in density of the particles bearing the hexokinase activity. They then made the further important observation that bound hexokinases could be fully solubilized by the addition of very low concentrations of glucose 6-phosphate. This effect is highly specific, and appears to be due to interaction at the same regulatory site (*100*) at which glucose 6-phosphate acts as an inhibitor. Related sugar phosphates which are products of

120. H. M. Katzen, D. D. Soderman, and C. E. Wiley, *JBC* **245**, 4081 (1970).
121. J. E. Wilson, *JBC* **243**, 3640 (1968).
122. J. E. Wilson, *ABB* **150**, 96 (1972).
123. J. H. Southard and H. O. Hultin, *FEBS Lett.* **19**, 96 (1972).
124. R. Tanaka and L. G. Abood, *J. Neurochem.* **10**, 571 (1963).
125. O. H. Lowry, N. R. Roberts, D. W. Schulz, J. E. Clow, and J. R. Clark, *JBC* **236**, 2813 (1961).
126. D. E. Green, E. Murer, H. O. Hultin, S. H. Richardson, B. Salmon, G. P. Brierly, and H. Baum, *ABB* **112**, 635 (1965).
127. P. Emmelot and C. J. Bos, *BBA* **121**, 434 (1966).
128. I. A. Rose and J. V. B. Warms, *JBC* **242**, 1635 (1967).

the hexokinase reaction (mannose 6-phosphate, 2-deoxyglucose 6-phosphate) are not highly effective either as inhibitors or as solubilizers of the enzyme, whereas 1, 5 anhydroglucitol 6-phosphate, a strong inhibitor, was very effective at solubilizing the enzyme.

In the same paper (128) they showed that the dissociation by glucose 6-phosphate requires measurable time to reach equilibrium ($t_{0.5}$ = 18 sec at 35°). The rate of dissociation by glucose 6-phosphate was inhibited by low concentrations of phosphate, which alone had a weak dissociating effect. The dissociated enzyme could be rebound with a similar time course by adding low concentrations of Mg^{2+}, Mn^{2+} or Ca^{2+}, and the equilibrium reached suggested formation of a complex in which a single metal ion was involved in binding each enzyme molecule to the particle.

In the absence of added Mg^{2+}, ATP has a strong but nonspecific dissociating effect, shared by other nucleosidetriphosphates as well as inorganic pyrophosphate (128). The enzyme in tumor mitochondria can also be eluted by high salt concentrations at low pH (128), as first shown by Hernandez and Crane (129) with heart particles.

Rose and Warms showed that there was no tissue specificity for the interaction between hexokinase and mitochondria. Brain, liver, or tumor mitochondria were all similarly susceptible to the dissociating and associating agents. Furthermore, the soluble enzyme from brain or tumor could interact equally well with mitochondria from all three sources. The only exception was the soluble hexokinase fraction of liver, which failed to bind to mitochondria from any source. This is apparently due to a lysosomal cathepsin (iodoacetate-sensitive) that, like trypsin and chymotrypsin, converts soluble hexokinases to a form that cannot be bound by mitochondria (128).

The question of the location of the bound hexokinase in the mitochondrial structure has been approached in several ways. One approach has been through metabolic studies designed to test whether the ATP generated by oxidative phosphorylation is in the same compartment as the mitochondrial hexokinase. This is apparently not the case, since (1) an externally added glycerol kinase competes effectively with mitochondrial hexokinase for the generated ATP (128), (2) atractyloside, which blocks transport of ATP, prevented completely the phosphorylation of sugar by the generated ATP, but had no effect on phosphorylation when ATP was added rather than generated (128, 130), and (3) the ^{32}P incorporated into ATP by mitochondria can be diluted by external ATP prior to phosphorylation of the sugar (131).

129. A. Hernandez and R. K. Crane, *ABB* **113**, 223 (1966).
130. C. G. Vallejo, R. Marco, and J. Sebastian, *Eur. J. Biochem.* **14**, 478 (1970).
131. P. Siekevitz and V. R. Potter, *JBC* **215**, 237 (1955).

In agreement with the above metabolic studies, direct measurements of the location of the bound enzyme show that it is not within the inner mitochondrial space. Most investigators have concluded that the enzyme is associated with the outer membrane, based on (1) hexokinase content of mitochondrial fragments (*132, 133*), (2) binding of added hexokinase by the isolated fragments (*128*), or (3) use of antibodies against the pure brain enzyme for the localization of the enzyme in intact mitochondria (*134, 135*). Vallejo et al. (*130*) report that the brain hexokinase is concentrated in the inner membrane fragments, but agree that the enzyme is outside of the atractyloside barrier, thus concluding that it is on the outer surface of the inner membrane. However, in view of the ease of dissociation and rebinding of the enzyme to intact mitochondria, it seems more likely that the enzyme is on the outer membrane. Its association with fragments of the inner membrane may be an artifact resulting from exposure of new sites for hexokinase binding after the mitochondria are disrupted (*128*).

Craven and Basford (*136*) have obtained evidence for an intimate association of the bound hexokinase with certain lipids in brain mitochondria. They report that the solubilization of the enzyme by glucose 6-phosphate is accompanied by the release of phosphatidylethanolamine and other lipids from the mitochondria, and conclude that the enzyme is released as a lipoprotein complex. They report further that when mitochondria are rendered lipid-deficient by treatment with 10% acetone, the bound hexokinase becomes inactive, but can be reactivated by addition of phosphatidylethanolamine.

Wilson (*137*) has obtained further evidence for a role of phospholipid in the interaction of brain hexokinases with mitochondria. He finds that the enzyme, after solubilization by glucose 6-phosphate, appears to contain bound phospholipid which is necessary for reattachment to the mitochondria. Treatment of the solubilized enzyme with phospholipase C does not affect its activity but abolishes its ability to bind to mitochondria. When the solubilized enzyme is purified (see below), the ability to bind to mitochondria is lost, presumably due to loss of the essential phospholipid. There are discrepancies between the results of Craven and Basford (*136*) and those of Wilson (*137*) with respect

132. P. A. Craven, P. J. Goldblatt, and R. E. Basford, *Biochemistry* **8**, 3525 (1969).
133. E. S. Kropp and J. E. Wilson, *BBRC* **38**, 74 (1970).
134. P. A. Craven and R. E. Basford, *Biochemistry* **8**, 3520 (1969).
135. J. F. Jemionek, Ph.D. Thesis, Univ. Pittsburgh, Pittsburgh, Pennsylvania, 1972.
136. P. A. Craven and R. E. Basford, *BBA* **255**, 620 (1972).
137. J. E. Wilson, *ABB* **154**, 332 (1973).

to the molecular weight and lipid content of the hexokinase–lipid complex, but the main findings, taken together, suggest that the solubilized enzyme can maintain an active conformation when phospholipid is removed, but that the phospholipid plays a role not only for the binding of the enzyme to the mitochondria but for maintenance of the active conformation in the bound state.

C. PURIFICATION AND MOLECULAR PROPERTIES

1. *Purification Procedures*

Partial purification of the types I, II, and III hexokinases was achieved by Grossbard and Schimke (*117*) using the soluble fraction of homogenates from various tissues. The specific activities reached were about 9, 13, and 1.3 units/mg for the respective types. Partial purification of the type IV enzyme (glucokinase) has been effected from the soluble fraction of liver homogenates (*138–140*). The highest specific activity reached was about 30 units/mg, which is believed to represent about 80% purity for the rat liver enzyme (*140*).

Highly purified hexokinase type I has been prepared, using brain mitochondria as starting material. There is an obvious advantage to starting with enzyme that has been separated from the soluble proteins of the homogenate. Chou and Wilson (*141*) have recently reported a remarkably simple method for obtaining the rat brain enzyme from the mitochondria in a homogeneous state. There are three steps: (1) Freezing and thawing, which is believed to disrupt the synaptosomal vesicles and release the enclosed mitochondria. About one half of the total mitochondrial hexokinase is thought to be present within synaptic vesicles that must be disrupted to convert this "latent" activity to "overt" activity. (2) Solubilization of hexokinase by means of glucose 6-phosphate. (3) Chromatography on a DEAE-cellulose column.

The Chou-Wilson procedure seems more convenient and less drastic than some of the earlier methods for isolation of brain hexokinase. However, it may lack convenience for large-scale preparations, as from beef brain, because of the high dilutions of mitochondria needed for efficient solubilization by glucose 6-phosphate (*136*). In the procedure

138. T. Salas, M. Salas, E. Vinuela, and A. Sols, *JBC* **240**, 1014 (1965).
139. D. G. Walker and M. J. Parry, "Methods in Enzymology," Vol. 9, p. 381, 1967.
140. S. J. Pilkis, *ABB* **149**, 349 (1972).
141. A. C. Chou and J. E. Wilson, *ABB* **151**, 48 (1972).

of Jagganathan (*142*), the beef brain enzyme is solubilized by a combination of elastase and freeze-thaw treatments. In the procedure of Schwartz and Basford (*143*) solubilization is achieved by a combination of detergent and protease treatment. A milder procedure for large-scale isolation of the beef brain enzyme has recently been described by Redkar and Kenkare (*144*), who avoid enzymic digestion by using high salt concentrations at low pH (*129*) for solubilization. The latter procedure may now be the method of choice for large-scale preparation of the beef brain enzyme, since the specific activity, 83 units/mg is as good as that of the Schwartz-Basford preparation, and the yield is higher. The Chou-Wilson preparation from rat brain has a lower specific activity, 60 units/mg, but this may be due to species difference rather than impurity.

Procedures have recently been described for isolation of highly purified hexokinase from heart tissue (*145, 146*). Easterby (*145*) reports the isolation of the type I hexokinase from porcine heart, with a specific activity of at least 60 units per mg, but details of the method are not given. Paranjpe and Jagganathan (*146*) report 500-fold purification of the ox heart enzyme (final specific activity 48 units/mg) following solubilization by their elastase technique. The enzyme exhibits catalytic properties resembling those of type I, e.g., a K_M for glucose of $5 \times 10^{-5} M$, and an attenuation of the glucose 6-phosphate inhibition by P_i (see Section III,E). However, it shows an anomalous lack of mobility in gel electrophoresis, which leads these workers to conclude that the isolated enzyme is different from the two isoenzymes normally seen in the soluble fraction of the heart.

There has thus far been no report of isolation of types II or III in a homogeneous state. The type II enzyme can be partially purified (*147*) from tumor mitochondria, after solubilization by freezing and thawing, by means of the usual chromatographic procedures, but the specific activity, 15 units per mg, is not better than that of type II enzyme previously purified from the soluble fraction of muscle or fat pad (*117*). Since types III and IV are not known to be bound to mitochondria, purification attempts for these must be limited to the soluble fraction.

142. V. Jagganathan, *Indian J. Chem.* **1**, 192 (1963).
143. G. P. Schwartz and R. E. Basford, *Biochemistry* **6**, 1070 (1967).
144. V. D. Redkar and U. W. Kenkare, *JBC* **247**, 7576 (1972).
145. J. S. Easterby, *FEBS Lett.* **18**, 23 (1971).
146. S. V. Paranjpe and V. Jagganathan, *Indian J. Biochem. and Biophys.* **8**, 227 (1971).
147. D. P. Kosow and I. A. Rose, *JBC* **243**, 3623 (1968).

2. Molecular Weight and Subunit Structure

The native type I enzyme preparations obtained in the pure state show a molecular weight close to 100,000, whether obtained from rat brain (141), beef brain (144), or porcine heart (145). The ox heart enzyme (146) is approximately the same size. These values agree with those estimated earlier with partially purified preparations (117) of types I, II and III.

A very surprising recent finding is that the molecular weight of the type I enzyme under denaturing conditions is the same as that of the native enzyme. That is, in contrast to the yeast enzymes, which exist as dimers with subunit molecular weights of 50,000, the type I mammalian enzyme from rat brain (141) or porcine heart (145) appears to have no subunits but rather to consist of a single chain of molecular weight 100,000. This conclusion is based on the following evidence. (1) The molecular weight by gel electrophoresis in sodium dodecyl sulfate is close to 100,000 for both preparations (141, 145). (2) The sedimentation velocity of the porcine enzyme (145) is independent of the presence of glucose or salts, both of which are known to cause dissociation of the yeast enzymes into subunits. (3) The sedimentation velocity is also independent of the presence of thiols, suggesting that disulfide bond formation is not responsible for the failure to dissociate. (4) N-Terminal analysis of the rat brain enzyme (141) indicated the presence of a single glycine residue per 100,000 MW.

Kenkare (personal communication) has similarly found that the pure beef brain enzyme has the same MW, near 100,000, in denaturing solvents as in the native state.

It is interesting to recall that Rose and Warms (128), on the basis of the effect of dilution on the equilibrium for the association of bound brain hexokinase with mitochondria, had concluded that the brain enzyme did not undergo dissociation into subunits under their conditions.

These observations all suggest that type I hexokinase differs drastically in quaternary structure from the yeast enzymes and may have arisen during evolution by gene duplication and fusion. Easterby (145) favors this view and suggests that each molecule may have two catalytic sites. The possibility must still be considered, however, that separate subunits do exist but are more difficult than usual to separate. The data of Easterby (145) on the maleylated enzyme showed a decrease in sedimentation rate, which was interpreted as being due to unfolding of the molecule, but might also represent dissociation into subunits.

No direct information is available with respect to the subunit structure of hexokinases II and III. The fact that type II is much more susceptible than type I to inactivation by proteases and heat (*117*) suggests that type II may have a looser structure and may dissociate more readily into subunits, The effect of glucose in protecting type II against inactivation by proteases, but permitting the conversion to active forms with altered electrophoretic mobility (*117*), is very analogous to the effects seen with the yeast enzymes. It will be of interest to determine whether the altered forms of type II have undergone dissociation into subunits. Although type I is not readily inactivated by proteases, it undergoes slow modification to a new active form with increased mobility (*117*). Type III is neither inactivated nor modified by proteases (*117*) and so may possibly be an even more compact structure than type I.

Of all the mammalian enzymes that phosphorylate glucose, only type IV has been reported (*113, 140*) to have a molecular weight significantly below 100,000. Values of 68,000 and 50,000 were found at low and high salt concentrations, suggesting a loose quaternary structure similar to that of the yeast hexokinases (*140*). This view is supported by the finding (*140*) that glucokinase is even more susceptible than type II enzyme to inactivation by trypsin or urea. Inactivation by trypsin or urea can be prevented by glucose. It has not been reported whether or not trypsin treatment in the presence of glucose yields a modified active form of the type IV enzyme. If such modification occurs, it may help to explain the splitting of type IV into two bands during electrophoresis under some conditions (*116*).

Thus, of the four mammalian hexokinases, the so-called "glucokinase" seems to bear the closest resemblance to the yeast hexokinases in terms of quaternary structure. It will be of interest to determine whether the N-terminal sequences found essential for association of the yeast hexokinase subunits have been retained in the evolution of the liver glucokinase.

The molecular basis for the high K_m of glucokinase for all sugar substrates remains unknown. It would be of interest to know whether conditions that lead to dissociation of liver glucokinase into subunits can result in a greater affinity for sugars, as is the case with the yeast enzymes.

3. *Aggregation Phenomenon*

A slight tendency of the purified type I enzyme to undergo aggregation was noted by Easterby (*145*) during sedimentation equilibrium ultracentrifugation of the porcine heart enzyme. A marked aggregating effect of glucose 6-phosphate on the pure beef brain enzyme has been noted by Redkar and Kenkare (*144*). They report an increase in sedimentation constant from 5.9 S to about 8.0 S. This effect may be very significant in explaining the regulatory effect of glucose 6-phosphate. The observation may be related to the apparent aggregation reported by Craven and Basford (*136*) upon solubilization of the brain enzyme with glucose 6-phosphate, a phenomenon that was not, however, confirmed by Wilson (*137*).

4. Amino Acid Composition and Essential Groups

The amino acid composition of the rat brain enzyme is reported by Chou and Wilson (*141*) to have many analogies with that of the beef brain enzyme (*143*). The content of aromatic acids is relatively low, accounting for the low extinction coefficient at 280 nm (*141*).

There are about 12 sulfhydryl groups per mole in the beef brain enzyme (*144*), which are accessible for reaction with Ellman's reagent at varying rates. The reagent causes rapid inactivation by reacting with one or two of the relatively reactive groups. This inactivation is prevented completely by glucose 6-phosphate, partially by glucose and not at all by Mg·ATP, Mg·ADP, or P_i. Glucose 6-phosphate also has a striking effect in protecting about 8 of the 12 SH groups from reaction. This suggests a profound effect of glucose 6-phosphate on the conformation of the protein, in agreement with the aggregating effect described in the preceding section.

5. Isoelectric Point

The isoelectric points for the pure type I enzymes from rat brain (*141*) and porcine heart (*145*) have been reported to be 6 and 6.3, respectively. The values for types II, III, and IV are likely to be progressively lower on the basis of their known behavior in gel electrophoresis and column chromatography.

D. Enzyme Mechanism

1. Evidence from Kinetic Studies

Although initial rate studies with ATP as the variable substrate and glucose at different fixed levels had indicated a "ping-pong" mechanism for the brain enzyme (*148, 149*), later studies with fructose instead of

148. H. J. Fromm and V. Zewe, *JBC* **237**, 1661 (1962).
149. M. Copley and H. J. Fromm, *Biochemistry* **6**, 3503 (1967).

glucose as well as with sugar analogs have shown a "sequential" mechanism for the same enzyme (*150–152*). The physical reasons for different results with glucose and fructose are not clear, but in mathematical terms it can be shown that a "sequential" mechanism can give apparent "ping-pong" kinetics for certain values of the kinetic constants (*150–152*). In the case of rat muscle enzymes type II, this anomalous effect of glucose was not seen (*153*) so that a "sequential" mechanism was clearly established. Another rat muscle preparation, undefined as to type, behaved like the brain enzyme (type I) in giving apparent "ping-pong" kinetics with glucose (*154*).

Thus although the apparent "ping-pong" kinetics with glucose had at first led to the view that there was a fundamental difference between the mechanism of mammalian and yeast enzymes, it is now generally agreed that they both involve a "sequential" mechanism.

Furthermore, it now seems likely that the mammalian and yeast enzymes are similar with respect to the order of addition of substrates and the order of release of products.

The evidence for an ordered release of products from the ternary product complex comes from the work of Kosow and Rose (*42*) and is quite analogous to their results with the yeast enzymes, described above. They have shown for both type I and II tumor enzymes a preequilibrium exchange of added ADP into ATP when only enzyme-bound G6P was available for reaction. On the other hand, there was no evidence for preequilibrium exchange of G6P into ATP when only bound ADP could have been available for reaction. These experiments thus established unequivocally, as for the yeast enzymes, that the primary path for product release is via an E·G6P complex that dissociates slowly.

Kosow and Rose (*42*) point out that the ability of ADP to react with E·G6P provides an adequate explanation for its ability to act as a noncompetitive inhibitor versus ATP, so that it becomes unnecessary to postulate a separate regulatory binding site for ADP and other purine nucleotides to account for this type of inhibition. They point out further that this unusual noncompetitive component of ADP inhibition is characteristic not only of yeast isoenzymes P-I and P-II and mammalian enzymes I and II (*42, 117, 148*), but also of the types III (*117*) and IV (*138, 155*), and suggest that preequilibrium measurements

150. H. J. Fromm and J. Ning, *BBRC* **32**, 672 (1968).
151. J. Ning, D. L. Purich, and H. J. Fromm, *JBC* **244**, 3840 (1969).
152. D. Purich and H. J. Fromm, *JBC* **246**, 3456 (1971).
153. T. L. Hanson and H. J. Fromm, *JBC* **242**, 501 (1967).
154. T. L. Hanson and H. J. Fromm, *JBC* **240**, 4133 (1965).
155. M. J. Parry and D. G. Walker, *BJ* **99**, 266 (1966).

of ADP incorporation into ATP might help to demonstrate an evolutionary relationship among the hexokinases.

They also point out that the noncompetitive component of ADP inhibition disappears at high ADP concentrations in the case of the type IV enzyme (*138*, *155*), just as for the yeast enzymes (*42*). This would simply mean, as discussed above, that the alternative pathway of dissociation of products via E·ADP becomes prominent at high ADP concentrations, where the "normal" pathway is suppressed.

Another striking analogy between the yeast enzymes and the mammalian enzymes (types I and II) is seen in the enhancing effect of Mg on the ADP inhibition (*142*). The intercept effect also becomes more apparent at high Mg concentrations.

As to the order of addition of substrates to the enzyme, the conclusion has been drawn that this is random, partly on the basis that certain structural analogs of ATP, such as AMP, cause noncompetitive rather than uncompetitive inhibition with respect to glucose (*147*, *153*). However, as pointed out in the section on kinetics of the yeast enzymes, the fact that glucose has no effect on the binding of certain structural analogs of ATP to the enzyme does not at all rule out the possibility that glucose has a strong enhancing effect on the binding of ATP itself or of other structural analogs more closely related to ATP. It would therefore be of great interest to test the effect of the glucose-containing analog of ATP, which showed uncompetitive inhibition with yeast hexokinase (*53*), to see whether similar evidence for an ordered addition of substrates could then be shown with the brain enzyme.

It would also be extremely useful, for the purpose of determining the order of addition of substrates, to apply the flux ratio technique of Britton and Clark (*51*), which has clearly established an ordered addition, with glucose as leading substrate, for yeast hexokinase.

2. *Evidence from Studies on Enzyme–Substrate Interaction*

Evidence is accumulating that the mammalian enzymes, like the yeast enzymes, can form binary complexes with glucose much more readily than with ATP. As mentioned above, glucose can prevent inactivation of mammalian enzymes type II (*117*) and type IV (*140*) by proteolytic enzymes, and of type I by Ellman's reagent (*144*). In the latter case, Mg·ATP had no protective effect. It appears likely from these studies that, as in the case of the yeast enzymes, Mg·ATP does not bind very readily to the mammalian enzymes unless glucose is present. These results therefore support the view that substrate addition is probably ordered, with glucose addition preceding ATP addition in the reaction sequence.

Direct binding measurements on nucleotides and sugars have not yet been reported, but would be highly desirable. It would also be of interest to determine whether the pure mammalian enzymes have an intrinsic ATPase function, as the yeast enzymes do. If so, this would provide another means of assessing the role of sugars in nucleotide binding.

From the meager information available at present, it seems reasonable to conclude tentatively that there are probably no essential differences between the yeast and mammalian systems with respect to order of addition of substrates or release of products.

3. *On the Question of a Phosphoenzyme Intermediate*

The early evidence for a "ping-pong" mechanism with certain mammalian hexokinases (see above) had strongly suggested the possibility of a phosphoenzyme intermediate. However, no evidence was found to support this possibility when partial exchange reactions were studied. Neither ADP:ATP nor G6P:G exchanges were appreciable in the absence of the other substrate pair (*42, 150, 154*). Since it is now well established that the mechanism is not "ping-pong" but "sequential," there is no longer a strong kinetic argument in favor of a phosphoenzyme.

However, the possibility that a phosphoenzyme intermediate is formed during interconversion of the central ternary complexes cannot be ruled out. It would be of interest to determine whether the treatment of mammalian enzymes with ATP and D-xylose can lead to formation of an inactive phosphoenzyme, as has been found for the yeast hexokinases (*62*). If this occurred, a powerful tool for comparing the catalytic sites of the various hexokinases would be available.

E. Regulation of Enzyme Activity

The unique feature in the regulation of mammalian hexokinases of types I, II, and III is the striking inhibition by G6P (*100, 101*). This inhibition has long been thought to be due to action at a regulatory site rather than at the catalytic site (*100*). This view was among the earliest expressions of the concept of allosteric control.

However, the question of whether the control is truly exerted at an allosteric site is not yet fully resolved. Fromm and his colleagues (*151, 152*) have presented kinetic arguments that this inhibition is due to binding at the catalytic site, but they invoke a regulatory site to explain inhibition by ADP and related purine nucleotides. Kosow and Rose (*42*) take the inverse position, arguing that G6P action is at a regulatory site but that ADP inhibition is fully explainable by action at the catalytic site. A compelling nonkinetic argument in favor of a special site for regulation by G6P is that its inhibitory action is much greater than that of mannose or 2-deoxyglucose 6-phosphate, although the corresponding hexoses are equally good substrates with equally good affinities for the enzyme (*99, 100*). The special effects of G6P as a solubilizer of the enzyme (*128*) and in causing marked conformational changes as measured by sedimentation or by protection of the enzyme against Ellman's reagent (*144*) also tend to favor action at a regulatory site.

The fact that analogous inhibition is not seen with the yeast enzyme also speaks for the evolution of a special regulatory site. Since the mammalian enzyme (type I) seems to have evolved by gene duplication and fusion, perhaps one of the two catalytic sites has evolved into a regulatory site.

Regardless of the question of the site of action of G6P, its importance as a regulator of the action of the mammalian enzymes remains undeniable. It is generally accepted that metabolic control of phosphofructokinase activity serves to regulate the G6P level that in turn regulates the hexokinase activity (*156*). It is also known that the inhibition of hexokinase by G6P can be attenuated by P_i (*157–159*). Thus P_i can regulate glucose metabolism at both the hexokinase and phosphofructokinase levels.

In a recent report of Kosow and Rose (*160*), a second way to attenuate the inhibitory effect of G6P was described. It was shown that there was a distinct lag period for the onset of inhibition when G6P was added to the ascites tumor hexokinase type II. The half-time for onset of inhibition was 12 seconds for the soluble enzyme and 130 seconds for the enzyme bound to mitochondria. The discoveries of this "hysteretic" response has helped to explain the phenomenon of delayed feedback control of hexokinase activity in intact ascites tumor cells. It is known that there is a sudden burst of hexokinase activity when glucose is added to such cells, so that the G6P reaches a very high peak level of 0.5 to 1 mM in the cell within 3 seconds. However, the hexokinase activity in the cell remains high for 20 to 60 seconds before succumbing to feedback inhibition by G6P. This delayed feedback control appears to have the function of permitting the glycolytic intermediates G6P, F6P, etc., to rise rapidly to a new steady-state level before the hexokinase is turned off. That is, the delayed feedback permits a more rapid metabolic response.

156. O. H. Lowry and J. V. Passoneau, *JBC* **239**, 31 (1964).
157. H. Tiedemann and J. Born, *Z. Physiol. Chem.* **321**, 205 (1959).
158. I. A. Rose, J. V. B. Warms, and E. L. O'Connell, *BBRC* **15**, 33 (1964).
159. K. Uyeda and E. Racker, *JBC* **240**, 4682 (1965).
160. D. P. Kosow and I. A. Rose, *BBRC* **48**, 376 (1972).

In the latest report of Kosow et al. (161) on this subject, they show for the first time that the type I and type II hexokinases differ in their mode of regulation by G6P. It turns out that only the type I enzyme shows the P_i effect on sensitivity to G6P inhibition, and only the type II enzyme shows the delayed onset of inhibition by G6P. The site of action of P_i in modulating G6P inhibition is different from that for its interference with elution of hexokinase from mitochondria by G6P, since the latter effect of P_i is not type-specific.

The different types of regulation are thought to be geared to the different metabolic characteristics of the different tissues. Tumor and muscle tissue, which are rich in type II enzyme, are thought to require adjustment to sudden changes of availability of intracellular glucose, whereas brain and erythrocytes, which are rich in type I enzyme, are likely to have a constant supply of intracellular glucose and to regulate their glucose utilization through alterations in the level of intracellular P_i.

An additional important finding in this paper is that type I enzyme but not type II is stimulated by citrate (161) or catecholamines (162) under the low pH conditions that are similar to those required to observe stimulation of the yeast enzyme by these compounds (43, 92a). Since the regulatory effects of ADP and citrate in the case of yeast hexokinases were much more pronounced with form P-II than P-I (91, 92), it may be deduced that there is an evolutionary relationship between yeast hexokinase P-II and mammalian hexokinase type I. The stimulatory effect of catecholamine on the yeast enzyme was noted by Harrison and Gray (92a) to vary with the enzyme source. It seems likely that only those preparations rich in P-II gave a good response.

Kosow et al. (161) have proposed that the physiological significance of citrate stimulation is to signal that there is an adequate supply of ATP in the cell. By simultaneously inhibiting phosphofructokinase and stimulating hexokinase, the citrate would promote the utilization of G6P for both glycogen storage and fat deposition.

IV. Comparative Aspects

This survey of the yeast and mammalian hexokinases shows that although there are striking evolutionary differences, especially with re-

161. D. P. Kosow, F. A. Oski, J. V. B. Warms, and I. A. Rose, *ABB* (1973) (in press).
162. W. H. Harrison and R. M. Gray, *BBA* **237**, 391 (1971).

TABLE III
COMPARISON OF YEAST AND MAMMALIAN HEXOKINASES[a,b]

	Yeast hexokinases		Mammalian hexokinases			
Property	P-I	P-II	I	II	III	IV
MW × 10^{-3} (native)	50–100	50–100	100	100	100	50–68
MW × 10^{-3} (denatured)	50	50	100	n.d.	n.d.	n.d.
Glucose K_m (mM)	0.1	0.1	0.05	0.2	0.007	12
Fructose K_m (mM)	1	1	3	3	3	>800
V (glucose) μmoles/min/mg[c]	200	800	80	n.d.	n.d.	40
V (fructose)/V (glucose)	3.0	1.0	1.1	1.1	1.2	0.9[d]
Strong inhibition by G6P	No	No	Yes	Yes	Yes	No
Attenuation of by P_i	n.a.	n.a.	Yes	No	n.d.	n.a.
Delayed onset of	n.a.	n.a.	No	Yes	n.d.	n.a.
Marked regulatory effects at low pH (ADP inhibition, citrate activation)	No	Yes	Yes	No	n.d.	n.d.

[a] Values shown are approximate and taken from references cited in the text.

[b] Abbreviations: n.d., not determined; n.a., not applicable.

[c] Values are maximum velocities for pure enzymes.

[d] Calculated from observed value of 0.08 for 100 mM sugars.

spect to molecular weight in denaturing solvents and regulation by glucose 6-phosphate, there are also some striking homologies in specificity, structure, mechanism, and regulation that will make future comparative studies of considerable interest (see Table III). In addition to the features shown in the Table III in which certain differences in the properties of the isoenzymes are illustrated, there are some properties that all isoenzymes have in common. These include the properties of "mixed inhibition" by ADP versus ATP, strong inhibition by N-acylated derivatives of glucosamine, broad sugar specificity [with the exception that type IV fails to use glucosamine as substrate (108)], and narrow nucleotide specificity toward ATP among the natural nucleoside-triphosphates. To the extent investigated, all seem to function by the same "sequential" mechanism, which involves a highly ordered but not compulsory sequence of addition of substrates and removal of products.

This article has been largely limited to the yeast and mammalian hexokinases because these have been the best characterized to date. Similar hexokinases seem to be present in *Aspergillus oryzae* (163) and

163. M. Ruiz-Amil and A. Sols, *Biochem. Z.* **334**, 168 (1961).

in wheat germ (164). The molecular properties of the latter show striking similarities to the yeast enzymes. The bacterial hexokinases, although having a fairly broad sugar specificity, have high affinities for particular sugars and are named accordingly. The mannofructokinase of *Leuconostoc mesenteroides* (165) and the mannokinase *of E. coli* (166) both act preferentially on mannose and fructose. Others act preferentially on individual sugars, e.g., the glucokinase of *Aerobacter aerogenes* (167) or *Entamoeba histolytica* (168), and the fructokinase (169) and mannokinase (170) of *Streptomyces violaceoruber*. Enzymes with high specificity for glucose, fructose, mannose, and glucosamine, respectively, were separated from extracts of *Schistosoma mansoni* (171).

This article has not included other systems for phosphorylating sugars, e.g., in which phosphorylation of a sugar by ATP occurs in the 1-position rather than the 6-position (liver fructokinase, galactokinase) or in which the phosphate donor is not ATP. The latter systems include those in which phosphoenolpyruvate (172) and acetyl phosphate (173), phosphoramidate (174), and pyrophosphate or carbamyl phosphate (175, 176) serve as donors.

The results with the glucokinase of *Entamoeba histolytica* are of special interest here because of the analogy with results described above for the yeast system. Two isoenzymes were separated by electrophoresis. With the slower-migrating form, there was a strong stimulation by phosphate at pH values of 6.5 to 7.4, while at pH values around 8, which are optimal for activity in imidazole or tris buffer, there was little stimulation by phosphate. The faster-migrating isoenzyme showed similar but much less pronounced effects. The analogy with the behavior of the yeast isoenzymes P-I and P-II is striking. The *E. histolytica* enzymes (prior to separation) also showed unusually strong inhibition by ADP in measurements at pH 7.

164. J. C. Meunier, J. Buc, and J. Ricard, *FEBS Lett.* **14,** 25 (1971).
165. V. Sapico and R. L. Anderson, *JBC* **242,** 5086 (1967).
166. J. Sebastian and C. Asensio, *ABB* **151,** 227 (1972).
167. M. Y. Kamel, D. P. Allison, and R. L. Anderson, *JBC* **241,** 690 (1966).
168. R. E. Reeves, F. Montalvo, and A. Sillero, *Biochemistry* **6,** 1752 (1967).
169. B. Sabater, J. Sebastian, and C. Asensio, *BBA* **284,** 414 (1972).
170. B. Sabater, J. Sebastian, and C. Asensio, *BBA,* **284,** 406 (1972).
171. E. Bueding and J. A. MacKinnon, *JBC* **215,** 495 (1955).
172. W. Kundig and S. Roseman, *JBC* **246,** 1407 (1971).
173. M. Y. Kamel and R. L. Anderson, *ABB* **120,** 322 (1967).
174. R. A. Smith and M. C. Theisen, "Methods in Enzymology" Vol. 9, p. 403, 1966.
175. J. L. Herrman and R. C. Nordlie, *ABB* **152,** 180 (1972).
176. J. D. Lueck, J. C. Herrman, and R. C. Nordlie, *Biochemistry* **11,** 2792 (1972).

2
Nucleoside and Nucleotide Kinases

ELIZABETH P. ANDERSON

I. Introduction 49
II. Nucleoside Kinases 51
    A. Adenosine Kinase 51
    B. Inosine-Guanosine Kinase 54
    C. Uridine-Cytidine Kinase 56
    D. Pseudouridine Kinase 62
    E. Deoxycytidine-Deoxyadenosine-Deoxyguanosine Kinases . 62
    F. Deoxythymidine Kinase 69
    G. Riboflavin Kinase 74
    H. Nicotinamide-Adenine Dinucleotide Kinase 76
III. Nucleotide Kinases 82
    A. GMP-dGMP Kinase 82
    B. dAMP Kinase 86
    C. dCMP-CMP-UMP Kinase(s) 87
    D. dTMP Kinase 91
    E. Other Deoxynucleotide Kinases 94

I. Introduction

The scope of this chapter includes kinases that phosphorylate nucleosides to nucleoside monophosphates and kinases that phosphorylate these nucleoside monophosphates to the nucleoside diphosphates. Adenylate kinase is, however, covered by Noda in Chapter 8 of Volume VIII and is therefore omitted here. Also, a separate chapter is devoted to nucleoside diphosphokinase (Parks and Agarwal, Chapter 9, Volume VIII).

Nucleoside kinases catalyze a phosphoryl transfer from a nucleoside triphosphate to an R–OH acceptor, which is typically the 5′-hydroxyl group in the sugar moiety of the nucleoside. The next section of this chapter considers these enzymes. Nicotinamide-adenine dinucleotide kinase likewise catalyzes phosphoryl transfer to an R–OH group, the 2′-hydroxyl of a ribose, although the molecule phosphorylated is a dinucleotide rather than a nucleoside. In accord with traditional enzyme classification (1, 2), this enzyme is discussed next, following the nucleoside kinases, since the type of phosphoryl transfer is analogous. Nucleoside monophosphokinases, on the other hand, transfer phosphate from a nucleoside triphosphate to an acceptor

$$R-O-\overset{\overset{O}{\|}}{\underset{\underset{OH}{|}}{P}}-OH$$

with the formation of a new phosphoanhydride bond. These enzymes are covered in the final section of the chapter.

The enzymes within each group have been subdivided in the usual manner (1) on the basis of present information about their specificity with regard to the phosphoryl acceptor. It will become clear from the discussion that these specificities are not always clear-cut, and that they vary for enzymes from different sources, and even sometimes with the experimental conditions under which substrate specificity is examined. The present organization is therefore somewhat arbitrary and may have to be revised as more information is acquired about these enzymes. In general, however, the substrate specificity of the nucleoside kinases appears to depend on the the sugar moiety of the nucleoside, as well as on its purine, as distinct from pyrimidine, base. In contrast, the substrate specificity of the nucleoside monophosphokinases, although it depends on the base constituent, is apparently not so strict with regard to the carbohydrate moiety.

Both nucleoside kinases (3) and nucleoside monophosphokinases (4) were covered by chapters in the previous edition of "The Enzymes." The difference between the relatively small number of these enzymes on which information was then available, and the number that can now reasonably be discussed here, bears witness to the vast amount of literature that has appeared in the intervening years on the purification and characterization of these enzymes. Apologies are offered for the omissions in this coverage

1. "Enzyme Nomenclature," Recommendations of the Commission of Biochemical Nomenclature of the International Union of Pure and Applied Chemistry and the International Union of Biochemistry on the Nomenclature and Classification of Enzymes. Elsevier, Amsterdam, 1973 (in press).
2. R. Nordlie and H. Lardy, "The Enzymes," 2nd ed., Vol. 6, p. 3, 1962.
3. R. Caputto, "The Enzymes," 2nd ed., Vol. 6, p. 133, 1962.
4. L. Noda, "The Enzymes," 2nd ed., Vol. 6, p. 139, 1962.

that were necessary in the face of this quantity of material. The reader is also referred to other recent and more general reviews (5, 6).

II. Nucleoside Kinases

A. ADENOSINE KINASE

1. *Distribution and Purification*

Adenosine kinase (ATP:adenosine 5′-phosphotransferase, EC 2.7.1.20) catalyzes the phosphorylation of adenosine to AMP according to the following reaction:

$$\text{Adenosine} + \text{NTP} \rightarrow \text{AMP} + \text{NDP}$$

Early work on this enzyme was discussed in the previous edition of "The Enzymes" (3). The enzyme is widely distributed. It was first discovered in yeast (7, 8) and in mammalian tissues (7) and partially purified from these sources (3, 7, 8). It has since been further purified from a variety of mammalian tumor cells (9–12), from rabbit liver (10, 11), and most recently from brewer's yeast (13). Successful purification procedures have included Sephadex gel filtration (9, 10, 13), chromatography on DEAE-cellulose (9, 12, 14), DEAE-Sephadex (13) or hydroxyapatite (9, 13), and fractionation at pH 5 (10) and with ammonium sulfate (13).

2. *Assay*

The enzyme has been assayed in several ways. Estimation of the conversion of radioactive substrate into product has the advantage of specificity in the presence of other enzyme reactions and has been pre-

5. S. J. Benkovic and K. J. Schray, "The Enzymes," 3rd ed., Vol. 8, p. 201, 1973.
6. J. F. Morrison and E. Heyde, *Annu. Rev. Biochem.* **41**, 29 (1972).
7. A. Kornberg and W. E. Pricer, Jr., *JBC* **193**, 481 (1959).
8. R. Caputto, *JBC* **189**, 801 (1951).
9. H. P. Schnebli, D. L. Hill, and L. L. Bennett, Jr., *JBC* **242**, 997 (1967).
10. B. Lindberg, H. Klenow, and K. Hansen, *JBC* **242**, 350 (1967).
11. B. Lindberg, *BBA* **185**, 245 (1969).
12. A. Y. Divekar and M. T. Hakala, *Mol. Pharmacol.* **7**, 663 (1971).
13. T. K. Leibach, G. I. Spiess, T. J. Neudecker, G. J. Peschke, G. Puchwein, and G. R. Hartmann, *Hoppe-Seyler's Z. Physiol. Chem.* **352**, 328 (1971).
14. In addition to standard abbreviations accepted for the *Journal of Biological Chemistry,* the following abbreviations have been used in this chapter: in nomenclature and equations, NTP and NDP for general nucleoside triphosphate and nucleoside diphosphate, respectively; and PEI for polyethyleneimino.

ferred by various investigators, especially in early stages of purification (9, 10, 12). On the other hand, the reaction can be continuously monitored spectrophotometrically by (a) coupling ADP formation to the pyruvate kinase-lactate dehydrogenase system and following NADH oxidation (7, 10, 11), or (b) phosphorylating adenosine all the way to the triphosphate in the presence of myokinase and an ATP-regenerating system and assaying the ATP spectrophotometrically with glucose, hexokinase, glucose-6-phosphate dehydrogenase, and NADP$^+$ (14a) or assaying the ADP formed as in method (a) (7, 10). The enzyme reaction can also be assayed by following the disappearance of adenosine, either by precipitating the phosphorylated products with barium hydroxide-zinc sulfate and spectrophotometrically determining the adenosine remaining (3, 8) or by exposing the products of the reaction to adenosine deaminase to estimate the nucleoside (7).

3. Kinetic and Molecular Properties

The properties of partially purified adenosine kinase preparations from various sources are quite similar. Apparent K_m values for adenosine are reported to be in the range of 0.5–$5.8 \times 10^{-6}\,M$ for several different preparations (9, 10, 12, 15, 16), while apparent K_m values of 2–$4 \times 10^{-4}\,M$ have been recorded for ATP (12, 15, 16). Some partially purified preparations of the animal enzyme have exhibited a pH optimum near 5 (8, 12), but most of the more highly purified preparations have shown a higher pH optimum in the range of 6–7 (9, 10). The yeast enzyme has been reported to have a pH optimum near 7 (8) and to be unstable below pH 5.5 (13). The enzyme requires a divalent cation, usually Mg^{2+} or sometimes Mn^{2+} (9, 10, 12, 13), and there is some evidence that Mg^{2+} can be bound to the enzyme (9, 15). This may be relevant to the high requirement for ATP relative to Mg^{2+} in some preparations (12).

Purification procedures for adenosine kinase have been hampered by the general instability of the enzyme. The animal enzyme was stabilized by 20% glycerol (10), and the yeast enzyme was strongly stabilized during purification by adenosine (5 mM), especially in the presence of Mg^{2+}, and by dithiothreitol (10 mM, 13). Thiols also protected this enzyme against heat or urea denaturation; however, thiols could reversibly inhibit to some extent. This purified yeast enzyme exhibited one binding site for p-mercuribenzoate, and the mercurial destroyed

14a. A. Kornberg, "Methods in Enzymology," Vol. 2, p. 497, 1955.
15. A. W. Murray, *BJ* **106**, 549 (1968).
16. F. L. Meyskens and H. E. Williams, *BBA* **240**, 170 (1971).

enzymic activity; preparations inactivated by removal of adenosine no longer bound p-mercuribenzoate (13). These data strongly suggest the presence of one thiol group essential for activity.

Adenosine kinase appears to be of low molecular weight, in the range of 25,000–40,000 (10, 13); the most highly purified preparation reported to date, purified to homogeneity as judged by disc gel electrophoresis (13), behaved like a single polypeptide chain. In the absence of adenosine, the purified enzyme aggregated to an inactive enzyme of much higher molecular weight which could be dissociated again on sodium dodecyl sulfate gels (13). The inactive aggregate form could be partially (75%) reactivated by treatment with 8 M urea in the presence of dithiothreitol and adenosine, followed by dialysis. During this reactivation the aggregate forms of the enzyme were converted to the monomer, based on behavior in disc gel electrophoresis (13).

4. Substrate Specificity

Adenosine kinase has a fairly broad specificity for the naturally occurring nucleoside triphosphates. In addition to ATP, dATP could serve as phosphate donor with various preparations of the enzyme (9, 10, 13); GTP, ITP, dGTP, and UTP have also been found to donate phosphate in this reaction (9, 13). The enzyme is somewhat more restricted with regard to the nucleoside substrate; it phosphorylated guanosine or inosine poorly if at all (9, 10, 12, 13) and did not phosphorylate pyrimidine ribo- or deoxyribonucleosides (11–13, 17). 2′-Deoxyadenosine was weakly phosphorylated, even by highly purified preparations (10, 13), but deoxyinosine was not (13). However, a number of adenosine analogs were found to be well phosphorylated by the enzyme, some of them better than adenosine itself (9, 10, 12).

The specificity of the enzyme allowed a wide variety of substitutions at the 6 position of the purine ring, a methyl or =O substituent at the N-1 of the ring, modification of the imidazole ring of the purine to a pyrazole, pyrrole, or triazole ring and linkage of the pentose at positions other than N-2 (9–12). The enzyme showed greater specificity with regard to the 2 position of the ring, but it did phosphorylate 2-fluoroadenosine, and to a lesser extent 2-amino- and 2-hydrazinoadenosine (9, 10). Most analogs in which the ribofuranosyl moiety was replaced by arabinofuranosyl, xylofuranosyl, or deoxyribofuranosyl were poorly phosphorylated (9). However, 3′-deoxyadenosine (cordycepin) and some of its derivatives were phosphorylated by partially purified enzyme from

17. H. L. A. Tarr, Can. J. Biochem. **42**, 1535 (1964).

mammalian sources (*10, 11, 18–20*). Among the compounds which were phosphorylated better than adenosine were purine riboside, 6-methylpurine riboside, 6-methylthiopurine riboside, N^6-methyladenosine, adenoside-N^1-oxide, and 7-deazaadenosine (tubercidin) and its derivative toyocamycin (*9, 10, 12*).

Analogs of particular interest which were not phosphorylated by the mammalian enzyme included 6-mercaptopurine riboside (*9*) and 4-amino-5-imidazolecarboxamide riboside (*9, 12*), although the latter compound is apparently phosphorylated by a kinase present in yeast (*21*). This broad specificity of the enzyme with regard to nucleoside analogs has attracted attention because there is considerable correlation between cytotoxic or antitumor activity of these compounds and their capacity to undergo phosphorylation by cellular enzymes, presumably to more active derivatives (*9–12, 18–27*). This is also supported by the findings that cells resistant to certain adenosine analogs apparently lacked adenosine kinase activity and could not phosphorylate the analogs (*28, 29*).

B. Inosine-Guanosine Kinase

The picture on the distribution of inosine-guanosine kinase is considerably less clear; apparently this enzymic activity is distributed much less widely than is adenosine kinase or is much less active in most tissues (cf. *30*), and a highly purified preparation catalyzing these phosphorylations has not been reported. The phosphorylation of inosine and/or

18. H. Klenow, *BBA* **76**, 347 (1963).
19. J. T. Truman and H. Klenow, *Mol. Pharmacol.* **4**, 77 (1968).
20. H. T. Shigeura, G. E. Boxer, M. L. Meloni, and S. D. Sampson, *Biochemistry* **5**, 994 (1966).
21. H. T. Huang, *Biochemistry* **4**, 58 (1965).
22. G. Acs, E. Reich, and M. Mori, *Proc. Nat. Acad. Sci. U. S.* **52**, 493 (1964).
23. H. T. Shigeura, G. E. Boxer, S. D. Sampson, and M. L. Meloni, *ABB* **111**, 713 (1965).
24. I. C. Caldwell, J. F. Henderson, and A. R. P. Paterson, *Can. J. Biochem.* **44**, 229 (1966).
25. S. Frederiksen, A. Ø. Jørgensen, A. H. Rasmussen, and T. Tønnesen, *Mol. Pharmacol.* **4**, 358 (1968).
26. A. Tavitian, S. C. Uretsky, and G. Acs, *BBA* **157**, 33 (1968).
27. I. C. Caldwell, J. F. Henderson, and A. R. P. Paterson, *Can. J. Biochem.* **47**, 901 (1969).
28. L. L. Bennett, Jr., H. P. Schnebli, M. H. Vail, P. W. Allan, and J. A. Montgomery, *Mol. Pharmacol.* **2**, 432 (1966).
29. D. H. W. Ho, J. K. Luce, and E. Frei, *Biochem. Pharmacol.* **17**, 1025 (1968).
30. M. F. Utter, "The Enzymes," 2nd ed., Vol. 2, Part A, p. 75, 1960.

guanosine has been recorded in *Salmonella typhimurium* (*31, 32*), *Escherichia coli* (*33, 34*), *Streptococcus faecalis* (*35*), immature salmon milts (*17*), and chick fibroblasts (*36–38*). With regard to mammalian tissues, inosine-guanosine kinase activity has also been reported in human erythrocytes (*39, 40*) and leukocytes (*41*) and in mammalian tumor cells (*42–44*), but other workers have looked for such activity in similar mammalian cells and been unable to detect it (*45, 46*).

Since all of these studies have been with crude systems, one problem is that adenosine kinase may have some weak activity with at least inosine. A more difficult problem is that the nucleosides are readily broken down to the free bases, which can then react with phosphoribosyl pyrophosphate to form the nucleotides by this route rather than by direct phosphorylation. When the nucleoside cleavage is phosphorolytic, it also produces ribose 1-phosphate, which can be further converted through ribose 5-phosphate to phosphoribosyl pyrophosphate, thus favoring this pathway. This mechanism therefore has to be ruled out before incorporation of labeled nucleoside into nucleotide can be attributed to kinase activity, and it is difficult to do this rigorously. Investigators have tried to do this by using mutant cell systems that supposedly lack all IMP-GMP pyrophosphorylase (*31, 35, 41–44, 46*) or the nucleoside phosphorylase (*32*), or they have inferred that this pathway was not the one responsible for the results because the conversion of the nucleoside to the nucleotide was independent of phosphate and K^+ and was not inhibited by arsenate (the reverse of the characteristics of the phosphorylase-pyrophosphorylase pathway) (*40*). However, with crude preparations, the

31. E. F. Zimmerman and B. Magasanik, *JBC* **239**, 293 (1964).
32. J. Hoffmeyer and J. Neuhard, *J. Bacteriol.* **106**, 14 (1971).
33. A. J. Tomisek, A. P. V. Hoskins, and M. R. Reid, *Cancer Res.* **25**, 1925 (1965).
34. P. W. Allan and L. L. Bennett, Jr., *Fed. Proc., Fed. Amer. Soc. Exp. Biol.* **30**, 1169 (1971).
35. R. W. Brockman, *Cancer Res.* **23**, 1191 (1963).
36. C. Scholtissek, *BBA* **155**, 14 (1968).
37. C. Scholtissek, *BBA* **158**, 435 (1968).
38. J. Rau and C. Scholtissek, *Z. Naturforsch. B* **25**, 292 (1970).
39. H. Banaschak, *Acta Biol. Med. Ger.* **17**, 261 (1966).
40. H. Banaschak, *Acta Biol. Med. Ger.* **18**, 581 (1967).
41. M. R. Payne, J. Dancis, P. H. Berman, and M. E. Balis, *Exp. Cell Res.* **59**, 489 (1970).
42. G. A. LePage, I. G. Junga, and B. Bowman, *Cancer Res.* **24**, 835 (1964).
43. K. J. Pierre and G. A. LePage, *Proc. Soc. Exp. Biol. Med.* **127**, 432 (1968).
44. R. W. Brockman and S. Chumley, *BBA* **95**, 365 (1965).
45. A. W. Meikle, A. M. Gotto, and O. Touster, *BBA* **138**, 445 (1967).
46. T. Friedmann, J. E. Seegmiller, and J. H. Subak-Sharpe, *Exp. Cell Res.* **56**, 425 (1969).

question remains an open one, and the answer will have to await purification of this enzymic activity from these various sources, if it indeed exists there. In any case, it appears at this point that such an activity does not play a significant physiological role in mammalian cells, although it may do so in certain bacterial or other cells.

C. URIDINE-CYTIDINE KINASE

1. Distribution and Purification

Uridine kinase (ATP:uridine 5'-phosphotransferase, EC 2.7.1.48) was included among the nucleoside kinases reviewed in the previous edition of "The Enzymes" (3). It catalyzes the reaction:

$$\genfrac{}{}{0pt}{}{\text{Uridine}}{\text{Cytidine}} + \text{NTP} \rightarrow \genfrac{}{}{0pt}{}{\text{UMP}}{\text{CMP}} + \text{NDP}$$

This enzyme was first observed in mammalian liver (47) and in Ehrlich ascites tumor (48); for mammalian cells, the activity is especially well developed in cells exhibiting a high growth rate (49–51). Uridine kinase has now been reported in a variety of cells including *Streptococcus faecalis*, *Staphylococcus aureus* and *Escherichia coli* (52), *Salmonella typhimurium* (53, 54), *Saccharomyces cerevisiae* (55), *Tetrahymena pyriformis* (56, 57), *Chlorella pyrenoidosa* (58), sea urchin embryos (59), cultured chick (36, 37) and mouse (60) fibroblasts, human lymphocytes (61, 62), and a variety of mammalian tumors (51, 63–68). Wherever tested, this enzymic activity was also found to phosphorylate cytidine.

47. E. S. Canellakis, *JBC* **227**, 329 (1957).
48. P. Reichard and O. Sköld, *Acta Chem. Scand.* **11**, 17 (1957).
49. P. Reichard and O. Sköld, *BBA* **28**, 376 (1958).
50. O. Sköld, *BBA* **44**, 1 (1960).
51. P. G. W. Plagemann, G. A. Ward, B. W. J. Mahy, and M. Korbecki, *J. Cell. Physiol.* **73**, 233 (1969).
52. E. P. Anderson and R. W. Brockman, *BBA* **91**, 380 (1964).
53. C. F. Beck, J. L. Ingraham, J. Neuhard, and E. Thomassen, *J. Bacteriol.* **110**, 219 (1972).
54. G. A. O'Donovan and J. Neuhard, *Bacteriol. Rev.* **34**, 278 (1970).
55. M. Grenson, *Eur. J. Biochem.* **11**, 249 (1969).
56. R. L. Heinrikson and E. Goldwasser, *JBC* **239**, 1177 (1964).
57. W. Plunkett and J. G. Moner, *BBA* **250**, 92 (1971).
58. F. Wanka and C. L. M. Poels, *Eur. J. Biochem.* **9**, 478 (1969).
59. A. Orengo, *Exp. Cell Res.* **41**, 338 (1966).
60. S. Kit, Y. Valladares, and D. R. Dubbs, *Exp. Cell Res.* **34**, 257 (1964).
61. Z. J. Lucas, *Science* **156**, 1237 (1967).
62. P. Hausen and H. Stein, *Eur. J. Biochem.* **4**, 401 (1968).

The enzyme was first purified from Ehrlich ascites tumors cells (*69, 70*), and has since been purified from several other mammalian tumors (*68, 71–76*), from sea urchin embryos (*59*), and from *T. pyriformis* (*57*). The purification procedures independently developed for uridine kinase from different sources have been similar in many ways; generally successful procedures have included fractionation with ammonium sulfate (*57, 59, 68–76*), DEAE-cellulose chromatography (*57, 69, 71, 75*), and Sephadex or Sepharose gel filtration (*57, 68, 71–74, 76*). The enzyme has also been adsorbed onto calcium phosphate gel (*68*) and hydroxyapatite (*75*) and purified by zone electrophoresis on agarose gel (*69, 70*).

2. *Assay*

Like other nucleoside kinases, the enzyme has usually been assayed in one of two ways:

1. In less pure preparations a specific assay has often been based on the conversion of radioactive substrate to product, with chromatographic isolation of the products formed (*57, 59, 68–76*); a particularly satisfactory separation procedure involves thin-layer chromatography on PEI-cellulose (*53, 75*). Radioactive 6-azauridine (*73, 74*) or 5-fluorouridine (*67*) has sometimes been used instead of uridine or cytidine as assay substrate. Azauridine was found preferable to uridine in crude preparations because azaUMP is apparently not phosphorylated to higher derivatives, and azauridine and azaUMP are also less extensively degraded than the normal compounds (*65, 77*). Other investigators have found advantages in the use of GTP instead of ATP as phosphate donor (*53*).

---

63. C. A. Pasternak, G. A. Fischer, and R. E. Handschumacher, *Cancer Res.* **21**, 110 (1961).
64. E. P. Anderson, *Cancer Res.* **23**, 1270 (1963).
65. J. Veselý, A. Čihák, and F. Šorm, *Int. J. Cancer* **2**, 639 (1967).
66. J. Veselý, A. Čihák, and F. Šorm, *Cancer Res.* **30**, 2180 (1970).
67. D. Kessel, R. Bruns, and T. C. Hall, *Mol. Pharmacol.* **7**, 117 (1971).
68. A. Orengo, *JBC* **244**, 2204 (1969).
69. O. Sköld, *JBC* **235**, 3273 (1960).
70. P. Reichard and O. Sköld, "Methods in Enzymology," Vol. 6, p. 194, 1963.
71. J. E. Ciardi, Ph.D. Thesis, George Washington University, Washington, D. C., 1968.
72. G. Krystal and T. E. Webb, *BJ* **124**, 943 (1971).
73. J. Veselý, M. Dvořák, A. Čihák, and F. Šorm, *Int. J. Cancer* **8**, 310 (1971).
74. J. Veselý, A. Čihák, and F. Šorm, *Eur. J. Biochem.* **22**, 551 (1971).
75. A. S. Liacouras and E. P. Anderson, *Fed. Proc., Fed. Amer. Soc. Exp. Biol.* **32**, 554 (1973).
76. G. Krystal and P. G. Scholefield, *Can. J. Biochem.* **51**, 379 (1973).
77. J. Kara, F. Šorm, and A. Winkler, *Neoplasma* **10**, 3 (1963).

2. With more highly purified preparations, it has been possible to assay the enzymic activity spectrophotometrically by coupling ADP formation to the pyruvate kinase-lactate dehydrogenase system and monitoring, at 340 nm, the oxidation of NADH (*69, 70, 75, 78*). Since the reaction can then be followed continuously, this method has been especially useful for initial velocity studies (*75, 78*).

3. Substrate Specificity

Uridine-cytidine kinase resembles other nucleoside kinases in exhibiting a broad specificity with regard to phosphate donor. With preparations from various sources, ATP, GTP, ITP, dATP, dGTP, dUTP, dCTP, and dTTP have been observed to donate phosphate in this reaction (*57, 68–71, 75, 76*); dGTP was not active with the enzyme from *Tetrahymena* (*57*). However, with regard to the phosphoryl acceptor, only uridine and cytidine, among the normal, physiological nucleosides, appear to be phosphorylated by this enzyme. Specificity of a single nucleoside kinase for uridine and cytidine is not limited to mammalian species and was also evident in bacteria (*53*), yeast (*55*), *Tetrahymena* (*57*), sea urchin (*59*), and chick fibroblasts (*36, 37*). The activity for uridine has generally been higher than that for cytidine (*59, 69, 71, 75, 76, 79*). The ratio of activities for these two substrates was essentially constant during different purifications (*57, 68–70, 75, 79*), and the two substrates were competitive with each other (*57, 68, 71*); in the preparations from the P815 mast cell tumor, uridine inhibited cytidine phosphorylation only mildly, while cytidine was a potent inhibitor of uridine phosphorylation (*71*).

A number of pyrimidine nucleoside analogs were also phosphorylated by this enzyme to the respective nucleoside monophosphates; these included 5-fluorouridine, 5-fluorocytidine, 5-azacytidine, and 6-azauridine (*69–71, 75*). The enzyme is, however, apparently specific for the ribofuranosyl moiety, since deoxyribonucleosides and xylofuranosyl nucleosides tested were not phosphorylated (*57, 69*). The 5-methyl derivatives of uridine and cytidine were also inactive as substrates (*57, 69*). Activity of the enzyme in phosphorylating the analogs is of interest because of the relation of this anabolism to the cytotoxic and antitumor properties of these compounds (see Section II,A; see also *53, 80–83*). Thus, a decrease

78. A. S. Liacouras, T. Q. Garvey, III, F. K. Millar, and E. P. Anderson, unpublished experiments.
79. J. M. Glick and E. P. Anderson, unpublished experiments.
80. E. P. Anderson and L. W. Law, *Annu. Rev. Biochem.* **29**, 577 (1960).
81. R. W. Brockman and E. P. Anderson, *Annu. Rev. Biochem.* **32**, 463 (1963).
82. R. W. Brockman and E. P. Anderson, *in* "Metabolic Inhibitors" (R. M.

in uridine kinase activity has been observed to be associated with resistance of tumor cells to azauridine (*63*), fluorouracil and fluorouridine (*64, 67, 71, 79, 84, 85*), and 5-azacytidine (*65, 66, 73, 74, 86*); when examined, cytidine and azacytidine phosphorylation was found to be likewise decreased (*65, 79, 86*).

Resistance of tumor sublines to anticancer agents has generally proven to be a stable, heritable characteristic and has been considered to be a mutational phenomenon (*80–83*). Patterns of resistance are therefore of interest as genetic evidence with regard to the substrate specificity of the enzyme altered or deleted. For example, mouse leukemia cells resistant to azacytidine exhibited decreased uridine-cytidine kinase activity, while cells resistant to 5-aza-2'-deoxycytidine showed low deoxycytidine kinase activity while uridine-cytidine kinase remained high (*66, 86*). In yeast and bacteria, resistance to fluorouridine has been used as a genetic marker to search for mutants deficient in uridine kinase (*55, 87*). Definitive genetic evidence for the substrate specificity of uridine-cytidine kinase has come from experiments with *S. typhimurium*, which have shown that both uridine and cytidine are phosphorylated by a single enzyme protein encoded by a single gene (*53, 87*). A mutant of *S. cerevisiae* deficient in uridine kinase has apparently not yet been tested for metabolism of cytidine (*55*).

4. Kinetic and Molecular Properties

Uridine kinase has a requirement for Mg^{2+} which was partially fulfilled by Mn^{2+} (*57, 59, 69–71, 75*), Fe^{2+} (*57, 69, 70, 75*) or Co^{2+} (*75*), and to a much smaller extent by Ni^{2+} (*75*) or Ca^{2+} (*59, 76*). Various pH optima, mostly between 6.5 and 8.5, have been reported for different preparations of the enzyme (*57, 68–71, 75, 76*). Apparent K_m values of the enzyme have been reported to be in the range of $10^{-5} M$ for the nucleoside substrates and 10^{-3} to $10^{-4} M$ for ATP (*57, 69, 70, 76*). In a study of the initial velocity pattern with the purified enzyme from P815 ascites tumor cells (*75*), K_m values based on replots of the intercepts were $1.5 \times 10^{-4} M$ for uridine, $4.5 \times 10^{-5} M$ for cytidine, and $3.9 \times 10^{-3} M$ and $2.8 \times 10^{-3} M$ for ATP with uridine and cytidine, respectively (*78*).

Hochster and J. H. Quastel, eds.), Vol. 1, p. 239. Academic Press, New York, 1963.
   83. R. W. Brockman, *Advan. Cancer Res.* **7**, 129 (1963).
   84. P. Reichard, O. Sköld, and G. Klein, *Nature (London)* **183**, 939 (1959).
   85. P. Reichard, O. Sköld, and G. Klein, L. Révész, and P.-H. Magnusson, *Cancer Res.* **22**, 235 (1962).
   86. J. Veselý, A. Čihák, and F. Šorm, *Cancer Res.* **28**, 1995 (1968).
   87. C. F. Beck, J. L. Ingraham, and J. Neuhard, *Mol. Gen. Genet.* **115**, 208 (1972).

The molecular weight of uridine-cytidine kinase has generally been observed to be quite high, in the region of 150,000–200,000, as judged by gel filtration (*57, 71, 75*). However, uridine kinase from rat liver (*72*) or Ehrlich ascites (*76*) has been reported to exist as two species of enzyme protein, one of molecular weight around 120,000 and the other around 30,000. The two species apparently did not differ in kinetic properties (*76*), but the high molecular weight species appeared to be more labile (*72, 76*); there was some evidence of possible interconversion of the two (*76*). The data suggested that the high molecular weight species predominated in adult rat liver and increased in regenerating liver, while the small molecular weight species was low in these tissues but was predominant in fetal rat liver and in rapidly growing ascites hepatoma (*72*).

Some evidence has been obtained for sulfhydryl involvement in the activity of the enzyme; the purified preparation from Ehrlich ascites cells was inhibited by p-mercuribenzoate and this was reversed by glutathione (*69*), and activity lost in aged preparations of the *Tetrahymena* enzyme could be restored by dithiothreitol or mercaptoethanol (*57*). Uridine-cytidine kinase purified from the azacytidine-resistant subline of mouse leukemia was observed to be less sensitive to p-mercuribenzoate inhibition than was the kinase purified in a parallel fashion from the parent sensitive line (*74*). This "mutant" enzyme was also more stable than the sensitive-line enzyme to inactivation by mild heat. The kinetic properties of the two enzymes appeared to be the same except that the reaction velocity with the resistant-line enzyme was only half that with the sensitive-line preparation. Behavior on Sephadex and on polyacrylamide gel was also the same for both enzymes, and neither of these characteristics was altered by either p-mercuribenzoate or the mild heat (*74*). In contrast to this, uridine kinase purified from fluorouracil-resistant sublines of Ehrlich ascites tumor appeared to be more labile to heat, as compared to the enzyme from the sensitive line. In this case, the resistant-line enzyme(s) also fractionated somewhat differently than did that from the parent line with ammonium sulfate and in DEAE chromatography (*88*).

5. Reaction Mechanism

With purified preparations of the enzyme from Novikoff hepatoma or Ehrlich ascites cells, the uridine kinase reaction has been reported to proceed by a ping-pong mechanism, based on velocity studies with dis-

88. O. Sköld, *BBA* **76**, 160 (1963).

continuous time-point assays that measured conversion of radioactive substrate to product (*68, 76*). However, with the purified enzyme from the P815 mast cell tumor, studies in this laboratory of the initial velocity of the reaction, using the continuous spectrophotometric assay, have indicated that with either uridine or cytidine as substrate the initial velocity pattern was one of intersecting rather than parallel lines, indicating that the reaction mechanism was not ping-pong but involved instead sequential addition of both substrates to the enzyme to form a ternary complex (*75, 78*). The reasons for the discrepancy are not, at this point, clear. A sequential mechanism would be in agreement with results for many other kinases (*6*).

6. Regulatory Properties

Various lines of evidence have indicated that the uridine kinase reaction is the rate-limiting step in the anabolism of the nucleoside to its triphosphate (*52, 59–62*). Feedback regulation of this enzyme was reported in early studies by Anderson and Brockman (*52*); the phosphorylation of either uridine or cytidine was strikingly inhibited by CTP and less strongly by UTP, the nucleoside triphosphate end products of the anabolic pathway. This regulatory inhibition was evident in preparations from two mammalian tumors and from several species of bacteria (*52*). The observation has since been confirmed for uridine-cytidine kinase preparations from several other sources, with both crude and purified preparations of the enzyme (*54, 57, 59, 68, 74, 76*), and it seems evident that this represents a mechanism for regulation of this salvage pathway in a variety of cells.

Both 6-aza-UTP and 5-fluoro-UTP mimicked the normal nucleoside triphosphates in exerting inhibition (*52*). The initial studies with crude preparations showed that the end product inhibition could be reversed by either substrate and that both ATP and GTP were effective in this reversal (*52*). With a partially purified kinase from sea urchin embryos, dGTP as well as GTP reversed the inhibition (*59*). These observations are of interest with regard to the particular effectiveness of GTP as a phosphate donor in some preparations of the enzyme (*53*). Subsequent studies with purified kinase from various sources have indicated that the inhibition by CTP is competitive with the phosphate donor and noncompetitive with the nucleoside substrate (*57, 68, 76*). Values of K_i reported for CTP and UTP have varied from $10^{-4}\,M$ to $10^{-6}\,M$ for preparations of the enzyme from different sources (*57, 74*). It has been reported that the sensitivity to inhibition by CTP could be enhanced by p-mercuribenzoate (*74*) and that in some cases the sensitivity disappeared upon aging (*68*). Under such conditions, CTP had some activity as a phosphate donor (*68*).

D. PSEUDOURIDINE KINASE

A kinase phosphorylating pseudouridine to pseudoUMP with ATP has been reported in extracts from strains of *Escherichia coli* that grow on pseudouridine as a sole pyrimidine source (*89*). This activity was not detected in preparations from a pyrimidine auxotroph that does not grow on pseudouridine. The activity appeared to be distinct from uridine kinase or deoxythymidine kinase activities. An ammonium sulfate fraction containing the pseudouridine kinase activity exhibited kinetics indicating apparent K_m values for pseudouridine of $1.8 \times 10^{-4} M$ and for ATP of $2.9 \times 10^{-3} M$. The reaction also required K^+ or NH_4^+ ion; Li^+ or Na^+ could not substitute.

E. DEOXYCYTIDINE-DEOXYADENOSINE-DEOXYGUANOSINE KINASES

Enzymes phosphorylating these three nucleosides will be considered together since there is overlap among them in the specificity for the nucleoside substrate and some discrepancies between the results obtained with different preparations of these kinases.

1. *Deoxycytidine Kinase*

   a. *Distribution, Purification, and Assay.* Deoxycytidine kinase, classified as NTP:deoxycytidine 5'-phosphotransferase (EC 2.7.1.74), catalyzes the phosphorylation of deoxycytidine to dCMP, using a variety of nucleoside triphosphates as phosphoryl donor (*1*). Enzymic activity phosphorylating deoxycytidine was reported in early studies with regenerating rat liver (*89a*) and rat embryo (*90*) extracts, and its inhibition by dCTP was noted (*90*). The enzymic activity is particularly high in lymphoid tissues (*91*) and has been purified notably from calf thymus (*92–95*). It has also been purified from L1210 mouse leukemia cells (*96*),

89. L. R. Solomon and T. R. Breitman, *BBRC* **44**, 299 (1971).
89a. R. E. Beltz, *ABB* **99**, 304 (1962).
90. F. Maley and G. F. Maley, *Biochemistry* **1**, 847 (1962).
91. J. P. Durham and D. H. Ives, *Mol. Pharmacol.* **5**, 358 (1969).
92. Y. Sugino, S. Kobayashi, H. Nakamura, H. Shimono, S. Sonoda, and K. Tsukamoto, *Annu. Rep. Inst. Virus Res., Kyoto Univ.* **8**, 91 (1965).
93. R. L. Momparler and G. A. Fischer, *JBC* **243**, 4298 (1968).
94. J. P. Durham and D. H. Ives, *JBC* **245**, 2276 (1970).

but it was not found to be especially high in several other tumors (*91*). The activity has usually been assayed as conversion of labeled nucleoside to nucleotide, separating the substrate and product, frequently with DEAE-cellulose (*91, 93–96*). Successful purification procedures included fractionation with ammonium sulfate (*92–95*), adsorption onto protamine (*93*), DEAE-cellulose (*92, 94, 95*), or DEAE-Sephadex (*95, 96*), gel filtration (*93, 95*), and adsorption onto calcium phosphate gel (*94, 96*) or hydroxyapatite (*92*) and onto alumina C_γ gel (*96*).

b. Substrate Specificity. The specificity of this enzyme with regard to the nucleoside acceptor is not clear-cut. Most of the preparations purified from thymus were also active in phosphorylating deoxyguanosine and deoxyadenosine, and it seemed clear that these activities purified together with the deoxycytidine phosphorylating activity (*91–95*). The deoxyguanosine kinase activity, in particular, could not be separated from deoxycytidine kinase by gel filtration, isoelectric focusing, or sucrose density gradient centrifugation (*94*). However, the ratio of these activities to deoxycytidine phosphorylation was not constant throughout purification, and there was more variability in the activity toward the purine deoxyribonucleosides (*91, 94*). One possible explanation is that this activity was apparently considerably more labile than the activity for deoxycytidine, especially to the absence of thiol protection (*91, 94*). It has been suggested that the enzyme may have a stable site or conformation for deoxycytidine phosphorylation and less stable site(s) or conformation for deoxyguanosine and deoxyadenosine kinase activities as well (*94*). Deoxycytidine was by far the preferred substrate kinetically, with K_m values much lower than those for the other nucleoside substrates (*91, 93, 95, 97*); high concentrations of the purine nucleosides were therefore necessary to demonstrate competitive inhibition of deoxycytidine phosphorylation by these alternate substrates (*91, 93–95*).

All preparations of the enzyme that were tested were also found to be active in phosphorylating cytosine arabinoside [1-β-D-arabinofuranosylcytosine (araC)], and it seems clear that this antitumor agent is phosphorylated to its active nucleotide forms by deoxycytidine kinase (*91, 93, 94, 96–104*). Tumor cell sublines resistant to araC were found

---

95. Y. Kozai, S. Sonoda, S. Kobayashi, and Y. Sugino, *J. Biochem. (Tokyo)* **71**, 485 (1972).
96. D. Kessel, *JBC* **243**, 4739 (1968).
97. D. H. Ives and J. P. Durham, *JBC* **245**, 2285 (1970).
98. M. Y. Chu and G. A. Fischer, *Biochem. Pharmacol.* **14**, 333 (1965).
99. D. Kessel, T. C. Hall, and I. Wodinsky, *Science* **156**, 1240 (1967).
100. A. W. Schrecker and M. J. Urshel, *Cancer Res.* **28**, 793 (1968).

to have lost most or all of their capacity to phosphorylate araC, and deoxycytidine phosphorylation was lost concurrently (*98, 100, 102*). The capacity to phosphorylate deoxyguanosine was almost completely lost, but the phosphorylation of deoxyadenosine was only slightly decreased (*102*). The activity phosphorylating araC purified together with deoxycytidine kinase in a parallel fashion (*94*), and the two substrates competed with each other (*91, 93, 96, 97, 100, 101*). Competition was also evident between araC and deoxyguanosine or deoxyadenosine (*91, 97, 102*). Deoxycytidine was still the preferred substrate, activity of the enzyme with araC being intermediate between that with deoxycytidine and that with deoxyguanosine (*91, 93, 102, 104*).

One preparation of deoxycytidine kinase, 400-fold purified from thymus, also phosphorylated cytidine, although the usual uridine-cytidine kinase activity (see above) was separated from the deoxycytidine kinase during the purification (*95*). The phosphorylation of cytidine was strongly inhibited by deoxycytidine; the apparent K_m for cytidine was two orders of magnitude higher than that for deoxycytidine (*95*).

Deoxycytidine kinase has a broad specificity with regard to the phosphate donor. Of the naturally occurring nucleoside triphosphates, only dCTP was unable to donate phosphate in the reaction (*91–96*). However, the capacity of the different triphosphates to serve as phosphate donor varied both with the phosphate acceptor and with the presence of a thiol compound (*91, 94*).

c. Kinetic, Molecular, and Allosteric Properties. The requirement of deoxycytidine kinase for a divalent cation could be fulfilled by Mg^{2+} and, to a lesser extent, by a number of other metals. Different preparations exhibited activity with Mn^{2+}, Ca^{2+}, Fe^{2+}, Co^{2+}, Zn^{2+}, Ni^{2+}, or Sr^{2+} (*93–96*). The molecular weight of two purified preparations of the enzyme was estimated to be, for one, around 60,000 as judged by gel filtration (*96*), and for the other, 56,000, calculated from the results with gel filtration, sucrose density gradient centrifugation, and isodensity centrifugation in cesium chloride (*94*). The protein is therefore not a large molecule and has given no indication of a subunit structure that aggregated or disaggregated in the presence of substrates or effectors (*94*) or surface active agents (*96*). Nevertheless, the kinetic behavior

---

101. D. Kessel, *Mol. Pharmacol.* **4**, 402 (1968).
102. A. W. Schrecker, *Cancer Res.* **30**, 632 (1970).
103. R. L. Momparler, T. P. Brent, A. Labitan, and V. Krygier, *Mol. Pharmacol.* **7**, 413 (1971).
104. Y. Kozai and Y. Sugino, *Cancer Res.* **31**, 1376 (1971).

and regulatory properties of the enzyme have been found to be exceedingly complex (*94–97, 105*).

The enzyme has been variously reported to exhibit a pH optimum at 5–7 (*96*), at 8 (*95*), and between 6 and 10 (*94*), and pH was observed to affect the inhibition of the enzyme by various nucleotides (*94*). Although the preparation from leukemia L1210 cells was not found to be sensitive to sulfhydryl reagents (*96*), that from calf thymus was affected (*93, 94*). The effect of thiols in modifying substrate specificity has already been noted, and they have been observed to affect also the regulatory inhibition by nucleotides (*94*).

Feedback regulation of deoxycytidine kinase by the end product dCTP was reported in early studies on the enzyme (*90, 92*), and this has been confirmed with the various purified preparations (*93–97*). The inhibition by dCTP was competitive with the phosphate donor and not with the phosphate acceptor (*93, 95*); the inhibition was, however, different with various phosphate acceptors, especially in its response to thiol (*94*). The regulatory effects of other nucleotides have also been studied; dCDP, dCMP, UDP, and dTDP have all been found to inhibit the enzyme (*93–95, 97, 106*). These inhibitions could be reversed by various triphosphates that could also serve as phosphate donors in the reaction—dTTP, UTP, and dUTP (*91, 94, 96*). The inhibition by dCTP and reversal by dTTP have been related to the location of dCMP at the metabolic branch point leading to either of these triphosphates (*94, 95*). In a similar manner, several enzymes that lead to the formation of dTTP are inhibited by dTTP and activated by dCTP (*94*).

The complexity of deoxycytidine kinase in its kinetic behavior and allosteric regulatory properties is indicated by a recent study with the calf thymus enzyme (*97*). This detailed kinetic analysis revealed that the double reciprocal plot for varying concentrations of ATP was bimodal (cf. *91, 95*), with a rather abrupt increase in the apparent K_m for $[\text{ATP} \cdot \text{Mg}]^{2-}$ ($3 \times 10^{-5} M$ and $9 \times 10^{-5} M$, both lower than earlier estimates of this value, *91, 93, 96*). Double reciprocal plots for varying concentrations of deoxycytidine were even more complex and suggested negative cooperativity in the binding of this substrate (*97*). Apparent K_m values calculated from the two different slopes were $5.3 \times 10^{-6} M$ and $1.6 \times 10^{-5} M$, the latter similar to those reported in other, earlier studies (*93, 96*). In contrast to these results, the plots with varying deoxyguanosine or deoxyadenosine as phosphate acceptors were linear.

105. J. Seifert and J. Riman, *Folia Biol. (Prague)* **11**, 156 (1965).
106. Y. Sugino, S. Kobayashi, H. Nakamura, K. Tsukamoto, H. Shimono, K. Shimada, S. Sonoda, and Y. Kozai, *Annu. Rep. Inst. Virus Res., Kyoto Univ.* **9**, 148 (1966).

Inhibition by dCTP was also complex; however, it appeared to be noncompetitive with deoxycytidine at low concentrations of this substrate and competitive with [ATP·Mg]$^{2-}$ at high concentrations of this substrate (97). Reversal of this inhibition by dTTP was also evident only at low concentrations of deoxycytidine; at high nucleoside concentrations dTTP became an inhibitor. With limiting deoxycytidine, dCTP would then inhibit and dTTP stimulate this reacton, while at higher concentrations of deoxycytidine both triphosphates would inhibit (97).

2. Deoxyadenosine Kinase

a. Distribution, Purification, and Assay. An enzyme preparation has also been purified from calf thymus for capacity to phosphorylate deoxyadenosine (107, 108). This enzyme, classified as ATP:deoxyadenosine 5'-phosphotransferase (EC 2.7.1.76), catalyzes the phosphorylation of deoxyadenosine and deoxyguanosine to their respective 5'-monophosphates, using a variety of nucleoside triphosphates as phosphate donor (1). The presence of such an enzyme has been suggested in a variety of mammalian cells (109–111). Deoxyadenosine was shown in early studies to inhibit DNA synthesis in several different cell systems (110–114), and this has been attributed to its phosphorylation to nucleotide derivatives and perhaps inhibition, at that level, of ribonucleotide reductase (108). It was this property of the deoxynucleoside that led to the exploration of the kinases responsible for its phosphorylation.

Deoxyadenosine kinase has been purified 140-fold from calf thymus (107, 108). The purification involved fractionation with protamine and ammonium sulfate, Sephadex gel filtration, and DEAE-cellulose chromatography; mercaptoethanol was used to stabilize the enzyme during purification (108). The kinase activity was assayed as conversion of radioactive nucleoside to nucleotide.

b. Substrate Specificity. This preparation catalyzed the phosphorylation of both deoxyadenosine and deoxyguanosine with comparable facility (K_m for deoxyadenosine $0.7 \times 10^{-3} M$, K_m for deoxyguanosine $1.1 \times 10^{-3} M$, K_i for deoxyguanosine with deoxyadenosine as substrate

107. V. Krygier and R. L. Momparler, *BBA* **161**, 578 (1968).
108. V. Krygier and R. L. Momparler, *JBC* **246**, 2745 (1971).
109. A. Munch-Petersen, *BBRC* **3**, 392 (1960).
110. G. F. Maley and F. Maley, *JBC* **235**, 2964 (1960).
111. H. Klenow, *BBA* **61**, 885 (1962).
112. H. Klenow, *BBA* **35**, 412 (1959).
113. W. H. Prusoff, *Biochem. Pharmacol.* **2**, 221 (1960).
114. K. G. Lark, *BBA* **45**, 121 (1960).

$1.1 \times 10^{-3} M$, K_i for deoxyadenosine with deoxyguanosine as substrate $0.8 \times 10^{-3} M$, V_{max} for both nucleoside substrates also comparable) (115). In addition it apparently catalyzed the phosphorylation of cytidine very well (K_m for cytidine $0.6 \times 10^{-3} M$), and this phosphorylation was also decreased by either deoxyadenosine or deoxyguanosine (K_i for deoxyadenosine with cytidine as substrate $0.35 \times 10^{-3} M$) (115). This phosphorylation of cytidine did not appear to result from contamination with uridine-cytidine kinase since the K_m for cytidine was comparable to that for the purine deoxyribonucleosides, and since an excess of uridine had no effect on the phosphorylation. The physiological significance of this phosphorylation is not clear since cytidine is believed to be phosphorylated usually by uridine-cytidine kinase (see above). However, the latter enzyme is apparently very low in thymus (115); thus, deoxyadenosine kinase could conceivably be important for the phosphorylation of cytidine in this tissue.

The deoxyadenosine-deoxyguanosine kinase also phosphorylated deoxycytidine to a smaller extent, but this activity differed from the purine deoxynucleoside and cytidine kinase activity in being stable to dialysis and in being undiminished by the addition of excess deoxyadenosine or deoxyguanosine as an alternate substrate (108). Nucleosides which were inactive as phosphate acceptors were adenosine, guanosine, uridine, and deoxythymidine (108). It is therefore not completely clear to what extent this enzymic activity is distinct from deoxycytidine kinase, although the two activities were apparently separated from each other to some degree during purification (93, 103, 108).

Of the nucleotides tested as possible phosphate donors for this enzyme, with either deoxyadenosine or deoxyguanosine as phosphate acceptor, ATP and GTP were the most active, followed by UTP and dTTP (108); CTP, dATP, and dGTP were much less active, and dCTP was inactive.

c. Kinetic and Molecular Properties; Reaction Mechanism. The partially purified deoxyadenosine kinase exhibited a broad pH optimum between 6.5 and 8.5, and a requirement for Mg^{2+} that could be partially replaced by Mn^{2+} or, to a lesser extent, Ca^{2+} (108). The activity was sensitive to Hg^{2+} and to dialysis in the absence of a sulfhydryl-protecting agent. The molecular weight of the partially purified kinase was 63,000 as judged by Sephadex gel filtration (108).

The enzymic activity was inhibited by deoxyribonucleotides containing adenine, guanine, or cytosine (107, 115). Of these the most potent inhibitors were the deoxyribonucleotides of cytidine, followed by those

115. V. Krygier and R. L. Momparler, *JBC* **246**, 2752 (1971).

of adenosine and then by the guanosine deoxynucleotides (*107*). In each case the relative inhibitory capacity of the nucleotides, with respect to the number of phosphate groups, was tri > di > mono. The potency of nucleotide inhibitor was also dependent on the nature of its pentose moiety, with deoxyribose > arabinose > ribose; cytosine arabinoside triphosphate did, however, exert appreciable inhibition (*103, 115*). Kinetic data suggested that the deoxyribonucleotide inhibition was competitive with respect to the phosphate donor (ATP) and noncompetitive with respect to the phosphate acceptor (deoxyadenosine) (*103, 115*). The apparent K_m for ATP was $2 \times 10^{-4} M$ and the apparent K_i values for dCTP and dATP were $1.5 \times 10^{-7} M$ and $2.7 \times 10^{-5} M$, respectively. The potent inhibition by the deoxynucleoside triphosphates and the relatively high K_m of this enzyme for its substrates, especially the nucleoside substrates, make it potentially very susceptible to end product regulation. The initial velocity pattern evident from the kinetic studies suggested that the kinase catalyzes the phosphorylation of deoxyadenosine by a ping-pong type of mechanism (*115*).

3. *Deoxyguanosine Kinase*

It seems clear that the phosphorylation of deoxyguanosine can be catalyzed both by the preparation purified as deoxycytidine kinase and by that purified as deoxyadenosine kinase. If these are indeed distinct enzymes, it is then an open question which plays the major role in the physiological phosphorylation of this nucleoside. Schrecker's (*102*) genetic evidence with the araC-resistant subline of mouse leukemia L1210 indicated that when deoxycytidine kinase activity was lost, most, but not all, of the deoxyguanosine kinase activity was lost concurrently. Since deoxyadenosine kinase activity was not lost, the deoxyguanosine phosphorylation remaining could have been catalyzed by this enzyme. This would suggest two separate enzymes, both able to phosphorylate deoxyguanosine, with deoxycytidine kinase carrying the major responsibility for deoxyguanosine phosphorylation, at least in these cells. However, it must be borne in mind that the detection and quantitation of these phosphorylations have been found to be very much affected by the conditions of assay, and conclusions of this nature must be regarded as tentative. It is certainly not ruled out that all of these phosphorylations are carried out by a single enzyme protein, or by closely related proteins, with the activity for each particular nucleoside substrate regulated by the environmental conditions with regard to thiol concentration or the concentration of specific nucleoside triphosphates. To the author's knowledge, an enzyme has not yet been purified for deoxyguanosine phosphorylating capacity. It therefore remains to be seen what the substrate specificity of such a preparation would be and what light this might shed on the overlap of these kinases.

F. DEOXYTHYMIDINE KINASE

Deoxythymidine kinase (ATP:deoxythymidine 5′-phosphotransferase, EC 2.7.1.75) catalyzes the reaction:

$$\text{Deoxythymidine} + \text{NTP} \rightarrow \text{dTMP} + \text{NDP}$$

This enzyme has attracted a great deal of attention in the past decade in several contexts. These include its susceptibility to end product feedback regulation, its dramatic increase in activity in cells during rapid growth and DNA synthesis, and its induction in cells infected with any one of several DNA viruses. These various physiological and biochemical aspects of deoxythymidine kinase have been extensively covered in other recent reviews (e.g., [116–118]), to which the reader is referred. The present discussion will be confined primarily to a consideration of the properties of purified preparations of the enzyme and to the current state of our knowledge about the enzyme molecules in these preparations.

1. *Distribution, Purification, and Assay*

Deoxythymidine kinase is widely distributed; its activity and contribution to DNA synthesis may vary depending on the organism and the physiological state of the cell, but it is apparently present in most species. Some mutant strains of cells and viruses selected for resistance to the 5-halogenated analogs, such as 5-bromodeoxyuridine and 5-iododeoxyuridine, have been found to be deficient in deoxythymidine kinase, but its presence certainly appears to be nearly universal ([54, 116, 117]). Recent studies on the intracellular localization of the enzyme in rat thymus have indicated that most of the enzyme was in the cytoplasmic supernatant but that significant activity was also located in the nuclei and low levels in the mitochondria ([119], cf. [117]).

The enzyme has been purified from *E. coli* ([120, 121]), from T4-infected

116. J. E. Cleaver, *Front. Biol.* **6**, 43 (1967).
117. J. S. Roth, *in* "Protein Metabolism and Biological Function" (C. P. Bianchi and R. Hilf, eds.), p. 141. Rutgers Univ. Press, New Brunswick, New Jersey, 1970.
118. H. M. Keir, *Symp. Soc. Gen. Microbiol.* **18**, 67 (1968).
119. J. P. Durham and D. H. Ives, *Exp. Cell Res.* **70**, 97 (1972).
120. R. Okazaki and A. Kornberg, *JBC* **239**, 269 (1964).
121. P. Voytek, P. K. Chang, and W. H. Prusoff, *JBC* **246**, 1432 (1971).

coli (122), from certain mammalian tumors *(123–125)*, from regenerating rat liver *(126)*, and from calf thymus *(127, 127a)*. Purification procedures have employed fractionation with ammonium sulfate *(120–122, 124–127a)*, heat *(120, 124, 126)*, acid *(123, 127a)*, protamine *(127)*, calcium phosphate gel or hydroxyapatite *(124–126)*, DEAE-cellulose or DEAE-Sephadex *(120, 122, 125, 127a)*, phosphocellulose *(122)*, gel filtration *(123, 125, 127, 127a)*, and preparative disc gel electrophoresis *(121)*. Enzymic activity has usually been assayed as conversion of labeled deoxythymidine to the mononucleotide, with separation of the substrate and product by electrophoresis or with anion exchange cellulose.

2. Substrate Specificity

Deoxythymidine kinase purified from either bacterial or mammalian sources was found to be specific for the phosphorylation of deoxythymidine, deoxyuridine, and their 5-halogenated analogs (5-bromodeoxyuridine, 5-iododeoxyuridine, 5-fluorodeoxyuridine, etc.) *(120, 124–127)*, and alternate substrates were competitive with each other *(124)*. Aza-deoxythymidine was phosphorylated to a smaller extent *(126)*, and the enzyme also apparently phosphorylated 5-mercaptodeoxyuridine *(128)* and 5-trifluoromethyldeoxyuridine *(128a)*. Specificity of this enzyme for deoxythymidine, deoxyuridine, and their analogs has also been observed for *Salmonella typhimurium* *(53, 54)*, and the structural gene for deoxythymidine kinase has been mapped in both *Salmonella (87)* and *E. coli (129)*.

Like other nucleoside kinases, deoxythymidine kinase has a broad specificity with regard to the phosphate donor; most nucleotide triphosphates with the exception of dTTP have been found able to donate phosphate in the reaction *(120, 127, 127a, 130)*. With the *E. coli* enzyme, however, the kinetics were different with different donors and depended on their capacity to activate the enzyme as well, in an allosteric fashion *(130*, see below).

122. R. Okazaki, personal communication.
123. H. J. Grav and R. M. S. Smellie, *BJ* **94**, 518 (1965).
124. E. Bresnick and U. B. Thompson, *JBC* **240**, 3967 (1965).
125. T. Hashimoto, T. Arima, H. Okuda, and S. Fujii, *Cancer Res.* **32**, 67 (1972).
126. E. Bresnick, K. D. Mainigi, R. Buccino, and S. S. Burleson, *Cancer Res.* **30**, 2502 (1970).
127. M. O. Her and R. L. Momparler, *JBC* **246**, 6152 (1971).
127a. K. Shimada, *Annu. Rep. Inst. Virus Res., Kyoto Univ.* **12**, 41 (1969).
128. T. I. Kalman and T. J. Bardos, *Mol. Pharmacol.* **6**, 621 (1970).
128a. E. Bresnick and S. S. Williams, *Biochem. Pharmacol.* **16**, 503 (1967).
129. K. Igarashi, S. Hiraga, and T. Yura, *Genetics* **57**, 643 (1967).
130. R. Okazaki and A. Kornberg, *JBC* **239**, 275 (1964).

3. Kinetic, Molecular, and Allosteric Properties

The pH optimum of deoxythymidine kinase has been found to be between 7.5 and 8.5 for purified preparations from several sources (*116, 120, 124, 125, 127, 127a*). The requirement for Mg^{2+} could be partially fulfilled by Co^{2+}, Mn^{2+}, or Ca^{2+}; but in the presence of Mg^{2+}, higher concentrations of these metals were sometimes inhibitory (*122, 124*). With the highly purified *E. coli* enzyme, maximum activity was obtained in the presence of both Mg^{2+} and Mn^{2+} (*120*).

Deoxythymidine kinase was observed in early studies to be susceptible to end product feedback inhibition by dTTP (*90, 116–118, 131, 132*). A careful study of the response of the enzyme to allosteric effectors was first made by Okazaki and Kornberg (*130*) with the enzyme from *E. coli*. They observed that this enzyme was not only inhibited by the end product dTTP, but it was also activated by a number of nucleoside di- and triphosphates, such as dCDP, that might be expected to accumulate in the cell when the dTTP concentration was low. The activators functioned by decreasing the K_m of the enzyme for the substrate deoxythymidine (*130*). The enzyme was saturated with the substrate ATP only at quite high concentrations, and the saturation curve did not follow classic Michaelis–Menten kinetics but was sigmoid rather than hyperbolic in shape. In the presence of dCDP, the kinetics were "normalized." Moreover, with dATP rather than ATP as phosphate donor, classic kinetics were observed. It was concluded that two molecules of ATP were utilized by the enzyme molecule for two different functions, one as an allosteric effector to activate the enzyme, and one as a substrate to donate phosphate in the reaction. The alternate substrate dATP was assumed to function similarly, but to be a much more efficient activator than ATP so that the kinetics resembled those in the presence of dCDP. The feedback inhibition by dTTP appeared to be competitive with the phosphate acceptor deoxythymidine but not with the phosphate donor or the activator nucleotide (*130*).

Subsequent studies with the *E. coli* enzyme, using sucrose density gradient centrifugation and gel filtration, indicated that the addition of effectors, whether negative (dTTP) or positive (dCDP, dCTP, and dADP), approximately doubled the sedimentation coefficient and the apparent molecular weight of the enzyme, as if the enzyme "dimerized" in the presence of these molecules (*133*). The molecular weights of the "monomer" and "dimer" were 42,000 and about 90,000, respectively.

131. D. H. Ives, P. A. Morse, Jr., and V. R. Potter, *JBC* **238**, 1467 (1963).
132. T. R. Breitman, *BBA* **67**, 153 (1963).
133. N. Iwatsuki and R. Okazaki, *JMB* **29**, 139 and 155 (1967).

The sedimentation coefficient increased from 3.4 S to around 5.4 S on addition of an activator deoxynucleotide, and to near 6.0 S in the presence of dTTP; this small difference could reflect differences in the conformation of the "dimer" formed on addition of activator or inhibitor (133). The "monomer" form appeared to be extremely temperature-sensitive, maximum reaction rates being obtained at temperatures below 30°; the lower velocities at higher temperatures were, however, linear, and reflected reduced affinity of the enzyme for both substrates rather than enzymic inactivation (133). The temperature effects were reversible. In the presence of effectors the temperature sensitivity was abolished, and presumably the "dimer" is insensitive to this effect (133).

With the preparation of deoxythymidine kinase purified from E. coli by Prusoff and co-workers (121), the halogen analogs 5-iodo-dCTP and 5-bromo-dCTP were more potent allosteric activators than the natural dCTP, and they produced similar alterations in the sedimentation coefficient of the enzyme. The dependence of the effect on pH was, however, complex.

These observations do not necessarily carry over to other systems and organisms. Deoxythymidine kinase purified from regenerating rat liver was compared with that purified from E. coli (126). Whereas the E. coli enzyme had a sedimentation coefficient of 3.6 S that was increased by dTTP, the enzyme from regenerating liver had a coefficient of 5.0 S, and this was not altered by the feedback inhibitor. The kinase purified from calf thymus had a molecular weight of 55,000 as judged by gel filtration and density gradient centrifugation (127). Inhibition of this enzyme by dTTP appeared to be competitive with deoxythymidine but bore a more complex relationship to ATP, being noncompetitive with this substrate at low concentrations of ATP and competitive at higher concentrations. This preparation showed no evidence of activation by any nucleotides.

Deoxythymidine kinase purified from mammalian tumors also did not appear to be activated by dCDP (124). The molecular weight of the kinase from some mammalian tumors was considerably higher than that from E. coli, and the enzyme appeared to exist in different states of aggregation (116, 117, 124, 134). In dilute solution it aggregated to a molecular weight in excess of 600,000, while in solutions of higher ionic strength it appeared to disaggregate to a molecular weight near 100,000 (134). The disaggregated form was more sensitive to inhibition by dTTP, and the inhibition was of a different type (competitive with deoxythymidine with the aggregated form and noncompetitive with the

134. E. Bresnick, U. B. Thompson, and K. Lyman, ABB 114, 352 (1966).

disaggregated form). Results with the kinase from Yoshida sarcoma also indicated that the molecular weight of the enzyme species was affected by ionic strength (125, 135); however, it was observed that after treatment with RNase, the enzyme behaved like the "disaggregated" form even in dilute solutions, and it was suggested that the "aggregated" species might be a complex between the enzyme and some modifier containing RNA. Multiple forms of deoxythymidine kinase, as judged by behavior in DEAE-cellulose chromatography, were, however, observed in a variety of rat tumors, but not in normal rat tissues (135). Two of the forms from Yoshida sarcoma were examined separately; they resembled each other in pH optima, substrate specificity, and response to inhibitors, and differed from each other primarily in K_m for deoxythymidine and in molecular weight (125).

The significance of these different forms of the enzyme is not yet clear. Different forms of deoxythymidine kinase have also been reported in human adult tissues compared with fetal tissues (135a), and differences have likewise been found between the kinases from fetal and adult rat liver (136). Deoxythymidine kinase activity stimulated by adrenocorticotropic hormone administration appeared to have different properties and subcellular distribution as compared with the control enzyme (137). Two different forms of the kinase have been suggested in germinating plant seedlings (138), but recent studies have indicated that these activities may be, not deoxythymidine kinase, but a nucleoside phosphotransferase that catalyzes phosphate transfer between AMP and deoxythymidine and does not require Mg^{2+} (139).

The feedback inhibition of deoxythymidine kinase by dTTP appears to be general, however (116, 117). The induction of deoxythymidine kinase in cells infected with DNA viruses has been reviewed by Keir (118); in some cases the induced enzyme could be shown to be virus-specific and immunologically distinct from that of the host cell (117, 118), and this induced enzyme has sometimes been observed to be relatively insensitive to dTTP inhibition (140). The deoxythymidine kinase of *Tetrahymena pyriformis* has also been observed not to be inhibited by dTTP (141), but these situations are exceptional.

135. H. Okuda, T. Arima, T. Hashimoto, and S. Fujii, *Cancer Res.* **32**, 791 (1972).
135a. A. T. Taylor, M. A. Stafford, and O. W. Jones, *JBC* **247**, 1930 (1972).
136. H. G. Klemperer and G. R. Haynes, *BJ* **108**, 541 (1968).
137. H. Masui and L. D. Garren, *JBC* **246**, 5407 (1971).
138. F. Wanka, I. K. Vasil, and H. Stern, *BBA* **85**, 50 (1964).
139. Q. Deng and D. H. Ives, *BBA* **277**, 235 (1972).
140. H. G. Klemperer, G. R. Haynes, W. I. H. Shedden, and D. H. Watson, *Virology* **31**, 120 (1967).
141. G. D. Shoup, D. M. Prescott, and J. R. Wykes, *J. Cell Biol.* **31**, 295 (1966).

The increase in deoxythymidine kinase activity in cells during rapid growth and DNA synthesis has been considered to be an example of derepression (see *116*). The decline of the enzymic activity at the end of such periods of DNA synthesis has also been explored; this process may be more variable from one system to another and several explanations have been considered, including catabolism of the enzyme or of its messenger RNA, or possible alteration of the enzyme to some less active form (*116, 117, 142, 143*). Deoxythymidine kinase is stabilized by its substrate, and in some systems the enzymic activity could be maintained by the addition of deoxythymidine to the cell cultures (*116*).

4. Reaction Mechanism; Active Site

Deoxythymidine kinase purified from calf thymus exhibited classic rather than allosteric kinetics, and this preparation has been used for initial velocity studies to explore the mechanism of the reaction (*127*). The pattern observed was one of intersecting lines, indicating that the reaction involved the sequential addition of both substrates to the enzyme to form a ternary complex.

Studies with deoxythymidine kinase purified from *E. coli* have shown that the enzyme could be inactivated by ultraviolet irradiation and that protection against this inactivation was afforded by substrates as well as by the allosteric effectors dCDP or dTTP. In contrast, the analog and alternate substrate, 5-iododeoxyuridine, enhanced the inactivation. Since iododeoxyuridine is photolabile and undergoes dehalogenation upon ultraviolet irradiation, it was suggested that the photolysis of iododeoxyuridine bound at the catalytic site might result in modification of amino acids at the active site and that this active site-directed sensitization might be used to label and identify such amino acids (*144*).

G. RIBOFLAVIN KINASE

Riboflavin kinase (ATP:riboflavin 5′-phosphotransferase, EC 2.7.1.26) catalyzes the reaction:

$$\text{Riboflavin} + \text{ATP} \rightarrow \text{FMN} + \text{ADP}$$

It is therefore fundamental in the conversion of the vitamin to its coenzyme forms, both FMN and FAD. As reviewed in the previous

142. E. Bresnick and S. S. Burleson, *Cancer Res.* **30**, 1060 (1970).
143. S. Kit, D. R. Dubbs, and P. M. Frearson, *JBC* **240**, 2565 (1965).
144. R. Cysyk and W. H. Prusoff, *JBC* **247**, 2522 (1972).

edition of this series (*3*), the enzyme was first partially purified from yeast (*144a*) and from plant tissues (*145*), and the properties and substrate specificity of the yeast enzyme were explored (*144a*, *146*). It has since been shown that this enzyme is also present in a variety of animal tissues (rat liver, kidney, brain, spleen, and heart), although its presence in extracts may often be masked by acid phosphatases that hydrolyze FMN (*147–150*).

The kinase has been partially purified from rat liver by ammonium sulfate fractionation and DEAE-cellulose chromatography (*149*); it is localized in the cell supernatant. The purified preparation exhibited a temperature optimum near 50°, a pH optimum near 8, and activation by divalent cations, with Zn^{2+} being more effective than Mg^{2+}. However, the preparation was not free of phosphatase, and this fact was thought to account for the effect of Zn^{2+}, which inhibits the contaminating phosphatase (*148*, cf. also *144a*). This may also have contributed to the apparent pH and temperature optima.

The substrate specificity of the liver flavokinase was also studied (*149, 150*) and appeared to resemble that of the yeast enzyme (*146*). Flavins phosphorylated were riboflavin and dichloroflavin and, to a lesser extent, 2-iminoriboflavin and D-araboflavin. Based on the structures of the flavins phosphorylated, those inhibitory, and those inactive as either substrate or inhibitor, it was concluded that, for reaction with the kinase, the isoalloxazine ring must be substituted at position 6 or 7 and preferably both, that position 5 and nitrogen 3 should be unsubstituted, and that the 5-carbon polyol side chain needed hydroxyl groups in the D configuration at carbons 3' and 4'. With side chains less than three carbons in length, attachment depended primarily on the isoalloxazine portion of the molecule. Certain of the biological properties of flavin analogs could be correlated with their behavior in the flavokinase system (*150*).

With the partially purified liver enzyme, ATP was the most effective phosphate donor, with GTP and CTP exhibiting less activity, and ITP and UTP ineffective (*149*). The apparent K_m values were $2 \times 10^{-4} M$ for ATP and $1.2 \times 10^{-5} M$ for riboflavin (*149*).

144a. E. B. Kearney and S. Englard, *JBC* **193**, 821 (1951).
145. K. V. Giri, P. R. Krishnaswamy, and N. Appaji Rao, *BJ* **70**, 66 (1958).
146. E. B. Kearney, *JBC* **194**, 747 (1952).
147. D. B. McCormick, *Proc. Soc. Exp. Biol. Med.* **107**, 784 (1961).
148. D. B. McCormick and M. Russell, *Comp. Biochem. Physiol.* **5**, 113 (1962).
149. D. B. McCormick, *JBC* **237**, 959 (1962).
150. D. B. McCormick and R. C. Butler, *BBA* **65**, 326 (1962).

H. NICOTINAMIDE-ADENINE DINUCLEOTIDE KINASE

Nicotinamide-adenine dinucleotide (NAD⁺) kinase (ATP:NAD⁺ 2′-phosphotransferase, EC 2.7.1.23) catalyzes a phosphoryl transfer to an R–OH acceptor, a reaction analogous to those catalyzed by the nucleoside kinases discussed above, even though all of them are 5′-phosphotransferases. In accord with traditional enzyme classification (*1, 2*), NAD⁺ kinase is therefore discussed here together with the nucleoside kinases. The reaction it catalyzes is

$$NAD^+ + ATP \rightarrow NADP^+ + ADP$$

in which a phosphoryl is transferred from ATP to the 2′-OH of the ribose in the adenosine portion of NAD⁺.

1. *Distribution, Purification, and Stability*

Early studies reported the presence of enzymes capable of phosphorylating NAD⁺ to NADP⁺ in yeast (*151, 152*), pigeon liver (*153*), and rabbit liver and kidney (*154*). The first purifications of the enzyme from brewer's yeast (*152*) and from pigeon liver (*155*) were discussed in the previous edition of "The Enzymes" (*2*). In the last decade a great deal more literature on NAD⁺ kinase has appeared, including reports on its further purification from pigeon liver (*156, 157*), rat liver (*158–160*), rabbit liver (*161*), spinach leaves (*162*), bacteria (*Azotobacter vinelandii*) (*163*), baker's yeast (*164*), and insects (*165*). The enzyme has been quite highly purified from acetone powders of pigeon liver

---

151. H. von Euler and E. Z. Adler, *Hoppe-Seyler's Z. Physiol. Chem.* **252**, 41 (1938).
152. A. Kornberg, *JBC* **182**, 805 (1950).
153. A. H. Mehler, A. Kornberg, S. Grisolia, and S. Ochoa, *JBC* **174**, 961 (1948).
154. B. Katchman, J. J. Betheil, A. I. Schepartz, and D. R. Sanadi, *ABB* **34**, 437 (1951).
155. T. P. Wang and N. O. Kaplan, *JBC* **206**, 311 (1954).
156. V. L. Nemchinskaya, V. P. Kushner, V. M. Bozhkov, E. I. Turchenko, and S. E. Tukachinsky, *Biokhimiya* **31**, 306 (1966).
157. D. K. Apps, *Eur. J. Biochem.* **5**, 444 (1968).
158. V. L. Nemchinskaya and T. B. Smirnova, *Biokhimiya* **32**, 854 (1967).
159. H. Oka and J. B. Field, *JBC* **243**, 815 (1968).
160. I. L. Yero, B. Farinas, and L. S. Dietrich, *JBC* **243**, 4885 (1968).
161. A. E. Chung, "Methods in Enzymology," Vol. 18, p. 149, 1971.
162. Y. Yamamoto, *Plant Physiol.* **41**, 523 (1966).
163. A. E. Chung, *JBC* **242**, 1182 (1967).
164. D. K. Apps, *Eur. J. Biochem.* **13**, 223 (1970).
165. M. Agosin, J. Ilivicky, and S. Litvak, *Can. J. Biochem.* **45**, 619 (1967).

(*156, 157*), and equal or greater degrees of purification have been achieved starting with homogenates of rat liver (*158, 160*) or insects (*165*) or with sonicates of *Azotobacter* (*163, 166*).

Purifications of the enzyme from various sources have made use of its capacity to be precipitated by protamine (*155–158, 160, 162*) or to be adsorbed by DEAE-cellulose or DEAE-Sephadex (*156, 159, 161, 163–165*), and electrophoretic data have also indicated that the enzyme is a negatively charged protein (*156*). The enzyme was also adsorbed by calcium phosphate gel or hydroxyapatite (*157, 160–165*) and by alumina $C\gamma$ gel (*155, 156*). Purification procedures have also taken advantage of the relative heat stability of the enzyme protein (*156–158, 160, 163*). Other methods that have proven fruitful have been ammonium sulfate precipitation (*155–160, 163, 165*), gel filtration (*159, 166*), and precipitation with polyethylene glycol (*164*). Rabbit antiserum to NAD^+ kinase has been prepared using a highly purified preparation of the enzyme from insects (*165*).

Except for the preparation from rabbit liver (*161*), the enzyme has not been characterized by marked instability, but some preparations have been found to be stabilized by cysteine or mercaptoethanol (*157–160, 163, 166*). Chung (*163, 166*) found that the 500- to 1500-fold purified NAD^+ kinase from *Azotobacter* was also stabilized by substrates. From his data he concluded that the enzyme could exist in several forms and that the active form required intact sulfhydryl groups and had a larger molecular weight, perhaps a dimer form, that was maintained by NAD^+ (*166*). Based on alkylation studies with 3-(bromoacetyl)-pyridine, Apps (*167*) concluded that inactivation of the enzyme by this reagent occurred in three independent stages and that the final stage, which totally inactivated the enzyme, probably involved alkylation of a sulfhydryl group with abolition of NAD^+ binding. The earlier stages produced a less active enzyme with modified properties, and apparently the first of these may also have involved sulfhydryl groups (*167*).

2. Assay

The enzyme has usually been assayed by a time-point method that then determined the amount of $NADP^+$ formed. This can be done enzymically with an enzyme specific for $NADP^+$ such as pig heart isocitric dehydrogenase (*152, 155, 156, 160, 161, 163*), or chemically as by the method of Slater and Sawyer (*168*). In this, the $NADP^+$ is the rate-

166. A. E. Chung, *BBA* **159**, 490 (1968).
167. D. K. Apps, *Eur. J. Biochem.* **19**, 301 (1971).
168. T. F. Slater and B. Sawyer, *Nature (London)* **193**, 454 (1962).

limiting carrier in the reduction of the blue dye, 2,6-dichlorophenol-indophenol, by glucose 6-phosphate; the coenzyme is cyclically reduced by glucose-6-phosphate dehydrogenase and reoxidized by phenazine methosulfate. This is linked to reduction of the dye, and the rate of this reduction is directly proportional to the NADP$^+$ concentration (*157*). The isocitric dehydrogenase assay of NADP$^+$ has also been used as a continuous assay for NAD$^+$ kinase, the NADP$^+$ being reduced as it is formed, and the reaction being monitored at 340 nm (*161, 163*); a similar continuous assay used glucose-6-phosphate dehydrogenase to reduce the NADP$^+$ (*169*). Other assay methods have estimated the conversion of ^{14}C-labeled NAD$^+$ to radioactive product or have measured the NADP$^+$ formed enzymically with 6-phosphogluconate dehydrogenase and [^{14}C]6-phosphogluconate and collected $^{14}CO_2$ (*159*). Crude fractions from animal tissues have often shown interference with some assays at 340 nm, perhaps as a result of contaminating enzymes that reduced NADP$^+$ (*160*); this has sometimes been circumvented by reducing all NADP$^+$ formed and assaying NADPH$^+$.

3. *Kinetic and Molecular Properties*

The pH optimum for NAD$^+$ kinase has been reported to be in the vicinity of 6.6–7.6 for preparations from a variety of sources (*157–160, 162*); however, the pH optimum for yeast NAD$^+$ kinases was found to be 8.3–9.4 (*164*). The enzyme requires a divalent cation, and this requirement could be satisfied by a variety of metals including Mg^{2+}, Mn^{2+}, Co^{2+}, and Zn^{2+} (*170*); in addition, the yeast enzyme was activated by Ca^{2+} (*164*). The optimum metal concentration was different for different cations, e.g., the optimum concentration of Mn^{2+} proved to be appreciably lower than the optimum level of Mg^{2+} for NAD$^+$ kinase from pigeon or rat liver (*152, 155, 157, 160*) or from yeast (*164*). The enzyme from rat liver has also been reported to be activated by KCl (*158*). For the divalent cation, the substrate is believed to be the ATP·M^{2-} complex, and a careful study by Apps (*170*) has determined the various K_m values of the pigeon liver enzyme for ATP·M^{2-} complexes to be $2.1 \times 10^{-3} M$ for ATP·Mg^{2-}, $2.0 \times 10^{-3} M$ for ATP·Co^{2-}, $1.4 \times 10^{-4} M$ for ATP·Mn^{2-}, and $9 \times 10^{-5} M$ for ATP·Zn^{2-}. The value for ATP·Mg^{2-} agrees well with the K_m of NAD$^+$ kinase for ATP determined in other preparations of the enzyme (*156, 159, 161, 162*). The enzyme from different sources also has a comparable K_m for NAD$^+$, 3–$8 \times 10^{-4} M$ (*155, 157–159, 163*), although a somewhat higher value (2–$2.6 \times 10^{-3} M$)

169. D. K. Apps, *FEBS Lett.* **15**, 277 (1971).
170. D. K. Apps, *Eur. J. Biochem.* **7**, 260 (1969).

has been reported for NAD⁺ kinase from rabbit liver (*161*) and from plant tissues (*162*). It is worthy of note that these K_m values have been considered to be in the region of physiological concentrations for the nucleotides (e.g., *160*).

The molecular weight of the enzyme protein has been reported to be near 270,000 for the pigeon liver enzyme, based on gel filtration (*157*), and near 125,000–136,000 for the kinase from *Azotobacter* or from rabbit liver, based on sucrose density gradient sedimentation (*161, 166*). It is not clear whether this difference represents different proteins or different forms of a similar enzyme protein in these two organisms.

4. Reaction Mechanism

Based on isotope exchange results and the effect of ATP concentration on the double reciprocal plot of initial velocity vs. NAD⁺ concentration, Orringer and Chung (*171*) concluded that the reaction with the purified *Azotobacter* enzyme suggested a ping-pong mechanism and a phosphorylated enzyme intermediate. However, Apps (*157*), from a kinetic analysis of the reaction with the purified NAD⁺ kinase from pigeon liver, concluded that the enzyme acts by a random addition, rapid equilibrium mechanism with interconversion of the two ternary complexes being the rate-limiting step of the reaction. This latter conclusion was based on the pattern of both double reciprocal plots for 1/v against 1/[S] at different concentrations of the second substrate, which indicated that the K_m values were independent of the second substrate at all concentrations studied, and on the independent variation of V_{max} and the two K_m values with pH (*157*). The kinetic coefficients for the rate equation were calculated (*157*). Furthermore, substitution of acetyl-pyridine-adenine dinucleotide for NAD⁺ did not alter the K_m for ATP·Mg²⁻, and the K_m for NAD⁺ was the same with all the various effective nucleoside triphosphates (*170*) confirming the independence of substrate binding sites. As mentioned above, replacement of Mg²⁺ with other divalent cations markedly affected the K_m of the enzyme for ATP, but it had no effect on the K_m for NAD⁺ (*170*). Data suggested that the rate-limiting step in the reaction might be different when metals other than Mg²⁺ activated the enzyme (*172*). The enzyme could, reproducibly, be partially inactivated by reaction with the alkylating agent, 3-(bromoacetyl)-pyridine, and either NAD⁺ or the ATP·M²⁻ complexes protected against this attack, again suggesting that either substrate could

171. B. P. Orringer and A. E. Chung, *BBA* **250**, 86 (1971).
172. D. K. Apps and A. Marsh, *Eur. J. Biochem.* **28**, 12 (1972).

bind independently (173). For the enzyme modified with this reagent, the V_{max} and the K_m values for both $ATP \cdot M^{2-}$ and NAD^+ were reduced (173). NAD^+ kinase partially purified from autolysates of baker's yeast appeared to proceed by a similar reaction mechanism (164).

5. Substrate Specificity

The substrate specificity has been explored for NAD^+ kinase from a variety of sources (155, 158, 160, 161, 163, 170). Broad specificity for the nucleoside triphosphate phosphoryl donor has not been reported for many preparations of the enzyme (155, 158, 160–162). However, Chung observed that ATP, GTP, ITP, UTP, and CTP were all effective phosphate donors with the Azotobacter enzyme (163), and the very complete study by Apps (170) showed that the same was true for the pigeon liver kinase, although the K_m and V_{max} values differed with the various triphosphates. These latter results were obtained using Mn^{2+} or Zn^{2+} as the activating divalent cation since both of these metals lowered the K_m of the enzyme for ATP as compared with Mg^{2+}. Since Mg^{2+} has usually been used as the activating divalent cation, it was pointed out by Apps (170) that some reports of inactivity with particular triphosphates might have resulted therefore from triphosphate concentrations below the K_s for the $ATP \cdot M^{2-}$ substrate. However, it is not clear that all of the triphosphates would be important as physiological substrates for NAD^+ kinase (170).

The specificity for the pyridine nucleotide substrate has also been investigated in a number of studies (155, 160, 170). The 3-acetyl-pyridine-adenine dinucleotide has been found to serve as phosphate acceptor with the pigeon liver enzyme (170, 174), with a K_m of $1.6 \times 10^{-3} M$; this was derived by extrapolation to zero concentration of the second substrate since the K_m was not independent of this concentration (170). Thionicotinamide and 6-amino-nicotinamide analogs of NAD^+ have also been reported to be phosphorylated by NAD^+ kinase from rat liver (160). Suzuki et al. (175) have reported that α-NAD^+ was slowly phosphorylated by pigeon liver NAD^+ kinase and that a crude enzyme preparation from Azotobacter vinelandii phosphorylated the α-isomer of NAD^+ to α-$NADP^+$. It is not clear whether this activity represents a separate kinase or simply low activity of β-NAD^+ kinase with this substrate. Other pyridine nucleotides tested with β-NAD^+ kinase have been found to be inactive. However, it must be remembered that substrate specificity studies are often open to some question when they are based on assays

173. D. K. Apps, FEBS Lett. **5**, 96 (1969).
174. T. P. Wang, N. O. Kaplan, and F. E. Stolzenbach, JBC **211**, 465 (1954).
175. K. Suzuki, H. Nakano, and S. Suzuki, JBC **242**, 3319 (1967).

that depend on other coupled enzyme reactions, since in many cases the substrate specificity of the coupled enzymes may not have been adequately explored. Thus, acetyl-pyridine-adenine dinucleotide phosphate is reduced by pig heart isocitric dehydrogenase, while the phosphate derivatives of some other NAD^+ analogs are not (176).

Several NAD^+ analogs have been found to inhibit the kinase competitively with NAD^+; these include 3-acetyl-pyridine-adenine dinucleotide (K_i near $1.5 \times 10^{-3} M$), thionicotinamide-adenine dinucleotide (K_i near $1.4 \times 10^{-4} M$), nicotinamide-2'-deoxyadenosine dinucleotide, which cannot be phosphorylated in the 2' position (K_i near $3.9 \times 10^{-4} M$), and the reduced form of the last of these (K_i near $10^{-4} M$) (169, 170). These inhibitions were noncompetitive with $ATP \cdot Mg^{2-}$ (169). The enzyme has also been reported to be inhibited by ADP (157, 158), which was competitive with $ATP \cdot Mg^{2-}$ and noncompetitive with NAD^+ (K_i approximately $10^{-2} M$) (157), by high concentrations of substrate $ATP \cdot Mg^{2-}$ and by free ATP^{-4} and free Mg^{2+}, which were considered to be competitive with the $ATP \cdot Mg^{2-}$ complex (K_i about $2 \times 10^{-2} M$) (157). NAD^+ kinase was also inhibited by a number of other adenine nucleotide species (172); the inhibition by free ATP, $ADP \cdot Mg^-$, or AMP^{2-} was considered to be competitive with respect to both $ATP \cdot Mg^{2-}$ and NAD^+. ATP appeared to bind equally well to the free enzyme and to the binary enzyme–substrate complexes, but the other adenine nucleotide species had greater affinity for the enzyme–substrate complexes than for the free enzyme.

Of particular interest is the behavior of NAD^+ kinase with NADH. Neither the animal nor plant enzyme phosphorylated the reduced form of the nucleotide (155, 157, 159, 162, 163, 170), but NADH was an effective inhibitor, acting competitively with NAD^+ and noncompetitively with $ATP \cdot Mg^{2-}$ (K_i reported near $10^{-4} M$) (157–159, 162). The modes of inhibition by NADH and other NAD^+ analogs and by ADP also supported the independence of substrate binding sites. NADPH was also an inhibitor, competing with both NAD^+ and ATP (K_i 4–$5 \times 10^{-5} M$), and it has been proposed that these inhibitions by the reduced pyridine nucleotide could be of importance in the physiological regulation of the enzyme (157, 159). In this connection it is of interest that most studies have indicated that NAD^+ kinase was localized intracellularly in the cytoplasmic supernatant where nucleotides would also be located (155, 160–162, 177, 178) although there have been some reports of its presence

176. G. Plaut, "The Enzymes," 2nd ed., Vol. 7, p. 105, 1963.
177. J. B. Clark, A. L. Greenbaum, P. McLean, and E. Reid, *Nature (London)* **201,** 1131 (1964).
178. B. Middleton and D. K. Apps, *BBA* **177,** 276 (1969).

in nuclei (*179, 180*) or in cell particles (*154, 179*), and it seemed to be clearly present in yeast mitochondria (*164, 178*).

6. *NADH Kinase*

The substrate specificity of NAD$^+$ kinase with regard to NADH does not hold true for the enzyme from yeast. The NAD$^+$ kinase first purified from brewer's yeast (*152*) phosphorylated NADH, although somewhat less readily than NAD$^+$. However, an NAD$^+$ kinase recently found in yeast mitochondria preferentially phosphorylated NADH and had only slight activity for NAD$^+$ (*181*). It is not clear whether these are distinct enzymes. Apps (*164*) concluded that *Saccharomyces cerevisiae* contained three NAD$^+$ kinase activities differing in their pH optima and kinetic properties and in their stability to heat and to 3-(bromoacetyl)-pyridine. One of these, located in the cytoplasm, was specific for NAD$^+$, while the mitochondria contained both NAD$^+$ and NADH phosphorylating enzymes. However, Griffiths and Bernofsky (*182*) have reported that yeast mitochondrial NADH kinase was activated either by a particulate fraction from mitochondria or, more effectively, by high concentrations of certain salts such as sodium acetate. Only the phosphorylation of NADH was influenced; the slower rate of phosphorylation of NAD$^+$ was unaffected. NADH kinase activity may therefore depend upon the conditions of assay and may have been overlooked in some preparations of NAD$^+$ kinase.

Apps (*164*) reported that the mitochondrial NADH kinase had a K_m for NADH of $5.2 \times 10^{-5} M$ and was inhibited by higher concentrations of this substrate; this could also affect the detection of this activity. The K_m for ATP·Mg^{2-} was approximated ($\sim 2.5 \times 10^{-4} M$) since the double reciprocal plot was nonlinear (and the Michaelis plot sigmoid). Similar kinetics were observed for the yeast mitochondrial NAD$^+$ kinase (*164*).

III. Nucleotide Kinases

A. GMP-dGMP Kinase

1. *Distribution and Purification*

A kinase that specifically phosphorylates nucleoside monophosphates containing the guanine base [ATP:(d)GMP phosphotransferase, EC

179. V. L. Nemchinskaya, *Biokhimiya* **28**, 951 (1963).

2.7.4.8] was recognized in early studies with *Escherichia coli* (*183, 184*) and with mammalian tissues (*4, 185*) and partially purified from *E. coli* (*183*), from roundworms (*Ascaris lumbricoides*) (*186*), and from mouse fibroblasts (*187*). The reaction catalyzed is

$$(d)GMP + (d)ATP \rightarrow (d)GDP + (d)ADP$$

Enzymic activity with this specificity has since been more highly purified from hog brain (*188*), rat liver (*189*), mouse and rat tumors (*189–191*), calf thymus (*192*), human erythrocytes (*193*), and *E. coli* (*194*), and appears to have a wide distribution in different organs of the rat (*192*) and in erythrocytes from a variety of species (*195, 196*). The purified preparations have proven to be specific for nucleoside monophosphates containing the guanine moiety without regard for the sugar; i.e., they phosphorylated specifically GMP, dGMP, and a few base analogs of GMP such as 8-aza-GMP (*188–190, 192, 194*). A highly purified preparation from *E. coli* was homogeneous as judged by DEAE-cellulose chromatography, disc gel electrophoresis, and sedimentation equilibrium pattern; it was also free of contaminating tryptophan fluorescence (*194*).

A distinction has been made between the enzyme in normal *E. coli* that phosphorylates GMP or dGMP (*183, 194*) and enzymes in phage-infected *E. coli* that phosphorylate a broader group of deoxynucleotides. These are presently classified as separate enzymes (*1*) and are discussed as such in a subsequent section.

---

180. J. B. Field, S. M. Epstein, A. K. Remer, and C. Boyle, *BBA* **121**, 241 (1966).
181. C. Bernofsky and M. F. Utter, *Science* **159**, 1362 (1968).
182. M. M. Griffiths and C. Bernofsky, *FEBS Lett.* **10**, 97 (1970).
183. L. J. Bello, M. J. Van Bibber, and M. J. Bessman, *JBC* **236**, 1467 (1961).
184. S. Hiraga and Y. Sugino, *BBA* **114**, 416 (1966).
185. J. L. Strominger, L. A. Heppel, and E. S. Maxwell, *BBA* **32**, 412 (1959).
186. N. Entner and C. Gonzales, *BBA* **47**, 52 (1961).
187. T. J. Griffith and C. W. Helleiner, *BBA* **108**, 114 (1965).
188. R. P. Miech and R. E. Parks, Jr., *JBC* **240**, 351 (1965).
189. R. J. Buccino, Jr. and J. S. Roth, *ABB* **132**, 49 (1969).
190. R. P. Miech, R. York, and R. E. Parks, Jr., *Mol. Pharmacol.* **5**, 30 (1969).
191. K. C. Agarwal and R. E. Parks, Jr., *Mol. Pharmacol.* **8**, 128 (1972).
192. H. Shimono and Y. Sugino, *Eur. J. Biochem.* **19**, 256 (1971).
193. R. P. Agarwal, E. M. Scholar, K. C. Agarwal, and R. E. Parks, Jr., *Biochem. Pharmacol.* **20**, 1341 (1971).
194. M. P. Oeschger and M. J. Bessman, *JBC* **241**, 5452 (1966).
195. P. R. Brown, R. P. Agarwal, J. Gell, and R. E. Parks, Jr., *Comp. Biochem. Physiol.* **40B**, 777 (1972).
196. R. E. Parks, Jr., P. R. Brown, Y.-C. Cheng, K. C. Agarwal, C. M. Kong, R. P. Agarwal, and C. C. Parks, *Comp. Biochem. Physiol.* **45B**, 355 (1973).

Purification procedures for GMP-dGMP kinase have included fractionation with ammonium sulfate (*186–189, 192, 194*), K$_2$HPO$_4$ (*194*), calcium phosphate gel or hydroxyapatite (*187–189, 192*), and alumina Cγ gel (*189, 192*), chromatography on DEAE-cellulose (*186–189, 192, 194*), and gel filtration on Sephadex (*189, 192, 194*). The enzyme from different sources was also stable to fractionation at pH 5 (*189, 192*), to precipitation with acetone (*186, 194*), to some degree of heat, especially at more purified stages (*186, 187, 192*), and to storage for long periods of time in the frozen state (*186, 189, 194*) or at 4° in the presence of KCl (*192*). A procedure has been reported for the large-scale purification of the enzyme from human erythrocytes (*193*). During various purification procedures the ratio of GMP to dGMP kinase activities could be shown to remain relatively constant (*186, 189, 192, 194*), and the two activities exhibited similar chromatographic behavior (*184, 189, 192*).

2. Assay

Like other nucleoside monophosphokinases, GMP-dGMP kinase has most often been assayed as (1) conversion of radioactive substrate to product with subsequent chromatographic separation of the two (*187, 189*); (2) conversion of the monophosphate to phosphomonoesterase-resistant di- and triphosphate, subjecting the incubation products to the monoesterase and adsorbing the resistant nucleotides onto charcoal (*183, 186, 187, 192, 194, 197*); or (3) formation of diphosphate, linking the reaction to pyruvate kinase and lactate dehydrogenase and continuously monitoring the oxidation of NADH at 340 nm (*186, 188, 194*). Although the phosphorylation of GMP by ATP results in the oxidation of 2 moles of NADH per mole of GMP phosphorylated, since both GDP and ADP are substrates of the pyruvate kinase reaction, this last assay can be made proportional to the GMP kinase concentration in the presence of an excess of the coupled enzymes because the V_{max} of the pyruvate kinase reaction was found to be the same with either ADP or GDP as substrate (*188*).

3. Kinetic and Molecular Properties

GMP-dGMP kinase from several sources exhibited a broad pH optimum in the vicinity of pH 7 (*186, 187, 192, 194*), and with the hog brain enzyme the V_{max} of the reaction (using the coupled spectrophotometric assay), and the K_m for ATP were independent of pH over a range

197. I. R. Lehman, M. J. Bessman, E. S. Simms, and A. Kornberg, *JBC* **233**, 163 (1958).

of two to three pH units *(188)*. The K_m for GMP, however, varied somewhat with pH. The requirement of the enzyme for a divalent cation could be satisfied by Mg^{2+} or Mn^{2+} *(186, 187, 192, 194)*; other metals which were reported to substitute for Mg^{2+} with enzyme preparations from different sources were Fe^{2+} *(187)*, Co^{2+} *(192, 194)*, and Zn^{2+} *(192)*; Zn^{2+} produced more activation with GMP as substrate than with dGMP. The kinase was inhibited by various sulfhydryl reagents *(189)* but did not require the presence of a thiol for stability *(189, 192)*. The molecular weight of the enzyme from rat liver, mouse tumor, or hog brain was observed to be 19,000–20,000, based on gel filtration *(189, 190)*, but that of the highly purified *E. coli* enzyme was 88,000, calculated from the sedimentation equilibrium pattern *(194)*. With agarose gel electrophoresis and isoelectric focusing, six isozymes of GMP–dGMP kinase have been isolated, four from human erythrocytes and two from mouse sarcoma 180 *(191)*. These differed in isoelectric points, and the isozymes from erythrocytes ranged in molecular weight from 18,500 to 24,000, but the substrate specificity and kinetic parameters were similar for all isozymes.

4. Substrate Specificity; Reaction Mechanism

In general, for these various preparations of the enzyme, velocity of the kinase reaction has been greater with the GMP substrate than with dGMP *(186–188, 192, 194)*, while reaction rate with analogs of GMP has been even less *(188–190)*. However, reversibility of the reaction has been demonstrated *(194)*, and equilibrium constants were similar for the reaction of ATP with either GMP or dGMP *(192)*. Apparent K_m values for dGMP have been reported to be $1\text{–}4 \times 10^{-4} M$ with different preparations from normal mammalian tissues *(188, 192)*; a value of $3 \times 10^{-5} M$ was observed for the enzyme from sarcoma 180 *(190)*. In careful kinetic studies with the enzyme from hog brain *(188)* and from erythrocytes *(193)*, the intercepts of the double reciprocal plots were extrapolated to give K_m values for each substrate at saturating levels of the second substrate; these K_m values were, for GMP, $6 \times 10^{-6} M$ and $1.8 \times 10^{-5} M$ for the two enzymes, respectively, and for ATP, $1.2 \times 10^{-4} M$ and $1.9 \times 10^{-4} M$. Activities with GMP and dGMP as substrate were not additive in any of several assays *(189, 194)*; heat inactivation of both activities was similar *(186, 194)*, and each of these substrates served as a competitive inhibitor of the other with a K_i near its K_m value *(192)*. These facts again indicated that the same enzyme protein catalyzes the phosphorylation of both substrates.

Various analogs of GMP also served as inhibitors (8-aza-GMP, 8-bromo-GMP, and 6-thio-GMP) and were more potent against the reaction with dGMP as substrate (*189*). IMP, AMP, dXMP, guanosine, GDP, GTP, and dGTP were also inhibitors (*194*). Inhibition by 6-thio-GMP was shown to be competitive with GMP and noncompetitive with ATP (*188, 190*), and this analog was a potent inhibitor of the enzyme from tumor cells (*190*) as well as that from normal tissues (*188, 189, 193*) (K_i 5–8 × 10^{-5} M). With large amounts of the hog brain enzyme, it could be shown that this resulted from capacity of the analog to serve as an alternate substrate with a very low reaction velocity (*190*). Presumably this, plus subsequent anabolism, accounts for the small amount of incorporation of 6-thioguanine into nucleic acids.

The enzyme also exhibited a high degree of specificity with regard to the phosphoryl donor; with various preparations only ATP or dATP served as effective donor among several nucleoside triphosphates tested (*186, 188, 189, 192, 194*).

The kinase activity has also been found to be stimulated by K$^+$ ions, but for most preparations K$^+$ enhanced primarily the dGMP kinase activity, while the phosphorylation of GMP was unaffected or increased only slightly (*187, 189, 192*). Other monovalent cations that could stimulate activity to a smaller extent were NH$_4^+$ and Rb$^+$, while Cs$^+$ and Li$^+$ were ineffective or inhibitory (*192, 194*). With the *E. coli* enzyme, K$^+$ greatly increased the K_m for dGMP; NH$_4^+$, however, had no such effect on K_m and appeared to activate by a different mechanism (*194*). The effect of K$^+$ has been cited as one of the characteristics distinguishing the dGMP kinase in phage-infected *E. coli* (see below) from the enzyme in uninfected cells; K$^+$ inhibited the dGMP kinase activity in extracts from *E. coli* infected with three different T-even phages (*183*).

Initial velocity studies with the hog brain (*188*) and erythrocyte (*193*) enzymes indicated that the pattern, in the double reciprocal plots of 1/v against 1/[S], was one of intersecting lines, indicating that the reaction proceeded by a sequential addition of both substrates to the enzyme rather than by a ping-pong mechanism of substrate addition and product release.

B. dAMP Kinase

A kinase that phosphorylates dAMP to dADP was also recognized in early studies on the substrate specificity of nucleoside monophosphokinases (*184–187, 198*), and this enzymic activity was partially purified from *Ascaris lumbricoides* (*186*) and from mouse fibroblasts (*187*) as an entity distinct from dGMP kinase and other nucleoside monophospho-

198. Y. Sugino, H. Teraoka, and H. Shimono, *JBC* **241**, 961 (1966).

kinases [ATP: (d)AMP phosphotransferase, EC 2.7.4.11]. It was recently reported that dAMP kinase had been purified 2500-fold from calf thymus (192). The substrate specificity of this enzyme has been found to be analogous to that of GMP-dGMP kinase, i.e., it phosphorylated both AMP and dAMP but had a high degree of specificity with regard to the base constituent of the nucleotide (184, 186, 187). This enzyme therefore overlaps with adenylate kinase(s) which are covered in a separate chapter (Noda, Chapter 8, Volume VIII) and will not therefore be discussed here.

C. dCMP-CMP-UMP KINASE(S)

Early studies on pyrimidine nucleoside monophosphokinases were reviewed in the previous edition of "The Enzymes" (4). An enzyme purified 800-fold from *Azotobacter vinelandii* phosphorylated both dCMP and CMP, and the ratio of the two activities remained constant through the purification (UMP not reported) (199). This enzymic activity is classified as ATP: (d)CMP phosphotransferase, EC 2.7.3.14 (1). Meanwhile, phosphotransferase activities for ATP:UMP and ATP:CMP followed together during several steps of purification from calf liver acetone powder, while ATP:GMP and ATP:AMP kinase activities were fractionated away (dCMP not tested) (185). Ruffner and Anderson (200) purified ATP:UMP-CMP kinase 300-fold from *Tetrahymena pyriformis*, and the ratio of these two activities again was constant during the purification. The purified enzyme also retained appreciable kinase activity for dCMP. Similarly, a preparation of dCMP kinase purified 150-fold from calf thymus by Sugino *et al.* (198) also catalyzed the phosphorylation of CMP and UMP, and the activities for the ribonucleotide substrates bore a nearly constant ratio to that for dCMP throughout the purification. On the other hand, with *E. coli* extracts, CMP and dCMP kinase activities were inseparable during DEAE-cellulose chromatography, but they could be separated from UMP kinase activity (184). CMP and dCMP kinase activities also purified together from *E. coli* (UMP not tested) (201). The structural gene encoding UMP kinase in *Salmonella typhimurium* has recently been mapped, and mutations in this gene did not affect CMP kinase activity (202). Thus, there is at present conflicting evidence on the substrate specificity of this group of enzymes which could have various explanations. Among these are possible dif-

199. F. Maley and S. Ochoa, *JBC* **233**, 1538 (1958).
200. B. W. Ruffner, Jr. and E. P. Anderson, *JBC* **244**, 5994 (1969).
201. J. Hurwitz, *JBC* **234**, 2351 (1959).
202. J. L. Ingraham and J. Neuhard, *JBC* **247**, 6259 (1972).

ferences in the specificity of the enzyme proteins from various organisms, differences in the lability during purification of kinase activity for particular substrates (see, e.g., *198*), or complexities at the catalytic site(s) on the enzyme that could affect substrate specificity. Further purifications that explore these possibilities, as well as more detailed studies with the purified preparations, will be necessary to clarify these substrate specificities and their explanations.

1. *dCMP Kinase Activity*

The dCMP kinase from calf thymus was purified by fractionation at pH 5 and with ammonium sulfate, and by alumina Cγ gel, DEAE-cellulose, and DEAE-Sephadex (*198*). The assay was a modification of that of Lehman *et al.* (*197*) in which ^{32}P-labeled substrate sensitive to phosphomonoesterase (dCMP) is converted to product(s) resistant to the monoesterase (di- and triphosphates), and the esterase is then used to cleave only the monophosphate to nucleoside and ^{32}P$_i$. In Sugino's work, however, the subsequent adsorption of the di- and triphosphate products onto charcoal was eliminated and, instead, the ^{32}P$_i$ released by the monoesterase was precipitated with triethylammonium molybdate and counted; the assay therefore measured, by difference, the decrease in monophosphate substrate (*198*).

In the purification, dCMP kinase was separated from dAMP, dGMP, and dTMP kinases, while CMP and UMP kinase activities were retained and were eluted together with dCMP kinase on the DEAE-cellulose and DEAE-Sephadex columns. The relative rates of activity of the purified preparation with the three substrates were CMP > dCMP > UMP. Kinetic studies also indicated that the three nucleotides were alternate substrates for the enzyme since CMP and UMP were competitive inhibitors of dCMP kinase activity, dCMP and UMP competed with CMP, and dCMP and CMP inhibited UMP kinase activity. The respective apparent K_m values were $1.4 \times 10^{-3} M$ for dCMP, $8.5 \times 10^{-5} M$ for CMP, and $1.7 \times 10^{-4} M$ for UMP. Stoichiometry was also demonstrated for the phosphorylation of each of these three pyrimidine nucleotides by ATP. The reaction with dCMP was also inhibited to a smaller extent by ψUMP and dUMP; dUMP was, however, inactive as a phosphate acceptor in the reaction with this preparation and a dUMP kinase activity was recovered in another fraction separate from the dCMP kinase activity (*198*).

The preparation could use only ATP or dATP as phosphoryl donor, and dATP was about 80% as active as ATP under the assay conditions used. The pH optimum of the dCMP kinase activity was near 9. The requirement for divalent cation could be satisfied equally well by Mg^{2+} or Mn^{2+}, while Co^{2+} and Ni^{2+} could partially substitute.

The dCMP kinase activity of the preparation was lost rapidly, even at 4° or −10°, and it was discovered that the stability of this activity was highly dependent on the presence of a thiol compound such as β-mercaptoethanol, glutathione, cysteine, or reduced lipoic acid (*106, 198*). Furthermore, activity once lost could be completely restored by incubation of the enzyme with a thiol for 1–2 hr at 37°. Thus, the inactivation and reactivation were quite reversible. To measure full activity of dCMP kinase, it was therefore found advisable to preincubate the enzyme with mercaptoethanol prior to assay (*198*). Similar reversible inactivation of dCMP kinase activity was observed in hepatoma extracts (*203*), and it was thought that the low levels of dCMP kinase activity found in certain tumor tissues (*203, 204*) might result from low concentrations of thiol compounds in these tissues (*203*). In *E. coli* extracts and in rat liver extracts dCMP kinase activity was stable in the absence of added thiols. This may again reflect thiol levels in these cells; if liver extracts were treated with Sephadex G-25 to remove low molecular weight compounds, dCMP kinase activity in these extracts was then dependent on added thiol (*203*).

The active and inactive forms of dCMP kinase were compared in sucrose density gradient centrifugation (*106*); it was concluded that conversion to the inactive form did not involve subunit dissociation or association, but that it was accompanied by a slight but distinct change in S value as if there were a conformational change in the molecule. It was found that the inactive dCMP kinase could be reactivated, not only by various thiol compounds but also, and more rapidly, by purified thioredoxin$(SH)_2$ or by catalytic levels of thioredoxin and its reductase in the presence of substrate levels of NADPH (*106, 205*). It has been suggested that a thioredoxin-like protein might function in the regulation of these enzymic activities *in vivo* (*205*).

With the calf thymus enzyme, inactivation was selective for dCMP kinase activity, and UMP and CMP kinase activities of the preparation were not lost in the absence of thiol (*106, 198*). In addition, it appeared that these enzymic activities were affected differently by the concentration of oxidized forms of thiols (*205*). These facts might well be relevant to the substrate specificities recorded for different preparations of these pyrimidine nucleoside monophosphokinases.

203. H. Nakamura and Y. Sugino, *Cancer Res.* **26**, 1425 (1966).
204. R. K. Kielley, *Cancer Res.* **23**, 801 (1963).
205. S. Kobayashi, H. Shimono, and Y. Sugino, *Abstr. Int. Congr. Biochem., 7th, 1967* **4**, 773 (1968).

2. UMP-CMP Kinase Activity

The UMP-CMP kinase purified from Tetrahymena pyriformis (200) also remained in the supernatant upon fractionation at pH 5 and fractionated, with ammonium sulfate, in a fashion resembling that of the dCMP kinase. This enzyme was further purified, to 300-fold, by gel filtration on Sephadex. The purified enzyme was free of UDP kinase, uridine triphosphatase, and adenosine triphosphatase, and was stable for at least 2 weeks at 4° when stored at a concentration of 0.2 mg protein per milliliter.

The assay used measured conversion of radioactive substrate to product, with chromatographic separation of substrate and products. Two assays of this enzyme are presently in use in the author's laboratory: one, a modification of that employed in the earlier studies (200), employs chromatographic separation on thin layers of PEI-cellulose; the second is a continuous spectrophotometric assay that links ADP formation to the pyruvate kinase-lactate dehydrogenase system and follows the oxidation of NADH at 340 nm. The latter assay has proven feasible for studies of CMP phosphorylation since CDP is a very poor substrate of pyruvate kinase; it is not so satisfactory for the assay of UMP phosphorylation, at least for initial velocity studies, because UDP is an appreciable substrate and, unlike the situation with GDP (188), the V_{max} with UDP is very different from that with ADP (206).

The purified Tetrahymena enzyme phosphorylated CMP, UMP, and dCMP at rates decreasing in that order (200). The ratio of CMP to UMP kinase activity remained constant throughout the purification and also in fractions eluted from DEAE-cellulose or CM-cellulose. Both activities were also lost together in a parallel manner during inactivation. The activities were unstable to dilution and were stabilized by bovine serum albumin; activity was also unstable in the presence of tris buffer rather than phosphate. The effect of thiol on dCMP kinase activity was not investigated; the preparations did not contain added thiol since both CMP and UMP kinase activities were very labile to 1 mM dithiothreitol.

CMP was a potent inhibitor of UMP kinase, while UMP was a less effective inhibitor of CMP phosphorylation. Other nucleotides also inhibited the reaction; at levels one-fifth that of monophosphate substrate, CDP, CTP, dCDP, and dCTP were modest inhibitors of CMP kinase and considerably more potent against UMP kinase. However, UDP and UTP also inhibited phosphorylation of UMP more than that of CMP. The substrate concentration giving half-maximal velocity was estimated

206. T. Q. Garvey, III, F. K. Millar, and E. P. Anderson, BBA 302, 38 (1973): Fed. Proc., Fed. Amer. Soc. Exp. Biol. 31, 859 (1972); unpublished experiments.

to be higher for UMP than for CMP, and the differences in the effects of inhibitors were attributed to this dissimilarity in the affinity of the enzyme for the two substrates. The kinetics of the reaction also differed with the two monophosphates; CMP phosphorylation followed Michaelis–Menten kinetics, but UMP phosphorylation did not, and it was suggested that there might be two binding sites on the enzyme for UMP with CMP binding at only one of them (*200*).

As with the dCMP kinase from calf thymus, only ATP or dATP served as phosphoryl donor for the UMP-CMP kinase, but in this case dATP was only one-tenth as active as ATP under the assay conditions used. The pH optimum of UMP kinase activity was between 7 and 8. Inhibition was observed in the presence of high levels of ATP·Mg. The reaction was also inhibited by ADP and dADP, and the ADP inhibition appeared to be competitive with ATP (*200*).

More recent work with the *Tetrahymena* enzyme has explored the mechanism of the reaction through initial velocity studies and isotope exchange in the partial reactions (*206*). The data on the isotope exchange indicated that the enzyme catalyzed neither half-reaction, i.e., exchange of phosphate between one substrate-product pair in the absence of the second substrate. This ruled out a ping-pong reaction mechanism of addition of one substrate and release of its product prior to addition of the second substrate. Initial velocity studies with CMP confirmed this conclusion. Double reciprocal plots of 1/v against 1/[S] at different fixed concentrations of the second substrate gave a pattern of intersecting lines, indicating again that the reaction did not proceed by a ping-pong mechanism but rather by the sequential addition of both substrates to the enzyme to form a ternary complex. Based on replots of the intercepts, the K_m values were found to be $0.5 \times 10^{-3}\,M$ for ATP, $0.8 \times 10^{-3}\,M$ for CMP, and near $1.25 \times 10^{-3}\,M$ for UMP. The kinetic constants for the rate equation were calculated. CDP inhibited the reaction in a manner that was competitive with CMP. The reaction was found to be reversible, and the initial velocity pattern was again intersecting, indicating that the mechanism of the reverse reaction was likewise sequential (*206*).

D. dTMP KINASE

1. *Distribution and Purification; Assay; Stability*

Thymidylate kinase (deoxythymidine 5'-monophosphate kinase, ATP: dTMP phosphotransferase, EC 2.7.4.9) catalyzes the reaction:

$$\text{dTMP} + (\text{d})\text{ATP} \rightarrow \text{dTDP} + (\text{d})\text{ADP}$$

This enzyme is widely distributed. In early studies it was partially purified from *Escherichia coli* (*197, 201*), mouse fibroblast L cells (*187*), calf thymus (*198, 207*), and mammalian tumors (*123, 208*). Highly purified preparations have recently been isolated from *E. coli* (*209*) and mouse hepatoma (*210*). Purification procedures have involved fractionation with acid (*123, 198, 209, 210*), ammonium sulfate (*198, 209, 210*), and DEAE-cellulose or DEAE-Sephadex (*209, 210*), adsorption onto hydroxyapatite (*209*) or alumina Cγ gel (*123*), gel filtration (*123, 209, 210*), and polyacrylamide gel electrophoresis (*209*). The preparation purified 5000-fold from *E. coli* appeared homogeneous in analytical disc gel electrophoresis (*209*).

Like other kinases, this enzyme has usually been assayed either by (1) converting labeled substrate to product and separating the two on ion exchange cellulose (*209, 210*) or on DEAE-Sephadex after destroying unreacted monophosphate with phosphatase (*197, 210*), or (2) coupling the ADP formation to the pyruvate kinase-lactate dehydrogenase system and monitoring NADH oxidation at 340 nm (*209*).

Early work with mammalian dTMP kinase indicated its instability and the fact that it was stabilized by substrate (*211*). The activity of the hepatoma enzyme was maintained through 7500-fold purification by the constant presence of substrate (dTMP) and mercaptoethanol, and by elimination of substrate-destroying phosphatase activity (*210*). Thiol was found to be essential for the protective action of dTMP; dTDP and dTTP were also effective stabilizers (*210*). In contrast to these results with the mammalian preparation, the *E. coli* enzyme did not require the presence of substrate during purification, although mercaptoethanol was added (*209*). The *E. coli* preparation also was not inhibited by sulfhydryl reagents (*209*). Kinase purified from mouse leukemia cells, but not from normal mouse tissues or from *E. coli*, was irreversibly inhibited by diethylstilbestrol, and this inhibition was prevented by sulfhydryl compounds such as dithiothreitol (*212*).

2. *Substrate Specificity*

Purified dTMP kinase from either the bacterial or mammalian source was specific for the phosphorylation of dTMP and dUMP (*209, 210*);

207. Y. Sugino, S. Kobayashi, K. Shimada, Y. Kozai, and K. Hamada, *Annu. Rep. Inst. Virus Res., Kyoto Univ.* **10**, 105 (1967).
208. S. M. Weissman, R. M. S. Smellie, and J. Paul, *BBA* **45**, 101 (1960).
209. D. J. Nelson and C. E. Carter, *JBC* **244**, 5254 (1969).
210. R. K. Kielley, *JBC* **245**, 4204 (1970).
211. T. B. Bojarski and H. H. Hiatt, *Nature* (*London*) **188**, 1112 (1960).
212. D. J. Nelson and C. E. Carter, *Mol. Pharmacol.* **3**, 341 (1967).

ribo-TMP was tested with the *E. coli* enzyme and was not phosphorylated, but 5-iodo-dUMP was active as a phosphate acceptor *(209)*, and presumably other analogs of dUMP or dTMP might likewise be alternate substrates. Both dUMP and 5-iodo-dUMP were competitive inhibitors of the phosphorylation of dTMP; the apparent K_m for 5-iodo-dUMP was 4 times, and the apparent K_m for dUMP 50 times, that for dTMP ($2.4 \times 10^{-4} M$) *(209)*. With the hepatoma enzyme the apparent K_m for dTMP was likewise $1.9 \times 10^{-4} M$, and dUMP was again a much less active substrate *(210)*.

Specificity with regard to the phosphate donor was also evident. With the hepatoma enzyme only ATP and dATP were active as donors, dATP being 70% as active as ATP *(210)*. ATP and dATP were also the most active donors with the *E. coli* enzyme; in this case other triphosphates were also active but considerably less effective *(209)*. Sugino *et al.* *(207)* found that a preparation of dTMP kinase 500-fold purified from calf thymus was specific for ATP or dATP with dUMP as phosphate acceptor but that with dTMP as substrate, GTP and dGTP were also highly active donors, and several other nucleoside triphosphates exhibited less activity.

3. *Kinetic and Molecular Properties*

The dTMP kinase reaction was shown to be stoichiometric and reversible *(209, 210)*. The enzyme from different sources exhibited a pH optimum between 7 and 7.8 *(123, 209, 210)*. The requirement for Mg^{2+} could be partially satisfied by Mn^{2+} and to a smaller extent by Co^{2+} but not by Ni^{2+}, Zn^{2+}, Cd^{2+}, or Ca^{2+} *(209, 210)*. Monovalent cations (Na^+, K^+, or NH_4^+) inhibited the *E. coli* enzyme *(209)*.

The molecular weight of the homogeneous enzyme from *E. coli* was estimated from gel filtration and sucrose density gradient centrifugation to be around 65,000 *(209)*. However, there were indications from gel filtration and sedimentation equilibrium experiments that the purified hepatoma enzyme had a molecular weight closer to 35,000 *(210)*.

Nucleoside triphosphates were not found to be effective inhibitors of dTMP kinase *(209, 210)*. However, Sugino *et al.* *(207)* have reported that, with the purified calf thymus enzyme, plots of reaction velocity against ATP concentration gave sigmoid curves that were "normalized" by the presence of dTTP.

The apparent K_m of the *E. coli* enzyme for ATP was $1.2 \times 10^{-3} M$. With the hepatoma enzyme, the apparent K_m for ATP·2Mg was $1.4 \times 10^{-3} M$ and for ATP·1.5Mg $1.25 \times 10^{-3} M$. The hepatoma enzyme was strikingly inhibited by product ADP or dADP, and to a smaller extent by dTDP *(210)*. Kinetic analysis of the ADP inhibition indicated that it was of a mixed type. Thus, regulation of the enzymic activity might be inherent in the metabolic state of the cell. Mouse liver mitochondria were found to contain large amounts of inactive dTMP kinase which could be activated by disrupting the particles (*213*); the enzyme in mitochondria might exist there in an inhibited state as a result of the high concentrations of ADP in the particles (*210*).

E. OTHER DEOXYNUCLEOTIDE KINASES

1. *5-Hydroxymethyl-dCMP–dGMP–dTMP Kinase*

The DNA of the T-even bacteriophages of *E. coli* (T2, T4, and T6) contains 5-hydroxymethyldeoxycytidylate instead of the deoxycytidylate present in DNA from other sources, including the host cell DNA. An enzyme able to phosphorylate hydroxymethyldeoxycytidylate (dHMP) is induced in phage-infected cells (*214*) and has been studied in detail by Bessman and co-workers (*215, 216*). In uninfected *E. coli* separate kinases phosphorylate the various deoxynucleotides with specificity that depends on the base constituent without regard for the sugar moiety. In T2-infected *E. coli*, dGMP and dTMP kinase activities increase markedly and dHMP kinase appears (*214–217*). It was found that all three of these kinase activities were associated with what appeared to be a single protein (*215, 218*) that was separable in DEAE-cellulose chromatography from all of the four kinase activities of the uninfected cells (*183*). This kinase also exhibited some activity toward dUMP (K_m about $10^{-2} M$), 5-bromo-dUMP, and 5-methyl-dCMP; it was not observed to phosphorylate dAMP or the ribonucleotides (*215, 216*). This enzyme is classified as ATP:(d)NMP phosphotransferase, EC 2.7.4.12 (*1*).

An enzyme with this specificity for the nucleoside monophosphate substrate has been highly purified from both T2 and T4 phage-infected *E. coli* by DEAE-cellulose chromatography, adsorption onto calcium phosphate gel and alumina Cγ gel, fractionation at acid pH and with ammonium sulfate, and chromatography on hydroxyapatite (*215, 216*). Both preparations could use only ATP or dATP as phosphoryl donor;

213. R. K. Kielley, *BBRC* **10**, 249 (1963).
214. A. Kornberg, S. B. Zimmerman, S. R. Kornberg, and J. Josse, *Proc. Nat. Acad. Sci. U. S.* **45**, 772 (1959).
215. L. J. Bello and M. J. Bessman, *JBC* **238**, 1777 (1963).
216. D. H. Duckworth and M. J. Bessman, *JBC* **242**, 2877 (1967).
217. M. J. Bessman, *JBC* **234**, 2735 (1959).
218. M. J. Bessman and L. J. Bello, *JBC* **236**, PC72 (1961).

stoichiometric formation of diphosphate products could be demonstrated, and the reaction was reversible. The T2-induced enzyme had a broad pH optimum for dGMP and dTMP kinase activities near pH 8 and for dHMP kinase activity around 8.6. Mg^{2+}, Mn^{2+}, and, to a lesser extent, Ca^{2+} could fulfill the divalent cation requirement (215). For the T2-induced kinase, the apparent K_m values were 0.56, 0.85, and 2.78 $\times 10^{-4} M$ for dHMP, dGMP, and dTMP, respectively, and 0.82, 1.25, and $5 \times 10^{-3} M$ for ATP with dTMP, dGMP, and dHMP, respectively.

Several lines of evidence suggested that one enzyme catalyzes the phosphorylation of all three substrates.

1. All three activities fractionated together in essentially constant ratio during the purification with either T2- or T4-infected *E. coli* (215, 216). In both cases activities for dGMP and dTMP were about equal, and that for dHMP was somewhat lower.

2. The three activities were lost in a parallel fashion during inactivation by heat (216, 218), low pH (215), 8 M urea (216), or tryptic digestion (215).

3. The different activities were not additive (218), and each of the three substrates served as a competitive inhibitor against the other two with a K_i similar to its K_m value (215).

Genetic evidence also supports the conclusion that only one enzyme protein is involved (216). Four amber mutants of T4 phage that were defective in gene 1 did not induce any of these three kinase activities when grown in the nonpermissive host, *E. coli* B, in which the genetic defect would be evident. Spontaneous revertants of two of these mutants regained all three kinase activities simultaneously (216). These results strongly suggest that a single gene in the phage is responsible for the synthesis of a single protein that catalyzes the phosphorylation of all three deoxynucleotides. A "mutant" enzyme protein was induced by one of these phage mutants in the presence of suppressor gene in the permissive host *E. coli* CR 63; this enzyme was much more heat labile than the normal protein (216).

2. *dAMP-dTMP-dGMP-dCMP Kinase*

The DNA of *E. coli* T5 bacteriophage contains the normal nucleotide dCMP rather than dHMP. Infection with this phage does not induce dHMP kinase; however, kinase activities toward dTMP, dGMP, dCMP, and dAMP all increase after T5 infection (214, 217, 219). Again the data have suggested that, as in the case of the T-even phages, these activities are separable from the corresponding kinases in uninfected

219. M. J. Bessman, S. T. Herriott, and M. J. V. B. Orr, *JBC* **240**, 439 (1965).

E. coli, and that all four are likely associated with a single protein. This enzyme has been classified as ATP:deoxynucleoside monophosphate phosphotransferase, EC 2.7.4.13 (*1*).

This enzyme was purified 500-fold (for dTMP kinase activity) from T5-infected *E. coli*; the purified enzyme phosphorylated the deoxynucleotides with the following relative velocities: dAMP > dTMP = dGMP > 5-bromo-dUMP > dCMP (*219*). In contrast to the nucleoside monophosphokinases in the normal cell, this enzyme did not phosphorylate ribonucleoside monophosphates; this makes it analogous to the enzyme induced by T-even phages. Again only ATP or dATP served as phosphoryl donor. The pH optima for all activities were near 7.0–7.5. Mg^{2+}, Mn^{2+}, and, to a lesser extent, Ca^{2+} or Fe^{2+} activated all four kinase activities. The apparent K_m values were 2.2, 2.2, 1.73, and $0.34 \times 10^{-4}\ M$ for dAMP, dGMP, dTMP, and dCMP, respectively, and $3.6 \times 10^{-4}\ M$ for ATP.

Kinase activities for all four substrates were purified together in nearly constant ratio. Also, each substrate served as competitive inhibitor of the phosphorylation of the other three with a K_i near its K_m value. Thus, data again support the involvement of a single protein; this is then a second instance in which a single enzyme induced by viral infection catalyzes reactions requiring separate enzymes in the uninfected cell.

3

Carbamate Kinase

L. RAIJMAN • M. E. JONES

I. Introduction	97
A. Historical Background	97
B. Distribution	101
II. Molecular Properties	102
A. Purification	102
B. Composition, Size, and Subunit Structure	103
C. Other General Properties	104
III. Catalytic Properties	105
A. Reactions Catalyzed	105
B. Specificity and Cofactors	107
C. Assays	107
D. Thermodynamics, Kinetics, and Catalytic Mechanism	110
IV. Function and Relation to Other Enzymes; Metabolite Control of Activity	115

I. Introduction

A. HISTORICAL BACKGROUND

The discovery that carbamyl phosphate (1) is the universal energy-rich carbamyl donor for two major biosyntheses (uridylic acid and arginine) and a carbamyl or energy-rich phosphate donor in a few more restricted enzyme systems (see Section IV and Chapter 7; 1a) resulted

---

1. The following abbreviations are used: CKase, carbamate kinase; CAP, carbamyl phosphate.

1a. P. P. Cohen, "The Enzymes," 2nd ed., Vol. 6, p. 477, 1962.

from studies on the phosphorolysis of citrulline in *Streptococcus faecalis* (*2*). This organism cannot grow in the absence of arginine for it lacks the enzymes necessary to convert citrulline to arginine (*3*). However, several microbiological laboratories had observed that if this or other microorganisms were provided with arginine, this amino acid could be dissimilated to two moles of ammonia and one mole each of ornithine and CO_2 (*4*).

Cell-free extracts from cells grown on arginine catalyzed the conversion of arginine or citrulline to ornithine, ammonia, and CO_2, if orthophosphate, ADP, and magnesium were added, in which case 1 mole of ATP was formed for every mole of arginine or citrulline utilized. The dissimilation of arginine, therefore, was an alternate source of energy made available by the induction of at least two enzymes after all the glucose was utilized. Indeed, Bauchop and Elsden (*5*) have observed that each mole of arginine dissimilated *in vivo* provides these cells with one-half the amount of energy that they obtain from the normal degradation of glucose to lactic acid.

The cell-free extracts also catalyzed an arsenolysis of citrulline which did not require ADP to yield 1 mole each of ornithine, ammonia, and bicarbonate. The arsenolysis reaction strongly suggested that the phosphorolysis might be a two-step reaction, the first step of which yielded an unknown phosphate compound which could react with ADP in a second step to form ATP.

The phosphorolysis reaction was a poor reaction to study for the identification of a phosphate intermediate, for optimal conditions required nearly 100 mM orthophosphate and citrulline to form about 1 μmole of the unknown intermediate, which was presumably carbamyl phosphate (CAP), δ-N-phosphoornithine, or the N-carboxyphosphate of ornithine. It was obvious, therefore, that to identify the intermediate it would be necessary to reverse the physiological reaction and to form citrulline from ATP, ornithine, ammonia, bicarbonate, and magnesium. In considering the three possible products, a reaction which involved an attack of citrulline by orthophosphate at the secondary amide N atom of citrulline seemed unlikely; a more likely possibility was that an acylphosphate was formed, either by displacement of ammonia to yield an N-carboxyphosphate of ornithine or by displacement of ornithine to yield CAP. This means that if one reverses the physiological reaction and tries to form the unknown intermediate from ATP, conditions ought to favor

2. M. E. Jones, L. Spector, and F. Lipmann, *JACS* **77**, 819 (1955).
3. B. E. Volcani and E. E. Snell, *JBC* **174**, 893 (1948).
4. M. E. Jones and F. Lipmann, *Proc. Nat. Acad. Sci. U. S.* **46**, 1194 (1960).
5. T. Bauchop and S. R. Elsden, *J. Gen. Microbiol.* **23**, 457 (1960).

the formation of a carbamate. Despite the known instability of carbamates in solution, the literature (6) indicated that they were more stable at slightly alkaline pH values. Therefore, the synthesis of the unknown phosphate derivative might occur at a more alkaline pH than pH 6.1, which was optimal for the phosphorolysis (7–10). Indeed, the whole reaction could be reversed to yield citrulline from ATP, ornithine, ammonium carbonate, and magnesium, and as predicted, the pH optimum for citrulline formation from ATP was more alkaline, between pH 8 and 9 (2, 11).

However, the search for the intermediate required additional refinements. The first attempt to identify a new acylphosphate was to see whether a hydroxamate was formed when ATP, carbonate, magnesium, and enzyme were incubated with either ornithine or ammonia, with hydroxylamine either added at the end of the reaction or present during the reaction (12). This search was in vain and, indeed, we now know it was destined to be so since (1) carbamyl phosphate does not react readily with hydroxylamine at 30°, the temperature of the incubation; and (2) even if the hydroxamate of carbamyl phosphate had been formed, it would not give the typical purple color with ferric chloride at micromolar concentrations (although it yields a fading deep blue color if the solution of hydroxyurea tested is near 50 mM) (13).

It was obvious one had to search directly for a labile phosphate compound.

Amounts of enzyme that converted 10 μmoles of ATP to ADP and P_i with the formation of 10 μmoles of citrulline when ammonium carbonate and ornithine were present gave no release of orthophosphate unless all substrates were added. It seemed possible that the new intermediate might be more "energy-rich" than ATP and that we should, therefore, give a thermodynamic "push" to its accumulation by using phosphoenolpyruvate as the initial energy donor with only a small amount of ADP added and an excess of pyruvate kinase. In this system, little ADP would be present to back-react with the energy-rich intermediate, and

6. C. Faurholt, *Kgl. Dan. Vidensk. Selsk., Mat.-Fys. Medd.* **3**, 20 (1921).
7. M. Korzenovsky, *in* "Amino Acid Metabolism" (W. D. McElroy and B. Glass, eds.), p. 309. Johns Hopkins Press, Baltimore, Maryland, 1955.
8. V. A. Knivett, *BJ* **56**, 602 (1954).
9. E. L. Oginsky, *in* "Amino Acid Metabolism" (W. D. McElroy and B. Glass, eds.), p. 300. Johns Hopkins Press, Baltimore, Maryland, 1955.
10. H. D. Slade, *in* "Amino Acid Metabolism" (W. D. McElroy and B. Glass, eds.), p. 321. Johns Hopkins Press, Baltimore, Maryland, 1955.
11. M. E. Jones, "Methods in Enzymology," Vol. 5, p. 903, 1962.
12. F. Lipmann and L. C. Tuttle, *JBC* **159**, 21 (1945).
13. M. Grassl, S. McKinley, and M. E. Jones, *ABB* **129**, 98 (1969).

since phosphoenolpyruvate has a free energy of hydrolysis of -12.8 kcal mole^{-1} (*14*), while ATP has one of about -9 kcal mole^{-1} at pH 8.5, both circumstances should favor the accumulation of an intermediate more energy-rich than ATP. In addition, fluoride ion had to be added to prevent a release of orthophosphate from the phosphoenolpyruvate and ADP which was observed in the absence of the other substrates for carbamate kinase (CKase). Using two phosphate assays, the Lowry-Lopez (*15*) for "true" orthophosphate and the Fiske-SubbaRow (*16*) assay for "labile" phosphate plus orthophosphate, we found the accumulation of an acid-labile phosphate when only ATP, ammonia, and bicarbonate were added: ATP with either ornithine or ornithine and bicarbonate led to no labile phosphate compound. The product, therefore, was CAP.

At this point Dr. Leonard Spector joined Jones and Lipmann to synthesize carbamyl phosphate. He conceived of the incubation of KCNO and KH$_2$PO$_4$ (*2, 17*), and the very first incubation of molar amounts of these two substances at 30° yielded an amount of CAP equal to half of the KH$_2$PO$_4$ originally present; these conditions were later shown to be optimal (*4*). The synthesis allowed the determination of the acid- and base-stability characteristics of CAP which were essential to measure it quantitatively. Finally, the new intermediate was isolated in gram quantities. It reacted with ADP to form ATP (*2*), and with ornithine in the presence of the crude extract to form citrulline (*2*), or with the same extract and aspartate to form carbamyl aspartate (*18*).

Perhaps one of the most exciting aspects of this particular discovery was the speed with which it progressed. The whole period from initial growth of the arginine-adapted cells to the isolation and characterization of grams of CAP was a mere 2 months. It is obvious in retrospect that reasonably pure CAP had already been isolated by Grisolia and Cohen as "compound X" (see Grisolia and Marshall, *19*). However, they thought the material contained carbamyl glutamate. They measured for the latter compound by a method which is much more sensitive to urea, a common contaminant of aged CAP preparations, than to carbamyl glutamate.

14. W. P. Jencks, in "Handbook of Biochemistry" (H. A. Sober, ed.), p. J-148. Chemical Rubber Publ. Co., Cleveland, Ohio, 1968.

15. O. H. Lowry and J. A. Lopez, *JBC* **162**, 421 (1946).

16. C. H. Fiske and Y. SubbaRow, *JBC* **66**, 375 (1925).

17. L. Spector, M. E. Jones, and F. Lipmann, "Methods in Enzymology," Vol. 3, p. 653, 1957.

18. M. E. Jones, L. Spector, and F. Lipmann, *Proc. Int. Congr. Biochem. 3rd, 1955*, p. 278 (1956).

19. S. Grisolia and R. O. Marshall, in "Amino Acid Metabolism" (W. D. McElroy and B. Glass, eds.), p. 258. Johns Hopkins Press, Baltimore, Maryland, 1955.

B. Distribution

Carbamate kinase activity has been detected in a number of microorganisms: Streptococcus faecalis R (2), Serratia marcescens (20), Streptococcus lactis (21), Streptococcus strain D_{10}, group D (22), Mycoplasma hominis type II, strain 07 (23), and perhaps Neurospora crassa (24) and Bacillus subtilis (25). However, the subject of the distribution of this enzyme contains some uncertainties at present.

The synthesis of CAP when ATP, ammonium carbamate or carbonate, and magnesium are the only substrates utilized is frequently called CKase activity; the synthesis of CAP (in the presence or absence of an acylglutamate) from glutamine or ammonia, HCO_3^-, and 2 moles of MgATP per mole of CAP formed is referred to as CAP synthetase activity (26-29). The major glutamine-utilizing synthetase (28) can also form CAP from carbamate or ammonium carbonate; therefore, the simple measurement of the product CAP, without the determination of the amount of ATP utilized, and of ADP and orthophosphate formed, does not establish whether the activity observed is the result of a CKase or of a CAP synthetase.

This problem is compounded because the specificity of the synthetase for glutamine is given by a subunit which can be easily separated from a catalytic subunit in the course of purifications (30, 31); the modified enzyme can then synthesize CAP only from ammonia, and would appear to be a CKase if the stoichiometry were not determined. One of the enzymes of Escherichia coli previously referred to as CKase (32-34) may be a synthetase, and the presumed kinase of Neurospora, which provides

20. K. T. Glasziou, Aust. J. Biol. Sci. 9, 253 (1956).
21. J. M. Ravel, J. S. Humphreys, and W. Shive, ABB 92, 525 (1961).
22. L. C. Mokrasch, J. Caravaca, and S. Grisolia, BBA 37, 442 (1960).
23. R. T. Schimke, C. M. Berlin, E. W. Sweeney, and W. R. Carroll, JBC 241, 2228 (1966).
24. M. F. Lou and R. L. Herrmann, BBA 139, 199 (1967).
25. I. M. Issaly, A. S. Issaly, and J. L. Reissig, BBA 198, 482 (1970).
26. S. Grisolia and P. P. Cohen, JBC 204, 753 (1953).
27. R. L. Metzenberg, L. M. Hall, M. Marshall, and P. P. Cohen, JBC 229, 1019 (1957).
28. B. Levenberg, JBC 237, 2590 (1962).
29. P. R. Tramell and J. W. Campbell, JBC 245, 6634 (1970).
30. P. M. Anderson and A. Meister, Biochemistry 4, 2803 (1965).
31. P. P. Trotta, M. E. Burt, R. H. Haschemeyer, and A. Meister, Proc. Nat. Acad. Sci. U. S. 68, 2599 (1971).
32. L. Gorini and S. M. Kalman, BBA 69, 355 (1963).
33. J. Yashphe and L. Gorini, JBC 240, 1681 (1965).
34. S. M. Kalman, P. H. Duffield, and T. Brzozowski, BBRC 18, 530 (1965).

CAP for arginine biosynthesis (*35*), appears now to be a CAP synthetase (*36*). The same comment may apply to the pyrimidine-specific CAP synthetase (*37*) and presumed CKase (*24*) of that organism.

Cell extracts might contain any combination of these activities (CKase, CAP synthetase, and disrupted CAP synthetase), the distinction among which is obviously in their stoichiometry. Since the latter is difficult to determine unless the enzymes are purified, that information has not always been obtained, and the identity of certain enzymes able to synthesize CAP from ammonia remains open to question.

In other cases, apparent CKase activity is the result of the utilization of carbamate by an acetate kinase. This seems to apply to some *E. coli* mutants which are double auxotrophs for arginine and uracil, yet contain apparently normal levels of "carbamate kinase" (*33, 38, 39*). These acetate kinases have low affinity for carbamate, and their ability to synthesize CAP can be completely inhibited by the addition of acetate (*25, 33*); they are easily distinguishable from CKases.

II. Molecular Properties

A. Purification

Carbamate kinase is generally obtained from microorganisms adapted to growth in an arginine-rich medium (*8, 11*); such adaptation results in a 10- to 40-fold increase in CKase activity in crude extracts of *S. faecalis* (*11*).

The enzyme has been partially purified from many sources and has been obtained in the crystalline state from extracts of *Streptococcus* D_{10} (*40–42*) and of *Streptococcus faecalis* ATCC 11,420 (*43*).

The crystalline enzyme from *Streptococcus* D_{10} sedimentated in a centrifugal field, and migrated on acrylamide gel disc electrophoresis, as a

35. R. H. Davis, *BBA* **107**, 44 (1965).
36. L. G. Williams and R. H. Davis, *J. Bacteriol.* **103**, 335 (1970).
37. L. G. Williams, P. H. Bernhardt, and R. H. Davis, *Biochemistry* **9**, 4329 (1970).
38. S. Grisolia, P. Harmon, and L. Raijman, *BBA* **62**, 293 (1962).
39. K. J. I. Thorne and M. E. Jones, *JBC* **238**, 2992 (1963).
40. This organism, previously referred to as *Streptococcus* strain D_{10}, group D (*22*), would appear to be a strain of *S. faecium*. See Bishop and Grisolia (*41*) and Deibel (*42*).
41. S. H. Bishop and S. Grisolia, *BBA* **118**, 211 (1966).
42. R. H. Deibel, *J. Bacteriol.* **87**, 988 (1964).
43. M. Marshall and P. P. Cohen, *JBC* **241**, 4197 (1966).

homogeneous protein. This preparation synthesized 3100 μmoles of ATP min^{-1} mg^{-1} (38°, pH 7.4) from CAP and ADP. It also synthesized ATP from acetyl phosphate and ADP at a rate 30 times slower than from CAP; the ratio of CKase to acetate kinase, however, remained constant throughout the last stages of purification, suggesting that the ability to utilize acetyl phosphate may be intrinsic to the enzyme rather than a result of the presence of an impurity (*42*).

Crystalline CKase from *S. faecalis* sedimented as a single species in the ultracentrifuge, and only traces of two other components were noticeable on starch gel electrophoresis. The activity of the crystalline enzyme in the direction of ATP synthesis was 2260 μmoles min^{-1} mg^{-1} (37°, pH 7.54) (*43*).

B. Composition, Size, and Subunit Structure

The amino acid composition of crystalline CKase from *S. faecalis* ATCC 11,420 has been determined (*43*); the enzyme is rich in aspartic and glutamic acids, which is in agreement with its migratory characteristics in an electrophoretic field. Treatment with dithiothreitol did not change the number of sulfhydryl groups of the protein.

The minimum molecular weight calculated from the amino acid composition was 31,000. Studies on the binding of ADP to the enzyme indicated that the latter has one binding site for ADP per 33,000 g of protein, a finding which is consistent with the value obtained from the amino acid composition. The authors chose to use 31,000 as the equivalent weight of the enzyme (*43*).

Sedimentation data obtained at a protein concentration of 8.8 mg/ml gave an $s_{20,w}$ of 4.1 S for the crystalline enzyme from *S. faecalis*. This value is similar to the one obtained with the purified enzyme from *Streptococcus* D$_{10}$, group D ($s_{20,w} = 3.9$ S) at a concentration of 3 mg of protein per milliliter (*38*); the molecular weight of this enzyme was estimated to be approximately 66,000 (*44*). On the basis of these data and of the minimum molecular weight obtained from the amino acid composition, Marshall and Cohen (*43*) suggested that the enzyme of *S. faecalis* (and possibly that of *Streptococcus* D$_{10}$) may be a dimer.

The molecular weight of a homogeneous preparation of CKase from *S. faecalis* R, ATCC 8043, was determined (*45*) at a concentration of 10 mg/ml by the method of sedimentation equilibrium. Values of 45,300 and 46,900 were obtained for the weight-average molecular weight and

44. S. H. Bishop, *Fed. Proc., Fed. Amer. Soc. Exp. Biol.* **25**, 523 (1966).
45. S. M. Kalman and P. H. Duffield, *BBA* **92**, 498 (1964).

the z-average molecular weight, respectively. The molecular weight obtained by the method of Martin and Ames (46), using lysozyme as a standard, was approximately 40,000 (45). This molecular weight would seem to suggest that the CKase of this *S. faecalis* is quite different in size from the CKase isolated from *S. faecalis* ATCC 11,420 or *Streptococcus* D_{10} (41, 43). However, since each group of investigators carried out their analyses at vastly different pH values and used different salts as buffers, perhaps no comparison should be attempted at this time.

The molecular weight of the CKase from *M. hominis* type II was estimated by the sucrose density gradient technique (46), using arginine deiminase as a standard; a value of 61,000 was obtained with a kinase preparation of approximately 50% purity (23).

C. OTHER GENERAL PROPERTIES

1. *Stability*

Extensively purified CKases are unstable when in solution in dilute buffers, even at very low temperatures (21, 47). The enzyme from *S. lactis*, for example, underwent an 80% loss of activity on dialysis against 0.04 M tris, pH 8.5, at 4°, for 18 hr. Storage at −20° for 1 week resulted in loss of 50% of the activity, and heating at 60° for 2 min in 85% loss; however, the enzyme was stable under any of those conditions if the solutions contained 0.5 M ammonium sulfate (47). After inactivation, most of the activity was restored by incubation of the enzyme with ammonium sulfate at pH 8.5, at 4°, for 1 hr. Sodium and potassium sulfate were as effective as ammonium sulfate, but other ammonium salts and potassium phosphate were ineffective.

The degree of reactivation depended on the concentration of sulfate present during the incubation for 1 hr, and not on the concentration of sulfate present during assay of the enzyme (21). Salt solutions which did not contain sulfate were ineffective even though their ionic strength might be the same as those of the sulfate solutions; furthermore, high concentrations of other salts did not diminish the amount of sulfate required for reactivation.

The effect of sulfate showed a remarkable temperature dependence, the rate of reactivation increasing approximately 3-fold with each 10° rise in temperature in the range from 0° to 20°.

Purified preparations of enzyme from *S. faecalis* have also been re-

46. R. G. Martin and B. N. Ames, *JBC* **236**, 1372 (1961).
47. J. M. Ravel, R. F. Sund, and W. Shive, *BBRC* **1**, 186 (1959).

ported to lose activity upon dialysis; this was prevented by glutathione (45). Protection from heat denaturation of the enzyme from *M. hominis* was achieved with 2-mercaptoethanol (23). Several purification procedures for different CKases include 2-mercaptoethanol, dithioerythritol, EDTA and glycerol, and glutathione as components of the reagents used (24, 25, 41).

2. Effect of Some Sulfhydryl Reagents

p-Hydroxymercuribenzoate inhibits the CKase from *S. faecalis*; at 10 μM concentration, almost 90% inhibition of the forward and backward reactions was observed. This effect was partially reversed by a 50-M excess of cysteine with respect to *p*-hydroxymercuribenzoate. Silvertris at 10 μM concentration inhibited the reverse reaction 95%. The enzyme was little affected by iodoacetic acid or iodoacetamide; 0.01 M iodoacetamide was required to produce a 42% inhibition of the reverse reaction (45).

III. Catalytic Properties

A. Reactions Catalyzed

Carbamate kinase catalyzes the reversible reaction in Eq. (1). The

$$NH_2\text{—}COO^- + ATP^{4-} \underset{}{\overset{Mg^{2+}}{\rightleftharpoons}} \underset{NH_2}{O=C\text{—}O\text{—}\overset{O^-}{\underset{O_-}{P}}=O} + ADP^{3-} \quad (1)$$

carbamyl donor is carbamate (4). In practice, ammonium bicarbonate or other salts of ammonium and bicarbonate or carbonate have been used as reactants for the synthesis of CAP since carbamic acid is readily formed from such mixtures by reactions (2)–(5).

$$H_2N\text{—}COOH \rightleftharpoons NH_3 + CO_2 \quad (2)$$
$$NH_3 + H^+ \rightleftharpoons NH_4^+ \quad (3)$$
$$CO_2 + H_2O \rightleftharpoons HCO_3^- + H^+ \quad (4)$$
$$\overline{H_2N\text{—}COOH + H_2O \rightleftharpoons (NH_4)HCO_3} \quad (5)$$

Carbamic acid has a pK of 5.8 and would therefore be ionized above pH 7.

The rate at which overall equilibrium is reached is increased by carbonic anhydrase, which enhances the velocity of the reaction described by Eq. (4). This enzyme is present in crude extracts of some bacteria (48) but not in extracts of S. faecalis (39).

Working with purified CKase from the latter organism, Jones and Lipmann (4) showed that during the first 10 min of incubation at 10° the velocity of CAP synthesis from carbamate was at least three times greater than that obtained with equimolar concentrations of freshly prepared ammonium carbonate. After 10 min the CAP that had accumulated when carbamate was the substrate, began to decrease, while that formed from ammonium carbonate continued to increase slightly. These observations could be understood to mean that carbamate was the substrate of CKase in terms of the reaction described by Eq. (5); since equilibrium was approached in opposite directions, the concentration of carbamate decreased with time when it was the initial substrate, but it increased when ammonium carbonate was the substrate. The rate of synthesis of CAP will respond to the changes in concentration of the true substrate. When equilibrium was established between carbamate and ammonium carbonate, the amount of CAP in the incubation mixtures was the same whether carbamate or ammonium carbonate was used as the initial substrate since the CKase was then also at equilibrium. In the experiments of Jones and Lipmann, equilibrium was achieved after 80 min of incubation at 10°. The addition of carbonic anhydrase to incubation mixtures of CKase, MgATP, and freshly prepared carbamate or ammonium carbonate should accelerate the attainment of equilibrium between the last two reagents, and therefore could yield the same rate of CAP synthesis from either substrate. This was in fact observed by Jones and Lipmann (4), and it confirmed their proposal that carbamate is the true substrate of CKases.

Carbamate kinase also catalyzes the reaction described by Eq. (6).

$$CH_3-COO^- + ATP^{4-} \underset{}{\overset{Mg^{2+}}{\rightleftarrows}} CH_3-\overset{O}{\underset{}{C}}-O-\overset{O^-}{\underset{O_-}{P}}=O + ADP^{3-} \quad (6)$$

Early experiments with E. coli mutants requiring arginine and uracil for growth showed that extracts of these cells catalyzed the synthesis of ATP from CAP and ADP (38). Since these mutants were assumed to contain little or no CKase, it was postulated that the CKase activity might result from acetate kinase, reflecting a broader substrate specificity of the latter enzyme (38). A comparative study (39) of the CKase and acetate kinase activities of S. faecalis R (ATCC 8043), E. coli W (ATCC 9637), and E. coli R 185-482 (an arginine- and uracil-requiring mutant) established that a kinase from E. coli W could indeed synthesize CAP or

---

48. W. T. Shoaf, III and M. E. Jones, Arch. Biochem. Biophys. 139, 130 (1970).

acetyl phosphate from ATP and carbamate or acetate, respectively. The authors could not separate these activities, which by several criteria appeared to result from a single protein with very low affinity for acetate ($K_m = 0.1\ M$) and even lower affinity for carbamate ($K_m = 0.17\ M$).

After partial purification of extracts from *S. faecalis* R, two distinct fractions were recovered (*39*). One of them (CKase fraction) contained most of the CKase activity and 25% of the acetate kinase activity detected in crude extracts. The second fraction (acetate kinase fraction of *S. faecalis*) contained 66% of the total acetate kinase activity in crude extracts and had no CKase activity. The CKase fraction could not be resolved into two separate activities; the enzyme utilized acetate although its affinity for this substrate was very low ($K_m = 1.7\ M$) as compared to that for carbamate ($K_m = 2\ \text{m}M$).

It was therefore established that some carbamate kinases also utilize acetate, and some, but not all, acetate kinases can utilize carbamate.

B. SPECIFICITY AND COFACTORS

The utilization of acetate (*39*) and of acetyl phosphate (*41*) by CKase has not been studied exhaustively. The K_m for acetate of the enzyme from *S. faecalis* is $1.7\ M$ (*39*), which fairly well excludes a physiological function for this activity.

Using this enzyme, maximal synthesis of both CAP and acetyl phosphate is achieved with ATP as the phosphoryl donor; equal concentrations of GTP (10 mM) result in little synthesis of carbamyl-P. In view of the high concentration of GTP used, this residual activity might have resulted from contamination of the GTP with ATP. Activity with ITP is negligible (*39*). The enzyme from *S. lactis* cannot utilize GTP, ITP, UTP, or CTP (*21*).

A bivalent cation is required, and Mg^{2+} is the most commonly used. Carbamate kinase from *S. lactis* is also fully active with Mn^{2+} and less so with Fe^{2+} and Co^{2+}; Cd^{2+}, Cu^{2+}, and Pb^{2+} inhibit it (*21*).

An important negative finding is that biotin is not involved in this system, either as a cofactor or as a component of the enzyme (*21, 38*). Furthermore, biotin appears to play no role in the synthesis of CKase (*21*).

C. ASSAYS

A wide variety of conditions has been used with the purpose of overcoming or minimizing certain difficulties inherent to the CKase system, mainly as a result of the relative instability of carbamate and CAP, or of utilizing certain of its advantages such as the broad pH optimum of the reverse reaction and the thermodynamically and kinetically favored synthesis of ATP. The enzyme can be assayed in the direction of CAP synthesis (forward reaction) or of ATP synthesis (reverse reaction).

1. Design of Assays for the Forward Reaction

The major problems are concerned with the relative instability of carbamate and of CAP. The interconversion of carbamate and ammonium bicarbonate or carbonate has already been discussed [see Section III,A and Eq. (5)]. Either ammonium carbamate or carbonate can be used as a source of substrate; preequilibration of these reactants by incubation in the absence of enzyme is essential in order to obtain constant and defined concentrations of substrate (and therefore reliable rates of CAP synthesis).

The stability of CAP under different conditions of pH and temperature has been thoroughly studied (*4, 49*). Knowledge of this is essential since the pattern of decomposition of CAP varies with conditions and is complex; the reader is referred to the original publications for discussion of this subject.

Two general types of assay are used for the measurement of CKase activity in the direction of CAP synthesis. If the experimental conditions are such that the decomposition of the product CAP during incubation can be expected to be negligible (for example, short incubations, low temperature, pH between 2.5 and 8.5), CAP can be allowed to accumulate as such in the reaction mixture; the product is then measured by one of several procedures outlined below. If, on the other hand, the decomposition of CAP can be expected to be significant (long incubations at higher temperatures and pH above 8.5), a coupling system is included in the assay mixture to trap the CAP in the form of a stable compound. The ornithine and aspartate transcarbamylating systems are commonly used; the products are citrulline and carbamyl aspartate, respectively, compounds which can be accurately measured by many methods based on Fearon's reaction (*50–53*).

Accumulated CAP can be measured by differential phosphate analysis. This procedure is based on the fact that, although molybdate enhances

49. C. M. Allen and M. E. Jones, *Biochemistry* **3**, 1238 (1964).
50. W. R. Fearon, *BJ* **33**, 902 (1939).
51. R. M. Archibald, *JBC* **156**, 121 (1944).
52. L. M. Prescott and M. E. Jones, *Anal. Biochem.* **32**, 408 (1969).
53. G. Guthöhrlein and J. Knappe, *Anal. Biochem.* **26**, 188 (1968).

the decomposition of CAP, it does so only slowly (*49*); the method of Fiske and SubbaRow (*16*), with minor modifications (*17*), can be used, or that of Lowry and Lopez (*15*), which is carried out at a higher pH. Alternatively, CAP can be selectively hydrolyzed at alkaline pH (*40*), and the resulting inorganic phosphate measured under conditions which do not cause significant hydrolysis of ATP or ADP (*2, 45, 54*).

At pH above 7, CAP decomposes quantitatively to cyanate (*2, 49*). Since this is a base-catalyzed reaction (*49*), the rate of decomposition is markedly increased by more alkaline conditions, and the reaction is usually carried out at pH about 11 (*17*). The cyanate can be converted to urea by the classic Wohler synthesis by heating in the presence of excess ammonia at or near pH 8.5 (*33, 49*). The resulting urea can be measured by the Fearon reaction mentioned above.

The ADP formed in the forward reaction can be measured by incorporating the pyruvate kinase and lactic dehydrogenase systems into the assay; ATP is formed, and NADH is oxidized. The decrease in NADH can be followed spectrophotometrically (*43*) or fluorimetrically. This method offers the advantage that a continuous record of the reaction rate is obtained.

Finally, ^{14}C-labeled bicarbonate and ^{32}P-labeled ATP have been used: ^{14}C was measured in either citrulline (*24*) or urea (*25*), and ^{32}P incorporated into CAP was measured after removal of the remaining AT^{32}P with Norit (*43*).

2. *Design of Assays for the Reverse Reaction*

The synthesis of ATP can be measured indirectly by determining the residual amount of easily hydrolyzable phosphate (i.e., of CAP) after heating at or near pH 2 (*2, 22*); this type of assay usually includes an ATP-utilizing system, such as hexokinase and glucose, which serves the double purpose of regenerating ADP and synthesizing a more stable phosphate compound from ATP. The disadvantages of this method, those common to assays which measure residual substrate rather than product formed, can be overcome by including a second coupling system. The glucose 6-phosphate which would be synthesized in the basic system described above is linked to the conversion of NADP to NADPH by means of glucose-6-phosphate dehydrogenase; the appearance of NADPH is followed spectrophotometrically at 340 nm (*23*) or fluorimetrically.

The transfer of ^{32}P from labeled CAP to glucose 6-phosphate via ATP has also been used (*43*); ^{32}P from residual CA^{32}P is precipitated with the

54. J. B. Martin and D. M. Doty, *Anal. Chem.* **21**, 965 (1949).

reagent of Sugino and Miyoshi (55), and the radioactive glucose 6-phosphate in the supernatant is counted.

If the reverse reaction takes place in an acid medium, it can be followed by the resulting evolution of CO_2, which is determined manometrically (20, 45).

The substrate and cofactor concentrations which have been used vary greatly, as do the pH, temperature, and length of incubation. The reader is referred to the original publications for detailed descriptions of the assays.

D. Thermodynamics, Kinetics, and Catalytic Mechanism

The apparent equilibrium constant of the reaction [Eq. (7)] at

$$K = \frac{[\text{CAP}][\text{ADP}]}{[\text{carbamate}][\text{ATP}]} \quad (7)$$

10° and pH 9.5 is between 0.037 and 0.048 (4). At 25° and pH 8.14, values between 0.021 and 0.033 have been reported (43). The reaction proceeds without uptake or production of H^+ because the pK_3 and pK_4 of ADP and the pK_4 and pK_5 of ATP are nearly 4 and 7 (56), while the pK_2 of CAP is 4.9 (49) and the pK of carbamic acid is 5.8 (57). The equilibrium constant is independent of pH above 6.8. The rather large difference between the two sets of values reported and the variations within each set most likely result from the different temperatures at which the constant was determined and from experimental difficulties related to the instability of CAP and carbamate. Jones and Lipmann (4) determined the initial concentration of carbamate in their experiments after allowing a mixture of ammonium carbamate and ammonium carbonate to reach equilibrium, then estimating carbamate by the method of Faurholt (6). Marshall and Cohen calculated the initial concentration of carbamate on the basis of the following equilibrium constant (43, 58), measured at 25° and ionic strength 0.2 [Eq. (8)]:

55. Y. Sugino and Y. Miyoshi, *JBC* **239**, 2360 (1964).
56. R. M. Izatt and J. J. Christensen, in "Handbook of Biochemistry" (H. A. Sober, ed.), pp. J-49 to J-139 (see especially pp. J-56 and J-57). Chemical Rubber Publ. Co., Cleveland, Ohio, 1968.
57. J. T. Edsall and J. Wyman, "Biophysical Chemistry," Vol. 1, p. 572. Academic Press, New York, 1958.
58. The constant defined by Eq. (8) has a dimension, molar. Marshall and Cohen (43) did not designate this dimension, perhaps because water is a reactant in this system. A balanced equation for the reaction is given by Eq. (9):

$$HCO_3^- + NH_3 \rightleftharpoons NH_2CO_2^- + H_2O \quad (9)$$

$$K = \frac{(\mathrm{NH_3})(\mathrm{HCO_3^-})}{(\mathrm{NH_2COO^-})} = 0.53 \tag{8}$$

The concentration of carbamate was determined by the method of Faurholt as adapted by Yashphe and Gorini (33). On the basis of an average value of 0.04 for the equilibrium constant of CKase at 10°, the phosphoryl potential of CAP is approximately 1.8 kcal mole^{-1} more negative than that of ATP at pH 9.5; the synthesis of CAP from ATP, therefore, proceeds uphill. If CO_2 and NH_3 are used as substrates, the resulting equilibrium between ammonium carbamate and ammonium carbonate adds a further 0.8 kcal mole^{-1}, making the synthesis of CAP even more uphill. Conversely, ATP can be readily synthesized from CAP (4). This property is at the basis of procedures for the synthesis of ATP labeled with ^{32}P at the γ,β, or β and γ positions; of ADP labeled in the β position (22, 59, 60); and of ^{18}O-labeled ATP (61).

Reported pH optima and K_m values for the substrates of carbamate kinases of different sources are given in Table I. In the direction of CAP synthesis, all of those kinases have alkaline pH optima which occur between pH 8.5 and 9.5. The enzyme from *S. faecalis* ATCC 11,420 appears to have a broad optimum between pH 7.6 and 8.9, at which range the K_m for carbamate remains nearly constant (43). In the reverse direction, maximal activity is constant between pH 4.5 and 8 for the two *Streptococci* studied (40, 43).

It is apparent in Table I that there are enormous variations in the reported K_m values for any one substrate, even if only the kinases from *Streptococci* or, more specifically, from *S. faecalis*, are considered. The K_m for ATP of the enzyme from *S. lactis*, calculated from Lineweaver-Burk plots, was reported to be 3.4 mM in the presence of 10 mM MgCl$_2$ (at pH 8.7) but also to be dependent on the specific concentrations of metal ion and ATP (21). For the kinase from *S. faecalis* R, the K_m for ATP in the presence of 20 mM MgCl$_2$ (at pH 8.5), also calculated from double-reciprocal plots, was reported to be 4.8 mM (39), while at 5 mM Mg^{2+} and pH 8.5 a value of 0.7 mM has been reported (45). In both cases, an assay coupled to the synthesis of citrulline was used. A K_m value for ATP of 8 μM has been reported for the crystalline enzyme from *S. faecalis* ATCC 11,420 (43). Beyond the variations which might be attributable to the relative concentrations of ATP and Mg^{2+}, Marshall and Cohen (43) reported that at 10 mM Mg^{2+}, 0.99 mM carbamate, and pH 8.15, the

59. R. Tanaka, Y. Mano, and N. Shimazono, *BBA* **36**, 262 (1959).
60. M. R. Hokin and L. E. Hokin, *JBC* **234**, 1381 (1959).
61. R. L. Metzenberg, M. Marshall, P. P. Cohen, and W. G. Miller, *JBC* **234**, 1534 (1959).

TABLE I
REPORTED K_m VALUES AND pH OPTIMA OF CKASES OF VARIOUS SOURCES

Enzyme source	Ammonium carbamate (mM)	ATP (mM)	CAP (mM)	ADP (mM)	Mg (mM)	pH optimum		Ref.
						CAP synthesis	ATP synthesis	
S. lactis 8039	10[a]	3.4[b]			2.2[b]	8.9–9.1		21
Streptococcus, strain D_{10}, group D		1[a]			0.5			38
S. faecalis R, ATCC 8043	2	4.8				9.5	4.5–8	39, 41
S. faecalis R, ATCC 8043		0.7		1[c]		8.4		45
S. faecalis ATCC 11,420	0.08	0.008	0.1	0.05		7.6–8.9	5–8	43
M. hominis type II, strain 07		1.4	0.02	0.2		8.3–8.5	8.3–8.5	23
B. subtilis try C_2	8	15			1.7	8.5		25

[a] Approximate value.
[b] Dependent on the specific concentration of ATP and magnesium.
[c] At pH 5.

velocity of the reaction of CKase from *S. faecalis* continues to increase slowly up to concentrations of ATP at least as high as 10 mM; that is, approximately 800 times the apparent K_m for ATP. The same authors calculated K_m values of 80 μM for carbamate and CAP, respectively, using Lineweaver-Burk plots, while Kalman and Duffield (*45*), using an enzyme preparation from *S. faecalis* R, found nonlinear relationships for either of these substrates. There is no explanation for the great differences in the reported K_m values for ATP of the CKases from *S. faecalis*.

Kinetic studies indicate (*43*) that the association constant of MgADP and the enzyme is only four times greater than that of ADP and the enzyme. In this respect, CKase resembles creatine kinase (*62*) and hexokinase in its interaction with ATP (*63*); it differs from carbamyl phosphate synthetase, which requires a large molar excess of Mg^{2+} in order to bind ATP (*64*). The association constant of MgADP and the enzyme is 250 times greater than that of adenine and the kinase, indicating that the phosphate groups may be important in the binding of the nucleotide (*43*); no data are available on the effect of the ribose moiety. The third phosphate of ATP appears to interfere with binding since the association constant of MgATP and the enzyme is 10 times smaller than that of MgADP and the kinase.

Marshall and Cohen (*43*) studied the kinetic and binding properties of the crystalline CKase from *S. faecalis*, and proposed that the reaction catalyzed by CKase proceeds by the sequential mechanism described by the following reactions:

$$E + MgATP \underset{k_{-1}}{\overset{k_{+1}}{\rightleftharpoons}} EMgATP$$

$$EMgATP + CA \underset{k_{-2}}{\overset{k_{+2}}{\rightleftharpoons}} X \underset{k_{-3}}{\overset{k_{+3}}{\rightleftharpoons}} EMgADP + CAP$$

$$EMgADP \underset{k_{-4}}{\overset{k_{+4}}{\rightleftharpoons}} E + MgADP$$

in which E represents the enzyme, CA represents carbamate, and X a ternary complex; Mg^{2+} was present in their experiments at concentrations such that the nucleotides existed almost entirely as the magnesium complexes.

62. S. A. Kuby, T. A. Mahowald, and E. A. Noltmann, *Biochemistry* **1**, 748 (1962).
63. G. G. Hammes and D. Kochavi, *JACS* **84**, 2073 (1962).
64. M. E. Marshall, R. L. Metzenberg, and P. P. Cohen, *JBC* **236**, 2229 (1961).

For the forward and reverse reactions, double reciprocal plots of v and substrate concentration at various fixed concentrations of the other substrate had intercepting patterns suggestive of a sequential mechanism.

Kinetic constants obtained by fitting the data to the initial rate equation for an ordered reaction indicate that the rate-limiting steps are the dissociations of the second products from the enzyme.

Product inhibition analysis identified ADP as the substrate of the reverse reaction which binds to the enzyme first. Carbamate was shown to be noncompetitive with CAP in the reverse reaction; the competition between ATP and ADP was shown in both directions.

On the basis of these data, the authors suggested that the reaction proceeds by an obligatory sequential mechanism (43).

The binding of MgADP to CKase was studied by a modification of the method of Hummel and Dryer (65); the dissociation constant of the enzyme–MgADP complex measured by this method was 6.6 μM. This value is in good agreement with that calculated from kinetic data which was 7.6 μM.

On the basis of the kinetic data reported above, the binding of CAP would be expected to be dependent on ADP. Marshall and Cohen (43) studied the binding of CAP to CKase, alone and in the presence of adenine, which is a competitive inhibitor with ADP and must therefore bind to the free enzyme. The authors found that CAP was bound equally well in the absence or in the presence of adenine. The dissociation constant of the enzyme–CAP complex, measured at a single concentration of CAP, was 0.15 mM. These findings cannot be simply explained by the reaction mechanism postulated by the authors.

Difficulties encountered in studies on the binding of MgATP to the enzyme, and a discrepancy in the value of the equilibrium constant obtained directly and that calculated from kinetic data, leave some questions concerning the mechanism of the reaction unanswered. Despite these problems, the kinetic data favor the interpretation that at substrate concentrations of the order of their Michaelis constants, the reaction takes place according to the proposed obligatory sequential mechanism (43).

Studies with CKase from *Streptococcus* D_{10} were carried out in an attempt to learn whether the enzyme is phosphorylated in the course of the reaction (38). The amount of ^{32}P which remained attached to the enzyme after incubation with CA^{32}P under different conditions, followed by repeated precipitations of the protein, was measured. From 0.02 to 0.03 mole of ^{32}P per mole of enzyme remained bound; it was concluded that CKase is not phosphorylated.

65. J. P. Hummel and W. J. Dryer, *BBA* 63, 530 (1962).

In recent studies on acetate kinase from *E. coli*, Anthony and Spector (*66*) have shown that this enzyme can be phosphorylated by ATP or acetyl phosphate, and that the isolated phosphoenzyme reacts with ADP or acetate to yield ATP or acetyl phosphate, respectively. Since this enzyme can utilize CAP and some CKases can utilize acetyl phosphate (see Section III,A), the question whether CKase is phosphorylated should be reexamined with the aid of the more sensitive methods now available.

IV. Function and Relation to Other Enzymes; Metabolite Control of Activity

In early studies on the phosphorolysis of citrulline, it was recognized that this reaction might be part of a major pathway of energy derivation catalyzed by the arginine dihydrolase system (*7–10*), which can now be represented as follows:

$$\text{Arginine} + H_2O \xrightarrow{\text{arginine deiminase}} \text{citrulline} + NH_3$$

$$\text{Citrulline} + P_i \xleftrightarrow{\text{ornithine transcarbamylase}} \text{CAP} + \text{ornithine}$$

$$\text{CAP} + \text{ADP} \xleftrightarrow{\text{carbamate kinase}} \text{ATP} + CO_2 + NH_3$$

The discovery by Jones *et al.* (*2*) that CAP is the product of the phosphorolytic cleavage of citrulline by ornithine transcarbamylase (as well as the carbamyl donor for the synthesis of citrulline), and the demonstration by these authors of the presence of CKase in extracts of *S. faecalis*, initiated the systematic study of this enzyme in microorganisms. The reversibility of the reaction was established, and the finding that it proceeds more readily in the direction of ATP synthesis lent support to its proposed role as a source of ATP in certain organisms. Since the spontaneous condensation of cyanate and phosphate to form CAP occurs readily, it has been speculated that CAP could have been formed in the preorganismic period and might have provided phosphoryl groups for energy derivation (*4*).

In *S. faecalis* strains grown in media containing some glucose and high concentrations of arginine (*42*), CKase adaptively increases, and the organisms appear to grow on the CAP arising from the phosphorolysis of citrulline as the major, and perhaps the only, source of energy (*4*) after the glucose is utilized.

In studies on the classic arginine dihydrolase, Glasziou (*20*) used

---

66. R. S. Anthony and L. B. Spector, *JBC* **247**, 2120 (1972).

arginine-requiring mutants of *Serratia marcescens*. These cells grow on arginine or citrulline, but not ornithine or CAP. The inability to grow on CAP probably resulted from its spontaneous decomposition in the culture medium as well as from the presence in that organism of a CAP phosphatase; furthermore, it is not certain how permeable the cells were to CAP. The stoichiometry and requirements of CKase from *S. marcescens* were similar to those reported for the enzyme from *S. faecalis*.

The enzymes of the arginine dihydrolase pathway have been shown to exist in several species of nonfermenting *Mycoplasma*. In *Mycoplasma hominis* type II, strain 07, arginine deiminase, ornithine transcarbamylase, and CKase occur in high concentrations, amounting to 10 and 4% of the soluble protein of cell extracts for the first two enzymes, respectively (*23*).

A direct relationship between growth of these cells and the availability of arginine has been demonstrated. With initial arginine concentrations of 2 and 10 mM, growth ceases when the medium is depleted of arginine; it resumes at the same rate as in the initial culture on addition of arginine to the depleted medium to give a 50 mM concentration. Eventually, a new plateau is reached which is not the result of complete arginine depletion but rather of an undetermined factor or factors.

Compounds such as the L-isomers of ornithine, citrulline, lysine, glutamine, glutamic acid, homoarginine, and other analogs of arginine do not stimulate growth. Since ornithine transcarbamylase and CKase are present in the cells, L-citrulline should substitute for arginine, but does not because it does not enter the cells appreciably (*67, 68*). A different group of compounds which might provide energy for growth, including glycerol, succinate, acetate, D,L-β-hydroxybutyrate, and pyruvate, was tested. None of these compounds promoted growth beyond that resulting from the arginine present. In experiments using uniformly labeled [^{14}C]-arginine, Schimke *et al.* (*23*) have shown that the amount of ATP produced by the degradation of arginine is most likely sufficient to provide for the energy requirements of *M. hominis* type II, strain 07, during growth.

Whether the enzymes of the arginine dihydrolase pathway are constitutive or inducible in this organism is not known, since the latter does not grow in the absence of arginine. No data are available on the regulation of the pathway. However, it is known that in resting cultures, citrulline rather than ornithine is the main product of the breakdown of arginine (*67*). This is probably a result of the equilibrium constant of the

67. P. F. Smith, *J. Bacteriol.* **70**, 552 (1955).
68. R. T. Schimke and M. F. Barile, *J. Bacteriol.* **86**, 195 (1963).

ornithine transcarbamylation reaction, which strongly favors citrulline synthesis. In such a system, the reaction would proceed significantly in the reverse direction only if one or both of the products were removed from the medium. This can be assumed to occur during the growth of *M. hominis* (and other organisms utilizing this pathway) when ATP is continually utilized for synthetic purposes. It is interesting in this context that the K_m values of the CKase of this organism for CAP and ADP are low (0.02 and 0.2 mM, respectively) and significantly smaller than the values for the corresponding substrates for the forward reaction, CAP synthesis.

Unlike in other microorganisms in which CKase increases adaptively under certain conditions of growth, providing an alternate source of energy, this enzyme may be part of the main system for energy derivation of *M. hominis* type II [which contains only certain enzymes of the tricarboxylic acid cycle at low levels (*69*)] and other species of *Mycoplasma*. These bacteria are extremely small and contain only limited amounts of DNA. It has been suggested (*23*) that if the physiological conditions of growth of *Mycoplasma* determine that small size and low DNA content are necessary for survival, then an organism with a functioning arginine dihydrolase pathway would have a greater chance to survive, having to code for only three enzymes for its energy supply.

An activity which appears to result from CKase has been described in *Neurospora crassa* (*24*). The enzyme was obtained from extracts of a pyrimidine-requiring strain of *Neurospora* and was partially purified in the presence of 20% glycerol. Activity is induced by growth on limiting uracil: a 9-fold increase in activity was found in crude extracts of acetone powders, and a similar increase (5-fold) was observed in the purified fractions from those extracts. The enzyme requires ATP and Mg^{2+}; it utilizes NH_4^+ and glutamine, but the velocity appears to be four times slower with the latter substrate. This kinase would appear to differ from other CKases, occupying an intermediate position between the latter and the glutamine-dependent CAP synthetases. The stoichiometry of the reaction has not been determined, however, and it is possible that the diminished activity with glutamine might result from alterations of the native protein during purification. The authors suggested that the enzyme is specific for pyrimidine synthesis; this is again unlike other CKases which appear to function predominantly in the direction of ATP synthesis from CAP arising from the degradation of arginine.

Some interesting and important questions are raised by the work of Issaly *et al.* (*25*). These authors have partially purified two CAP-related

69. P. J. van Demark and P. F. Smith, *J. Bacteriol.* **88**, 1602 (1964).

enzymes from *Bacillus subtilis* try C_2. They differ in their pattern of fractionation, in substrate specificity, pH optimum, stability, K_m values for ATP and Mg^{2+}, and the effect of inhibitors. While these differences could be artifacts resulting from alterations of the protein structure in the course of purification, the authors noted that the ratio of these two activities is very reproducibly dependent on the conditions of growth of their cultures. It seems certain that the two enzymic activities reflect the presence of two different proteins. One is a glutamine-dependent CAP synthetase, which is inhibited by UTP and arginine and repressed by arginine and uracil. This appears to be the only CAP synthetase present in this strain of *B. subtilis* since single-step mutations often result in the appearance of auxotrophs with a double requirement for arginine and pyrimidine. The second enzyme utilizes ammonium carbamate, and is in fact inhibited by glutamine. Millimolar arginine, UMP, and UTP are also inhibitory, in decreasing order. This enzyme, apparently a carbamate kinase, is induced by arginine but unaffected by uracil alone; the latter, however, abolishes the induction resulting from arginine. On the basis of the patterns of inhibition and repression, the authors proposed that the synthetase fulfills a biosynthetic role, while the CKase is concerned mostly with the catabolism of arginine, and possibly the synthesis of ATP by a path resembling the arginine dihydrolase system.

However, the fact that the kinase is inhibited by arginine is not consistent with the proposed role of the enzyme, nor is the abolition by uracil of the induction of the kinase by arginine. The latter observation might be "rationalized as a mechanism to salvage carbamyl phosphate from excessive depletion by the kinase when the synthetase is subject to concerted repression by uracil and arginine" (*25*). Further questions are posed by the fact that the first enzyme in the arginine dihydrolase pathway, arginine deiminase, which catalyzes the conversion of arginine to citrulline, has not been demonstrated in this organism, while an alternative degradative pathway for arginine exists (*70, 71*). Moreover, when the standard strain of *B. subtilis* is grown in the presence of arginine, and CKase activity is induced, ornithine transcarbamylase is significantly repressed. This finding is not coherent with the proposed role of the kinase.

Finally Issaly *et al.* (*25*) found that the single mutations resulting in double auxotrophy for arginine and uracil also cause a deficiency in the CKase. The synthetase and kinase have different regulatory patterns and would appear not to be a part of the same operon. The authors have no

70. E. J. Laishley and R. W. Bernlohr, *BBA* **167**, 547 (1968).
71. G. De Hauwer, R. Lavalle, and J. M. Wiame, *BBA* **81**, 257 (1964).

explanation for this puzzling effect; they proposed, as a working hypothesis, that the two enzymes may share structural or regulatory elements.

Although the different carbamate kinases so far studied are activated or inhibited by related metabolites, it is not known how important the regulation by metabolites is *in vivo*. Reliable data on the intracellular content of metabolites in bacteria are mostly unavailable; however, a technique has been developed which allows the measurement of different compounds under conditions such that the values obtained should closely approach those *in vivo* (*72*). That information would permit a more complete evaluation of the physiological role of carbamate kinases and of the mechanisms by which they are regulated.

72. O. H. Lowry, J. Carter, J. B. Ward, and L. Glaser, *JBC* **246,** 6511 (1971).

4

N^5-Methyltetrahydrofolate-Homocysteine Methyltransferases

ROBERT T. TAYLOR • HERBERT WEISSBACH

I. Introduction 121
II. B_{12} Methyltransferase from *Escherichia coli* 122
   A. Assay and Purification 122
   B. Physical Properties 124
   C. Catalytic Properties 127
   D. Alkylation Studies with Radioactive N^5-Methyl-H_4-folate
     and the Light Stability of a Methyl-B_{12} Enzyme . . . 137
   E. Studies on the Role of S-Adenosyl-L-methionine . . . 143
   F. Mechanism of N^5-Methyltetrahydrofolate-Homocysteine
     Transmethylation 151
III. Non-B_{12} Methyltransferase from *Escherichia coli* 154
   A. Assay and Purification 154
   B. Physical Properties 155
   C. Catalytic Properties and Binding of Folate Substrate . . 156
   D. Repression of Enzyme Synthesis 158
IV. Other Sources of Non-B_{12} and B_{12} N^5-Methyltetrahydrofolate-
Homocysteine Methyltransferases 160
   A. Non-B_{12} Methyltransferases 160
   B. B_{12} Methyltransferases 162

I. Introduction

The terminal reaction in *de novo* methionine biosynthesis involves a methyl group transfer from 5-methyl-H_4-folate (*1*) to homocysteine.

1. Abbreviations: 5-methyl-H_4-folate, l,N^5-methyltetrahydrofolate (L-monoglutamate); 5-methyl-H_4-folate (Glu$_3$), l,N^5-methyltetrahydrofolate (γ-L-triglutamate);

In *Escherichia coli* there are two enzymes with quite different characteristics that catalyze this conversion (*1a–3*). The two reactions are shown below:

5-Methyl-H$_4$-folate (Glu$_1$, Glu$_3$, etc.) + homocysteine $\xrightarrow[\text{AMe}]{\text{reducing system}}$

methionine + H$_4$-folate (Glu$_1$, Glu$_3$, etc.) (1)

5-Methyl-H$_4$-folate (Glu$_2$, Glu$_3$, etc.) + homocysteine $\xrightarrow[\text{phosphate}]{\text{Mg}^{2+}}$

methionine + H$_4$-folate (Glu$_2$, Glu$_3$, etc.) (2)

Reaction (1), which is catalyzed by a cobalamin-containing protein, also requires catalytic levels of a reducing system and AMe (*1*; reviews, references *4–6*). Methyl-H$_4$-folates containing one or more glutamates can function as the methyl donor in this reaction. The enzyme that catalyzes reaction (2) does not contain a B$_{12}$ (*1*) prosthetic group, and the only requirement is Mg^{2+} ions, although a stimulation by inorganic phosphate has been observed. Poly-L-glutamate forms of 5-methyl-H$_4$-folate, but not the monoglutamate derivative, can function as substrates in this reaction (*6, 7*). This report will emphasize the characteristics of both types of methyltransferase isolated from *E. coli* and the mechanism by which the B$_{12}$-dependent enzyme catalyzes reaction (1).

II. B$_{12}$ Methyltransferase from *Escherichia coli*

A. Assay and Purification

1. *Methods of Assay*

Catalysis of reaction (1) can be determined by measuring either the formation of H$_4$-folate or methionine. H$_4$-Folate is formed stoichiometri-

---

AMe, *S*-adenosyl-L-methionine; AH, *S*-adenosyl-L-homocysteine; B$_{12}$ is used to denote various cobalamins, e.g., cyano-B$_{12}$, cyanocobalamin; methyl-B$_{12}$, methylcobalamin; and propyl-B$_{12}$, propylcobalamin; B$_{12r}$, a one-electron reduced derivative of aquo-B$_{12}$ containing Co^{2+}; B$_{12s}$, a two-electron reduced derivative of aquo-B$_{12}$ containing Co^{1+}; K$_2$HPO$_4$, potassium phosphate buffer, pH 7.4.

1a. D. D. Woods, M. A. Foster, and J. R. Guest, *in* "Transmethylation and Methionine Biosynthesis" (S. K. Shapiro and F. Schlenk, eds.), p. 138. Univ. of Chicago Press, Chicago, Illinois, 1965.

2. M. A. Foster, G. Tejerina, J. R. Guest, and D. D. Woods, *BJ* **92**, 476 (1964).

3. J. M. Buchanan, H. L. Elford, R. E. Loughlin, B. M. McDougall, and S. Rosenthal, *Ann. N. Y. Acad. Sci.* **112**, 756 (1964).

cally with methionine (8, 9) and can be converted quantitatively with formic acid to 5,10-methenyl-H_4-folate. The latter is measured by its absorption at 350 nm (9). Measurement of the methionine produced is a more sensitive method of assay. Initially, it was determined by microbiological assay (10–12); however, the chemical synthesis of dl-[5-^{14}C]-methyl-H_4-folate from dl-H_4-folate and [^{14}C]formaldehyde (13) led to the development of a simple tracer assay in which [^{14}C]methylmethionine is separated from the unreacted dl-[5-^{14}C]methyl-H_4-folate by an ion exchange procedure (14).

The absolute amount of B_{12} methyltransferase activity observed depends markedly on the reducing system. But a convenient system for routine assays contains in a total volume of 0.2 ml: K_2HPO_4 (1), 20 μmoles; dl-[5-^{14}C]methyl-H_4-folate (2000 cpm/mμmole), 30 mμmoles; L-homocysteine, 50 mμmoles; AMe, 10 mμmoles; 2-mercaptoethanol, 40 μmoles; cyano-B_{12}, 10 mμmoles; and enzyme (15). After a 15-min incubation at 37°, catalysis is terminated with 0.8 ml of ice cold water. The diluted mixture is placed on a 0.5 × 3.0 cm column of Bio-Rad AG1-X8 (chloride) (15, 16) and the effluent fluid containing [^{14}C]methylmethionine is assayed for radioactivity.

2. Purification

The B_{12} protein has been partially purified from extracts of *E. coli* B grown commercially in cyano-B_{12} supplemented media (15). A sum-

---

4. H. Weissbach and R. Taylor, *Fed. Proc., Fed. Amer. Soc. Exp. Biol.* **25**, 1649 (1966).
5. H. A. Barker, *BJ* **105**, 1 (1967).
6. H. Weissbach and R. T. Taylor, *Vitam. Horm. (New York)* **28**, 415 (1970).
7. R. L. Blakley, in "Frontiers of Biology: The Biochemistry of Folic Acid and Related Pteridines" (A. Neuberger and E. L. Tatum, eds.), p. 332. North-Holland Publ., Amsterdam, 1969.
8. A. R. Larrabee, S. Rosenthal, R. E. Cathou, and J. M. Buchanan, *JBC* **238**, 1025 (1963).
9. S. Rosenthal, L. C. Smith, and J. M. Buchanan, *JBC* **240**, 836 (1965).
10. B. F. Steel, H. E. Sauberlich, M. S. Reynolds, and C. A. Baumann, *JBC* **177**, 533 (1949).
11. F. Gibson and D. D. Woods, *BJ* **74**, 160 (1960).
12. F. T. Hatch, A. R. Larrabee, R. E. Cathou, and J. M. Buchanan, *JBC* **236**, 1095 (1961).
13. J. C. Keresztesy and K. O. Donaldson, *BBRC* **5**, 286 (1961).
14. H. Weissbach, A. Peterkofsky, B. G. Redfield, and H. Dickerman, *JBC* **238**, 3318 (1963).
15. R. T. Taylor and H. Weissbach, *JBC* **242**, 1502 (1967).
16. R. T. Taylor, *ABB* **144**, 352 (1971).

TABLE I
PURIFICATION OF B_{12} TRANSMETHYLASE FROM 2 KG OF *E. coli* B[a]

Fraction	Volume (ml)	Total activity (k-units)	Protein (mg/ml)	Specific activity (units/mg)	Yield (%)
1. Extract	9,400	10.0	38.3	26.5	100
2. Manganese chloride supernatant solution	10,200	8.9	20.0	43.0	89
3. Ammonium sulfate and protamine sulfate	600	4.6	67.0	115	46
4. pH 4.4 precipitation	345	3.9	56.0	200	39
5. First DEAE-Sephadex	1,000	2.9	5.3	555	29
6. Hydroxylapatite	105	1.9	10.0	1,800	19
7. Second DEAE-Sephadex	145	1.45	2.4	4,200	14.5
8. Sephadex G-200	60	1.1	3.3	5,500	11
9. Third DEAE-Sephadex	102	0.8	1.0	7,600	8

[a] From Taylor and Weissbach (*15*).

mary of this purification is given in Table I. One unit of activity is defined as the amount of enzyme that catalyzes the formation of 1 mμmole of [^{14}C]methylmethionine per 15 min at 37° in a 2-mercaptoethanol + cyano-B_{12} reducing system (*15*). The details of the purification and a recent modification of the early step sequences have been published (*15, 17*). B_{12} Enzyme that has been subjected to the steps in Table I sediments as a single major species in the ultracentrifuge, yet it is definitely not homogeneous. Cellulose-polyacetate and polyacrylamide-gel electrophoresis further separate these enzyme preparations into several closely spaced protein bands, only one of which is active (*16*).

B. PHYSICAL PROPERTIES

1. *Absorption Spectrum and B_{12} Chromophore*

Partially purified preparations of B_{12} methyltransferase (*15, 17*) contain 1–2.5 mμmoles of firmly bound cobalamin per milligram of protein. Because of this bound B_{12}, the enzyme preparations are distinctly salmon-colored and in the visible region display absorption maxima at 355, 405, and 475 nm plus a shoulder at 530 nm (*15*). Their absorption spectra closely mimic that of B_{12r} (*1, 18*). However, unlike B_{12r}, the enzyme-bound cobalamin does not yield an electron spin resonance (ESR) spec-

17. R. T. Taylor, *ABB* **137**, 529 (1970).
18. H. Diehl and R. Murie, *Iowa State Coll. J. Sci.* **26**, 555 (1952).

trum, and it is not oxygen labile (15). It has been pointed out (19, 20) that the spectrum of the B_{12} protein also resembles that of a cobalamin at low pH values, where the 5,6-dimethylbenzimidazolyl nucleotide is not coordinated to the cobalt (21).

Attempts to identify the cobalamin chromophore by treatment with hot ethanol in the dark have yielded sulfito-B_{12} as the major corrinoid in the extracts (19, 22). It remains questionable, though, whether the enzyme as usually prepared (15, 17) actually does contain sulfito-B_{12} or whether the sulfito-B_{12} is an artifact of the alcohol stripping and subsequent identification procedures (19, 22). Sulfito-B_{12} is a light-labile cobalamin which forms nonenzymically by exposing aquo-B_{12} to bisulfite (23). The salmon-colored B_{12} protein will react with alkaline cyanide to give a completely different absorption spectrum that is typical for the dicyano derivative of vitamin B_{12} (15). Subsequent extraction of the chromophore from the protein then yields only dicyano-B_{12} (19).

2. Resolution–Reconstitution and Molecular Weight

Selective treatment of the B_{12} protein with $6 M$ urea + 1,4-dithiothreitol resolves it into a colorless apoprotein plus B_{12r} (17). Release of the bound cobalamin as B_{12r}, rather than sulfito-B_{12}, results from its reduction by the 1,4-dithiothreitol which is added to stabilize the much more labile apoenzyme (23a). Upon Sephadex G-25 filtration, apoenzyme retaining only about 5% of the original B_{12} is separated from the resolution mixture (17). Incubation of the resolved apoenzyme with methyl-B_{12} results in the spontaneous formation of a methyl-B_{12} holoenzyme. The binding of methyl-B_{12} resulting in the transformation of apoenzyme into holoenzyme requires no ancillary protein fractions or cofactors, and it is temperature dependent (17). From Table II it can be seen that reconstitution requires a complete corrinoid, i.e., a cobalamin. The ligand at the sixth coordinating site is equally important since methyl-B_{12} yields predominantly holoenzyme but deoxyadenosyl-B_{12} and cyano-B_{12} promote no conversion. Sulfito-B_{12} is bound, but its sulfonate group is reductively cleaved when one tests for holoenzyme activity in a thiol-containing reducing system (17). From these resolution–reconstitution studies it was concluded that all the bonds between the B_{12} and the apoprotein must be noncovalent in nature.

19. R. Ertel, N. Brot, R. Taylor, and H. Weissbach, *ABB* **126**, 353 (1968).
20. H. P. C. Hogenkamp, *Annu. Rev. Biochem.* **37**, 225 (1968).
21. J. A. Hill, J. M. Pratt, and R. J. P. Williams, *J. Theor. Biol.* **3**, 423 (1962).
22. S. Takeyama and J. M. Buchanan, *J. Biochem. (Tokyo)* **49**, 578 (1961).
23. F. Wagner, *Annu. Rev. Biochem.* **35**, 405 (1966).
23a. Various forms of the B_{12} protein are defined as in reference 17.

TABLE II
CONVERSION OF UREA-RESOLVED APOENZYME TO HOLOENZYME
WITH VARIOUS CORRINOIDS[a,b]

Compound present in preliminary incubation	Holoenzyme (%)
None	3
Methyl-B_{12} (or propyl-B_{12} and subsequent photolysis)	77–80
Deoxyadenosyl-B_{12}	4
Sulfito-B_{12}	48–52
Photolyzed sulfito-B_{12}	15–18
Hydroxy-B_{12}, 10–70 mμmoles	8–13
Cyano-B_{12}	6
B_{12r}[c]	11
Diaquocobinamide	4
Methylcobinamide, 10–50 mμmoles	3–4

[a] From Taylor (17).

[b] Mixtures (0.2 ml) containing apoenzyme, 1.0 mg; the indicated corrinoid compounds, 10 mμmoles of each except where noted; and K_2HPO_4, pH 7.4, 20 μmoles, were incubated in the dark 5 min at 37°. Unbound corrinoids were removed with Norite cellulose columns, and the percentages of holoenzyme were assayed (17, 24) with identical aliquots of the column effluent solutions, both before and after they were illuminated (100 W, 15 min, 20 cm, 0°).

[c] B_{12r} was generated by photolysis of a duplicate deoxyadenosyl-B_{12} system under an H_2 atmosphere and was incubated under H_2 gas.

In vitro, the conversion of initial B_{12} holoenzyme (23a) into apoenzyme and back to a methyl-B_{12} holoenzyme is accompanied by a gross, but reversible, change in the shape of the protein (17, 25). Initial (purified) B_{12} holoenzyme has a sedimentation coefficient of 7.0 S (15, 17). Apoenzyme has a sedimentation coefficient of only 6.2 S (17) but a larger Stokes radius than either the initial or the reconstituted forms (23a) of holoenzyme (Table III). Sephadex G-200 chromatography of apoenzyme in 6 M urea + 1,4-dithiothreitol gives no indication that loss of the B_{12} chromophore causes the apoenzyme to dissociate into subunits (25). Based on their sedimentation coefficients in sucrose gradients and their elution from a calibrated Sephadex G-200 column, all forms of B_{12} holoenzyme have a molecular weight of 140,000–150,000 (15, 17, 25). A 9-Å decrease in the Stokes radius of apoenzyme (Table III) cannot be effected with methylcobinamide (25) which also cannot transform apoenzyme into active holoenzyme (Table II). An attachment between the 5,6-dimethylbenzimidazolyl nucleotide and the apoprotein is appar-

24. H. Weissbach, B. Redfield, and H. Dickerman, *BBRC* **17**, 17 (1964).
25. R. T. Taylor, *BBA* **242**, 355 (1971).

TABLE III
Stokes Radii of Urea-Resolved Apoenzyme and Three Holoenzymes Determined by Sephadex G-200 Chromatography[a]

Form of enzyme	No. of experiments	K_D	Stokes radius[b] (Å)	f/f_0
Initial B_{12} holoenzyme	6	0.215	54.2 ± 1.1	1.55
Urea-resolved apoenzyme	4	0.156	62.9 ± 0.3	1.79
Reconstituted methyl-B_{12}-^{3}H holoenzyme	4	0.219	53.6 ± 0.4	1.53
Propyl-B_{12} enzyme[c]	4	0.209	55.1 ± 0.3	1.57

[a] From Taylor (25).

[b] Values are the mean ± S.D.

[c] Prepared by cobalt alkylation with propyl iodide and 90% propylated (27). It was chromatographed in the dark and then column fractions were assayed after photolysis of the cobalt–propyl bond (27).

ently necessary in order to return the loosely folded apoenzyme to a more compact structure.

Using a molecular weight of 140,000, purified enzyme with 2.5 mμmoles/mg of bound B_{12} (Table I, fraction 9) would contain 0.35 mole of cobalamin per mole of protein. The estimated purity of such a preparation is, therefore, only about 33%, assuming that there is only one molecule of bound B_{12} per molecule of enzyme. It was reported by Stavrianopoulos and Jaenicke (26) that a homogeneous preparation of B_{12} enzyme was obtained from *E. coli* (strain unspecified) by a series of steps similar to those in Table I. Its molecular weight, too, was about 140,000, although its B_{12} content of 0.5 mole/mole of protein (26) would not be compatible with a homogeneous protein.

C. Catalytic Properties

1. Specific Inhibition by Propyl Iodide

Although the cobalamin chromophore is tightly bound to the apoprotein, its cobalt atom still possesses the chemical reactivity (23) of unbound aquo-B_{12} or cyano-B_{12}. In particular, the cobalt can be reductively alkylated with propyl iodide (27, 28) to give an inhibited, propyl-B_{12} enzyme [reaction (3)].

26. J. Stavrianopoulos and L. Jaenicke, *Eur. J. Biochem.* **3**, 95 (1967).
27. R. T. Taylor and H. Weissbach, *JBC* **242**, 1509 (1967).
28. N. Brot and H. Weissbach, *JBC* **240**, 3064 (1965).

$$\underset{\text{Initial}}{\underset{\text{B}_{12} \text{ enzyme}}{\overset{|}{\underset{|}{\text{Co}}}\text{Enz}}} + CH_3-CH_2-CH_2-I \xrightarrow[\text{dark}]{\text{reducing system}} \underset{\text{Inactive propyl-B}_{12} \text{ enzyme}}{\overset{CH_2-CH_2-CH_3}{\underset{|}{\overset{|}{\text{Co}}}\text{Enz}}} + I \quad (3)$$

Reversal of the inhibition by a short exposure to light (28) provided a strong clue that the site of blockage in unpurified enzyme preparations is the bound B_{12} group. Alkyl-B_{12} compounds are extremely photosensitive because of the ease with which a carbon-to-cobalt bond is cleaved by light under aerobic conditions (23). When millimicromole quantities of purified B_{12} enzyme became available (15), direct proof was obtained (27, 29) for reaction (3) as well as reaction (4).

$$\underset{\text{Inactive propyl-B}_{12} \text{ enzyme}}{\overset{CH_2-CH_2-CH_3}{\underset{|}{\overset{|}{\text{Co}}}\text{Enz}}} \xrightarrow{\text{light}} \underset{\text{Active aquo-B}_{12} \text{ enzyme}}{\overset{|}{\underset{|}{\text{Co}}}\text{Enz}} + [CH_3-CH_2-CH_2] \quad (4)$$

Treatment of purified enzyme with propyl iodide in a $FMNH_2$ + 1,4-dithiothreitol reducing system yielded an inactive B_{12} protein with an altered chromophore spectrum. Light restored activity and simultaneously generated an absorption spectrum resembling that of aquo-B_{12} (27). When [1-^{14}C]propylbromide was the alkylating agent, the inhibited B_{12} protein was radioactive, but most of the ^{14}C was reversibly lost upon photolysis (27). By extracting the cobalamin from propyl iodide-treated enzyme in the dark, it was possible to identify the prosthetic group by spectral and paper chromatographic procedures as propyl-B_{12} (29). The ability to inhibit selectively the B_{12} methyltransferase with propyl iodide and reverse the inhibition with light has been a powerful aid in establishing whether or not this enzyme is involved in the catalysis of other reactions.

If the reader refers to either of the two initial papers (27, 28) on chemical propylation of *E. coli* B_{12} methyltransferase, an obvious discrepancy will be noted regarding the incubation conditions needed. S-Adenosyl-L-methionine was required in the earlier study using unpurified B_{12} enzyme (28), but it actually prevented propylation of the purified B_{12} protein in a $FMNH_2$ + 1,4-dithiothreitol reducing system (27). This anomaly has been ascribed to the presence of contaminating enzymes in the unpurified

29. R. T. Taylor and H. Weissbach, *ABB* **123**, 109 (1968).

B_{12} enzyme preparation (30) which degraded AMe to homocysteine. It will become apparent later in the discussion why the inhibition of propylation by AMe (27, 30) provided an important clue as to the role of AMe in reaction (1).

2. *Methyl Group Transfer Reactions Catalyzed by the Enzyme*

The biosynthesis of methionine via reaction (1) is the important catalytic function of the *E. coli* B_{12} methyltransferase. Table IV illustrates its unique dependencies on both a reducing system and AMe. While 2-mer-

TABLE IV
REQUIREMENTS FOR METHYL GROUP TRANSFER FROM
[5-^{14}C]METHYL-H$_4$-FOLATE TO HOMOCYSTEINE[a]

Reaction mixture	Methionine formed (mμmoles)
Experiment A[b]	
Complete system (H$_2$)	6.4
− Enzyme	0
− 2-Mercaptoethanol	0.06
− Cyano-B$_{12}$	1.2
− AMe	0.08
− Homocysteine	0.8[c]
Complete system (aerobic)	4.4
Experiment B[d]	
Complete system (H$_2$)	4.7
− Enzyme	0
− 1,4-dithiothreitol	2.9
− AMe	0.02
− Homocysteine	0.14
− FMNH$_2$ and platinum	0.14

[a] From Taylor and Weissbach (15).

[b] The reaction conditions for experiment A are described in the text in Section II,A,1, except that an H$_2$ gas phase was used unless indicated otherwise. Mixtures (0.2 ml) contained 2.4 μg of B$_{12}$ enzyme and were incubated for 15 min at 37°.

[c] In the absence of homocysteine, 2-mercaptoethanol (0.2 M) functions also as a methyl group acceptor. In the complete system, however, saturation of the B$_{12}$ enzyme with homocysteine (0.25 mM) completely eliminates any methyl group transfer to 2-mercaptoethanol and renders this assay method specific for radioactive methionine.

[d] For experiment B the complete system (0.2 ml) contained the same amounts of buffer, AMe, homocysteine, and *dl*-[5-^{14}C]methyl-H$_4$-folate as in experiment A, but only 0.7 μg of B$_{12}$ enzyme. The reducing system consisted of FMNH$_2$, 50 mμmoles; platinum, 0.1 mg; and 1,4-dithiothreitol, 5.0 μmoles under a H$_2$ gas phase.

30. R. T. Taylor, C. Whitfield, and H. Weissbach, *ABB* **125**, 240 (1968).

captoethanol + cyano-B_{12} can be used as a convenient artificial reducing system (15), $FMNH_2$ + 1,4-dithiothreitol (Table IV) has consistently been observed (15, 31–33) to sustain the highest rate of catalysis. Most investigators have used $FADH_2$, generated either enzymically (26, 34, 35) or chemically (3, 35, 36). However, in the presence of 1,4-dithiothreitol and NADH, reduced flavin bound to diaphorases and lipoamide dehydrogenases will substantially replace the need for free $FMNH_2$ or $FADH_2$ (33). The coproduct of reaction (1) (H_4-folate) will also serve, at high concentrations, one-third as well as $FMNH_2$ as a reducing agent (31). Both AMe and reduced flavin function catalytically in reaction (1) (14, 34, 35), the latter even when it is bound in a flavoprotein (33). When one uses a 2-mercaptoethanol + cyano-B_{12} reducing system (Table IV), the vitamin only functions nonenzymically to accelerate reduction by the thiol (3) as described by Peel (37).

Galivan and Huennekens (38) reported that a 3000 molecular weight "S" protein derived from extracts of E. coli K-12 is required for the B_{12} protein to catalyze reaction (1). They concluded that the "S" protein is an essential subunit of the latter, but exogenous methyl-B_{12} (or aquo-B_{12}) plus a dithiol compound or an NADH-dependent flavoprotein fraction were also essential in their system (38). It seems probable that the "S" protein is merely a component of the supplementary reducing system since no evidence for an essential "S" protein has been found by other investigators (9, 15, 25, 34, 39) who have used purified E. coli B_{12} enzyme.

In the presence of an excess of B_{12} enzyme, reaction (1) is catalyzed essentially to completion (8), and for all practical purposes it is unidirectional. Rüdiger and Jaenicke (40) found it to be slightly reversible and published an equilibrium constant of 1.4×10^5 in the forward direction. 5-Methyl-H_4-folate containing one or more L-glutamates can serve as a substrate in reaction (1) (8); however, 5-methyl-H_4-folate (Glu_3)

31. R. T. Taylor and H. Weissbach, *ABB* **129**, 745 (1969).
32. R. T. Taylor and M. L. Hanna, *ABB* **137**, 453 (1970).
33. R. T. Taylor and M. L. Hanna, *ABB* **139**, 149 (1970).
34. S. Rosenthal and J. M. Buchanan, *Acta Chem. Scand.* **17**, Suppl. 1, 288 (1963).
35. M. A. Foster, M. J. Dilworth, and D. D. Woods, *Nature (London)* **201**, 39 (1964).
36. H. L. Elford, H. M. Katzen, S. Rosenthal, L. C. Smith, and J. M. Buchanan, *in* "Transmethylation and Methionine Biosynthesis" (S. K. Shapiro and F. Schlenk, eds.), p. 157. Univ. of Chicago Press, Chicago, Illinois, 1965.
37. J. L. Peel, *JBC* **237**, PC263 (1962).
38. J. Galivan and F. M. Huennekens, *BBRC* **38**, 46 (1970).
39. L. Jaenicke and H. Rüdiger, *Fed. Proc., Fed. Amer. Soc. Exp. Biol.* **30**, 160 (1971).
40. H. Rüdiger and L. Jaenicke, *FEBS Lett.* **4**, 316 (1969).

was reported to be only one-half as active as 5-methyl-H_4-folate (Glu_1) (*26*).

Chemically synthesized *dl*-5-methyl-H_4-folate (*13*, *41*) contains an asymmetric center at carbon 6 in the pteridine ring. It is introduced when folic acid is chemically reduced to H_4-folate (*42*). B_{12} Methyltransferases distinguish between these isomers and stereospecifically remove methyl groups from only *l*-5-methyl-H_4-folate (*8*, *43*, *44*). It must be stressed that designating the active isomer as *l* (*8*, *43*, *44*) does not refer to the direction that it rotates light. It is merely based on the fact that the active isomer of 5-methyl-H_4-folate is derived enzymically from *l*-H_4-folate which is (−) or levorotary. Despite a recent paper (*45*) deducing that *d*- and *l*-5-methyl-H_4-folate rotate light in different directions, direct measurements have, in fact, yielded specific rotations of +12.9° and +4.0° for the *l* (active) and *d* (inactive) isomers, respectively (*46*).

Under saturating conditions with respect to the other components the K_m values for *l*-[5-^{14}C]methyl-H_4-folate, homocysteine, and AMe were 30, 16, and 1.6 μM (*47*). These K_m values were obtained when purified enzyme was used to catalyze reaction (1) aerobically in a 2-mercaptoethanol + cyano-B_{12} reducing system. In a $FMNH_2$ + 1,4-dithiothreitol reducing system (*15*), the K_m for *l*-[^{14}C]methyl-H_4-folate, 35 μM, does not change significantly (*48*), but the apparent K_m for AMe decreased about 10–20-fold (*32*).

Rosenthal *et al.* (*9*) first reported that 2-mercaptoethanol would substitute, though poorly, for homocysteine as a folate methyl group acceptor [reaction (5)].

5-Methyl-H_4-folate + 2-mercaptoethanol $\xrightarrow[\text{AMe}]{\text{reducing system}}$

S-methylmercaptoethanol + H_4-folate (5)

At an optimal concentration of 0.1 M 2-mercaptoethanol the rate of transmethylation was 8-fold slower than when a saturating level of homocysteine (0.25 mM) was present (*9*, *15*). At equal concentrations of 10 mM, 2-mercaptoethylamine, thioglycolic acid, cysteine, glutathione, and 4-mercaptobutyric acid all displayed less than 10% of the acceptor

41. W. Sakami and I. Ukstins, *JBC* **236**, PC50 (1961).
42. C. K. Mathews and F. M. Huennekens, *JBC* **235**, 3304 (1960).
43. K. O. Donaldson and J. C. Keresztesy, *JBC* **237**, 3815 (1962).
44. B. T. Kaufman, K. O. Donaldson, and J. C. Keresztesy, *JBC* **238**, 1498 (1963).
45. H. Rüdiger, *FEBS Lett.* **11**, 265 (1970).
46. J. C. Keresztesy and K. O. Donaldson, *Iowa State Coll. J. Sci.* **38**, 41 (1963).
47. R. T. Taylor and H. Weissbach, "Methods in Enzymology," Vol. 17B, p. 379, 1971.
48. R. T. Taylor and M. L. Hanna, *ABB* **151**, 401 (1972).

activity of 2-mercaptoethanol (9). Dithiols such as 2,3-dimercaptoethanol (9) and 1,4-dithiothreitol (Table IV, experiment B) have negligible methyl acceptor activity.

In addition to reactions (1) and (5), preparations of E. coli B_{12} enzyme also catalyze reactions (6), (7), and (8).

$$\text{AMe} + \text{homocysteine} \xrightarrow{\text{FMNH}_2 + 1,4\text{-dithiothreitol}} \text{methionine} + \text{AH} \quad (6)$$

$$\text{AMe} + \text{H}_4\text{-folate} \xrightarrow{\text{FMNH}_2 + 1,4\text{-dithiothreitol}} 5\text{-methyl-H}_4\text{-folate} + \text{AH} \quad (7)$$

$$\text{Methyl-B}_{12} + \text{homocysteine} \xrightarrow[\text{dark}]{\text{aerobic}} \text{methionine} + \text{aquo-B}_{12} \quad (8)$$

Reaction (6) was also first detected by Rosenthal et al. (9) who demonstrated that the coproduct was AH (1). Catalysis requires a reduced flavin and like reaction (1) (Table IV) it is stimulated by 1,4-dithiothreitol (15). Even in the presence of saturating levels of both substrates, it is a very slow enzymic reaction. Reaction (7) was first observed qualitatively by Stavrianopoulos and Jaenicke (26). It was subsequently studied in detail by Taylor and Weissbach (31, 49). The folate reaction product is the active l isomer of 5-methyl-H$_4$-folate (49) and the K_m for dl-H$_4$-folate is 0.17 mM (31). Table V summarizes the relative rates (in terms of the bound B_{12}) and the apparent K_m values of AMe for reactions (1), (6), and (7), respectively. Strikingly, 5-methyl-H$_4$-folate is the preferred methyl donor in spite of the low K_m of AMe both as a cofactor [reaction (1)] and as a substrate [reactions (6) and (7)].

Guest et al. (50) first observed the catalysis of reaction (8). It occurs in the dark under both aerobic and anerobic conditions and requires no cofactors (14, 15, 50). Under anaerobic conditions, the cobalamin product that accumulates stoichiometrically to the methionine is B_{12r} (51–53). Under aerobic conditions aquo-B_{12} accumulates (51) as a result of the oxidation of B_{12r}. It is not possible to designate B_{12r} as the initial cobalamin product of reaction (8) because it may be formed by the anaerobic, spontaneous reaction of aquo-B_{12} with the homocysteine substrate (50, 51). Alternatively, B_{12r} may arise as a result of the rapid oxidation of B_{12s} (1) to B_{12r} or the reaction of B_{12s} with water (54). Reac-

49. R. T. Taylor and H. Weissbach, ABB **129**, 728 (1969).
50. J. R. Guest, S. Friedman, D. D. Woods, and E. L. Smith, Nature (London) **195**, 340 (1962).
51. H. Weissbach, B. G. Redfield, and H. Dickerman, JBC **239**, 1942 (1964).
52. S. S. Kerwar, J. H. Mangum, K. G. Scrimgeour, J. D. Brodie, and F. M. Huennekens, ABB **116**, 305 (1966).
53. J. R. Guest, S. Friedman, M. J. Dilworth, and D. D. Woods, Ann. N. Y. Acad. Sci. **112**, 774 (1964).
54. S. L. Tackett, J. W. Collat, and J. C. Abbott, Biochemistry **2**, 919 (1963).

TABLE V
APPARENT K_m (OR 50% REACTION CONCENTRATION) FOR AMe[a]

Reaction	Apparent K_m	B_{12} enzyme concn.	Apparent K_m [Bound B_{12}]	Turnover number[b]	Ref.
(1) [5-^{14}C]Methyl-H$_4$-folate + homocysteine $\rightarrow$ [^{14}C]methylmethionine + H$_4$-folate	$5.2 \times 10^{-8} M$	$3.0 \times 10^{-9} M$	17	500–780	32
(6) [^{14}C]Methyl-AMe + homocysteine $\rightarrow$ [^{14}C]methylmethionine + AH	$5.0 \times 10^{-6} M$	$2.6 \times 10^{-7} M$	19	2.2	15
(7) [^{14}C]Methyl-AMe + H$_4$-folate $\rightarrow$ [5-^{14}C]methyl-H$_4$-folate + AH	$3.7 \times 10^{-6} M$	$1.8 \times 10^{-7} M$	21	1.3	31

[a] Each apparent K_m was determined in the FMNH$_2$(Pt) + 1,4-dithiothreitol reducing system cited in the legend to Table IV and described in detail in reference 15. All reaction mixtures (0.2 ml) contained 20 μmoles of K$_2$HPO$_4$, pH 7.4, and all incubations were for 15 min at 37° with continuous H$_2$ gassing (15).

[b] Millimicromoles of [^{14}C]methyl transferred/min/mμmole of bound B$_{12}$ in the presence of a saturating concentration of AMe (or [^{14}C]-methyl-AMe), 50 μM.

tion (8) can be balanced if one depicts the immediate cobalamin product as B_{12s} [reaction (9)]

$$\text{Methyl-B}_{12} + \text{homocysteine} \xrightarrow[\text{dark}]{} \text{methionine} + B_{12s} + H^+ \qquad (9)$$

as has been suggested (7, 16, 51, 52).

In view of the fact that a completely homogeneous B_{12} protein preparation has not been obtained, one might question whether reactions (1), and (5)–(8) are indeed catalyzed by the same enzyme. Over a 7-fold purification of enzyme from E. coli 113-3, a constant ratio was observed between the activities for reactions (1) and (5) (9). Likewise, for over a 150–300-fold enrichment of enzyme from E. coli B, constant ratios were found between the activities for reactions (1), (6), and (8) (15). While these findings were strongly suggestive of a single enzyme, the most compelling evidence was provided by chemical propylation data. Table VI shows that, except for reaction (8), all of these reactions were inhibited in a light-reversible manner and to the same extent by propyl iodide. Since propyl iodide specifically alkylates the B_{12}-cobalt (27), one must conclude that the cobalamin enzyme is the catalyst for at least reactions (1) and (5)–(7).

The insensitivity of reaction (8) to propylation (27, 55), despite the

TABLE VI
Effect of Propylation on Five Reactions Catalyzed by a Purified Preparation of E. coli B Methyltransferase[a]

[^{14}C]Methyl group transfer reaction	Propylation or inhibition in the dark (%)
(1) [5-^{14}C]Methyl-H$_4$-folate + homocysteine → [^{14}C]methylmethionine + H$_4$-folate	92
(5) [5-^{14}C]Methyl-H$_4$-folate + 2-mercaptoethanol → S-[^{14}C]methyl-mercaptoethanol + H$_4$-folate	86
(6) [^{14}C]Methyl-AMe + homocysteine → [^{14}C]methylmethionine + AH	87
(7) [^{14}C]Methyl-AMe + H$_4$-folate → [5-^{14}C]methyl-H$_4$-folate + AH	90
(8) [^{14}C]Methyl-B$_{12}$ + homocysteine → [^{14}C]methylmethionine + aquo-B$_{12}$	0

[a] Separate samples of B_{12} enzyme were incubated in the dark with propyl iodide in a FMNH$_2$ (Pt) + 1,4-dithiothreitol reducing system (27). Appropriate aliquots from these mixtures were then assayed (15, 31) for the indicated transmethylase activities before and after being exposed to light (27). For each reaction, the percentage of propylation was estimated from the ratio of activity given by an aliquot of enzyme in the dark relative to an identical aliquot which had been photolyzed (27).

copurification of this activity (15), suggested that two distinct sites on the B_{12} protein might catalyze reactions (1) [plus (5)–(7)] and (8), respectively. Recently, additional evidence for this interpretation was obtained. A 200-fold purified preparation of the B_{12} enzyme was further fractionated to give a small amount of material which appeared to be predominantly a single protein upon disc gel electrophoresis (16). This protein retained the ability to catalyze both reactions (1) and (8) with no change in the ratio of the activities. Moreover, a propyl-B_{12} enzyme was shown to catalyze reaction (8) with no loss of the propyl group. Also, reconstituted holoenzyme containing tritium in the cobalamin group catalyzed this reaction with little loss of its radioactive chromophore (16). The apparent K_m of apoenzyme for methyl-B_{12} as a prosthetic group [reaction (1)] is 2.0 μM, while the K_m of holoenzyme for methyl-B_{12} as a substrate is >2.5 mM. These results and other kinetic data (16) indicate that [in the catalysis of reaction (8)] methyl-B_{12} and homocysteine form a ternary complex with a site on the enzyme separate from where the tightly bound B_{12} is attached. After the ternary complex has reacted to yield methionine and B_{12s} (or its oxidation product B_{12r}), this loosely bound cobalamin product must be replaced by a new molecule of methyl-B_{12} before catalysis [reaction (8)] can continue.

The site for reaction (8) can apparently use exogenous methyl-B_{12} as a substrate because of its lower affinity for cobalamins compared to the site for reaction (1). In contrast to the reaction (1) site, the cobalamin site for reaction (8) probably lacks specific attachment points for the 5,6-dimethylbenzimidazolyl nucleotide. Consequently, methylcobinamide is reported to function equally as well as methyl-B_{12} as a substrate in reaction (8) (36, 53). It will be recalled from Table II, however, that methylcobinamide will not bind tightly with apoenzyme at the reaction (1) site and form holoenzyme. Irrespective, the cobalamin site for reaction (8) does have in common with the site for reaction (1) a high degree of specificity for transferring cobalt-methyl groups. Neither ethyl-B_{12} (53, 56) nor propyl-B_{12} (16) can function as alkyl donors to homocysteine in reaction (8). Similarly, a propyl-B_{12} enzyme is inhibited with respect to reaction (1) because the cobalt-propyl group is totally unreactive with homocysteine (16, 17). Reaction (8) is catalyzed over a broad pH range (6–10) with optima at pH 7.5 and 9.5 in phosphate and carbonate buffer, respectively (16). It is an irreversible reaction (51, 53). Reaction (1) is catalyzed optimally only near pH 7 (9, 16).

---

55. N. Brot, R. T. Taylor, and H. Weissbach, *ABB* **114**, 256 (1966).
56. H. Weissbach, *in* "Transmethylation and Methionine Biosynthesis" (S. K. Shapiro and F. Schlenk, eds.), p. 179. Univ. of Chicago Press, Chicago, Illinois, 1965.

3. Binding of Radioactive Folate Substrate

When purified B_{12} protein is incubated with dl-[5-^{14}C]methyl-H_4-folate in the absence of both AMe and reduced flavin, an enzyme-^{14}C complex is formed (48). While the amount of complex formed is independent of the temperature between 0° and 37°, its subsequent recovery is quite temperature-dependent. Sephadex G-25 filtration at 22° yields virtually no complex, but filtration at 2°–4° results in the isolation of 0.1–0.15 mµmoles of bound ^{14}C per millimicromole of bound B_{12} (Fig. 1). The low yield of complex resulted from its reversible dissociation upon passage through the column. Since the same amount of complex was obtained with dl-5-methyl-H_4-[2-^{14}C]folate and the ^{14}C stripped from heat-denatured enzyme chromatographed like authentic dl-5-methyl-H_4-folate, it was concluded that the folate substrate can bind intact to the B_{12} protein. Moreover, only the active l isomer of [5-^{14}C]methyl-H_4-folate was bound to the enzyme. By using substrate equilibrated G-25 columns, the maximum amount of binding observed was 0.81 mµmole/mµmole of bound B_{12}. From a study of the equilibrium binding versus the folate substrate concentration, a dissociation constant of 6.2 μM was calculated, and the

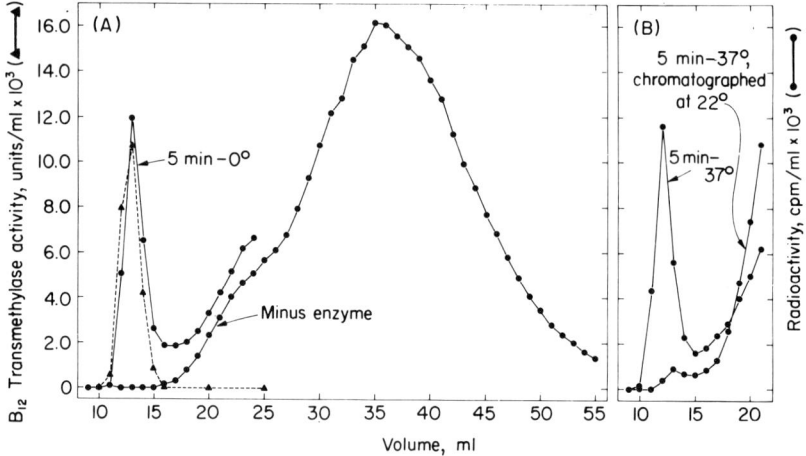

Fig. 1. Sephadex G-25 filtration of preformed initial B_{12} [^{14}C]enzyme complex. (A), Systems (0.2 ml) containing K_2HPO_4, 20 µmoles; initial B_{12} enzyme, 7.7 mµmoles; and dl-[5-^{14}C]methyl-H_4-folate (34,000 cpm/mµmole), 10 mµmoles, were incubated 5 min at 0°. K_2HPO_4 buffer, 0.8 ml of 10 mM, was added and the entire mixture was chromatographed at 2°–4°. Aliquots were counted for radioactivity and assayed for transmethylase activity. Enzyme was omitted for the control filtration. (B), Procedure as in (A) except that the preliminary incubations were 5 min at 37°, and one filtration was made at 22° rather than 2°–4°. Only the ^{14}C elution profile was determined. From Taylor and Hanna (48).

theoretical number of binding sites was 0.92 per molecule of bound B_{12} (48). Folate has a much lower affinity for the B_{12} protein than l-5-methyl-H_4-folate.

Interestingly, urea-resolved apoenzyme possessed 90% as much [5-^{14}C]methyl-H_4-folate binding activity as the original B_{12} holoenzyme from which it was prepared (48). It therefore appears that the folate substrate attaches initially to a special protein site rather than directly to the B_{12} chromophore. At all pH values where the B_{12} enzyme catalyzes reaction (1), 5-methyl-H_4-folate (isoelectric pH = 3.89 ± 0.04 S.D.) binds with its N^5-amino group in its free base rather than its protonated form (48).

D. ALKYLATION STUDIES WITH RADIOACTIVE 5-METHYL-H_4-FOLATE AND THE LIGHT STABILITY OF A METHYL-B_{12} ENZYME

Since the B_{12} enzyme catalyzed methyl transfer from methyl-B_{12}, it naturally followed that a methyl-B_{12} enzyme could be an intermediate in 5-methyl-H_4-folate transmethylation [reaction (1)]. The fact that propylation of the B_{12}-cobalt blocked reaction (1) also suggested this possibility, but direct evidence was needed. [^{14}C]Methyl-H_4-folate was, therefore, incubated with stoichiometric amounts of B_{12} enzyme in a $FMNH_2$ + 1,4-dithiothreitol reducing system containing AMe (Table VII). The amount of radioactive enzyme was determined by precipitating the protein with cold 10% trichloroacetic acid in the dark (57). ^{14}C-Labeled precipitates were then collected on a Millipore filter for liquid scintillation counting. Increasing the concentration of [5-^{14}C]methyl-H_4-folate to 30 μM (not shown) gave maximal precipitable counts, namely, 0.5 mμmole of ^{14}C per millimicromole of B_{12} protein (58). As seen in Table VII, the use of acid precipitation and filtration in the dark (57) removes both the unreacted folate substrate as well as any that is reversibly bound intact (Fig. 1).

Experiments using variously labeled 5-methyl-H_4-folate showed that only the methyl-labeled substrate reacted with the enzyme to form an acid-stable radioactive species (Table VIII). Although the results in Tables VII and VIII prove that the folate methyl group can be transferred to an acid-stable position in the presence of the cofactors, these findings did not establish either the location of the methyl group on the enzyme or its reactivity with homocysteine. For these purposes, enzyme was first methylated with [5-^{14}C]methyl-H_4-folate and then separated,

57. R. T. Taylor and H. Weissbach, *ABB* **119**, 572 (1967).
58. R. T. Taylor and H. Weissbach, *ABB* **123**, 109 (1968).

TABLE VII
REQUIREMENTS FOR THE FORMATION OF TRICHLOROACETIC ACID–PRECIPITABLE
[^{14}C]B_{12}-TRANSMETHYLASE WITH [5-^{14}C]METHYL-H_4-FOLATE[a]

Reaction mixture	Trichloroacetic acid precipitate (cpm)	^{14}C bound (mµmole)
Complete system, 0°[b]	187	<0.01
Complete system, 37°	6219	0.51
−Enzyme[c]	147	—
−FMNH$_2$ and platinum	1278	0.09
−1,4-dithiothreitol	2369	0.18
−FMNH$_2$, platinum, and 1,4-dithiothreitol	147	—
−AMe	276	0.01
+Homocysteine	180	<0.01

[a] From Taylor and Weissbach (57).

[b] Complete system (0.2 ml) contained K_2HPO_4, pH 7.4, 20 µmoles; dl-[5-^{14}C]methyl-H_4-folate (12,300 cpm/mµmole), 3 mµmoles; AMe, 10 mµmoles; 1,4-dithiothreitol, 5 µmoles; FMNH$_2$, 50 mµmoles; platinum, 0.1 mg; and B_{12} transmethylase, 0.6 mg. Homocysteine, 0.1 µmole, was added to the complete system as indicated. Incubations were for 10 min at 37° under continuous H_2 gassing.

[c] The amount of B_{12} enzyme added contained about 1.4 mµmoles of bound B_{12} chromophore.

without denaturation, by passage through a Sephadex G-25 column at 22° (Fig. 2). A large peak of radioactivity is associated with the enzyme (peak I), and the reaction is dependent on AMe. Peak II in the figure is an oxidized degradation product of the folate substrate, and peak III is unreacted dl-[5-^{14}C]methyl-H_4-folate. Methylated enzyme isolated as

TABLE VIII
FORMATION OF TRICHLOROACETIC ACID–PRECIPITABLE COUNTS PER MINUTE
WITH THREE DIFFERENTLY LABELED 5-METHYL-H_4-FOLATES[a,b]

Folate in complete system	Trichloroacetic acid precipitate (cpm)	Isotope bound (mµmole)
(1) [5-^{14}C]Methyl-H_4-folate 16,500 cpm/mµmole	6222	0.37
−Enzyme	128	0
(2) 5-Methyl-H_4-[2-^{14}C]folate 9300 cpm/mµmole	374	<0.01
−Enzyme	352	0
(3) 5-Methyl-^{3}H$_4$-folate 22,000 cpm/mµmole	235	<0.01
−Enzyme	179	0

[a] From Taylor and Weissbach (58).

[b] Incubation conditions were the same as in Table VII except that all the systems contained 10 mµmoles of labeled dl-5-methyl-H_4-folate and 0.52 mg of enzyme.

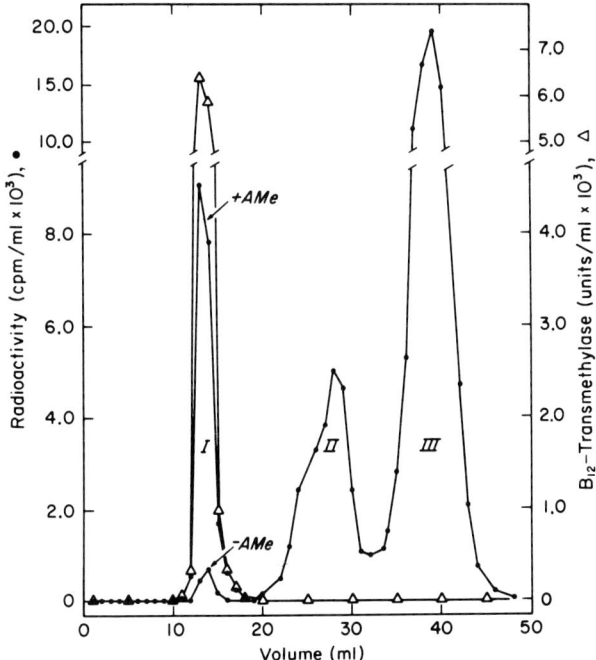

FIG. 2. Isolation of trichloroacetic acid–precipitable [^{14}C]B$_{12}$-transmethylase by Sephadex filtration. A reaction mixture (0.28 ml) containing 2.0 mg of B$_{12}$ enzyme, 6 μmoles of 1,4-dithiothreitol, 10 mμmoles of dl-[5-^{14}C]methyl-H$_4$-folate (12,300 cpm/mμmole), and all the other components of the complete system in Table VII were incubated 10 min at 37° under H$_2$. Then 0.3 ml of water was added, and the entire mixture was applied to a Sephadex G-25 column (for details, see reference 57). A second identical reaction mixture minus AMe was treated in the same manner.

in Fig. 2 readily transferred most of its [^{14}C]methyl to homocysteine without any further requirement for AMe or a reducing system (Table IX). Transfer was not impaired by either oxygen or light and is actually a very rapid reaction. The 15-min incubation used initially in Table IX is not essential because transfer is largely complete in <5 sec (31).

Other experiments firmly established that the homocysteine reactive enzyme (Table IX) contained a [^{14}C]methyl-B$_{12}$ chromophore. Substrate methylation shifted the major absorption peak in the visible portion of the spectrum from 475 to 520 nm (Fig. 3). This new maximum plus the double peaks at 340 and 380 nm are characteristic of methyl-B$_{12}$ (59) whose structure is shown in Fig. 4. Addition of homocysteine essentially regenerated the spectrum of the original B$_{12}$ enzyme prior to methylation

59. O. Müller and G. Müller, Biochem. Z. 336, 299 (1962).

TABLE IX
REQUIREMENT FOR [^{14}C]METHYLMETHIONINE FORMATION FROM
[^{14}C]B$_{12}$-TRANSMETHYLASE[a]

Experiment	Reaction mixture[b]	(cpm)	Methionine formed (mµmole)
1. Dark (anaerobic)	Complete system ([^{14}C]enzyme + homocysteine)	2000	0.16
	−Homocysteine	160	0.01
	Complete system + AMe + reducing system	1900	0.15
2. Light (aerobic)	Complete system ([^{14}C]enzyme + homocysteine)	2650	0.21
	−Homocysteine	230	0.02

[a] From Taylor and Weissbach (57).
[b] Complete systems (0.3 ml) contained [^{14}C]enzyme, isolated as described in Fig. 2, 2600 cpm (Exp. 1) or 2730 cpm (Exp. 2) and homocysteine, 0.2 µmole. In Exp. 1, AMe, 10 µmoles, and a reducing system were included in the reaction mixture as indicated. The reducing system consisted of FMNH$_2$, 50 mµmoles; platinum, 0.1 mg; and 1,4-dithiothreitol, 5 µmoles. Incubations were for 15 min at 37° under H$_2$ gas in the dark (Exp. 1) or aerobic for 15 min at 37° in room light (Exp. 2).

(Fig. 3). Direct confirmation that reaction with [5-^{14}C]methyl-H$_4$-folate yielded a [^{14}C]methyl-B$_{12}$ enzyme was obtained by extracting the chromophore from the protein with hot ethanol in the dark. It was subsequently identified as [^{14}C]methyl-B$_{12}$ by paper chromatography, paper electrophoresis, and its absorption spectrum before and after photolysis (58).

The relative light stability of a [^{14}C]methyl-B$_{12}$ enzyme (Table IX and Fig. 3) was unexpected in view of the previously observed lability of a propyl-B$_{12}$ enzyme (27). Free methyl-B$_{12}$ is rapidly degraded by light under aerobic conditions, the major cleavage product being formaldehyde (60). Therefore, a variety of solvent conditions were tested to determine if the cobalt–methyl group could be made light sensitive. Only acidification to below pH 2.5, which also precipitates the enzyme, resulted in complete photolysis (58). Photolysis at acidic pH values is not the result of removal of the entire cobalamin from the protein, followed by light cleavage of the freed [^{14}C]methyl-B$_{12}$; it is only the subsequent exposure of acid-precipitated [^{14}C]methyl-B$_{12}$ enzyme to light which causes a 90–100% loss of ^{14}C from the precipitated protein (57, 58, 61). The light promoted loss of ^{14}C from the precipitated B$_{12}$ protein was correlated with

60. H. P. C. Hogenkamp, *Biochemistry* **5**, 417 (1966).
61. R. T. Taylor and H. Weissbach, *BBRC* **27**, 398 (1967).

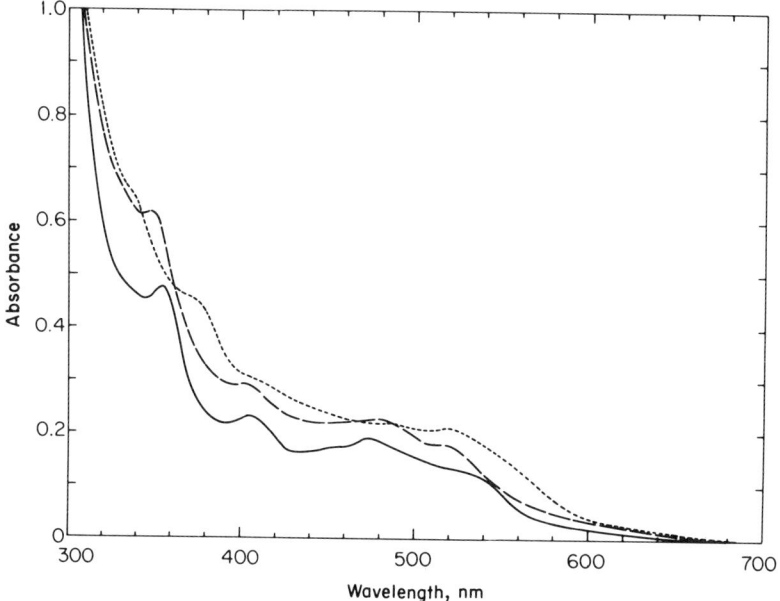

FIG. 3. Effect of methylation with [5-^{14}C]methyl-H$_4$-folate on the absorption spectrum of vitamin B$_{12}$ transmethylase. (———) Vitamin B$_{12}$ enzyme: initial enzyme, 22 mg/ml, in 0.05 M K$_2$HPO$_4$, pH 7.4; (- - -) [^{14}C]methyl-B$_{12}$ enzyme: same concentration of enzyme in K$_2$HPO$_4$ buffer after methylation with [5-^{14}C]methyl-H$_4$-folate (plus unlabeled AMe) followed by Sephadex G-25 filtration and lyophilization in the dark; same tracing was obtained after illumination for 20 min with a 100-W tungsten lamp at 10 cm and 0°; (– –) [^{14}C]methyl-B$_{12}$ enzyme + homocysteine: spectrum immediately after the addition of 0.2 μmole of homocysteine at 22°, same tracing was obtained after an additional 30-min aerobic incubation at 37°. From Taylor and Weissbach (58).

the release of ^{14}C-formaldehyde into the supernatant fluid (58). An increased light stability of bound methyl-B$_{12}$ vs. free methyl-B$_{12}$ has also been observed when a [^{14}C]methyl-B$_{12}$ enzyme is prepared from apoenzyme + [^{14}C]methyl-B$_{12}$ (17).

Shortly after a [^{14}C]methyl-B$_{12}$ enzyme was first reported to be relatively light stable (57), Stavrianopoulos and Jaenicke (26) offered a different explanation. These investigators also prepared a [^{14}C]-methyl-B$_{12}$ enzyme from radioactive folate substrate and a B$_{12}$ protein purified from *E. coli*. They attributed the retention of 65–76% of its ^{14}C after lighting (30 min, 10 cm, 160 W, 22°) to a secondary reaction between the photolytically produced methyl radicals and amino acid side chains in the protein (26). Such an explanation, however, cannot account for the findings made by Taylor and Weissbach (57, 58, 61).

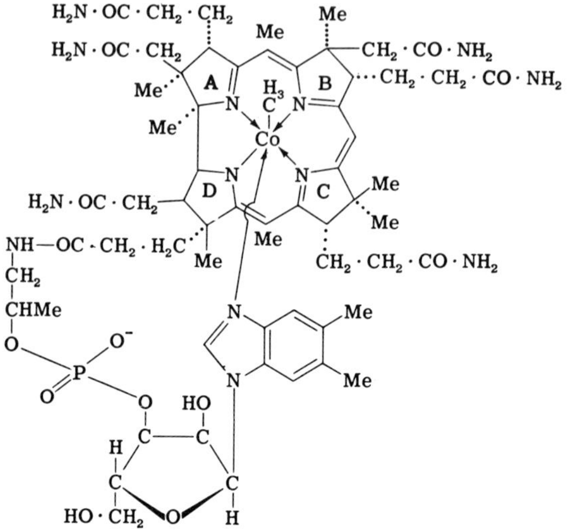

FIG. 4. Structure of methyl-B_{12} (5,6-dimethylbenzimidazolyl-cobamide methyl).

First, photolysis did not alter the reactivity of their [^{14}C]methyl-B_{12} enzyme with homocysteine (57, 58). Second, light did not alter its absorption spectrum (Fig. 3) which occurs when the [^{14}C]methyl–cobalt bond is irreversibly broken. Third, lighting the [^{14}C]methyl-B_{12} enzyme did not decrease the yield of free [^{14}C]methyl-B_{12} which was subsequently extracted off the protein (58). Rüdiger (62) has since confirmed independently that enzyme-bound [^{14}C]methyl-B_{12} is indeed less light sensitive than free [^{14}C]methyl-B_{12}.

There are other examples in the literature of an increased carbon–cobalt bond stability to light. Sheep liver methylmalonyl mutase was purified as a light-stable 5′-deoxyadenosyl-B_{12}–protein complex (63) and δ-(9-adenyl)butyl-B_{12} forms a light-stable complex with ethanolamine deaminase (64). In a free state, both of these alkylcorrinoids are as photolabile as methyl-B_{12}. Pailes and Hogenkamp (65) found that the first-order rate constant at pH 7 for the photolysis of methylcobinamide was about 15% less than that for methyl-B_{12}. In the presence of imidazole, the cobalt in methylcobinamide appears to become even more electrophilic because the photolysis rate constant decreases an additional twofold. The light stabilizing influence of imidazole is lost at low pH

62. H. Rüdiger, *Eur. J. Biochem.* **21**, 264 (1971).
63. J. J. B. Cannata, A. Focesi, Jr., R. Mazumder, R. D. Warner, and S. Ochoa, *JBC* **240**, 3249 (1965).
64. B. Babior, H. Kon, and H. Lecar, *Biochemistry* **8**, 2662 (1969).
65. W. H. Pailes and H. P. C. Hogenkamp, *Biochemistry* **7**, 4160 (1968).

values. From such model experiments, it has been suggested that in the methylated B_{12} holoenzyme, the cobalt-5,6-dimethylbenzimidazolyl link is broken and a new bond between a histidine side chain and the cobalt is formed (*65*). In contrast, in the propylated B_{12} holoenzyme, the histidine is not able to bind to the cobalt as a result of the added inductive effect of the extra ethyl group (*65*). While these are certainly plausible suggestions, it would seem that other possible contributions by the protein cannot be overlooked. Photolysis involves a homolytic split of the carbon–cobalt bond, and it requires oxygen (*23, 60*). Consequently, apoenzyme binding may merely protect the methyl–cobalt bond from oxygen, and, in addition, stabilize the methyl radical, thereby favoring its recombination with cobalt (*17, 58*). Experimental evidence obtained with the use of methyl-B_{12} enzyme for any of these suggestions is presently lacking.

Regardless of the light stability considerations, the substrate methylation experiments (Tables VII–IX and Figs. 2 and 3), *a priori*, indicated that reaction (1) had been separated into two partial reactions (10) and (11).

$$[5\text{-}^{14}\text{C}]\text{Methyl-H}_4\text{-folate} + B_{12}\text{ enzyme} \xrightarrow[\text{AMe}]{\text{FMNH}_2 + 1,4\text{-dithiothreitol}} [^{14}\text{C}]\text{methyl-}B_{12}\text{ enzyme} + H_4\text{-folate} \quad (10)$$

$$[^{14}\text{C}]\text{Methyl-}B_{12}\text{ enzyme} + \text{homocysteine} \xrightarrow{\text{aerobic}} [^{14}\text{C}]\text{methylmethionine} + B_{12}\text{ enzyme} \quad (11)$$

In reaction (10), the enzyme acquires a cobalt–methyl group from the folate substrate, this step being dependent on AMe and a reducing system. In reaction (11) the methyl group is transferred to homocysteine, this step being closely analogous to reaction (8), although reaction (8) is catalyzed at a separate site (*16*) on the B_{12} protein. However, it became apparent that the AMe-dependent accumulation of a [^{14}C]methyl-B_{12} enzyme could not simply be described by reaction (10) when the process was examined stoichiometrically. It is seen in Table X that too little l-H$_4$-folate is formed in the absence of homocysteine (complete system) to account for the amount of [^{14}C]methyl-B_{12} protein that was produced in these same incubations. This finding and a better understanding of reaction (10) became clear when the function of AMe in reaction (1) was further elucidated. These points are discussed together in Section II,E.

E. STUDIES ON THE ROLE OF *S*-ADENOSYL-L-METHIONINE

Early studies demonstrating the need for catalytic amounts of AMe in reaction (1) relied solely upon the use of relatively crude enzyme (*14,*

TABLE X
FORMATION OF H_4-FOLATE UPON METHYLATION WITH
[5-^{14}C]METHYL-H_4-FOLATE[a]

Reaction mixture	Trichloroacetic acid precipitate (cpm)	^{14}C Bound (mμmole)	H_4-Folate formed (mμmole)
Complete[b]	13,695	0.18	0.038
Complete[b]	14,495	0.20	0.038
−FMNH$_2$ and platinum	2,147	0.03	0.016
−B$_{12}$ Enzyme	245	0	0
−AMe	4,388	0.06	0.02
+Homocysteine	240	0	0.43

[a] From Taylor and Weissbach (49).
[b] Complete systems (0.2 ml) contained K$_2$HPO$_4$, pH 7.4, 20 μmoles; [5-^{14}C]methyl-H_4-folate (73,000 cpm/mμmole), 1.0 mμmole; AMe, 1.0 mμmole; B$_{12}$ enzyme, 0.92 mμmole; 1,4-dithiothreitol, 5 μmoles; FMNH$_2$, 50 mμmoles; and platinum, 0.1 mg. After 5-min incubations in the dark at 37° under H$_2$ gas, 0.3 ml of cold water were added to each mixture. Samples of 0.2 ml were then removed for trichloroacetic acid precipitation and Millipore filtration. To the remaining 0.3 ml of the diluted mixtures was added 0.1 ml of potassium ascorbate, pH 6.0, 100 mg/ml. These ascorbate-protected samples were stored at −15° in the dark and then assayed microbiologically for H_4-folate.

34, 35). They provided no information as to how AMe functions catalytically except that it was shown (9, 34) that AH could not be substituted for AMe. Prompted by the observation that both AMe and methyl iodide could prevent chemical propylation (27), methyl iodide was tested and found to satisfy partially the requirement for AMe in reaction (1) (66). Methyl iodide was also shown to function catalytically in the system (67). Since radioactive folate substrate was used in these experiments, any direct nonenzymic reaction between the unlabeled methyl iodide and homocysteine did not affect the estimation of [5-^{14}C]methyl-H_4-folate-homocysteine transmethylation. These observations with methyl iodide indicated that AMe could only be serving as a methyl group donor. Since the slow catalysis (Table V) of AMe-homocysteine transmethylation [reaction (6)] was inhibited by chemical propylation (27), a possible site for methylation by AMe (or methyl iodide) was the B$_{12}$ cobalt. It is known that methyl iodide (59, 68) and AMe (69) will react rapidly with free B$_{12s}$ to form methyl-B$_{12}$.

66. R. T. Taylor and H. Weissbach, JBC 241, 3641 (1966).
67. R. T. Taylor and H. Weissbach, JBC 242, 1517 (1967).
68. E. L. Smith, L. Mervyn, A. W. Johnson, and N. Shaw, Nature (London) 194, 1175 (1962).
69. W. Friedrich and E. Königk, Biochem. Z. 336, 444 (1962).

Direct evidence for AMe methylation of the cobalt was obtained by using [^{14}C]methyl-AMe (Table XI). A reducing system was required and the properties of the resulting [^{14}C]methyl-B$_{12}$ enzyme (light stability and homocysteine reactivity) were indistinguishable from those of enzyme that had been methylated with [5-^{14}C]methyl-H$_4$-folate (plus unlabeled AMe) (49, 61). Consequently, it was anticipated that both unlabeled 5-methyl-H$_4$-folate and homocysteine should markedly decrease the accumulation of a [^{14}C]methyl-B$_{12}$ enzyme from [^{14}C]methyl-AMe. This, in fact, did happen as seen in Table XI. Homocysteine very likely functioned as a methyl group acceptor and demethylated most of the [^{14}C]methyl-B$_{12}$ enzyme to form [^{14}C]methylmethionine, but the effect of unlabeled 5-methyl-H$_4$-folate on the methylation of the enzyme by [^{14}C]methyl-AMe was not clear and required further study. Short-time enzyme labeling experiments (Fig. 5) revealed that methylation by [^{14}C]methyl-AMe occurred within the first 30 sec of incubation. In the absence of unlabeled 5-methyl-H$_4$-folate the [^{14}C]methyl from [^{14}C]-methyl-AMe remained on the enzyme. But in the presence of unlabeled folate substrate, it was removed from 30 sec to 6 min after the start of the incubation. In a correlative experiment (Fig. 5), the AMe-dependent formation of [^{14}C]methyl-B$_{12}$ enzyme with [5-^{14}C]methyl-H$_4$-folate showed a lag during the first 30 sec and required 5–6 min to reach completion.

The data in Fig. 5 indicated that in the absence of homocysteine a [^{14}C]methyl-B$_{12}$ enzyme accumulated from [5-^{14}C]methyl-H$_4$-folate by an exchange with a cobalt–methyl group derived initially from unlabeled

TABLE XI

REQUIREMENTS FOR THE FORMATION OF TRICHLOROACETIC ACID–PRECIPITABLE ^{14}C WITH [^{14}C]METHYL-AMe[a]

Reaction mixture	Trichloroacetic acid precipitate (cpm)	^{14}C Bound (mµmole)
Complete (37°)[b]	12,800	0.61
−B$_{12}$ Enzyme	1,480	0
−FMNH$_2$ and platinum	2,074	0.03
+5-Methyl-H$_4$-folate	2,600	0.06
+Homocysteine	2,470	0.05

[a] From Taylor and Weissbach (49).
[b] Complete systems (0.2 ml) contained K$_2$HPO$_4$, pH 7.4, 20 µmoles; [^{14}C]methyl-AMe (18,500 cpm/mµmole), 10 mµmoles; 1,4-dithiothreitol, 5 µmoles; B$_{12}$ enzyme, 0.54 mµmole; FMNH$_2$, 50 mµmoles; and platinum, 0.1 mg. Where indicated dl-5-methyl-H$_4$-folate, 30 mµmoles, or homocysteine 0.2 µmole, was added to the complete system. Incubations were for 15 min in the dark under H$_2$ gas.

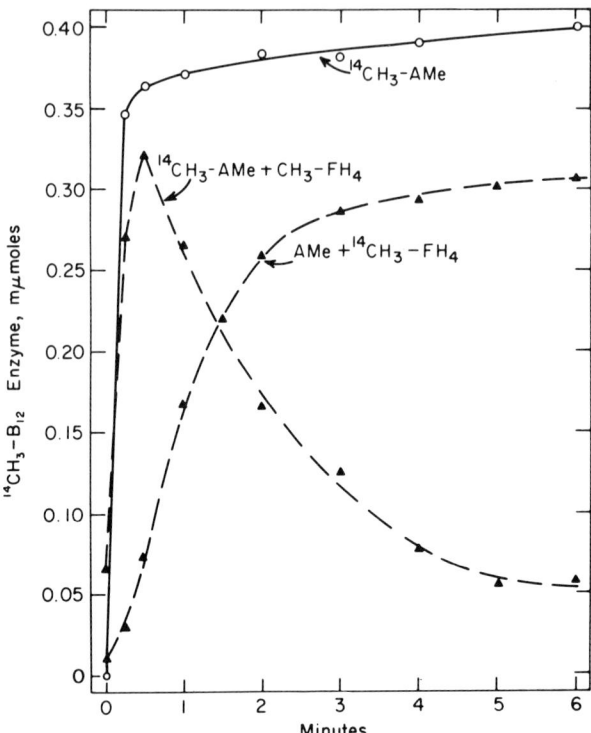

FIG. 5. Time dependence of [^{14}C]methyl-B_{12} enzyme formation with [^{14}C]methyl-AMe and [5-^{14}C]methyl-H_4-folate. All reaction mixtures (0.2) ml contained 0.46 mμmole of B_{12} enzyme; 10 mμmoles of AMe either unlabeled or [^{14}C]methyl (18,000 cpm/mμmole); and, where indicated, 10 mμmoles of dl-5-methyl-H_4-folate either unlabeled or [^{14}C]methyl (73,000 cpm/mμmole). Incubations were under H_2 gas at 37° for the times indicated. They were initiated within 3 min after the injection at 0° of reduced flavin. From Taylor and Weissbach (49).

AMe. Thus, the dependency on AMe for methyl transfer from [5-^{14}C]-methyl-H_4-folate to the enzyme (Table VII) and the formation of insufficient H_4-folate (Table X) would be explained by this exchange reaction. The mechanism for this exchange was realized when analysis showed (49) that the dl-[5-^{14}C]methyl-H_4-folate prepared by the method of Keresztesy and Donaldson (13) contained traces of dl-H_4-folate. Enzyme that had been methylated with [^{14}C]methyl-AMe was shown to react with H_4-folate to produce the active isomer of [5-^{14}C]-methyl-H_4-folate (49). Based on these findings the mechanism for the results in Fig. 5 is depicted in Fig. 6. H_4-Folate accepts a methyl group (derived initially from AMe) from methyl-B_{12} on the enzyme to form 5-methyl-H_4-folate and a postulated Co^{1+} enzyme species (B_{12s} enzyme)

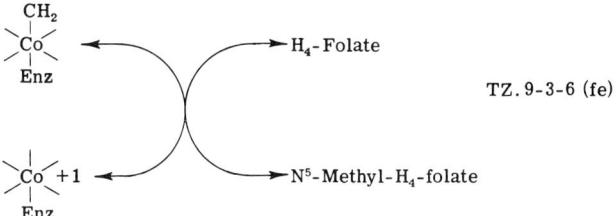

FIG. 6. Role of H_4-folate in methyl group exchange between 5-methyl-H_4-folate and the methyl-B_{12} enzyme. From Taylor and Weissbach (*31, 49*).

The latter then reacts reversibly with exogenously added [5-^{14}C]methyl-H_4-folate to reform a [^{14}C]methyl-B_{12} enzyme with the methyl group now having been derived from the radioactive folate substrate (*31*). Figure 6 predicts (a) that small amounts of H_4-folate should accelerate the AMe-dependent rate of [^{14}C]methyl-B_{12} enzyme accumulation from an excess of [5-^{14}C]methyl-H_4-folate, (b) the B_{12} enzyme should catalyze AMe-H_4-folate transmethylation [reaction (7)], and (c) the B_{12} enzyme should catalyze an AMe-dependent exchange reaction between pteridine ring labeled 5-methyl-H_4-folate and unlabeled H_4-folate to give pteridine ring labeled H_4-folate. All of these predictions have been demonstrated experimentally (*31*).

The above results on exchange transmethylation (*31, 49*) suggested a mechanism for the participation of AMe in reaction (1). Inactive B_{12} protein is first primed or activated by B_{12}-cobalt methylation. This priming can be facilitated by AMe or methyl iodide. Once the enzyme has been converted to one of its functional forms by AMe methylation, the enzyme should catalyze reaction (1). The key to testing for AMe activation depended on devising an experimental system that would permit one to detect even a few cyclic turnovers by the enzyme. Table XII summarizes the results of such an experiment. Millimicromole amounts of B_{12} enzyme were first incubated with AMe (or methyl iodide) in the presence of an $FMNH_2$ reducing system. The reaction mixtures were then shaken aerobically to oxidize the flavin and reincubated with the two substrates, 5-methyl-H_4-folate and homocysteine. The activated enzyme (methylated in the first incubation) catalyzed the aerobic synthesis of about 10 equivalents of [^{14}C]methylmethionine (Table XII). The catalysis of reaction (1) was complete in 30 sec because the enzyme was approximately two-thirds inactivated within 5 sec after adding the substrates (*31*). Requirements for both AMe and $FMNH_2$ during the first incubation were also observed when the activated enzyme was separated from the smaller components in the preliminary reaction mixture, prior to being challenged with the two substrates (*31*). The important point

TABLE XII
REQUIREMENTS FOR AEROBIC METHYL[^{14}C] GROUP TRANSFER FROM
[5-^{14}C]METHYL-H$_4$-FOLATE TO HOMOCYSTEINE[a]

First incubation reaction mixture	15 min, 37°[b] Gas phase	[^{14}C]Methylmethionine formed after a second aerobic incubation, 5 min, 37°[c] (mμmole)
Complete[b]	H$_2$	3.4–3.8
Complete (FMN)	Air	0.07
−FMNH$_2$ and platinum	H$_2$	0.01
−AMe	H$_2$	0.10
−AMe	H$_2$	0.13 (+AMe)[d]
−AMe + CH$_3$I[e]	H$_2$	3.40
−B$_{12}$ Enzyme	H$_2$	0
Complete	H$_2$	0.09 (−homocysteine)[f]

[a] From Taylor and Weissbach (*31*).

[b] Complete systems (0.2 ml) for the first incubation contained B$_{12}$ enzyme, 0.39 mμmole; K$_2$HPO$_4$, pH 7.4, 20 μmoles; AMe, 10 mμmoles; FMNH$_2$, 50 mμmoles, and platinum 0.1 mg.

[c] At the end of the first incubation the reaction mixtures were chilled to 0° in the dark and shaken for several minutes to oxidize the FMNH$_2$. They were then equilibrated at 37° for 5 min and tested for their transmethylase activity by the addition of 0.06 ml of a pH 7.4 substrate solution containing 0.5 m*M* dl-[5-^{14}C]methyl-H$_4$-folate (14,400 cpm/mμmole), 2.5 m*M* homocysteine, 0.05 *M* 1,4-dithiothreitol, and 0.05 *M* K$_2$HPO$_4$, pH 7.4. [^{14}C]Methylmethionine synthesis was determined at the end of a 5-min incubation (*15*).

[d] AMe (10 mμmoles) was added at the end of the first incubation after the FMNH$_2$ had been oxidized.

[e] Methyl iodide (0.15 μmole in 2 μl of ethanol) was substituted for AMe and the complete system (0.2 ml) also contained 1,4-dithiothreitol (5 μmoles) during the first incubation.

[f] Homocysteine was omitted from the second incubation system.

in these experiments was that limited [^{14}C]methylmethionine synthesis occurred only when the first incubation conditions were such as to permit the preliminary formation of a methyl-B$_{12}$ enzyme (*49, 58, 61*). Thus, both AMe (or methyl iodide) and reduced flavin were always essential (*31*) to preform a catalytically active B$_{12}$ protein.

It has been verified with the use of apoenzyme that only cobalt methylation is involved in AMe activation. Purified B$_{12}$ protein was resolved with urea into apoenzyme (*17*) as discussed earlier (Section II,B). The apoprotein was incubated with [^{14}C]methyl-B$_{12}$ and then freed of unbound [^{14}C]methyl-B$_{12}$ with a charcoal column. Since this [^{14}C]methyl group reacts quantitatively with homocysteine (*17*), one has an accurate estimate of the amount of reconstituted [^{14}C]methyl-B$_{12}$ enzyme added to a system. The importance of testing its catalytic activity resided in the fact that it was a methylated B$_{12}$ enzyme that had been prepared by circumventing a preliminary exposure (Table XII) to AMe and $FMNH_2$. Hence, aerobic turnover (Table XII) resulting from activation by any reduced flavin or by nonspecifically bound AMe (*49*) could be eliminated. Table XIII demonstrates that reconstituted [^{14}C]methyl-B_{12} enzyme (0.25 mµmole) synthesized aerobically 18 equivalents (4.5 mµmoles) of [^{14}C]methylmethionine in the absence of both AMe and $FMNH_2$. The activation of the enzyme was not lost if the folate substrate was added prior to homocysteine, but the catalytic activity was lost by demethylation with homocysteine for 15 sec prior to addition of the [5-^{14}C]methyl-H_4-folate. A similar observation was made using enzyme that had been activated by AMe methylation (*31*), i.e., an activated enzyme was a methylated enzyme, and any procedure that removed the methyl group resulted in complete loss of activation. These findings were indicative of an extremely oxygen-sensitive form of the cobalamin (such as B_{12s}, Fig. 6) being formed in the catalytic cycle. Consistent with this view, it was observed that the total catalytic turnover of both AMe premethylated (*31*) and reconstituted (*32*) methyl-B_{12} enzyme was increased as much as 10–30-fold by carrying out reaction (1) under anaerobic conditions (e.g., H_2 gas phase, Fig. 7). Under H_2 gas, the catalytic life of reconstituted [^{14}C]methyl-B_{12} enzyme lasted sufficiently longer to increase the total number of enzyme turnovers to 530-fold. A turnover rate (after 5 sec) of 910 molecules of [^{14}C]methylmethionine formed per minute per molecule of added methyl-B_{12} enzyme was calculated from the data in Fig. 7. Like an AMe-premethylated enzyme (*31*), the reconstituted enzyme in Fig. 7 could not maintain its initial rate of synthesis unless both AMe and $FMNH_2$ were included in the system (*32*). Irrespective, the initial turnover rate in Fig. 7 is comparable to the steady-state turnover

TABLE XIII
EFFECT OF ORDER OF SUBSTRATE ADDITION ON THE AEROBIC CATALYSIS BY A RECONSTITUTED [^{14}C]METHYL-B_{12} ENZYME[a,b]

Order of addition	[^{14}C]Methylmethionine formed (mµmole)
[5-^{14}C]Methyl-H_4-folate + homocysteine (5 min)	4.5
[5-^{14}C]Methyl-H_4-folate (15 sec), then homocysteine (5 min)	4.0
Homocysteine (15 sec), then [5-^{14}C]methyl-H_4-folate (5 min)	0.4

[a] From Taylor and Hanna (*32*).
[b] [^{14}C]Methyl-B_{12} enzyme, 0.92 mg (0.25 mµmole) in 0.2 ml of 0.1 M K_2HPO_4, pH 7.4, was equilibrated at 37° for 5 min. A 1.0 mM solution of dl-[5-^{14}C]methyl-H_4-folate (29,000 cpm/mµmole), a 5.0 mM solution of homocysteine, and a 1:1 mixture of the substrate solutions were also equilibrated separately at 37° for 5 min. Then 0.06 ml of the 1:1 mixture or 0.03 ml of each substrate was added and incubated at 37° as shown.

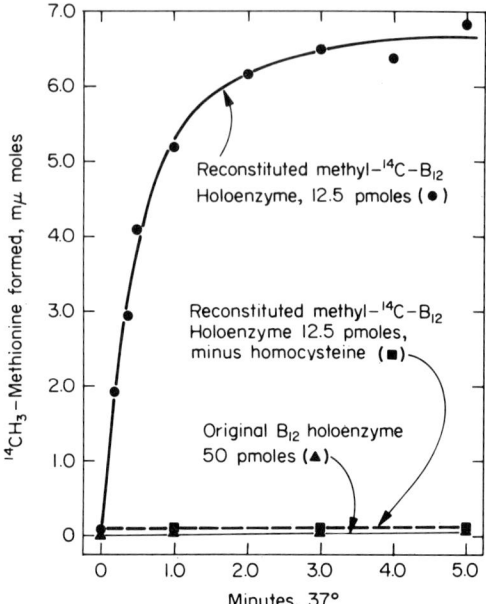

FIG. 7. [5-^{14}C]Methyl-H$_4$-folate-homocysteine transmethylation under H$_2$ gas by a reconstituted [^{14}C]methyl-B$_{12}$ holoenzyme. Final reaction mixtures (0.26 ml) contained K$_2$HPO$_4$, pH 7.4, 20 μmoles; dl-[5-^{14}C]methyl-H$_4$-folate (29,000 cpm/mμmole-, 30 mμmoles; homocysteine, 50 mμmoles; and enzyme, either 46 μg of reconstituted [^{14}C]methyl-B$_{12}$ enzyme (12.5 pmoles of bound [^{14}C]methyl-B$_{12}$) or 50 μg of the original B$_{12}$ holoenzyme (50 pmoles of bound B$_{12}$). Enzyme in 0.1 M K$_2$HPO$_4$ buffer (0.2 ml) and a pH 7.4 substrate solution containing 0.5 mM dl-[5-^{14}C]-methyl-H$_4$-folate plus 2.5 mM homocysteine were pregassed separately with H$_2$ for 5 min at 0° and with constant agitation. Then the buffered enzyme and substrate mixture were equilibrated separately at 37° for 5 min. Reactions were initiated by the injection of 0.06 ml of substrate mixture and were terminated at the times indicated with 0.8 ml of ice cold water. From Taylor and Hanna (32).

rate of 860 which was obtained using the reconstitued enzyme and conditions of Fig. 7 except with the supplementation of AMe + FMNH$_2$ (32). It was, therefore, concluded that a methyl-B$_{12}$ protein is a fully active enzyme species which, at least transiently, requires only the two substrates to catalyze reaction (1); AMe and reduced flavin are only involved in the preliminary formation of methylated enzyme (32).

Rüdiger and Jaenicke (70, 71) have since reported the partial purification of a methyl-B$_{12}$ holoenzyme directly from extracts of E. coli. It is active in the absence of exogenously supplied AMe but only in a reducing

70. H. Rüdiger and L. Jaenicke, Eur. J. Biochem. 10, 557 (1969).
71. H. Rüdiger and L. Jaenicke, Eur. J. Biochem. 16, 92 (1970).

system formed by adding 1,4-dithiothreitol + aquo-B_{12} + FMN. In an enzymic $FADH_2$ generating system, AMe is still essential (70). Also, in the AMe-independent reducing system, their methyl-B_{12} enzyme regularly shows an unexplained lag in the catalysis of reaction (1) during the first 15 min of incubation (70). It should be noted that the absorption spectra of the latter enzyme preparations (70, 71) differ significantly from that of free methyl-B_{12} (59) or that of the methyl-B_{12} enzyme prepared *in vitro* by other investigators (17, 58). Thus, before assessing their data, one should wait until the cobalamin is removed from the protein and adequately characterized as methyl-B_{12}. It is possible that by avoiding all precipitation-type steps (39, 70, 71), their new enzyme may show AMe-independent activity as a result of the presence of ionically bound, endogenous AMe (49). The extended lag in AMe-independent catalysis (70) might reflect the slow release of this endogenous AMe in a 1,4-dithiothreitol-containing reducing system.

F. Mechanism of N^5-Methyltetrahydrofolate-Homocysteine Transmethylation

A schematic mechanism for reaction (1) that is compatible with the data presented (Sections II,C,D, and E) is shown in Fig. 8. It is not intended as a detailed kinetic scheme to account for all of the possible substrate and product complexes with the enzyme that may exist. Hence, even the now-known complex (48) between B_{12} enzyme and the intact l-5-methyl-H_4-folate substrate has been deleted. The usefulness of Fig. 8 for discussion purposes is that it shows the relationship between the three forms of bound B_{12} that have been identified with the enzyme. Figure 8 also accounts for the equal sensitivity of reactions (1) and (5)–(7) to chemical propylation and the catalysis of exchange transmethylation. The salient feature when this scheme was first proposed (31) was that it

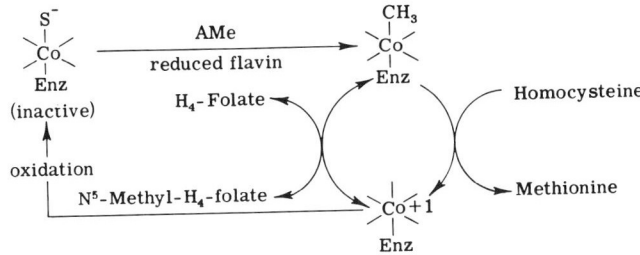

Fig. 8. Schematic reaction mechanism for AMe-dependent methionine synthesis catalyzed by *E. coli* B cobalamin transmethylase. From Taylor and Weissbach (31).

Unfortunately, for both the B_{12}-dependent [reaction (1)] and the B_{12}-independent [reaction (2)] synthesis of methionine, no concrete information is available as to how the N^5-methyl group is activated for enzymic transfer. 5-Methyl-H_4-folate is not a high energy onium compound like AMe (74). Instead, its N^5-methyl group is chemically quite unreactive at pH 7, even with free B_{12s} (58, 78). Recently, it was reported that a slight amount of nonenzymic reaction takes place in acidic media between 5-methyl-H_4-folate and B_{12s} to yield a trace of methyl-B_{12} (62).

III. Non-B_{12} Methyltransferase from *Escherichia coli*

A. Assay and Purification

Non-B_{12} transmethylase activity can be routinely assayed by measuring the formation of [^{14}C]methylmethionine using [5-^{14}C]methyl-H_4-folate (Glu$_3$) as a substrate. The radioactive methionine can be separated from the folate substrate with an anion exchange resin (14). A typical reaction mixture would contain in a total volume of 50 µl: dl-[5-^{14}C]methyl-H_4-folate (Glu$_3$), 3 mµmoles, 13,300 cpm/mµmole; L-homocysteine, 50 mµmoles; Na$_2$HPO$_4$ buffer, pH 7.8, 500 mµmoles; magnesium acetate, 5 mµmoles; 1,4-dithiothreitol, 500 mµmoles; and enzyme. After incubation for 15 min at 37°, the reaction is stopped by

TABLE XIV
Purification of Non-B_{12} Transmethylase from *E. coli* K_{12}[a]

Fraction	Volume (ml)	Total activity (k-units)	Protein (mg/ml)	Specific activity (units/mg)	Yield (%)
1. Extract	319	1220	30.4	126	100
2. Ammonium sulfate (0–65%)	154	1080	50.0	140	89
3. Protamine sulfate	306	1080	15.4	232	89
4. Ammonium sulfate (50–65%)	31	664	48.4	463	54.5
5. First DEAE-Sephadex	102	625	4.28	1430	51
6. Hydroxylapatite	61	376	3.58	1720	31
7. Second DEAE-Sephadex	85	362	2.03	2100	29.7
8. Sephadex G-100	47	320	2.7	2520	26.2

[a] From Whitfield *et al.* (79).

78. G. N. Schrauzer and R. J. Windgassen, *J. Amer. Chem. Soc.* **89**, 3607 (1967).
79. C. D. Whitfield, E. J. Steers, Jr., and H. Weissbach, *JBC* **245**, 390 (1970).

system formed by adding 1,4-dithiothreitol + aquo-B_{12} + FMN. In an enzymic $FADH_2$ generating system, AMe is still essential (70). Also, in the AMe-independent reducing system, their methyl-B_{12} enzyme regularly shows an unexplained lag in the catalysis of reaction (1) during the first 15 min of incubation (70). It should be noted that the absorption spectra of the latter enzyme preparations (70, 71) differ significantly from that of free methyl-B_{12} (59) or that of the methyl-B_{12} enzyme prepared *in vitro* by other investigators (17, 58). Thus, before assessing their data, one should wait until the cobalamin is removed from the protein and adequately characterized as methyl-B_{12}. It is possible that by avoiding all precipitation-type steps (39, 70, 71), their new enzyme may show AMe-independent activity as a result of the presence of ionically bound, endogenous AMe (49). The extended lag in AMe-independent catalysis (70) might reflect the slow release of this endogenous AMe in a 1,4-dithiothreitol-containing reducing system.

F. Mechanism of N^5-Methyltetrahydrofolate-Homocysteine Transmethylation

A schematic mechanism for reaction (1) that is compatible with the data presented (Sections II,C,D, and E) is shown in Fig. 8. It is not intended as a detailed kinetic scheme to account for all of the possible substrate and product complexes with the enzyme that may exist. Hence, even the now-known complex (48) between B_{12} enzyme and the intact *l*-5-methyl-H_4-folate substrate has been deleted. The usefulness of Fig. 8 for discussion purposes is that it shows the relationship between the three forms of bound B_{12} that have been identified with the enzyme. Figure 8 also accounts for the equal sensitivity of reactions (1) and (5)–(7) to chemical propylation and the catalysis of exchange transmethylation. The salient feature when this scheme was first proposed (31) was that it

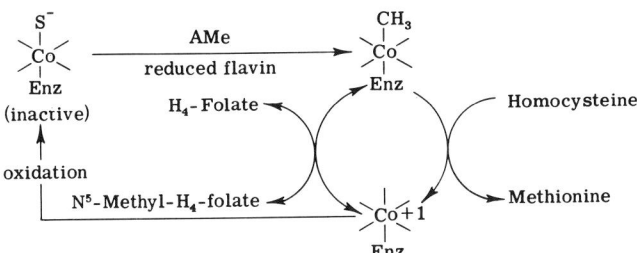

Fig. 8. Schematic reaction mechanism for AMe-dependent methionine synthesis catalyzed by *E. coli* B cobalamin transmethylase. From Taylor and Weissbach (31).

assigned a cobalt methylation role to AMe. Yet, it satisfied the fact that unlabeled 5-methyl-H_4-folate does not inhibit (9, 15) the slow catalysis of reaction (6).

In Fig. 8, the notation

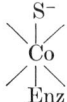

represents collectively any or all of the nonmethylated but inactive forms of the B_{12} chromophore that may exist, including the salmon-colored derivative that is customarily associated with purified preparations (15, 22). Although the precise structure of the salmon-colored cobalamin is uncertain, the above symbol does denote the fact that a sulfur-containing B_{12} compound has been recovered (19, 22) from most of these preparations. Irrespective of its exact structure, the salmon-colored protein isolated by typical purification methods (15, 22) clearly does not contain methyl-B_{12} (19, 22).

The first step in the sequence is methylation of the inactive enzyme by AMe (or CH_3I) in the presence of a reducing system. No attempt is made (Fig. 8) to indicate the number of steps involved in the conversion of the initial B_{12} protein to a methyl-B_{12} enzyme. It may be worthwhile, however, to comment on the role of the reducing system in this reaction. Reduced flavin has the redox potential to reduce aquo-B_{12} or cyano-B_{12} by at least one electron to B_{12r} (72). It also undoubtedly reacts with traces of oxygen to improve the anaerobiosis of a system. Reduction of the bound B_{12} to the Co^{2+} (B_{12r}) stage followed by disproportionation (73) might produce traces of B_{12s} enzyme + aquo-B_{12} enzyme. Although the equilibrium of disproportionation greatly favors B_{12r} (73), it has been suggested (63) that AMe (or CH_3I) with its high methyl group transfer potential (74) can trap such traces of B_{12s} enzyme. Disproportionation is attractive in that the bound B_{12} could be methylated initially without a direct two-electron reduction of the cobalt. This would avoid overcoming the E'_0 of -1.07 V (72) required to reduce B_{12r} to B_{12s}. However, disproportionation (73) requires that the chromophores on two separate molecules of B_{12r} protein must react. Conclusive spectral evidence showing that $FMNH_2$ or $FADH_2$ per se (in the absence of thiols) will reduce an inactive B_{12} protein to a B_{12r} protein has not yet been obtained. It is clear, though, that whatever the form of this reduced B_{12} protein prepared from the inactive enzyme, it has a preference for AMe, not 5-methyl-H_4-folate.

72. H. P. C. Hogenkamp and S. Holmes, *Biochemistry* **9**, 1886 (1970).
73. R. Yamada, S. Shimizu, and S. Fukui, *Biochemistry* **7**, 1713 (1968).
74. S. H. Mudd, W. A. Klee, and P. D. Ross, *Biochemistry* **5**, 1653 (1966).

In the next step in Fig. 8, the methyl-B_{12} enzyme formed by AMe methylation of the inactive enzyme transfers its methyl group to homocysteine to form methionine plus a Co^{1+} enzyme. The latter species is then methylated by the folate substrate, and all subsequent methyl groups appearing in methionine come from the folate substrate. Thus, the methyl group from AMe is used only to activate the enzyme. Although AMe and 5-methyl-H_4-folate can both methylate the B_{12} cobalt, these reactions take place sequentially on different species of the B_{12} protein. In 5-methyl-H_4-folate transmethylation, two forms of the bound cobalamin (methyl-B_{12} and B_{12s}) function cyclically as prosthetic groups analogous to pyridoxamine phosphate and pyridoxal phosphate in enzymic transamination.

While the existence of a methyl-B_{12} enzyme is well-documented (Sections II,D and E), the evidence for a Co^{1+} enzyme (B_{12s} enzyme) pictured in Fig. 8 has been largely circumstantial or indirect. The initial evidence was based on the extreme oxygen lability of the B_{12} enzyme during transmethylation (*31, 32*) in comparison to the known nucleophilicity and lability of free B_{12s} (*21, 23, 54, 75*). Recently, however, spectral evidence for a Co^{1+} enzyme (B_{12s} enzyme) was obtained in an anaerobic system containing methyl-^{14}C-AMe, homocysteine, H_4-folate, 1,4-dithiothreitol, and the salmon-colored B_{12} protein (*76*). During transmethylation from methyl-^{14}C-AMe to homocysteine [reaction (6)], the B_{12r}-like spectrum of the salmon-colored protein shifted to give new absorption maxima at 385 and 460 nm. These maxima are highly indicative of B_{12s} formation (*77*). They were transient, however, even under H_2 gas, since the B_{12s} spectrum reverted back to a B_{12r} spectrum as soon as the methyl-^{14}C-AMe had been metabolized (*76*). Therefore, as seen in Fig. 8, the Co^{1+} enzyme is pictured as continually oxidizing, even in a highly anaerobic system, to an inactive B_{12} protein that must again be primed (methylated) by AMe. One can regard the slow methylation of homocysteine by AMe [reaction (6)] via a methyl-B_{12} enzyme (*27, 31, 49*) as a manifestation of its cofactor role in reaction (1). In support of these views is the observation that the apparent K_m for AMe in reaction (1) increased from $5.2 \times 10^{-8} M$ to $5.0 \times 10^{-7} M$ (*32*) upon switching from an $FMNH_2$ + 1,4-dithiothreitol reducing system to a 2.5-fold-less active reducing system (*15*). This is what one would predict from Fig. 8 since the apparent K_m for AMe should largely reflect the rate of Co^{1+} enzyme inactivation in a particular system.

75. G. N. Schrauzer, E. Deutsch, and R. J. Windgassen, *J. Amer. Chem. Soc.* **90**, 2441 (1968).
76. R. T. Taylor and M. L. Hanna, *BBRC* **38**, 758 (1970).
77. R. Bonnet, *Chem. Rev.* **63**, 573 (1963).

Unfortunately, for both the B_{12}-dependent [reaction (1)] and the B_{12}-independent [reaction (2)] synthesis of methionine, no concrete information is available as to how the N^5-methyl group is activated for enzymic transfer. 5-Methyl-H_4-folate is not a high energy onium compound like AMe (74). Instead, its N^5-methyl group is chemically quite unreactive at pH 7, even with free B_{12s} (58, 78). Recently, it was reported that a slight amount of nonenzymic reaction takes place in acidic media between 5-methyl-H_4-folate and B_{12s} to yield a trace of methyl-B_{12} (62).

III. Non-B_{12} Methyltransferase from *Escherichia coli*

A. Assay and Purification

Non-B_{12} transmethylase activity can be routinely assayed by measuring the formation of [^{14}C]methylmethionine using [5-^{14}C]methyl-H_4-folate (Glu$_3$) as a substrate. The radioactive methionine can be separated from the folate substrate with an anion exchange resin (14). A typical reaction mixture would contain in a total volume of 50 μl: dl-[5-^{14}C]methyl-H_4-folate (Glu$_3$), 3 mμmoles, 13,300 cpm/mμmole; L-homocysteine, 50 mμmoles; Na$_2$HPO$_4$ buffer, pH 7.8, 500 mμmoles; magnesium acetate, 5 mμmoles; 1,4-dithiothreitol, 500 mμmoles; and enzyme. After incubation for 15 min at 37°, the reaction is stopped by

TABLE XIV
Purification of Non-B_{12} Transmethylase from *E. coli* K_{12}[a]

Fraction	Volume (ml)	Total activity (k-units)	Protein (mg/ml)	Specific activity (units/mg)	Yield (%)
1. Extract	319	1220	30.4	126	100
2. Ammonium sulfate (0–65%)	154	1080	50.0	140	89
3. Protamine sulfate	306	1080	15.4	232	89
4. Ammonium sulfate (50–65%)	31	664	48.4	463	54.5
5. First DEAE-Sephadex	102	625	4.28	1430	51
6. Hydroxylapatite	61	376	3.58	1720	31
7. Second DEAE-Sephadex	85	362	2.03	2100	29.7
8. Sephadex G-100	47	320	2.7	2520	26.2

[a] From Whitfield *et al.* (79).

78. G. N. Schrauzer and R. J. Windgassen, *J. Amer. Chem. Soc.* **89**, 3607 (1967).
79. C. D. Whitfield, E. J. Steers, Jr., and H. Weissbach, *JBC* **245**, 390 (1970).

the addition of 0.9 ml of cold water. The [^{14}C]methylmethionine is then separated from the unreacted [5-^{14}C]methyl-H$_4$-folate (Glu$_3$) by chromatography on a Dowex 1 (Cl$^-$) column (0.5 by 3 cm) and assayed for radioactivity (79).

The enzyme has been purified from *E. coli* AB1909, a methionine auxotroph defective in $N^{5,10}$-methylenetetrahydrofolate reductase by Whitfield *et al.* (79). Under derepressed conditions the enzyme represents about 5% of the soluble protein of the cell. A summary of its purification is seen in Table XIV. One unit of non-B$_{12}$ transmethylase activity is defined as the amount of enzyme catalyzing the formation of 1 mμmole of [^{14}C]methylmethionine per 15 min at 37° under standard assay conditions. A homogeneous material has been obtained, and the protein has been partially characterized.

B. PHYSICAL PROPERTIES

The sedimentation coefficient, $s_{20,w}$, of the enzyme was determined from velocity ultracentrifugation at a protein concentration of 10 mg/ml. It was calculated to be 4.7 S (79) by the method of Schachman (80). The molecular weight of the enzyme was determined to be 84,000 by equilibrium sedimentation according to the meniscus depletion method of Yphantis (81). Similar experiments employing a carboxymethylated transmethylase in the denaturing agent, 5 M guanidine·HCl, revealed a lower molecular weight of 50,800. The lower molecular weight of the denatured, carboxymethylated enzyme suggests that the native transmethylase is composed of two subunits.

The percentage of nitrogen in the transmethylase was found to be 16.7, and the gram per liter extinction coefficient, $E_{280}^{1\%}$, of the enzyme at 280 nm in $1 \times 10^{-2} M$, Na$_2$HPO$_4$, pH 7.8, was 1.62 liters g^{-1} cm^{-1}. In 0.1 M NaOH, the $E_{280}^{1\%}$ was 1.6 liters g^{-1} cm^{-1} at 280 nm, and the $E_{294}^{1\%}$ was 1.18 liters g^{-1} cm^{-1} at 294 nm. The latter values were used to calculate the tryptophan and tyrosine content of the protein. Analysis of amino acid composition revealed that the most frequently occurring amino acids were aspartic acid, glutamic acid, alanine, and leucine. Cysteine and methionine were present in the lowest amounts. Using an enzyme molecular weight of 84,000 and a protein concentration based on dry weight and nitrogen analysis, the tyrosine residues per molecule were 23.2 from spectrophotometric determination, as compared to 18.9 from amino acid analysis (79). The number of tryptophan residues per

80. H. K. Schachman, "Methods in Enzymology," Vol. 4, p. 32, 1957.
81. D. A. Yphantis, *Biochemistry* 3, 297 (1964).

molecule by spectrophorometric determination and spectrophotofluorometric analysis was 18.4 to 20.4.

C. CATALYTIC PROPERTIES AND BINDING OF FOLATE SUBSTRATE

Characteristics of the overall catalytic reaction [reaction (2)] were examined using a homogeneous preparation of the purified enzyme (79). The formation of [^{14}C]methylmethionine from [5-^{14}C]methyl-H$_4$-folate (Glu$_3$) was dependent on the presence of homocysteine, enzyme, and phosphate and was stimulated by Mg^{2+} and 1,4-dithiothreitol as seen in Table XV. The Na$_2$HPO$_4$ requirement was specific for the phosphate ion since K$_2$HPO$_4$ could replace Na$_2$HPO$_4$, but Cl$^-$ and sulfate ion could not. Magnesium ions could be replaced by manganese ions and less effectively by calcium ions. The enzyme was nonspecifically inhibited by a high ionic strength medium.

TABLE XV
REQUIREMENTS OF NON-B$_{12}$ TRANSMETHYLASE FOR METHIONINE SYNTHESIS[a]

Reaction mixture	Methionine formed	
	(cpm)	(pmoles)
Complete[b]	2715	202
−Homocysteine	0	0
−Enzyme	0	0
−Na$_2$HPO$_4$ buffer	328	24.3
−Na$_2$HPO$_4$ buffer + K$_2$HPO$_4$ buffer	2685	199
−Na$_2$HPO$_4$ buffer + NaCl	280	20.8
−Na$_2$HPO$_4$ buffer + tris-Cl buffer	220	16.3
−Mg^{2+}	1331	98.6
−[5-^{14}C]Methyl-H$_4$-folate (Glu$_3$) + [5-^{14}C]Methyl-H$_4$-folate (Glu$_1$) or [5-^{14}C]Methyl-H$_4$-folate (α-Glu$_2$)	0	0
−Homocysteine + cysteine, 2-mercaptoethanol, or 1,4-dithiothreitol	0	0
+1,4-Dithiothreitol	3017	224

[a] From Whitfield et al. (79).
[b] The complete system contained dl-[5-^{14}C]methyl-H$_4$-folate (Glu$_3$), 3 mμmoles, 13,500 cpm/mμmole; L-homocysteine, 50 mμmoles; Na$_2$HPO$_4$ buffer, pH 8.2, 500 mμmoles; magnesium acetate, 5 mμmoles; and purified enzyme, 0.12 μg, in a total volume of 50 μl. The following were added where indicated: 23 mμmoles of dl-[5-^{14}C]methyl-H$_4$-folate (Glu$_1$), 54,000 cpm/mμmole; 30 mμmoles of dl-[5-^{14}C]methyl-H$_4$-folate (α-Glu$_2$), 2340 cpm/mμmole; and 500 mμmoles of NaCl, K$_2$HPO$_4$ buffer, pH 8.2; tris-Cl buffer, pH 8.2; 1,4-dithiothreitol, 2-mercaptoethanol, and L-cysteine. The pH (8.1) of the reaction mixture did not change when Na$_2$HPO$_4$ buffer was omitted because of the buffering capacity of the other reaction components.

The substrate specificity of the enzyme was investigated by the use of derivatives or analogs of the two substrates (*79*). [5-^{14}C]Methyl-H$_4$-folate (Glu$_1$) and [5-^{14}C]methyl-H$_4$-folate (α-Glu$_2$) could not replace the triglutamate folate derivative as a methyl donor. In addition, cysteine, 2-mercaptoethanol, and 1,4-dithiothreitol could not replace homocysteine as a methyl acceptor. At pH 7.8 the K_m for l-[5-^{14}C]methyl-H$_4$-folate (Glu$_3$) was 2.35 μM (*79*). Under saturating conditions and based on one catalytic site per enzyme molecule (*82*), the turnover number of purified transmethylase [reaction (2)] is only 14 mμmoles of methionine formed per minute per millimicromole of enzyme. Thus, compared to the *E. coli* B$_{12}$ enzyme [reaction (1) turnover number about 800, Table V], the non-B$_{12}$ transmethylase is a considerably less efficient catalyst.

The pH optimum of the enzyme was determined in phosphate buffer only, since phosphate is required for the reaction. Other buffers such as tris, *N*-tris, (hydroxymethyl)methylglycine, barbital, and imidazole plus optimal levels of phosphate resulted in an inhibition of the reaction as a result of the high ionic strength created. The enzyme was active between pH 6.0 and 8.5, with optimal activity at pH 7.5–7.8.

Although no evidence for a substrate-methylated, non-B$_{12}$ enzyme has been obtained, [5-^{14}C]methyl-H$_4$-folate (Glu$_3$) does form a specific complex with this enzyme in the absence of homocysteine (*82*). As seen in Fig. 9, incubation of the enzyme and *dl*-[5-^{14}C]methyl-H$_4$-folate (Glu$_3$) at 0° followed by chromatography on Sephadex G-50 yielded a peak of radioactivity which eluted with the protein. No radioactivity was seen in this region when the ^{14}C substrate was chromatographed alone in a similar manner. Figure 9 also shows the highest level of enzyme eluted in fraction 12, whereas the highest concentration of complexed radioactivity occurred in fraction 13. The trailing of radioactivity in fractions 16–19 behind the enzyme–substrate complex peak, as well as the slight displacement of the radioactive peak from the enzyme peak, suggested that the complex dissociated during chromatography. Studies on the nature of the bound radioactivity using enzymic and paper chromatographic procedures (*82*) showed that the active isomer of [5-^{14}C]-methyl-H$_4$-folate (Glu$_3$) was present intact in the complex.

A Sephadex G-50 column in which the enzyme and substrate were constantly in equilibrium was used to study the stoichiometry of the binding of substrate to enzyme as well as to ascertain the dissociation equilibrium constant of the enzyme–substrate complex (*82*). The maximum amount of substrate bound corresponded to 0.76 mμmole of [5-^{14}C]-

---

82. C. D. Whitfield and H. Weissbach, *JBC* **245**, 402 (1970).

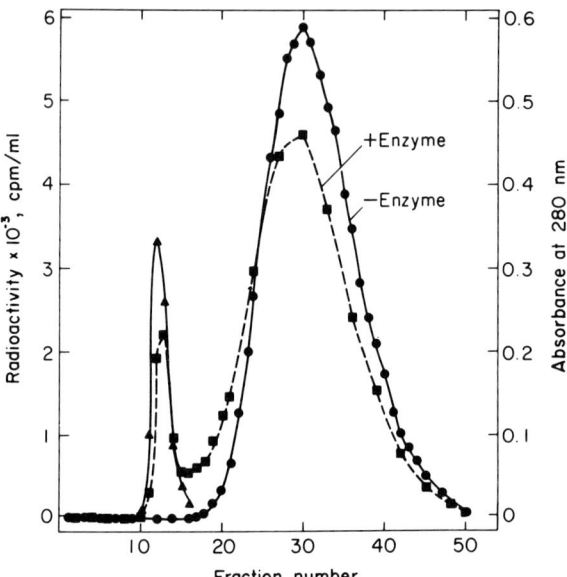

FIG. 9. Isolation of preformed [^{14}C]enzyme complex by Sephadex G-50 chromatography. After incubation of 0.5 mg of non-B_{12} transmethylase with 5 mμmoles of dl-[5-^{14}C]methyl-H_4-folate (Glu$_3$), 75,000 cpm, at 0°, the mixture was applied to a Sephadex G-50 column: (○) radioactivity in the absence of enzyme, (□) radioactivity after incubation with enzyme, and (△) absorbance of the enzyme at 280 nm. From Whitfield and Weissbach (82).

methyl-H_4-folate (Glu$_3$)/mμmole of transmethylase, using a molecular weight of 84,000 (79) for the enzyme.

D. REPRESSION OF ENZYME SYNTHESIS

In *E. coli* it is well established that methionine represses all of the enzymes involved in its biosynthesis including the non-B_{12} transmethylase (79, 83–85). In addition, it has been shown that cyano-B_{12} also represses the non-B_{12} transmethylase (86) and $N^{5,10}$-methylene-H_4-folate reductase (84). The available data (86) indicate that the repression of the non-B_{12} transmethylase observed with the vitamin does not result from excess synthesis of methionine as might be expected since cyano-B_{12} *in vivo* is metabolized to cobalamins which convert the apoenzyme form of the B_{12}

83. R. J. Rowbury and D. D. Woods, *J. Gen. Microbiol.* **24**, 129 (1961).
84. H. M. Katzen and J. M. Buchanan, *JBC* **240**, 825 (1965).
85. R. J. Rowbury and D. D. Woods, *J. Gen. Microbiol.* **35**, 145 (1964).
86. L. Milner, C. Whitfield, and H. Weissbach, *ABB* **133**, 413 (1969).

TABLE XVI
EFFECT OF VARIOUS COBAMIDE COMPOUNDS AND L-METHIONINE
ON THE LEVELS OF THE Non-B_{12} TRANSMETHYLASE[a]

Addition to the growth medium	Specific activity	Repression (%)
None	13.5	0
L-Methionine 10^{-2} M	2.2	83
L-Methionine 10^{-3} M	2.6	80
L-Methionine 5×10^{-5} M	8.7	35
L-Methionine 10^{-5} M	12.1	10
Cyano-B_{12} 10^{-9} M	0.1	99
Cyano-B_{12} 10^{-10} M	12.8	5
Factor B 10^{-7} M	10.1	5
Factor B 10^{-8} M	10.2	23
Factor B 10^{-9} M	13.4	0
Cyano-B_{12}-anilide 10^{-7} M	12.2	11
Cyano-B_{12}-anilide 10^{-8} M	12.3	10
Cyano-B_{12}-anilide 10^{-9} M	13.6	0

[a] From Milner et al. (86).

transmethylase to the active holoenzyme (87). The effect of L-methionine and several cobamides on the repression of the non-B_{12} transmethylase in *E. coli* K_{12} is seen in Table XVI. Although both L-methionine and cyano-B_{12} repress synthesis of the non-B_{12} transmethylase, the vitamin is more effective, and the repression by the various cobamides tested appears to correlate with the ability of the cobamide to form the B_{12} transmethylase holoenzyme.

More recent studies using methionine auxotrophs of *E. coli* have provided further evidence that the mechanism of non-B_{12} enzyme repression by L-methionine and cyano-B_{12} are different (88). In a methionine regulatory mutant (Met J$^-$), L-methionine is not able to repress the enzymes involved in the biosynthetic pathway although cyano-B_{12} is still an effective repressor. In contrast, studies with an *E. coli* auxotroph lacking the B_{12} transmethylase (Met H$^-$) have shown that the non-B_{12} transmethylase is not repressed by cyano-B_{12}, but it can be repressed by L-methionine (88). The above results indicate that the vitamin and other cobamides capable of forming the B_{12} transmethylase holoenzyme can specifically repress the synthesis of the non-B_{12} transmethylase. The B_{12} enzyme itself appears to be part of the repressor system, but the regulatory gene required for repression of all the biosynthetic enzymes

87. H. Weissbach, B. G. Redfield, H. Dickerman, and N. Brot, *JBC* **240**, 856 (1965).
88. H. F. Kung, C. Spears, and H. Weissbach, *ABB* **150**, 23 (1972).

in the pathway by methionine does not appear to be involved in the cyano-B_{12} response. The finding that organisms with low levels of S-adenosyl-L-methionine synthetase have depressed levels of the methionine biosynthetic enzymes (*89*) also indicates that methionine repression may be mediated by S-adenosylmethionine. These results are summarized below:

Methionine → S-adenosylmethionine → repression of all methionine biosynthetic enzymes

Cyano-B_{12} → B_{12} enzyme → specific repression of non-B_{12} transmethylase and, perhaps, $N^{5,10}$-methylene-H_4-folate reductase

IV. Other Sources of Non-B_{12} and B_{12} N^5-Methyltetrahydrofolate-Homocysteine Methyltransferases

A. Non-B_{12} Methyltransferases

1. *Occurrence*

Except for the methionine-cobalamin auxotrophs, all strains of *Escherichia coli* that have been studied (PA15, K-12, 518-W, and B) utilize only the non-B_{12} enzyme for growth in cobalamin-free minimal media (*1a, 3*). In addition to *E. coli*, the presence of a non-B_{12} methyltransferase has also been documented for *Aerobacter aerogenes* (*90*) and *Salmonella typhimurium* (*91*). Both of these enteric non-B_{12} methyltransferases have substrate and ion requirements that are very similar to the *E. coli* non-B_{12} enzyme (*90, 91*). A key difference between *E. coli* (*2*) and these other two enteric bacteria (*90, 91*), however, is that the latter can synthesize enough corrinoids during growth on minimal media to convert partially their B_{12} requiring apoenzyme into a B_{12} holoenzyme. *Saccaromyces cerevisiae* (*92, 93*), *Neurospora crassa* (*94*), *Chlorella pyrenoidosa* (*95*), and higher plants (*1a, 36, 53, 96, 97*) only contain

89. R. C. Greene, C. H. Su, and C. T. Holloway, *BBRC* **38**, 1120 (1970).
90. J. F. Morningstar and R. L. Kisliuk, *J. Gen. Microbiol.* **39**, 43 (1965).
91. S. E. Cauthen, M. A. Foster, and D. D. Woods, *BJ* **98**, 630 (1966).
92. J. D. Botsford and L. W. Parks, *J. Bacteriol.* **94**, 966 (1967).
93. E. G. Burton and W. Sakami, *Fed. Proc., Fed. Amer. Soc. Exp. Biol.* **26**, 387 (1967).
94. E. Burton, J. Selhub, and W. Sakami, *BJ* **111**, 793 (1969).
95. E. G. Burton, Dissertation, Microfilm No. 71-1667, University Microfilms, Ann Arbor, Michigan, 1970.
96. E. G. Burton and W. Sakami, *BBRC* **36**, 228 (1969).
97. W. A. Dodd and E. A. Cossins, *ABB* **133**, 216 (1969).

non-B_{12} methyltransferases. An absence of the B_{12} enzyme from higher plant cell extracts is consistent with the finding that corrinoids are not metabolites in their tissues except in the root nodules of those species which carry out symbiotic nitrogen fixation (98).

Several years ago considerable interest was generated by a report that rat liver also contained a non-B_{12} 5-methyl-H_4-folate (Glu$_3$) transmethylase (99). This claim could not be verified by Burton and Sakami (100). Using [5-^{14}C]methyl-H_4-folate (Glu$_3$) (100) instead of unlabeled folate substrate (99), Burton and Sakami carefully determined that non-B_{12} transmethylation was insignificant compared to the unlabeled methionine formed by proteolysis and methyl transfer from the endogenous betaine. Therefore, it still appears safe to conclude that mammalian tissues do not contain a non-B_{12} methyltransferase for folates.

2. Yeast Non-B_{12} Enzyme

Other than from *E. coli* (Table XIV), a non-B_{12} folate methyltransferase has been highly purified only from *S. cerevisiae* (101). The yeast enzyme preparation, estimated to be 85% pure, has a sedimentation coefficient of 5.25 S and a molecular weight of 75,000. It has a catalytic [reaction (2)] specific activity of 3600 mµmoles of methionine formed per milligram per 15 min relative to 2520 mµmoles (Table XIV) for the *E. coli* enzyme. Yeast non-B_{12} enzyme is also specific for L-homocysteine ($K_m = 22$ µM) as a methyl acceptor, but it utilizes 5-methyl-H_4-folate (Glu$_3$) ($K_m = 380$ µM) and 5-methyl-H_4-folate (Glu$_2$) ($K_m = 430$ µM) equally well as methyl donors (94). 5-Methyl-H_4-folate (Glu$_1$) is inactive as a substrate (94, 101). The optimum pH region for the yeast enzyme is 6.6–7.6. It displays an absolute requirement for phosphate, but unlike the *E. coli* enzyme (Table XV) it is not stimulated by Mg^{2+} (101). A partial inhibition of its activity by EDTA (95, 101) is suggestive that a bound metal may also be required by this enzyme.

3. Folate Substrates of Non-B_{12} Enzymes

Until recently it was generally held (7, 94) that the inability to use 5-methyl-H_4-folate (Glu$_1$) as a substrate was a distinguishing feature of all non-B_{12} folate methyltransferases. However, extracts of pea seeds (97), green beans, spinach, and barley sprouts (96) were found subsequently to catalyze 5-methyl-H_4-folate (Glu$_1$)-homocysteine trans-

98. H. J. Evans and M. Kliewer, *Ann. N. Y. Acad. Sci.* **112**, 735 (1964).
99. F. K. Wang, J. Koch, and E. L. R. Stokstad, *Biochem. Z.* **346**, 458 (1966).
100. E. Burton and W. Sakami, *Eur. J. Biochem.* **7**, 1 (1968).
101. E. Burton and W. Sakami, "Methods in Enzymology," Vol. 17B, p. 388, 1971.

methylation in the presence of Mg^{2+} and phosphate buffer. Catalysis resulting from the presence of a B_{12} enzyme in these plant extracts was excluded by the lack of any stimulation with AMe or reduced flavin and the absence of any inhibition by oxygen (96). Upon subjecting extracts of green beans to Sephadex G-100 chromatography, the activities for 5-methyl-H_4-folate (Glu_1) and (Glu_3) eluted in the exact same fractions. The monoglutamate substrate was one-seventh as active as the triglutamate (96). Thus, higher plants apparently contain a second type of non-B_{12} methyltransferase. It resembles the bacterial non-B_{12} enzyme in requiring only Mg^{2+} and phosphate, but it resembles the bacterial B_{12} enzyme in utilizing 5-methyl-H_4-folate (Glu_1) as a substrate. Detection of this second type of non-B_{12} enzyme suggests that the extra L-glutamate groups are required only for binding the folate substrate to the active site of the *E. coli* (82) and the yeast (94) non-B_{12} enzymes. It argues against an essential participation of the α-carboxyl group (on the second glutamate residue) in the catalysis per se of reaction (2). At this time, it would seem that the influence of oxygen as opposed to AMe plus a reducing system (3, 96) provides the best criteria as to whether one is dealing with a B_{12} or a non-B_{12} enzyme in crude systems.

B. B_{12} METHYLTRANSFERASES

1. *Occurrence and Methyl-B_{12}-Homocysteine Transmethylation*

Although B_{12} methyltransferases have been widely reported in both microorganisms and mammalian tissues, in none of these instances has the enzyme been purified and studied to the same extent as the enzyme from *E. coli* B (Section II). In most cases, evidence for the B_{12} enzyme rests on the presence of reaction (1) activity which shows the unique dependencies on a reducing system, anaerobiosis, AMe, and a B_{12} compound if the cells were cultured in the absence of a cobalamin. In contrast to *E. coli* (1a, 3), *A. aerogenes* (90), *S. typhimurium* (91), and *C. pyrenoidosa* (95) which contain both types of methyltransferases, only the B_{12} enzyme has been found in *Rhodopseudomonas spheroides* (102), *Ochromonas malhamensis* (103), and *Streptomyces olivaceus* (104). Similarly, only the B_{12} enzyme is found in mammalian cells. Thus far,

102. S. E. Cauthen, J. R. Pattison, and J. Lascelles, *BJ* **102**, 774 (1967).
103. J. M. Griffiths and L. J. Daniel, *ABB* **134**, 463 (1969).
104. H. Ohmori, K. Sato, S. Shimizu, and S. Fukui, *Agr. Biol. Chem.* **35**, 338 (1971).

it has been detected in cell-free extracts of pig liver (*3, 14, 41, 52, 105, 106*), beef liver (*52*), chicken liver (*107*), rat liver (*108, 109*), human liver (*110, 111*), pig kidney (*112*), human kidney (*110, 111*), rat brain (*113*), beef brain (*114*), human brain (*115*), and a variety of mammalian cell lines grown in tissue culture (*116–120*). Studies of the subcellular distribution of B_{12} methyltransferase in rat liver (*99, 108*) have shown it to be located predominantly in the cytoplasmic and mitochondrial fractions. Brown *et al.* (*108*) found 50% in the cytoplasm and 36% in the mitochondria. A survey of the organ distribution of B_{12} enzyme in bovine tissues showed the pancreas and brain to contain the highest levels of activity (*114*). Arranged in the order of their enzyme activities per milligram of protein, the following pattern was observed: pancreas > brain > liver > adrenals > heart > kidney.

It will be recalled from Section II,C that the *E. coli* B_{12} enzyme catalyzes free methyl-B_{12}-homocysteine transmethylation [reaction (8)] in addition to reaction (1). It is noteworthy that in every case where it has been examined, other sources of the B_{12} enzyme have also displayed this subsidiary activity. This holds true for extracts or partially purified enzyme from *R. sphaeroides* (*102*), *S. olivaceus* (*104*), *A. aerogenes* (*53*), *S. typhimurium* (*53*), chicken liver (*14, 107*), pig liver (*14, 36*), and rat liver (*14*). While in none of these systems has the

105. J. H. Mangum and K. G. Scrimgeour, *Fed. Proc., Fed. Amer. Soc. Exp. Biol.* **21**, 242 (1962).
106. R. E. Loughlin, H. L. Elford, and J. M. Buchanan, *JBC* **239**, 2888 (1964).
107. H. Dickerman, B. G. Redfield, J. G. Bieri, and H. Weissbach, *JBC* **239**, 2545 (1964).
108. S. S. Brown, G. E. Neal, and D. C. Williams, *BJ* **97**, 34c (1965).
109. C. Kutzbach, E. Galloway, and E. L. R. Stokstad, *Proc. Soc. Exp. Biol. Med.* **124**, 801 (1967).
110. S. H. Mudd, H. L. Levy, and R. H. Abeles, *BBRC* **35**, 121 (1969).
111. S. H. Mudd, H. L. Levy, and G. Morrow, *Biochem. Med.* **4**, 193 (1970).
112. J. H. Mangum and J. A. North, *Biochemistry* **10**, 3765 (1971).
113. T. Nakazawa, K. Yoshiba, and M. Takasugi, *Bitamin* **41**, 333 (1970); **42**, 193 (1970).
114. J. H. Mangum, B. W. Stewart, and J. A. North, *ABB* **148**, 63 (1972).
115. H. L. Levy, S. H. Mudd, J. D. Schulman, P. M. Dreyfus, and R. H. Abeles, *Amer. J. Med.* **48**, 390 (1970).
116. J. H. Mangum and J. A. North, *BBRC* **32**, 105 (1968).
117. J. H. Mangum, B. K. Murray, and J. A. North, *Biochemistry* **8**, 3496 (1969).
118. S. H. Mudd, B. W. Uhlendorf, K. R. Hinds, and H. L. Levy, *Biochem. Med.* **4**, 215 (1970).
119. S. S. Kerwar, C. Spears, B. McAuslan, and H. Weissbach, *ABB* **142**, 231 (1971).
120. M. J. Mahoney, L. E. Rosenberg, S. H. Mudd, and B. W. Uhlendorf, *BBRC* **44**, 375 (1971).

relationship of reaction (8) activity to reaction (1) activity been studied in as much detail as for the $E.$ $coli$ B enzyme (16), the apparent widespread coexistence of both activities is certainly suggestive that they are properties of a single enzyme. Over a 30-fold enrichment from extracts of chicken liver, both activities copurify together along with the cobalt-60 label from previously injected cyano-^{60}Co-B_{12} (107). It would be of interest to learn whether any organisms which are reported to contain only the non-B_{12} methyltransferase also possess any appreciable free methyl-B_{12}-homocysteine transmethylase activity.

2. *Pig Liver and Pig Kidney B_{12} Enzymes*

The most purified and best characterized mammalian enzyme preparations have been those from pig liver and pig kidney. Loughlin et $al.$ (106) purified the enzyme 250-fold from pig liver with an overall yield of 10%. Their preparation showed a nearly complete dependency on reduced flavin and AMe, and its specific catalytic activity [reaction (1)] was 1.0 μmole of methionine synthesized per hour per milligram. The cofactor requirements (106) were, therefore, quite analogous to those of the $E.$ $coli$ enzyme (Table IV). Both the mono- and the triglutamate forms of 5-methyl-H_4-folate were equally active as substrates for the pig liver enzyme (106). Most significant, however, was their demonstration (106) that a bound cobalamin is an essential component of a mammalian transmethylase. A constant ratio was observed between the B_{12} content and the activity content of the fractions at each step over the 250-fold purification procedure. The final preparations contained about 21 pmoles of cobalamin per milligram of protein (106).

Very recently, Mangum and North (112) published in detail a purification scheme to enrich the pig kidney B_{12} enzyme 1800-fold over crude extracts. The final preparation catalyzed the formation of 54.4 μmoles of methionine per hour per milligram, but it was obtained in an overall yield of only 0.8%. Homocysteine was used to stabilize the protein during its isolation. An absorption spectrum of their best preparation (10 mg/ml) shows (112) that it is probably rather similar in the visible region to the $E.$ $coli$ B enzyme (15). The best pig kidney enzyme is still contaminated by cytochrome absorption in the 410 nm region, however (112). No attempt was made to identify the protein-bound B_{12}.

The only catalytic property reported (112) for 1800-fold purified pig kidney enzyme is that it has a near absolute dependency on reduced flavin, but is stimulated only 1.3–1.8-fold by AMe. It is suggested by Mangum and North (112) that only a slight dependency on AMe represents a fundamental difference in the mechanism of reaction (1) between the pig kidney enzyme and the *E. coli* B enzyme (depicted in Fig. 8). However, since the B_{12} enzyme that was partially purified from pig brain by Mangum *et al.* (*114*) showed an absolute requirement for both reduced flavin and AMe, the possibility of nonspecifically bound AMe (*49*) in the pig kidney enzyme preparation (*112*) should be considered.

From the combined data in two communications (*121, 122*), Burke *et al.* concluded that the mechanism depicted in Fig. 8 essentially accommodates most of the observations made using pig kidney B_{12} methyltransferase. A key part of the experimental evidence for this conclusion was the accumulation of a [^{14}C]methyl-B_{12} enzyme from [5-^{14}C]methyl-H_4-folate. The Sephadex G-25 filtered [^{14}C]methyl-B_{12} enzyme then transferred its [^{14}C]methyl group to homocysteine (*121*) under the same conditions as for the *E. coli* enzyme (Table IX). Folate substrate methylation required AMe, reduced flavin, and 0.15 M levels of methionine (*121*). The high level of methionine required has not been explained. It is possible that high levels of methionine were needed merely to inhibit (*122*) the transfer of [^{14}C]methyl from the [^{14}C]methyl-B_{12} protein to the contaminating homocysteine in the enzyme preparation. Partially purified pig kidney methyltransferase can also be methylated with [^{14}C]methyl-AMe, and it catalyzes reaction (6) at a slow rate compared to reaction (1) (*121*).

121. G. T. Burke, J. H. Mangum, and J. D. Brodie, *Biochemistry* **9**, 4297 (1970).
122. G. T. Burke, J. H. Mangum, and J. D. Brodie, *Biochemistry* **10**, 3079 (1971).

5
Enzymic Methylation of Natural Polynucleotides

SYLVIA J. KERR • ERNEST BOREK

I. Introduction 167
II. Transfer RNA Methyltransferases 168
   A. Occurrence 168
   B. Purification and Properties 169
   C. Regulation 176
   D. Biological Significance 184
III. Ribosomal RNA Methyltransferases 187
   A. Occurrence 187
   B. Isolation and Properties 187
   C. Biological Significance 189
IV. DNA Methyltransferases 190
   A. Occurrence 190
   B. Properties 190
   C. Regulation 193
   D. Biological Significance 194

I. Introduction

Consider the problem faced by evolution. A unique invention, without parallel in the inorganic world, information storage of amino acid sequence, was created via a new language: that of permutations of nucleotides in DNA. Retrieval of information was also effected by a universal code of nucleotides in mRNA. Expression of information was achieved by the nucleotide-composed macromolecules rRNA and tRNA. But these macromolecules were essentially identical in all species. The DNA's of even contemporary organisms are identical, except for permutations and fluctuations in the abundance of the four bases. How was the species identity of these macromolecules to be achieved? This is no small solipsist problem. How is the DNA of an invading parasite to be prevented from integrating into the DNA of the host? The answer, in the hindsight of ten years experience, is obvious: A species-specific imprint on all of these macromolecules—except mRNA—was developed. A summary of our current knowledge of these species-specific modifying enzymes of nucleic acids is the subject of this article.

II. Transfer RNA Methyltransferases

A. OCCURRENCE

The existence of enzymes which methylate transfer RNA were first demonstrated in extracts of *Escherichia coli* in 1962 (*1*). Since then, tRNA methyltransferases have been detected in every organism examined including yeast, insects, plants, and mammals (*2*). The enzymes are base, species, as well as organ specific (*3, 4*).

The clearest evidence for organ specificity of the enzymes has been presented by Turkington (*4*), who studied the relative distribution of six different enzymes in seven different organs of mice. The large variations in the extracts from the different organs are readily apparent from the data in Fig. 1. Even greater organ specific variation was found with respect to the enzyme 7-methylguanine methyltransferase. This enzymic activity, which is high in extracts of liver, is absent from those of the heart, brain, and mammary gland of normal mice.

In most methods of extraction of the methyltransferases homogenization of tissues in aqueous media is used. Under these conditions the enzymes are found in the high-speed supernatant fraction after centrifugation and thus were at first thought to be cytoplasmic components of the mammalian cell. However, Kahle *et al.* (*5*) have reported that strictly nonaqueous extraction techniques indicate that the tRNA methyltransferases are located in the nuclei of rat liver cells.

Liau *et al.* (*6*) have found a class of tRNA methyltransferases in the nucleoli of the Novikoff ascites cells. These enzymes formed a distinct subgroup of the total cellular methyltransferases since the pattern of

1. E. Fleissner and E. Borek, *Proc. Nat. Acad. Sci. U. S.* **48**, 1199 (1962).
2. E. Borek and P. R. Srinivasan, *Annu. Rev. Biochem.* **35**, 275 (1966).
3. P. R. Srinivasan and E. Borek, *Proc. Nat. Acad. Sci. U. S.* **49**, 529 (1963).
4. R. W. Turkington and M. Riddle, *Cancer Res.* **30**, 650 (1970).
5. P. Kahle, P. Hoppe-Seyler, and H. Kroeger, *BBA* **240**, 384 (1971).
6. M. C. Liau, C. M. O'Rourke, and R. B. Hurlbert, *Biochemistry* **11**, 629 (1972).

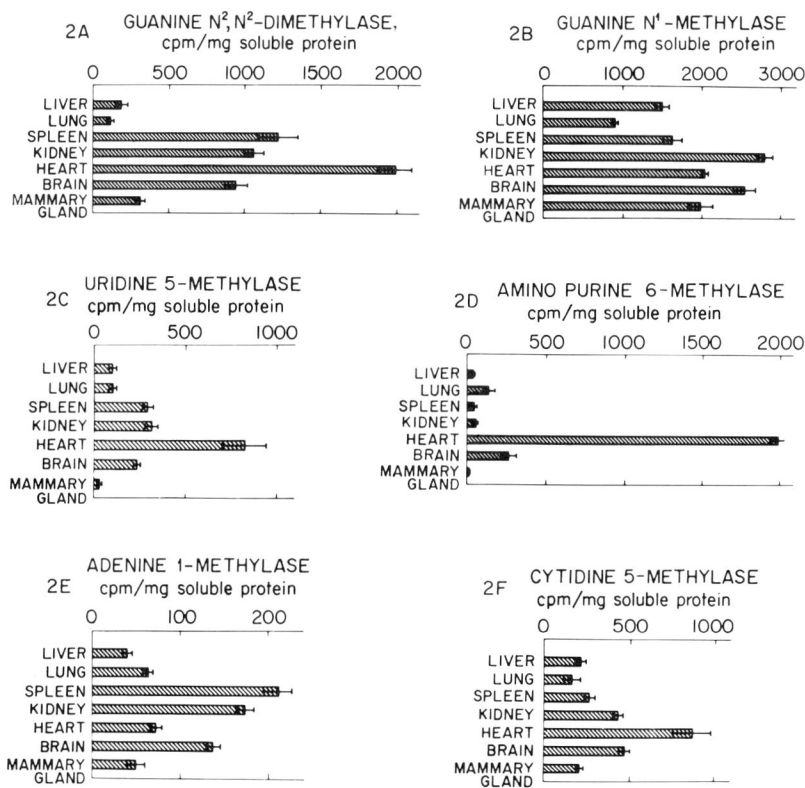

Fig. 1. Profiles of specific activities of tRNA methyltransfer enzymes in various tissues of C3H mice. Each value represents the mean in 3 female mice studied (4).

methylated bases formed by the nucleolar enzymes differed from the pattern yielded by the soluble enzymes.

An even greater example of sequestration was found by Gantt et al. (7), who showed that a specific tRNA methyltransferase is located within the virion of the avian myeloblastosis virus. This would suggest that the enzyme is a requisite ancillary in the viral life cycle.

B. Purification and Properties

1. *Purification*

The tRNA methyltransferases modify the structure of preformed tRNA by the addition of methyl groups at the macromolecular level. The methyl donor is the high energy compound S-adenosylmethionine.

7. R. R. Gantt, K. J. Stromberg, and F. Montes de Oca, *Nature* (*London*) **234**, 35 (1971).

The first tRNA methyltransferases to be partially purified were those from E. coli (1, 8). These enzymes were assayed using methyl-deficient tRNA obtained from relaxed methionine auxotrophs as a substrate. Hurwitz et al. (8, 9) separated six distinct methylating activities which yielded methylated derivatives of all four major bases. These enzymic activities had pH optima in the range 8.0–9.0 and exhibited varying requirements for divalent cations such as Mg^{2+}. The enzyme fractions were quite specific for tRNA and would not methylate synthetic polynucleotides, natural viral RNA's, or rRNA.

Although only six enzymic activities were separated, this is undoubtedly only a lower limit of detection. The adenine-methylating fraction alone was capable of forming three different methylated adenine derivatives, which in view of the known specificity of these enzymes strongly suggests three separate enzymes. Indeed, recently, workers in the laboratory of Kersten have been able to subdivide this same fraction derived by Hurwitz's procedure into eight separate enzyme fractions (10).

Evidence for the multiplicity of the tRNA methyltransferases also comes from fractionation of yeast enzymes. Bjork and Svensson (11) separated eight different methyltransferase activities from Saccharomyces cerevisiae, including three which all methylated uracil in the 5 position. Sequence analysis of the products of these enzymes revealed that they were distinctly different in sequence specificity (12). The partially purified methyltransferases were relatively unstable, losing activity at either 0° or −15°. Genetic studies on yeast have also provided evidence that the enzymes which monomethylate guanine at the N^2 position are distinct from those which dimethylate guanine in the same position (13). This conclusion stems from finding mutants which lack the guanine dimethylating enzyme but still possess the monomethylating enzyme.

The instability of mammalian tRNA methyltransferases has been a hindrance in their isolation and purification. The first attempts to partially characterize mammalian tRNA methyltransferases were carried out by Rodeh et al. (14) on rat liver and Simon et al. (15) on rat brain. These workers found evidence for the existence of at least five different

8. J. Hurwitz, M. Gold, and M. Anders, JBC **239**, 3462 (1964).
9. J. Hurwitz, M. Gold, and M. Anders, JBC **239**, 3474 (1964).
10. W. Kersten and H. Kersten, personal communication (1972).
11. G. R. Bjork and I. Svensson, Eur. J. Biochem. **9**, 207 (1969).
12. I. Svensson, G. R. Bjork, and P. Lundahl, Eur. J. Biochem. **9**, 216 (1969).
13. J. H. Phillips and K. Kjellin-Straby, JMB **26**, 509 (1967).
14. R. Rodeh, M. Feldman, and U. Z. Littauer, Biochemistry **6**, 451 (1967).
15. L. N. Simon, A. J. Glasky, and T. H. Rejal, BBA **142**, 99 (1967).

TABLE I
Methyl Acceptor Capacities of Individual *E. coli*
tRNA's with Liver Methylases *in Vitro*[a]

tRNA	Extent of methylation (cpm $^{14}CH_3$)		
	Methylase I	Methylase II	Methylase III
tRNA^{Asp}	210	290	150
tRNA$_2^{Glu}$	440	530	140
tRNA$_2^{Leu}$	1790	2340	170
tRNA^{fMet}	4580	4110	210
tRNA$_1^{Met}$	74	110	510
tRNA^{Phe}	33	240	770
tRNA$_1^{Ser}$	32	220	520
tRNA$_3^{Ser}$	2380	3270	110
tRNA$_2^{Tyr}$	260	480	210
tRNA^{Val}	64	350	710
E. coli tRNA	1440	1280	460
Methyl-deficient *E. coli* tRNA	1730	1600	540
Rat liver tRNA	45	84	34
Yeast tRNA	230	240	70

[a] From Kuchino and Nishimura (*17*).

methyltransferases in rat liver and three to four different activities in rat brain.

Baguley and Staehelin (*16*) separated an adenine-specific tRNA methyltransferase from rat liver and spleen as well as from leukemic rat spleen using DEAE-cellulose column chromatography and gel filtration on Sephadex G-200. This achieved an approximately 40-fold purification and separated the adenine-specific activity from other base specific activities. The enzyme methylated adenine only in a particular sequence, although, as will be discussed later, tertiary structure of the tRNA also plays a part in enzyme recognition.

Kuchino and Nishimura (*17*) have separated three distinct guanine-specific methylating activities from rat liver using gradient elution from hydroxylapatite columns. They were able to distinguish the enzyme fractions by their elution profiles and on the basis of their reactivity toward individual purified tRNA species. Their results are shown in Table I where it can be seen that fractions I, II, and III have quite distinct specificities toward purified tRNA's. This was further confirmed by actual sequence analysis of the methylated bases in the puri-

16. B. C. Baguley and M. Staehelin, *Eur. J. Biochem.* **6**, 1 (1968).
17. Y. Kuchino and S. Nishimura, *BBRC* **40**, 306 (1970).

fied tRNA's. Figure 2 shows the sites of methylation by fraction II in tRNA^{fMet} and by fraction III in tRNA$_\text{I}^{\text{Val}}$. Again, as in the case of the adenine-specific methyltransferase from rat liver (16), sequence alone was not sufficient for enzyme recognition since other tRNA species containing the same trinucleotide sequences were bypassed by the enzyme.

To date (May, 1972) no one has prepared a pure, homogeneous tRNA methyltransferase from any source.

2. Substrate Specificity

All tRNA methyltransferases use S-adenosylmethionine as a methyl donor, but their reactions with tRNA are governed by their species specificity. In the *in vitro* assay of the methyltransferases, tRNA from a heterologous source must be used as a methyl acceptor since the homologous tRNA has already been exposed *in vivo* to the indigenous methyltransferases and will no longer act as a substrate for them. The only exception is methyl-deficient tRNA from *E. coli* which can be used as a substrate for the homologous methyltransferases.

The interaction with tRNA can be characterized in two ways. The first is the standard enzymological method of determining rate (moles product/mg protein/time), which yields the specific activity of a given enzyme preparation. The second method takes advantage of the limited number of introducible methyl groups, which are characteristic of each substrate and enzyme pair. Enzymes from a particular source are able to introduce only a fixed number of methyl groups into a unit amount of heterologous tRNA. This has been called capacity or extent of methylation. It can be defined as the absolute number of methyl groups per unit of tRNA at infinite time and maximum protein concentration. This number is a characteristic parameter of any given enzyme source. It is a sensitive probe for changes in enzyme specificities.

The tRNA methyltransferases will not react with DNA, ribosomal RNA, viral RNA's, synthetic polynucleotides, or the monomeric ribonucleotides. They also require an intact tRNA molecule. Shershneva *et al.* (18) have shown that separated half molecules of yeast tRNA^{Val} would not serve as a substrate for a crude mixture of tRNA methyltransferases from rat liver or enzymes from Novikoff hepatoma. When the halves were recombined the level of methylation returned to 90% of that achieved in the intact molecule.

Kuchino *et al.* (19) found some slightly different results with fragments of *E. coli* tRNA^{fMet}. As mentioned above, they had isolated a

18. L. P. Shershneva, T. V. Venkstern, and A. A. Baer, *FEBS Lett.* **14**, 297 (1971).
19. Y. Kuchino, T. Seno, and S. Nishimura, *BBRC* **43**, 476 (1971).

5. ENZYMIC METHYLATION OF NATURAL POLYNUCLEOTIDES

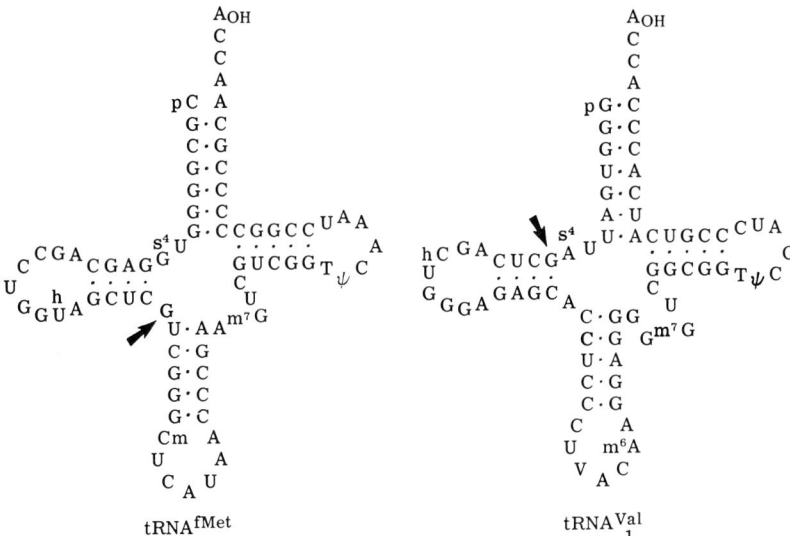

FIG. 2. Cloverleaf structure of *E. coli* tRNA's, indicating the site of methylation in tRNA^{fMet} by methyltransferase II and in tRNA$_1^{Val}$ by methyltransferase III. Enzyme II did not accept valyl tRNA as a substrate while enzyme III would not act on formyl-methionyl tRNA. V stands for uridine-5-oxyacetic acid (17).

tRNA methyltransferase fraction from rat liver which methylated a specific guanine residue at the fifty-first position from the 3'-OH end of the tRNA^{fMet} (Fig. 2) (17).

When the tRNA^{fMet} was split into two fragments comprising approximately three quarters and one quarter of the molecule, the three-quarter fragment could be methylated to a slight extent and the product was not N^2-methylguanine but 1-methyladenine at the nineteenth position from the 3'-OH end. When the fragments were recombined the methyltransferase fraction now methylated adenine and guanine. Thus it is clear that the conformation of the tRNA is essential for its recognition by tRNA methyltransferases, and alteration of the conformation can modify the specificity of recognition.

Further proof of specificity rooted in conformation has been adduced by Baguley *et al.* (20) from studies of purified yeast tRNA$_1^{Ser}$. The 1-methyladenine methyltransferase derived from leukemic rat spleen was able to methylate an adenine at the nineteenth position from the 3'-OH end of yeast tRNA$_1^{Ser}$. 1-Methyladenine occurs in this position normally in rat liver tRNA^{Ser}. Furthermore, other yeast tRNA species, tRNA^{Tyr}, tRNA^{Phe}, and tRNA^{Val}, also have a 1-methyladenine naturally

20. B. C. Baguley, W. Wehrli, and M. Staehelin, *Biochemistry* **9**, 1645 (1970).

occurring at that position. Thus, yeast must have a methyltransferase capable of recognizing that sequence in the nineteenth position from the 3'-OH end in certain tRNA species but not in the yeast $tRNA_1^{Ser}$. However, the rat spleen enzyme can react at that site in the heterologous tRNA as well. Baguley et al. (20) suggested that the yeast $tRNA_1^{Ser}$ exists in a closed conformation unavailable to the yeast adenine methyltransferase for reaction and that the mammalian adenine methyltransferase can induce a conformational change in the molecule or that the high pH (9.25) required for *in vitro* methylation induces the conformational change.

Thus, we can conclude that the tRNA methyltransferases are base-specific, sequence-specific, and conformation-specific in their requirements for reaction with tRNA and that these specificities are characteristics of the enzymes from various organisms.

3. *Ionic Stimulation*

The tRNA methyltransferases all require the presence of cations in some form—monovalent or divalent inorganic cations or organic cations—for activity. Early workers in the field generally used Mg^{2+} (*1*, *8*). Hurwitz et al. (*9*) tested the effect of a series of divalent inorganic cations as well as spermine on the activity of the various base-specific enzyme fractions they had separated from *E. coli*: Mg^{2+} stimulated all the fractions, Ni^{2+} and Zn^{2+} were unable to stimulate the reaction, Ca^{2+} was almost as effective as Mg^{2+}, spermine stimulated only a few of the enzyme fractions, and spermine and Mg^{2+} combined exhibited an antagonistic effect.

Rodeh et al. (*14*) first reported that NH_4^+ would stimulate the activity of the tRNA methyltransferases from rat liver. The optimal concentration was 0.25 *M*. Kaye and Leboy (*21*) extended the work of Rodeh et al. (*14*) on the effect of NH_4^+ on the methyltransferases. In extracts of mouse organs and tumors, they found an NH_4^+ optimal concentration of 0.36 *M* and also found an alteration in the pattern of bases methylated by the extracts compared to patterns obtained using Mg^{2+}. In particular, the cytosine-specific methyltransferases were stimulated 9-fold, indicating that different methyltransferases have different responses to ionic variations. However, the very high concentration of NH_4^+ (0.25–0.36 *M*) required for maximal stimulation of the methyltransferases makes it improbable that such conditions would be encountered *in vivo*. Also, as mentioned above, the individual tRNA methyltransferase species undoubtedly vary in their ion requirements, and several workers have

21. A. M. Kaye and P. S. Leboy, *BBA* **157**, 289 (1968).

reported tRNA methyltransferase systems which do not respond to ammonium ion (*22–25*).

Several workers have studied the effect of diamines and polyamines on the methyltransferases (*26–28*). Leboy has shown that the tRNA methyltransferases can be stimulated both in rate of reaction and extent of reaction by physiological concentrations of spermine, spermidine, or putrescine (*26*). Again the pattern of bases methylated by the extracts was different in the presence of Mg^{2+} or polyamine, with cytosine-specific and the adenosine-specific methyltransferases showing a large response to the polyamines (*29*). Pegg (*27*) has confirmed and extended the work of Leboy (*26, 29*) presenting some evidence that the effect of the polyamines on the methylation of tRNA resulted from their combination with the tRNA substrate rather than a direct effect on the methyltransferases.

Young and Srinivasan (*28*) have studied the effects of polyamines and inorganic cations on the homologous methylation of *E. coli* methyl-deficient tRNA by *E. coli* methyltransferases. In this system the polyamines give a greater stimulation of the rate of *in vitro* methylation than does Mg^{2+}, but they do not alter the pattern of bases methylated nor do they cause hypermethylation.

The monovalent cations NH_4^+, Na^+, K^+, and Li^+ also stimulate the rate of tRNA methylation in this system. It was found that Mg^{2+} and the polyamines are antagonistic; that is, in combination they caused a reduction in rate of *in vitro* methylation. This was interpreted as competition for binding sites on the tRNA molecules. On the other hand, the monovalent cations were not antagonistic toward either Mg^{2+} or the polyamines; their effects were additive.

It is clear that in such a complex multienzyme system as that which comprises the family of tRNA methyltransferases existing in bacterial or mammalian cells, the exact pH and ionic optima for each enzyme cannot be achieved in a single *in vitro* reaction mixture. The control of these enzymes *in vivo* will then depend on the localization of the enzymes and the availability in the cell of the various ionic components.

22. R. W. Turkington, *JBC* **244**, 5140 (1969).
23. E. S. McFarlane, *Can. J. Microbiol.* **15**, 189 (1969).
24. S. J. Kerr, *Biochemistry* **9**, 690 (1970).
25. R. E. Gallagher, R. C. Y. Ting, and R. C. Gallo, *Proc. Soc. Exp. Biol. Med.* **136**, 819 (1971).
26. P. S. Leboy, *Biochemistry* **9**, 1577 (1970).
27. A. E. Pegg, *BBA* **232**, 630 (1971).
28. D. V. Young and P. R. Srinivasan, *BBA* **238**, 447 (1971).
29. P. S. Leboy, *FEBS Lett.* **16**, 117 (1971).

C. Regulation

1. *Inhibitors*

A number of inhibitors of the tRNA methyltransferases, both synthetic and natural, have been described.

Hurwitz et al. (*9*) showed that S-adenosylhomocysteine, which is the product derived from S-adenosylmethionine after methyl transfer, is a competitive inhibitor of the *E. coli* tRNA methyltransferases. They found adenosine to be a much weaker inhibitor.

Wainfan and Borek (*30*) found that while adenosine did inhibit the *E. coli* guanine-specific methyltransferases, it was ineffective toward mammalian enzymes. Adenine was an inhibitor of the guanine-specific methyltransferases from both bacterial and mammalian sources. Wainfan also found that several analogs of these compounds such as 7-deazaadenosine and a number of cytokinins—kinetin riboside, zeatin riboside, 6-benzylamino purine riboside, and N^6-(Δ^2-isopentenyl) adenosine—were active as inhibitors of the methyltransferases (*30, 31*). These findings are particularly interesting since cytokinins are known to occur naturally as modifications of tRNA molecules (*32*).

Rodeh et al. (*14*) showed that the tRNA methyltransferases from rat liver were inhibited by *Micrococcus lysodeikticus* DNA and by a number of synthetic ribopolynucleotides, including polyadenylic acid, polycytidylic acid, polyinosinic acid, polyuridylic acid, and polyguanylic acid. Double-stranded copolymers such as poly UA did not inhibit the enzymes.

Young and Srinivasan (*28*) found that anions such as Cl^-, Br^-, ClO_4^-, and SCN^- inhibited the methyltransferases from *E. coli* with increasing inhibition in that order.

Moore and Smith (*33*) investigated the effects of S-adenosylethionine and ethylthioadenosine on the methyltransferases from *E. coli* and rat liver and observed competitive inhibition. Moore (*34*) also showed that the inhibition by S-adenosylethionine was specific for the *E. coli* tRNA-cytosine and tRNA-adenine methyltransferases.

Pegg (*35*) has confirmed that the methyltransferases from rat liver

30. E. Wainfan and E. Borek, *Mol. Pharmacol.* **3**, 595 (1967).
31. E. Wainfan and B. Landsberg, *FEBS Lett.* **19**, 144 (1971).
32. D. J. Armstrong, W. J. Burrows, F. Skoog, K. L. Roy, and D. Söll, *Proc. Nat. Acad. Sci. U. S.* **63**, 834 (1969).
33. B. G. Moore and R. C. Smith, *Can. J. Biochem.* **47**, 561 (1969).
34. B. G. Moore, *Can. J. Biochem.* **48**, 702 (1970).
35. A. E. Pegg, *FEBS Lett.* **16**, 13 (1971).

and rat kidney are inhibited by S-adenosylethionine as well as by S-adenosylhomocysteine. Further, he showed that these mammalian enzymes are also strongly inhibited by ethidium bromide, acridine orange, and proflavin.

Halpern et al. (36) have reported that nicotinamide is a naturally occurring inhibitor of the methyltransferases from rat liver and Walker-256 carcinosarcoma and that it was in competition with S-adenosylmethionine. Buch et al. (37) found that nicotinamide also inhibited the tRNA methyltransferases of a human tumor cell line grown in tissue culture (KB cells) and that several structural analogs of nicotinamide were capable of inhibiting the KB cell enzymes. They further showed that nicotinamide was able to inhibit the tRNA methyltransferase activity in extracts of several human malignancies. However, nicotinamide had no effect on the methyltransferases from normal human tissues.

Several macromolecular inhibitors of the methyltransferases have been reported (38–40). Sheid and Wilson (38) have identified an inhibitor of mammalian tRNA methyltransferases present in bull sperm and seminal plasma as a high molecular weight polysaccharide.

Sharma and Borek (39) have shown that 8 hr after the onset of morphogenesis an inhibitor of the methyltransferases appears in differentiating *Dictyostelium discoideum*. The inhibitor was also present in spores and appeared to be a protein since it was heat sensitive, nondialyzable, and trypsin sensitive.

Sharma and Borek (40) have also reported the presence of an inhibitor of the methyltransferases in extracts of pig uterus, which seemed to be a protein.

Several investigators have reported elevated levels of the tRNA methyltransferases in organs of fetal and newborn animals compared to their adult counterparts (15, 41, 42). Kerr (24) has shown that this is in part the result of the presence of inhibitors of the methyltransferases in extracts of adult tissues which are absent from fetal tissues and several tumor tissues, all of which exhibit elevated levels of methyltransferase activity. In the case of liver, kidney, and pancreas, the inhibitor has been

36. R. M. Halpern, S. Chaney, B. C. Halpern, and R. A. Smith, *BBRC* **42**, 602 (1971).
37. L. Buch, D. Streeter, R. M. Halpern, L. N. Simon, M. G. Stout, and R. A. Smith, *Biochemistry* **11**, 393 (1972).
38. B. Sheid and S. M. Wilson, *BBA* **224**, 382 (1970).
39. O. K. Sharma and E. Borek, *J. Bacteriol.* **101**, 705 (1970).
40. O. K. Sharma and E. Borek, *Biochemistry* **9**, 2507 (1970).
41. A. M. Kaye, B. Fridlender, and R. Salomon, *Isr. J. Chem.* **3**, 78 (1966).
42. R. L. Hancock, P. McFarland, and P. R. Fox, *Experientia* **23**, 806 (1967).

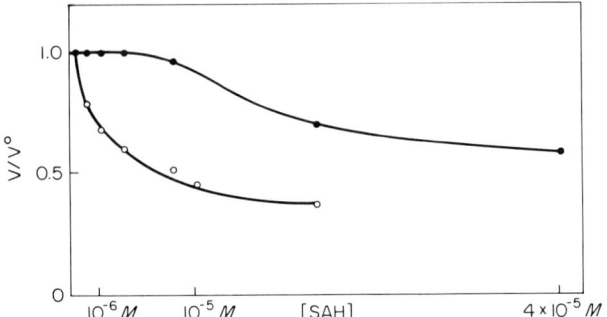

FIG. 3. Inhibition of the tRNA methyltransferase and the glycine methyltransferase from rabbit liver by S-adenosylhomocysteine: (●) glycine methyltransferase activity and (○) tRNA methyltransferase activity. Both enzymes were assayed in the presence of 10^{-5} M S-adenosylmethionine and varying concentrations of S-adenosylhomocysteine. Rates are expressed as percent of the initial reaction velocity ($V°$) in the absence of S-adenosylhomocysteine (43).

identified as a competing enzyme system which methylates glycine to yield sarcosine (N-methylglycine) (43). In tissue extracts of the abovementioned normal adult organs the specific activity of the glycine methyltransferase is orders of magnitude higher than the specific activity of the tRNA methyltransferases; thus, in *in vitro* enzymic reactions using undialyzed extracts, the glycine methyltransferase competes for the common substrate S-adenosylmethionine very effectively. Further examination of the system revealed another facet of this regulation which is a differential inhibition of the tRNA methyltransferases by S-adenosylhomocysteine. This is shown in Fig. 3 where it can be seen the tRNA methyltransferases are much more sensitive to inhibition by S-adenosylhomocysteine than is the glycine methyltransferase. Thus, even in the presence of low concentrations of the glycine methyltransferases, such that all the available S-adenosylmethionine is not exhausted, the tRNA methyltransferases are still inhibited as a result of the production of low levels of S-adenosylhomocysteine. There is an interesting correlation between the presence of glycine methyltransferase in liver, kidney, and pancreas, the organs where elevated levels of the S-adenosylmethione-synthesizing enzyme exist (44).

The report by Murai *et al.* (45) of the inhibition of tRNA methyltransferase in rat liver by nicotinamide and a nondialyzable component

43. S. J. Kerr, *JBC* **247**, 4248 (1972).
44. J. B. Lombardini and P. Tallalay, *Advan. Enzyme Regul.* **9**, 349 (1971).
45. J. T. Murai, P. Jenkinson, R. M. Halpern, and R. A. Smith, *BBRC* **46**, 999 (1972).

may be analogous to the glycine methyltransferase system: The enzyme may methylate nicotinamide thus producing inhibitory levels of S-adenosylhomocysteine.

While the competing glycine methyltransferase is found only in adult liver, kidney, and pancreas, inhibitors of the tRNA methyltransferases do exist in other adult organs. The inhibitor in pig uterus mentioned above (40) is one example and Swiatek et al. (46) have reported the presence of an inhibitor in developing pig brain which appears as early as 12 hr after birth and persists in the adult animal. The nature of this inhibitor is not yet clear. Its appearance can be delayed up to 48 hr by fasting the newborn animals and it is inactivated by freezing, by standing at 2° for 18 hr, or dialysis at 2° for 18 hr.

The regulation of tRNA methyltransferases by naturally occurring inhibitors, either small molecules or macromolecules, is just beginning to be understood and should prove to be a very useful tool in the investigation of control mechanisms in higher organisms.

2. *Hormones*

Alterations in the levels and capacities of the tRNA methyltransferases have been observed in several hormonally regulated systems (22, 40, 47–50). Hacker (47) found that in immature oviduct of chicks which had been maintained on a diet supplemented with diethylstilbesterol the tRNA methyltransferase levels rose several-fold. Turkington (22) showed that the tRNA methyltransferase levels were altered during the cell differentiation which occurs in mouse mammary gland during pregnancy. Similar alterations could be simulated by treatment of mammary gland explants in organ culture with the hormones insulin, hydrocortisone, and prolactin. All three hormones were required for maximal stimulation of the enzymes and the effect was blocked by the addition of either actinomycin D or cycloheximide, indicating that the hormone action depended on the synthesis of RNA and protein.

Sharma and Borek (40) found that the level of tRNA methyltransferase activity in the uterus of an ovariectomized rat or pig fell to one-half or less of the level of activity in the normal mature uterus. Administration of physiological levels of estradiol to the ovariectomized animals returned the enzyme levels to normal. At these low levels of estrogen

46. K. R. Swiatek, D. G. Streeter, and L. N. Simon, *Biochemistry* **10**, 2563 (1971).
47. B. Hacker, *BBA* **190**, 38 (1969).
48. D. J. Pillinger, E. Borek, and W. K. Paik, *J. Endocrinol.* **49**, 547 (1971).
49. B. Sheid, E. Bilik, and L. Biempica, *ABB* **140**, 437 (1970).
50. L. L. Mays and E. Borek, *Biochemistry* **10**, 4612 (1971).

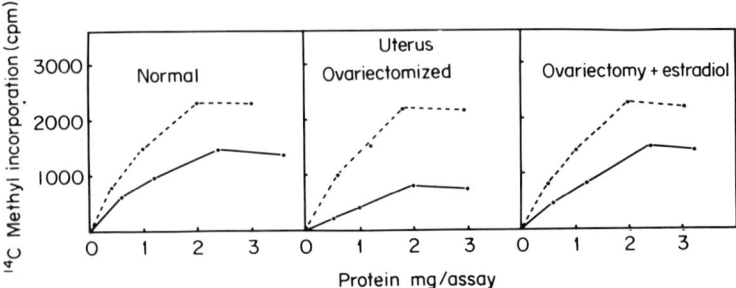

FIG. 4. Effect of pH 5 precipitation on tRNA methyltransferase activity in extracts of normal and ovariectomized pig uterus: (———) enzymic activity in the 100,000 × g supernatant extracts and (- - -) activity in the pH 5 precipitates derived from those extracts (40).

no other organs except the target organ, uterus, exhibited any response to the hormone.

The effect of the hormone appears to be the modulation of inhibitor levels in the uterus rather than a direct effect on the methyltransferases. Evidence for this is presented in Fig. 4. It had been shown earlier that by bringing a high-speed supernatant extract of mammalian tissues to pH 5, it is often possible to separate the tRNA methyltransferases from inhibitors (24). The methyltransferases are precipitated at pH 5 while the inhibitors remain in the supernatant. Figure 4 shows the tRNA methyltransferase capacity in crude extracts of normal uterus, ovariectomized uterus, and ovariectomized uterus treated with estradiol as well as the capacity of the pH 5 precipitated methyltransferases from these same tissues. The capacity of the pH 5 precipitated enzymes to methylate tRNA is identical in all three samples, and the difference in activity observed in the crude extracts appears to result from differences in inhibitor levels. As mentioned above, in pig uterus the inhibition appears to involve a protein (40) and evidence from studies with enzyme extracts from uterus indicate that some of the base-specific methyltransferases may be more sensitive to the inhibitor than others (51). Table II shows the pattern of base methylation by extracts of rat uteri, and it can be seen that the N^2,N^2-dimethylguanine-specific methyltransferase activity is lowered to a relatively greater extent than the N^2-monomethylguanine-specific methyltransferase in crude extracts of ovariectomized uteri.

If massive doses of estrogen are given to animals, changes in the tRNA methyltransferases are found in tissues other than the natural target organs (49, 50). Sheid et al. (49) showed that high doses of mestranol

51. O. K. Sharma, S. J. Kerr, R. Lipshitz-Wiesner, and E. Borek, Fed. Proc., Fed. Amer. Soc. Exp. Biol. **30**, 167 (1971).

TABLE II
PATTERN OF BASE METHYLATION BY EXTRACTS OF RAT UTERI[a,b]

Methylated derivative	Isolated bases (cpm)							
	Normal			Ovariectomized			Ovariectomized plus estradiol	
	Exp 1	Exp 2	Exp 3	Exp 1	Exp 2	Exp 3	Exp 1	Exp 2
1. N$_2$diMeG	3,790	3,550	4,059	630	857	632	4,940	4,220
2. N$_2$MeG	3,330	3,325	3,579	1,310	3,313	1,514	5,020	4,984
Ratio $\frac{1}{2}$	1.17	1.08	1.13	0.48	0.26	0.41	0.98	0.85

[a] From Sharma et al. (51).
[b] tRNA labeled with ^{14}C-methyl was prepared using 100,000 × g extracts from normal rat uterus, ovariectomized rat uterus, and estradiol-treated ovariectomized rat uterus. The tRNA was purified and the methylated bases isolated and identified by acid hydrolysis and two-dimensional chromatography. N$_2$diMeG = N^2,N^2-dimethylguanine and N$_2$MeG = N^2-methylguanine.

caused a 60% enhancement of the tRNA methyltransferase capacity in rat liver. The uracil-specific methyltransferase appeared to be the most altered, showing a relative increase of 130%.

The synthesis of phosvitin, an egg yolk protein can be induced in the liver of the rooster by massive doses of estrogen (52). The synthesis of this protein is normally restricted to hens. Mays and Borek (50) have examined the tRNA methyltransferases in this system. They found that after administration of estradiol (10 mg/kg body weight) or diethylstilbesterol (25 mg/kg body weight) the tRNA methyltransferase capacity of the rooster liver was decreased. Analysis of hydrolyzed tRNA methylated *in vitro* by enzyme extracts showed that the different base-specific enzymes were not uniformly affected. The enzymic activities producing N^2-methylguanine were higher in overall capacity after estrogen treatment, while those producing a number of other methylated bases were either constant or diminished after hormone treatment.

Hormonal influence on the tRNA methyltransferases has been shown in another system by Pillinger et al. (48), who studied the enzymes in the giant bullfrog, *Rana catesbeiana*, during thyroxine-induced metamorphosis. Three days after the administration of physiological levels of thyroxine into the environment of the bullfrog tadpole there was a diminution in methyltransferase capacity of extracts of both the liver and tail to one-half of that in the untreated animals. Four days after the administration, there was a beginning of return to normal in the capacity of

52. O. Greengard, M. Gordon, M. A. Smith, and G. Acs, *JBC* **239**, 2079 (1964).

the enzymes. Eight days after the administration of the hormone the enzyme capacity was almost that of the untreated animals.

Hormones, then, regulate the activity of the tRNA methyltransferases in a complex manner probably involving the induction or repression of their synthesis as well as modulating their activity through regulation of the levels of inhibitors of the methyltransferases.

3. Bacteriophage Induction and Infection

Wainfan et al. (53) have studied the tRNA methyltransferases in lysogenic E. coli at various intervals after ultraviolet light induction and observed a profound reduction in total enzyme capacity between 10 and 15 min after induction. Thirty minutes after the induction the enzyme capacity returned to the level found in extracts from noninduced bacteria. The changes among the various base-specific enzymes were not uniform, with the reduction in the capacity of the uracil-specific methyltransferases accounting for most of the diminution of the total methyltransferase capacity. The differential effect on the uracil-specific methyltransferase was traced to the appearance of a dialyzable inhibitor specific for this enzyme (54).

When the λ prophage was induced by heat in a heat-inducible strain of E. coli, there was also a transient decrease in total tRNA methyltransferase capacity, but in this case caused by a direct preferential heat inactivation of uracil-specific and adenine-specific methyltransferases. The methyltransferase then returned to preinduction levels following the development of new relatively more heat stable enzymes (55).

In the case of bacteriophage infection, Wainfan et al. (53) found that, after infection of E. coli B with the virulent bacteriophage T2, major shifts in the relative levels of base-specific methyltransferases occurred. After T1 infection no changes in the enzymes were observed.

When E. coli B was infected with bacteriophage T3, Gefter et al. (56) showed that the tRNA methyltransferases were completely prevented from expressing their activity as a result of the appearance of a new enzyme which destroyed the methyl donor, S-adenosylmethionine.

4. Non-oncogenic and Oncogenic Virus Infection

No direct measurements of the tRNA methyltransferases themselves in cells infected with non-oncogenic viruses have been reported. However,

53. E. Wainfan, P. R. Srinivasan, and E. Borek, *Biochemistry* **4**, 2845 (1965).
54. E. Wainfan, P. R. Srinivasan, and E. Borek, *JMB* **22**, 349 (1966).
55. E. Wainfan, *Virology* **35**, 282 (1968).
56. M. Gefter, R. Hausmann, M. Gold, and J. Hurwitz, *JBC* **241**, 1995 (1966).

at least three cases of alterations in the methyl content of tRNA extracted from virus infected cells have been reported. These include foot-and-mouth disease virus (FDMV) infected baby hamster kidney cells, where an inhibition of methylation of tRNA was noted (57); a poliovirus infected human cell line (HEp 2), in which incorporation of methyl groups into tRNA was also inhibited (58); and vaccinia virus infected HeLa cells, where a shift to a more extensively methylated form of tRNA was observed (59).

Kit et al. (60) have studied the effect of the oncogenic virus SV40 on the tRNA methyltransferases during productive infection, abortive infection, and transformation of cells. SV40 yields a productive infection with African green monkey kidney cells (CV-1). In this case neither the specific activity nor the capacity of tRNA methyltransferase appeared to be altered.

With primary mouse kidney cells, SV40 gives primarily an abortive infection and a small percentage of transformed cells. No changes in the methyltransferases were noted in the abortively infected cells. However, in mouse kidney cell lines transformed by SV40, 2- to 4-fold increases in both the specific activity and in the capacity of the enzymes were found.

Gallagher et al. (25) have examined the tRNA methyltransferases after transformation of a rat embryo cell line by the oncogenic polyoma virus. They found a 3- to 6-fold increase in the rate of methylation and a 7-fold increase in the capacity or extent of methylation in extracts from the transformed cell line, using *E. coli* tRNA as a substrate. The differences were even more dramatic when yeast tRNA was used as a substrate.

Thus, the changes in the tRNA methyltransferases seen after viral transformation would appear to be much more extensive than those seen after viral infection.

5. *In Tumor Tissues*

It has been found that in crude extracts of over 35 different neoplastic tissues the tRNA methyltransferases exhibit an abnormally high capacity, from a 2- to 10-fold increase, for methylation as compared to normal tissue counterparts (61). This in part results from a lack of the inhibitors of the methyltransferases found in normal adult tissue (24) as well as

57. G. F. Vande Woude, J. Polatnick, and R. Ascione, *J. Virol.* **5**, 458 (1970).
58. C. Grado, B. Friedlender, M. Ihl, and G. Contreras, *Virology* **35**, 339 (1968).
59. M. Klagsbrun, *Virology* **44**, 153 (1971).
60. S. Kit, K. Nakajima, and D. R. Dubbs, *Cancer Res.* **30**, 528 (1970).
61. E. Borek and S. J. Kerr, *Advan. Cancer Res.* **15**, 163 (1972).

from the appearance of enzymes with altered specificities (*62*). For an extensive review of this subject see Borek and Kerr (*61*).

D. BIOLOGICAL SIGNIFICANCE

Proper modification of the tRNA molecule has been implicated as a requirement for a number of the functions of tRNA. These include repression of enzyme synthesis (*63*), reaction with the amino-acyl synthetases (*64*), codon response (*65*), prevention of wobble (*66*), as well as ribosomal binding (*67–69*).

The tRNA methyltransferases have been observed to undergo profound alterations, both qualitative and quantitative, in a number of biological systems undergoing changes in regulatory mechanisms (Table III) (*70–75*). The changes in the tRNA modifying enzymes imply a change in the population of the tRNA molecules themselves; in fact, this has been confirmed in some of the systems listed in Table III (*40, 61, 76–78*).

The variegated, species-specific structures of transfer RNA and the species specificity of the enzymes which achieve those modifications combined with the alterations in the enzymes observed in many differentiating systems have prompted the suggestion that these modulations of structure, achieved at great cost in energy, serve as some recognition sites for regulatory mechanisms (*79*). Control of protein synthesis at the

62. A. Mittelman, R. H. Hall, D. S. Yohn, and J. T. Grace, Jr., *Cancer Res.* **27**, 1409 (1967).
63. M. Brenner and B. Ames, *JBC* **247**, 1080 (1972).
64. L. Shugart, G. D. Novelli, and M. P. Stulberg, *BBA* **157**, 83 (1968).
65. J. D. Capra and A. Peterkofsky, *JMB* **33**, 591 (1968).
66. M. Yoshida, K. Takeishi, and T. Ukita, *BBRC* **39**, 852 (1970).
67. F. Fittler and R. H. Hall, *BBRC* **25**, 441 (1966).
68. R. Thiebe and H. G. Zachau, *Eur. J. Biochem.* **5**, 546 (1968).
69. M. Gefter and R. L. Russell, *JMB* **39**, 145 (1969).
70. B. S. Baliga, P. R. Srinivasan, and E. Borek, *Nature (London)* **208**, 555 (1965).
71. D. Pillinger and E. Borek, *Proc. Nat. Acad. Sci. U. S.* **62**, 1145 (1969).
72. S. J. Kerr and Z. Dische, *Invest. Ophthalmol.* **9**, 286 (1970).
73. O. K. Sharma, L. Loeb, and E. Borek, *BBA* **240**, 558 (1971).
74. R. S. L. Wong, G. A. Scarborough, and E. Borek, *J. Bacteriol.* **108**, 446 (1971).
75. D. H. Riddick and R. C. Gallo, *Blood* **37**, 282 (1971).
76. J. A. Boezi, R. L. Armstrong, and M. DeBacker, *BBRC* **29**, 281 (1967).
77. S. J. Kerr, *Fed. Proc., Fed. Amer. Soc. Exp. Biol.* **29**, 893 (1970) (abstr.).
78. P. H. Mäenpää and M. R. Bernfield, *Biochemistry* **8**, 4926 (1969).
79. E. Borek, *Cold Spring Harbor Symp. Quant. Biol.* **28**, 139 (1963).

TABLE III

MODULATIONS OF tRNA METHYLTRANSFERASES IN BIOLOGICAL SYSTEMS

System	Ref.
Bacteriophage infection	53
Bacteriophage induction	53, 54
Insect metamorphosis	70
Embryonic vs. neonatal tissue	15, 24, 41, 42
Colonizing slime mold	71
Differentiating lens tissue	72
Mammary epithelial cell differentiation	22
Ovariectomized uterus	40, 51
Thyroxine-induced morphogenesis in the tadpole	48
Sea urchin embryogenesis	73
Germination of spores	74
Hormone-induced phosvitin synthesis	50
Phytohemagglutinin induction of lymphocytes	75
Viral transformation	25, 60

translation level has been suggested frequently (79, 80). If tRNA's do have a regulatory role for specific protein synthesis, the mechanism cannot depend on variation among the anticodons alone; the possible permutations are too few. Modulations of the structure of tRNA's by the addition of a variety of modifying moieties offers a much more abundant system of variability.

Evidence for participation of methylation in translational control comes from two *in vivo* systems (81, 82). As mentioned earlier, bacteriophage T3 infection of *E. coli* induces the production of an enzyme which destroys S-adenosylmethionine, thus essentially eliminating methylation (56). Bacteriophage T3 can be inactivated by ultraviolet light, but the S-adenosylmethionine hydrolase is still inducible upon infection (83). This provides a useful tool for studying the effect of methylation on processes within the infected cell.

Siersma and Lederberg (81) have shown that infection with inacti-

80. B. N. Ames and P. E. Hartman, *Cold Spring Harbor Symp. Quant. Biol.* 28, 349 (1963).
81. P. W. Siersma and S. Lederberg, *J. Bacteriol.* 101, 398 (1970).
82. W. Sauerbier, M. Schweiger, and P. Herrlich, *J. Virol.* 8, 613 (1971).
83. M. Gold, M. Gefter, R. Hausmann, and J. Hurwitz, *J. Gen. Physiol.* 49, Part 2, 5 (1966).

vated T3 of an alkaline phosphatase *amber* mutant of *E. coli* caused an increase in phosphatase production, indicating an *in vivo* suppression of the *amber* mutation. This correlates with the observed effect of hypomethylation on codon response of tRNA species in *in vitro* assays (*65*). Early communications reported that co-infection of bacteria with inactivated T3 and either T2 or T4 led to normal production of bacteriophage (*83*). This was interpreted as indicating that methylation of tRNA or DNA was not necessary for the successful production of these bacteriophages, even though changes in the tRNA methyltransferases had been observed in T2 and T4 infection (*53, 76*). However, mere observation of the gross, overall production of bacteriophage was not a subtle enough measure of the effect of methylation. Sauerbier et al. (*82*) have shown that co-infection with T4 and inactivated T3 does have an effect upon the course of bacteriophage production.

Synthesis of early T4 proteins starts shortly after infection and stops at about the middle of the latent period (*84*), even though the messenger RNA homologous to the early genes is present throughout the whole cycle (*85*). Since the messenger RNA is present and not translated, this might be an indication that some form of translational control is operating. Sauerbier *et al.* (*82*) demonstrated that, upon elimination of the methylating capacity in the T4 infected cells by co-infection with inactivated T3, synthesis of early T4 proteins was not shut off at 10 min but continued until lysis. Thus, methylation *is* involved in the normal control of bacteriophage T4 production, and presumably it exerts its effect at the translation level.

Another role for tRNA, unrelated to protein synthesis, has emerged, which is direct enzyme inhibition by specific tRNA species. Feedback inhibition of enzymes in the pathways of amino acid syntheses by the cognate tRNA's has been observed in microorganisms (*86, 87*). An even more provocative situation has been uncovered by Jacobson (*88*), who observed that in a mutant of *Drosophila melanogaster* a tyrosyl tRNA is capable of inhibiting the enzyme tryptophan pyrrolase, which is, of course, totally unrelated to the synthesis of tyrosine. The potential variation of the structure of tRNA's for this kind of role is also enormous.

84. R. Somerville, K. Ebisuzaki, and G. R. Greenberg, *Proc. Nat. Acad. Sci. U. S.* **45**, 1240 (1959).
85. J. D. Friesen, B. Dale, and W. Bode, *JMB* **28**, 413 (1967).
86. E. Duda, M. Staub, P. Venetianer, and G. Denes, *BBRC* **32**, 992 (1968).
87. G. W. Hatfield and R. O. Burns, *Proc. Nat. Acad. Sci. U. S.* **66**, 1027 (1970).
88. K. B. Jacobson, *Nature (London), New Biol.* **231**, 17 (1971).

III. Ribosomal RNA Methyltransferases

A. Occurrence

Enzymes which catalyze the methylation of ribosomal RNA *in vitro* have been observed in extracts of both bacteria and mammalian cells (*89–92*).

In bacteria the enzymes appear to be associated to some extent with the ribosomal particles (*89, 90*). In mammalian cells, they are found in the nucleolus (*91, 92*) where presumably they act on ribosomal RNA concomitantly with or very soon after the synthesis of high molecular weight ribosomal precursor RNA (*93, 94*). Both the ribose moieties as well as the bases are methylated, but we will limit our discussion to base-specific methylation.

As with the tRNA methyltransferases, the rRNA methyltransferases are species specific and act at the macromolecular level, with an interesting exception in bacteria (*90*).

B. Isolation and Properties

The rRNA methyltransferases have not been as well characterized as the tRNA methyltransferases. The most extensive studies have been carried out in bacterial systems.

Hurwitz *et al.* (*89*) isolated four fractions from *E. coli* W which methylated rRNA, two specific for adenine and two specific for cytosine. A fifth activity specific for guanine was observed in *E. coli* K12, strain 735. The enzymic activities in extracts of *E. coli* W were distributed between the high-speed supernatant solution and the ribosomal pellet and probably actually represent two enzymic activities artificially distributed in such a manner by the procedure used to prepare cell-free extracts. The guanine-specific activity from *E. coli* strain 735 was isolated solely from the ribosomal pellet.

89. J. Hurwitz, M. Anders, M. Gold, and I. Smith, *JBC* **240**, 1256 (1965).
90. J. E. Sipe, W. M. Anderson, Jr., C. N. Remy, and S. H. Love, *J. Bacteriol.* **110**, 81 (1972).
91. L. A. Culp and G. M. Brown, *ABB* **137**, 222 (1970).
92. M. C. Liau, N. C. Flatt, and R. B. Hurlbert, *BBA* **224**, 282 (1970).
93. H. Greenberg and S. Penman, *JMB* **21**, 527 (1966).
94. E. F. Zimmerman and B. W. Holler, *JMB* **23**, 149 (1967).

The enzyme fractions catalyzed the methylation of both tRNA and rRNA, but this probably resulted from contamination with tRNA methyltransferases since a highly purified preparation reported recently did not act on tRNA (90). The methyltransferase fractions would not act on whole ribosomal particles, synthetic polynucleotides, or viral RNA's. They are species and strain specific. The enzymes from *E. coli* W methylated rRNA from a variety of different species such as *Micrococcus lysodeikticus*, but they were inactive with rRNA isolated from the same strain of *E. coli* W or from *E. coli* B. Conversely, the guanine-specific methyltransferase activity from *E. coli* K12, strain 735, could methylate the rRNA from either of those two strains.

The adenine-specific methyltransferase activity catalyzed the formation of both N^6-methyladenine and N^6,N^6-dimethyladenine in rRNA. It did not appear to require ionic stimulation. The cytosine-specific activity yielded 5-methylcytosine as its product in rRNA. It was stimulated by both Mg^{2+} and NH_4^+. The guanine-specific enzyme fraction from strain 735 formed 7-methylguanine and was only slightly stimulated by Mg^{2+}.

An adenine-specific rRNA methyltransferase [S-adenosylmethionine: ribosomal ribonucleic acid-adenine (N^6) methyltransferase] from *E. coli* B has now been extensively purified (1500-fold) by Sipe et al. (90). As indicated by its name, this enzyme is responsible for the formation of N^6-methyladenine moieties in rRNA. It is species-specific and methylates a limited number of adenine residues in heterologous rRNA of *M. lysodeikticus* and *Bacillus subtilis* and methyl-deficient homologous rRNA. The site recognition mechanism does not require intact 16 S or 23 S rRNA. The enzyme will not use tRNA as a methyl acceptor and does not synthesize any other methylated adenine derivatives in rRNA, such as N^6,N^6-dimethyladenine. Monovalent or divalent cations increase the reaction rate but not the extent of methylation, and high levels of cations inhibit the enzyme.

The purified enzyme also uses 9-β-ribosyl-2,6-diaminopurine as a methyl acceptor yielding the 2-methyl derivative. The relative specific activities for the methylation of rRNA and of the purine remained constant throughout the purification procedure, indicating that the two activities reside in one enzyme. This is a very unusual lack of discrimination for such an enzyme.

The methyltransferase activity was localized in the ribosomal fraction of crude cell extracts prepared in buffers of low ionic strength, but this may in part represent an artifact of cell disruption rather than a specific functional relationship. The enzyme was inhibited by adenine, polyvinylsulfate, and an unidentified macromolecular inhibitor from

M. lysodeikticus. Evidence was found for the existence of a second, less cationic rRNA-adenine (N^6) methyltransferase, which did not react with synthetic purines.

The isolation of rRNA base-specific methyltransferases from mammalian cells has not been accomplished to date. *In vitro* systems for the methylation of the ribose in rRNA have been described using nucleolar preparations from rat liver (*91*) and Novikoff ascites tumor (*92*), and there is one report of a cytoplasmic fraction from HeLa cells which was capable of methylating methyl-deficient rRNA from *E. coli* (*95*). However, no mammalian rRNA methyltransferase specific for base methylation has yet been characterized although there is ample *in vivo* evidence that these enzymes must exist. This may result from difficulties in solubilizing the enzymes as well as finding the proper RNA substrate.

C. Biological Significance

As mentioned above, in animal cells ribosomal RNA is synthesized in the nucleolus in the form of a larger precursor molecule which is methylated as it is being synthesized (*93, 94*). During further processing into the smaller ribosomal subunits, all the methyl groups incorporated into the 45 S molecule are conserved even though about half the 45 S molecule is lost during maturation (*96*). Thus, the methyl groups may be markers in the maturation process. There is evidence for a secondary methylation of adenine in the 18 S ribosomal subunit (*97, 98*).

Elucidation of another aspect of methylation of rRNA has come from drug resistance studies in microorganisms. Certain antibiotics appear to interact with the RNA of ribosomal subunits, and two cases of drug resistance associated with a change in methylation of the rRNA have been noted (*99, 100*). Lai and Weisblum (*99*) have found that the 23 S rRNA obtained from induced- and constitutively erythromycin resistant cells contained an N^6,N^6-dimethyladenine moiety which was absent from the 23 S rRNA of erythromycin-sensitive cells of *Staphylococcus aureus*. Helser *et al.* (*100*) showed that in *E. coli* a mutation to resistance to the drug kasugamycin was associated with a change in the methylation

95. L. A. Culp and G. M. Brown, *ABB* **124**, 483 (1968).
96. E. Wagner, S. Penman, and V. Ingram, *JMB* **29**, 371 (1967).
97. E. F. Zimmerman, *Biochemistry* **7**, 3156 (1968).
98. E. H. Maden, M. Salim, and D. F. Summers, *Nature (London), New Biol.* **237**, 5 (1972).
99. C. J. Lai and B. Weisblum, *Proc. Nat. Acad. Sci. U. S.* **68**, 856 (1971).
100. T. L. Helser, J. E. Davies, and J. E. Dahlberg, *Nature (London), New Biol.* **233**, 12 (1971).

of a specific nucleotide sequence in the 16 S rRNA subunit. In the resistant mutant a normally occurring N^6,N^6-dimethyladenine was absent, with an unmodified adenine in its place.

IV. DNA Methyltransferases

A. Occurrence

Methylated bases occur in the DNA of every organism examined with the exception of some bacteriophage (83), animal viruses (101–103), and a few strains of bacteria (104, 105). The main methylated derivatives are N^6-methyladenine and 5-methylcytosine. N^6-Methyladenine is found only in the DNA of bacteriophage and certain bacterial species (106), sometimes as the only derivative and sometimes in combination with 5-methylcytosine. The DNA of higher plants and animals contains solely 5-methylcytosine (107), although there has been a report that tumor DNA contains a very small percentage of other methylated derivatives (108).

The recent observation that M. radiodurans, an extraordinarily radiation-resistant microorganism, lacks methylated bases from its DNA provides a useful tool for the standardization of methodology for the in vitro studies of DNA methyltransferases. DNA from this organism may serve as a standard "methyl-deficient" substrate in various laboratories (105).

B. Properties

1. Bacteria

The DNA methyltransferases, like the other polynucleotide modifying enzymes, are species- and strain-specific (83). They act on the preformed macromolecule and use S-adenosylmethionine as the methyl

101. A. M. Kaye and E. Winocur, JMB 24, 475 (1967).
102. M. Low, J. Hay, and H. M. Keir, JMB 46, 205 (1969).
103. M. Low, M. J. Mechie, and J. Hay, BJ 124, 63P (1971).
104. J. T. Wachsman and V. Irwin, J. Bacteriol. 104, 814 (1970).
105. A. Schein, B. J. Berdahl, M. Low, and E. Borek, BBA 272, 481 (1972).
106. D. B. Dunn and J. D. Smith, BJ 68, 627 (1958).
107. G. R. Wyatt, BJ 48, 581 (1951).
108. L. A. Culp, E. Dore, and G. M. Brown, ABB 136, 73 (1970).

donor. Kornberg suggested that the methyl groups of DNA may be acquired, as are the glucose moieties, via some enzymic reaction on the polymer (*109*).

Gold and Hurwitz (*110*) demonstrated enzymic activity in extracts of *E. coli* W capable of transferring methyl groups to DNA from a variety of bacterial and mammalian sources. The enzyme extract catalyzed the formation of both N^6-methyladenine and 5-methylcytosine. A partial purification of the extracts did not separate the activities, and it was not clear whether one or two enzymes were involved. However, biological evidence has been adduced which indicates that two distinct enzymes exist (*111*).

The DNA methyltransferase activity from *E. coli* W required no metal ions and had a pH optimum of 8. It required double-stranded DNA as a template, but the reaction was not substantially altered by treatment of the DNA by either sonic oscillation or exposure of the DNA to small quantities of nuclease. Heat denaturation of the DNA abolished its methyl acceptance. The methyltransferase reaction was inhibited by *S*-adenosylhomocysteine and the DNA binding agents, actinomycin D and proflavin. Actinomycin D was more effective in inhibiting the methylation of adenine residues and proflavin the methylation of cytosine residues.

Oda and Marmur (*112*) have isolated a 5-methylcytosine-specific DNA methyltransferase from *Bacillus subtilis* and the enzyme differs significantly from the *E. coli* enzymes. The methyltransferase was produced only in the early exponential phase of growth and disappeared rapidly upon further cellular growth. It catalyzed the methylation of both native and heat-denatured DNA. The extent of methylation of both types of DNA was approximately proportional to the guanine plus cytosine (GC) content. The higher the percent of GC in a particular DNA, the greater was the extent of methylation. DNA from the *B. subtilis* bacteriophage 2C was denatured and the two strands separated; both strands were equally methylated by the enzyme *in vitro*, and the results were identical with *in vivo* methylation of the phage DNA.

As early as 1962, Arber and Dussoix (*113*) postulated that methylation of bacteriophage λ DNA was involved in the phenomenon of host-induced modification and restriction of bacteriophage infectivity. Several

109. A. Kornberg, S. B. Zimmerman, S. R. Kornberg, and J. Josse, *Proc. Nat. Acad. Sci. U. S.* **45**, 772 (1959).
110. M. Gold and J. Hurwitz. *JBC* **239**, 3858 (1964).
111. D. Fujimoto, P. R. Srinivasan, and E. Borek, *Biochemistry* **4**, 2849 (1965).
112. K. Oda and J. Marmur, *Biochemistry* **5**, 761 (1966).
113. W. Arber and D. Dussoix, *JMB* **5**, 18 (1962).

of the early workers in the field of DNA methyltransferases tried to test this hypothesis and came to the conclusion that methylation had no role in this mechanism (*83, 112, 114*). However, the methods of analysis, as it turned out, were simply too crude.

With great perseverance and ingenuity Arber and his co-workers have isolated and characterized the B-specific modification enzyme from *E. coli* B and demonstrated that it is indeed a DNA methyltransferase with an extremely restricted specificity (*115–117*). This enzyme is specific to *E. coli* B and catalyzes the formation of N^6-methyladenine. It uses S-adenosylmethionine and requires no ionic stimulation. Its extreme specificity is demonstrated by its reaction with bacteriophage *fd* DNA. This particular phage DNA is a relatively small molecule (6600 nucleotides) and has a low overall methyl content, making detection of slight differences in methylation feasible. The reaction of the B-specific DNA N^6-adenine methyltransferase followed first-order kinetics and reached saturation when four adenine moieties per replicative form were methylated. DNA from phage *fd* grown in *E. coli* B, and thus already exposed to the modifying enzyme *in vivo*, would not accept any further methyl groups *in vitro*. The number of N^6-methyladenines per fully modified double-stranded DNA molecule is two. Thus, two methylated bases per 6600 nucleotides are sufficient to protect this DNA from the restricting endonuclease of *E. coli* B. It is not too surprising that such a subtle change went undetected with the crude earlier experiments.

Arber's work illuminates brilliantly the subtleties of macromolecular modifications and refutes the hastily drawn early conclusions on the lack of biological role for modifications of nucleic acids by methylation.

2. *Eucaryotes*

DNA methyltransferases from mammalian sources are refractory to isolation because of their status as membrane-bound components of the nucleus. Sheid *et al.* (*118*) demonstrated DNA methyltransferase in a particulate fraction derived from rat liver nuclei. The enzyme yielded 5-methylcytosine as the only derivative and was species-specific. Different organs of the same animal appeared to have different levels of activity. The pH optimum was between 8.0 and 8.4 and no metal requirement was observed, nor was the activity inhibited by the pres-

114. W. B. Wood, *Pathol. Microbiol.* **28**, 73 (1965).
115. U. Kuhnlein, S. Linn, and W. Arber, *Proc. Nat. Acad. Sci. U. S.* **63**, 556 (1969).
116. J. D. Smith, W. Arber, and U. Kuhnlein, *JMB* **63**, 1 (1972).
117. U. Kuhnlein and W. Arber, *JMB* **63**, 9 (1972).
118. B. Sheid, P. R. Srinivasan, and E. Borek, *Biochemistry* **7**, 280 (1968).

ence of EDTA. Heat-denatured DNA did not serve as a substrate. The enzyme activity was unstable at 4° but was protected by storage at −20°.

Burdon et al. (119) have described a DNA methyltransferase activity associated with the chromatin fraction of Krebs 2 mouse ascites tumor cells.

One laboratory has reported the solubilization of DNA methyltransferase from mammalian tissue (120, 121). The variability of the technique was evident in the first report which claimed that the enzyme existed only in rat and mouse spleen and could not be detected in any other organ or any other animal (120). In a subsequent communication the same group did detect the enzyme in rat liver (121). Further studies on the liver enzyme showed that it is bound tightly to either single-stranded or double-stranded *E. coli* DNA and could use both as methyl acceptors, with a preference for the single-stranded form (122). The authors interpret some kinetic data with anthropomorphic terminology implying that the enzyme "walks" along the DNA helix. Presumably it is not detached from the DNA molecule between each separate methylation. What relation this has to the *in vivo* action of the enzyme is unclear for two reasons: The enzyme did not act on denatured mammalian DNA (120) and, since *in vivo* it is particulate-bound, it is not obvious how it would accomplish a "walk" along a DNA molecule. DNA methyltransferase activity in developing embryos of loach, *Misgurnus fossilis*, has been reported to be in the soluble cytoplasmic fraction (supernatant, 105,000 × *g*, 60 min) (123).

C. REGULATION

In bacteria (124, 125) and in mammals (126, 127) methylation of DNA appears to be closely coupled with its synthesis. How this control is brought about is completely unclear at present. Methylation of DNA has also been shown to be altered in a number of cases of bacteriophage

119. R. H. Burdon, B. T. Martin, and B. M. Lal, *JMB* **28**, 357 (1967).
120. F. Kalousek and N. R. Morris, *JBC* **244**, 1157 (1969).
121. N. R. Morris and K. D. Pih, *Cancer Res.* **31**, 433 (1971).
122. D. Drahovský and N. R. Morris, *JMB* **57**, 475 (1971).
123. B. F. Vanyushin, G. I. Kiryanov, I. B. Kudryashova, and A. N. Belozersky, *FEBS Lett.* **15**, 313 (1971).
124. C. Lark, *JMB* **31**, 389 (1968).
125. D. Billen, *JMB* **31**, 477 (1968).
126. R. W. Turkington and R. L. Spielvogel, *JBC* **246**, 3835 (1971).
127. J. W. Kappler, *J. Cell. Physiol.* **75**, 21 (1970).

infection (*83*) or induction (*128*). Bacteriophages T2, T4, and T1 cause increases in DNA methyltransferase upon infection; infection with T7 or λ produced very little change, T5 and T6 gave a slight decrease in activity, and T3 provoked a rapid loss of methyltransferase activity (*83*). The reason for this last phenomenon was mentioned earlier: T3 causes the production of an enzyme which destroys S-adenosylmethionine (*56*). In contrast to productive infection by λ, where little change in methyltransferase activity was seen, induction of λ leads to a marked rise in DNA methyltransferase activity (*128*).

What modulates these variations of the methyltransferase is not resolved. In the case of T2 infection a new bacteriophage-coded DNA methyltransferase appears to be responsible for the rise in activity (*129*). Inhibitors are possible agents for the control of the enzymes. Falaschi and Kornberg (*130*) have characterized a potent inhibitor of DNA methyltransferases as a lipopolysaccharide normally associated with the bacterial cell wall.

Evidence has been presented that a drug resistance transfer factor, N-3, controls a specific restriction–modification system which involves a 5-methylcytosine-specific DNA methyltransferase (*131*). This would be the first instance where host specificity is imparted by a methylated cytosine residue. All the others studied have been dependent on adenine methylation. Whether the DNA methyltransferase activity determined by the transfer factor is also involved in the drug resistance conferred by this factor is unknown.

Ultraviolet irradiation of *E. coli* causes aberrant methylation of the DNA (*132, 133*) indicating that this process is normally under some form of control in the cell. The exact mechanisms of these controls await elucidation.

D. BIOLOGICAL SIGNIFICANCE

One function that can definitely be assigned to methylation of DNA is the achievement of species individuality. In bacteria it is clear that

128. G. Medoff, *Virology* **44**, 642 (1971).
129. M. Gold, R. Hausmann, U. Maitra, and J. Hurwitz, *Proc. Nat. Acad. Sci. U. S.* **52**, 292 (1964).
130. A. Falaschi and A. Kornberg, *Proc. Nat. Acad. Sci. U. S.* **54**, 1713 (1965).
131. S. Hattman, E. Gold, and A. Plotnick, *Proc. Nat. Acad. Sci. U. S.* **68**, 187 (1972).
132. A. M. Ryan and E. Borek, *BBA* **240**, 203 (1971).
133. B. L. Whitfield and D. Billen, *JMB* **63**, 363 (1972).

this forms the basis of the restriction–modification systems. Whether similar phenomena exist in higher organisms is unknown.

Scarano has postulated that in developing sea urchin embryos a small fraction of the DNA thymine is synthesized at the polymer level, possibly by the deamination of 5-methylcytosine. He postulates that the synthesis of this minor thymine is related to the stages of embryonic development and that a mechanism of base changes in DNA in this manner may underlie the stepwise transitions by which cell differentiation and embryogenesis are realized (*134*). Sneider and Potter (*135*) have also reported deamination of 5-methylcytosine in the DNA of Novikoff hepatoma cells.

Other possible functions of DNA methylation will undoubtedly emerge as experimental techniques become more refined.

134. E. Scarano, M. Iaccarino, P. Grippo, and E. Parisi, *Proc. Nat. Acad. Sci. U. S.* **57**, 1394 (1967).
135. T. W. Sneider and V. R. Potter, *JMB* **42**, 271 (1969).

6

Folate Coenzyme-Mediated Transfer of One-Carbon Groups

JEANNE I. RADER • F. M. HUENNEKENS

I. Introduction	197
II. Transfer of Formate and Its Congeners	198
A. 10-Formyltetrahydrofolate Synthetase	198
B. 10-Formyltetrahydrofolate Deacylase	200
C. 5,10-Methenyltetrahydrofolate Cyclohydrolase	201
D. 5-Formyltetrahydrofolate Cyclodehydrase	201
E. 5-Formiminotetrahydrofolate Cyclodeaminase	202
F. Glycinamide Ribonucleotide Transformylase	203
G. 5-Amino-4-Imidazole Carboxamide Ribonucleotide Transformylase	204
H. Formiminoglycine Formiminotransferase	206
I. Formiminoglutamate Formiminotransferase	206
J. N-Formylglutamate Transformylase	207
K. Methionyl-tRNA Transformylase	208
III. Transfer of Formaldehyde and Its Congeners	209
A. Deoxycytidylate Hydroxymethyltransferase	209
B. Deoxyuridylate Hydroxymethyltransferase	210
C. Thymidylate Synthetase	210
D. Serine Hydroxymethyltransferase	215
E. Oxidative Decarboxylation of Glycine	221

I. Introduction

Enzymic reactions involving the transfer of one-carbon groups (1) at the oxidation states of formate, formaldehyde, and methanol occur as

1. Abbreviations: C_1, one-carbon group; FH_4, tetrahydrofolate; CH_2–FH_4, methylenetetrahydrofolate; FIG, formiminoglycine; FIGLU, formiminoglutamate; and DTNB, 5,5'-dithiobis(2-nitrobenzoic acid).

steps in various metabolic sequences for the synthesis or degradation of purines, pyrimidines, and amino acids. These transfer reactions are mediated by the coenzyme form of folic acid, tetrahydrofolate, whose structural features (I) allow it to form adducts with the one-carbon groups by means

(I)

of covalent linkages at N-5 and N-10. This property provides the basis for the ability of the coenzyme to accept and donate one-carbon groups from metabolites, i.e.,

$$C_1\text{-metabolite} + FH_4 \rightleftharpoons C_1\text{-}FH_4 + \text{metabolite} \tag{1}$$

These concepts have been discussed extensively in previous reviews (*2–8*).

Interrelationships of the individual reactions involving tetrahydrofolate-mediated transfer of one-carbon groups are shown in Fig. 1. The present review covers all of the enzymes responsible for these reactions with the exception of: (a) the B_{12}-dependent, 5-methyltetrahydrofolate-homocysteine transmethylase (enzyme 19 in Fig. 1) which is treated elsewhere in this volume (*9*); and (b) methylenetetrahydrofolate dehydrogenase (enzyme 17), and methylenetetrahydrofolate reductase (enzyme 18), both of which catalyze reactions involving oxidoreduction, rather than transfer, of the one-carbon group.

II. Transfer of Formate and Its Congeners

A. 10-Formyltetrahydrofolate Synthetase

Formyltetrahydrofolate synthetase [formate:tetrahydrofolate ligase (ADP)] (EC 6.3.4.3), also called "formate-activating enzyme," catalyzes

2. F. M. Huennekens and M. J. Osborn, *Advan. Enzymol.* **21**, 369 (1959).
3. J. C. Rabinowitz, "The Enzymes," 2nd ed., Vol. 2, Part A, p. 185, 1960.
4. L. Jaenicke, *Ciba Found. Study Group* [*Pap.*] **11**, 38 (1962).
5. M. Friedkin, *Annu. Rev. Biochem.* **32**, 185 (1963).
6. F. M. Huennekens and K. G. Scrimgeour, in "Pteridine Chemistry" (W. Pfleiderer and E. C. Taylor, eds.), p. 355. Pergamon, Oxford, 1964.
7. F. M. Huennekens, in "Biological Oxidations" (T. P. Singer, ed.), p. 439. Wiley, New York, 1968.

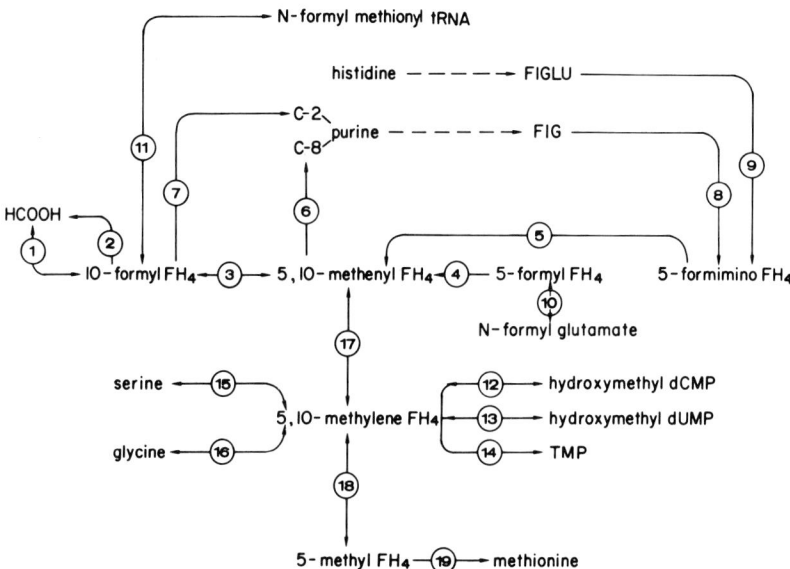

FIG. 1. Tetrahydrofolate-mediated transfer reactions involving one-carbon groups. Arrows indicate reversibility or irreversibility of reactions. Numbers refer to the following enzymes: 1, 10-formyltetrahydrofolate synthetase; 2, 10-formyltetrahydrofolate deacylase; 3, 5,10-methenyltetrahydrofolate cyclohydrolase; 4, 5-formyltetrahydrofolate cyclodehydrase; 5, 5-formiminotetrahydrofolate cyclodeaminase; 6, glycinamide ribonucleotide transformylase; 7, 5-amino-4-imidazole carboxamide ribonucleotide transformylase; 8, formiminoglycine formiminotransferase; 9, formiminoglutamate formiminotransferase; 10, N-formylglutamate transformylase; 11, methionyl-tRNA transformylase; 12, deoxycytidylate hydroxymethyltransferase; 13, deoxyuridylate hydroxymethyltransferase; 14, thymidylate synthetase; 15, serine hydroxymethyltransferase; 16, enzyme system responsible for oxidative decarboxylation of glycine; 17, 5,10-methylenetetrahydrofolate dehydrogenase; 18, 5,10-methylenetetrahydrofolate reductase; and 19, 5-methyltetrahydrofolate-homocysteine transmethylase.

the ATP-dependent formylation of tetrahydrofolate (at the N-10 position) according to Eq. (2):

$$\text{HCOOH} + \text{tetrahydrofolate} + \text{ATP} \xrightleftharpoons{\text{Mg}^{2+},\ \text{K}^+} \text{10-formyltetrahydrofolate} + \text{ADP} + \text{P}_i \quad (2)$$

This reaction was first observed in pigeon liver extracts (10–13) and the

---

8. R. L. Blakley, "The Biochemistry of Folic Acid and Related Pteridines," Chapters 6, 7, 8, and 9. North-Holland Publ., Amsterdam, 1969.
9. R. T. Taylor and H. Weissbach, Chapter 4 this volume.
10. G. R. Greenberg, Fed. Proc., Fed. Amer. Soc. Exp. Biol. **13**, 221 (1954).
11. G. R. Greenberg, Fed. Proc., Fed. Amer. Soc. Exp. Biol. **13**, 745 (1954).
12. G. R. Greenberg, L. Jaenicke, and M. Silverman, BBA **17**, 589 (1955).
13. L. Jaenicke, BBA **17**, 588 (1955).

enzyme is now known to be widely distributed in nature. Properties of formyltetrahydrofolate synthetases of plant, bacterial, and mammalian origin have been reviewed elsewhere (*3, 7, 8, 14*). The enzyme is the subject of a separate chapter scheduled for a later volume of this series, and thus will not be discussed further here.

B. 10-Formyltetrahydrofolate Deacylase

Osborn *et al.* (*15*) reported the partial purification from acetone powders of beef liver of an enzyme catalyzing the conversion of 10-formyltetrahydrofolate to tetrahydrofolate. The latter was identified by its absorption spectrum and by its reactivity in the formyltetrahydrofolate synthetase system (enzyme 1 in Fig. 1) and in the coupled serine hydroxymethyltransferase-methylenetetrahydrofolate dehydrogenase system (enzymes 15 and 17 in Fig. 1). Some indication that this process might involve more than a simple deformylation of the substrate [Eq. (3)] was provided by the observation that the enzyme preparation was

$$\text{10-Formyltetrahydrofolate} + H_2O \rightarrow \text{tetrahydrofolate} + HCOOH \qquad (3)$$

stimulated by catalytic amounts of TPNH or TPN. An explanation for this pyridine nucleotide requirement was provided subsequently by Kutzbach and Stokstad (*16*), who found that a similar preparation from pig liver catalyzed the oxidative deformylation of 10-formyltetrahydrofolate:

$$\text{10-Formyltetrahydrofolate} + TPN^+ + H_2O \rightarrow$$
$$\text{tetrahydrofolate} + CO_2 + TPNH + H^+ \qquad (4)$$

In this latter system, CO_2 production was demonstrated manometrically and by the formation of $^{14}CO_2$ from [^{14}C]formyltetrahydrofolate. Free formate could not be detected when 10-formyltetrahydrofolate and TPN were incubated with the purified enzyme. Aged preparations of the enzyme, however, were able to carry out the deformylation reaction [Eq. (3)] in the absence of TPN. This suggests that deacylation of 10-formyltetrahydrofolate [Eq. (4)] proceeds via a formyl intermediate which is oxidized before being released as CO_2. These results also help to explain why CO_2 production from formate is depressed when animals are maintained on a folate-deficient diet (*17, 18*).

14. H. R. Whiteley, *Comp. Biochem. Physiol.* **1**, 227 (1960).
15. M. J. Osborn, Y. Hatefi, L. D. Kay, and F. M. Huennekens, *BBA* **26**, 208 (1957).
16. C. Kutzbach and E. L. R. Stokstad, *BBRC* **30**, 111 (1968).
17. G. W. E. Plaut, J. J. Betheil, and H. A. Lardy, *JBC* **184**, 795 (1950).
18. B. Friedmann, H. I. Nakada, and S. Weinhouse, *JBC* **210**, 413 (1954).

C. 5,10-METHENYLTETRAHYDROFOLATE CYCLOHYDROLASE

The interconversion of 10-formyltetrahydrofolate and 5,10-methenyltetrahydrofolate [Eq. (5)] occurs readily in the absence of enzymes.

10-Formyltetrahydrofolate + H$^+$ ⇌ [5,10-methenyltetrahydrofolate]$^+$ + H$_2$O (5)

At pH 7, the equilibrium constant for this reaction is near unity. Acidification drives the reaction to the right, and this procedure, which results in formation of an oxygen-stable product with an absorbance maximum in the 350-nm region, has been used for quantitation of the more labile 10-formyltetrahydrofolate. The rate of reaction (5) depends upon the nature and concentration of anions present. In phosphate, pyrophosphate, or arsenate at pH 7.0, the half-life of methenyltetrahydrofolate is 5–6 min, while in maleate this is increased to 40 min (*3*).

Reaction (5) is catalyzed by 5,10-methenyltetrahydrofolate 5-hydrolase (decyclizing) (EC 3.5.4.9), also called "cyclohydrolase." This enzyme was first detected in extracts of *Clostridium cylindrosporum* (*19*), *Clostridium acidi-urici* (*20*), and pig kidney (*21*) and has been purified subsequently from rabbit liver (*22*) and beef liver (*23*).

D. 5-FORMYLTETRAHYDROFOLATE CYCLODEHYDRASE

Greenberg (*10, 24*) reported that ATP-supplemented pigeon liver extracts were able to utilize 5-formyltetrahydrofolate (folinic acid) in the conversion of the purine precursor, 5-amino-4-imidazole carboxamide-5′-phosphoribotide, to inosine-5′-phosphate. Subsequent studies (*25*) showed that the actual donor of the one-carbon group was 10-formyltetrahydrofolate and that the earlier results were referable to an ATP-dependent isomerization:

5-Formyltetrahydrofolate + ATP $\xrightarrow{\text{Mg}^{2+}}$ 10-formyltetrahydrofolate + ADP + P$_i$ (6)

The enzyme system responsible for reaction (6) has been partially purified from sheep liver (*26–28*), chicken liver (*29*), and *Micrococcus aerogenes* (*29*).

19. J. C. Rabinowitz and W. E. Pricer, Jr., *JACS* **78**, 5702 (1956).
20. J. C. Rabinowitz and W. E. Pricer, Jr., *JACS* **78**, 4176 (1956).
21. H. Tabor and J. C. Rabinowitz, *JACS* **78**, 5705 (1956).
22. H. Tabor and L. Wyngarden, *JBC* **234**, 1830 (1959).
23. L. Lombrozo and D. M. Greenberg, *ABB* **118**, 297 (1967).
24. G. R. Greenberg, *JACS* **76**, 1458 (1954).
25. S. C. Hartman and J. M. Buchanan, *JBC* **243**, 1812 (1959).
26. J. M. Peters and D. M. Greenberg, *JBC* **226**, 329 (1957).

Reaction (6) is further separable into two steps (28). Initially, the formyl group is cyclized from the 5 to the 5,10 position of tetrahydrofolate [reaction (7)], and this is followed by an opening of the bridge compound via reaction (5) to yield 10-formyltetrahydrofolate.

5-Formyltetrahydrofolate + ATP →
$$\text{5,10-methenyltetrahydrofolate} + ADP + P_i \quad (7)$$

Reaction (7) is catalyzed by 5-formyltetrahydrofolate cyclodehydrase, while reaction (5) proceeds nonenzymically or is catalyzed by cyclohydrolase (see Section II,C). 5-Formylcyclodehydrase partially purified from sheep liver had a pH optimum of 4.8; K_m values of $1.4 \times 10^{-4} M$ (5-formyltetrahydrofolate) and $4.5 \times 10^{-4} M$ (ATP) have been reported (27). The enzyme was inhibited by p-mercuribenzoate and N-ethylmaleimide (27).

E. 5-FORMIMINOTETRAHYDROFOLATE CYCLODEAMINASE

Rabinowitz and Pricer (19, 20) identified the 5-formimino derivative of tetrahydrofolate as a product of purine fermentation by extracts of clostridia (see Section II,H). The enzyme 5-formiminotetrahydrofolate ammonia-lyase (cyclizing) (EC 4.3.1.4), also called "cyclodeaminase," catalyzes the conversion of 5-formiminotetrahydrofolate to 5,10-methenyltetrahydrofolate:

5-Formiminotetrahydrofolate + 2 H$^+$ ⇌
$$[\text{5,10-methenyltetrahydrofolate}]^+ + NH_4^+ \quad (8)$$

A mechanism has been proposed for the enzymic and nonenzymic catalysis of this reaction (3). Cyclodeaminase activity was first identified in extracts of *C. cylindrosporum* (19, 20) and has been purified from this source (30) as well as from rabbit liver (21) and pig liver (22). Tetrahydrofolate was a competitive inhibitor of the cyclodeaminase from *C. cylindrosporum* (30) and pig liver (22).

Cyclodeaminase purified 600-fold from pig liver (22) was markedly stimulated by K$^+$ and NH$_4^+$; K_m values for 5-formiminotetrahydrofolate of $17.5 \times 10^{-5} M$ (without KCl) and $2.6 \times 10^{-5} M$ (with KCl) have been reported (22). Cyclodeaminase purified 40-fold from *C. cylindro-*

---

27. J. M. Peters and D. M. Greenberg, *JACS* **80**, 2719 (1958).
28. D. M. Greenberg, L. K. Wynston, and A. Nagabhushanan, *Biochemistry* **4**, 1872 (1965).
29. L. D. Kay, M. J. Osborn, Y. Hatefi, and F. M. Huennekens, *JBC* **235**, 195 (1960).
30. K. Uyeda and J. C. Rabinowitz, *JBC* **242**, 24 (1967).

sporum (*30*) had a pH optimum of 7.2 and a K_m value for 5-formiminotetrahydrofolate of $2.8 \times 10^{-5} M$. The latter preparation gave a single homogeneous peak during ultracentrifugation (MW $\simeq$ 38,000), but it also had 5,10-methenyltetrahydrofolate cyclohydrolase activity. Attempts to separate the cyclodeaminase and cyclohydrolase activities were unsuccessful (*30*). After starch gel electrophoresis in various buffers ranging from pH 5.0 to 8.4, both activities were found at the same R_f. This result, as well as the sedimentation behavior mentioned above, suggests that both enzymic activities are associated with a single protein (*30*).

F. GLYCINAMIDE RIBONUCLEOTIDE TRANSFORMYLASE

Two steps in the *de novo* synthesis of the purine ring involve addition of one-carbon groups catalyzed by tetrahydrofolate-dependent transformylases [reviewed by Greenberg and Jaenicke (*31*), Buchanan and Hartman (*32*), and Hartman (*33*)]. Mechanistic aspects of these transformylation reactions have been described elsewhere (*7, 34*).

The enzyme 5'-phosphoribosyl-*N*-formylglycinamide:tetrahydrofolate-5,10-methyltransferase (EC 2.1.2.2), also called "glycinamide ribonucleotide transformylase," mediates an irreversible transfer of the one-carbon unit from 5,10-methenyltetrahydrofolate to the terminal amino group of glycinamide ribonucleotide [Eq. (9)]. After several additional steps, the final product in purine biosynthesis, inosine-5'-phosphate, contains this one-carbon group at C-8.

Goldthwait *et al.* (*35, 36*) and Hartman *et al.* (*37*) first studied this transformylase activity in extracts of pigeon liver. Formylation of glycinamide ribonucleotide required formate, ATP, and tetrahydrofolate (*35, 36*). Warren and Buchanan (*38*) demonstrated that 5,10-methenyltetrahydrofolate, rather than the 10-formyl derivative, was the reactant in Eq. (9). In order to obtain this result, it was necessary to use an enzyme preparation that had been partially purified from chicken liver

31. G. R. Greenberg and L. Jaenicke, *Chem. Biol. Purines, Ciba Found. Symp. 1956* p. 204 (1957).
32. J. M. Buchanan and S. C. Hartman, *Advan. Enzymol.* **21**, 200 (1959).
33. S. C. Hartman, *in* "Metabolic Pathways" (D. M. Greenberg, ed.), 3rd ed., Vol. 4, pp. 1–68. Academic Press, New York, 1970.
34. J. M. Buchanan, S. C. Hartman, R. L. Herrmann, and R. A. Day, *J. Cell. Comp. Physiol.* **54**, Suppl. 1, 139 (1959).
35. D. A. Goldthwait, R. A. Peabody, and G. R. Greenberg, *JBC* **221**, 553 (1956).
36. D. A. Goldthwait, R. A. Peabody, and G. R. Greenberg, *JBC* **221**, 569 (1956).
37. S. C. Hartman, B. Levenberg, and J. M. Buchanan, *JBC* **221**, 1057 (1956).
38. L. Warren and J. M. Buchanan, *JBC* **229**, 613 (1957).

$$\text{Glycinamide ribonucleotide} + \text{5,10-Methenyltetrahydrofolate} \longrightarrow \text{Formylglycinamide ribonucleotide} + \text{Tetrahydrofolate} \quad (9)$$

and then stored for several months to inactivate any cyclohydrolase (see Section II,C) still present, and to carry out the reaction in maleate buffer to minimize nonenzymic catalysis of reaction (5). The pH optimum of the partially purified transformylase was found to be 7.8, and the K_m values of $5.8 \times 10^{-5} M$ for 5,10-methenyltetrahydrofolate and $5.2 \times 10^{-5} M$ for glycinamide ribonucleotide were reported (38).

G. 5-Amino-4-Imidazole Carboxamide Ribonucleotide Transformylase

C-2 of the purine ring originates from the interaction of 10-formyl-tetrahydrofolate with 5-amino-4-imidazole carboxamide ribonucleotide [Eq. (10)]. The product, 5-formamido-4-imidazole carboxamide ribonucleotide, is converted, via ring closure, to hypoxanthine ribonucleotide (IMP); inosinicase is the enzyme responsible for this latter step. Reaction (10), which is reversible, is catalyzed by 5′-phosphoribosyl-5-formamido-4-imidazole carboxamide:tetrahydrofolate 10-formyltransferase (EC 2.1.2.2), also called "aminoimidazole carboxamide ribonucleotide transformylase." Both the transformylase and inosinicase have been

$$\begin{array}{c}\text{5-Amino-4-imidazole}\\\text{carboxamide ribonucleotide}\end{array} + \text{10-Formyltetrahydrofolate} \rightleftarrows$$

$$\begin{array}{c}\text{5-Formamido-4-imidazole}\\\text{carboxamide ribonucleotide}\end{array} + \text{Tetrahydrofolate} \quad (10)$$

partially purified from chicken liver (39). The transformylase, but not inosinicase, was markedly stimulated by K^+. The pH optimum for both activities was 7.4. Although the formylated product was not isolated in reaction (10), the chemically synthesized material was converted to IMP by inosinicase (40). Crude preparations of the transformylase utilized both 5,10-methenyl- and 10-formyltetrahydrofolate as formyl donors, but when the reaction was run in maleate buffer (see Section II,C), it was specific for the latter (41).

It is still not certain whether one protein is responsible for both the transformylase and inosinicase activities. The existence of two separate enzymes is suggested by the K^+ dependence of only the transformylase. On the other hand, both activities are lost as the result of a single mutation in certain *Enterobacteriaceae* (42).

39. J. G. Flaks, M. J. Erwin, and J. M. Buchanan, *JBC* **229**, 603 (1957).
40. L. Warren, J. G. Flaks, and J. M. Buchanan, *JBC* **229**, 627 (1957).
41. S. C. Hartman and J. M. Buchanan, *JBC* **234**, 1812 (1959).
42. A. P. Levin and B. Magasanik, *JBC* **236**, 184 (1961).

H. FORMIMINOGLYCINE FORMIMINOTRANSFERASE

Cell-free extracts of certain *Clostridia* ferment purines via a sequence of reactions (*3*) involving formiminoglycine as an intermediate. The formimino group is transferred from glycine to tetrahydrofolate by reaction (11). *N*-Formiminoglycine:tetrahydrofolate 5-formiminotransferase

$$\underset{\text{Formimino-glycine}}{\overset{\text{H}}{\underset{\text{O HN}}{\overset{\text{H}_2\text{C}-\text{N}}{\text{O}-\text{C}}}}\text{CH}} + \text{Tetrahydrofolate} \rightleftharpoons \text{Glycine} + \underset{\text{tetrahydrofolate}}{\text{5-Formimino-}} \quad (11)$$

(EC 2.1.2.4) was partially purified from *C. cylindrosporum* and shown to have a molecular weight of about 180,000 (*43*). Reaction (11) can be assayed directly by taking advantage of the spectral difference between tetrahydrofolate and its 5-formimino derivative at 280 nm. Using this procedure, Uyeda and Rabinowitz (*43*) reported K_m values of $4.3 \times 10^{-3} M$ and $4.8 \times 10^{-5} M$ for formiminoglycine and *dl*,L-tetrahydrofolate, respectively. The pH optimum for this reaction was 7.1. Formimino-D,L-alanine and formiminoglycine methyl ester also served as substrates while formimino-D,L-aspartate, formimino-L-glutamate, and formylglycine were inactive. The following metal ions inhibited the enzyme: $Zn^{2+} = Fe^{2+} > Cu^{2+} > Mn^{2+}$.

I. FORMIMINOGLUTAMATE FORMIMINOTRANSFERASE

The degradation of histidine in *Pseudomonas fluorescens* (*44, 45*), *Aerobacter aerogenes* (*45*), and mammalian liver (*44, 46*) leads to the formation of *N*-formimino-L-glutamate. In liver, the formimino group is transferred to tetrahydrofolate (*22*) [Eq. (12)]. This reaction is catalyzed by *N*-formimino-L-glutamate:tetrahydrofolate 5-formiminotransferase (EC 2.1.2.5), which has been purified extensively from pig liver (*22*). The enzyme had a broad optimum between pH 7 and 8 and was markedly inhibited by Ca^{2+}, Mn^{2+}, Zn^{2+}, and Ba^{2+}. The K_m values were found to be $1.1 \times 10^{-2} M$ for formimino-L-glutamate and $1.0 \times 10^{-4} M$ for tetrahydrofolate. Formiminoglycine was inactive as a sub-

43. K. Uyeda and J. C. Rabinowitz, *JBC* **240**, 1701 (1965).
44. H. Tabor and A. Mehler, *JBC* **210**, 559 (1954).
45. B. Magasanik and H. R. Bowser, *JBC* **213**, 571 (1955).
46. B. A. Borek and H. Waelsch, *JBC* **205**, 459 (1953).

$$\begin{array}{c}\text{COOH}\\|\\\text{HC}-\underset{H}{N}-\underset{H}{C}=\text{NH}\\|\\\text{CH}_2\\|\\\text{CH}_2\\|\\\text{COOH}\end{array} + \text{Tetrahydro-folate} \longrightarrow \text{Glutamic acid} + \text{5-Formimino-tetrahydrofolate} \quad (12)$$

Formimino-
glutamic acid

strate, thereby distinguishing the present enzyme from the formiminoglycinetransferase (see Section II,H).

As discussed previously (Section II,E), the cyclodeaminase from *C. cylindrosporum* could not be separated from cyclohydrolase (*30*). Similarly, the formiminoglutamate formiminotransferase and cyclodeaminase activities from pig liver remained together during a 700-fold purification procedure. Ultracentrifugal analysis showed no evidence for the separation of these activities or for the presence of more than one protein. Treatment of the preparation with chymotrypsin, however, caused more destruction of the cyclodeaminase activity than of the transferase activity (*22*).

J. N-Formylglutamate Transformylase

Following an earlier observation (*47*) that 5-formyltetrahydrofolate disappeared rapidly when incubated with liver extracts, Silverman et al. (*48, 49*) described the purification from pig liver of an enzyme, N-formyl-L-glutamate:tetrahydrofolate 5-formyltransferase (EC 2.1.2.6), that catalyzed the reversible transfer of the formyl group from 5-formyltetrahydrofolate (folinic acid) to glutamic acid:

5-Formyltetrahydrofolate + L-glutamate $\rightleftharpoons$
$$\text{tetrahydrofolate} + N\text{-formyl-L-glutamate} \quad (13)$$

This reaction and that catalyzed by 5-formyltetrahydrofolate cyclodehydrase (Section II,D) are the only ones known in which 5-formyltetrahydrofolate serves as a substrate. Either or both of these reactions is presumably responsible for the growth-promoting activity of this folate derivative.

10-Formylfolate, 10-formyltetrahydrofolate, and 5,10-methenyltetrahydrofolate were found to be inactive as formyl donors in reaction (13).

47. J. C. Keresztesy and M. Silverman, *JACS* **75**, 1512 (1953).
48. M. Silverman, J. C. Keresztesy, G. J. Koval, and R. C. Gardiner, *JBC* **226**, 83 (1957).
49. M. Silverman, "Methods in Enzymology," Vol. 5, p. 790, 1962.

D-Glutamic acid and 19 other amino acids were inactive as formyl acceptors; K_m values of $1 \times 10^{-5} M$ (5-formyltetrahydrofolate) and $3 \times 10^{-4} M$ (L-glutamate) were determined (49).

The purified formyltransferase preparation also contained formiminoglutamate formiminotransferase activity (Section II,I) (49). Similarly, these two activities were copurified in a preparation (50) from calf liver.

K. METHIONYL-tRNA TRANSFORMYLASE

The role of N-formylmethionine in the initiation of protein synthesis has been discussed elsewhere (51–53). Methionine occurs as the N-terminal amino acid in about 40% of the proteins in $E.\ coli$ (54), and this organism contains two species of methionine-accepting transfer RNA ($tRNA_F^{met}$ and $tRNA_M^{met}$). Only the former species can be formylated (55, 56) in a reaction catalyzed by 10-formyltetrahydrofolate:methionyl-$tRNA_F$ transformylase:

10-Formyltetrahydrofolate + methionyl-$tRNA_F^{met} \rightarrow$
$\qquad N$-formylmethionyl-$tRNA_F^{met}$ + tetrahydrofolate $\qquad$ (14)

The formyl donor in this reaction was identified as 10-formyltetrahydrofolate (57–59). Transformylase activity has been detected in $E.\ coli$ B (59–63), $Saccharomyces\ cerevisiae$ (64), $Bacillus\ subtilis$ (65, 66), and Ehrlich ascites cells (67). The most highly purified preparations of

50. A. Miller and H. Waelsch, JBC **228**, 397 (1957).
51. P. Lengyel and D. Söll, Bacteriol. Rev. **33**, 264 (1969).
52. C. Baglioni and B. Colombo, in "Metabolic Pathways" (D. M. Greenberg, ed.), 3rd ed., Vol. 4, pp. 277–351. Academic Press, New York, 1970.
53. J. Lucas-Lenard and F. Lipmann, Annu. Rev. Biochem. **40**, 409 (1971).
54. J. Waller, JMB **7**, 483 (1963).
55. B. F. C. Clark and K. A. Marcker, Nature (London) **207**, 1038 (1965).
56. B. F. C. Clark and K. A. Marcker, JMB **17**, 394 (1966).
57. K. A. Marcker, JMB **14**, 63 (1965).
58. J. M. Adams and M. R. Capecchi, Proc. Nat. Acad. Sci. U. S. **55**, 147 (1966).
59. H. Dickerman, E. Steers, B. G. Redfield, and H. Weissbach, Cold Spring Harbor Symp. Quant. Biol. **31**, 287 (1966).
60. H. W. Dickerman and B. C. Smith, ABB **122**, 105 (1967).
61. H. W. Dickerman and H. Weissbach, "Methods in Enzymology," Vol. 12, Part B, p. 681, 1968.
62. H. W. Dickerman and B. C. Smith, Biochemistry **9**, 1247 (1970).
63. H. W. Dickerman and B. C. Smith, JMB **59**, 425 (1971).
64. A. Halbreich and M. Rabinowitz, Proc. Nat. Acad. Sci. U. S. **68**, 294 (1971).
65. K. Horikoshi and R. H. Doi, ABB **122**, 685 (1967).
66. L. Migita and R. H. Doi, ABB **138**, 457 (1970).
67. C-C. Li and C-T. Yu, BBA **182**, 440 (1969).

the transformylase have been obtained from *E. coli* B (*61*, *68*). Dickerman and Weissbach (*61*) have shown that the enzyme from this latter source has a molecular weight of 25,000, a pH optimum of 7.5, and a requirement for Mg^{2+}; K_m values were found to be $1.3 \times 10^{-5} M$ for 10-formyltetrahydrofolate and $1 \times 10^{-5} M$ for methionyl-tRNA. Formylation of methionyl-tRNA was competitively inhibited by tetrahydrofolate, tetrahydropteroate, tetrahydropteroyl triglutamate, 5-formyltetrahydrofolate, tetrahydrohomofolate, and tetrahydrohomopteroate (*62*). 5-Methyltetrahydrofolate was a noncompetitive inhibitor of the reaction, while folate, dihydrofolate, and aminopterin were not inhibitory. Aminoacylated methionyl-tRNA, but not 10-formyltetrahydrofolate, protected the enzyme against heat and trypsin inactivation (*60*).

III. Transfer of Formaldehyde and Its Congeners

A. Deoxycytidylate Hydroxymethyltransferase

5-Hydroxymethylcytosine was first identified in the DNA of T-even phages that infect *E. coli* (*69*). Introduction of the hydroxymethyl group was found to result from the presence of deoxycytidylate hydroxymethyltransferase (*70*, *71*), which catalyzes the following reaction:

dCMP + 5,10-methylenetetrahydrofolate + H_2O ⇌
5-hydroxymethyl dCMP + tetrahydrofolate   (15)

The enzyme, purified from phage-infected *E. coli* (*72*, *73*), had a molecular weight of approximately 68,000, as determined by gel filtration (*73*) and ultracentrifugation (*73*). In the latter procedure, the enzymic activity was associated with the major protein component. The pH optimum was 6.5–8.5 and a K_m value of $6 \times 10^{-4} M$ was reported for dCMP. A K_m value was not determined directly for the other substrate, 5,10-methylenetetrahydrofolate. Instead, the latter was generated *in situ* from HCHO and tetrahydrofolate (see Section III,C); under these conditions, apparent K_m values of $1.5 \times 10^{-3} M$ and $1 \times 10^{-4} M$ were reported for these components.

68. H. W. Dickerman, E. Steers, B. G. Redfield, and H. Weissbach, *JBC* **242**, 1522 (1967).
69. G. R. Wyatt and S. S. Cohen, *BJ* **55**, 774 (1953).
70. J. G. Flaks and S. S. Cohen, *BBA* **25**, 667 (1957).
71. J. G. Flaks and S. S. Cohen, *JBC* **234**, 1501 (1959).
72. L. I. Pizer and S. S. Cohen, *JBC* **237**, 1251 (1962).
73. C. K. Mathews, F. Brown, and S. S. Cohen, *JBC* **239**, 2957 (1964).

Greenberg and Yeh (74) have shown that this enzyme catalyzes an exchange of the hydrogen atom at C-5 of dCMP with the medium. Tetrahydrofolate, but not HCHO, was required for this process. The K_m value for tetrahydrofolate in the exchange reaction was estimated to be $1.3 \times 10^{-4} M$ which is similar to that reported for the overall reaction (see above). These observations were interpreted to mean that formation of the carbanion at C-5 is the rate-limiting step in the hydroxymethyl transfer reaction.

B. Deoxyuridylate Hydroxymethyltransferase

Another hydroxymethylated pyrimidine, 5-hydroxymethyluracil, was identified in the DNA of the *B. subtilis* phage SP8 (75). This component has also been identified in the DNA of phages Φe (76), SPO-1 (77), and SP5C (78). Deoxyuridylate hydroxymethyltransferase, which catalyzes a reaction analogous to that shown in Eq. (15), viz.,

dUMP + 5,10-methylenetetrahydrofolate + $H_2O \rightarrow$
$$\text{5-hydroxymethyl dUMP + tetrahydrofolate} \quad (16)$$

has been detected in extracts of phage-infected *B. subtilis* (76, 79). Of interest is the fact that the hydroxymethylase appeared within 10 min following infection and increased steadily over the next 15 min; thymidylate synthetase activity (see Section III,C), on the other hand, declined 10 min after infection (80).

Unlike dCMP hydroxymethyltransferase, the dUMP-dependent enzyme has not yet been purified appreciably. The known properties of the enzyme have been determined with relatively crude preparations. The enzyme had a pH optimum of 7–8, and it appeared to show some activity with dCMP (80).

C. Thymidylate Synthetase

Because of its importance in DNA synthesis, thymidylate synthetase is one of the better-characterized tetrahydrofolate-dependent transfer enzymes (2, 3, 5, 8). In an early and definitive investigation of this

74. G. R. Greenberg and Y. C. Yeh, *JBC* **242**, 1307 (1967).
75. R. G. Kallen, M. Simon, and J. Marmur, *JMB* **5**, 248 (1962).
76. D. H. Roscoe and R. G. Tucker, *BBRC* **16**, 106 (1964).
77. S. Okubo, B. Strauss, and M. Stodolsky, *Virology* **24**, 552 (1964).
78. H. V. Aposhian, *BBRC* **18**, 230 (1965).
79. F. Kahan, E. Kahan, and B. Riddle, *Fed. Proc., Fed. Amer. Soc. Exp. Biol.* **23**, 318 (1964).
80. D. H. Roscoe and R. G. Tucker, *Virology* **29**, 157 (1966).

enzyme, Friedkin (81) demonstrated that cell-free extracts of E. coli, supplemented with tetrahydrofolate and serine or formaldehyde as a one-carbon source, converted ^{32}P-labeled dUMP to TMP. Soluble preparations from rabbit (82) and rat (83) thymus catalyzed the same reaction. Subsequent studies showed that the actual substrate was 5,10-methylene tetrahydrofolate, which could be generated by the nonenzymic interaction (84–87) of formaldehyde and tetrahydrofolate [Eq. (17)]

$$\text{HCHO} + \text{tetrahydrofolate} \rightleftharpoons \text{5,10-methylenetetrahydrofolate} \qquad (17)$$

or by the reaction catalyzed by serine hydroxymethyltransferase (see Section III,D).

From the studies cited above, it was evident that some reductive process was necessary for conversion of the one-carbon group from the methylene to the methyl level. Humphreys and Greenberg (88) observed that DPNH had no effect upon thymidylate synthesis in crude preparations when high concentrations of tetrahydrofolate were present. However, when dihydrofolate replaced tetrahydrofolate, there was a marked stimulation by DPNH. McDougall and Blakley (89, 90) found in crude preparations that synthesis of thymidylate was inhibited by aminopterin (4-amino-4-deoxyfolate) when the system contained DPNH and catalytic amounts of tetrahydrofolate. Aminopterin had no effect on the reaction when excess tetrahydrofolate was present. Based upon these results, both groups suggested that tetrahydrofolate (via oxidation to dihydrofolate) was the source of reducing power. Regeneration of tetrahydrofolate was attributed to the pyridine nucleotide-dependent, aminopterin-sensitive dihydrofolate reductase (6–8). Corroboration of this hypothesis was obtained from labeling experiments. Pastore and Friedkin (91) demonstrated that, when tetrahydrofolate labeled with ^{3}H in the 6 and 7 positions was used, the label appeared without dilution by the solvent in the methyl group of thymidylate. These results were extended by Blakley et al. (92) and Lorenson et al. (93), who showed that the

81. M. Friedkin, Fed. Proc., Fed. Amer. Soc. Exp. Biol. **16**, 183 (1957).
82. R. L. Blakley, BBA **24**, 224 (1957).
83. E. A. Phear and D. M. Greenberg, JACS **79**, 3737 (1957).
84. R. L. Kisliuk, JBC **227**, 805 (1957).
85. R. L. Blakley, Nature (London) **182**, 1719 (1958).
86. M. J. Osborn, P. T. Talbert, and F. M. Huennekens, JACS **82**, 4921 (1960).
87. R. G. Kallen and W. P. Jencks, JBC **241**, 5851 (1966).
88. G. K. Humphreys and D. M. Greenberg, ABB **78**, 275 (1958).
89. B. M. McDougall and R. L. Blakley, BBA **39**, 176 (1960).
90. B. M. McDougall and R. L. Blakley, JBC **236**, 832 (1961).
91. E. J. Pastore and M. Friedkin, JBC **237**, 3802 (1962).
92. R. L. Blakley, B. V. Ramasastri, and B. M. McDougall, JBC **238**, 3075 (1963).
93. M. Y. Lorenson, G. F. Maley, and F. Maley, JBC **242**, 3332 (1967).

hydrogen transferred originates specifically from C-6 of the reduced pyrazine ring of tetrahydrofolate.

These results established that the primary reaction, catalyzed by thymidylate synthetase, involves both a transfer and reduction of the one-carbon group:

$$\text{dUMP} + \text{5,10-methylenetetrahydrofolate} \rightarrow \text{TMP} + \text{dihydrofolate} \qquad (18)$$

Dihydrofolate is reduced to tetrahydrofolate [Eq. (19)] by

$$\text{Dihydrofolate} + \text{TPNH} + \text{H}^+ \rightarrow \text{tetrahydrofolate} + \text{TPN}^+ \qquad (19)$$

dihydrofolate reductase (EC 1.5.1.3), a TPNH-dependent enzyme whose properties have been reviewed elsewhere (*6, 7, 8, 94, 95*).

Thymidylate synthetase is present at rather low levels in most tissues, and this has led to the development of special methods for its assay [reviewed by Friedkin (*96*)]. These include: (a) conversion of ^{32}P- or ^{3}H-labeled dUMP to TMP (*97, 98*); (b) transfer of ^{3}H from tetrahydrofolate to the methyl group of TMP; (c) transformation of 5,10-methylenetetrahydrofolate to dihydrofolate, measured by changes at 340 nm (*99–101*); and (d) release into the solvent of ^{3}H from the C-5 position of dUMP (*102*).

Thymidylate synthetase has been identified in a variety of tissues and purified extensively from some of these sources (Table I) (*93, 100, 103–107*). Purification is generally hampered by the low level of the enzyme in the starting material, but in some instances this difficulty has been ameliorated by selecting a cell subline that is resistant to an

94. G. H. Hitchings and J. J. Burchall, *Advan. Enzymol.* **27**, 417 (1965).
95. F. M. Huennekens, R. B. Dunlap, J. H. Freisheim, L. E. Gundersen, N. G. L. Harding, S. A. Levison, and G. P. Mell, *Ann. N. Y. Acad. Sci.* **186**, 85 (1971).
96. M. Friedkin, "Methods in Enzymology," Vol. 6, p. 124, 1963.
97. M. Friedkin and A. Kornberg, in "The Chemical Basis of Heredity" (W. D. McElroy and B. Glass, eds.), p. 609. Johns Hopkins Press, Baltimore, Maryland, 1957.
98. K.-U. Hartmann and C. Heidelberger, *JBC* **236**, 3006 (1961).
99. A. J. Wahba and M. Friedkin, *JBC* **237**, 3794 (1962).
100. R. L. Blakley and B. M. McDougall, *JBC* **237**, 812 (1962).
101. P. Reyes and C. Heidelberger, *Mol. Pharmacol.* **1**, 14 (1965).
102. M. Smith and G. R. Greenberg, *Fed. Proc., Fed. Amer. Soc. Exp. Biol.* **23**, 271 (1964).
103. E. Jenny and D. M. Greenberg, *JBC* **238**, 3378 (1963).
104. A. Fridland and C. Heidelberger, *Fed. Proc., Fed. Amer. Soc. Exp. Biol.* **29**, 878 (1970).
105. R. L. Blakley, *JBC* **238**, 2113 (1963).
106. R. B. Dunlap, N. G. L. Harding, and F. M. Huennekens, *Biochemistry* **10**, 88 (1971).
107. T. Crusberg, R. Leary, and R. L. Kisliuk, *JBC* **245**, 5292 (1970).

TABLE I
THYMIDYLATE SYNTHETASES

Source	Purification (fold)	Specific activity (μmole/hr/mg)	pH optimum	MW	K_m dUMP (mM)	K_m CH_2-FH_4 (mM)	Activation Mg^{2+}	Activation Thiols	Ref.
I. Mammalian									
Calf thymus	205	0.98	7.1	n.d.[a]	0.02	0.038–0.045	No	Yes	103
Chick embryo	600–800	3.8	6.5	58,000	0.0075	0.014	No	Yes	93
Ehrlich ascites	1500	3.48	6.7	67,000	0.0085	0.0125	n.d.	Yes	104
II. Bacterial									
S. faecium R	100–600	54	7.8	n.d.	0.0057	0.037	Yes	Yes	100, 105
L. casei (amethopterin-resistant)	38	150	6.5–6.8	70,000	0.00068	0.032	Yes	Yes	106
L. casei (dichloroamethopterin-resistant)	17	30	6.9–7.8	67,000	0.0052	0.045	Yes	Yes	107

[a] n.d. stands for not determined.

antifolate such as amethopterin or aminopterin. These cells are usually characterized by an elevated level of dihydrofolate reductase [reviewed in Blakley (8)] and thymidylate synthetase (106–109).

The molecular weights of the purified thymidylate synthetases range from 60,000 to 70,000; the L. casei enzyme has been shown to consist of two subunits (106). pH optima fall between 6.5 and 8.0, and K_m values are in reasonably good agreement (i.e., about $10^{-6} M$ for dUMP and about $10^{-5} M$ for methylenetetrahydrofolate); Mg^{2+} is required for the bacterial, but not the mammalian, thymidylate synthetases.

All of the thymidylate synthetases are sensitive to mercurials and other reagents that react with sulfhydryl groups. The crystalline enzyme from L. casei (106) contains four sulfhydryl groups per molecule; one of these appears to be required for catalytic activity (108). One of the most striking properties of thymidylate synthetases is their extreme sensitivity to 5-fluorodeoxyuridylate (98, 110, 111). This substrate analog inhibits the enzyme with a K_i value of about $10^{-8} M$. Inhibition of the enzyme by analogs of the other substrate, 5,10-methylenetetrahydrofolate, has been discussed by Baker (112).

The mechanism of the reaction catalyzed by thymidylate synthetase can be considered formally in two parts: (1) loss to the solvent of hydrogen at C-5 of deoxyuridylate; and (2) detachment of the one-carbon unit from the 5 and 10 positions of tetrahydrofolate, its attachment to C-5 of deoxyuridylate, and transfer of the C-6 hydrogen of tetrahydrofolate to the one-carbon unit. The loss (or exchange) of the hydrogen at C-5 of uridine derivatives has been investigated with the enzyme (113, 114) and in model systems (108, 115, 116). The results from these studies indicate that labilization of the hydrogen at C-5 of the pyrimidine is facilitated by a nucleophilic attack (probably by a sulfhydryl group on the enzyme) at the adjacent C-6 position. Transfer of the methylene group from tetrahydrofolate to deoxyuridylate and

108. R. B. Dunlap, N. G. L. Harding, and F. M. Huennekens, *Ann. N. Y. Acad. Sci.* **186**, 153 (1971).

109. A. M. Albrecht, F. K. Pearce, and D. J. Hutchison, *JBC* **241**, 1036 (1966).

110. S. S. Cohen, J. G. Flaks, H. D. Barner, J. R. Loeb, and J. Lichtenstein, *Proc. Nat. Acad. Sci. U. S.* **44**, 1004 (1958).

111. C. Heidelberger, G. Kaldor, K. L. Mukherjee, and P. B. Danneberg, *Cancer Res.* **20**, 903 (1960).

112. B. R. Baker, "Design of Active-Site-Directed Irreversible Enzyme Inhibitors," pp. 269–284. Wiley, New York, 1967.

113. H. O. Kammen, *Anal. Biochem.* **17**, 553 (1966).

114. M. I. S. Lomax and G. R. Greenberg, *JBC* **242**, 1302 (1967).

115. D. V. Santi and C. F. Brewer, *JACS* **90**, 6236 (1968).

116. T. I. Kalman, *Biochemistry* **10**, 2567 (1971).

reduction of the group via hydrogen emanating from C-6 of tetrahydrofolate are less well-understood. Rupture of the N-10 bond, followed by the attack of deoxyuridylate on the methylene group (or vice versa), would result in formation of the "Friedkin intermediate," i.e., a hypothetical structure in which the methylene group is linked covalently to N-5 of tetrahydrofolate and C-5 of deoxyuridylate (117). In a subsequent step, the hydrogen at C-6 could be envisioned as being transferred directly to the methylene group to form the required methyl group, or it might be transferred via some residue on the enzyme. Irreversible inhibition of the enzyme by 5-fluorodeoxyuridylate [which requires the presence of methylene tetrahydrofolate (118)] is envisioned as resulting from formation of a covalent bond between the enzyme and the inhibitor (118, 119).

D. Serine Hydroxymethyltransferase

L-Serine:tetrahydrofolate 5,10-methylenetransferase (EC 2.1.2.1), also called "serine hydroxymethyltransferase" or "serine aldolase," catalyzes the reversible transfer of a one-carbon group between glycine and serine:

$$\text{Glycine} + \text{HCHO} \rightleftharpoons \text{L-serine} \qquad (20)$$

On the basis of the participation of tetrahydrofolate (2, 3, 8) in the overall reaction, Blakley (120) suggested that the coenzyme combined with HCHO to form 5,10-methylenetetrahydrofolate [reaction (17)] and that the latter reacted with glycine:

$$5{,}10\text{-Methylenetetrahydrofolate} + \text{glycine} + \text{H}_2\text{O} \rightleftharpoons$$
$$\text{L-serine} + \text{tetrahydrofolate} \qquad (21)$$

Subsequent studies (121) confirmed the stoichiometry of these reactions and showed that the hydroxymethyltransferase was stereospecific for one isomer of methylenetetrahydrofolate. Threonine and allothreonine also served as substrates for the enzyme (122).

Various assay methods have been devised for this enzyme [reviewed in Blakley (8)]. These include: (a) analyses for serine or glycine; (b) formation of methylenetetrahydrofolate or tetrahydrofolate, measured by

117. M. Friedkin, in "The Kinetics of Cellular Proliferation" (F. Stohlman, Jr., ed.), p. 99. Grune & Stratton, New York, 1959.
118. R. J. Langenbach, P. V. Danenberg, and C. Heidelberger, BBRC 48, 1565 (1972).
119. D. V. Santi and T. T. Sakai, Biochemistry 10, 3598 (1971).
120. R. L. Blakley, BJ 58, 448 (1954).
121. R. L. Blakley, BJ 77, 459 (1960).
122. L. Schirch and T. Gross, JBC 243, 5651 (1968).

absorbance changes at the appropriate wavelengths; and (c) coupling of the hydroxymethyltransferase reaction with that catalyzed by methylenetetrahydrofolate dehydrogenase [Eq. (22)] and measuring TPNH formation at 340 nm.

$$5,10\text{-Methylenetetrahydrofolate} + TPN^+ \rightleftharpoons [5,10\text{-methenyltetrahydrofolate}]^+ + TPNH \quad (22)$$

Serine hydroxymethyltransferase has a specific requirement for tetrahydrofolate (*123–125*). Blakley (*124, 125*) reported that the enzyme from rabbit liver can utilize, in addition to this coenzyme, the tetrahydro derivatives of folyldiglutamate (pteroyltriglutamate) and folylhexaglutamate (pteroylheptaglutamate). The activity of serine hydroxymethyltransferases obtained from mammalian (*126–129*), avian (*130, 131*), bacterial (*132, 133*), and plant sources (*134*) was stimulated by pyridoxal phosphate. As isolated, however, most of these enzymes contained a sufficient amount of bound pyridoxal phosphate to make this requirement only partial. This was paralleled by the observation (*135, 136*) that the purified rabbit liver enzyme contained 4 moles of pyridoxal phosphate per mole of protein (MW 331,000), as estimated by the absorbance at 430 nm; when fully activated, however, the content of this coenzyme was increased to 6 moles/mole of protein.

The occurrence and properties of serine hydroxymethyltransferases from various sources have been reviewed elsewhere (*2, 3, 5, 7, 8, 137*). Parameters for some of these enzymes are listed in Table II (*138–141*).

123. R. L. Kisliuk and W. Sakami, *JBC* **214**, 47 (1955).
124. R. L. Blakley, *BJ* **65**, 331 (1957).
125. R. L. Blakley, *BJ* **65**, 342 (1957).
126. N. Alexander and D. M. Greenberg, *JBC* **220**, 775 (1956).
127. R. L. Blakley, *BJ* **61**, 315 (1955).
128. R. L. Blakley, *Nature (London)* **174**, 652 (1954).
129. F. M. Huennekens, Y. Hatefi, and L. D. Kay, *JBC* **224**, 435 (1957).
130. S. Deodar and W. Sakami, *Fed. Proc., Fed. Amer. Soc. Exp. Biol.* **12**, 195 (1953).
131. W. Sakami, in "Amino Acid Metabolism" (W. D. McElroy and H. B. Glass, eds.), p. 658. Johns Hopkins Press, Baltimore, Maryland, 1955.
132. B. E. Wright and T. C. Stadtman, *JBC* **219**, 863 (1956).
133. B. E. Wright, *JBC* **219**, 873 (1956).
134. A. P. Wilkinson and D. D. Davies, *Nature (London)* **181**, 1070 (1958).
135. L. Schirch and M. Mason, *JBC* **238**, 1032 (1963).
136. L. Schirch and W. T. Jenkins, *JBC* **239**, 3797 (1964).
137. D. M. Greenberg, in "Metabolic Pathways," (D. M. Greenberg, ed.) 3rd ed., Vol. 3, pp. 95–190. Academic Press, New York, 1969.
138. Y. Nakano, M. Fujioka, and H. Wada, *BBA* **159**, 19 (1968).
139. M. Fujioka, *BBA* **185**, 338 (1969).
140. L. Schirch, "Methods in Enzymology," Vol. 17B, pp. 335–340, 1971.
141. K. Uyeda and J. C. Rabinowitz, *ABB* **123**, 271 (1968).

TABLE II
Serine Hydroxymethyltransferases

Source	Purification (fold)	Specific activity (μmole/min/mg)	pH optimum	MW	K_m L-Serine (mM)	K_m FH_4 (mM)	K_m Glycine (mM)	K_m CH_2-FH_4 (mM)	Ref.
Rat liver									
Supernatant	600	4.3	8.0	n.d.[a]	0.54	0.0072	1.2	0.13	138
Mitochondria	800	5.5	7.0–7.7	n.d.	0.54	0.010	1.8	0.13	
Rabbit liver									139
Supernatant	180	10.0	7.3	185,000	1.3	n.d.	n.d.	n.d.	
Mitochondria	240	15.5	7.3	170,000	1.0	n.d.	n.d.	n.d.	
Rabbit liver	320	11.7	6.6–7.3	330,000	1.0	0.017 (+glycine)	1.4 (+FH_4)	n.d.	140
C. cylindrosporum	13	5.7	6.8–7.4	n.d.	0.086	0.037	19	0.034	141

[a] n.d. stands for not determined.

The most highly purified serine hydroxymethyltransferases are those from rat and rabbit liver (138–140). Partially purified isozymes from the soluble and mitochondrial fractions of rat liver differed in pH optima, stability, affinity for DEAE-cellulose, and electrophoretic mobility (138). In contrast, serine hydroxymethyltransferase isozymes purified to apparent homogeneity from soluble and mitochondrial fractions of rabbit liver had similar pH optima, stabilities, and electrophoretic mobilities (139). These isozymes were distinguishable, however, on the basis of their immunological reactivities.

Detailed studies on the mechanism of serine hydroxymethyltransferase from rabbit liver have been made by Schirch and co-workers (122, 135, 136, 140, 142–145). The highly purified enzyme catalyzed a variety of reactions [in addition to (21)]:

α-Methylserine + tetrahydrofolate $\rightleftharpoons$
$\qquad$ D-alanine + 5,10-methylenetetrahydrofolate $\qquad$ (23)

$\qquad$ L-Threonine $\rightleftharpoons$ glycine + acetaldehyde $\qquad$ (24)

$\qquad$ Allothreonine $\rightleftharpoons$ glycine + acetaldehyde $\qquad$ (25)

D-Alanine + hydroxymethylase $\rightleftharpoons$
$\qquad$ (holoenzyme)
$\qquad\qquad$ pyruvate + pyridoxamine phosphate + hydroxymethylase $\qquad$ (26)
$\qquad\qquad\qquad$ (apoenzyme)

Tetrahydrofolate was not required for reactions (24), (25), and (26) (143). The K_m value for α-methylserine in reaction (23) was $7 \times 10^{-3} M$, and the reaction proceeded at a much slower rate than the normal reaction (135). α-Methylserine was a competitive inhibitor in the conversion of serine to glycine. The enzyme also catalyzed a novel, but probably nonphysiological, transamination reaction between pyridoxal phosphate and D-alanine (135, 136). This process, however, was about three orders of magnitude slower than catalysis of the serine–glycine interconversion (136). Addition of D-alanine to the enzyme resulted in a decrease in the 430-nm absorbance and the appearance of a new maximum at 327 nm, characteristic of free pyridoxamine. The enzyme was inactivated by this treatment, but could be reactivated by the addition of pyridoxal phosphate (135).

By analogy with other pyridoxal-dependent reactions, it has been

142. L. Schirch and M. Mason, JBC 237, 2578 (1962).
143. L. Schirch and A. Diller, JBC 246, 3961 (1971).
144. L. Schirch and W. T. Jenkins, JBC 239, 3801 (1964).
145. L. Schirch and M. Ropp, Biochemistry 6, 253 (1967).

suggested (*146–148*) that the first step in the conversion of glycine to serine is the formation of a Schiff base intermediate:

$$\underset{\text{Glycine}}{\overset{H}{\underset{NH_2}{\overset{|}{C}}}\!\!-\!COOH} + \underset{\text{Pyridoxal phosphate}}{\begin{array}{c}HC=O\\ HO\diagup\!\!\diagdown CH_2OPO_3H_2\\ H_3C\diagdown_N\diagup\end{array}} \rightleftarrows \underset{\text{Schiff base intermediate}}{\begin{array}{c}H\diagdown\!\!\diagup H\\ C-COOH\\ HC=N\\ HO\diagup\!\!\diagdown CH_2OPO_3H_2\\ H_3C\diagdown_N\diagup\end{array}} \quad (27)$$

In support of this hypothesis, Wilson and Snell (*149*) showed that α-methylserine hydroxymethyltransferase could interconvert D-alanine and (+)-2-methylserine (α-methylserine). In this reaction the entering group occupied the position from which the proton had been released:

$$\underset{\text{D-Alanine}}{\overset{H^S}{\underset{NH_2}{\overset{|}{C}}}\!\!\!\diagup^{CH_3}_{\!-COOH}} \rightleftarrows \underset{\text{2-Methylserine}}{\overset{HO-C}{\underset{NH_2}{\overset{H}{\underset{|}{\overset{|}{C}}}}}\!\!\!\diagup^{CH_3}_{\!-COOH}} \quad (28)$$

Similarly, during the cleavage of hydroxymethylserine to D-serine and formaldehyde, the enzyme distinguished between "identical" hydroxymethyl groups (i.e., the $-CH_2OH$ removed during cleavage was the same one added when the reaction was run in the synthetic direction). Schirch and Jenkins (*144*) showed that serine hydroxymethyltransferase catalyzed an exchange of the α-proton of D-alanine or glycine with solvent protons. The exchange was accelerated by the presence of tetrahydrofolate, which suggested that this coenzyme also played a role (along with pyridoxal phosphate) in the activation of the amino acid.

Akhtar and Jordan (*150–152*) recently provided evidence that the pro-S proton of glycine [see (II)] is lost stereospecifically during the conversion of glycine to L-serine or to L-threonine. Samples of asymmetrically tritiated glycine were prepared by incubating randomly

146. H. C. Dunathan, *Advan. Enzymol.* **35**, 79 (1971).
147. T. C. Bruice and S. J. Benkovic, "Bioorganic Mechanisms," Vol. II, p. 181–300. Benjamin, New York, 1966.
148. A. E. Braunstein, "The Enzymes," 2nd ed., Vol. 2, Part A, p. 113, 1960.
149. E. M. Wilson and E. E. Snell, *JBC* **237**, 3180 (1962).
150. M. Akhtar and P. M. Jordan, *Chem. Commun.* No. 24, p. 1691 (1968).
151. M. Akhtar and P. M. Jordan, *Tetrahedron Lett.* No. 11, p. 875 (1969).
152. P. M. Jordan and M. Akhtar, *BJ* **116**, 277 (1970).

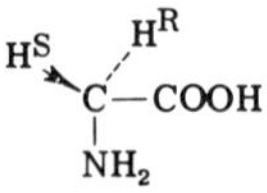

$$\begin{array}{c} H^S \diagdown \diagup H^R \\ C-COOH \\ | \\ NH_2 \end{array}$$

(II)

labeled glycine ([2 RS-^{3}H$_2$]glycine) with serine hydroxymethyltransferase or by incubating glycine with the enzyme in the presence of [^{3}H]water [reactions (29) and (30)].

$$[2\ RS\text{-}^3H_2]\text{glycine} \xrightarrow{\text{enzyme}} [2\ R\text{-}^3H]\text{glycine} \qquad (29)$$

$$\text{Glycine} \xrightarrow[\text{[}^3\text{H]water}]{\text{enzyme}} [2\ S\text{-}^3H]\text{glycine} \qquad (30)$$

The assignment of absolute configuration was made by measuring the loss or retention of label when these glycines were incubated with D-amino acid oxidase. Samples of glycine prepared in the presence of serine hydroxymethyltransferase released more than 90% of their tritium when treated with D-amino acid oxidase (153). This hydrogen [see Eq. (31)] corresponds in configuration to the α-hydrogen of D-amino acids.

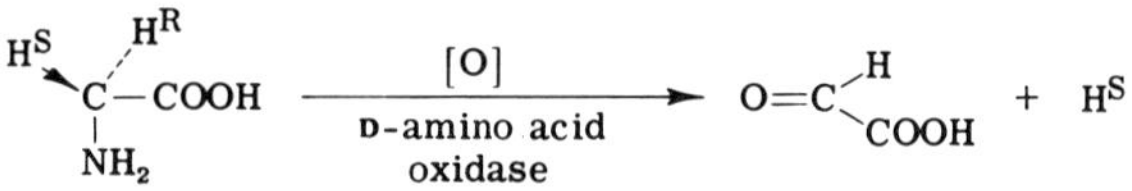

$$\underset{\text{Glycine}}{\begin{array}{c} H^S \diagdown \diagup H^R \\ C-COOH \\ | \\ NH_2 \end{array}} \xrightarrow[\text{oxidase}]{[O] \atop \text{D-amino acid}} \underset{\text{Glyoxylate}}{O=C \diagup^H _{\diagdown COOH}} + H^S \qquad (31)$$

Samples of the stereospecifically labeled glycines were used in the hydroxymethyltransferase reaction, and radioactivity in the serine or threonine was measured (152). The conversion of (2 R-^{3}H)glycine to serine or threonine resulted in retention of the label, indicating that the R-hydrogen atom remained unchanged during the reaction.

Although certain serine hydroxymethylases can also use α-methylserine as a substrate [see Eq. (23)], there is a specific α-methylserine hydroxymethylase that occurs in *Pseudomonas* MS. This soil microorganism utilizes α-methylserine or α-hydroxymethylserine as its sole source of carbon and nitrogen. The enzyme has been partially purified (149, 154, 155) and shown to catalyze the reversible cleavage of three α-substituted serines:

153. D. Wellner, *Biochemistry* **9**, 2307 (1970).
154. E. M. Wilson and E. E. Snell, *BJ* **83**, 1P (1962).
155. E. M. Wilson and E. E. Snell, *JBC* **237**, 3171 (1962).

(+)-α-Methylserine + tetrahydrofolate ⇌
  5,10-methylenetetrahydrofolate + D-alanine (32)
(−)-α-Ethylserine + tetrahydrofolate ⇌
  5,10-methylenetetrahydrofolate + D-α-aminobutyrate (33)
α-Hydroxymethylserine + tetrahydrofolate ⇌
  5,10-methylenetetrahydrofolate + D-serine (34)

The partially purified enzyme lost all activity after prolonged dialysis at pH 8.0; full activity was restored by incubation with pyridoxal phosphate. The enzyme was protected against heat inactivation by the presence of tetrahydrofolate, pyridoxal phosphate, and hydroxymethylserine. The pH optimum for reaction (32) was 9.0. D- and L-cycloserine competitively inhibited the enzyme; Zn^{2+} and Cu^{2+} were markedly inhibitory, while Fe^{2+}, Fe^{3+}, and pCMB inhibited to a lesser extent. The following K_m values have been reported (156): hydroxymethylserine, 0.8 mM; α-methyl-D,L-serine, 1.5 mM; α-ethyl-D,L-serine, 6.2 mM; dl,L-tetrahydrofolate, 0.25 mM; and pyridoxal phosphate, 0.1 mM.

E. OXIDATIVE DECARBOXYLATION OF GLYCINE

Cell-free extracts of *Peptococcus glycinophilus* (157) and avian liver (158) catalyze the following reaction:

Glycine + tetrahydrofolate + DPN^+ →
  5,10-methylenetetrahydrofolate + CO_2 + NH_4^+ + DPNH (35)

The enzyme system responsible for this complex process was fractionated (159–161) into four separate proteins (P_1, P_2, P_3, and P_4): P_1, a pyridoxal phosphate-containing protein, has a molecular weight of about 125,000 (162); P_2 is an acidic, heat-stable protein (MW about 10,000) that contains a vicinal dithiol group (160); P_3 is an FAD-containing flavoprotein (163); and no functional group has yet been detected on P_4.

Although all four proteins were required for catalysis of reaction (35), an exchange of $^{14}CO_2$ with the carboxyl group of glycine was achieved

156. E. W. Miles, "Methods in Enzymology," Vol. 17B, p. 341, 1971.
157. R. D. Sagers and I. C. Gunsalus, *J. Bacteriol.* **81**, 541 (1961).
158. D. A. Richert, R. Amberg, and M. Wilson, *JBC* **237**, 99 (1962).
159. R. D. Sagers and S. M. Klein, *Fed. Proc., Fed. Amer. Soc. Exp. Biol.* **24**, 219 (1965).
160. M. L. Baginsky and F. M. Huennekens, *BBRC* **23**, 600 (1966).
161. S. M. Klein and R. D. Sagers, *JBC* **242**, 297 (1967).
162. S. M. Klein and R. D. Sagers, *JBC* **241**, 206 (1966).
163. M. L. Baginsky and F. M. Huennekens, *ABB* **120**, 703 (1967).

in the presence of only P_1 and P_2 (*164*). With the aid of a thiol such as DTNB as a source of reducing power, the sequence of electron transfer between P_2, P_3, and DPN was demonstrated to be the following (*160*):

$$\text{DTNB} \rightarrow P_2 \rightarrow P_3 \rightarrow \text{DPN} \qquad (36)$$

Based upon these observations, a tentative mechanism has been proposed (*7*) for the overall reaction [Eq. (35)]. Glycine is assumed to form a Schiff base with the protein-bound pyridoxal phosphate (P_1). Interaction of this complex with P_2 leads to the release of the carboxyl carbon of glycine as CO_2. The resulting intermediate then reacts with P_3, P_4, and tetrahydrofolate in a complex step or steps in which: (a) reducing power from the α-carbon is transferred to DPN [cf. Eq. (36)], (b) ammonia is released, and (c) the α-carbon appears as 5,10-methylene tetrahydrofolate.

A new route for glycine biosynthesis was proposed recently (*165, 166*). The overall reaction proceeds according to the following equation:

$$CO_2 + NH_3 + \text{serine} + 2\,[H] \rightleftharpoons 2\,\text{glycine} + H_2O \qquad (37)$$

The enzyme system, which has been partially purified from rat liver mitochondria (*167–171*), required tetrahydrofolate and pyridoxal phosphate. One of the components of this multienzyme system has been identified as serine hydroxymethyltransferase (*167*). The remainder of the system appears to be the glycine decarboxylase described above. Two of the protein components in this system have been isolated from mitochondria (*170*). One of these, termed the "carboxylation enzyme," contained pyridoxal phosphate and had serine hydroxymethyltransferase as well as methylenetetrahydrofolate dehydrogenase activities. The second component, termed "hydrogen carrier protein," was electrophoretically homogeneous and contained one disulfide group per molecule; the latter group could be reduced by dihydrolipoamide dehydrogenase (*170*). This fraction may be analogous to the P_2 protein of the *P. glycinophilus* system (see above). Incubation of glycine-2-^{14}C with these protein fractions resulted in the formation of a complex (separable by chromatog-

164. S. M. Klein and R. D. Sagers, *JBC* **241**, 197 (1966).
165. H. Kawasaki, T. Sato, and G. Kikuchi, *BBRC* **23**, 227 (1966).
166. T. Sato, Y. Motokawa, H. Kochi, and G. Kikuchi, *BBRC* **28**, 495 (1967).
167. T. Sato, H. Kochi, Y. Motokawa, H. Kawasaki, and G. Kikuchi, *J. Biochem.* (*Tokyo*) **65**, 63 (1969).
168. Y. Motokawa and G. Kikuchi, *J. Biochem.* (*Tokyo*) **65**, 71 (1969).
169. T. Sato, H. Kochi, N. Sato, and G. Kikuchi, *J. Biochem.* (*Tokyo*) **65**, 77 (1969).
170. Y. Motokawa and G. Kikuchi, *ABB* **135**, 402 (1969).
171. T. Yoshida and G. Kikuchi, *ABB* **139**, 380 (1970).

raphy on Sephadex G-100) in which the α-carbon of glycine was linked to the hydrogen carrier protein (172). Treatment of the reduced form of the hydrogen carrier protein with N-ethylmaleimide inhibited formation of this complex. Based upon these observations, the authors (172) proposed a reaction scheme in which the pyridoxal phosphate–containing protein is detached from an initial complex and the remaining steps are carried out by the hydrogen carrier protein–glycine complex, tetrahydrofolate, and the dihydrolipoamide dehydrogenase.

172. Y. Motokawa, K. Hiraga, H. Kochi, and G. Kikuchi, *BBRC* **38**, 771 (1970).

7
Aspartate Transcarbamylases

GARY R. JACOBSON • GEORGE R. STARK

    I. Introduction 226
   II. Isolation and Characterization of the Native Enzyme from
       Escherichia coli 227
       A. Properties Associated with Regulation 227
       B. Purification 228
       C. Assay Procedures 228
       D. Size and Subunit Composition 230
  III. Isolation and Characterization of Subunits 231
       A. The Catalytic Subunit 231
       B. The Regulatory Subunit 234
  IV. Reconstitution of Native ATCase from the Isolated Subunits . 237
       A. Methods of Reassociation 237
       B. Importance of Metals in Reassociation 237
       C. Substitution of Other Metals for Zinc 238
   V. Detailed Subunit Structure of Native ATCase 239
  VI. Mechanism of Catalysis by the *E. coli* Enzyme 243
       A. Detailed Mechanism 243
       B. Functional Groups at the Active Site 262
 VII. Cooperative Properties of the *E. coli* Enzyme 268
       A. Cooperative Substrate Binding 268
       B. Allosteric Effectors 269
       C. Stoichiometry of Ligand Binding 273
       D. Conformational Changes Induced by Ligands . . . 275
       E. A Comparison of the Properties of Native ATCase
          with Those of Its Subunits 277
       F. Cooperative Properties of Native ATCase after
          Structural Modifications 278
       G. Kinetics of Ligand Binding 285
       H. Possible Mechanisms for Cooperativity 287
VIII. Biosynthesis and Genetics of Bacterial Aspartate
       Transcarbamylases 292
       A. Location of the Genes 292

  B. Are the Two Kinds of Chain in a Single Operon? . . 293
  C. Control of Enzyme Biosynthesis 295
IX. Aspartate Transcarbamylases from Other Organisms . . . 297
  A. Bacterial ATCases 297
  B. ATCases from *Neurospora crassa* and *Saccharomyces cerevisiae* 302
  C. Mammalian ATCases 306
  D. Plant ATCases 307

I. Introduction

Aspartate transcarbamylase catalyzes the formation of carbamyl-L-aspartate, the first compound unique to the biosynthetic pathway for pyrimidine nucleotides. In *Escherichia coli*, ATCase (*1*) is specifically inhibited in feedback fashion by CTP, one of the end products of the pathway (Fig. 1). Interest in understanding the molecular basis of the allosteric feedback regulation and details of catalysis has led to an extensive investigation of the *E. coli* enzyme by a variety of approaches. Thus, the primary structure of the enzyme is largely known, the crystallographic structure has proceeded to the stage of 5.5 Å resolution, detailed views of the enzyme have been obtained by electron microscopy after negative staining, extensive studies by steady state and rapid reaction kinetics have been carried out, and the enzyme in solution has been prodded and probed by a substantial fraction of the chemical and physical tools available today. In the last few years the accumulated information has prompted several authors to generate reviews, which have usually been focused rather strongly on allosteric properties and on allosteric models (*2–4*). The review by O'Donovan and Neuhard (*5*) has in addition much information on the genetics of the enzyme in *E. coli* and in *Salmonella typhimurium*. The most extensive review is the one by Gerhart (*6*), in which several different aspects of the enzyme are discussed in depth and a detailed allosteric model is proposed.

  In this review, we have attempted to discuss the *E. coli* enzyme in

 1. The following abbreviations are used: ATCase, aspartate transcarbamylase; C subunit, catalytic subunit; R subunit, regulatory subunit; CPS-pyr, pyrimidine-specific carbamyl-P synthetase; NMR, nuclear magnetic resonance; DTNB, 5,5'-dithiobis(2-nitrobenzoate); PALA, *N*-(phosphonacetyl)-L-aspartate; and BrCTP, 5-bromocytidine triphosphate.
 2. E. Whitehead, *Progr. Biophys. Mol. Biol.* **21**, 321 (1970).
 3. G. G. Hammes and C.-W. Wu, *Science* **172**, 1205 (1971).
 4. K. Kirschner, *Curr. Top. Cell. Regul.* **4**, 167 (1971).
 5. G. A. O'Donovan and J. Neuhard, *Bacteriol. Rev.* **34**, 278 (1970).
 6. J. C. Gerhart, *Curr. Top. Cell. Regul.* **2**, 275 (1970).

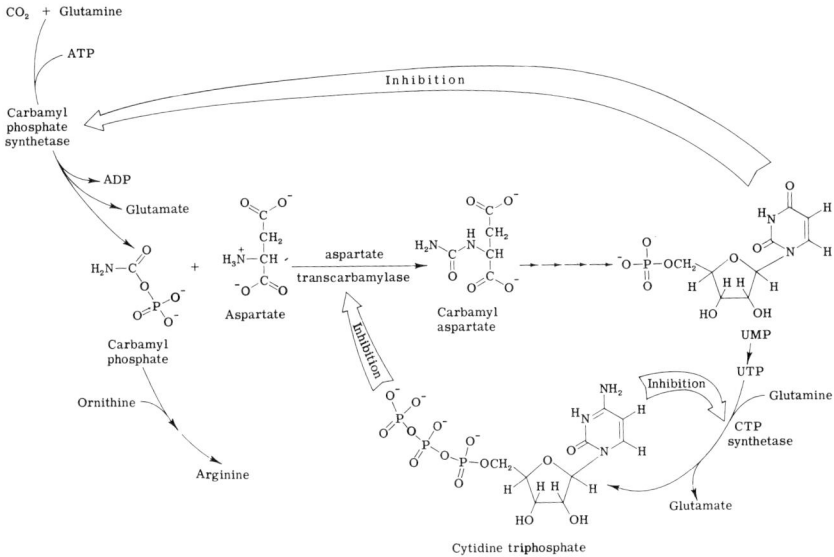

FIG. 1. Regulatory circuits governing pyrimidine biosynthesis at the level of enzymic activity in *E. coli*. Heavy open arrows signify the inhibitory action of a specific metabolite on one of the enzymes of the pathway. From Gerhart (*6*).

breadth as well as in depth but have concentrated most on those facts not dealt with extensively by others, especially a detailed discussion of the catalytic mechanism. The enzyme from *E. coli* is the most studied ATCase, primarily because it is readily available in large amounts as a homogeneous protein and because of its interesting subunit structure. Much more limited information is available for ATCases from other organisms, which are discussed briefly in Section IX of this review.

II. Isolation and Characterization of the Native Enzyme from *Escherichia coli*

A. PROPERTIES ASSOCIATED WITH REGULATION

Much of the early work concerning the regulation of ATCase has been extensively reviewed by Gerhart (*6*) and will only be outlined briefly here. Yates and Pardee (*7, 8*) were the first to show that ATCase is under both feedback control by derivatives of cytidine and repression

7. R. A. Yates and A. B. Pardee, *JBC* **221**, 757 (1956).
8. R. A. Yates and A. B. Pardee, *JBC* **227**, 677 (1957).

of its biosynthesis by uracil (or a metabolite of uracil). Purification of the enzyme from *E. coli* by Shepherdson and Pardee (*9*), utilizing a pyrimidine auxotroph capable of synthesizing up to 7% of its protein as ATCase, permitted a more extensive examination of the regulatory properties of the enzyme by Gerhart and Pardee (*10*). Purified ATCase was inhibited by a variety of pyrimidine nucleotides, but the most striking inhibition was exhibited by CTP. Adenosine triphosphate antagonized the CTP effect and, in the absence of CTP, ATP could actually activate the enzyme. In addition, when the activity of the enzyme was tested at various concentrations of L-aspartate, a sigmoidal curve was obtained, indicating cooperative binding of this substrate. Treatment with heat, urea, or mercurials abolished both the CTP effect and the sigmoidal binding of L-aspartate simultaneously without impairing catalysis. It was concluded that the binding site for CTP was probably distinct from the active site, and thus ATCase became a classic (although, as we shall see, unusual) example of an allosteric enzyme.

B. PURIFICATION

After Shepherdson and Pardee had purified ATCase to homogeneity (*9*), Gerhart and Holoubek (*11*) devised a new purification procedure which gives enzyme of slightly higher specific activity in much better yield. A special strain of *E. coli* was selected which has only limited ability to synthesize pyrimidines because of a mutational defect in the gene coding for orotidylate decarboxylase. The strain is also diploid for the region containing the genes for ATCase. Under conditions of depression (absence of exogenous pyrimidines), this strain produces 8% of its protein as ATCase. The purification procedure yields approximately 5 g of native ATCase from 700 g of cells and has been used widely to produce quantities of the enzyme sufficiently large to support the broad series of investigations currently under way. Large quantities of the enzyme have also been purified at the New England Enzyme Center using essentially the Gerhart and Holoubek procedure.

C. ASSAY PROCEDURES

In the early work, detection of ATCase activity depended upon colorimetric determination of carbamyl aspartate, using the procedure of

9. M. Shepherdson and A. B. Pardee, *JBC* **235**, 3233 (1960).
10. J. C. Gerhart and A. B. Pardee, *JBC* **237**, 891 (1962).
11. J. C. Gerhart and H. Holoubek, *JBC* **242**, 2886 (1967).

Koritz and Cohen (*12*) for determining carbamyl amino acids, modified (*13*) for increased sensitivity toward carbamyl aspartate. More recently, a colorimetric procedure has been developed which is sensitive to nanomole amounts of carbamyl aspartate and which seems superior to earlier methods both in ease and reproducibility (*14*). The method employs a color reagent containing antipyrine (1,5-dimethyl-2-phenyl-3-pyrazolone) and diacetylmonoxime in H_2SO_4 solution into which aliquots may be pipetted directly. Mercaptoethanol interferes and must be diluted out before the determination. Care must also be taken to avoid doing too many assays at once (unpublished observations of G. Jacobson) since some color development takes place even before the incubation at 60° and since the developed samples are light-sensitive (*14*).

Two radioisotope assays have been developed for ATCase (*15, 16*). Porter et al. (*15*) modified a method originally used by Yates and Pardee (*17*) in which ^{14}C-labeled L-aspartate is used. The product, ^{14}C-labeled carbamyl aspartate, passes through a small Dowex-50 column, while unreacted substrate is bound. This method, like the succeeding one, can be made more or less sensitive, depending upon the specific activity of the ^{14}C-labeled substrate. It suffers somewhat from the necessity of running a small Dowex-50 column for each assay and drying the eluates overnight before counting them. However, with crude bacterial extracts, this assay is complicated by the presence of an interfering activity which results, even in the absence of carbamyl-P, in the formation of radioactive products which are not retained by the column (*18*). A modification which avoids the inconvenience of running columns has been proposed recently by Ong and Jackson (*19*). [^{14}C]Carbamyl aspartate is separated by high voltage paper electrophoresis and then counted.

The second radioisotope assay employs carbamyl-P as the labeled substrate (*16, 20, 21*). In this procedure, aliquots are pipetted directly into scintillation vials containing acetic acid; the vials are then placed in a warm air oven for a few hours to remove unreacted label as $^{14}CO_2$

12. S. B. Koritz and P. P. Cohen, *JBC* **209**, 145 (1954).
13. R. Crokaert and E. Schram, *Bull. Soc. Chim. Biol.* **40**, 1093 (1958).
14. L. M. Prescott and M. E. Jones, *Anal. Biochem.* **32**, 408 (1969).
15. R. W. Porter, M. O. Modebe, and G. R. Stark, *JBC* **244**, 1846 (1969).
16. G. E. Davies, T. C. Vanaman, and G. R. Stark, *JBC* **245**, 1175 (1970).
17. R. A. Yates and A. B. Pardee, *JBC* **221**, 743 (1956).
18. M. Syvanen, Ph.D. Thesis, University of California, Berkeley, 1972.
19. B. L. Ong and J. F. Jackson, *Anal. Biochem.* **42**, 289 (1971).
20. J. E. Young, M. D. Prager, and I. C. Atkins, *Proc. Soc. Exp. Biol. Med.* **125**, 860 (1967).
21. M. R. Bethell, K. E. Smith, J. S. White, and M. E. Jones, *Proc. Nat. Acad. Sci. U. S.* **60**, 1442 (1968).

and then counted. The only remaining isotope is in the form of [^{14}C] carbamyl aspartate. In our experience, this method has been the most convenient for routine assays and is at least as reproducible as any of the others.

A continuous assay, utilizing a pH stat to monitor proton release during the ATCase reaction, has also been used [(*22*); F. Quiocho, unpublished observations; C-W. Wu and G. G. Hammes, unpublished observations] and shows promise of filling the need for a fast method for routine assays. The method is somewhat laborious, since only one sample can be assayed at a time, and is restricted to a narrow range of pH, near pH 8.3.

D. Size and Subunit Composition

The molecular weight of native ATCase has been studied by a variety of methods (*23*, *24*). Gerhart and Schachman (*23*) found that purified ATCase sediments as a single component with an $s_{20,w}$ of 11.7. A molecular weight of 310,000 was calculated from sedimentation equilibrium studies. Similar experiments by Rosenbusch and Weber (*24*) have confirmed the value of 11.7 S and placed the molecular weight between 290,000 and 310,000. As we shall see, these values are consistent with the subunit structure of the enzyme.

As mentioned above, Gerhart and Pardee (*10*) noticed that certain treatments were able to eliminate the sensitivity of ATCase to CTP without eliminating catalytic activity. In 1965 Gerhart and Schachman (*23*) were able to show that one of these treatments, reaction of native ATCase with *p*-mercuribenzoate, dissociates the enzyme into two distinct kinds of separable subunits: one which retains catalytic activity but is insensitive to CTP, and one which retains the ability to bind the CTP analog BrCTP but is inactive. The larger, catalytically active subunit (C subunit) has a molecular weight of about 100,000 and comprises 63% of the total weight of ATCase; the smaller, BrCTP, binding subunit (R subunit) has a molecular weight of about 34,000 and comprises 37% of the total weight of ATCase. If the two subunits are mixed together under appropriate conditions. The catalytic activity of the mixture is again sensitive to CTP, indicating that the subunits can

22. J. C. Gerhart, Ph.D. Thesis, Princeton University, Princeton, New Jersey, 1962.
23. J. C. Gerhart and H. K. Schachman, *Biochemistry* **4**, 1054 (1965).
24. J. P. Rosenbusch and K. Weber, *JBC* **246**, 1644 (1971).

reassociate into a structure like that of the native enzyme and further confirming a regulatory role for the R subunit.

Gerhart and Holoubek (11) have provided a procedure for separating the subunits, utilizing p-mercuribenzoate and a DEAE-Sephadex column, and have tabulated various physical characteristics both of native ATCase and of the subunits isolated by this procedure. More recently, other treatments have been used to obtain subunits in pure form, including heat (25, 26), which selectively precipitates R subunit, and treatment with another mercurial, neohydrin (1-(3-chloromercuri-2-methoxypropyl) urea) which, in conjunction with a DEAE-cellulose step, is faster than the p-mercuribenzoate procedure and gives better yields of R subunit (27). In addition, the p-mercuribenzoate procedure has recently been modified to give a more homogeneous preparation of R subunit, free of mercury and containing zinc (26, 28).

III. Isolation and Characterization of Subunits

A. THE CATALYTIC SUBUNIT

1. Size and Substructure

Early evidence seemed to support a dimeric structure for the C subunit. Sedimentation equilibrium studies in $8 M$ urea (29), N-terminal analyses (29, 30), and equilibrium dialysis studies of the binding of the aspartate analog succinate to the C subunit in the presence of carbamyl-P (31) were all consistent with a subunit composed of two polypeptide chains, each with a molecular weight of 45,000–50,000. More recent evidence, however, is overwhelmingly in favor of a trimeric structure. Possible explanations for the earlier erroneous conclusions have been discussed by Gerhart (6).

In 1968, Weber (32) reinvestigated the molecular weight of the poly-

---

25. P. D. J. Weitzman and I. B. Wilson, *JBC* **241**, 5481 (1966).
26. J. P. Rosenbusch and K. Weber, *Proc. Nat. Acad. Sci. U. S.* **68**, 1019 (1971).
27. J. A. Cohlberg, V. P. Pigiet, and H. K. Schachman, *Biochemistry* **11**, 3396 (1972).
28. M. E. Nelbach, V. P. Pigiet, J. C. Gerhart, and H. K. Schachman, *Biochemistry* **11**, 315 (1972).
29. K. Weber, *JBC* **243**, 543 (1968).
30. G. L. Hervé and G. R. Stark, *Biochemistry* **6**, 3743 (1967).
31. J.-P. Changeux, J. C. Gerhart, and H. K. Schachman, *Biochemistry* **7**, 531 (1968).
32. K. Weber, *Nature (London)* **218**, 1116 (1968).

peptide chains of the C subunit, using sodium dodecyl sulfate–polyacrylamide gel electrophoresis. He found the molecular weight of one chain to be 33,000, a value which agreed very well with the amount of leucine released from the C subunit upon digestion with carboxypeptidase A. A molecular weight of 33,000 has been reconfirmed by Rosenbusch and Weber (26) by a variety of techniques. Therefore, each C subunit of molecular weight 100,000 is a trimer.

Other independent evidence supports this conclusion. When denatured C subunit and denatured succinyl C subunit are mixed and the denaturant is removed, four species are found upon electrophoresis (33). Such a pattern would be expected for a trimeric structure (one unmodified species, two hybrids, and one fully modified species). The hybrids have one-third and two-thirds of the specific activity of unmodified C subunit, whereas the fully modified derivative is inactive, consistent with hybrids containing one or two "good" C chains per trimer, respectively.

Cross-linking of the C subunit with dimethyl suberimidate yields a banding pattern on sodium dodecyl sulfate–polyacrylamide gels consistent with a trimeric structure (34). Titration of the C subunit with thiol reagents yields a value of one cysteinyl group per 33,000 daltons (35), and titration of the C subunit with the transition state analog N-(phosphonacetyl)-L-aspartate (PALA) also reveals one binding site per 33,000 daltons (36). The C subunit binds one molecule of carbamyl-P per 33,000 daltons in the presence (37) or absence (38) of succinate. Finally, electron microscopy of isolated C subunit shows a triangular structure, again consistent with the trimeric model for C subunit (39).

2. Primary Structure

The amino acid composition of C subunit has been published (29, 40, 41), but the sequence is not yet known although work on it is well along (W. Konigsberg, personal communication). End group analysis revealed

33. E. A. Meighen, V. Pigiet, and H. K. Schachman, *Proc. Nat. Acad. Sci. U. S.* **65**, 234 (1970).
34. G. E. Davies and G. R. Stark, *Proc. Nat. Acad. Sci. U. S.* **66**, 651 (1970).
35. T. C. Vanaman and G. R. Stark, *JBC* **245**, 3565 (1970).
36. K. D. Collins and G. R. Stark, *JBC* **246**, 6599 (1971).
37. G. G. Hammes, R. W. Porter, and C.-W. Wu, *Biochemistry* **9**, 2992 (1970).
38. J. Rosenbusch and J. H. Griffin, personal communication (1973).
39. K. E. Richards and R. C. Williams, *Biochemistry* **11**, 3393 (1972).
40. J.-P. Changeux and J. C. Gerhart, *in* "The Regulation of Enzyme Activity and Allosteric Interactions" (E. Kvamme and A. Pihl, eds.), p. 13. Academic Press, New York, 1968.
41. W. F. Benisek, *JBC* **246**, 3151 (1971).

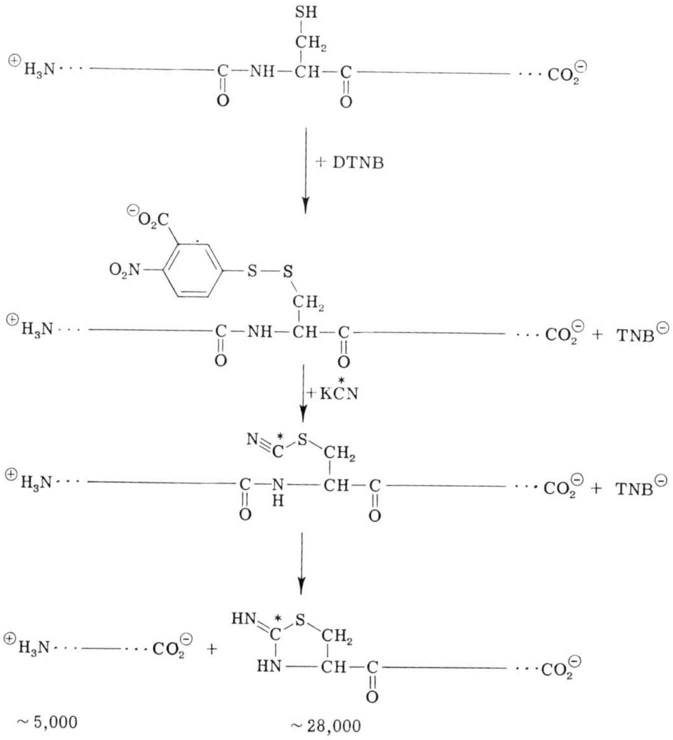

7. ASPARTATE TRANSCARBAMYLASES

FIG. 2. Specific chemical cleavage of the C chain of ATCase at the cysteinyl residue, according to Vanaman and Stark (*35*). See text for details.

only 2 NH_2-terminal alanine residues per 100,000 daltons (*29, 30*). No other NH_2-terminal or NH_2-blocked residues have been detected. The NH_2-terminal residues of the R subunit are also low, and the principal residue identified is either methionine or threonine, depending on the preparation of enzyme (*29, 30, 32*). Perhaps under conditions for derepression when a large amount of a single protein is being synthesized, as in the *E. coli* mutant used for production of ATCase, the enzyme systems responsible for removing formylmethionine from the nascent chains (*42*) are overwhelmed, resulting in variable amounts of methionine and formylmethionine in place of the NH_2-terminal group which would have been found in the wild type cell. Such a possibility has not yet been carefully investigated for ATCase.

The C subunit contains one free SH group per chain, and various derivatives of this group have been made (*35*). One such derivative is

42. D. Housman, D. Gillespie, and H. F. Lodish, *JMB* **65**, 163 (1972).

the thiocyanate, made with DTNB and ^{14}C-labeled KCN, as shown in the first two steps of Fig. 2. Under denaturing conditions, the thiocyanate cleaves nearly quantitatively (step 3 of Fig. 2) to yield an unlabeled NH$_2$-terminal peptide of molecular weight about 5000 and a radioactive COOH-terminal peptide of molecular weight about 28,000. This result places the SH group about 50 residues in from the NH$_2$-terminal alanine. Specific chemical cleavage at cysteinyl residues with KCN has been studied previously by Wood and Catsimpoolas (43, 44). The method illustrated in Fig. 2 has been extended to several other proteins including the R subunit (45).

B. THE REGULATORY SUBUNIT

1. *Size and Substructure*

The initial experiments of Gerhart and Schachman (23) indicated that R-subunit sediments with an $s_{20,w}$ of 2.8 and has a molecular weight of 30,000, as judged by sedimentation-equilibrium analysis. However, Weber (32) determined the amino acid sequence of the R subunit and found that it corresponds to a molecular weight of 17,000 daltons per R chain. This value was confirmed by sodium dodecyl sulfate–polyacrylamide gel electrophoresis and carboxypeptidase analysis of the COOH-terminus (32) and by sedimentation equilibrium studies in guanidine hydrochloride (24).

Recently, Nelbach et al. (28) and Rosenbusch and Weber (26) have examined the metal content of the R subunit after using various procedures to dissociate native ATCase. It has been clear for some time the enzyme contains tightly bound zinc (6, 28). Two independent studies have shown that native ATCase contains six molecules of Zn^{2+} per 300,000 daltons (26, 28) and that under appropriate conditions the isolated R subunit can be shown to bind zinc. No metal is required for the catalytic activity since fully active C subunit prepared by the method of Gerhart and Holoubek (11) contains much less than stoichiometric quantities of zinc, mercury, or other metals (26, 28, 46). Nelbach et al. (28) have also shown that the R subunit, isolated by the p-mercuribenzoate procedure (11), contains variable amounts of mercury, accounting for the variable values of the extinction coefficient at 280 nm for

43. J. L. Wood and N. Catsimpoolas, *JBC* **238**, PC2887 (1963).
44. N. Catsimpoolas and J. L. Wood, *JBC* **241**, 1790 (1966).
45. G. R. Jacobson, M. H. Schaffer, G. R. Stark, and T. C. Vanaman, *JBC* (1973) (in press).
46. P. G. Schmidt, G. R. Stark, and J. D. Baldeschwieler, *JBC* **244**, 1860 (1969).

different R-subunit preparations. The mercury–carbon bond in the mercaptide initially formed with p-mercuribenzoate is labile, and the aromatic ring is easily lost, leaving only mercury attached to the protein. Both Nelbach et al. (28) and Rosenbusch and Weber (26) have worked out procedures for isolating R subunit completely free of mercury and containing two atoms of zinc per 34,000 daltons. Cohlberg et al. (27) have shown that such a preparation of R subunit sediments as a single species with an $s_{20,w}$ of 2.84 and a molecular weight of 34,000 and that the form of the R subunit studied by Gerhart and Schachman (23), which contains mercury, is in a monomer–dimer equilibrium with a monomer molecular weight of 17,000. They also were able to cross-link the zinc R subunit by the method of Davies and Stark (34) and obtained as the predominant species a diffuse band of molecular weight 34,000, as judged by sodium dodecyl sulfate–polyacrylamide gel electrophoresis.

Metals have been removed completely from isolated R-subunit either by treating the p-mercuribenzoate-dissociated subunit with a chelating agent (26, 28) or by denaturation, gel filtration, and renaturation of the R chains (26). Metal-free R subunit, like the mercury-containing R subunit obtained directly from p-mercuribenzoate dissociation, appears to be much less stable than R subunit which contains a full complement of zinc (26, 28). In addition, removal of the zinc to form apo R subunit has the effect of destabilizing the subunit dimers to a mixture of monomers and dimers (27). Metal appears to be important in stabilizing the dimeric form of the isolated R subunit and perhaps in making it more resistant to denaturation.

2. Primary Structure

The amino acid sequence of the R subunit, as reported by Weber (32), is

NH$_2$-Met-Thr-His-Asn-Asp-Lys-Leu-Gln-Val-Ala-Glu-Ile -Lys-Arg-Gly-
Thr-Val-Ile -Asn-His-Ile -Pro-Ala-Glu-Ile -Gly-Phe-Lys-Leu-Leu-Ser-
Leu-Phe-Lys-Leu-Thr-Glu-Thr-Gln-Asp-Arg-Ile -Thr-Ile -Gly-Leu-Asn-
Leu-Pro-Ser-Gly-Glu-Met-Gly-Arg-Lys-Asp-Leu-Ile -Lys-Ile -Glu-Asn-
Thr-Phe-Leu-Ser-Glu-Asx-Glx-Val-Asx-Glx-Leu-Ala-Leu-Tyr-Ala-Pro-
Gln-Ala-Thr-Val-Asn-Arg-Ile -Asn-Asp-Tyr-Glu-Val-Val-Gly-Lys-Ser-
Arg-Pro-Ser-Leu-Pro-Glu-Arg-Asn-Ile -Asp-Val-Leu-Val-Cys-Pro-Asp-
Ser-Asn-Cys-Ile -Ser-His-Ala-Glu-Pro-Val-Ser-Ser-Ser-Phe-Ala-Val-
Arg-Arg-Ala-Asx-Asx-Ile -Ala-Leu-Lys-Cys-Lys-Tyr-Cys-Glu-Lys-Glu-
Phe-Ser-His-Asn-Val-Val -Leu-Ala-Asn-CO$_2$H

As noted earlier, the NH$_2$-terminal residue can be methionine, threonine, or a combination of the two, depending on the preparation (32), and quantitative estimates of the NH$_2$-terminus have been low.

3. The Metal Binding Site

The studies of Nelbach et al. (28) and Rosenbusch and Weber (26) have provided some clues about the nature of the metal binding site in the R subunit. Spectral evidence strongly implicates one or more of the 4 cysteinyl residues as a ligand for the bound metal. R subunit, which has been completely carboxymethylated on its SH groups, fails to bind zinc at all (26, 27). Furthermore, p-mercuribenzoate dissociation, which presumably occurs in an all or none fashion by reaction with the R-subunit SH groups (47), produces zinc-free R subunit which contains mercury (28). R subunit containing 1.1 mercury atoms per 17,000 daltons, made from metal-free apo R subunit, has spectral properties almost identical with those of p-mercuribenzoate-dissociated R subunit (28).

The SH groups of R subunit are present as two groups of two cysteinyl residues each, close together in the sequence. Either or both of these dithiol segments is a good candidate for a metal binding site since dithiols complexed with Hg^{2+}, in contrast to monothiol-Hg^{2+} complexes, show absorption spectra very similar to those of the R subunit–Hg complex (28). Similar suggestions have been made by Rosenbusch and Weber (26). All the data are consistent with a model in which the Zn^{2+} of the R subunit in native ATCase is displaced by p-mercuribenzoate, with substitution of Hg^{2+} as the metal to which the SH groups are liganded. Both the Zn^{2+} and Hg^{2+} derivatives of the R subunit are able to interact in a similar way with isolated C subunit to form functional ATCase.

4. R Subunit as a Regulatory Protein

Gerhart and Schachman (23) were the first to show that the ability of native ATCase to bind the CTP analog BrCTP could be completely accounted for by the isolated R subunit, and Changeux et al. (31) further studied the binding of BrCTP and CTP to the R subunit by equilibrium dialysis. Winlund and Chamberlin (48) have found two classes of sites for the binding of CTP to native ATCase at 4°. The homogeneity of these sites with respect to the dissociation constant for CTP for both the R subunit and native ATCase is discussed in more detail in Section VII.

Using the method of continuous variation, Hammes et al. (37) showed that the R subunit binds one BrCTP per 17,000 daltons and that native ATCase binds approximately six molecules of BrCTP per 310,000 daltons in the presence of carbamyl-P. Using a different procedure, Rosenbusch and Weber (26) also obtained one CTP site per 17,000 daltons for various

---

47. J. C. Gerhart and H. K. Schachman, *Biochemistry* **7**, 538 (1968).
48. C. C. Winlund and M. J. Chamberlin, *BBRC* **40**, 43 (1970).

preparations of the R subunit. The experiments of Cohlberg et al. (27) suggest that the dimeric form of R subunit binds CTP more strongly than the monomer. Thus, it appears that the R subunit probably contains one CTP binding site per peptide chain of 17,000 daltons.

Metal does not seem to be necessary for the binding of CTP to the isolated R subunit. Cohlberg et al. (27) found that removal of all metal to form apo R subunit increased the dissociation constant for CTP only about 10-fold. Using apo R subunit prepared somewhat differently, Rosenbusch and Weber (26) found 0.8 binding site per 17,000 daltons with approximately unchanged affinity for CTP. However, their preparation was somewhat contaminated with metal ions, making an exact evaluation of these numbers difficult. The metal itself is probably not necessary for CTP binding, but its absence may reduce the affinity for CTP. This idea is supported by the finding of Cohlberg et al. (27) that the R subunit dimer, which is the predominant form in zinc-containing R subunit but is only one of several components in apo R subunit, binds CTP more strongly than monomeric R subunit.

IV. Reconstitution of Native ATCase from the Isolated Subunits

A. METHODS OF REASSOCIATION

Spontaneous reassociation of the C and R subunits after dissociation with p-mercuribenzoate was first accomplished by Gerhart and Schachman (23) by treating the unfractionated reaction mixture with β-mercaptoethanol. The reassociated enzyme had a sedimentation coefficient and catalytic activity close to those of native ATCase and was inhibited by CTP. Only 80–90% of the dissociated protein recombined. Some faster sedimenting species were detected and were assumed to be aggregates. More recently, Rosenbusch and Weber (24) have been able to renature both C subunit and native ATCase from solutions of each which have been denatured with guanidine hydrochloride. The renatured native enzyme retains all of the normal allosteric properties of untreated ATCase, indicating a high degree of fidelity during the refolding and recombination steps.

B. IMPORTANCE OF METALS IN REASSOCIATION

Metal ions bound to the R subunit or present in the reassociation mixture appear to be essential for reconstitution of native ATCase from its separated subunits (26, 28). Preparations of metal-free R subunit fail to recombine with C subunit, whereas addition of zinc or other metals (see below) to apo R subunit restores its ability to recombine. Preparations of R subunit derived from p-mercuribenzoate-dissociated native ATCase contain varying amounts of mercury and no zinc and also exhibit varying efficiencies of recombination with C subunit (49).

The metal has the ability both to stabilize the R-subunit dimer and to promote its recombination with C subunit. It is not yet possible, however, to draw firm conclusions about the location of the metal site relative to the site of interaction between the R chains (R:R bonding domain) in the dimeric R subunit or the R:C bonding domains in native ATCase. As discussed below, the dimeric R subunit is probably the form which recombines with C subunit, and this dimer seems to be retained in the native enzyme that is formed. It is quite clear, however, that the form of accessibility of the metal binding site differs considerably in native enzyme and in dimeric R subunit. Attempts to remove zinc from native ATCase under conditions which completely remove the metal from R subunit have failed (28, 49).

C. Substitution of Other Metals for Zinc

Zinc may be replaced by Hg^{2+} or Cd^{2+} in isolated R subunit simply by adding mercury or cadmium ions to a solution of freshly prepared apo R subunit (26, 28). The derivatives contain stoichiometric equivalents of the metals and are fully capable of recombining with C subunit. The enzyme thus obtained is very similar in physical, catalytic, and regulatory properties to native ATCase containing zinc. In addition, ATCase containing Cd^{2+} instead of Zn^{2+} may be obtained by growing the bacteria from which the enzyme is to be purified in a medium containing cadmium instead of zinc. Such an enzyme is identical to enzyme prepared from cadmium-containing R subunit and contains six atoms of Cd^{2+} per 300,000 daltons (26).

Nelbach et al. (28) have reported that Co^{2+}, Ni^{2+}, and Cu^{2+} also promote aggregation of apo R subunit with C subunit to form an enzyme with the same electrophoretic mobility as native ATCase. However, these recombined species were not tested for actual incorporation of the respective metals. These is some evidence that recombination may occur in the absence of intentionally added metals, although the possibility of metal contamination in such an experiment is difficult to rule out. When a preparation of apo R subunit which recombined efficiently with Zn^{2+} and C subunit was incubated with Mn^{2+} and C subunit under the same

49. M. E. Nelbach, Ph.D. Thesis, University of California, Berkeley, 1972.

conditions, a low yield of a product electrophoretically indistinguishable from native ATCase was obtained. However, the product contained less than 10% of the expected amount of Mn^{2+} by atomic absorption and little Zn^{2+} as well (S. Seaver, unpublished results). Confirmation that derivatives of native ATCase containing metals other than Zn^{2+}, Cd^{2+}, or Hg^{2+} have been prepared must await their full characterization. Derivatives containing any paramagnetic ion (such as Co^{2+}) would be extremely useful in a study by electron spin resonance spectroscopy of the structure of the metal site and the changes produced when various ligands become bound.

V. Detailed Subunit Structure of Native ATCase

Given the molecular weights of native ATCase (300,000 ± 10,000) and of each polypeptide chain (R = 17,000, C = 33,000) (*32*) and the weight fraction of each type of chain in the native enzyme (*23*), one can calculate that the native structure should contain six chains of each type, i.e., $(RC)_6$. If the C subunits are present as trimers and the R subunits as dimers, then the native structure can be written as $(C_3)_2(R_2)_3$, i.e., as a dimer of C subunits and a trimer of R subunits.

The most detailed information about the arrangement of R-subunit dimers and C-subunit trimers in native ATCase comes from the X-ray crystal structure of native ATCase at 5.5 Å resolution (*50*) and from electron micrographs of native ATCase and the C subunit (*39*). Both methods lead to a model like the one shown in Fig. 3, which has been reproduced from the paper of Cohlberg *et al.* (*27*), although the model derived from the most recent crystallographic data does differ in some details (see Fig. 20). The C-subunit trimers are eclipsed and are connected by R-subunit dimers which join to C chains displaced by 120° in different trimers.

The electron micrographs show for native ATCase (Fig. 4A) a solid triangular structure, 95 Å on a side, inscribed within a larger equilateral triangle, 145 Å on a side, with the outer triangle rotated 60° with respect to the inner one. Another method of preparation (Fig. 4B) shows what is taken to be a "side" view of the molecule with two equivalent, somewhat crescent-shaped segments separated by 20–40 Å. The picture of the C subunit (Fig. 4C) is almost exactly like that of the same view

50. D. C. Wiley, D. R. Evans, S. G. Warren, C. H. McMurray, B. F. P. Edwards, W. A. Franks, and W. N Lipscomb, *Cold Spring Harbor Symp. Quant. Biol.* **36**, 285 (1971).

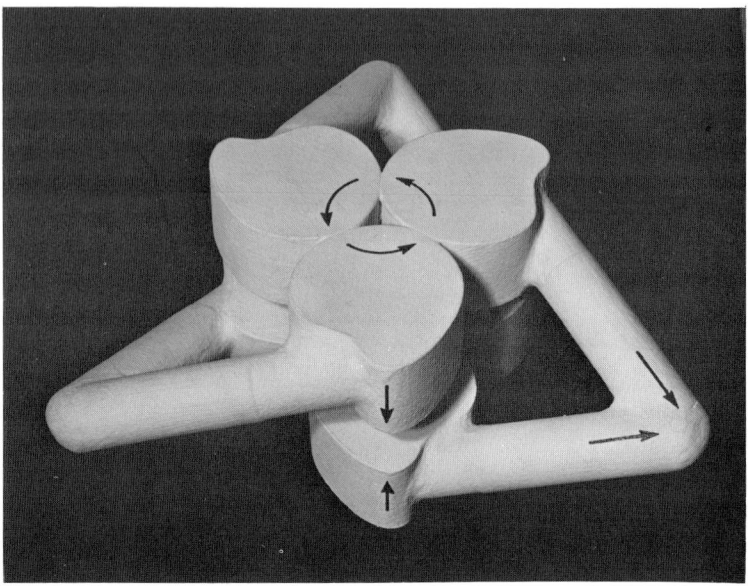

Fig. 3. A model for the arrangement of polypeptide chains in ATCase, courtesy of H. K. Schachman [see Cohlberg et al. (27)]. Each catalytic subunit is shown as a triangular array of three asymmetrically shaped C chains. Arrows on the upper face indicate the heterologous association of the C:C bonding domains; arrows on the side faces of the C chains indicate the isologous relationship of the C-chain trimers. Each R subunit is shown as a pair of cylindrical R chains oriented to one another at an acute angle; arrows on the cylinders illustrate the isologous association of the R:R bonding domains.

of native ATCase in dimensions and shape but *without* the outer, staggered triangular structure. Thus, this latter structure has been taken to represent the R subunit within the native enzyme. Because the C-subunit portion of the native structure is identical in appearance to isolated C subunit, the C-subunit trimers are presumed to lie on top of one another (shown also by the side view in Fig. 4B), each chain presumably located at the vertex of the smaller triangle. The R-subunit dimers presumably do not connect C chains which lie directly above one another, but rather they connect C chains which are 120° apart on different trimers. Whether or not the dimers connect the centers of these chains (as indicated in the model) is impossible to tell, but the fact that the R subunits show up as vertices of a triangular structure which is rotated by 60° with respect to the C subunit triangle indicates that the regulatory dimers probably do not sandwich in between the catalytic trimers as proposed in the model of Rosenbusch and Weber (*24*).

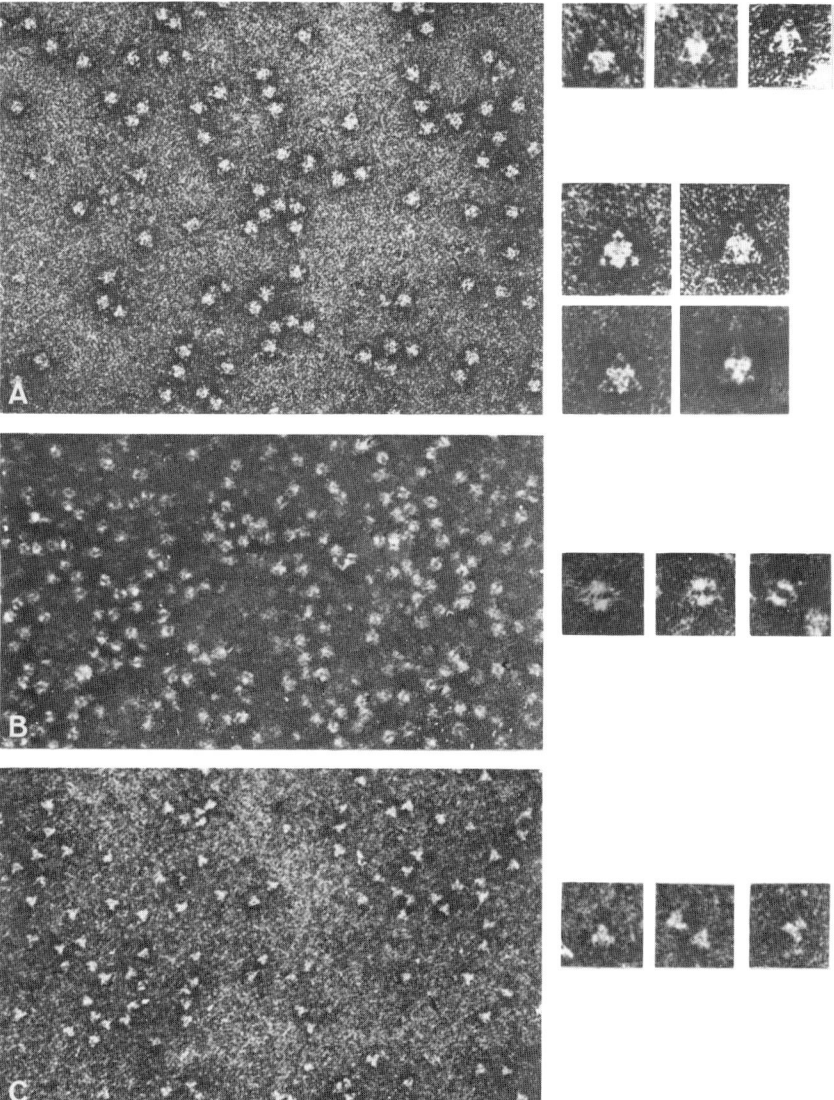

Fig. 4. Electron micrographs of ATCase and the C subunit, courtesy of Richards and Williams (39). (A) ATCase molecules, deposited by method I of Richards and Williams (39). 1. Field at ×200,000. 2. Particles selected to show a solid triangular structure within a larger triangle mainly delineated by its edges. The former shows evidence of a closed trimeric substructure. ×400,000. 3. Composite views. Each view is a photographic montage of the images of five particles selected from the same electron micrograph. ×400,000. (B) ATCase molecules deposited by method II of Richards and Williams (39). 1. Field at ×200,000. 2. Particles selected for their appearance of twofold symmetry. Faintly evident is material extending beyond the limits of the dense portions. ×400,000. (C) The C subunit of ATCase. 1. Field at ×200,000. 2. Particles selected to show triangular form and evidence of closed trimeric substructure. ×400,000.

Aspartate transcarbamylase has been crystallized in a variety of forms including tetragonal (9, 51) trigonal (51), octahedral (11, 51), and canoe-shaped crystals (11). More recently, a form suitable for X-ray crystallography has been found (50). Previous crystallographic data (51) had established both a 3-fold and a 2-fold axis of symmetry in the molecule. These results are inconsistent with any model of the enzyme which does not contain three or a multiple of three of each kind of subunit and, along with the independent results of Weber (32), were the first to suggest a C_6R_6 structure for native ATCase.

The X-ray data at 5.5 Å (50) reveal an approximately triangular molecule when viewed down the 3-fold axis, with one side of the triangle having a length of 105 ± 10 Å. The length of the molecule along this 3-fold axis is 92 ± 10 Å. Further analysis of the data reveals a large hole at the center of the molecule (3,2 symmetry position). The three mercury atoms of a heavy metal derivative are probably located near the active site of each C chain (see Section VI,B). The mercury atoms are placed at 120° intervals around the center, about 22 Å apart. The R-chain dimers are seen as connecting the catalytic trimers, as depicted in Fig. 3, and appear above and below the plane of the twofold axes of the molecule. A model based on the most recent crystallographic data is depicted in Fig. 20 in Section VII,H.

Crystallographic data at higher resolution will be required to reveal the detailed structures. Wiley et al. (50) commented that the two C-subunit trimers are probably not in contact. The electron microscope side views indicate a space of 20–40 Å between the trimers, but the uncertainties of micrograph interpretation, resulting especially from changes in structure as artifacts of sample preparation, make a definite conclusion very uncertain. It should be emphasized that both the crystallographic data and the electron micrographs were obtained with unliganded enzyme. Since major conformational changes occur in the presence of ligands (see below), many structural features of the native enzyme may change dramatically when ligands are added.

Recent chemical evidence indicates that the substructure of each isolated subunit is preserved in the native enzyme. When an equal mixture of fully succinylated C subunits (C_S) and unmodified C subunits (C_N) was allowed to reassociate with R subunits, three electrophoretically distinct species were found, corresponding to $(C_S)_2(R)_3$, $(C_S)(C_N)(R)_3$, and $(C_N)_2(R)_3$ (33). If rearrangement of each C-subunit trimer had occurred upon incorporation into the native structure, many more bands would have been evident. This result argues strongly that the C-subunit

51. D. C. Wiley and W. N. Lipscomb, *Nature* (London) **218**, 1119 (1968).

trimer is not produced from some other arrangement in native ATCase upon dissociation. Other logic for this argument has been discussed by Rosenbusch and Weber (24). Cohlberg et al. (27) have cross-linked R-subunit containing one zinc molecule per 17,000 daltons, fractionated the reaction mixture to obtain dimers, and then recombined the cross-linked dimers with C subunit. A high yield of molecules like native ATCase was obtained. Cross-linking of native ATCase, followed by sodium dodecyl sulfate–polyacrylamide gel electrophoresis reveals a major band corresponding to R-subunit dimers (G. E. Davies and G. R. Stark, unpublished observations). Finally, native and succinylated R subunits, when recombined with C subunits, form a hybrid set of native molecules consistent with three R-subunit combining units which do not dissociate into their component R chains during recombination (G. Nagel and H. K. Schachman, personal communication). The chemical evidence indicates that the dimeric R subunit isolated from native ATCase is present as a similar dimeric structure in the native enzyme.

VI. Mechanism of Catalysis for the *E. coli* Enzyme

A. DETAILED MECHANISM

The way in which the carbamyl group of carbamyl-P is thought to be transferred to the amino group of L-aspartate by ATCase is illustrated in outline form in Fig. 5. Individual steps in this process are described below, together with the supporting experimental evidence. It should be emphasized that, as usual, the mechanism has been arrived at by excluding alternatives rather than by complete and unambiguous proof of a single possibility. It seems likely that at least some of the details discussed below will be modified as more experimental evidence becomes available.

1. The carbamyl group is transferred directly from carbamyl-P to L-aspartate without the intermediate formation either of cyanate or of a carbamyl enzyme.

Carbamyl-P decomposes at neutral pH via the intermediate formation of cyanate and phosphate (52). Therefore, it seemed possible that ATCase might catalyze the conversion of carbamyl-P to cyanate in the presence of L-aspartate (53). This possibility is now rendered highly improbable since it is likely that the catalytic mechanism for acetyl-P,

52. C. M. Allen, Jr. and M. E. Jones, *Biochemistry* 3, 1238 (1964).
53. G. R. Stark, *Biochemistry* 4, 1030 (1965).

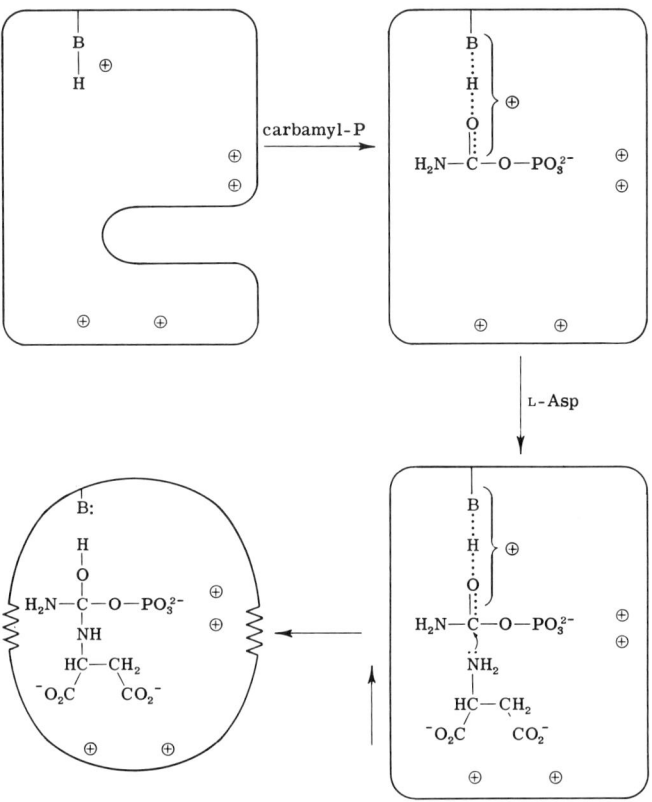

FIG. 5. Schematic representation of the compression mechanism for ATCase. (1) Unliganded enzyme with positively charged sites for substrate binding, the acid BH⁺ and a steric constraint on the binding site for L-aspartate. (2) Carbamyl-P binds and interacts with BH⁺ through its carbonyl oxygen. The steric constraint upon L-aspartate binding is relieved. (3) L-aspartate binds. Its α-amino group is in a position to react with the bound and activated carbamyl-P. (4) Compression of the two substrates together by a conformational change of the enzyme aids reaction. There is virtually no evidence for the tetrahedral intermediate; it is shown only for purposes of illustration.

a good substrate for ATCase (see below), is the same as for carbamyl-P. Yet acetyl-P cannot yield ketene by a mechanism analogous to any by which cyanate can be formed from carbamyl-P. Furthermore, M. O. Modebe and G. R. Stark (unpublished experiments) found that the C subunit fails to catalyze decomposition of carbamyl-P (measured by phosphate release) either in the presence or absence of succinate, a competitive inhibitor of L-aspartate. The intermediate formation of cyanate probably does not occur in the reaction catalyzed by frog liver

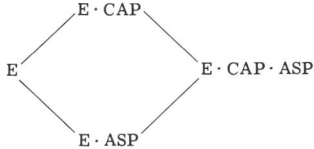

FIG. 6. Possible pathways leading to the enzyme–carbamyl-P–L-aspartate complex. A *random* mechanism would have significant contributions from both pathways. An *ordered* mechanism, with carbamyl-P binding first (upper pathway), is the predominant pathway for ATCase (see text).

ornithine transcarbamylase either, since added cyanate does not dilute the specific radioactivity of citrulline formed from ^{14}C-labeled carbamyl-P (54).

Reichard and Hanshoff (55) first showed that ATCase failed to catalyze exchange between $^{32}P_i$ and carbamyl-P unless L-aspartate was also present, and that exchange of [^{14}C]-L-Aspartate with carbamyl-L-aspartate was also dependent on the presence of carbamyl-P. On the basis of these results, they proposed direct transfer of the carbamyl group without the formation of a carbamyl enzyme intermediate. Porter et al. (15) repeated the experiment with $^{32}P_i$ and carbamyl-P with the C subunit in the presence and absence of succinate, and they achieved the same result. Furthermore, Schaffer and Stark (56) have recently shown that the C subunit is not labeled by [^{14}C]carbamyl-P in the presence of succinate and that the kinetic evidence of Porter et al. (15) is inconsistent with a ping pong mechanism. The evidence appears to rule out conclusively an intermediate carbamyl enzyme except if it is formed only in the ternary C-subunit–carbamyl-P–L-aspartate complex and not in the C-subunit–carbamyl-P–succinate complex.

2. In the major reaction path, carbamyl-P binds to the C subunit first, followed by L-aspartate; very little, if any, of the reaction can flow through the alternative path in which L-aspartate binds first.

A simple random mechanism for ATCase is illustrated in Fig. 6. Clearly, the two ordered mechanisms are special cases of the random mechanism in which one or the other of the two possible paths to the ternary complex is not used appreciably. Porter et al. (15) postulated that the mechanism is ordered, with carbamyl-P binding first, on the basis of the pattern of product inhibition of the C subunit. Although this method is capable in principle of distinguishing among the two ordered mechanisms and the random mechanism (57), the steady state

54. J. Carreras, A. Chabas, and S. Grisolia, *BBA* **250**, 456 (1971).
55. P. Reichard and G. Hanshoff, *Acta Chem. Scand.* **10**, 548 (1956).
56. M. H. Schaffer and G. R. Stark, *BBRC* **46**, 2082 (1972).
57. W. W. Cleland, "The Enzymes," 3rd ed., Vol. 2, p. 1, 1970.

kinetic experiments of Porter et al. (15), in which the concentration of carbamyl-P was varied, were carried out in the presence of rate-limiting concentrations of L-aspartate. (It was experimentally difficult to work with higher concentrations of this substrate in the presence of low concentrations of carbamyl-P since the Michaelis constants differ by about 1000-fold.) Under these conditions the reaction would have been forced through the upper branch of a random pathway (Fig. 6) and the distinction between this mechanism and the ordered one in which carbamyl-P binds first was obscured. However, an ordered mechanism in which L-aspartate binds first was definitely ruled out. The observation of Changeux et al. (31) that the binding of succinate to ATCase is greatly enhanced in the presence of carbamyl-P is also consistent either with a random mechanism or with an ordered mechanism in which carbamyl-P binds first, as these authors have pointed out. A clear decision between the two possibilities can be made with the use of an inhibitor which binds simultaneously to the sites for both substrates. N-(phosphonacetyl)-L-Aspartate is such an inhibitor (see Fig. 9), as shown by its tight binding, the ultraviolet difference spectrum of the C subunit in its presence, and its activation of native ATCase at low concentrations (36). Inhibition of the C subunit by PALA is competitive with carbamyl-P indicating that PALA and carbamyl-P compete for the same enzyme form (the free enzyme). Inhibition is noncompetitive with respect to L-aspartate, as predicted for an ordered mechanism in which carbamyl-P binds first. In the case of the random mechanism, saturation of the enzyme by either substrate (which occurs at the ordinate of the double reciprocal plots) eliminates the only form of the enzyme to which the inhibitor can bind (i.e., the free enzyme), and thus competitive inhibition is predicted for both substrates. Although we consider the random mechanism to be ruled out by these results with PALA, Drs. E. Heyde, A. Nagabhushanam, and J. F. Morrison have obtained kinetic evidence (unpublished) that they interpret in favor of a rapid equilibrium random addition of substrates.

The weak binding of succinate or L-aspartate to native ATCase or to the C subunit in the absence of carbamyl-P does appear to occur at the active site, on the basis of data concerning the immunologic reactivity and rate of p-mercuribenzoate-induced dissociation of native ATCase (58), enhancement of the rate of proteolysis of the R subunit in the native enzyme (59, 60), protection of the active site from chemical modification (41), and data from transient NMR measurements (61)

58. R. von Fellenberg, M. R. Bethell, M. E. Jones, and L. Levine, *Biochemistry* **7**, 4322 (1968).
59. D. K. McClintock and G. Markus, *JBC* **243**, 2855 (1968).
60. D. K. McClintock and G. Markus, *JBC* **244**, 36 (1969).

and difference spectroscopy (*62*). However, we argue that the kinetic pathway in which L-aspartate binds first makes an insignificant contribution to the observed rate. Other dicarboxylic acids can also bind weakly to the aspartate site, as shown, for example, by the dead-end complex formed with carbamyl-L-aspartate (*15*) and by the complexes with several dicarboxylic acids detected by difference spectroscopy (*62*).

3. Carbamyl-P binds to the C subunit rapidly through an initial interaction of its phosphate dianion with an easily accessible cationic site.

Porter (*15*) tested several phosphates and phosphonates as potential inhibitors competitive with carbamyl-P. No compound with the functional group $-PO_3^{2-}$ failed to inhibit the enzyme. The K_i values ranged from 0.09 mM (pyrophosphate) to 1.54 mM (N-methyl phosphonacetamide). Even CTP, a feedback inhibitor for native ATCase, is a competitive inhibitor of the C subunit ($K_i = .037$ mM). Therefore, a site (presumably cationic) which interacts with the phosphate dianion of carbamyl-P is easily accessible to a wide variety of compounds which bear a $-PO_3^{2-}$ group.

The rate constant for association of carbamyl-P with the C subunit, calculated from steady state kinetic data at pH 7.8 and 28°, is $\geq 5 \times 10^7$ M^{-1} sec^{-1} per binding site (*15*); the same constant calculated from temperature jump experiments is 2.4×10^8 M^{-1} sec^{-1} at pH 7.5 and 28° (*63*). These values are typical of rate constants for rapid association of other substrates and enzymes (*64*), and therefore they added to the impression that the carbamyl-P binding site is easily accessible.

4. A molecule of carbamyl-P bound to the C subunit is held rigidly through at least two interactions: one at the phosphate dianion and another at the carbonyl oxygen.

Schmidt et al. (*46*) used NMR spectroscopy to study the interaction of carbamyl-P analogs with the C subunit. The experimental conditions were selected so that the enzyme was saturated by a large excess of the ligand being studied. Exchange of the ligand between bound and free environments is fast enough to satisfy the limit $k_{off} \gg (1/T_{2\,EI})$ [where $(1/T_{2\,EI})$ is the NMR line width of the bound species], so that the line width observed in the presence of the C subunit is the average of the line widths for free and bound species, weighted by the relative amounts of each. Since the line widths for the bound ligands were large, major effects were observed even when there was a substantial excess of ligand

---

61. B. D. Sykes, P. G. Schmidt, and G. R. Stark, *JBC* **245**, 1180 (1970).
62. K. D. Collins and G. R. Stark, *JBC* **244**, 1869 (1969).
63. G. G. Hammes, R. W. Porter, and G. R. Stark, *Biochemistry* **10**, 1046 (1971).
64. M. Eigen and G. G. Hammes, *Advan. Enzymol.* **25**, 1 (1963).

TABLE I
NMR LINE WIDTHS FOR ANALOGS OF CARBAMYL-P BOUND TO
THE C SUBUNIT

Analog	Bound line width (Hz) per active site, 32° pH 6	pH 7
$\overset{O}{\underset{\|\|}{NH_2CCH_2PO_3}}$ (phosphonacetamide)	240 ± 22	67 ± 5
$CH_3PO_3^{2-}$ (methyl phosphonate)	0 ± 1	15 ± 2
$\overset{O}{\underset{\|\|}{CH_3NHCCH_2PO_3^{2-}}}$ (N-methyl phosphonacetamide)	(-CH$_2$-) 27 ± 5	32 ± 3
	(-CH$_3$) 23 ± 2	10 ± 2

over enzyme. An increase in line width in the presence of enzyme can result either from slow exchange, from immobilization of the ligand when bound, or from a combination of these effects. For the analogs of carbamyl-P studied, the exchange is rapid enough so that immobilization of the ligand when bound is probably responsible for a major portion of the observed line broadening. Some of the results obtained are given in Table I. Note particularly the large bound line width for phosphonacetamide, a close analog of carbamyl-P in which the anhydride oxygen is replaced by a methylene group, and the small bound line width for methyl phosphonate. The comparison indicates that, whereas the methyl group of $CH_3PO_3^{2-}$ is free to spin on the carbon–phosphorus bond when bound to the C subunit, the methylene group of $\overset{O}{\underset{\|\|}{NH_2CCH_2PO_3^{2-}}}$ is not. The most likely explanation is that some portion of the $NH_2C=O$ group of phosphonacetamide binds to the enzyme. By analogy, a similar interaction of the $NH_2C=O$ group of carbamyl-P is likely.

In carbamyl-P, the nonbonding electrons of the NH_2 group and anhydride oxygen are delocalized to such an extent that the carbonyl carbon atom is unreactive to nucleophiles at neutral pH (62). However, carbamyl-P does undergo acid-catalyzed hydrolysis by direct attack of water on a protonated species (52), a common mechanism for hydrolysis of carbonyl esters (65, 66). There is compelling evidence that carbonates (67), amides (68), and ureas (69) are all protonated predominantly on

65. M. L. Bender, *Chem. Rev.* **60**, 53 (1960).
66. W. P. Jencks, *Prog. Phys. Org. Chem.* **2**, 63 (1964).
67. G. A. Olah and M. Calin, *JACS* **90**, 401 (1968).
68. R. J. Gillespie and T. Birchall, *Can. J. Chem.* **41**, 148 (1963).
69. R. Stewart and L. J. Muenster, *Can. J. Chem.* **39**, 401 (1961).

the carbonyl oxygen; therefore, this site is the most probably one for protonation of carbamyl-P in acidic solution.

It seems highly probable on the basis of this sort of analogy and also on mechanistic grounds that the carbonyl group of carbamyl-P is the site of an interaction with an enzyme-bound acid (denoted BH⁺ in Fig. 5). Although it must be admitted that there is no *direct* evidence for such a hypothesis at present, the NH_2 group and anhydride oxygen of carbamyl-P, which might participate in a second interaction with the enzyme, can be eliminated from consideration since they are not essential for that interaction.

5. The NH_2 group of carbamyl-P is not required for activity. There is a severe steric restriction on the size of functional groups distal to the phosphate dianion.

Grisolia et al. (70) showed that extracts of *E. coli* could catalyze the concurrent disappearance of L-aspartate and acetyl-P and suggested that acetyl-P could replace carbamyl-P in the aspartate transcarbamylase reaction. Porter et al. (15) confirmed this suggestion with the purified C subunit. At pH 7.8 and 28°, acetyl-P is utilized 2.4% as fast as carbamyl-P when both are saturating. At lower pH, the relative maximum rate with acetyl-P is substantially increased, to about 25% of the maximum rate with carbamyl-P at pH 6.4. Therefore, the NH_2 of carbamyl-P is not required for activity and is an unlikely site for any crucial interaction with the C subunit.

N-Methyl carbamyl-P is an extremely poor substrate for the C subunit (V_{max} 0.03% of that with carbamyl-P at pH 7.8) and N,N-dimethyl carbamyl-P is not a substrate at all (15). Therefore, it appears that the *size* of the substituent adjacent to the carbonyl group is much more important than its chemical nature, implying that steric hindrance impedes full interaction of N-methyl carbamyl-P with the enzyme. This impression is strengthened by data presented in Table I. Since phosphonacetamide is larger than methyl phosphonate, it might be argued that the large bound line width for the former merely reflects steric interference to rotation. However, N-methyl phosphonacetamide, which is even larger, is much more free to rotate when bound. It is inferred that phosphonacetamide, acetyl-P and carbamyl-P all participate in a specific interaction with the enzyme, probably through the $-\overset{\overset{\displaystyle O}{\|}}{C}-$ group which they have in common. When a methyl group is added, as in N-methyl phosphonacetamide and N-methyl carbamyl-P, these interactions are largely prevented.

70. S. Grisolia, R. Amelunxen, and L. Raijman, *BBRC* **11**, 75 (1963).

6. A conformational change of the C subunit accompanies the binding of carbamyl-P.

Tight binding of analogs of L-aspartate to the C subunit occurs *only* if carbamyl-P (or an appropriate substitute) has been bound first. Logic requires that a conformational change of the enzyme occur upon the binding of carbamyl-P, since otherwise one would have to postulate that the carboxylate dianion binds *directly* to carbamyl-P in the ternary complex, which seems exceedingly unlikely.

This conformational change is not seen by relaxation measurements of the C-subunit–carbamyl-P system, monitored by changes in protonic equilibria, where only a single fast relaxation time is observed (*63*). Also, when the effect of phosphate on the sedimentation coefficient of the C subunit is compared with that of carbamyl-P by a sensitive difference method, only a small increase of sedimentation coefficient is seen with carbamyl-P, stemming principally from the added weight and density of the bound ligand (*71*).

The best direct evidence for the conformational change induced in the C subunit by carbamyl-P comes from the work of Pigiet (*72*) on optical rotatory dispersion and from the work of Griffin *et al.* (*73*) on circular dichroism in the presence of various ligands. It is particularly noteworthy that saturating carbamyl-P induces a large change in the optical rotatory dispersion spectrum (*72*) and in the circular dichroic spectrum (*73*) over and above any change induced by saturating phosphate alone. In 0.04 M phosphate buffer, the C subunit has a dispersion trough at 233 nm. In the presence of saturating carbamyl-P, the mean residue rotation of the C subunit becomes appreciably more negative at this wavelength ($\Delta = -120°$). Since the absorption maximum of carbamyl-P is far below 233 nm, the most likely explanation for this large change in the spectrum upon the binding of carbamyl-P is a conformational change of the protein, probably one affecting the amide bonds of the polypeptide backbone.

Somewhat more indirectly, it is highly probable that effects of carbamyl-P on properties of native ATCase stem from a conformational change of the C subunit which is propagated within the native enzyme by an alteration in quaternary structure. For example, saturing carbamyl-P induces an increase in the reactivity of SH groups of the R subunit toward *p*-mercuribenzoate of more than twofold, relative to their reactivity in the presence of saturating phosphate (*47*). Carbamyl-P

71. M. W. Kirschner and H. K. Schachman, *Biochemistry* **10**, 1919 (1971).
72. V. P. Pigiet, Jr., Ph.D. Thesis, University of California, Berkeley, 1971.
73. J. H. Griffin, J. P. Rosenbusch, K. K. Weber, and E. R. Blout, *JBC* **247**, 6482 (1972).

induces a small decrease (0.5%) in the sedimentation rate of native enzyme relative to the effect of phosphate (47). In this case, correction for the added weight and density of the bound carbamyl-P would make the difference even greater.

7. Tight binding of dicarboxylic acids to a C-subunit–first ligand complex occurs only if the first ligand has a carbonyl group and if it is not larger than carbamyl-P.

Changeux et al. (31) have shown by equilibrium dialysis that succinate does not bind strongly to the C subunit in the absence of carbamyl-P. Collins and Stark (62), using ultraviolet difference spectroscopy, confirmed this conclusion for succinate and showed in addition that L-malate, D-malate, and L-aspartate also bind poorly in the absence of carbamyl-P. The question of which structural features of carbamyl-P are important in inducing the C subunit to bind succinate more tightly was approached by Schmidt et al. (46) using NMR. It was demonstrated [and later confirmed in a more detailed study of Sykes et al. (61)] that broadening of the succinate line in the presence of the C-subunit–carbamyl-P complex resulted from slow exchange of succinate between the bound and unbound environments rather than from immobilization of succinate when bound. [This does not preclude the possibility, made probable on the basis of other data (61), that succinate is in fact bound rigidly to the C-subunit–carbamyl-P complex.]

A series of carbamyl-P analogs was then investigated for their ability to broaden the succinate line. Some of the results are seen in Table II. At pH 6, effects on the succinate line width in the presence of carbamyl-P, acetyl-P, phosphonacetamide, and phosphonacetate are appreciably larger than effects in the presence of the other phosphates and phosphonates. Thus, the ability to induce broadening of the succinate line under these conditions is specifically a property of those compounds which are not larger than carbamyl-P and which have a carbonyl group two atoms from the phosphorus atom. Interpretation of these results is complicated somewhat by the fact that the succinate line is also broad in the absence of a first ligand ("no analog" in Table II). Part of this effect results from nonspecific binding, since the line width of succinate is diminished appreciably by a noninhibitory concentration of the monoanion acetate. The remaining effect is probably the one observed by Sykes et al. (61).

In the presence of phosphate at pH 7, succinate exchanges rapidly between solution and the active site, a result to be contrasted with the slow exchange found in the presence of carbamyl-P. At pH 7, only carbamyl-P and perhaps acetyl-P induce slow exchange of succinate on the basis of line width measurements (Table II). The effect of pH is

TABLE II
Increase in Succinate Line Width Induced by the C Subunit in the Presence of Saturating Amounts of Carbamyl-P or Analogs[a]

First ligand	Increase in succinate line width (Hz)	
	pH 6	pH 7
Carbamyl-P	1.78	1.52
Acetyl-P	1.21	0.31
Phosphonacetamide	0.97	0.19
Phosphonacetate	0.54	
Pyrophosphate	0.18	
N,N-Dimethyl carbamyl-P	0.09	
N-Methyl phosphonacetamide	0.28	0.14
N-Methyl carbamyl-P	0.26	0.00
Methyl phosphonate	0.15	
(No carbamyl-P or analog)	1.97	0.25
Sodium acetate (70 mM)	0.98	

[a] The succinate concentration was 25 mM, the concentration of active sites was 0.6 mM and the temperature was 32° (46).

consistent with the finding that the affinity of succinate for the C-subunit–carbamyl-P complex increases with pH (15).

8. L-Aspartate binds to the C-subunit–carbamyl-P complex with its carboxylates approximately *cis* and is held rigidly by steric constraints.

Maleate is a potent competitive inhibitor of the C subunit (K_i = 4.3 mM at pH 7.8), whereas fumarate is an extremely poor inhibitor (K_i = 145 mM at pH 7.8) (15). Thus, the carboxylates of succinate (K_i = 3.5 mM at pH 7.9) must be oriented approximately *cis* when bound. The structure of L-aspartate, oriented with the carboxylates *cis*, is shown in Fig. 7. Davies et al. (16) showed that *erythro*-β-hydroxyl-L-aspartate (hydroxyl group in position H_C) is an excellent substrate, whereas *threo*-β-hydroxyl-L-aspartate is neither a substrate nor an inhibitor, i.e., a hydroxyl group in position H_B abolishes binding to the enzyme–carbamyl-P complex. Thus, L-aspartate is prevented from binding in an orientation rotated 180° from the one shown in Fig. 7 since its NH_2 group would then be in position H_B. By examining the ability of several substituted 4-carbon dicarboxylic acids to inhibit the enzyme, Davies et al. (16) were able to define a single set of restrictions on the four regions of the enzyme which interact with H_A, H_B, H_C, and NH_2 (Fig. 7), namely: Regions NH_2 and H_C can accept hydroxyl groups without substantial impairment of binding, whereas regions H_A and H_B cannot. An amino or methyl group in any of these four regions undergoes an unfavorable interaction with the C-subunit–carbamyl-P complex that

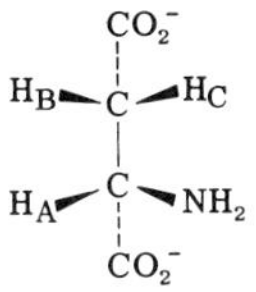

$$\begin{array}{c} CO_2^- \\ | \\ H_B \leftarrow C \rightarrow H_C \\ | \\ H_A \leftarrow C \rightarrow NH_2 \\ | \\ CO_2^- \end{array}$$

FIG. 7. Stereochemistry of the approximately *cis* conformation of L-aspartate which binds to ATCase. Constraints are placed by the enzyme on positions H_A, H_B, H_C, and NH_2 as discussed in the text.

prevents tight binding. The apparent paradox that the substrate L-aspartate is bound very poorly to the binary complex is an important aspect of the mechanism and will be discussed below.

9. A second conformational change of the enzyme accompanies the binding of 4-carbon dicarboxylic acids to the C-subunit–carbamyl-P complex.

Several properties related to protein conformation are dramatically changed when succinate binds to the C-subunit–carbamyl-P complex. Pigiet (*72*) has shown that the mean residue rotation centered at the 233-nm Cotton effect trough in the optical rotatory dispersion spectrum of the C subunit changes by $-30°$ when succinate binds to the enzyme–carbamyl-P complex but that no change is observed when succinate binds to the enzyme–phosphate complex. Similarly, Griffin *et al.* (*73*) have observed the appearance of a substantial circular dichroism difference spectrum upon addition of succinate to the C-subunit–carbamyl-P complex. These results indicate that the conformational change induced by the binding of carbamyl-P and succinate is not merely an enhancement of the change resulting from carbamyl-P alone since the difference spectra are opposite in sign. Kirschner and Schachman (*71*) found that the sedimentation coefficient of the C subunit increases by 1.05% upon addition of succinate plus carbamyl-P, more than three times the effect expected merely for the binding of the ligands. Either ligand alone produces an effect nearly equal to that expected from binding. Collins and Stark (*62*) showed that the ultraviolet difference spectrum of the C subunit in the presence of carbamyl-P plus succinate was far greater than the sum of the spectra obtained in the presence of each ligand singly (Fig. 8). Ultraviolet difference spectra arise from changes in the environment of chromophores and are not by themselves conclusive evidence for a conformational change. For example, it might be argued that the change observed when succinate binds to the C-subunit–carbamyl-P complex reflects merely the proximity of the carboxylate anions of succinate to the chromophores. However, saturation of the binary com-

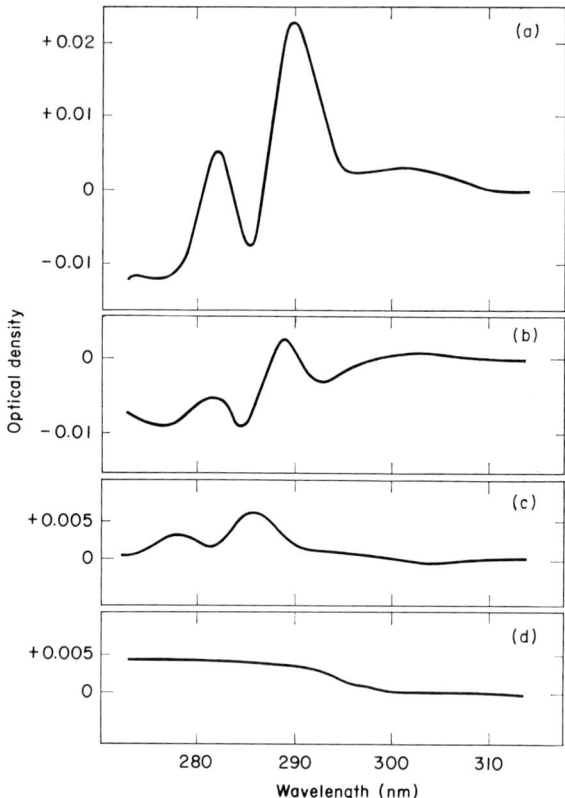

FIG. 8. Ultraviolet difference spectroscopy of the C subunit at pH 7 [from Collins and Stark (62)]. The absorbancy of one two-compartment cell, which contains enzyme alone in one compartment and ligands alone in the other, is subtracted from the absorbancy of a matched cell, which contains enzyme and ligands together in one compartment, and water in the other. Curve a, 4.9 mM carbamyl-P plus 0.1 M succinate, 3.54 mg/ml of enzyme; curve b, 4.9 mM carbamyl-P, 2.95 mg/ml of enzyme; curve c, 0.2 M succinate, 3.65 mg/ml of enzyme; and curve d, 0.4 M acetate, 3.58 mg/ml of enzyme.

plex with L-malate gives a difference spectrum which is much reduced in magnitude. Since L-malate is also a 4-carbon dicarboxylic acid, the difference between the maximum effects produced in the two ternary complexes is very likely to reflect a conformational change with succinate that occurs not at all or to a diminished extent with L-malate. Finally, the transition state analog PALA, which binds extremely tightly to the C subunit and combines in a single molecule most of the functional groups of carbamyl-P and succinate, induces a protein difference spectrum which is indistinguishable from that obtained when carbamyl-P and succinate are present simultaneously (*36*).

Effects of succinate on properties of native ATCase in the presence of saturating carbamyl-P must reflect indirectly conformational changes of the C subunit within the native structure. The reactivity of the SH groups of the R subunit toward p-mercuribenzoate is increased by more than 5-fold in the presence of carbamyl-P plus succinate. Carbamyl-P alone increases reactivity by 2.4-fold, and succinate alone has no effect (*47*). The sedimentation coefficient of the native enzyme *decreases* by 3.6% in the presence of both ligands, in contrast to a negligible effect of succinate alone and a much smaller effect of carbamyl-P alone (*47*). This result contrasts strongly with the effect of the same two ligands on the C subunit, in which the sedimentation coefficient is *increased* (*71*). The mean residue rotation at the Cotton effect trough at 233 nm increases by 65° in the case of the native enzyme in the presence of both ligands and is opposite in sign from the effect observed with the isolated C subunit. As with the C subunit, succinate alone has no effect (*72*) and the effect of carbamyl-P alone is much smaller (*74*). When circular dichroism difference spectra are considered (*73*), a similar situation is found; that is, in the presence of carbamyl-P and succinate, the spectrum for ATCase is substantially different from that of the C subunit, and succinate or phosphate alone have little effect. The complement-fixing activity of native ATCase in the presence of carbamyl-P plus succinate has been compared to that obtained in the absence of ligands by von Fellenberg *et al.* (*58*). With both anti-native ATCase serum and anti-C-subunit serum, the complement-fixing activity decreased in the presence of ligands. A decrease was also observed when the C subunit was assayed with anti-C subunit in the presence of these ligands. McClintock and Markus (*59*) examined the susceptibility of the native enzyme to three proteases. The R subunit is digested preferentially. A substantial increase in the rate of digestion was noted in the presence of both carbamyl-P and succinate, whereas the effect in the presence of either ligand alone was much smaller. The tendency of low concentrations of sodium dodecyl sulfate to cause dissociation of ATCase to its component subunits is *enhanced* by phosphate, carbamyl-P, L-aspartate plus phosphate, and succinate plus carbamyl-P, in order of increasing effectiveness (*75*). The same ligands protect against inactivation of the C subunit by the detergent.

Taken together, these data strongly indicate that a major conformational change accompanies the binding of succinate to the C-subunit–

---

74. E. A. Dratz and M. Calvin, *Nature (London)* **211**, 497 (1966).
75. P. D. Colman and G. Markus, *JBC* **247**, 3829 (1972).

carbamyl-P complex, both in the isolated subunit and when the C subunit is a part of native ATCase. Additional evidence for this conformational change is provided by data from temperature jump kinetic studies, which are discussed in the following section and in Section VII,G.

10. The amino group of L-aspartate and other substituents in the L-α positions of analogs encounter steric hindrance which weakens binding and impedes the completion of the second conformational change.

Comparison of the affinity of three inhibitors for the C-subunit–carbamyl-P complex indicates that the introduction of a substituent in the position occupied by the NH_2 group of L-aspartate reduces binding roughly in proportion to bulk, i.e., succinate (–H) is bound more tightly than L-malate (–OH), which is bound more tightly in turn than DL-α-methyl succinate (–CH_3) (Table III). What then of the binding of L-α-amino succinate (alias L-aspartate)? Of course, the affinity of this compound cannot be measured by equilibrium methods, but a lower limit for the dissociation constant can be obtained from steady state kinetics. In an ordered mechanism with carbamyl-P binding first, kinetic analysis predicts that inhibition by the product carbamyl-L-aspartate will be noncompetitive with L-aspartate (at constant carbamyl-P), i.e., the lines of a double reciprocal plot will intersect to the left of the ordinate. The amount of displacement from the ordinate is related to the affinity of L-aspartate for the binary complex (15). At pH 7.8, this displacement is undetectable, but an estimate of $K_i > 0.2\ M$ has been made (15). At pH 7.0, where inhibitors bind more tightly (Table III), the displacement is detectable and $K_i \geq 0.017\ M$ has been calculated

TABLE III
Equilibrium Constants at 28° for Dissociation of Aspartate Analogs from the C-Subunit–Carbamyl-P Complex

		Method for K_i(mM)		
Analog	pH	Steady state kinetics[a]	Temperature jump kinetics[b]	Difference spectroscopy[c]
Succinate	7.0	0.87		0.75
Succinate	7.4		1.0	
Succinate	7.9	3.5		
L-Malate	7.0	12		10
L-Malate	7.5		10	
L-Malate	7.9	61		
D,L-α-Methyl succinate	7.0	>150		

[a] Data from Porter et al. (15) and Davies et al. (16).
[b] Data from Hammes et al. (63).
[c] Data from Collins and Stark (62).

as a lower limit (*62*). (Note that the Michaelis constant for L-aspartate, 0.02 M at pH 7.8 and 28°, is not a measure of affinity.) Thus, it appears that the substrate L-aspartate is bound very poorly to the C-subunit–carbamyl-P complex, in accord with the trend shown by inhibitors with increasing bulk at the L-α-position.

The effect of an L-α substituent can also be seen by ultraviolet difference spectroscopy (*62*). If the spectrum of the C-subunit–carbamyl-P complex is assigned a magnitude of 1.0 (Fig. 8b), the spectrum of the ternary complex with succinate has a magnitude of 2.8. In the ternary complex with L-malate, the magnitude is 1.3. Therefore, if the difference spectrum measures this property, the extent of the conformational change with L-malate is much less than with succinate.

Strong support for the idea that binding is weakened because a conformational change is prevented is obtained from the data of Hammes *et al.* (*63*), who measured the relaxation spectra of the C-subunit–carbamyl-P complex in the presence of succinate and L-malate by the temperature-jump method. In both cases, binding to form a ternary complex was associated with two relaxation processes well separated on the time axis. The slower processes were analyzed quantitatively in terms of two-step binding mechanisms, i.e., rapid formation of initial complexes followed by a slow isomerization of those complexes (E + I $\rightleftharpoons$ EI $\rightleftharpoons$ E'I). All the kinetic parameters for succinate and L-malate were the same (within a factor of 2) *except* for the rate constant governing the conversion of E'I to EI, which was about ten times larger in the case of L-malate. This result suggests that the L-α-hydroxyl group encounters the most significant steric interference in E'I, the isomerized complex.

Estimates of the relative amount of the first ternary complex EI (at saturation) range from 12% for succinate and 60% for L-malate at pH 7.5 and 28°, measured by the temperature-jump method (*63*), to less than 2% for succinate at pH 7 and 33°, measured by NMR relaxation methods (*61*). More recent NMR measurements with succinate indicate that the relative amount of the first ternary complex (EI) increases substantially with increasing pH above pH 7 and 28° until, at pH 9.3, most of the complex is in this form (*76*). If a small difference spectrum (*62*) is associated with the state EI and a large one with the state E'I, then the ternary complex L-malate is predominantly in state EI and that with succinate is predominantly in state E'I at pH 7. Of course, it is possible that the ternary complex can exist in more than two states.

11. During catalysis, the activation energy is lowered by compression of the amino group of L-aspartate and the carbonyl carbon of car-

76. C. B. Beard and P. G. Schmidt, *Biochemistry* (1973) (in press).

bamyl-P. This compressional conformational change is driven by the potential binding energy of L-aspartate.

Jencks (77) has argued convincingly and at length that a single rigid structure for the active site of an enzyme cannot fit both the substrates and the products optimally. Therefore, it is reasonable to expect that the conformation of the enzyme will change during catalysis, and the evidence is good that such a conformational change in the C subunit closely accompanies catalysis of carbamyl transfer. A corollary to conformational change in the enzyme is the requirement that strain be induced concomitantly in the substrates. Such strain can be used productively to change the structures of the substrates toward the structure of the transition state, and hence to accelerate the reaction (77).

In the case of ATCase, both substrates are held quite rigidly by a combination of multiple ionic interaction and steric constraints. The conformational change which takes place when succinate binds to the C-subunit–carbamyl-P binary complex is in the direction of a more contracted or isometric conformation (71). This conformational change is opposed by succinate analogs which have an L-α substituent. Such a substituent increases the rate at which the conformationally changed ternary complex relaxes (63) and prevents tight binding of the analog. It is clear that amino group of L-aspartate must approach the carbonyl carbon of carbamyl-P closely in order to form a bond. Van der Waals repulsion between the nitrogen and carbon atoms opposes this close approach and contributes to the height of the activation energy barrier to reaction. Compression of the two substrates together along the reaction coordinate in the direction of the transition state clearly would help to lower the activation energy and hence would accelerate the reaction. Such a contribution to catalysis is probably an important part of the aspartate transcarbamylase mechanism (15, 62). In this case, an approximate measure of the potential energy available to push the two substrates together within the ternary complex is seen in the affinity of a dicarboxylic acid without an L-α substituent (e.g., succinate) for the C subunit. In an attempt to achieve the optimum binding interactions obtained by succinate, the conformational change drives the amino group of L-aspartate into the carbonyl carbon of carbamyl-P with the result *either* that bond formation begins to occur *or* that the ternary complex is pushed toward the un-isomerized (EI) form, where fast exchange is characteristic and binding is poor.

77. W. P. Jencks, "Catalysis in Chemistry and Enzymology," Chapter 5. McGraw-Hill, New York, 1969.

The binding of L-aspartate and L-malate is enhanced by inorganic phosphate, whereas the binding of succinate is not. Furthermore, the difference spectra observed in the first two cases are far larger than in the last (*62*). The conformational change should bring the amino group of L-aspartate or the hydroxyl group of L-malate close to the carbonyl carbon of carbamyl-P. When inorganic phosphate occupies the carbamyl-P site, the same conformational change brings the L-α substituent of the dicarboxylic acid close enough to the bound phosphate to interact, probably through a hydrogen bond. Succinate, having no L-α substituent, cannot form such a complex. As a further test, it was shown (*62*) that methyl phosphonate (in which a methyl group replaces that oxygen atom of phosphate which is most likely to be involved with the other ligand) does not increase the affinity of L-aspartate or L-malate for the C subunit and does not give a large difference spectrum in their presence. A C subunit (or native ATCase)–phosphate–L-aspartate complex has also been observed kinetically (*15, 78*) and in complement fixation assays (*58*).

There are several kinds of information which bear on the question of whether or not the rate of the conformational change associated with compression is rate determining. The turnover number of the C subunit at pH 7.5 and 28° is 390 sec^{-1} (*15*). From the temperature-jump work, the rate of the EI to E'I conversion is 4600 sec^{-1} for succinate and 3600 sec^{-1} for L-malate (*63*). If the rate of this process is similar for L-aspartate, then the isomerization is seen significantly in the overall rate of catalysis. Similar calculations for the isomerization have been presented by Sykes *et al.* (*61*), using data from transient NMR measurements.

^{14}C-Labeled and anhydride ^{18}O-labeled carbamyl-P have been used to study whether or not primary kinetic isotope effects are observed with the C subunit (*79*). At optimum concentrations of substrates and optimum pH, the primary kinetic isotope effects are 10% or less of the maximum effects predicted by theory. A likely explanation, but not the only one, is that kinetic steps in which the bonds to these atoms change are not important in determining the observed rate. Such an interpretation is consistent with a rate-determining conformational change of the enzyme. At pH 10, where the activity of the enzyme is low, a major effect (about 50% of the theoretical maximum) is seen for ^{14}C, but no substantial effect is seen for ^{18}O, suggesting that a step in which bonding to the carbonyl carbon of carbamyl-P changes but bonding to the anhydride oxygen does not change becomes important in determining the observed

78. K. Kleppe, *BBA* **122**, 450 (1966).
79. G. R. Stark, *JBC* **246**, 3064 (1971).

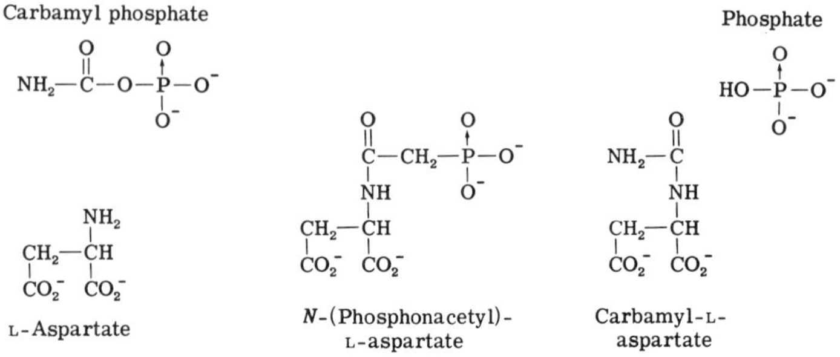

FIG. 9. The structures of PALA and the substrates and products of the ATCase reaction.

rate. Such a process might be the formation of a tetrahedral intermediate.

12. Some other aspects of the mechanism.

a. Is PALA an Analog of the Transition State? Wolfenden, in a recent review (*80*), has developed a quasi-thermodynamic treatment which reveals that the transition state for an enzyme-catalyzed reaction is bound much more tightly than substrates or products. In Wolfenden's words, "this conclusion . . . rests on no assumptions regarding the specific nature of the attractive forces involved, nor does it require that the enzyme be rigid or flexible. Conversely, evidence which may be obtained in favor of tight binding of the altered substrate in the transition state cannot by itself be regarded as evidence for or against particular forces of attraction or changes in conformation which may occur in a specific case." Therefore, the tight binding of PALA cannot be used as evidence for the compression mechanism, but the tightness of binding can be used to deduce whether or not PALA resembles the true transition state sufficiently to benefit from the increased affinity of the enzyme for that state.

N-(phosphonacetyl)-L-Aspartate has at least some of the structural features to be expected of a transition state for the ATCase reaction (Fig. 9). It certainly lacks some other features; for example, it is not unlikely that the carbonyl carbon of carbamyl-P is tetrahedral in the transition state, or nearly so. N-(phosphonacetyl)-L-Aspartate ($K_i = 2.7 \times 10^{-8} M$ at pH 7 and 28°) does bind somewhat more tightly to the C subunit than would be anticipated merely on the basis that it combines in one molecule most of the binding loci of the two substrates. The product of the dissociation constant for carbamyl-P ($2.7 \times 10^{-5} M$) and the Michaelis constant for L-aspartate ($2 \times 10^{-2} M$) (*15*) is $5.4 \times 10^{-7} M$, 20 times higher than K_i for PALA under the same conditions. Further-

80. R. Wolfenden, *Accounts Chem. Res.* **5**, 10 (1972).

more, the Michaelis constant of L-aspartate is very likely to be an appreciable underestimate of the dissociation constant for this substrate (see above); thus, the difference between PALA and the two substrates is probably much greater than 20-fold. Looking at the question another way, PALA can be considered to be a much closer analog of the inhibitors phosphonacetamide and succinate than of the substrates. For phosphonacetamide K_i is $6.6 \times 10^{-4} M$ (15) and K_i for dissociation of succinate from the C subunit–phosphonacetamide complex is $2.2 \times 10^{-2} M$ (62). Now the product is $1.45 \times 10^{-5} M$, 540 times greater than the dissociation constant for PALA. If L-aspartate is substituted for succinate, the difference is even greater. There is less translational entropy to overcome in the formation of the two body enzyme–PALA complex than in the formation of the three body complexes; thus, the difference between these cases must be reduced by an appropriate factor. The loss of entropy can be estimated to correspond very roughly to a factor of about 50 near room temperature (81). It might be anticipated that analogs with a tetrahedral carbon in place of the carbonyl would bind even more tightly than PALA.

b. *Is Base Catalysis Important?* The α-NH_3^+ group of L-aspartate must lose two protons in going to the –NH– group of carbamyl-L-aspartate, and base catalysis is therefore an attractive mechanistic possibility. Collins and Stark (62) observed an abnormally large difference spectrum but a weak affinity upon association of L-malate with the C subunit in the absence of a first ligand and postulated an interaction of the hydroxyl group with a putative base catalyst. It was later demonstrated (79) that the maximum velocity of the C subunit is the same in D_2O and in H_2O, putting into question the possibility of a rate-determining proton transfer step in the reaction. But as Jencks and Carriuolo (82) have emphasized, the lack of a solvent deuterium isotope effect is not unambiguous evidence for the lack of a proton transfer in the rate-determining step. The possibility of *preequilibrium* protonation of the carbonyl group of carbamyl-P is *not* ruled out by the lack of a solvent deuterium isotope effect and indeed is a likely mechanistic possibility (see above). Preequilibrium proton removal to convert α-NH_3^+ to α-NH_2 does not seem unlikely. The question of how the second proton is removed is also open. One possibility is that a proton is transferred from the NH_2 of L-aspartate to the anhydride oxygen of carbamyl-P since both L-aspartate and L-malate bind tightly to the enzyme–phosphate complex, with formation of a large

81. F. H. Westheimer and L. L. Ingraham, *J. Phys. Chem.* **60**, 1668 (1956).
82. W. P. Jencks and J. Carriuolo, *JACS* **82**, 675 (1960).

difference spectrum (*62*), suggesting a specific interaction between the L-α substituent and one of the oxygen atoms of the bound phosphate.

c. How Do 3- and 5-Carbon Dicarboxylic Acids Interact with the C subunit? L-Glutamate is not a substrate for the C subunit and neither it nor glutarate are good inhibitors (K_i's > 100 mM) (*15, 16*). Probably the tight fit of 4-carbon dicarboxylic acid into the L-aspartate site precludes the tight binding of larger molecules.

In contrast to glutarate, malonate is a reasonably good inhibitor of the C subunit (K_i 27 mM at pH 7.8, 28°) (*15*); and amino malonate, in contrast to L-glutamate, is a substrate (V_{max} one-sixtieth of that obtained with L-aspartate at pH 8.0 28°) (*16*). Malonate gives a difference spectrum equal in size to that given by succinate in the presence of carbamyl-P (*62*). In the presence of phosphonacetamide at pH 7, malonate also gives a large difference spectrum, in contrast to succinate, which does not (*62*). Presumably even the small change $-O-$ to $-CH_2-$ in going from carbamyl-P to phosphonacetamide is sufficient to prevent the conformational change in the presence of succinate but not in the presence of the smaller malonate. The same sort of effect is seen in NMR measurements of bound line widths. At pH 7, the succinate line is not appreciably broadened in the presence of the C-subunit–phosphonacetamide complex, in contrast to the appreciable effect seen at pH 6 (where exchange is expected to be slower) (*46*). However, the line width for malonate is broad at pH 7 in the presence either of carbamyl-P or of phosphonacetamide. Gregory and Wilson (*83*) have observed that malonate (but not glutarate) activates native ATCase at low concentrations of L-aspartate. If homotropic cooperativity results indirectly from the same conformational change associated with compression, then this change can take place with the 3-carbon analog.

Perhaps the conformational change with the smaller amino malonate bound to the C-subunit–carbamyl-P complex in place of L-aspartate can take place without much steric hindrance, and the 60-fold reduction in rate observed with the smaller compound may result at least in part from a loss of the compressional component of catalysis. An informative experiment in this regard would be to test the ability of methyl malonate to induce the full conformational change in the presence of carbamyl-P.

B. Functional Groups at the Active Site

At present, few amino acid residues have been placed specifically at the active site of ATCase by chemical modification. The technique of

83. D. S. Gregory and I. B. Wilson, *Biochemistry* **10**, 154 (1971).

affinity labeling has not been particularly fruitful, probably because the steric constraints on ligand binding are so severe that it is difficult to introduce a relative functional group without destroying the specificity of interaction. Exceptions seem to be some reagents for the SH groups of the C subunit and pyridoxal-P. Group specific reagents have yielded some useful information but, by comparison with other enzymes for which a comparable amount of mechanistic information is available, results with such reagents have been meager. Progress in chemical modification should be more rapid in the near future, when more information is available from the amino acid sequence and crystallographic studies now in progress.

1. *The Sulfhydryl Group*

The C subunit contains three SH groups, one per chain *(35, 47)*. In the milieu of the isolated C subunit, these groups are completely unreactive toward many reagents which ordinarily modify exposed SH groups in proteins very rapidly, including iodoacetate, N-ethylmaleimide, ethyleneimine, and ferricyanide *(35, 41)*. p-Mercuribenzoate, DTNB, and tetrathionate react sluggishly (much slower than with mercaptoethanol), whereas potassium permanganate reacts extremely rapidly and specifically *(35, 41)*. In the milieu of the native enzyme, the same SH groups are even more unreactive, since DTNB is no longer capable of modifying them *(35)*. However, potassium permanganate still reacts very rapidly and specifically *(84)*. Most mercurials (e.g., p-mercuribenzoate) react preferentially with the R-subunit SH groups, resulting in dissociation, followed by slow reaction with the C-subunit sulfhydryls. However, Wiley *et al.* *(50)* have reported that 2-chloromercuri-4-nitrophenol reacts preferentially and stoichiometrically with the C subunit in native ATCase. The product is a key heavy metal derivative for the crystallographic work. D. R. Evans (personal communication) has found that this mercurial reacts with $k = 3.2 \times 10^6 \, M^{-1} \, \text{min}^{-1}$ and that reaction results in complete inactivation without any dissociation into subunits. The pH-rate profiles and effects of anions indicate that in the reactive species the phenolic hydroxide is ionized and Cl⁻ is bound to the mercury. In the 5.5-Å map *(50)*, the mercury appears to be located in a relatively inaccessible region of the protein, largely hemmed in by areas of high electron density. C. McMurray (personal communication) has inferred from the spectral properties of the derivative that the SH group may be in a hydrophobic environment.

The SH group appears to react preferentially with derivatives which

---

84. G. R. Jacobson and G. R. Stark, *JBC* (1973) (submitted).

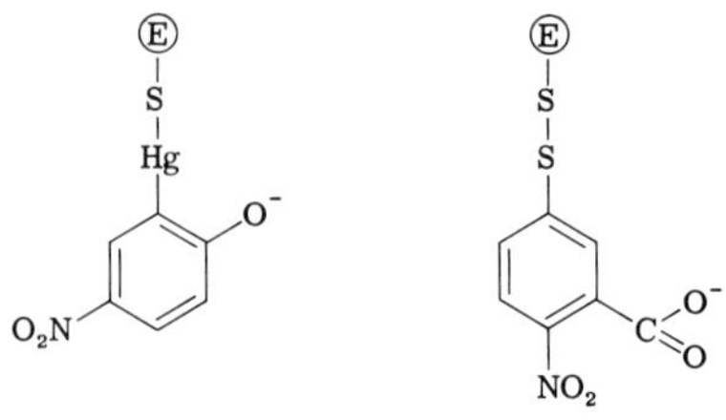

FIG. 10. (I) The 2-mercuri-4-nitrophenol and (II) the 5-thio-2-nitrobenzoate derivatives of the SH group of the C chain.

are (a) highly reactive and (b) negatively charged (see Fig. 10 for structures of the mercurial and thionitrobenzoate derivatives). Probably a positively charged group on the enzyme serves to bind and orient the inactivating reagents. Vanaman and Stark (35) have shown that an ionizable group of the enzyme with pK_a 7.9 is required for the reaction with DTNB. Potassium permanganate inactivates the enzyme by stoichiometrically oxidizing the SH groups to $-SO_3^-$ and appears to be an affinity label since reaction with the C subunit is 50-fold faster than reaction with mercaptoethanol (41). The SH group is very probably at the active site, on the grounds that substrates and substrate analogs protect against reaction with DTNB (35), permanganate (41), and 2-chloromercuri-4-nitrophenol (50) and that modification usually leads to inactivation. An alternative explanation, that modification leads to a structural alteration which results indirectly in inactivation, is less likely but cannot be ruled out rigorously at present.

Several derivatives of the C subunit modified on the SH groups are capable of recombining with R subunit with the same stoichiometry as is found in the native enzyme (84). The SH group is clearly not required for activity per se since three active modified derivatives are now known: A fully active –SCN derivative was prepared by reacting the TNB-C subunit with cyanide (35), and highly active $-SCH_3$ (84) and $-SCH_2CONH_2$ derivatives (J. P. Rosenbusch and K. Weber, personal communication) have been prepared by reacting the C subunit with methyl iodide or iodoacetamide in the presence of a denaturant, followed by renaturation. Inactive derivatives seem to be those in which a negative charge has been introduced or in which a bulky substituent has been added to the SH group. The bulky substituent probably inactivates by interacting with a binding site for L-aspartate in the C-subunit–carbamyl-P complex since succinate does not bind to the 2-mercuri-4-nitrophenol (85) or thionitrobenzoate (84) derivatives in the presence of

85. C. H. McMurray, D. R. Evans, and B. D. Sykes, BBRC 48, 572 (1972).

carbamyl-P on the basis of a complete lack of effect on the succinate line width in NMR experiments. However, acetyl-P (*85*) and carbamyl-P or PALA (*84*) do bind to the mercurial or thionitrobenzoate derivatives, respectively. Permanganate-oxidized ATCase does appear to bind succinate in the presence of carbamyl-P and yields an ultraviolet difference spectrum nearly identical to that obtained with unmodified native ATCase in the presence of these ligands. Thus, succinate cannot only bind to this oxidized derivative, but it can also induce most of the conformational change associated with catalysis (*84*). Therefore, the mechanism of inactivation upon conversion of the SH group to the negatively charged sulfonate remains unclear.

If a negative charge on a derivative of the SH group leads to inactivation, ionization of the SH group itself at high pH might be expected to result in an inactive form of the enzyme. The pH dependence of the $-SCH_3$ derivative is the same as that of the unmodified enzyme in the presence or absence of R subunit even at high pH, implying that the SH group may not be free to ionize in the complex with carbamyl-P (*84*).

The role of the SH group in the C subunit remains obscure. It does not appear to be a ligand for zinc in native ATCase, since the $-SCN$ and $-SCH_3$ derivatives combine normally with zinc R subunit (*84*). Furthermore, it does not appear to be involved in any allosteric transition since the same two derivatives of the native enzyme have unaltered homotropic and heterotropic allosteric interactions with succinate (at low concentrations of L-aspartate), with ATP, and with CTP (*84*).

The thionitrobenzoate derivative of the C subunit has been useful as a reactive intermediate in preparing other derivatives of the SH group which would have been difficult to make by another route (*35*). The $-SCN$ derivative undergoes a facile specific chemical cleavage in denaturing solvents as discussed in Section III,A.

2. *Amino Groups*

A highly schematic pictorial representation of the active site of ATCase is shown in Fig. 11. Five enzyme-bound positively charged groups are shown, and it seems highly probable that at least some of these are amino groups. Yet direct evidence for participation of such groups in catalysis is meager. Probably the best evidence at present comes from the inactivation of the C subunit by pyridoxal-P (*86*). Bulky phosphates inhibit the C subunit with K_i values of about $10^{-3} M$ (*15*), yet the value for pyridoxal-P is about $10^{-6} M$, implying that an interaction of the inhibitor at a second site is important. Pyridoxal-P gives rise to a differ-

86. P. Greenwell, S. L. Jewett, and G. R. Stark, *JBC* (1973) (in press).

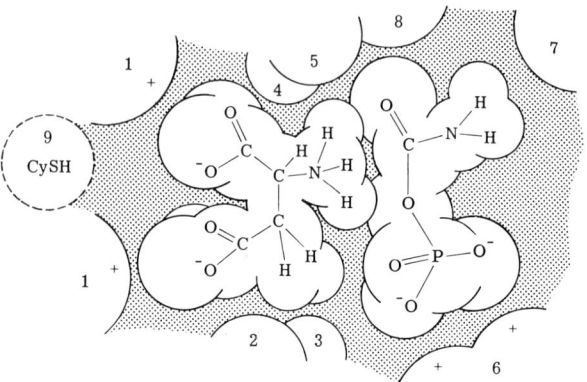

FIG. 11. Schematic representation of the active site of ATCase with carbamyl-P and L-aspartate bound showing positively charged residues interacting with L-aspartate (1) and carbamyl-P (6), steric constraints placed by the enzyme on the L-aspartate (2, 3, 4, 5) and carbamyl-P (7) sites, the acid catalyst –BH⁺ (8) and the nearby cysteinyl residue (9). See text for further details. Adapted from Fig. 5 of Gerhart (6).

ence spectrum when bound to the C subunit. Use can be made of this property in titrating the enzyme with the inhibitor. A dissociation constant of about 10^{-6} M and a stoichiometry of three pyridoxal-P molecules per C-subunit trimer were found. The difference spectrum disappears upon addition of PALA or carbamyl-P. The second interaction of pyridoxal-P with the C subunit is very likely to be formation of a Schiff's base in the active site (Fig. 12) since reaction of the complex with NaBH₄ leads to irreversible inactivation with incorporation of one pyridoxamine-P per active site. The reactive amino group has not yet been identified.

Reaction of the C subunit with 2.1 moles of succinic anhydride per mole of lysine residue leads to an inactive species which is homogeneous by sedimentation velocity analysis and by cellulose acetate electrophoresis (33). About 6 lysine residues per polypeptide chain were modified. Pigiet (72) has shown that the inactive product binds carbamyl-P but

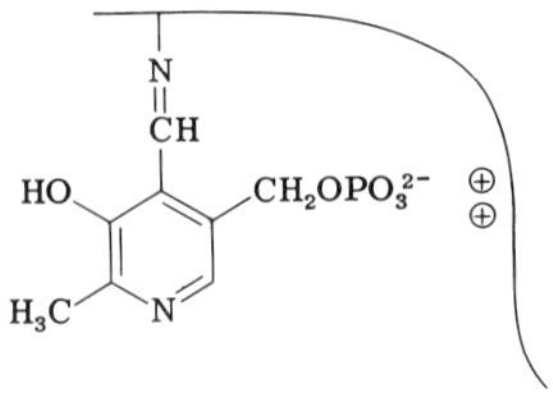

FIG. 12. Structure of the Schiff's base formed with pyridoxal-P and the C subunit; the phosphate moiety is shown interacting with the same positively charged residues that form ionic interactions with carbamyl-P.

not succinate. Hervé and Stark (unpublished observations) have also found that several reagents specific for amino groups inactivate the C subunit rapidly.

3. *Histidine Residues*

The possibility that histidine may be present in the active site is suggested by the observation that inhibition of the C subunit by succinate depends on a positively charged group with pK_a 7.1 (*15*). Gregory and Wilson (*83*) have shown that bromosuccinate inactivates native ATCase with a stoichiometry of one modifying group per catalytic chain; Hervé and Stark (unpublished results) observed a similar reaction with the C subunit and obtained preliminary evidence that histidine was modified.

When a photosensitizing dye is fixed to a discrete site on a protein, the highly reactive singlet state oxygen formed by light reacts preferentially with nearby residues because it will react with solvent before it has a chance to diffuse very far (*87*). The Schiff's base formed by pyridoxal-P with the C subunit is a convenient derivative to use for such affinity photoinactivation. Exposure of this derivative to light and air under controlled conditions leads to irreversible inactivation without a reduction in titratable SH groups or in the amount of any amino acid except histidine. The loss of approximately 2 histidines per C chain accompanies the inactivation (*86*).

4. *Aromatic Amino Acids*

One C chain has 2 tryptophan and 8 tyrosine residues (*29*). The difference spectrum induced by the binding of carbamyl-P to the C subunit or native enzyme probably reflects perturbation primarily of tryptophan, and the spectrum induced by the further addition of succinate is also the result of perturbation of tryptophan (*62, 72, 73*). Although the aromatic amino acid residues involved are not necessarily at the active site, large changes in the chemical shifts of the protons of methyl phosphonate upon binding at the active site do suggest ring current effects because of close proximity of aromatic residues (*46*). Kirschner (*88*) found that tetranitromethane modifies 1 tyrosine per chain and no other amino acid residue when the reaction is carried out in the presence of carbamyl-P and succinate at pH 6.7. The derivative, which is almost fully active, has an altered pH dependence which indicates that ionization of the nitrotyrosine residue results in loss of activity. The effect of this new negative charge is reminiscent of the effect of negatively

87. G. Jori, G. Gennari, C. Toniolo, and E. Scoffone, *JMB* **59**, 151 (1971).
88. M. W. Kirschner, Ph.D. Thesis, University of California, Berkeley, California, 1972.

charged modifications of the SH group, although a completely different phenomenon may be responsible (*89*). Nitration of additional residues of tyrosine leads to loss of activity.

VII. Cooperative Properties of the *E. coli* Enzyme

A. COOPERATIVE SUBSTRATE BINDING

When the activity of native ATCase is studied as a function of the concentration of L-aspartate in the presence of saturating carbamyl-P, the curve obtained is clearly sigmoidal (Fig. 13), indicating that the

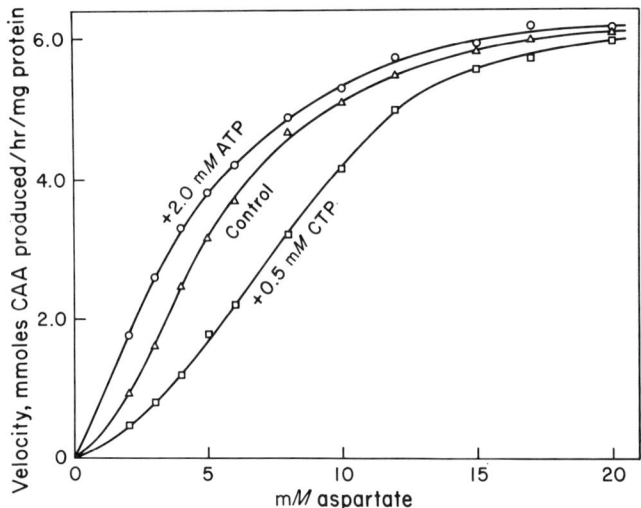

FIG. 13. The response of ATCase to the activator and inhibitor CTP. The reaction mixtures contained 3.6 mM dilithium carbamyl-P, 50 mM imidazole acetate, pH 7.0, 1.0 μg of ATCase per milliliter, CTP, ATP, and dipotassium L-aspartate as indicated. Incubation time was 5 min at 30°. These data closely resemble those previously reported (*10*, *91*) but are obtained from reactions catalyzed in the absence of phosphate ion and measured by the aspartate-^{14}C radioactive assay described by Porter *et al.* (*15*). The velocity of reaction is 7.6 mmoles/hr/mg protein at 120 mM aspartate, a value very close to the V_{max} extrapolated from the data shown. Data from V. Pigiet, Y. Yang, and H. K. Schachman, unpublished observations. Reprinted from Fig. 3 of Gerhart (*6*).

89. The ultraviolet difference spectrum of KMnO$_4$-oxidized ATCase vs. unmodified ATCase indicates a tyrosine perturbation. It is not known if this is the same one that is nitrated (G. R. Jacobson, unpublished).

binding of L-aspartate to the active sites of the oligomeric native enzyme is cooperative (10). The saturation curve for carbamyl-P at pH 7 in the presence of a relatively high concentration of L-aspartate (15 mM) is also sigmoidal (21). However, a kinetic study cannot prove that the *binding* of carbamyl-P is cooperative since the affinity of the enzyme for carbamyl-P should be affected by the binding of L-aspartate. Rosenbusch *et al.* (38) have shown that the curve for binding of carbamyl-P to native ATCase in the absence of succinate is hyperbolic. The kinetics of binding of carbamyl-P to the native enzyme have been analyzed by Hammes and Wu (90) using temperature-jump measurements. A simple bimolecular reaction was sufficient to explain the data, with an equilibrium dissociation constant of 4×10^{-5} M, identical to the value obtained for the isolated C subunit. No evidence of sigmoidal binding was found although the concentration range studied was small.

The binding of succinate to the enzyme–carbamyl-P complex has been studied by equilibrium dialysis, and the data clearly indicate that succinate is bound cooperatively under these conditions (31). A consequence of such positive homotropic cooperativity is that, at low concentrations of L-aspartate, analogs which inhibit the C subunit competitively can activate native ATCase by increasing the affinity of unoccupied sites for L-aspartate. Succinate, maleate, malate, and malonate have all been shown to activate native ATCase under such circumstances (83, 91). In addition, identical activation curves are obtained with succinate, maleate, or the transition state analog PALA (36). At pH and 28° in the presence of 4.9 mM carbamyl-P and 1 mM L-aspartate, the maximal activation of native ATCase is 3.3-fold with about 1 mM maleate or succinate or about 2 μM PALA. This result indicates that PALA alone is capable of inducing approximately the same conformational effect in a given site as carbamyl-P and succinate together. Higher concentrations of these activators will, of course, inhibit.

B. ALLOSTERIC EFFECTORS

1. *Properties of the Enzyme in the Presence of Effectors*

A series of nucleotides was studied by Gerhart and Pardee (10) for their effect on the activity of native ATCase. Cytidine triphosphate was the most potent inhibitor found, whereas ATP and dATP were the only compounds which stimulated ATCase activity. Other nucleotides were

90. G. G. Hammes and C-W. Wu, *Biochemistry* **10**, 2150 (1971).
91. J. C. Gerhart and A. B. Pardee, *Cold Spring Harbor Symp. Quant. Biol.* **28**, 491 (1963).

either much less effective as inhibitors than CTP or were neither inhibitors nor activators. Inhibition by CTP reaches a maximum of about 90% in the presence of low concentrations of L-aspartate (*92*). At high concentrations of both substrates, CTP has virtually no effect until its concentration is raised enough to permit it to act as a competitive inhibitor for carbamyl-P (*15*). As Gerhart (*6*) pointed out, the incomplete maximum inhibition by CTP is an indication that its effect is not exerted directly at the active site but is mediated through changes of protein conformation.

As shown in Fig. 13, CTP makes the saturation curve for L-aspartate more sigmoidal, while ATP makes it somewhat less sigmoidal, but neither effector changes the maximum velocity. As discussed in Section III,B, the binding site for CTP is within the R subunit. It seems probable (see below) that ATP is bound to the same site. If so, mediation of both inhibitory and stimulatory heterotropic allosteric effects in ATCase must involve interactions between the effector site on the R subunit and the active site on the C subunit, either by direct interaction of the two kinds of sites in native ATCase, or, much more likely, indirectly through conformational changes. Homotropic effects probably depend on changes in tertiary and quaternary structure as well, and since C subunit does not show cooperativity these effects must be mediated at least in part through C subunit:C subunit contacts.

2. *The Effector Binding Site*

Gerhart and Pardee (*10*) concluded that the entire CTP molecule is required for optimum inhibition. At 2 mM, cytosine does not inhibit, cytidine is a weak inhibitor, and each phosphate added to the 5' position of the nucleoside, up to CTP, increases the affinity for the enzyme markedly. Other ribonucleoside triphosphates are much weaker inhibitors, indicating that the nature of the base is important. Activation of the enzyme by ATP suggests that this purine nucleotide, if it interacts with the enzyme at the same site as CTP, must do so in a fundamentally different manner.

Recently, London and Schmidt (*93*) have developed a detailed model for the nucleotide binding site of ATCase, based on the following observations:

1. Cytidine inhibits, and the maximal inhibition produced by this compound approaches that obtainable with CTP (*92*). At con-

92. J. C. Gerhart and A. B. Pardee, *Fed. Proc., Fed. Amer. Soc. Exp. Biol.* **23**, 727 (1964).
93. R. E. London and P. G. Schmidt, *Biochemistry* **11**, 3136 (1972).

centrations higher than those tested by Gerhart and Pardee (10), cytosine also inhibits. However, 3-methyl cytidine is ineffective.
2. Inosine triphosphate inhibits weakly, in a manner qualitatively similar to inhibition by GTP. Inosine also inhibits, but 1-methyl inosine does not.
3. Adenosine monophosphate cannot produce the same maximal activation as ATP. Adenosine is a very weak activator, but its insolubility precludes studies at high concentrations.
4. 2-Thio UMP is a strong allosteric inhibitor of ATCase (94).
5. Uridine triphosphate is not a good inhibitor (10).

The development of the model is based on the assumption that CTP and ATP both interact with the same site on the enzyme. This assumption is consistent with the observations that ATP decreases the binding of CTP competitively (31) and that the rate of digestion by proteases (59), the protection of native ATCase against dissociation by low concentrations of sodium dodecyl sulfate (75), and the electron spin resonance spectrum of spin-labeled ATCase (95) are affected similarly by CTP and ATP. It is also reasonable that the identical ribose triphosphate portions of the two molecules should allow them to interact with the same site. However, direct confirmation of this hypothesis will probably have to await determination of the complete three-dimensional structure of the enzyme.

The model of London and Schmidt is represented in Fig. 14. All allosteric inhibitors have in common a basic ring nitrogen, located three atoms from the glycosidic bond, which is capable of binding to an electrophilic center in the protein (X in the figure). Thus, N-3 of CTP (Fig. 14a) and N-7 of GTP and ITP (Fig. 14b) could all interact with this electrophilic center. However, UTP (in which N-3 is protonated, Fig. 14c), 3-methyl cytidine, and 7-methyl inosine are poor inhibitors. Activation by ATP depends on a different kind of interaction. The most basic nitrogen of ATP is not N-7 but N-1. Thus, the enzyme interacts with a different conformation of ATP and, therefore, the structure of the binding site in the ATP complex is different than it is, for example, in the GTP complex. A hydrogen bond from residue Y in Fig. 14 to the base is suggested by other evidence (93) but is not essential to the model.

The most probable conformations of the various effectors in free solution are the ones which are proposed to bind optimally. For example, in the favored *anti* conformations, N-1 of ATP and N-3 of CTP are

94. M. E. Goodrich and P. Cardeilhac, *BBA* **222**, 621 (1970).
95. T. Buckman, *Biochemistry* **9**, 3255 (1970).

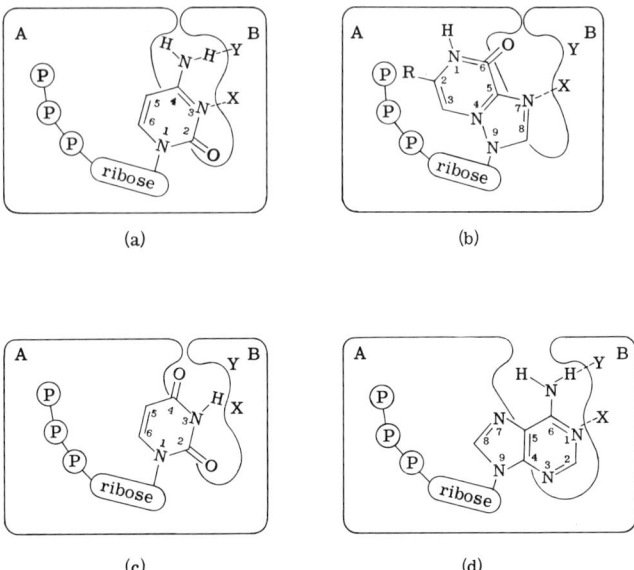

FIG. 14. The proposed interaction of several nucleotides with the regulatory site of ATCase, according to London and Schmidt (93). The site is considered to be composed of parts A and B, whose relative orientation is contracted when the enzyme is in a low energy state (a) and (b) and expanded when the enzyme is in a high energy state (d): (a) CTP bound to a contracted conformation of the site, (b) ITP (R=H) or GTP (R=NH$_2$) bound to the same conformation, (c) UTP unable to interact with part B of the site, and (d) ATP bound to an expanded conformation of the binding site.

placed in similar positions, facing away from the phosphate chain (93). What, then, is the basic difference in the binding of ATP which makes it an activator? From the observation that saturating ATP can induce more activation than saturating AMP, London and Schmidt postulated that the additional binding energy provided by the two additional phosphates of ATP allows it to bind to a minor, high energy conformation of the effector site in which the relative positions of the enzyme-bound groups which interact with the base and phosphate of the nucleotide are altered with respect to their positions in the major, low energy conformation to which inhibitors are bound. The fact that the maximum inhibition produced by cytidine is nearly the same as that produced by CTP, whereas the maximum effect of cytosine is considerably smaller, indicates that the ribose moiety is important in allowing cytidine to discriminate between the low and high energy states; this discrimination may depend on the relative position of the ribose and base binding sites in the two states. The recent analysis by Gerhart (6) is also consistent with such a hypothesis. In the high energy state of Fig. 14, with ATP bound, the base binding site is in a more expanded conformation relative to the ribose triphosphate site (i.e., the sites A and B are farther apart) than when CTP is bound. Consequently, the inhibitors ITP and GTP (Fig. 14b) must be in the *syn* conformation when bound to the more contracted state of the enzyme.

The strong inhibition of ATCase by 2-thio UMP is taken by London and Schmidt (*93*) as evidence that the electrophilic group associated with the binding of the basic ring nitrogens may in fact be the Zn^{2+} of the R subunit since Zn^{2+} is known to interact strongly with thiols as well as with basic ring nitrogens of nucleosides. Binding of 2-thio UMP in this manner would place it in a position favorable for interaction with the contracted form of the nucleotide binding site. The possibility of a direct metal–nucleotide interaction is made unlikely, however, by recent results of J. H. Griffin, J. P. Rosenbusch, K. Weber, and E. R. Blout (unpublished), who found that no change in the optical activity of the cadmium chromophore in cadmium ATCase or in the cadmium R subunit accompanies the binding of CTP.

If the model of London and Schmidt is essentially correct, then how are the different conformational states of the effector site sensed at the active site to produce either activation or inhibition? If the effector site is near the R:C domain, a conformational change at this site could affect the conformation of the active site on the C subunit directly (*93*). A more likely possibility is that conformational changes that occur within the R subunit as a result of effector binding are transmitted indirectly through an overall conformational change of the entire ATCase oligomer.

C. STOICHIOMETRY OF LIGAND BINDING

1. *Substrates and Substrate Analogs*

Equilibrium binding of succinate to native ATCase in the presence of carbamyl-P was studied by Changeux *et al.* (*31*). The binding curve is sigmoidal with a Hill coefficient of 1.6. A Scatchard plot of the data is complex, but the linear portion was extrapolated to 3.8 molecules of succinate bound per 300,000 daltons of enzyme. Based on the subunit structure of native ATCase, there should be 6 active sites per molecule. A difference spectrophotometric titration of native ATCase by PALA gives a stoichiometry of 6 PALA per 310,000 daltons and a difference spectral magnitude identical to that obtained with succinate plus carbamyl-P (*84*). Therefore, it now seems likely that there are 6 sites per

ATCase molecule capable of binding PALA. Evidence that there are 3 sites per C-subunit trimer has been discussed in Section III,A.

Rosenbusch and Griffin (38) have determined the stoichiometry of binding of carbamyl-P to ATCase. In the absence of succinate, only three molecules of carbamyl-P are bound per 300,000 daltons, whereas six molecules of carbamyl-P are bound in the presence of succinate. This exciting result has profound consequences for allosteric models of ATCase and is discussed at greater length in Section VII,H.

2. Nucleotides

The binding of the inhibitors CTP and BrCTP, both to native ATCase and to the isolated R subunit, has been studied under a variety of conditions by several different methods. Equilibrium dialysis and a phase-partition method carried out near 0° in phosphate buffer, pH 7, revealed two classes of binding sites, 1 of 3 strong sites and 1 of 3 weak sites per 310,000 daltons. However, the binding constants for two classes differ depending upon the method used (48, 96). Buckman (95) evaluated CTP binding to native ATCase at 23° (buffer and pH not specified) by measuring the concentration of unbound ligand in the supernatant of solutions from which the enzyme had been pelleted by ultracentrifugation. This study revealed 3 strong sites plus 6 weak sites per enzyme molecule. Changeux et al. (31) also obtained evidence for two classes of sites for CTP binding both to native ATCase at 21° and to isolated R subunit (dissociated by p-mercuribenzoate) at 4° in phosphate buffer at pH 7. However, experiments with zinc R subunit by Rosenbusch and Weber (26) in tris acetate buffer, pH 8.2, revealed only one class of sites for CTP over a limited concentration range; a stoichiometry of 1 site per 17,000 daltons was calculated. Cohlberg et al. (27) have shown that CTP binding to zinc R subunit is stronger in tris than in phosphate buffers. Hammes et al. (37) have shown that native ATCase binds approximately six molecules of BrCTP per 310,000 daltons in the presence of carbamyl-P and that isolated R subunit (dissociated by p-mercuribenzoate) binds 1 per 17,000, using the method of continuous variation.

Most recently, Matsumoto and Hammes (97) have found 6 binding sites for CTP in native ATCase at either 40° or 23° in the presence of saturating carbamyl-P with or without saturating succinate. The data quantitatively fit a model in which there are two sets of three sites, with binding constants for CTP which differ by one to two orders of

96. C. W. Gray and M. Chamberlin, *Anal. Biochem.* **41**, 83 (1971).
97. S. Matsumoto and G. G. Hammes, *Biochemistry* **12**, 1388 (1973).

magnitude. In the absence of carbamyl-P, additional weak binding sites for CTP are found, but the binding constants for the 6 regulatory sites are essentially unchanged. There is also evidence that the extent of activation or inhibition of the enzyme upon binding ATP or CTP, respectively, depends on the degree of saturation of all 6 sites rather than either of the two classes (97). These heterotropic interactions must be fitted to a model more complex than the simple, two-state model proposed by Monod et al. (98).

Cook (98a) has carefully examined the binding of CTP to ATCase at 4° and 23° by equilibrium dialysis. His data indicate six sites for CTP, but complete saturation of these sites is not achieved at 23°. The binding of CTP is complex, with evidence for both positive and negative cooperativity. Cook favors a ligand-induced sequential model for the cooperativity he observes; however, a model with dissimilar classes of sites is not ruled out.

A direct comparison of all of these data is difficult because different experimental conditions were used. In addition, exact determinations of binding constants are precluded in some of the studies discussed above because the phosphate buffer which was used would be expected to interact with the triphosphate portion of the nucleotide binding site. It does seem highly probable, however, that each R chain can bind one CTP molecule, both as an isolated dimer and when present in native ATCase. It is difficult to compare the data of Changeux et al. (27) on isolated R subunit with those of Rosenbusch and Weber (26), since the former preparation of R subunit probably contained mercury and no zinc and the latter contained primarily zinc. It is clear that, at least under some conditions, both isolated R subunit and native ATCase can be shown to contain heterogeneous sites for CTP. It is not clear whether this heterogeneity results from negative cooperativity between sites or represents two distinct classes of sites and what relationship heterogeneous binding might have to the allosteric mechanism. We must await more critical experiments with both the isolated zinc R subunit and native ATCase under comparable conditions.

D. Conformational Changes Induced by Ligands

1. Substrates and Substrate Analogs

Most of the evidence for conformational changes in ATCase upon binding carbamyl-P and succinate has already been reviewed in Section VI,A, points 6 and 9, respectively. Gerhart (6) has critically discussed

98. J. Monod, J. Wyman, and J.-P. Changeux, JMB 12, 18 (1965).
98a. R. A. Cook, Biochemistry 26, 3792 (1972).

much of the same evidence in terms of a model of ATCase in which homotropic cooperativity leads to a more open structure (R state). More recently, Pigiet (72) has shown that the spectral changes due to the binding of substrate ligands to native ATCase are qualitatively similar to the changes which occur with the isolated C subunit. Using circular dichroism difference spectroscopy, Griffin et al. (73) have shown that succinate plus carbamyl-P perturb tryptophan residues both in the C subunit and in native ATCase, but only in the case of native ATCase are there additional components in the difference spectra which suggest perturbation of buried tyrosyl residues.

Pigiet (72) has also shown that, in the presence of saturating carbamyl-P, some measures of conformational change are directly linked to the binding of succinate (as in the isolated C subunit), whereas others precede the binding of succinate and are complete at succinate concentrations which fail to saturate the first class. Similarly, Griffin et al. (99) have shown that in the presence of saturating carbamyl-P, the succinate concentration required for half-maximal change in the circular dichroism difference spectrum of cadmium ATCase is 0.2 mM for the cadmium chromophore of the R subunit but 0.4 mM for the tryptophan chromophore of the C subunit. (These difference spectra reflect conformational changes in the cadmium site of the R subunit upon the binding of ligands to the C subunit.) Results such as these strongly suggest the existence of hybrid conformational states, for example, a molecule of enzyme in the R state with only some of the catalytic sites in the "succinate binding" conformation. Since spectroscopic perturbations reflect changes in the immediate environment of a chromophore, it is not possible to suggest how extensive or how localized are the differences between conformational hybrids.

2. Allosteric Effectors

Changes in immunologic reactivity (58), sedimentation rates and reactivity toward p-mercuribenzoate (47), and susceptibility to proteases (59) upon the binding of ATP, CTP, or BrCTP to ATCase could reflect conformational changes primarily within the R subunit (tertiary changes) but may also reflect more extensive changes in the conformation of the whole ATCase molecule (quaternary changes). Other methods clearly reflect the changes in quaternary structure which must accompany the binding of allosteric effectors. Pigiet (72) has found that BrCTP causes a *decrease* in the optical rotation of native enzyme of

99. J. H. Griffin, J. P. Rosenbusch, K. Weber, and E. R. Blout, Fed. Proc., Fed. Amer. Soc. Exp. Biol. 31, 423 (1972) (abstr.).

50°, in contrast to the 65° *increase* caused by carbamyl-P plus succinate; UTP, which binds to the same site but does not modify the catalytic properties of the enzyme, had no effect. However, the effect of BrCTP on isolated R subunit was not studied. The binding of CTP to ATCase on to the isolated R subunit gives rise to circular dichroism difference spectra which are qualitatively similar but significantly different, indicating that there are some perturbations unique to the native enzyme (73). Buckman (95) has evaluated the effects of CTP and ATP on the mobility of a spin label localized in the C subunit within native ATCase. Both allosteric ligands caused broadening of the electron spin resonance signal, but UTP had no effect. This result again suggests a conformational change in the R subunit, which is transmitted in turn to the C subunit. Finally, Colman and Markus (75) have shown that concentrations of sodium dodecyl sulfate up to 0.5 mM cause dissociation of ATCase into subunits and that the integrity of the native enzyme is stabilized by both ATP and CTP. Adenosine triphosphate and CTP also protect against dissociation by heat.

The circular dichroism difference spectra resulting from the binding of CTP are the same for the zinc and cadmium forms of the R subunit. The spectra for the zinc and cadmium forms of native ATCase are also the same. Since no spectral change in the cadmium chromophore is observed, the conformation of the metal binding site is probably not altered by CTP and the metal is probably not involved either in CTP binding or in conformational changes caused by CTP (J. H. Griffin, personal communication). Since an intense cadmium difference spectrum *is* observed when carbamyl-P and succinate are bound (99), the conformational changes induced in these two instances cannot be the same. As will be discussed below, several other kinds of data also indicate that more than two states are probable.

E. A Comparison of the Properties of Native ATCase with Those of Its Subunits

Radioactive ATCase does not exchange with unlabeled C subunit, i.e., the half-time for such a reaction is greater than 4 months (49). Furthermore, zinc in the native enzyme cannot be removed by chelators and does not exchange with radioactive Zn^{2+}, even though Zn^{2+} is removed readily from the R subunit (28). Thus, the subunits are associated very tightly to form the native enzyme, and one may ask what changes in structural or chemical properties accompany this association. The reactivity of the SH group of the C subunit toward DTNB is lost in the native enzyme (*35*), but only small changes are seen when the optical rotatory dispersion spectrum of native ATCase is compared with the weighted average of the spectra of the two kinds of subunits (*27*). The ultraviolet spectrum of the native enzyme is not identical to the weighted average of the two subunit spectra and the change in absorbancy has been interpreted as resulting from transfer of tyrosyl residues to a more hydrophobic environment upon association (*73*). When the circular dichroic spectra are similarly compared, significant differences are observed which indicate that tryptophan residues in the C subunit and tyrosine residue in either or both kinds of subunits are perturbed upon aggregation to ATCase (*27, 73*). Kirschner (*88*) has found that the same specifically nitrated tyrosine in the C subunit which responds to the binding of ligands (see the discussion below) also responds in similar fashion to association with the R subunit. In summary, there are substantial changes in the properties of the subunits when they are combined to form ATCase, as might have been expected.

F. Cooperative Properties of Native ATCase after Structural Modifications

Modifications of native ATCase which alter expression of the normal cooperative properties are useful in studying the structural bases of cooperativity. Several such modifications will now be discussed.

1. *Effects of Urea and pH*

Relatively low concentrations of urea (0.8–1 M) have been shown to greatly lower or eliminate CTP inhibition of native ATCase (*10, 25*) and to largely abolish the sigmoidal saturation curve for L-aspartate (*25*). This effect is reversible: The enzyme reverts to its original fully sensitive cooperative behavior after removal of the urea (*25*).

Native ATCase has a pH dependence with two maxima, at pH 8.5 and at pH 10.2 (*25*). When assayed at pH 8.5, native ATCase shows fully sensitive cooperative behavior, while at pH 10.2 the L-aspartate saturation curve is more hyperbolic, and CTP at low concentrations (<0.25 mM) inhibits poorly when compared to its effect at pH 8.5 (*25*). The pH effect is reversible, too. Neither 1 M urea nor pH 10.2 cause native ATCase to dissociate into its subunits (*83*). The effects observed in these cases must be related to multiple changes in the interactions among the subunits (quaternary structure) caused by multiple changes in local conformation (tertiary structure) and thus cannot by themselves be used to pinpoint specific interactions crucial to mechanisms of cooperativity.

2. Allosteric Properties of ATCase Modified with Partial Specificity

Native ATCase reacts with cyanate generated *in situ* from carbamyl-P to yield a derivative with approximately 4–5 times the specific catalytic activity of the unmodified enzyme at low concentrations of L-aspartate [G. R. Stark, unpublished observations (*49*)]. Nelbach (*49*) has shown that the carbamyl derivative forms a sharp electrophoretically homogeneous boundary at several pH values and has about 25 stable carbamyl groups per C_6R_6 unit. Carbamyl groups are located on both kinds of polypeptide chain, the molecular weight of the modified enzyme is unchanged, and both homotropic and heterotropic cooperativity is lost. The L-aspartate saturation curve is similar to that of the C subunit, accounting for the higher activity at low concentrations of L-aspartate.

Limited succinylation of the C subunit with 2.1 moles of succinic anhydride per lysine residue yields a catalytically inactive product (C_S), which, however, retains the trimeric subunit structure of unmodified C subunit (*33*). Hybrid C subunits containing one or two succinylated C chains can be prepared (i.e., C_{nns} and C_{nss}) (*33*); $C_S(=C_{sss})$ and $C_N(=C_{nnn})$ can be recombined with R subunit to yield $(C_S)_2R_3$, $(C_SC_N)R_3$, and $(C_S)_2R_3$, which can be separated from one another (R represents the R-subunit dimer). Hybrid C subunits also recombine to form native-type molecules designed as $(C_{nns})_2R_3$ and $(C_{nss})_2R_3$ (*72*).

The allosteric properties of several of these recombined hybrid molecules have been studied by Pigiet (*72*). The hybrid $(C_NC_S)R_3$, consisting of one fully succinylated and one normal C subunit, shows reduced but significant heterotropic and homotropic interactions, as judged both by kinetic measurements and by difference sedimentation velocity in the presence of ligands. However, CTP binding to the R subunit of this hybrid is unaltered when compared to the binding to native ATCase.

Both $(C_{nns})_2R_3$ and $(C_{nss})_2R_3$ have maximal activities which are in proportion to the number of "good" C chains per native molecule: $(C_{nns})_2R_3$ has a sigmoidal L-aspartate saturation curve and is inhibited by CTP; $(C_{nss})_2R_3$ shows a small degree of homotropic cooperativity, as judged by the L-aspartate saturation curve, and also retains some sensitivity to CTP. The result with $(C_{nss})_2R_3$ suggests that the two active C chains, one on each trimer, can interact cooperatively with respect to L-aspartate binding through the R chains since the inactive succinylated C chains bind only carbamyl-P and not L-aspartate (*72*).

Bromosuccinate reacts with native ATCase in at least two fundamentally different ways (*83*). In the absence of substrates or substrate analogs, it inactivates the enzyme with incorporation of 0.8 molecule of succinate per C chain and 1.1 molecule per R chain. However, in the presence of maleate and carbamyl-P (which would produce the large conformational change associated with catalysis in the C subunit, see Section VI), bromosuccinate activates native ATCase approximately 50%, and the modified enzyme is not inhibited by CTP. Under these conditions of modification, 0.9 molecule of succinate are incorporated per C chain and 1.9 per R chain. The amino acid residues modified have not been identified. Apparently, maleate and carbamyl-P protect the active site from reaction with bromosuccinate, and modification of other residues on the C chains and R chains prevent the subunit interactions ordinarily responsible for normal cooperative effects. It is clear that activation is not associated with dissociation of the native enzyme into subunits.

Limited tryptic digestion of native ATCase in the presence of L-aspartate results in a loss of both homotropic and heterotropic interactions in the enzyme without destroying the catalytic activity (*60, 100*). The R subunit appears to be preferentially digested in native ATCase and the loss of cooperative properties is thus ascribed to selective cleavage of the R chains. A functional model for allosteric behavior is presented in which the R subunits play a central role. Such a model is consistent with models based on more recent evidence and discussed below.

3. *Specifically Modified ATCases*

C subunit reacts with tetranitromethane at pH 6.7 in the presence of carbamyl-P and succinate to yield a nitrated product containing 0.7–0.9 nitrotyrosine residues per C chain and retaining 75–90% of its original catalytic activity (*88*). No other amino acid residues appear to be affected significantly. It is likely that a unique tyrosine is involved, possibly one with an abnormally low pK, since the pK of the nitrated derivative is abnormally low. Only pairs of ligands which cause a major conformational change in the C subunit, as measured by difference spectroscopy (*62*) or change in sedimentation coefficient (*71*), produce a significant change in the spectrum of the nitrated C subunit.

This derivative recombines with R subunit to form a modified native ATCase which shows both homotropic and heterotropic cooperativity, but both effects are somewhat reduced when compared to those of the unmodified enzyme (*88*). As with the nitrated C subunit, pairs of substrate ligands which produce a major conformational change in C subunit change the spectrum of the nitrotyrosine chromophore, but unlike the nitrated C subunit, carbamyl-P alone also produces a small change. When the modified ATCase is titrated with succinate in the presence of

---

100. G. Markus, D. K. McClintock, and J. B. Bussel, *JBC* **246**, 762 (1971).

carbamyl-P, the saturation curve, as measured by the change of absorption of the nitrotyrosine residue, lags behind the curve obtained when the major conformational change is measured by difference sedimentation (88). This result is consistent with data obtained by Changeux et al. (31) and Gerhart and Schachman (47), who demonstrated that the saturation curve of succinate in the presence of carbamyl-P as measured by equilibrium dialysis is sigmoidal and lags behind the curve for the major conformational change, which appears to be nearly hyperbolic. Both sets of studies are therefore consistent with the idea that the major conformational change produced in ATCase in the presence of succinate and carbamyl-P is characterized by a saturation curve more hyperbolic than the curve which characterizes the binding of succinate and that a smaller concentration of succinate is required for half-saturation of the conformational change than is required to half fill the binding sites. Therefore, the extent of this major conformational change in native ATCase is *not* a linear function of succinate saturation.

As discussed in Section VI,B, a number of derivatives of the single SH group of the C chain have been made, using the intact C subunit. The active thiocyanate and S-methyl derivatives recombine with R subunit to form modified native ATCases (84). The recombined derivatives have homotropic and heterotropic cooperative interactions which are very similar to those of unmodified ATCase, suggesting that the SH group is *not* involved directly in relaying conformational information associated with cooperative transitions (for instance, as a ligand to the Zn^{2+} ion).

The thionitrobenzoate derivative of C subunit (TNB C subunit), prepared by reacting the SH groups of undenatured C subunit with DTNB, is inactive but recombines in the proper stoichiometry with R subunit to form a modified ATCase (84). Inactive TNB C subunit probably does not bind succinate (see above) but does bind carbamyl-P and the transition state analog PALA. The difference spectra obtained when the TNB C subunit is saturated with carbamyl-P or PALA are shown in Fig. 15. The two spectra are similar in the region of the TNB chromophore, with difference maxima at about 315 nm for carbamyl-P and about 320 nm for PALA. Addition of succinate alone does not give rise to a TNB difference spectrum. The similarity of these spectra in this region suggests that the spectral change of the TNB chromophore does *not* result from the major conformational change which is induced by PALA (over that induced by carbamyl-P alone) but rather results from perturbation of the TNB group resulting from an interaction with that part of the PALA molecule which corresponds to carbamyl-P. The protein difference spectrum induced by PALA (36) can be seen easily in the

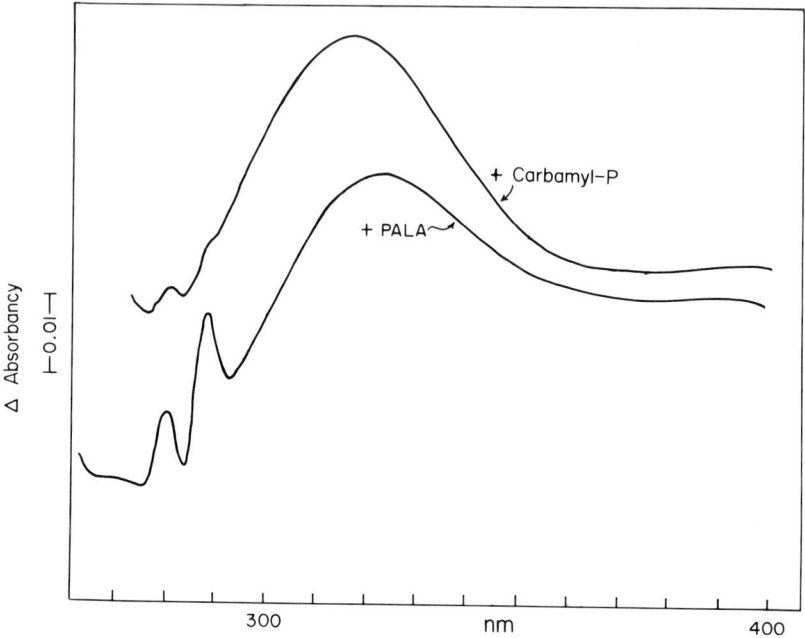

FIG. 15. Ultraviolet difference spectra of TNB C subunit in the presence of carbamyl-P (upper curve) or PALA (lower curve) vs. unliganded TNB C subunit. The curves have been displaced vertically by 0.005 absorbancy unit for clarity. Upper curve: The sample cell contained 6.8 mg/ml of TNB C subunit and 0.995 mM carbamyl-P in 0.2 M tris acetate buffer, 0.2 mM in EDTA, pH 8.0. The reference cell was the same, minus carbamyl-P. Light path, 0.2 cm. Lower curve: The sample cell contained 1.67 mg/ml of TNB C subunit and 12.4 µM PALA in 0.2 M tris acetate buffer, 0.2 mM in EDTA, pH 8.0. The reference cell was the same, minus PALA. Light path, 1.0 cm.

PALA curve of Fig. 15. It has approximately one-third the magnitude at saturation which would be expected if all three sites of the TNB C subunit were to bind PALA. To confirm this stoichiometry of binding, TNB C subunit has been titrated with PALA, using the TNB portion of the difference spectrum to monitor PALA binding (84). Only one molecule of PALA binds to this derivative with an affinity approximately like that for binding to unmodified C subunit, indicating that the other two sites of the trimer become inaccessible to PALA when one site is filled. Similarly, only one molecule of carbamyl-P is bound per trimer. This surprising result could be explained by one of the possibilities shown in Fig. 16: (I) The binding of PALA to one site may displace the TNB chromophore from that site into a position where it can physically block access of PALA to the other two sites or (II) PALA binding to one site may induce

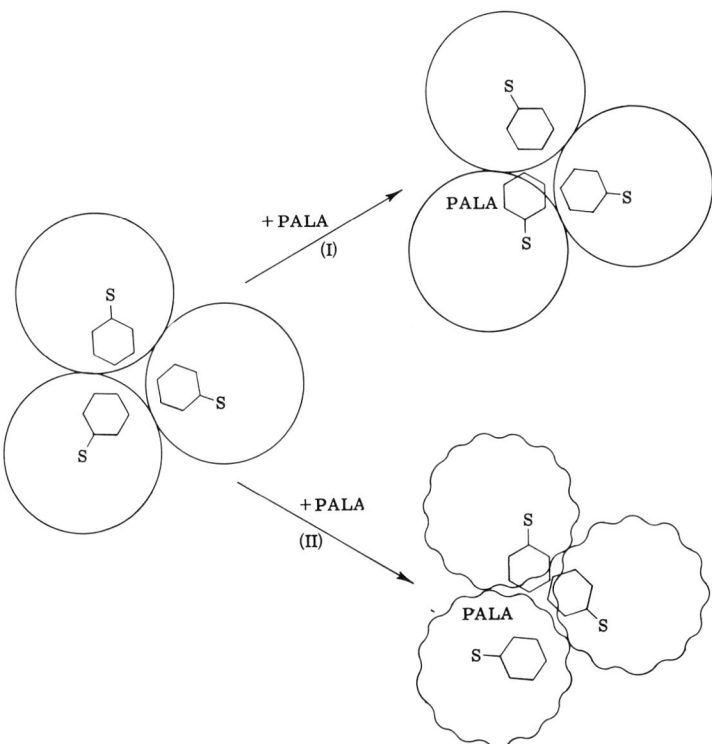

FIG. 16. Possible explanations for the stoichiometry with which PALA binds to the TNB C subunit. In (I) the TNB moiety (represented by S—◯) is displaced from the site to which PALA is bound and blocks access to the other two sites. In (II) the binding of one molecule of PALA induces a conformational change in the whole trimer distinct from the one associated with catalysis, which results in obstruction of the other two sites by their own TNB groups. Alternatively, there may be preexisting asymmetry in the sites.

a conformational change in the other two sites which, in combination with the TNB groups in these sites, prevents binding of PALA to them, or the three sites may be dissimilar before PALA is bound. Scheme (I) is not a reasonable possibility because, as shown in Fig. 17, mercury atoms attached to the SH groups of the C subunit lie about 22 Å apart (50). If Scheme (II) were correct, PALA would have to induce conformational changes in the two unliganded chains of the TNB C subunit trimer and yet, by all previous criteria, C subunit is not cooperative. The conformational changes detected in this way could be important to allosteric transitions in native ATCase and still have no cooperative effect in the isolated C subunit.

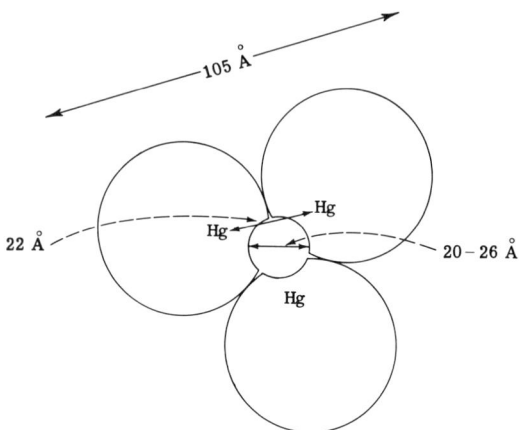

FIG. 17. Cross-sectional representation of the relative positions of mercury atoms bound to the three C-subunit SH groups. The view is down the threefold axis of native ATCase at the level of the active sites. No attempt has been made to represent the R subunits. The entire ATCase molecule, including the R chains, appears triangular when viewed down the 3-fold axis; each side of the triangle is 105 Å ± 10 Å long. All dimensions are from Wiley et al. (50).

The capacity of a modified ATCase composed of TNB C subunit and zinc R subunit to bind PALA has also been investigated. The recombined derivative binds only two PALA molecules per 300,000 daltons, as predicted on the basis of the stoichiometry with the TNB C subunit (84). This binding, in contrast to that with the isolated TNB C subunit, is *cooperative*, further confirming the conclusion of Pigiet (72) that homotropic interactions can take place between active sites on *different* C-subunit trimers within native ATCase.

4. *Other ATCases with Modified Allosteric Properties*

Beckwith et al. (101) first observed that when a uracil-requiring mutant of *E. coli* was derepressed for ATCase in the presence of 2-thiouracil, the enzyme produced had altered allosteric properties. Recently, Kerbiriou and Hervé (102) have investigated this enzyme further and found that it lacks homotropic interactions but is still sensitive to inhibition by CTP. The enzyme contains both types of subunits but has a slightly altered electrophoretic mobility compared to that of normal ATCase. The presence of 2-thiouracil during biosynthesis is required for produc-

101. J. R. Beckwith, A. B. Pardee, R. Austrian, and F. Jacob, *JMB* **5**, 618 (1962).
102. D. Kerbiriou and G. Hervé, *JMB* **64**, 397 (1972).

tion of the modified ATCase. The modified enzyme has many characteristics of the isolated C subunit, including a hyperbolic saturation curve for L-aspartate. The exact nature of the modification is unknown, but modification of SH groups may be involved (G. Hervé, personal communication).

A different altered ATCase with similar allosteric properties has been recognized as a product from recombination mixtures of the subunits in which the C subunit is in excess and as a minor component in normal preparations of ATCase (M. Syvanen, G. Nagel, and H. K. Schachman, personal communication; G. R. Jacobson and G. R. Stark, unpublished observations). This derivative has a slightly altered electrophoretic mobility and can be converted to a nativelike molecule by addition of more R subunit. It appears to lack one R subunit dimer (H. K. Schachman, personal communication). This R subunit deficient enzyme can still be inhibited by CTP but shows little activation by ATP or by succinate at low concentrations of L-aspartate and is considerably more active than native ATCase under these conditions, indicating a selective desensitization (G. R. Jacobson and G. R. Stark, unpublished observations).

G. Kinetics of Ligand Binding

Hammes and Wu (3) have summarized recently the extensive data they and their co-workers have gathered on the ATCase system, using temperature-jump relaxation techniques to measure very fast reactions. Most of the very interesting and extensive results have been discussed both by the principal authors (3) and by Kirschner (4) and hence they will only be presented very briefly here. Some more recent results are presented at greater length. In summary, the relaxation data of Hammes, Wu, and co-workers have added a great deal to our understanding of the cooperative processes in ATCase but have not succeeded in defining the cooperative mechanism uniquely. They pointed out that, in principle, this approach is capable of revealing elementary steps in the processes related to cooperativity, whereas averaging techniques such as steady state kinetics, spectroscopy, and ultracentrifugation may lack sufficient resolution. In practice, the temperature-jump technique may also be somewhat limited either because the elementary processes are too fast to measure at a convenient concentration of a ligand or because no change in an observable property of the system turns out to be coupled to the reaction of interest. Both sorts of complications were encountered in the work with ATCase. The latter difficulty is often overcome by coupling the processes of interest to rapid protonic equilibria involving a dye, and this has been a valuable tool in the work with ATCase. But it must be kept in mind that interesting and important processes may not result in uptake or liberation of protons and thus may be missed and, conversely, processes not central to the transformation of interest may complicate the relaxation spectra.

1. The binding of BrCTP was observed by utilizing changes in the absorption spectrum of this ligand. With isolated mercury R subunit or native ATCase, only a single relaxation time was observed, consistent with a bimolecular reaction.

2. In the presence of carbamyl-P and succinate, the binding of BrCTP to native ATCase reveals both the bimolecular step and a conformational change which can be interpreted to be quantitatively consistent with the concerted model of Monod et al. (98).

3. Utilizing pH coupling for observation, no relaxation spectra were observed for analogs of L-aspartate in the absence of carbamyl-P, and the binding of carbamyl-P either to the C subunit or to native enzyme is described satisfactorily by a simple bimolecular mechanism.

4. In the presence of saturating carbamyl-P, the binding of succinate or L-malate to the C subunit is consistent with a bimolecular reaction, followed by a conformational change. A detailed discussion of this process with respect to the catalytic mechanism can be found in Section VI,A. The same processes were observed with native ATCase and, in addition, another conformational change, presumably at the level of quaternary structure, was detected and found to be quantitatively consistent with the model of Monod et al. (98).

5. The concentrations of both BrCTP and aspartate analogs were varied at saturating carbamyl-P. Relaxations associated with the BrCTP and succinate-stimulated conformational changes were observed; furthermore, the value of the relaxation time associated with succinate binding was found to depend on the concentration of BrCTP and, conversely, that for BrCTP binding depends on the concentration of succinate. Although such synergistic effects are consistent with the idea that regulatory processes are being observed, the data are too complex to be accommodated by the simple two-state model of Monod et al. (98) since only one relaxation process should be observed in this case.

6. When the concentration of carbamyl-P is varied, yet another relaxation process is observed. There are at least three different conformational changes associated with the control mechanism, apart from the one associated with catalysis. All three show synergistic effects when BrCTP, succinate, and carbamyl-P are varied.

More recently, kinetic studies of the binding of CTP (103), BrCTP (104), and 6-mercapto-ATP (105) have revealed a single relaxation process with time constants in the range 0.1–1 msec, associated with binding to the regulatory site. In the presence of 2 mM carbamyl-P with or without 10 mM succinate, a conformational change is rate limiting in the binding process in all cases. In the absence of carbamyl-P, a conformational change is clearly rate limiting for 6-mercapto-ATP binding and probably also for BrCTP and CTP binding. For the binding of CTP and BrCTP to the isolated regulatory subunit, a simple bimolecular reaction mechanism is consistent with the data over a restricted concentration range, but a rate-limiting conformational change cannot be strictly excluded. If both BrCTP and mercapto-ATP are added to the enzyme, only a single relaxation process is seen, indicating that the conformational states involved in the observed transition are the same for both effectors.

A simple schematic mechanism which is consistent with all of the data is as follows:

$$ATP + E \rightleftharpoons X_1 \rightleftharpoons X_2$$
$$CTP + E \rightleftharpoons X_2 \rightleftharpoons X_1$$

In this mechanism, the effector–enzyme complexes for both CTP and ATP can exist in two different conformations, X_1 and X_2. However, the binding of ATP stabilizes the conformation X_2, while CTP binding stabilizes X_1. Carbamyl-P and succinate also stabilize the X_2 conformation for both effectors, as would be expected. Since there are two classes of binding sites for CTP and since complete inhibition appears to depend on binding to all of these regulatory sites (97), the transitions in the above mechanism cannot be concerted for the enzyme as a whole.

H. Possible Mechanisms for Cooperativity

1. Evidence for a Two-State Major Cooperative Transition and Indications of Further Complexities

The early data on cooperative properties of ATCase were quantitatively fitted to the Monod et al. model (98) by Changeux and Rubin

---

103. L. W. Harrison and G. G. Hammes, Biochemistry 12, 1395 (1973).
104. J. Eckfeldt, G. G. Hammes, S. C. Mohr, and C.-W. Wu, Biochemistry 9, 3353 (1970).
105. C.-W. Wu and G. G. Hammes, Biochemistry 12, 1400 (1973).

(*106*). The recent data of Kirschner (*88*), which indicate that the nitrotyrosine residue of an active derivative of the C subunit within native ATCase monitors a highly concerted conformational change, are also consistent with a simple two-state model. In an extreme situation for the native enzyme, in which both carbamyl-P and succinate are present at saturating levels, the rate constant and activation energy for dissociation of the first molecule of succinate from the saturated complex were found by Sykes et al. (*61*) to be very similar to the same parameters for the isolated C subunit, implying that in one limiting state for native ATCase (the R state) the active sites behave much as they do in the C subunit.

Gerhart (*6*) has discussed the conformational states of ATCase in terms of a more complex model in which there is a single highly concerted change of quaternary structure and more localized, more independent, changes in tertiary structure which give rise to "hybrid" forms of the enzyme. The data from temperature-jump relaxation kinetics (*3*) are clearly too complex for a simple two-state model, but they might be fitted to a model such as the one discussed by Gerhart.

In the discussion which ensues, "R state" and "T state" will be used for convenience to denote the two enzyme forms in the highly concerted major conformational change, but the reader should bear in mind that this conformational change alone is too simple to account for the overall regulatory mechanism.

Collins (*107*) has analyzed the data of Pigiet et al. (shown in Fig. 13), which describe the effect of varying the concentration of L-aspartate on reaction velocity in the presence of allosteric ligands, in terms of a two-state model. The data were converted to double reciprocal form as shown in Fig. 18. Two surprising observations are thereby revealed:

1. The double reciprocal plot is linear (i.e., the saturation curve is hyperbolic) above about 5 mM L-aspartate in the presence of 2 mM ATP, above about 8 mM L-aspartate in the absence of allosteric effectors, and above about 15 mM L-aspartate in the presence of 0.5 mM CTP, suggesting that the enzyme is wholly in the R state above these levels of L-aspartate.

2. The K_m of L-aspartate for the R state is significantly different in each case. It is 5.15 ± 0.1 mM in the presence of ATP, 5.75 ± 0.1 mM in the absence of any allosteric effectors, and 6.79 ± 0.1 mM in the presence of CTP, suggesting that these R states are significantly different,

---

106. J.-P. Changeux and M. M. Rubin, *Biochemistry* **7**, 553 (1968).
107. K. D. Collins, Ph.D. Thesis, Stanford University, Stanford, California, 1971.

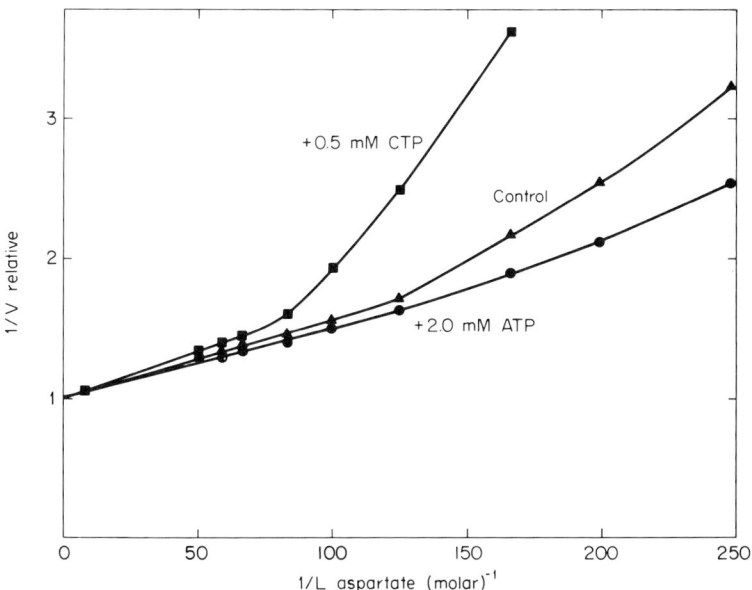

FIG. 18. The data of Fig. 13, replotted as a Lineweaver-Burk plot (*107*).

depending on the allosteric ligand. The K_m for L-aspartate with the C subunit is 20 mM under similar conditions (*15*).

Collins (*107*) has estimated the limiting K_m for L-aspartate for the T state of native ATCase from the activating effect of succinate as follows: The maximum activating effect, 3.3-fold, is observed at a succinate (or maleate) concentration which half-saturates the enzyme. If the enzyme is completely in the R state at this point, with K_m for aspartate equal to 5.75 mM (from Fig. 18), then it can be calculated that the limiting K_m for aspartate in the absence of succinate is at least 7.6 times larger than 5.75 mM. Changeux and Rubin (*106*) have estimated that about 20% of ATCase is in the R state in the absence of effectors and, since the presence of only 15% of the enzyme in the R state will explain the activity observed with 1 mM L-aspartate in the absence of succinate, the T state must make little or no contribution to the observed rate.

Collins (*107*) went on to show that a simple two-state analysis is consistent with the two curves of Fig. 13 which were obtained in the presence of 2 mM ATP and in the absence of allosteric effectors, except for small deviations near half-maximum velocity. The level of CTP used in obtaining the third curve of Fig. 13 was too low to allow analysis by the procedure used. It was also shown by the procedure of Blangy *et al.*

(108) that the data in the absence of allosteric effectors were consistent only with a concerted transition in which 6 sites participate, not 5 or 7.

The recent findings of Rosenbusch and Griffin (38) show that only three molecules of carbamyl-P bind to the 6 sites of native ATCase in the absence of succinate but that six bind in the presence of succinate. The Hill coefficient for the binding of carbamyl-P either in the presence or absence of succinate is near 1.0. These results are suggestive of a model for homotropic cooperativity such as the one represented in Scheme I.

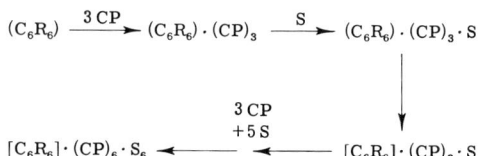

SCHEME I. A model for homotropic cooperativity in ATCase.

The binding of three molecules of carbamyl-P causes only minor conformational changes. These changes, or a preexisting asymmetry, prevent binding of three more molecules of carbamyl-P to the 3 unoccupied sites. Direct steric interference between bound carbamyl-P molecules is highly unlikely because the sites in unliganded ATCase are too far apart (see Fig. 20). When a single molecule of succinate binds, the major T → R transition is triggered, allowing full saturation with carbamyl-P and converting the sites unliganded by succinate to the high-affinity state.

2. *Structural Models for Cooperativity*

The rotational model of Gerhart (6) still seems to be the most attractive possibility for the major change in quaternary structure designated as the R → T transition. The newer evidence from electron microscopy and X-ray diffraction (Section V) indicates that Gerhart's model (Fig. 11 in reference 6) needs to be revised slightly: When the C subunits are eclipsed (viewed down the 3-fold axis), the R subunits are not (Fig. 19). In the absence of ligands, ATCase is predominantly in the T form as shown in Fig. 19A. The rotational model suggests that in the R form the two halves of the molecules are displaced relative to one another by a rotation of 60° (Fig. 19B) so that, viewed down the 3-fold axis, the C subunits are not eclipsed. The R chains are shown as eclipsed in Fig. 19B, but there is no evidence presently available to distinguish between this possibility, which would result from a counterclockwise rotation of the lower half of the enzyme of Fig. 19A relative to the upper half, and the alternative possibility that the R chains become

108. D. Blangy, H. Buc, and J. Monod, *JMB* **31**, 13 (1968).

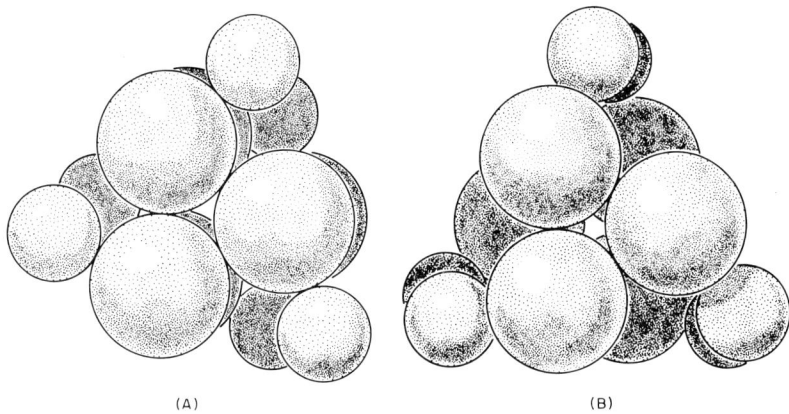

FIG. 19. A rotational model for the R → T transition of ATCase, modified from Fig. 11 of Gerhart (6). (A) The T state (absence of ligands). (B) The R state (one of two possibilities, see text).

less eclipsed as a result of a clockwise rotation of the lower half. A critical test of the rotational model would be to repeat the electron microscopy experiments of Richards and Williams (39) in the presence of a tightly bound ligand which favors the R state, such as PALA.

In addition to this major rotational motion, the overall R and T states must be modulated by additional, more minor, structural alterations which further modify the properties of the system. Such alterations are strongly suggested, for example, by the temperature-jump relaxation data. Thus, the R state with saturating PALA may be different in detail from the R state with saturating ATP. For example, although the R subunits might be eclipsed in both cases, the positions of the C subunits relative to the threefold axis might be different.

Very recently, Evans et al. (108a) and Warren et al. (108b) have presented a refinement of the previous crystallographic data at 5.5 Å resolution, leading to an interesting and novel possibility for allosteric regulation. The model now proposed is illustrated schematically in Fig. 20. A prominent feature is a large central cavity about $50 \times 50 \times 25$ Å in size. The location of the single SH group of each C chain suggests that the nearby active site is most probably accessible from the central cavity but not directly from the external solution. There are six channels, each about 15 Å in diameter, near the regulatory region which provide the most ob-

108a. D. R. Evans, S. G. Warren, B. F. P. Edwards, C. H. McMurray, P. H. Bethge, D. C. Wiley, and W. N. Lipscomb, Science 179, 683 (1973).
108b. S. G. Warren, B. F. P. Edwards, D. R. Evans, D. C. Wiley, and W. N. Lipscomb, Proc. Nat. Acad. Sci. U. S. 70, 1117 (1973).

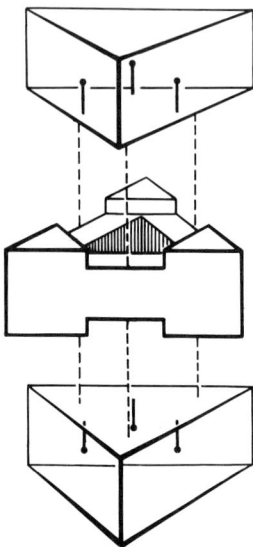

FIG. 20. An expanded view of ATCase based on crystallographic data, courtesy of W. N. Lipscomb [see Warren et al. (108b)]. A schematic representation of the major part of the trimeric C subunits is shown above and below, with a mostly regulatory region between. The assembled molecule has a central cavity (shaded region) about 25 Å along the 3-fold direction and about 50 Å along the 2-fold directions. The overall dimensions are about 90 Å along the 3-fold axis and 110 Å along the 2-fold axes. The largest openings from the outside to the central cavity are the six 15-Å channels in the regulatory region. Some of the central region near the ends of the dashed lines may be part of the C subunits, which may be nearly in contact with each other. Dots within the C subunits represent the mercury sites, which are easily accessible from the central cavity. The diagram is not meant to imply a sequential order of assembly of the subunits.

vious access to the central cavity. A component of the regulatory mechanism may be modulation of access of substrates through these channels. Critical evidence concerning this possibility will come from studies of liganded enzyme.

VIII. Biosynthesis and Genetics of Bacterial Aspartate Transcarbamylases

A. LOCATION OF THE GENES

The gene *pyrB* codes for the production of ATCase activity in *Escherichia coli* and *Salmonella typhimurium*. It is located at about 11 o'clock on the linkage map of both organisms and is unlinked to any of the other genes of the pyrimidine pathway. O'Donovan and Neuhard

(5) have reviewed much of the genetics of this locus, and only a brief summary of their conclusions will be presented here along with some more recent evidence concerning the structure and organization of the *pyrB* locus.

Mutants in both *E. coli* and *S. typhimurium* lacking ATCase activity have been isolated (*5, 18, 101, 109, 110*). In addition, regulatory mutants in *S. typhimurium* have been isolated by their overproduction of pyrimidines and pyrimidine intermediates (*5, 110*) and by their insensitivity to the pyrimidine analogs 5-fluorouracil and 5-fluorouridine (*110*). Of these strains, some apparently are linked to the *pyrB* locus and have lower sensitivity to the feedback inhibitor CTP, while most of the others are constituitive for enzymes of the pyrimidine pathway but unlinked to *pyrB* or loci for other enzymes of the pathway (*110*). All of these mutant enzymes show some inhibition by CTP (*110*); those mutant enzymes which are less sensitive to CTP may be altered in the C subunit (*18*). Recently, a deletion mutant in the *pyrB* locus has been isolated which has no detectable R subunit (*18*). Therefore, the possibility that R subunit has some other vital function in the cell appears to be ruled out. Syvanen (*18*) has mapped a series of *pyrB* mutants in *S. typhimurium* using a three-point mapping technique, with the closely linked *arg I* locus (*111*) as a reference point. All of these mutants have some detectable R subunit in a recombination assay based on the ability of the *S. typhimurium* R subunit to recombine with *E. coli* C subunit (see Section IX). The amount of R subunit produced by a selection of *pyrB* nonsense mutants correlated with the position of the mutations along the fine structure map of the *pyrB* locus. The mutations closest to the *arg I* marker make less R subunit than those further away with one exception, a leaky *amber* (*pyrB 641*) located distal from *arg I* on the map. All the mutants, however, produce less R subunit than the parent *pyrB*$^+$ strain. The effect of the position of a mutation in the C subunit in determining the amount of R subunit produced by the mutant strain has been explained by a polarity gradient through an ATCase operon where the C chain is synthesized first, followed by the R chain (*18*).

B. Are the Two Kinds of Chain in a Single Operon?

The fact that a mutant of *S. typhimurium* constitutive for enzymes of the pyrimidine pathway produces excess R subunit has been inter-

109. Y. Yan and M. Demerec, *Genetics* **52**, 643 (1965).
110. G. A. O'Donovan and J. C. Gerhart, *J. Bacteriol.* **109**, 1085 (1972).
111. M. Syvanen and J. R. Roth, *J. Bacteriol.* **110**, 66 (1972).

preted as evidence that the genes coding for the R subunit and C subunit may not be transcribed into a single species of mRNA, as with an operon (*18*). Other explanations are possible but have yet to be tested critically. Perbal and Hervé (*112*) have analyzed the proportion of ^{14}C-amino acids incorporated into the C subunit, compared to the R subunit, at various times after repression of ATCase synthesis (and, presumably, mRNA synthesis) by addition of uracil. They concluded that the synthesis of the C chains stops first, followed by cessation of R chain biosynthesis. They interpreted this result as evidence for a polycistronic mRNA containing the genes for both R chain and C chain, with the C chain gene being transcribed first. In their experiments, no excess C subunit was detected under any condition, but the techniques they used would not have detected excess R subunit. As Perbal and Hervé (*112*) suggested, the results could also be explained by a repressor which acts differentially on separate R chain and C chain operons or by C chain acting as an inducer of R chain biosynthesis.

Thus, it is not yet possible to conclude that the genes for R chain and C chain are directly adjacent to one another and constitute an operon in the classic sense. However, they are probably very close, because a *pyrB* deletion mutant produces no free R subunit (*18*). If the genes *are* adjacent, and if transcription proceeds sequentially through the region, and if the ordering of *arg I* and *pyrB* is the same for *S. typhimurium* as for *E. coli*, it can be concluded that the *pyrB* region is transcribed counterclockwise on the *E. coli* map with the R chain gene distal from *arg I* and transcribed last. Point mutants in the structural gene for the R chain and in the operator and promotor regions of *pyrB* will be necessary to test these hypotheses.

Legrain et al. (*113*) have recently raised the interesting possibility that the ATCase and ornithine transcarbamylase of *E. coli* may be evolutionarily related. The marker *arg I* codes for ornithine transcarbamylase in *E. coli* and is very close to the *pyrB* locus, and may be contiguous with it in some strains (*114*). The molecular weight of the *arg I* gene product is 35,000, close to that of the C chain of ATCase. Furthermore, the ornithine transcarbamylase polypeptide chains appear to form a trimer in the native structure as in the case of the C subunit of ATCase. Thus, there is good circumstantial evidence that the two activities may have arisen from a tandem duplication of an ancestral transcarbamylase gene.

112. B. Perbal and G. Hervé, *JMB* **70**, 511 (1972).
113. C. Legrain, P. Halleux, V. Stalon, and N. Glansdorff, *Eur. J. Biochem.* **27**, 93 (1972).
114. G. A. Jacoby, *J. Bacteriol.* **108**, 645 (1971).

C. Control of Enzyme Biosynthesis

The activity of ATCase in the bacterial cell is controlled both by feedback regulation and by repression of enzyme synthesis. Evidence has been obtained *in vivo*, using CTP synthetase mutants of *S. typhimurium*, that CTP is indeed an intracellular inhibitor of ATCase, just as it is with the purified enzyme (*115, 116*). A possible role *in vivo* for the activation of ATCase by ATP has also been discussed (*117*). In addition, the enzyme immediately preceding ATCase, carbamyl-P synthetase, is inhibited in *E. coli* by UMP (*117–121*). The interrelationship of these feedback control mechanisms in *E. coli* has been discussed by Gerhart (*6*) and by O'Donovan and Neuhard (*5*). Higher organisms have evolved a different and efficient mechanism for control of the pyrimidine pathway: an enzyme complex of ATCase and carbamyl-P synthetase (see Section IX).

Yates and Pardee (*8*) were the first to show that ATCase activity in *E. coli* could be inhibited by the addition of uracil to the growth medium and that uracil or an intracellular metabolite acts to decrease the amount of ATCase in the cell. Purification of large quantities of ATCase from *E. coli* has depended on the ability of certain pyrimidine mutants to increase greatly their production of the enzyme in the absence of exogenous pyrimidines (derepression) (*9, 11*). Under conditions of derepression, these mutants can produce up to 10% of their protein as ATCase. Beckwith *et al.* (*101*) found that ATCase synthesis was increased much more than the synthesis of other enzymes in the pathway upon derepression.

The recent experiments of Ingraham and Neuhard (*122*) suggest that the signal for repression of ATCase synthesis in *S. typhimurium* is UDP or UTP. Conditional mutants defective in UMP kinase are derepressed for ATCase synthesis. These mutants contain normal levels of CTP and UMP but are deficient in UDP and UTP. Thus it is likely that repression by uridine, a well-established characteristic of *E. coli* ATCase, is the direct result of UDP or UTP acting as a classic corepressor. The actual mechanism of the repression and how it is related to

115. J. Neuhard, *J. Bacteriol.* **96**, 1519 (1968).
116. J. Neuhard and J. L. Ingraham, *J. Bacteriol.* **95**, 2431 (1968).
117. A. Abd-El-Al and J. L. Ingraham, *JBC* **244**, 4033 (1969).
118. P. M. Anderson and A. Meister, *Biochemistry* **5**, 3164 (1966).
119. P. M. Anderson and S. V. Marvin, *Biochemistry* **9**, 171 (1970).
120. A. Pierard, N. Glansdorff, M. Mergeay, and J. M. Wiame, *JMB* **14**, 23 (1965).
121. A. Pierard, *Science* **154**, 1572 (1966).
122. J. L. Ingraham and J. Neuhard, *JBC* **247**, 6259 (1972).

regulation of the levels of other enzymes in the pyrimidine pathway in these organisms remains to be established.

Recently, Perbal and Hervé (112) have carried out an extensive investigation of the biosynthesis of ATCase in *E. coli*. Using a strain diploid in the *pyrB* region (11) and a haploid strain derived from it, they studied rates of ATCase biosynthesis under various conditions of repression and derepression. The repressed diploid strain produced twice as much ATCase (450 ± 50 molecules per cell) as the haploid strain (180 ± 50 molecules per cell), indicating that in the diploid strain the repression mechanism was probably acting on both copies of the ATCase gene to the same extent as in the haploid. No significant amount of free C subunit was detected in the diploid strain, indicating that this strain also carries the gene coding for the R subunit. Upon derepression, the production of ATCase was greatly increased, but a significant lag of 3.5 min was required before any new ATCase synthesis occurred. The rate of synthesis under depression conditions (minus uracil) was 1000 ± 300 molecules of ATCase per minute per cell in the diploid mutant. Upon the addition of uracil, the cultures stop production of ATCase within 3.5 min.

The half-life of the mRNA was calculated to be 1.5 min on the basis of its ability to synthesize enzyme. Using this value, it was calculated that if ATCase biosynthesis is directed by a bicistronic mRNA each mRNA should be translated by a maximum of 200 ribosomes. From the rate of ATCase production under derepression and from the half-life of the mRNA, ten molecules of the mRNA per ATCase operon were calculated to be present simultaneously (112).

Zinc seems to be required to produce normal amounts of native ATCase in the cell. When ATCase activity was examined in cells grown in a special low zinc medium, 70% of the activity corresponded to free C subunits and only 30% was present as native ATCase (28). In media containing zinc, less than 5% of the activity was present as free C subunit. This result agrees with the finding that zinc (or another metal) is required for association of R subunit with C subunit to form native ATCase (26, 28). A cadmium-containing ATCase has also been produced *in vivo* by replacement of Zn^{2+} by Cd^{2+} in the growth medium (26).

The biosynthesis of C chains and R chains appears to be tightly coupled. Little, if any, free C subunit is detectable during the course of biosynthesis of native ATCase under a variety of conditions (112), with the single exception described above. In addition, no active precursors of native ATCase which differ from it electrophoretically have been detected (112). In at least one *pyrB*⁺ organism, however, free R subunit has been detected (18). It would be interesting to know if other strains also contain excess free R subunit and under what conditions the amount might be changed. Information of this sort would be helpful in determining the mechanism(s) of control of C- and R-chain biosynthesis as well as the mechanism of assembly *in vivo* of the chains and subunits into a molecule of native ATCase.

IX. Aspartate Transcarbamylases from Other Organisms

A. BACTERIAL ATCASES

Aspartate transcarbamylases from bacteria other than *Escherichia coli* have been studied, primarily by M. E. Jones and co-workers. These are listed in Table IV (*5, 123–129*) according to resemblance in size and properties to the enzyme from *E. coli*. The enzymes from *Salmonella typhimurium, Citrobacter freundii, Streptococcus faecalis, Pseudomonas fluorescens,* and *Halobacterium cutirubrum* have been purified to various degrees and will be discussed below. Approximate sizes for the others were determined by Sephadex G-200 chromatography of crude extracts (*124*).

The ATCases listed in Table IV can be grouped according to their resemblance in size or properties to *E. coli* ATCase or to the C subunit (*124*). Enzymes of group 2 are approximately equal to or larger than native *E. coli* ATCase, and all those tested show some inhibition or activation by various nucleotides. Enzymes in group 3 resemble the *E. coli* C subunit in size and insensitivity to nucleotides. The *S. faecalis* and *Bacillus subtilis* enzymes are unaffected by a large number of nucleotides (*126, 128*) and thus are assumed to lack a feedback control mechanism. The *C. freundii* and *Proteus vulgaris* enzymes of group 3 may be subunits of a larger ATCase found with them in crude extracts: The group 3 *C. freundii* enzyme can be derived from its group 2 counterpart (*127*).

Bethell and Jones (*124*) have further subdivided groups 2 and 3 of Table IV into three groups, according to size and kinetic properties. On the basis of several properties of each organism which might be expected

123. G. A. O'Donovan, H. Holoubek, and J. C. Gerhart, *Nature (London) New Biology* **238**, 264 (1972).
124. M. R. Bethell and M. E. Jones, *ABB* **134**, 352 (1969).
125. L. B. Adair and M. E. Jones, *JBC* **247**, 2308 (1972).
126. J. Neumann and M. E. Jones, *ABB* **104**, 438 (1964).
127. M. S. Coleman and M. E. Jones, *Biochemistry* **10**, 3390 (1971).
128. L. M. Prescott and M. E. Jones, *Biochemistry* **9**, 3783 (1970).
129. T. Y. Hutson and M. Downing, *J. Bacteriol.* **96**, 1249 (1968).

TABLE IV
PROPERTIES OF BACTERIAL ATCases

Organism	Molecular weight	Subunits	Activators	Inhibitors	References
1. Very similar to *E. coli* native ATCase					
S. typhimurium	300,000	R and C	ATP	CTP	5, 123
2. Eluted with or ahead of *E. coli* native ATCase on Sephadex G-200					
P. aeruginosa	>300,000	nt[a]	nt	ATP, UTP, GTP, CTP	124
P. fluorescens	360,000	2 of 180,000	nf[b]	UTP, ATP, CTP	125
A. vinelandii	>300,000	nt	nt	nt	124
S. marcescens	~300,000	nt	nt	CMP, UMP	124, 126
R. spheroides	~300,000	nt	nt	nt	124
C. freundii[c]	~250,000	(see ref. 127)	ATP	Several nucleotides	124, 127
P. vulgaris[c]	~300,000	nt	CTP	nt	124
A. aerogenes	—	nt	nt	CMP, UMP	126
3. Resembling *E. coli* C subunit					
C. freundii[c]	~93,000	(see ref. 127)	nf	Several nucleotides	124, 127
P. vulgaris[c]	~100,000	nt	nt	nt	124
S. faecalis	120,000–140,000	3 or 4 of 33,000	nf	nf	128, M. E. Jones and T.-Y. Chang (unpublished)
B. subtilis	~100,000	nt	nf	nf	124, 126
L. leichmannii	—	nt	nf	nf	129

[a] Not tested.
[b] None found.
[c] Crude extracts contain both group 2 and 3 forms.

to reflect evolutionary relatedness, they concluded that the ATCases within each group were much closer to each other than to ATCases from the other groups. More extensive comparisons of the ATCases of these and other organisms will have to be made to test the generality and significance of this correlation.

Aspartate transcarbamylase has been purified from *S. typhimurium* (*5, 123*). Its properties are virtually identical to those of the *E. coli* enzyme. The protein has a molecular weight of about 300,000 and a sedimentation coefficient of about 12 S; it can be separated into C and R subunits with *p*-mercuribenzoate. The purified enzyme has a cooperative saturation curve for L-aspartate identical to that of the *E. coli* enzyme in the absence of nucleotides. The curve becomes much more sigmoidal than that of *E. coli* ATCase upon the addition of CTP (Fig. 21A), but the affinity of the two enzymes for CTP is approximately the same (*123*). The structures of the two ATCases must be very closely related since hybrid enzymes can be made from *E. coli* C subunit plus *S. typhimurium* R subunit and *S. typhimurium* C subunit plus *E. coli* R subunit. Both hybrids have cooperative aspartate saturation curves (Fig. 21B). More evidence is needed to interpret the slight differences in properties in terms of structural differences between subunits from the two organisms. It will be of mechanistic as well as evolutionary interest to compare the amino acid sequences of these two closely related enzymes.

Aspartate transcarbamylase has also been extensively purified from *S. faecalis* [(*128*); M. E. Jones and T.-Y. Chang, unpublished]. This enzyme belongs to group 3 of Table IV and resembles the *E. coli* C subunit in size and lack of nucleotide sensitivity. A 1500-fold purification leads to a preparation which is homogeneous by polyacrylamide gel electrophoresis, either in the presence or absence of sodium dodecyl sulfate. In a sucrose gradient a single peak of ATCase activity is observed. The homogeneous protein is a monomer of molecular weight 120,000 ± 10,000. The identical subunits of the enzyme have an apparent molecular weight of 33,000 ± 1000. No nucleotide tested was able to activate or inhibit the enzyme. Other properties, including many kinetic properties as well as the inability of the enzyme to cross-react with antibody to *E. coli* catalytic subunit, distinguish this enzyme from the *E. coli* C subunit. The *B. subtilis* enzyme, another of the noninhibitable ATCases of group 3, has been examined for coaggregation with the *E. coli* R subunit, but none was achieved (*5*). Thus, the possibility remains that the resemblance of the group 3 ATCases to the *E. coli* C subunit may be coincidental and may not reflect very similar structures.

The ATCase activity of *C. freundii* has been studied by Coleman and Jones (*127*). This organism contains two ATCases which belong to

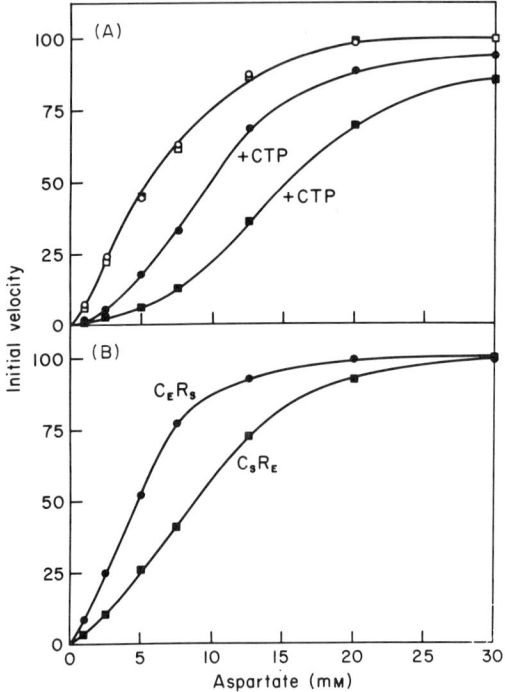

FIG. 21. Kinetic characteristics of purified ATCase from *E. coli* and *S. typhimurium*, and of intergeneric hybrid forms of ATCase [from O'Donovan et al. (123)]. Enzyme samples and assay mixtures were described by O'Donovan et al. (123). (A) Native ATCase. □ and ■ indicate the purified, undissociated, *S. typhimurium* enzyme minus and plus CTP, respectively, and ○ and ● indicate the undissociated *E. coli* enzyme minus and plus CTP. (B) Hybrid ATCase. The symbol C_SR_E designates the hybrid ATCase formed from *S. typhimurium* catalytic subunit and *E. coli* regulatory subunit. The reciprocal hybrid form is designated C_ER_S. The reconstituted parental (nonhybrid) forms, C_ER_E and C_SR_S, resembled the undissociated enzymes in their saturation kinetics as in (A).

groups 2 and 3. The enzymes have similar kinetic characteristics and both are inhibited by most nucleotides, by phosphate, and by pyrophosphate. However, the larger enzyme is activated by ATP, which inhibits the smaller one slightly. Dilution of the larger enzyme leads to formation of some of the smaller form, indicating that the two contain a subunit or subunits in common. The proportion of the two enzymes *in vivo* varies: The larger form increases during logarithmic growth until it is the sole component during the stationary phase. It will be interesting if the large form contains components analogous to the *E. coli* R subunit which are responsible for binding ATP during activation. Two ATCases also occur in *P. vulgaris* (*124*), but it is not known whether these are analogous to the two *C. freundii* forms except that the group 2 form is activated by CTP.

The ATCase of *P. fluorescens* has been purified to homogeneity by Adair and Jones (*125*). The enzyme is a dimer of two apparently identical subunits of molecular weight 180,000. Several nucleotides inhibit the enzyme (see Table IV), and the regulatory site appears distinct from the catalytic site. Mild heat treatment (*125*) does not separate these two sites as it does in the case of the *E. coli* enzyme.

Finally, partial purification of the ATCase of the extreme halophile *H. cutirubrum* has been achieved (*130, 131*). Activity depends upon a salt concentration of 3–4 M and is sensitive to CTP inhibition. Gel chromatography on agarose gives an estimated molecular weight of 160,000, and the chromatographed material is still fully sensitive to CTP. No evidence of cooperative kinetics was found with respect to substrates or inhibitor. Purification of the enzyme by precipitation with polyethylene glycol results in a decrease in molecular weight to 34,000. This form is no longer sensitive to CTP. Either substrate protects the enzyme against dissociation by polyethylene glycol. Thus, aggregation of the subunits is essential to regulatory but not catalytic function, as with the *E. coli* enzyme.

Little is known about the genetics or control of biosynthesis of ATCase in organisms other than *E. coli* and *S. typhimurium*. The pyrimidine pathways in both *Serratia marinorubra* (*132*) and *Serratia marcescens* (*133*) appear to coincide with the *E. coli* pathway, and at least some of the enzymes of the pathway in *S. marcescens* can be repressed by uracil (*133*). However, ATCase activity was not studied in these experiments. Four mutants in the pyrimidine pathway were found to be not closely linked in *Pseudomonas aeruginosa*, and no activity, including that of ATCase, could be repressed by uracil (*134*). The ATCase activity in *H. cutirubrum* also was not repressed by uracil. However, a fourfold to fivefold repression of ATCase was found when *Lactobacillus leichmannii* was grown in a medium containing uracil (*129*).

It is difficult to determine at this time the logic of the controls through which the activity of ATCases from many bacteria are regulated. In *E. coli* and *S. typhimurium*, ATCase, the first committed step in pyrimidine biosynthesis is feedback controlled by CTP and repressed by a

130. V. Liebl, J. G. Kaplan, and D. J. Kushner, *Can. J. Biochem.* **47**, 1095 (1969).
131. P. Norberg, J. G. Kaplan, and D. J. Kushner, *J. Bacteriol.* **113**, 680 (1973).
132. W. L. Belser, *BBRC* **4**, 56 (1961).
133. W. S. Hayward and W. L. Belser, *Proc. Nat. Acad. Sci. U. S.* **53**, 1483 (1965).
134. J. H. Isaac and B. W. Holloway, *J. Bacteriol.* **96**, 1732 (1968).

uridine nucleotide. *Escherichia coli* has one carbamyl-P synthetase for both pyrimidine and arginine biosynthesis. The enzyme is inhibited by UMP (*120, 121*), but the UMP inhibition can be relieved by ornithine, an intermediate in arginine biosynthesis (*120*); thus, the synthetase need not be turned off when the level of UMP is high if precursors for arginine biosynthesis are present. *Bacillus subtilis*, whose ATCase is not regulated by pyrimidine nucleotides, may have only one carbamyl-P synthetase as well (*135*), and it will be interesting to learn more of the control of both the pyrimidine and arginine pathways of this organism.

B. ATCases from *Neurospora crassa* and *Saccharomyces cerevisiae*

Donachie (*136, 137*) was the first to investigate extensively the mechanisms regulating ATCase activity in the fungus *Neurospora crassa*. He showed that uridine and some uridine derivatives are inhibitors of ATCase activity *in vitro* only at inhibitory concentrations of L-aspartate, and that these inhibitors affect either ATCase or a preceding enzyme in the pathway *in vivo*. He also showed that uridine or some metabolite of uridine probably acts as a repressor for ATCase synthesis (*136*). This repression of enzyme synthesis and end product inhibition of enzyme activity both appear to function during exponential growth of the wild type organism, and the total ATCase activity could theoretically be increased 3- to 4-fold under growth conditions which relieve these controls (*136*). Similar results concerning levels of derepression of ATCase in this organism were obtained by Caroline and Davis (*138*), who also studied the regulation of dihydro-orotic dehydrogenase. This enzyme is substrate-induced and a third enzyme of the pathway, dihydro-orotase, shows no clear regulatory behavior (*138*). Caroline has also isolated a mutant deficient in dihydro-orotase (*139*).

It is now known that *Neurospora* has two carbamyl-P synthetases, one specific for the biosynthesis of arginine and one specific for the pyrimidine pathway (CPS-pyr) (*140–142*). The two enzymes are coded for by

135. I. M. Issaly, A. S. Issaly, and J. L. Reissig, *BBA* **198**, 482 (1970).
136. W. D. Donachie, *BBA* **82**, 293 (1964).
137. W. D. Donachie, *BBA* **82**, 284 (1964).
138. D. F. Caroline and R. H. Davis, *J. Bacteriol.* **100**, 1378 (1969).
139. D. F. Caroline, *J. Bacteriol.* **100**, 1371 (1969).
140. R. H. Davis, *BBA* **107**, 54 (1965).
141. R. H. Davis, in "Organizational Biosynthesis" (H. J. Vogel, J. O. Lampen, and V. Bryson, eds.), p. 302. Academic Press, New York, 1967.
142. L. G. Williams and R. H. Davis, *J. Bacteriol.* **103**, 335 (1970).

separate genetic loci and differ in physical and chemical properties (*141*). The synthesis of both CPS-pyr and ATCase are coderepressed by pyrimidine starvation and CPS-pyr prefers L-glutamine to ammonia as the amino group donor (*142*). Uridine triphosphate inhibits the CPS-pyr completely but has no effect on ATCase activity under the same conditions (*142*). The genes for ATCase and CPS-pyr in *Neurospora* are closely linked and, in fact, single mutations at the *pyr-3* locus can eliminate either or both activities (*142, 143*). In addition, both enzymes may have a common mRNA (*144, 145*).

The fact that strains of *Neurospora* lacking CPS-pyr have a structurally altered ATCase (*146*) led to the search for a complex of the two enzymes. Williams *et al.* (*147*) have reported copurification of the two activities by designing the purification procedure for the more labile CPS-pyr. The molecular weight of the complex was estimated to be 650,000. Addition of UTP to the complex completely inhibits the CPS-pyr activity and may dissociate the complex, at least under certain conditions, into a form with lower molecular weight. The carbamyl-P produced *in vivo* by the pyrimidine complex is unavailable to ornithine transcarbamylase in the arginine pathway (*147*). This effect may be a consequence of the physical association of ATCase and CPS-pyr in *Neurospora*, so that carbamyl-P produced by CPS-pyr is immediately available for the ATCase reaction. In fact, this does seem to be the case for the complex in *Saccharomyces* (see below).

Lacroute (*148–150*) has studied the regulation of pyrimidine biosynthesis in baker's yeast, *Saccharomyces cerevisiae*. The ACTase activity of this organism is inhibited by UTP and repressed by uracil. Lacroute also found that the CPS-pyr was inhibited by UTP and that the synthesis of ATCase and CPS-pyr was controlled by the same genetic region (*150*). After these initial studies, Kaplan *et al.* (*151*) studied some of the physical and kinetic properties of yeast ATCase in crude extracts and in partially purified preparations. The kinetics of the enzyme were strictly Michaelian, and no cooperative effects were observed with either of the substrates or with the inhibitor.

A more intensive investigation into the genetics and regulation of

143. R. H. Davis and V. W. Woodward, *Genetics* **47**, 1075 (1962).
144. A. Radford, *Mutat. Res.* **8**, 537 (1969).
145. A. Radford, *Mol. Gen. Genet.* **104**, 288 (1969).
146. J. M. Hill and V. W. Woodward, *ABB* **125**, 1 (1968).
147. L. G. Williams, S. Bernhardt, and R. H. Davis, *Biochemistry* **9**, 4329 (1970).
148. F. Lacroute, *C. R. Acad. Sci.* **258**, 2884 (1964).
149. F. Lacroute and P. Slonimski, *C. R. Acad. Sci.* **258**, 2172 (1964).
150. F. Lacroute, *C. R. Acad. Sci.* **259**, 1357 (1964).
151. J. G. Kaplan, M. Duphil, and F. Lacroute, *ABB* **119**, 541 (1967).

pyrimidine biosynthesis in yeast by Lacroute (*152*) revealed that the genetic loci controlling the synthesis of the pyrimidine enzymes, including the locus for ATCase and CPS-pyr (*ura-2*), were unlinked. Unlike ATCase and CPS-pyr, however, the other enzymes of the pathway are induced in a sequential manner by intermediary products and are not controlled by pyrimidine repression. Mutants were also isolated which have lost simultaneously the sensitivity of both ATCase and CPS-pyr to UTP. The two activities exist as enzyme complex which retains sensitivity to the feedback inhibitor (*153–156*). The complex, recently purified from a derepressed diploid strain (*156*), has a molecular weight of 800,000 in a sucrose density gradient in the presence of UTP, Mg^{2+}, and glutamine. Omission of UTP in the gradient gives a peak of molecular weight 380,000 which contains both activities, but only the CPS-pyr activity is still highly sensitive to UTP. Omission of Mg^{2+} and glutamine from the gradient gives a distinct CPS-pyr peak of 250,000 molecular weight, retaining sensitivity to UTP. Heat-induced disaggregation of the complex leads to molecules of molecular weight 140,000 which retain ATCase activity but have no CPS-pyr activity or sensitivity to UTP (*155*). Gel electrophoresis of this component in the presence of sodium dodecyl sulfate reveals a single band of molecular weight 20,500. Thus, there seem to be six subunits in the ATCase protomer (*156a*). Recent evidence concerning the fine structure map of the *ura-2* locus of yeast (*157*) has led Lue and Kaplan (*156*) to propose that the two activities are coded for by a polycistronic region (*ura-2*) which contains cistrons for ATCase, CPSase, and possibly for a third, regulatory, protein containing the UTP site. The CPS-pyr subunit itself could contain the regulatory site since treatments which destroy the synthetase activity destroy the sensitivity to UTP. The ATCase subunit can be isolated free of sensitivity to UTP (*155*) and hence probably does not contain the binding site for the nucleotide.

In an elegant series of experiments, Lue and Kaplan (*158*) demonstrated that the carbamyl-P made by the ATCase–CPS-pyr complex of yeast is channeled very effectively from the CPS-pyr to the ATCase *in vitro*. Unlabeled carbamyl-P is unable to dilute the specific radio-

152. F. Lacroute, *J. Bacteriol.* **95**, 824 (1968).
153. P. F. Lue and J. G. Kaplan, *BBRC* **34**, 426 (1969).
154. J. G. Kaplan and I. Messmer, *Can J. Biochem.* **47**, 477 (1969).
155. P. F. Lue and J. G. Kaplan, *Can. J. Biochem.* **48**, 155 (1970).
156. P. F. Lue and J. G. Kaplan, *Can. J. Biochem.* **49**, 403 (1971).
156a. D. M. Aitken, A. R. Bhatti, and J. G. Kaplan, *BBA* **309**, 50 (1973).
157. M. Denis-Duphil and F. Lacroute, *Mol. Gen. Genet.* **112**, 354 (1971).
158. P. F. Lue and J. G. Kaplan, *BBA* **220**, 365 (1970).

activity of carbamyl-L-aspartate synthesized by the complex from ^{14}C-labeled bicarbonate and the other unlabeled substrates. In addition, the intermediate [^{14}C]carbamyl-P was utilized only inefficiently by *E. coli* ornithine transcarbamylase present in the assay mixture. In contrast, when the two activities were not associated in a complex, no channeling of the intermediate carbamyl-P was observed.

It is interesting to compare the carbamyl aspartate synthesizing systems of *Neurospora* and *Saccharomyces*. In both, the CPS-pyr and ATCase activities appear in the form of a complex within the cell. In both, the structural genes for the two enzymes are closely linked, probably within the same operon. The most striking difference between the two complexes is that UTP is an inhibitor of both the CPS-pyr and ATCase activities of *Saccharomyces*, but it inhibits only the CPS-pyr of *Neurospora*.

Both *Neurospora* (*159*, *160*) and *Saccharomyces* (*161*) contain two carbamyl-P synthetases under the regulatory control of the arginine and pyrimidine pathways, respectively. A deficiency of one of the two enzymes in *Neurospora* causes auxotrophy for either uracil or arginine (*141*), depending upon which enzyme is affected. In *Saccharomyces*, mutants in either of the two enzymes behave much like the wild type organisms (*161*). Thus, in *Neurospora*, effective channeling of carbamyl-P from each synthetase into its own pathway occurs, whereas in *Saccharomyces* there seems to be some overflow of carbamyl-P from either pathway into a common pool. The fact that UTP inhibits only CPS-pyr in the *Neurospora* complex makes sense since the ATCase activity need not be regulated in the absence of any other source of carbamyl-P, i.e., CPS-pyr catalyzes the first committed step of pyrimidine biosynthesis. In *Saccharomyces*, however, as pointed out by Lue and Kaplan (*156*), some of the carbamyl-P made by the arginine-specific synthetase is available to the pyrimidine complex, and thus both the CPS-pyr and ATCase activities of the complex are regulated, i.e., the ATCase reaction now becomes the first committed step. These and other aspects of regulation of the pyrimidine pathway of fungi have recently been discussed in detail by Jones (*162*, *163*). Further studies of the relationship between the active sites of the component enzymes in the two complexes, as well

159. R. H. Davis, *Science* **142**, 1652 (1963).
160. J. L. Reissig, *J. Gen. Microbiol.* **30**, 327 (1963).
161. F. Lacroute, A. Pierard, M. Grinson, and J. Wiame, *J. Gen. Microbiol.* **40**, 127 (1965).
162. M. E. Jones, *Advan. Enzyme Regul.* **9**, 19 (1971).
163. M. E. Jones, *Curr. Top. Cell. Regul.* **6**, 227 (1972).

as of the relationship of the UTP site to them, are awaited with great interest.

C. MAMMALIAN ATCASES

Lowenstein and Cohen investigated the ATCase activity in liver (164) and in various other tissues of the adult rat (165). The enzyme was detected in almost every tissue examined. Further work on the kinetic and physical properties of the partially purified enzyme from rat liver revealed hyperbolic saturation curves for both substrates and little, if any, inhibition of ATCase by pyrimidine nucleotides (166, 167). Feedback inhibition was also not detected in the ATCase of rabbit erythrocytes (168).

Sedimentation of rat liver homogenates revealed two ATCase peaks, corresponding to molecular weights 900,000 and 600,000 (169). The same forms also seem to be present in other rat tissues, in tissues from mouse and chicken, and in cultured cells (170). Rat liver extract treated with trypsin loses these higher molecular weight forms of ATCase completely and a new peak of activity appears with a molecular weight of 80,000. There is no loss of total ATCase activity in such experiments. None of these forms of ATCase is inhibited by a variety of nucleotides (170). Three forms of ATCase which differ in molecular weight have been detected in hematopoietic mouse spleen (171). One of these forms, comprising about 10% of the total activity, sediments with the microsomal fraction, and two are recovered in the supernatant. Again, these enzymes are inhibited only poorly by high concentrations of nucleotides and differ very little from one another in kinetic properties.

Evidence for at least two carbamyl-P synthetases in mammalian tissues has been summarized by Jones (162, 163). The glutamine-dependent pyrimidine specific enzymes from fetal rat liver (172) and from mouse spleen (173) both show sigmoidal saturation kinetics with respect to the substrate ATP, and the mouse spleen enzyme is inhibited by UTP (173,

164. J. M. Lowenstein and P. P. Cohen, *JBC* **213**, 689 (1955).
165. J. M. Lowenstein and P. P. Cohen, *JBC* **220**, 57 (1956).
166. E. Bresnick and H. Mosse, *Biochem. J.* **101**, 63 (1966).
167. E. Bresnick, *BBA* **67**, 425 (1963).
168. M. R. Curci and W. D. Donachie, *BBA* **85**, 338 (1964).
169. I. T. Oliver, O. Koskimies, R. Hurwitz, and N. Kretchmer, *BBRC* **37**, 505 (1969).
170. O. Koskimies, I. T. Oliver, R. Hurwitz, and N. Kretchmer, *BBRC* **42**, 1162 (1971).
171. A. Inagaki and M. Tatibana, *BBA* **220**, 491 (1970).
172. S. E. Hager and M. E. Jones, *JBC* **242**, 5674 (1967).
173. R. L. Levine, N. J. Hoogenraad, and N. Kretchmer, *Biochemistry* **10**, 3694 (1971).

174). Using the purification scheme of Tatibana and Ito (*175*) for the glutamine-dependent synthetase of mouse spleen, Hoogenraad et al. (*176*) were able to obtain a considerable copurification of ATCase activity. The complex has an approximate molecular weight of 600,000. The ATCase activity is not inhibited by several nucleotides and does not utilize carbamyl-P produced by the associated CPSase preferentially at a concentration of exogenous carbamyl-P of 1.6 mM. Shoaf and Jones (*177*) have demonstrated an enzyme complex of ATCase, CPS-pyr, and dihydroorotase in Ehrlich ascites carcinoma cells in the presence of dimethylsulfoxide, which stabilizes the CPS-pyr activity. The entire complex, in the presence of 30% dimethylsulfoxide, has a molecular weight of 750,000 to 850,000.

Thus, as in the fungi, ATCase and CPS-pyr are associated in some mammalian systems. All the ATCases studied in mammalian tissues appear to be at least partially associated into high molecular weight complexes, which makes it likely that most mammalian ATCases are associated with one or more of the other enzymes of the pyrimidine pathway. The situation appears to be similar to that found in *Neurospora* in that CPS-pyr rather than ATCase appears to be the target for feedback regulation of pyrimidine biosynthesis. The situation may be more complex, however, because of a recent report that the ATCase activity in rat liver tissue slices is inhibited by addition of uridine (*178*).

It is not yet clear whether channeling of carbamyl-P is an important consequence of association of the different carbamyl-P synthetases with the urea (arginine) and pyrimidine pathways (see references *162* and *163* for a discussion). However, if carbamyl-P is channeled, one would expect the CPS-pyr of mammals, rather than the ATCase, to be the site at which feedback control of the pyrimidine pathway is exerted, just as it appears to be in *Neurospora*. Other aspects of the control of pyrimidine biosynthesis in eukaryotes are discussed in detail by Jones (*162, 163*).

D. PLANT ATCASES

Aspartate transcarbamylases from lettuce seedlings (*126*), wheat germ (*179*), and mung bean (*180*) have been examined for their kinetic and

174. M. Tatibana and K. Ito, *BBRC* **26**, 221 (1967).
175. M. Tatibana and K. Ito, *JBC* **244**, 5403 (1969).
176. N. J. Hoogenraad, R. L. Levine, and N. Kretchmer, *BBRC* **44**, 981 (1971).
177. W. T. Shoaf and M. E. Jones, *BBRC* **45**, 796 (1971).
178. P. A. Bourget and G. C. Tremblay, *BBRC* **46**, 752 (1972).
179. R. J. Yon, *BJ* **128**, 311 (1972).
180. B. L. Ong, Ph.D. Thesis, University of Adelaide, South Australia, 1972.

allosteric properties. Little is known of the possible association of ATCase in plants with other enzymes of the pyrimidine pathway.

Acknowledgments

Work on ATCase has been supported by Public Health Service Research Grant GM-11788 from the National Institute of General Medical Sciences. The authors are indebted to many people for their comments and especially to John Gerhart and Kim Collins for critical readings of this manuscript.

8

Glycogen Synthesis from UDPG

W. STALMANS • H. G. HERS

I. Historical	310
II. Molecular Properties	311
A. Association with Glycogen	311
B. Purification	312
C. Physicochemical Properties	313
III. General Catalytic Properties	316
A. The Reaction	316
B. Assay Methods	317
C. Donor Specificity	318
D. The Glucosyl Acceptor	319
E. Reaction Mechanism	321
IV. The Two Forms of Glycogen Synthetase and Their Interconversion	322
A. Nomenclature	322
B. General Properties of the a and b Forms	324
C. The Basic System of Interconversion	325
D. Synthetase Kinase	326
E. Synthetase Phosphatase	327
F. An "Inactive" Form of Glycogen Synthetase	330
G. Proteolytic Inactivation	331
V. Glycogen Synthetase of Mammalian Muscle	332
A. Properties of the Two Forms	332
B. Control of Synthetase Activity in Muscle	336
VI. Glycogen Synthetase of Mammalian Heart	340
A. Properties	340
B. Control of Synthetase Activity in Heart	340
VII. Glycogen Synthetase of Mammalian Liver	341
A. Properties of the Two Forms	341
B. Control of Synthetase Activity in Liver	347
VIII. Glycogen Synthetase of Other Mammalian Tissues	353
A. Adrenals	353
B. Brain	353

C. Spleen 354
  D. Adipose Tissue 354
  E. Blood 354
  F. Tissues Sensitive to Sex Hormones 355
  G. Tumors 356
IX. Glycogen Synthetase of Nonmammalian Organisms 357
  A. Frog Muscle 357
  B. Tadpole Liver 357
  C. Fish 358
  D. Arthropoda 358
  E. Protozoa 359
  F. Molds and Yeast 359

I. Historical

The synthesis of glycogen from UDPG by a liver extract was described in 1957 by Leloir and Cardini (*1*). At that time, UDPG was known to act as a glucosyl donor in the biosynthesis of trehalose phosphate, sucrose, sucrose phosphate, and cellulose, whereas the only known mechanism for the synthesis of glycogen was the transfer of glucosyl from glucose 1-phosphate by phosphorylase. The glycogenolytic action of the hormones that activate liver phosphorylase as well as other experimental data had however given strong indication that another mechanism, possibly uridine-linked, was required for synthesis (*2*). Leloir and his co-workers (*3, 4*) also observed that the glycogen-forming enzyme (UDPG:α-1,4-glucan α-4-glucosyltransferase, EC 2.4.1.11, conveniently called glycogen synthetase) is stimulated by a heat-stable factor, which they identified as glucose 6-phosphate; the phosphoric ester was not used during the reaction, but the degree of activation was variable from one preparation to another.

The observation that a treatment of muscle with insulin decreases the requirement of the enzyme for glucose 6-phosphate allowed Larner and his co-workers (*5, 6*) to recognize the existence of two forms of glycogen synthetase, interconvertible through phosphorylation by a kinase and dephosphorylation by a phosphatase, similar in this respect to the *a* and *b* forms of glycogen phosphorylase. This discovery initiated a long

1. L. F. Leloir and C. E. Cardini, *JACS* **79**, 6340 (1957).
2. L. F. Leloir and C. E. Cardini, "The Enzymes," 2nd ed., Vol. 6, p. 317, 1962.
3. L. F. Leloir, J. M. Olavarría, S. H. Goldemberg, and H. Carminatti, *ABB* **81**, 508 (1959).
4. L. F. Leloir and S. H. Goldemberg, *JBC* **235**, 919 (1960).
5. C. Villar-Palasi and J. Larner, *BBA* **39**, 171 (1960).
6. D. L. Friedman and J. Larner, *Biochemistry* **2**, 669 (1963).

series of research on the hormonal and nonhormonal regulation of glycogen synthesis. The close interrelation of the antagonistic phosphorylase and synthetase systems became more and more apparent and has been solidified by the discovery that synthetase kinase and phosphorylase kinase kinase are identical to the cyclic AMP–stimulated protein kinase (7, 8), and that, in the liver, the activity of synthetase phosphatase is controlled by the level of phosphorylase a (9). The properties of glycogen synthetase have been reviewed by several authors (2, 10–12).

II. Molecular Properties

A. Association with Glycogen

Leloir and Goldemberg (4) have observed that when a liver homogenate is fractionated by differential centrifugation most of the enzyme is recovered with the particulate glycogen and that the amount of synthetase in the other fractions is roughly proportional to their glycogen content. Glycogen synthetase is also present as a sedimentable complex with glycogen in other mammalian tissues including skeletal muscle (13), brain (14), adrenal gland (15), hepatoma cells (16), and leukocytes (17); in kidney, which has a low glycogen content, the enzyme is soluble, but it associates readily with added glycogen (18). The enzyme from mammary gland was also reported as nonsedimentable, but the amount of glycogen in the tissue was not given (19). The synthetase of fish liver seems to be loosely bound to glycogen (20). Glycogen synthetase has also been isolated as a complex with glycogen from the fat body of

7. K. K. Schlender, S. H. Wei, and C. Villar-Palasi, *BBA* **191**, 272 (1969).
8. T. R. Soderling, J. P. Hickenbottom, E. M. Reimann, F. L. Hunkeler, D. A. Walsh, and E. G. Krebs, *JBC* **245**, 6317 (1970).
9. W. Stalmans, H. De Wulf, and H. G. Hers, *Eur. J. Biochem.* **18**, 582 (1971).
10. E. Helmreich, *Compr. Biochem.* **17**, 17 (1969).
11. B. E. Ryman and W. J. Whelan, *Advan. Enzymol.* **34**, 285 (1971).
12. J. Larner and C. Villar-Palasi, *Curr. Top. Cell. Regul.* **3**, 195 (1971).
13. P. W. Robbins, R. R. Traut, and F. Lipmann, *Proc. Nat. Acad. Sci. U. S.* **45**, 6 (1959).
14. B. M. Breckenridge and E. J. Crawford, *JBC* **235**, 3054 (1960).
15. M. M. Piras, E. Bindstein, and R. Piras, *ABB* **139**, 121 (1970).
16. R. Saheki, K. Sato, and S. Tsuiki, *BBA* **230**, 571 (1971).
17. L. Plesner, E. Salsas-Leroy, P. Wang, M. Rosell-Perez, and V. Esmann, *BBA* **268**, 344 (1972).
18. J. L. Hidalgo and M. Rosell-Perez, *Rev. Espan. Fisiol.* **27**, 343 (1971).
19. J. Mendicino and M. Pinjani, *BBA* **89**, 242 (1964).
20. P. Ingram, *Int. J. Biochem.* **1**, 263 (1970).

an insect (*21*) and from insect larvae (*22*). The amount of yeast enzyme that can be sedimented is proportional to the glycogen content of the colony (*23*).

The true association of glycogen synthetase with glycogen and not with elements of the endoplasmic reticulum, which often contaminate particulate glycogen isolated by centrifugation of tissue homogenates, has been confirmed by various types of experiments. These include separation of glycogen from microsomes by density gradient centrifugation (*24, 25*), ultrasonic treatment (*26*) or filtration (*27*), solubilization of the synthetase by digestion of glycogen *in vitro* (*8, 24, 27*), and reassociation of solubilized enzyme with added glycogen (*18, 24, 28, 29*).

Glycogen synthetase binds more strongly to particulate glycogen than do phosphorylase or other glycogen-metabolizing enzymes (*4, 8, 27*). It has preference for high molecular weight glycogen, whereas phosphorylase has a higher affinity for lighter glycogen (*30*). The high affinity of glycogen synthetase for glycogen explains why the enzyme is only partially soluble in the liver of a fasted animal (*31, 32*); complete solubilization is observed in the liver of fasted adrenalectomized animals (*33*) or after administration of glucagon to fasted mice (*34*).

B. PURIFICATION

Nearly all methods of purification of glycogen synthetase take advantage of the high affinity of the enzyme for glycogen and include, as a first step, the isolation of an enzyme–polysaccharide complex. This is usually performed by differential centrifugation but has also been

21. T. A. Murphy and G. R. Wyatt, *Nature (London)* **202**, 1112 (1964).
22. A. Vardanis, *JBC* **242**, 2306 (1967).
23. L. B. Rothman-Denes and E. Cabib, *Proc. Nat. Acad. Sci. U. S.* **66**, 967 (1970).
24. D. J. L. Luck, *J. Biophys. Biochem. Cytol.* **10**, 195 (1961).
25. J. C. Wanson and P. Drochmans, *J. Cell Biol.* **54**, 206 (1972).
26. S. Hizukuri and J. Larner, *Biochemistry* **3**, 1783 (1964).
27. S. DiMauro, W. Trojaborg, P. Gambetti, and L. P. Rowland, *ABB* **144**, 413 (1971).
28. H. J. Mersmann and H. L. Segal, *JBC* **244**, 1701 (1969).
29. W. Stalmans, *Abstr. Commun. 6th FEBS Meet.* p. 214 (1969).
30. A. A. Barber, S. A. Orrell, Jr., and E. Bueding, *JBC* **242**, 4040 (1967).
31. H. G. Sie, A. Hablanian, and W. H. Fishman, *Nature (London)* **201**, 393 (1964).
32. V. T. Maddaiah and N. B. Madsen, *Can. J. Biochem.* **46**, 521 (1968).
33. Y. Sanada and H. L. Segal, *BBRC* **45**, 1159 (1971).
34. Unpublished results from the authors' laboratory.

achieved by agglutination of glycogen with the phytoprotein, concanavalin A (*35, 36*).

The enzyme has been partially purified from skeletal muscle of different species (*37, 38*). Larner and his co-workers (*6, 39–42*) have obtained preparations that were enriched in the *a* or *b* form (*43*). Recently, homogeneous preparations of synthetase from rabbit muscle have been obtained consisting predominantly (*8*) or exclusively (*44*) of the *a* enzyme or being totally in the *b* form (*45*).

The enzyme from mammalian liver has not been purified to a similar extent. The synthetase can be isolated as an enzyme–glycogen complex, in which the proportion of the two enzyme forms is variable (*4, 26*); a method for the preparation of enzyme predominantly in the *a* or *b* form has been outlined (*29*). Glycogen-free preparations of the *a* (*46*) and *b* enzymes (*33*) have also been obtained. Glycogen synthetase mostly in the *a* or *b* form has been extensively purified from mammalian heart (*47*) and polymorphonuclear leukocytes (*17*), from frog liver (*48*), and from yeast (*49*).

C. Physicochemical Properties

1. *Muscle Glycogen Synthetase*

a. Molecular Weight. Molecular weights of 400,000 (*8*) and 250,000 (*45*) were found for homogeneous enzymes in the absence of ligands. A value of 195,000 was obtained by centrifugation of a partially purified preparation in a sucrose gradient containing NaF and $MgSO_4$ (*50*). Aggregation occurs at low temperature in the absence of glycogen (*45*),

35. A. Vardanis, *ABB* **130**, 408 (1969).
36. R. B. Scott and L. W. Cooper, *BBRC* **44**, 1071 (1971).
37. R. Kornfeld and D. H. Brown, *JBC* **237**, 1772 (1962).
38. R. R. Traut and F. Lipmann, *JBC* **238**, 1213 (1963).
39. M. Rosell-Perez, C. Villar-Palasi, and J. Larner, *Biochemistry* **1**, 763 (1962).
40. M. Rosell-Perez and J. Larner, *Biochemistry* **3**, 75 (1964).
41. M. Rosell-Perez and J. Larner, *Biochemistry* **3**, 81 (1964).
42. C. Villar-Palasi, M. Rosell-Perez, S. Hizukuri, F. Huijing, and J. Larner, "Methods in Enzymology," Vol. 8, p. 374, 1966.
43. See section IV A for terminology.
44. C. H. Smith, N. E. Brown, and J. Larner, *BBA* **242**, 81 (1971).
45. N. E. Brown and J. Larner, *BBA* **242**, 69 (1971).
46. D. F. Steiner, L. Younger, and J. King, *Biochemistry* **4**, 740 (1965).
47. J. Larner, C. Villar-Palasi, and N. E. Brown, *BBA* **178**, 470 (1969).
48. J. S. Sevall and K. H. Kim, *BBA* **206**, 359 (1970).
49. I. D. Algranati and E. Cabib, *JBC* **237**, 1007 (1962).
50. R. J. Staneloni and R. Piras, *BBRC* **42**, 237 (1971).

particularly with the *a* form (51). The presence of ligands can change the degree of aggregation: UDPG and UDP promote aggregation and the simultaneous presence of glucose 6-phosphate produces a whole series of heavy polymers, while ATP or an elevated ionic strength decrease the association; there is no correlation between molecular weight and activity (50). In the presence of sodium dodecyl sulfate the purified enzyme dissociates in subunits of molecular weight 90,000, of which the active synthetase is considered to be a trimer or a tetramer (8, 44). A small amount of subunits of about 15,000 daltons was also obtained upon gel filtration (45).

b. Phosphate Content. Pure synthetase *a* (43) contains a negligible amount of alkali-labile (serine-bound) phosphate (8, 44) and no pyridoxal phosphate (44). Pure phosphosynthetase contains 7.09 moles of alkali-labile phosphate per 100,000 g of protein, i.e., 6 for each 90,000 subunit (44). However, only 1.1 moles of phosphate per 100,000 g could be introduced by extensive reaction of synthetase *a* with ATP and protein kinase (8). The low value could result from an incomplete *a* to *b* transition; the discrepancy could also be explained by a dissociation between inactivation and phosphorylation of the synthetase. Such a dissociation is known to occur in the case of phosphorylase kinase (52, 53). Smith *et al.* (44) have proposed that the 90,000 daltons subunits are made of six smaller units of molecular weight around 15,000, on which the phosphate groups occupy identical sites. This suggestion is supported by the presence of six sulfhydryl groups in each 90,000 daltons unit. Other interpretations have also been considered (44).

c. Primary Structure. Tryptic digestion of synthetase *b* labeled with ^{32}P has enabled Larner and Sanger (54) to map the surrounding of the serine phosphate residue that is involved. The sequence of a hexapeptide, shown in Fig. 1 (54–56), is identical to the structure of the phosphorylated site in phosphorylase *a* (55). Subsequent work has, however, revealed important structural differences between the two enzymes; labeled

51. C. H. Smith and J. Larner, *BBA* **264**, 224 (1972).
52. R. J. DeLange, R. G. Kemp, W. D. Riley, R. A. Cooper, and E. G. Krebs, *JBC* **243**, 2200 (1968).
53. W. D. Riley, R. J. DeLange, G. E. Bratvold, and E. G. Krebs, *JBC* **243**, 2209 (1968).
54. J. Larner and F. Sanger, *JMB* **11**, 491 (1965).
55. C. Nolan, W. B. Novoa, E. G. Krebs, and E. H. Fischer, *Biochemistry* **3**, 542 (1964).
56. E. H. Fischer, P. Cohen, M. Fosset, L. W. Muir, and J. C. Saari, *in* "Metabolic Interconversion of Enzymes" (O. Wieland, E. Helmreich, and H. Holzer, eds.), p. 11. Springer-Verlag, Berlin and New York, 1972.

Glycogen phosphorylase a

```
        5                      10                    15
Ser – Asp – Gln – Glu – Lys – Arg – Lys – Gln – Ile – Ser – Val – Arg – Gly – Leu
                                                      |
                                                      P
```

Glycogen synthetase b $\quad\quad\quad\quad\quad \left.\begin{array}{l}\text{Arg}\\\text{Lys}\end{array}\right\}$ – Glu – Ile – Ser – Val – Arg
$\quad\quad\quad\quad\quad\quad\quad\quad\quad\quad\quad\quad\quad\quad\quad\quad\quad\quad\quad\quad$ |
$\quad\quad\quad\quad\quad\quad\quad\quad\quad\quad\quad\quad\quad\quad\quad\quad\quad\quad\quad\quad$ P

FIG. 1. Structure of the phosphorylated site in glycogen synthetase b (54) and in glycogen phosphorylase a (55) from rabbit muscle. In phosphorylase the seryl phosphate residue occupies position 14 (56).

peptides isolated from chymotryptic digests of phosphorylase are more basic than those of synthetase (57, 58).

d. *Electron Microscopy.* The homogeneous synthetase has been examined in the electron microscope in conditions that would minimize aggregation of the enzyme (58, 59). Three main figures with different dimensions have been observed; one of these is a flattened hexagon, with 160 Å as the long hexagonal distance (Fig. 2). A model relating morphological structure to subunit composition has been proposed (59).

2. *Glycogen Synthetase from Other Tissues*

The amount of phosphate incorporated into heart glycogen synthetase during a to b conversion is about 5 moles per 100,000 g of enzyme (47). Electrophoresis of phosphorylated peptides derived from the heart enzyme yields a pattern that is similar though not identical to that of the muscle enzyme.

No information is available at present concerning the subunit size of the liver enzyme, but evidence in favor of an oligomeric structure has been obtained with glycogen-free preparations from rat liver (33, 46). The enzyme can undergo dissociation and reassociation during reversible thermal inactivation (46, see Section IV,F). Synthetase b shows two major peaks upon centrifugation in a sucrose gradient (33). The heavier peak has an approximate molecular weight of 258,000–284,000 (11.5 S) and appears to be a dimer of the lighter enzyme (7.8 S). The addition of ligands (EDTA, Mg^{2+}, ATP, glucose-6-P, or UDPG) promotes the disappearance of the lighter component.

57. J. Larner, C. Villar-Palasi, N. D. Goldberg, J. S. Bishop, F. Huijing, J. I. Wenger, H. Sasko, and N. B. Brown, *in* "Control of Glycogen Metabolism" (W. J. Whelan, ed.), p. 1. Academic Press, New York, 1968.
58. J. Larner, *Diabetes* **21**, 428 (1972).
59. L. I. Rebhun, C. Smith, and J. Larner, *Mol. Cell. Biochem.* **1**, 55 (1973).

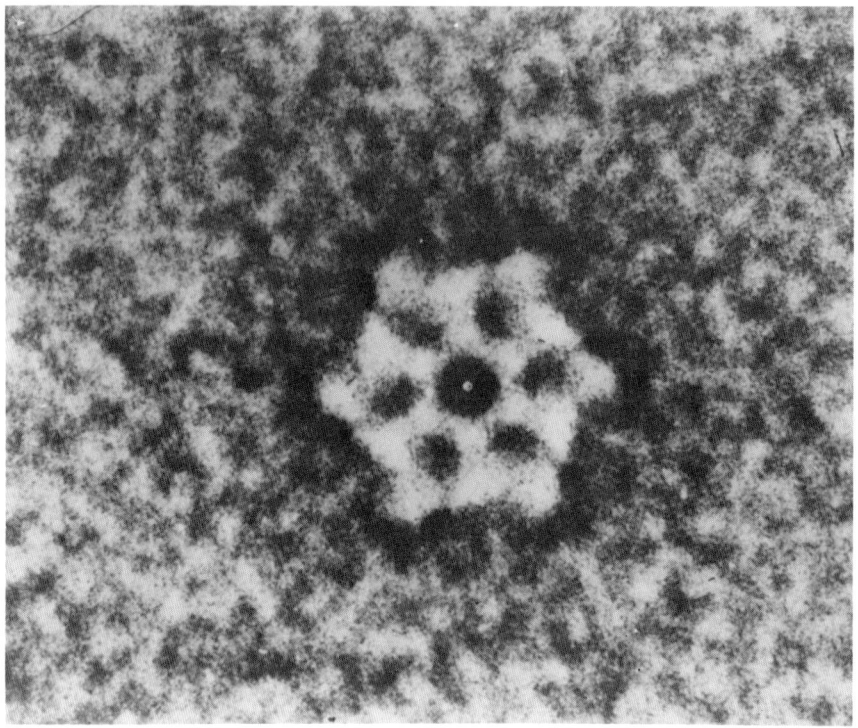

FIG. 2. Rotation micrograph (6-fold axis) prepared from a hexagonal figure of rabbit muscle glycogen synthetase. Uranyl oxalate stain. ×2,600,000. From Rebhun et al. (59).

III. General Catalytic Properties

A. THE REACTION

The reaction catalyzed by glycogen synthetase may be written as UDPG + α-primer $\rightarrow$ UDP + glucosyl-(α-1,4)-primer (3, 60). The reaction is only slightly reversible; Kornfeld and Brown (37) have observed the formation of 1.8 nmoles of UDPG in the presence of 7 μmoles of UDP at pH 7.5, allowing to estimate to 5 kcal mole^{-1} the change in standard free energy of the reaction.

This review is limited to the synthesis of glycogen from UDPG.

60. R. Hauk and D. H. Brown, BBA 33, 556 (1959).

Bacterial glycogen synthetase has originally been described using UDPG as a substrate (*61*). Since this enzyme utilizes preferentially ADPG as a glucosyl donor (*62*), it will not be considered here.

B. Assay Methods

Two methods are currently used for the determination of glycogen synthetase.

1. The amount of UDP produced is measured by coupling to the pyruvate kinase reaction. The pyruvate thus formed can be measured colorimetrically (*4, 63*) or spectrophotometrically (*64*). It is important to recall that ADP reacts like UDP in this procedure. Determination of inorganic phosphate after specific enzymic hydrolysis of UDP has also been used as an assay method (*14*).

2. The most popular method is the determination of the radioactivity incorporated into glycogen from UDPG labeled in the glucosyl moiety (*3*). Separation of unreacted UDPG from glycogen may be achieved by using charcoal (*38, 65*), paper chromatography (*66*), or ion exchangers (*67*); but generally precipitation of the polysaccharide by ethanol is used for this purpose: The glycogen is then collected by centrifugation (*42*), or by filtration on Millipore or glass filters (*68, 69*). An elegant method consists in spotting the reaction mixture on filter paper squares that are then washed in an ethanol bath (*70, 71*).

No assay method is completely satisfactory when crude enzyme preparations are used. UDPG may be converted to UDP in several ways; a striking example is encountered with insect tissues, which can synthesize trehalose phosphate from UDPG and glucose 6-phosphate (*21, 72*). Method 2 may underestimate the activity of glycogen synthetase if the radioactive glycogen is degraded by phosphorylase and α-amylase,

61. N. B. Madsen, *BBA* **50**, 194 (1961).
62. E. Greenberg and J. Preiss, *JBC* **239**, 4314 (1964).
63. E. Cabib and L. F. Leloir, *JBC* **231**, 259 (1958).
64. C. Villar-Palasi and J. Larner, *BBA* **30**, 449 (1958).
65. R. Schmid, P. W. Robbins, and R. R. Traut, *Proc. Nat. Acad. Sci. U. S.* **45**, 1236 (1959).
66. A. Vardanis. *BBA* **73**, 565 (1963).
67. H. De Wulf, W. Stalmans, and H. G. Hers, *Eur. J. Biochem.* **15**, 1 (1970).
68. R. Piras, L. B. Rothman, and E. Cabib, *Biochemistry* **7**, 56 (1968).
69. L. M. Blatt, J. O. Scamahorn, and K. H. Kim, *BBA* **177**, 553 (1969).
70. T. J. Kindt and H. E. Conrad, *Biochemistry* **6**, 3718 (1967).
71. J. A. Thomas, K. K. Schlender, and J. Larner, *Anal. Biochem.* **25**, 486 (1968).
72. J. C. Trivelloni, *ABB* **89**, 149 (1960).

which not only are present in crude tissue extracts but also are copurified with glycogen synthetase by adsorption on particulate glycogen. The error resulting from amylase is cancelled in a variant of the method, in which glycogen and oligosaccharides are first hydrolyzed to glucose before being separated from UDPG and counted (67).

C. Donor Specificity

The efficiency of different glucosyl donors has been studied with the synthetase from mammalian tissues (37, 73, 74); quantitative data may be found in Table I.

The transfer of other sugars or sugar derivatives from corresponding nucleotides has been investigated. UDP galactose, UDP-N-acetylglucosamine (37), and ADP maltose (73) are not accepted by muscle glycogen synthetase. UDP glucosamine is readily accepted, and this presumably explains the incorporation of glucosamine into glycogen in liver perfused with galactosamine (76, 77). UDP deoxyglucose is a substrate for glycogen synthetase from yeast (78), and deoxyglucose is indeed incorporated into glycogen not only by intact yeast cells (78) but also by hepatoma cells (79).

TABLE I
Specificity of the Glucosyl Donor

Glucosyl nucleotide[a]	UDPG	ADPG	TDPG	Pseudo-UDPG	CDPG	IDPG
Relative efficiency[b]	100	50	5	5–10	0	0
Enzyme source		Rat muscle	Rabbit muscle	Rat liver	Rat muscle	Rat muscle
Reference		73	37	74	73	73

[a] GDPG has been reported to be used at a low rate (75).
[b] Reaction rates as compared to UDPG at equimolar concentration.

73. S. H. Goldemberg, *BBA* **56**, 357 (1962).
74. M. Rabinowitz and I. H. Goldberg, *JBC* **238**, 1801 (1963).
75. S. H. Goldemberg, unpublished results; cited by L. F. Leloir, *Proc. Pan-Amer. Congr. Endocrinol., 6th, 1965* Excerpta Med. Found. Int. Congr. Ser. No. 112, p. 65 (1966).
76. F. Maley, J. F. McGarrahan, and R. DelGiacco, *BBRC* **23**, 85 (1966).
77. F. Maley, A. L. Tarentino, J. F. McGarrahan, and R. DelGiacco, *BJ* **107**, 637 (1968).
78. J. Zemek, V. Farkaš, P. Biely, and Š. Bauer, *BBA* **252**, 432 (1971).
79. V. N. Nigam, *ABB* **120**, 232 (1967).

D. THE GLUCOSYL ACCEPTOR

It has long been recognized that glucose cannot act as a glucosyl acceptor but that, like in the case of phosphorylase, a more complex primer is required (1). The acceptor may be a polysaccharide, an oligosaccharide, or even a protein.

1. *Polysaccharides*

The ability of a polysaccharide to serve as a glucosyl acceptor depends on several factors such as its degree of branching, its molecular size, and the length of its outer chains. Glycogen is by far the best acceptor for glycogen synthetase. Other polysaccharides, like native or solubilized starches, have a poor priming efficiency (3, 73). Small, KOH-treated glycogen is a better primer than cold water extracted glycogen (80). The same difference was noted for phosphorylase (81).

The apparent K_m of muscle glycogen synthetase b for glycogen is dependent on the concentration of UDPG (45, see Section III,E); values of 3.9 and 5.7 µg/ml were found at levels of UDPG equal to 0.67 and 3.3 mM, respectively; a similar value (82) as well as a 50-fold higher value (73) had previously been reported. The K_m of the enzyme from mammalian (80, 83) and fish liver (20) is expressed as milligram rather than microgram per milliliter.

Parodi et al. (84) have calculated that only 40% of the nonreducing ends of liver glycogen are available to transglucosylation. This value reaches 60% in small molecular weight, sonicated glycogen and falls to 20% for glycogen made *in vitro* by phosphorylase and branching enzyme. The suggestion (82, 85) that glycogen synthetase adds preferentially or exclusively to the main chains of glycogen has been disproved (84, 86).

The length of the outer chain is also an important factor since glycogen is a better acceptor than a phosphorylase limit dextrin which in turn is superior to a β-amylase dextrin (3, 82, 87); reduction of the

80. A. Vardanis, *JBC* **242**, 2312 (1967).
81. S. A. Orrell and E. Bueding, *JBC* **239**, 4021 (1964).
82. D. H. Brown, B. Illingworth, and R. Kornfeld, *Biochemistry* **4**, 486 (1965).
83. A. H. Gold and H. L. Segal, *ABB* **120**, 359 (1967).
84. A. J. Parodi, J. Mordoh, C. R. Krisman, and L. F. Leloir, *Eur. J. Biochem.* **16**, 499 (1970).
85. D. H. Brown, B. I. Brown, and C. F. Cori, *ABB* **116**, 479 (1966).
86. H. G. Hers and W. Verhue, *BJ* **100**, 3P (1966).
87. A. Vardanis, *Can. J. Biochem.* **46**, 579 (1968).

outer chain length decreases the V_{max} and increases the K_m value *(82)*. Glycogen can lose its primer ability by treatment with α-amylase *(3)*. A degradation of glycogen by liver α-amylase is known to affect initially the outer chains *(88)*. This presumably accounts for the observation that glycogen synthetase isolated from liver as an enzyme–glycogen complex is highly dependent on added glycogen for activity *(80, 87)*. This dependency also suggests that the tight association of the enzyme with particulate glycogen, described in Section II,A, does not necessarily involve its active site.

2. *Oligosaccharides*

Maltose and maltotriose at high concentrations act as acceptors but with very low efficiency. The affinity of the synthetase for larger linear or branched oligosaccharides is considerably better but still at least two orders of magnitude below that for glycogen when the concentration is calculated as end groups. In all cases, the reaction leads to the formation of the next higher homolog *(73, 82)*. The elongation of branched oligosaccharides occurs exclusively by addition to the main chains *(82)*.

The question whether new glycogen molecules could be formed from oligosaccharides has been discussed by Leloir *(89, 90)*. Efficient transglucosylation on oligosaccharides would require the virtual absence of glycogen. Furthermore, the oligosaccharides found in the liver *(91)* seem to be formed artifactually after death by α-amylase *(92, 93)*. There is therefore no reason to believe that the combined action of glycogen synthetase and branching enzyme on oligosaccharides could eventually lead to the formation of glycogen in cells.

3. *De novo Synthesis of Glycogen*

The study of glycogen synthesis *de novo*, that is, without preexisting primer, is a delicate enterprise because of the possible contamination of enzymes or substrates with a barely detectable amount of precursor. For instance, the early report of unprimed glycogen synthesis from glucose 1-phosphate by muscle phosphorylase *(94)* has been discounted on these grounds *(95)*.

88. J. M. Olavarría and H. N. Torres, *JBC* **237**, 1746 (1962).
89. L. F. Leloir, *Proc. Plenary Sess., Int. Congr. Biochem. 6th, 1964* p. 15 (1964).
90. L. F. Leloir, in "Control of Glycogen Metabolism" (W. J. Whelan and M. P. Cameron, eds.), p. 68. Churchill, London, 1964.
91. W. H. Fishman and H. G. Sie, *JACS* **80**, 121 (1958).
92. J. M. Olavarría, *JBC* **235**, 3058 (1960).
93. R. Sandruss, O. G. Gödeken, and J. M. Olavarría, *ABB* **116**, 69 (1966).

In recent years interest has been aroused again by the finding of possibly unprimed ADPG-dependent polysaccharide synthesis in *Aerobacter* (*96*) and in spinach leaves (*97*). A common feature in either case is a lag period before transglucosylation occurs, and the reduction of this latency by the addition of albumin. *De novo* synthesis with the plant enzyme also requires a high salt medium. Krisman (*98*) recently reported on the existence of a similar system in rat liver. Unprimed glycogen synthesis from UDPG is catalyzed by an enzyme preparation which is also capable of primer-dependent synthesis. Both types of activity can be distinguished by a clearly different pH optimum. The presence of a latency and the requirement of a high salt concentration are reminiscent of the plant system (*97*). An exciting feature is that the unprimed activity leads to the formation of a glucan that is covalently bound to protein. The finding that the addition of glycogen inhibits this unusual synthesis virtually eliminates the problem of primer contamination.

E. Reaction Mechanism

Brown and Larner (*45*) have applied Cleland's analysis (*99*) to the kinetics of the reaction catalyzed by purified glycogen-free muscle synthetase *b*. Parallel straight lines were obtained in double reciprocal plots of v against UDPG concentration at various glycogen concentrations and vice versa. These kinds of results indicate a ping-pong mechanism, in which a glucosyl–enzyme intermediate is presumably formed by transfer from UDPG and release of UDP, the glucosyl being in a second step transferred on the acceptor. Previous attempts to demonstrate the existence of a glucosyl–enzyme intermediate by an exchange between UDP and UDPG in the absence of glycogen had however given negative results (*37*). It has been suggested (*45*) that this failure could result from the presence of a small amount of glycogen in the enzyme preparation.

Several studies have been made of the "action pattern" of glycogen synthetase, i.e., the number of glucosyl units that are successively trans-

94. B. Illingworth, D. H. Brown, and C. F. Cori, *Proc. Nat. Acad. Sci. U. S.* **47**, 469 (1961); D. H. Brown, B. Illingworth, and C. F. Cori, *ibid.* p. 479.

95. M. Abdullah, E. H. Fischer, M. Y. Qureshi, K. N. Slessor, and W. J. Whelan, *BJ* **97**, 9P (1965).

96. L. C. Gahan and H. E. Conrad, *Biochemistry* **7**, 3979 (1968).

97. J. L. Ozbun, J. S. Hawker, and J. Preiss, *BBRC* **43**, 631 (1971).

98. C. R. Krisman, *BBRC* **46**, 1206 (1972).

99. W. W. Cleland, *BBA* **67**, 104 (1963).

ferred to a nonreducing terminal before the enzyme diffuses to another group. The two extreme possibilities to be considered are "single chain elongation" (successive addition to the same chain) or "multiple chain elongation" (random transglucosylation to all chains). The action of the synthetase on glycogen (*82, 84*) is by multirepetitive chain elongation, which is a combination of the two above mechanisms; the glucosyl residues are added randomly but in groups of more than one. The mean number of glucose units added successively to a nonreducing end of glycogen by the liver enzyme increased from 1.7 to 6.8 with the molecular weight of the polysaccharide (*84*). Multirepetitive elongation was also evident with yeast synthetase using UDPdeoxyglucose as a substrate (*78*).

As mentioned above, not all chains in a polysaccharide are equally good acceptors. When a poor substrate like a β-limit dextrin is used (*82, 84*), the elongation occurs by a nearly single chain mechanism, suggesting that the addition of one glucosyl to an outer chain greatly facilitates further transfer on the same chain until an optimum is reached. This process is probably also favored by the attachment of the enzyme to the polysaccharide at a site which, as suggested above, could be different from the active site. When linear (*73*) or branched (*82*) oligosaccharides are used as glucosyl acceptors with the muscle synthetase, the elongation follows an extreme multichain pattern. This is presumably explained by the fact that in this case all the acceptor chains have the same length and that the next higher homolog is not a markedly better substrate.

IV. The Two Forms of Glycogen Synthetase and Their Interconversion

A. Nomenclature

The two forms of glycogen synthetase have been recognized by Larner and his co-workers (*5, 39*) on the basis of a different degree of stimulation by glucose 6-phosphate. One form of the enzyme was largely active in the absence of the cofactor and was called I (glucose 6-phosphate independent), whereas the other showed a nearly complete dependency on glucose 6-phosphate and was called D (*39*).

The I and D nomenclature is still widely used, mostly by students of the muscle enzyme, whereas, at the suggestion of Mersmann and Segal (*100*), the a and b terminology, initially introduced by Cori and Green

100. H. J. Mersmann and H. L. Segal, *Proc. Nat. Acad. Sci. U. S.* **58**, 1688 (1967).

(*101*) for the two forms of phosphorylase, is often preferred by investigators of the liver synthetase. The main reason for this preference is that, in the ionic conditions prevailing in the cell, one form of the liver enzyme is nearly fully active and the other fully inactive, whereas neither form is significantly influenced by glucose 6-phosphate (*100, 102*). In this case the I and D nomenclature could be misleading. The fundamental meaning of an interconversion between two enzyme forms is indeed the change in activity and not the change in dependency on an effector as measured in artificial conditions. The less committal *a* and *b* terminology, which refers to the active (*a*) and less active (*b*) forms without specifying the actual factors that endow one form with activity and keep the other inactive, has therefore a much wider range of application. In tadpole liver, for instance, the two forms of glycogen synthetase are glucose 6-phosphate dependent and could adequately be called *a* and *b*, although not I and D. It must also be recalled that in several cells the I form is markedly stimulated by glucose 6-phosphate and is therefore partially "dependent," whereas the synthetase D may display some activity in the absence of the ligand. The I and D terminology should therefore only be an operational one, applicable to enzymic activity and not to enzyme forms.

Recently, Larner and Villar-Palasi (*12*) and Sols and Gancedo (*103*) have proposed adopting a terminology that would be general and applicable to other enzymic systems that are regulated by phosphorylation and dephosphorylation of the enzyme. These include phosphorylase, phosphorylase kinase, glycogen synthetase, pyruvate dehydrogenase, and lipase. One of the proposals was to call the two forms *phospho-* and *dephospho-* and to designate them as "P-O-enzyme" and "enzyme," respectively (*12*). Such a terminology was first used by Sutherland (*104*) for liver phosphorylase but, as recognized by the same author (*105*), it has never been very popular. Another proposal was to designate the two forms according to their degree of activity; the terminology is then also applicable to the two forms of glutamine synthetase which are interconverted by adenylylation and deadenylylation. The physiologically more active form and the physiologically less active form have

101. G. T. Cori and A. A. Green, *JBC* **151**, 31 (1943).
102. H. De Wulf, W. Stalmans, and H. G. Hers, *Eur. J. Biochem.* **6**, 545 (1968).
103. A. Sols and C. Gancedo, *in* "Biochemical Regulatory Mechanisms in Eukaryotic Cells" (E. Kun and S. Grisolía, eds.), p. 85. Wiley (Interscience), New York, 1972.
104. T. W. Rall, E. W. Sutherland, and W. D. Wosilait, *JBC* **218**, 483 (1956).
105. E. W. Sutherland, G. A. Robison, and R. W. Butcher, *Circulation* **37**, 279 (1968).

been designated "$\overset{*}{e}$" and "$\overset{o}{e}$" by Larner and Villar-Palasi (12) and a and b by Sols and Gancedo (103). The first system seems to be essentially a written notation and has not been further used by its promotors. A numerical classification (I and II) has been used for the two forms of glutamine synthetase (106); this system has the disadvantage of not clearly indicating which of the two forms is the active one and has not been used for other enzymes. It seems therefore that only the a and b terminology could be applied to all interconvertible enzymes; it has a historical value and has already been used for glycogen phosphorylase (101), glycogen synthetase (100, 106a), pyruvate dehydrogenase (107), lipase (108), and glutamine synthetase (109). Using it in the present review, we wish to make it clear that synthetase a and b are identical to the I and D forms, which do not necessarily correspond to the I and D activities.

B. General Properties of the a and b Forms

The total (I + D) and the I activities of glycogen synthetase are measured by assaying the enzyme with and without 10 mM glucose 6-phosphate. The specific determination of synthetase a is, however, better performed in more complex ionic conditions that may vary from tissue to tissue. This enzyme is only partially active in the absence of ligands and gains full activity in the presence of glucose 6-phosphate. The effect of glucose 6-phosphate on synthetase a is frequently shared by other anions such as phosphate and sulfate, although at higher concentrations. The K_a for glucose 6-phosphate is much smaller for synthetase a than for synthetase b; at saturating concentration of the ligand, the kinetic properties of the two forms are similar.

Several factors are responsible for the fact that synthetase b is poorly active in the absence of glucose 6-phosphate. For instance, in the muscle both a low V_{max} and a high K_m for UDPG are the most likely explanation. In mammalian liver, V_{max} is apparently comparable to that of synthetase a whereas K_m is high. In *Neurospora crassa*, on the contrary, K_m is similar to that of synthetase a whereas V_{max} is low. In the adrenals,

106. B. M. Shapiro, H. S. Kingdon, and E. R. Stadtman, *Proc. Nat. Acad. Sci. U. S.* **58**, 642 (1967).

106a. C. Villar-Palasi and J. Larner, *Abstr. 138th Amer. Chem. Soc. Meet.*, 78C (1960).

107. O. Wieland, E. Siess, F. H. Schulze-Wethmar, H. G. von Funcke, and B. Winton, *ABB* **143**, 593 (1971).

108. J. D. Corbin, E. M. Reimann, D. A. Walsh, and E. G. Krebs, *JBC* **245**, 4849 (1970).

109. D. Mecke, K. Wulff, and H. Holzer, *BBA* **128**, 559 (1966); H. Holzer, *Advan. Enzymol.* **32**, 297 (1969).

the b enzyme has a low V_{max} and paradoxically a higher affinity for UDPG than the a form. In all cell types, however, glycogen synthetase a and b are markedly inhibited by ATP, and this inhibition is released by a much lower concentration of glucose 6-phosphate in the case of synthetase a than in the case of synthetase b.

Some properties of glycogen synthetase appear to be fairly general. Inhibition by ATP is kinetically competitive with UDPG, although desensitization experiments indicate that binding occurs at an allosteric site. On the contrary, inhibition by UDP is truly competitive. If both glucose 6-phosphate and ATP are present simultaneously, cooperative effects are observed with one ligand, or with both of them, but usually not with each enzyme form. In spite of the oligomeric structure of several glycogen synthetases (see Section II,C), cooperative binding of UDPG has only been observed with the enzyme from mammalian and tadpole liver and from molds (*Dictyostelium* and *Blastocladiella*). Mg^{2+} positively affects the enzymic activity; at least part of its effect on the b form is the result of an increased affinity for glucose-6-P. The cation also decreases efficiently the inhibition by nucleotides. A detailed account of these properties will be found in Sections V to IX.

C. The Basic System of Interconversion

In most tissues of untreated animals glycogen synthetase is predominantly in the inactive b form. During incubation of a tissue extract there is a progressive activation of the enzyme, and the conversion into the a form is usually complete in about 1 hr at 20° or 30°. If at that time ATP and Mg^{2+} are added, a rapid inactivation occurs. This basic system of interconversion was first established by Friedman and Larner with a rat muscle extract (*6*). These authors demonstrated that the terminal phosphate of ATP is incorporated into the enzyme in the course of its inactivation and that this phosphate is removed during reactivation. This fundamental experiment established that, similar to what was known for a long time in the case of phosphorylase, the two forms of muscle glycogen synthetase are interconverted by phosphorylation and dephosphorylation with the main difference however that phosphosynthetase is inactive whereas phosphophosphorylase is the active form.

A relatively slow, spontaneous activation (*26, 83, 110*) and a more rapid ATP-Mg–dependent inactivation of glycogen synthetase also occur in liver preparations (*110, 111*). Moreover, muscle and liver synthetase kinases accept the heterologous as well as the homologous substrates

110. H. De Wulf and H. G. Hers, *Eur. J. Biochem.* **6**, 552 (1968).
111. J. S. Bishop and J. Larner, *BBA* **171**, 374 (1969).

(112) and conversion of liver synthetase *a* into *b* proceeds with phosphorylation of the enzyme *(58)*. In vitro interconversion of two forms also takes place in heart extracts *(113)*, and the changes in enzymic activity are coincident with phosphorylation and dephosphorylation *(47)*. Identical or similar interconversion reactions occur in extracts of mammalian brain *(114)*, spleen *(115)*, kidney *(18, 116)*, adrenal gland *(117)*, lymphocytes *(118)*, polymorphonuclear leukocytes *(119, 120)*, and hepatoma cells *(34, 121)*, as well as of frog muscle *(122, 123)* or of primitive organisms like *Neurospora crassa* *(124)* and yeast *(23, 125)*. Phosphorylation of the *a* enzyme from rabbit brain and kidney and from frog liver and muscle by purified rabbit muscle synthetase kinase has been reported *(58)*.

D. SYNTHETASE KINASE

Glycogen synthetase kinase has been purified about 300-fold from muscle *(7)*. The enzyme is different from phosphorylase kinase *(126)* but identical to the cyclic AMP–stimulated protein kinase, which also acts as a phosphorylase kinase kinase *(7, 8, 127)*.

A stimulation by cyclic AMP of the enzymic inactivation of muscle glycogen synthetase in the presence of ATP was first observed by Belocopitow *(128)* and was firmly established by Rosell-Perez and Larner *(41)* and by Appleman *et al.* *(129)*. Synthetase kinases of mammalian liver *(110, 111)*, heart *(130)*, brain *(114)*, kidney *(116)*, adrenal

112. A. T. Yip and J. Larner, *Physiol. Chem. Phys.* 1, 383 (1969).
113. O. Søvik, I. Øye, and M. Rosell-Perez, *BBA* 124, 26 (1966).
114. N. D. Goldberg and A. G. O'Toole, *JBC* 244, 3053 (1969).
115. S. Hizukuri and Y. Takeda, *BBA* 212, 179 (1970).
116. K. K. Schlender, *Fed. Proc., Fed. Amer. Soc. Exp. Biol.* 31, 594 (1972).
117. M. M. Piras and R. Piras, *ABB* 148, 581 (1972).
118. C. J. Hedeskov, V. Esmann, and M. Rosell-Perez, *BBA* 130, 393 (1966).
119. V. Esmann, C. J. Hedeskov, and M. Rosell-Perez, *Diabetologia* 4, 181 (1968).
120. M. Rosell-Perez, C. J. Hedeskov, and V. Esmann, *BBA* 156, 414 (1968).
121. K. Sato, N. Abe, and S. Tsuiki, *BBA* 268, 646 (1972).
122. M. Rosell-Perez and J. Larner, *Biochemistry* 1, 769 (1962).
123. J. L. Albert and M. Rosell-Perez, *Rev. Espan. Fisiol.* 26, 139 (1970).
124. M. T. Tellez-Iñon, H. Terenzi, and H. N. Torres, *BBA* 191, 765 (1969).
125. L. B. Rothman-Denes and E. Cabib, *Biochemistry* 10, 1236 (1971).
126. D. L. Friedman and J. Larner, *Biochemistry* 4, 2261 (1965).
127. C. Villar-Palasi, J. Larner, and L. C. Shen, *Ann. N. Y. Acad. Sci.* 185, 74 (1971).
128. E. Belocopitow, *ABB* 93, 457 (1961).
129. M. M. Appleman, E. Belocopitow, and H. N. Torres, *BBRC* 14, 550 (1964).
130. F. Huijing, F. Q. Nuttall, C. Villar-Palasi, and J. Larner, *BBA* 177, 204 (1969).

gland (117), and frog muscle (123) are all stimulated by cyclic AMP, albeit to a variable extent. No effect of the nucleotide was detected in yeast (125). Half-maximal stimulation of muscle synthetase kinase is obtained with cyclic AMP concentrations in the range of 6×10^{-8} to $10^{-7} M$ (7, 131–133); values of $2 \times 10^{-7} M$ (110, 134) and $4 \times 10^{-8} M$ (111) have been reported for the liver enzyme, and $5 \times 10^{-8} M$ for the kinase from heart (130). The same range has been found for the activation of phosphorylase kinase. One possible cause of variability is the presence of an inhibitor of the kinase in some commercial samples of cyclic AMP (34). The lowest K_m values appear then most probable. Other 3′,5′-cyclic nucleotides allow the same maximal activity of synthetase kinase although at much higher concentrations (7, 8, 133, 134). The available evidence indicates the following affinity of synthetase kinase for cyclic nucleotides: AMP > IMP > CMP ≃ dibutyryl AMP ≃ GMP ≃ UMP ≫ dAMP > TMP. No effect was observed with 3′-AMP, 5′AMP, 2′,3′-cyclic amp (7), cyclic dGMP, and cyclic dCMP (133).

The conversion of synthetase a into b in liver extracts is inhibited by high concentrations of glucose 6-phosphate ($K_i = 2$ mM) independently of the presence of cyclic AMP (110). Glycogen also inhibits the kinase, but only in the absence of cyclic AMP, and therefore amplifies the effect of the nucleotide.

Other properties of protein kinase are described by Walsh and Krebs (134a) in the preceding volume. We will only recall that two forms of the enzyme have been isolated. One of them, called C (catalytic subunit), is active in the absence of cyclic AMP; the other, made of the association of C with a regulatory subunit R, is dependent on the presence of cyclic AMP for activity. The two forms are assumed to be interconverted according to the reaction

$$CR + \text{cyclic AMP} \rightleftharpoons C + R\text{-cyclic AMP}$$

E. SYNTHETASE PHOSPHATASE

1. *Muscle*

Synthetase phosphatase has been purified 1000-fold from rabbit muscle (135). The enzyme seems identical to histone phosphatase (135) and to

131. M. M. Appleman, L. Birnbaumer, and H. N. Torres, *ABB* **116**, 39 (1966).
132. F. Huijing and J. Larner, *BBRC* **23**, 259 (1966).
133. O. Walaas, E. Walaas, and S. Osaki, in "Control of Glycogen Metabolism" (W. J. Whelan, ed.), p. 139. Academic Press, New York, 1968.
134. W. H. Glinsmann and E. P. Hern, *BBRC* **36**, 931 (1969).
134a. D. A. Walsh and E. G. Krebs, "The Enzymes," 3rd ed., Vol. 8, p. 555, 1973.
135. K. Kato and J. S. Bishop, *JBC* **247**, 7420 (1972).

phosphorylase kinase phosphatase (135a), and could thus be a more general protein phosphatase, acting antagonistically on the substrates of the cyclic AMP stimulated protein kinase. It is not specifically inhibited by phosphorylase a (135, 136) and is therefore presumably different from phosphorylase phosphatase.

The purified synthetase phosphatase is stabilized by Mn^{2+}, which also stimulates the enzyme, as do Ca^{2+} and Mg^{2+} to a lesser extent. It is inhibited by 10 mM NaF, as well as by millimolar concentrations of P_i, PP_i, and Na_2SO_3. Optimal activity is found between pH 7.0 and 7.4 (135). Some enzyme preparations require the presence of a reducing agent (39, 41). An apparent stimulation of the conversion of synthetase b into a by small amounts of glucose 6-phosphate has been noted (38, 135). In one case (135), however, the liberation of phosphate from the enzyme was measured and found to be unchanged. This indicates that the effect of glucose 6-phosphate at low concentration is to stimulate synthetase a more strongly than synthetase b. Alternatively, the hexose phosphate could change the action pattern of the phosphatase on synthetase b in order to produce partially phosphorylated enzyme that may behave kinetically like synthetase a (135). Such intermediate enzyme forms have been identified during interconversion of phosphorylase a and b (136a).

In a crude extract synthetase phosphatase is inhibited by glycogen at concentrations normally found in the tissue (136, 137), and this inhibition is believed to play an important feedback control of glycogen synthesis in muscle. The degree of inhibition seems to vary according to several factors such as the structure of the glycogen and the age of the animal as well as the age of the enzyme preparation (136). Purified synthetase phosphatase has optimal activity in the presence of about 0.1% glycogen (135).

2. *Liver*

Synthetase phosphatase from liver has not been purified. In a fresh liver extract (83) or gel filtrate (67) the enzyme is usually inactive and remains so for a period as long as 20 min at 20°; then it suddenly reaches full activity. This latency is markedly reduced by the addition of glucose or caffeine and also by treatment of animals with glucocorti-

---

135a. F. J. Zieve and W. H. Glinsmann, *BBRC* **50**, 872 (1973).
136. C. Villar-Palasi, *Ann. N. Y. Acad. Sci.* **166**, 719 (1969).
136a. S. S. Hurd, D. Teller, and E. H. Fischer, *BBRC* **24**, 79 (1966).
137. J. Larner, *Trans. N. Y. Acad. Sci.* [2] **29**, 192 (1966).

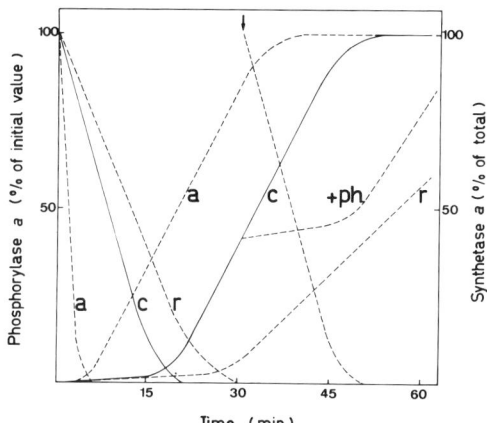

FIG. 3. Schematic representation of the inactivation of phosphorylase (descending lines) and of the activation of glycogen synthetase (ascending lines) as observed in a liver Sephadex filtrate incubated at 20°. Notations: c, control filtrate; a, control filtrate incubated in the presence of 0.5% glucose or 0.2 mM caffeine, or filtrate from mice treated with prednisolone; and r, control filtrate incubated in the presence of 6% glycogen. The arrow indicates the addition of an amount of purified liver phosphorylase a equivalent to that initially present (+ph). All filtrates contained 5 mM $(NH_4)_2SO_4$; similar results were obtained in the presence of 5 mM P_i (physiological concentration). The original data from which the scheme is constructed can be found in Stalmans et al. (9, 138) and De Wulf et al. (67). After Hers et al. (137a).

coids (67). These observations have been explained by the strong inhibition exerted by phosphorylase a on liver synthetase phosphatase (Fig. 3), while phosphorylase b is only slightly inhibitory (9). The latency in synthetase phosphatase activity is the time required for phosphorylase a to be inactivated by phosphorylase phosphatase. This time is shortened by caffeine and glucose, which bind to phosphorylase a, making it a better substrate for phosphorylase phosphatase, and also by a glucocorticoid treatment, which increases the activity of phosphorylase phosphatase in the liver (138, 139). The inhibition by phosphorylase a is cancelled by unphysiologically high concentrations of AMP, particularly when associated with Mg^{2+} (9, 67). In the presence of these agents, the synthetase phosphatase is active without latency.

137a. H. G. Hers, H. De Wulf, and W. Stalmans, FEBS Lett. 12, 73 (1970).
138. W. Stalmans, H. De Wulf, B. Lederer, and H. G. Hers, Eur. J. Biochem. 15, 9 (1970).
139. W. Stalmans, T. de Barsy, M. Laloux, H. De Wulf, and H. G. Hers, in "Metabolic Interconversion of Enzymes" (O. Wieland, E. Helmreich, and H. Holzer, eds.), p. 121. Springer-Verlag, Berlin and New York, 1972.

Synthetase phosphatase is different from phosphorylase phosphatase since purified liver phosphorylase phosphatase has no activity on synthetase b (34).

Some reports of altered synthetase phosphatase activity may have to be reconsidered with respect to the inhibition of the enzyme by phosphorylase a. The suggestion that the phosphatase might exist in two forms (67) had to be withdrawn on this basis (9). The inhibition of liver synthetase phosphatase by fluoride (110), at least in part, results from inhibition of phosphorylase phosphatase (34). Hormonal effects observed on synthetase phosphatase might also be mediated by a change in phosphorylase a content of the liver (see Sections VII,B,2 and 3). High doses of glycogen inhibit synthetase phosphatase (110), but this effect is inconstant (67) and may be at least partly explained by an inhibition of phosphorylase phosphatase (9). The low sensitivity of the liver system to glycogen, as compared to muscle (136, 137) or heart (130), could explain the much greater capacity of the liver for glycogen storage. However, a minimal amount of glycogen (0.3–0.5%) is required for the activity of liver synthetase phosphatase (9). This requirement may be an explanation for the lack of activity observed in conditions that deplete hepatic glycogen stores, like fasting and adrenalectomy (28) or intoxication with carbon tetrachloride (140).

F. AN "INACTIVE" FORM OF GLYCOGEN SYNTHETASE

Hizukuri and Larner (26) have observed that liver glycogen synthetase b, bound to particulate glycogen, becomes completely inactive when incubated at 30° for 10–15 min even when assayed in the presence of glucose 6-phosphate. The enzyme could be reactivated upon incubation in the presence of $MgCl_2$ and Na_2SO_3 and was recovered mostly in the a form. Steiner (141) has used a reversible inactivation at 37° as a means to separate the enzyme from glycogen. The loss of activity appears to result from a dissociation of the synthetase molecule into smaller inactive subunits which do not bind effectively to glycogen (46). Reactivation of the enzyme in the presence of glucose 6-phosphate and fluoride was characterized by reassociation of the fragments. Some uncertainty exists, however, whether the processes studied by both groups

140. R. S. Hickenbottom and K. R. Hornbrook, *J. Pharmacol. Exp. Ther.* **178**, 383 (1971).
141. D. F. Steiner, *BBA* **54**, 206 (1961).

(*26*, *46*) are identical. Furthermore, in the absence of glycogen and at 0°, synthetase *a* undergoes a partial inactivation which is reversed upon rewarming at 20° (*34*). This phenomenon might be related to the previously reported aggregation of muscle synthetase in the cold (see Section II,C,1).

The presence, in a fresh liver homogenate, of a third form of glycogen synthetase that is inactive even in the presence of glucose 6-phosphate could explain that in some experiments a marked increase in total activity during *b* to *a* conversion *in vitro* was observed (*142*, *143*). A similar increase in total activity has also been observed in several other tissues by Rosell-Perez and his co-workers (*18*, *119*, *144*, *145*). In addition these authors have reported a loss of total activity upon addition of ATP-Mg or UTP-Mg to the activated system. The phenomena have been interpreted as evidence for an "extra-phosphorylated," totally inactive form of glycogen synthetase (*145*).

G. Proteolytic Inactivation

A partially purified preparation of muscle synthetase *a* can be converted into a *b*-like form in the presence of trypsin (*129*) or of 1 mM Ca^{2+} plus a protein factor (*146*). These conversions are irreversible, do not require ATP, and are not stimulated by cyclic AMP (*146*, *147*). The protein factor seems identical (*146*) to that involved in the activation of phosphorylase kinase (*148*), and the latter factor has been recognized as a calcium-dependent proteolytic enzyme (*149*, *150*). The nonidentity of muscle phosphorylase kinase and glycogen synthetase has been demonstrated (*151*). The suggestion (*152*) that these two enzymes might be identical in liver has been disproved (*153*).

142. A. Vardanis, *ABB* **130**, 413 (1969).
143. L. M. Blatt and K. H. Kim, *JBC* **246**, 7256 (1971).
144. A. Sacristán and M. Rosell-Perez, *Rev. Espan. Fisiol.* **27**, 331 (1971).
145. M. Rosell-Perez, *Ital. J. Biochem.* **21**, 34 (1972).
146. E. Belocopitow, M. M. Appleman, and H. N. Torres, *JBC* **240**, 3473 (1965).
147. E. Belocopitow, M. C. Fernandez, L. Birnbaumer, and H. N. Torres, *JBC* **242**, 1227 (1967).
148. W. L. Meyer, E. H. Fischer, and E. G. Krebs, *Biochemistry* **3**, 1033 (1964).
149. G. I. Drummond and L. Duncan, *JBC* **243**, 5532 (1968).
150. R. B. Huston and E. G. Krebs, *Biochemistry* **7**, 2116 (1968).
151. F. Huijing, C. Villar-Palasi, and J. Larner, *BBRC* **20**, 380 (1965).
152. B. E. Ryman and W. J. Whelan, *FEBS Lett.* **13**, 1 (1971).
153. H. G. Hers, H. De Wulf, and W. Stalmans, *FEBS Lett.* **14**, 193 (1971).

V. Glycogen Synthetase of Mammalian Muscle

A. PROPERTIES OF THE TWO FORMS

1. *The Effect of Glucose 6-Phosphate and of Other Sugar Phosphates*

Data concerning the effect of glucose 6-phosphate on the kinetic constants of the two forms of muscle glycogen synthetase are collected in Table II. The effect of the ligand is to greatly increase the V_{max} of the b enzyme and to decrease the K_m for UDPG of both forms. At pH 6.6, however, the change in K_m of synthetase a is minimal. In the presence of glucose 6-phosphate, the a and b enzymes have the same K_m. They have apparently also the same V_{max}, since the activity of muscle synthetase remains usually unchanged during the *in vitro* interconversion of the two forms *(6, 8, 39, 41, 131)* as well as during activation by insulin in the isolated diaphragm *(157)*, at least if sufficiently high concentrations of substrate and ligand are present in the assay system.

The concentration of glucose 6-phosphate that allows half-maximal stimulation of synthetase b from various species has been measured at pH values between 7.2 and 8.5 and with UDPG concentrations from 0.75 to 5 mM. All values were comprised between 0.23 and 0.9 mM *(3, 121, 147, 155, 156, 158, 159)*. In the case of synthetase a, a K_a equal to 5 µM has been observed at pH 7.8 *(160)*.

Leloir *et al.* *(3)* found that the stimulatory effect of glucose 6-phosphate is shared by glucosamine 6-phosphate and by galactose 6-phosphate. Rosell-Perez and Larner *(158)* have extended this study to a large number of phosphate compounds. Among those that allow a stimulation of synthetase b comparable (75–96%) to that obtained with glucose 6-phosphate are the 6-phosphate esters of 1,5-sorbitan, galactose, glucosamine and allose, and sedoheptulose-7-P (all D-sugars); the K_a value of the enzyme for these substances increases in that order and reaches 2.5 mM for the last two compounds. A host of other phosphate esters, including triose phosphates and methyl phosphate but with the exception of erythrose 4-phosphate, produced smaller effects. Weak stimulators such

---

154. R. Piras, L. B. Rothman, and E. Cabib, *BBRC* **28**, 54 (1967).
155. O. Søvik, *Acta Physiol. Scand.* **68**, 246 (1966).
156. M. Rosell-Perez, *Rev. Espan. Fisiol.* **25**, 181 (1969).
157. J. W. Craig and J. Larner, *Nature (London)* **202**, 971 (1964).
158. M. Rosell-Perez and J. Larner, *Biochemistry* **3**, 773 (1964).
159. W. H. Danforth, *JBC* **240**, 588 (1965).
160. J. A. Thomas, K. K. Schlender, and J. Larner, *BBA* **193**, 84 (1973).

TABLE II
Effect of Glucose 6-Phosphate on Kinetic Constants of Muscle Synthetase b and a[a]

Species	pH	Synthetase b			Synthetase a			Ref.
		K_m UDPG − G6P (mM)	K_m UDPG + G6P (mM)	$\dfrac{V_{max} + G6P}{V_{max} - G6P}$	K_m UDPG − G6P (mM)	K_m UDPG + G6P (mM)	$\dfrac{V_{max} + G6P}{V_{max} - G6P}$	
Rat	7.8	0.6	0.4	6.7	1	0.2	1	39
Rat	6.6		0.25		0.34	0.25	1	154
Rat[b]	7.8				0.9			155
Rat	7.4		0.50			0.50		121
Rabbit	7.8	5	0.26	47	3.3	0.42	1	40
Dog	7.8	10	0.35	10	1.9	0.66	1	41
Man[b]	7.8		0.6					156

[a] Glucose-6-P, when added, was 10 mM. All assays were done at 30°.
[b] Crude muscle extracts were used. The other results were obtained with partially purified preparations.

as 2-deoxyglucose 6-phosphate inhibit the enzyme competitively with respect to glucose 6-phosphate.

2. The Effect of Inorganic Phosphate

A constituent of the muscle cell sap that stimulates synthetase a at pH 7.8 was identified as inorganic phosphate (*161*). Other anions such as sulfate (*40*), sulfite, arsenate, and pyrophosphate (*160*) have a similar effect. Inorganic phosphate also acts as a weak stimulator of synthetase b (*40*), behaving as a competitive inhibitor with respect to glucose 6-phosphate (*158*). At pH 6.6, inorganic phosphate inhibits both forms of synthetase (*68*); the inhibition of synthetase a is more easily reversed by glucose 6-phosphate.

3. The Effect of Nucleotides

With both forms of the synthetase, saturation curves with UDPG are hyperbolic (*68*). UDP, a reaction product, inhibits the enzyme (*3*). The inhibition is equally strong on the a and b forms, and is kinetically of the competitive type with respect to UDPG; a K_i value of 0.03 mM was found for the a enzyme (*68*). Reversion of the inhibitory effect by glucose-6-P was observed by some authors (*46*) but not by others (*68*).

ATP inhibits the b form in the presence of glucose 6-phosphate and at pH 7.8 (*158*). The absence of inhibition at pH 8.5 (*3*) and the very strong inhibition at pH 6.6 (*154*) are in agreement with an increasing inhibitory potency as pH decreases (*68*). UTP (*158*) and other adenosine, uridine, and guanosine nucleotides (*68*) are also strong inhibitors. Of considerable physiological importance is the demonstration that at low concentrations of glucose 6-phosphate, within the physiological range, the b form is much more strongly inhibited than the a form (*68*). ATP-Mg, presumably the prevalent form of the nucleotide, is less inhibitory since Mg^{2+} tends to reduce the inhibition by ATP.

A pronounced cooperative binding of glucose 6-phosphate by synthetase b is observed in the presence of ATP (without magnesium), whereas the cooperativity is only slight with synthetase a. Conversely, a cooperative effect with free ATP in the presence of glucose 6-phosphate is only observed with the a form (*68*). Rosell-Perez and Larner (*158*) found the inhibition of synthetase b by ATP and by UTP at pH 7.8 competitive with glucose 6-phosphate and not with UDPG. In contrast, Piras et al. (*68, 154*) found that at pH 6.6 the inhibition of synthetase a and b by ATP is of the competitive type with respect to UDPG. The inhibition

161. M. Rosell-Perez and V. Villar-Palasi, *Rev. Espan. Fisiol.* **20**, 131 (1964).

is, however, completely reversed by G6P without change in the K_m for UDPG. Furthermore, photooxidation in the presence of methylene blue desensitizes synthetase a toward ATP much more than toward UDP. The authors (68) concluded that UDPG and ATP bind to different sites.

4. *The Effect of Magnesium and of Other Ligands*

In the absence of glucose 6-phosphate, the affinity of synthetase a for UDPG is increased by 5–10 mM Mg^{2+} (39–41, 155), but the enzyme is inhibited by higher concentrations of the cation (40). The effect of magnesium on synthetase b is more complex: When glucose 6-phosphate is omitted, Mg^{2+} decreases the affinity for UDPG (39, 40), whereas in the presence of an excess of glucose 6-phosphate, the cation is without effect (39–41, 156); however, it increases the affinity for glucose 6-phosphate (158). Calcium seems to have an effect similar to that of magnesium (40).

Other inhibitors have been listed (3). Cinchona alkaloids have the unusual property of inhibiting synthetase a more than synthetase b (162).

5. *The Effect of pH*

Early experiments showed that glucose 6-phosphate shifted the pH optimum of the synthetase to a more alkaline range (37, 38). It is difficult to ascertain whether these changes result from an effect of the hexose phosphate on the a form, or from contribution of b enzyme with a different pH pattern, or both (147). Recent reevaluation of the pH-activity relationship of purified b and a enzymes showed a broad optimum in the pH range from 6.5 to 9 for both synthetase a and b without pronounced influence of glucose 6-phosphate (121, 160)

6. *Activity of the Two Forms in Physiological Conditions*

Piras et al. (68, 154) have tried to define the activity of the b and a enzymes in conditions resembling the intracellular environment of the muscle. At a rather acidic pH and a low concentration of UDPG they found that a "physiological mixture" of P_i, adenine nucleotides, creatine phosphate, and magnesium strongly inhibited either form of the enzyme (Fig. 4). The important feature is that low concentrations of glucose 6-phosphate efficiently reversed the inhibition of the a enzyme, whereas high amounts of the ligand were necessary to endow the b form with significant activity. It was concluded that the *in vivo* activity of muscle glycogen synthetase is determined by two mechanisms (68). One is the interconversion between b and a forms, which has full significance at the

162. L. Rossini and J. Larner, *Pharmacol. Res. Commun.* **3**, 21 (1971).

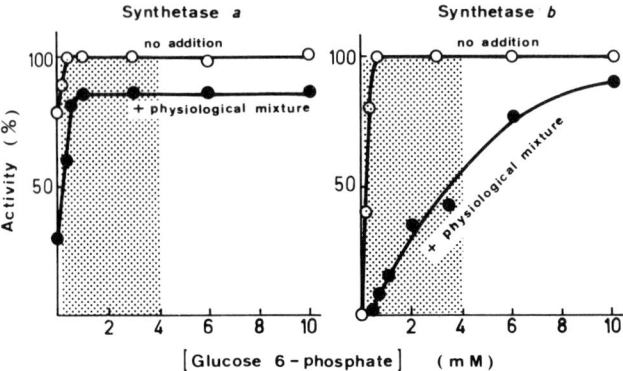

Fig. 4. The effect of glucose 6-phosphate concentration on the activity of muscle glycogen synthetase a and b, in the absence or in the presence of a "physiological mixture" containing 10 mM P_i, 7.3 mM ADP plus ATP, 14 mM creatine-P, and 11 mM $MgCl_2$. [UDPG] was 0.4 mM and pH 6.6. The concentration of glucose-6-P in resting muscle is 0.3 mM; the shaded area indicates the range observed in muscle during tetanic stimulation (173). After Piras et al. (68).

low glucose 6-phosphate level encountered in resting muscle. The other, the regulation by metabolites, is considered to operate during muscle contraction (see Section V,B,4).

B. Control of Synthetase Activity in Muscle

1. Control by Glycogen

Danforth (159) discovered that an inverse relationship exists between the amount of synthetase a and the concentration of glycogen in the muscle. This empirical relationship is probably explained by the inhibition of the synthetase phosphatase by glycogen (see Section IV,E,1). This feedback control by glycogen of its own synthesis superimposes upon other regulatory mechanisms. As illustrated in Fig. 5, the amount of synthetase found in the active form at a given concentration of glycogen was markedly decreased by epinephrine; in the isolated diaphragm it was increased by insulin.

2. Activation by Insulin

It was their study of the glycogenic effect of insulin on the isolated rat diaphragm which led Villar-Palasi and Larner (5) to propose the existence of interconvertible forms of glycogen synthetase. They found that the percentage of activity that can be measured in the absence of glucose

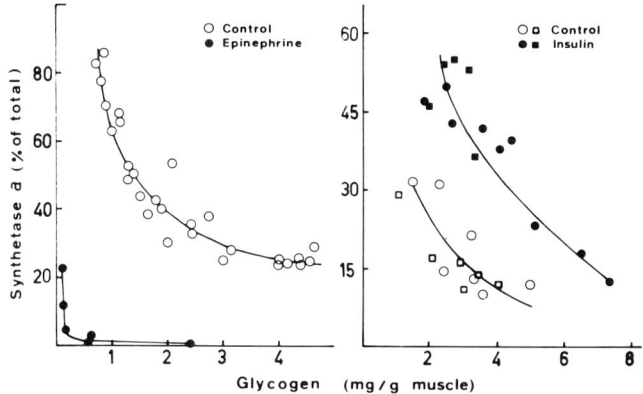

FIG. 5. The effect of epinephrine and of insulin on the relationship between muscle glycogen content and the level of glycogen synthetase a. Left: mouse skeletal muscle *in situ*. The glycogen level was altered by electrical stimulation and varying periods of rest. Measurements are reported for controls and for mice that received 10 μg epinephrine 5–10 min previously. Right: rat diaphragms were incubated with or without insulin (0.2 unit/ml) for 45 min at 37°. The glycogen content was lowered prior to death by hypoxia and by epinephrine, and varied by incubation of the diaphragms without (□■) or with (○●) 5 mM glucose. After Danforth (*159*).

6-phosphate increased from about 20% in the control tissue to about 30% in the presence of the hormone (0.1 unit/ml). It was readily recognized that this effect is unrelated to the hormonal facilitation of transmembrane transport of glucose (*163–165*; for a critical discussion, see Huijing et al., *130*). A similar activation was also observed in skeletal muscle of rats within 5 min after intraperitoneal injection of 2 units/kg of insulin (*166*). This speed of onset is hardly compatible with an effect on protein biosynthesis. The possibility that insulin acts by decreasing the concentration of cyclic AMP in the muscle was not supported by experimental evidence (*166, 167*). Although a previous incubation of diaphragm with the hormone reduces the increase in the level of the nucleotide in response to epinephrine, insulin alone has no effect; when administered *in vivo*, it produces a paradoxical rise in cyclic AMP content.

A stable change in the protein kinase (synthetase kinase) seems to be

163. C. Villar-Palasi and J. Larner, *ABB* **94**, 436 (1961).
164. O. Søvik, *Acta Physiol. Scand.* **63**, 325 (1965).
165. D. Eboué-Bonis, A. M. Chambaut, P. Volfin, and H. Clauser, *Bull. Soc. Chim. Biol.* **49**, 415 (1967).
166. N. D. Goldberg, C. Villar-Palasi, H. Sasko, and J. Larner, *BBA* **148**, 665 (1967).
167. J. W. Craig, T. W. Rall, and J. Larner, *BBA* **177**, 213 (1969).

the explanation of this insulin action. Indeed, it has been found that a greater proportion of the enzyme is cyclic AMP dependent in the muscle after insulin administration *in vivo* (*57, 168*) or after incubation of rat diaphragm with insulin (*169, 170*). This change presumably reflects a reassociation of the regulatory and the catalytic subunits of protein kinase. Contrary to expectation, this reduced efficiency of protein kinase, which also acts as a phosphorylase kinase kinase, is not accompanied by a diminution in the amount of phosphorylase *a* (*157, 167*).

3. *The Effect of Epinephrine*

Belocopitow (*128*) first showed that incubation of the isolated diaphragm in the presence of epinephrine decreases the activity of glycogen synthetase, measured in the presence of glucose 6-phosphate. This observation was confirmed by Craig and Larner (*157*) who demonstrated, in addition, a decrease in the level of the *a* form. The change in total enzymic activity, which is an unusual feature in muscle, has not been further studied. Inactivation of muscle synthetase has also been observed after epinephrine administration *in vivo* (*159*); the effect was already evident after 20 sec (*171*). The inactivation is concomitant with a large increase in the concentration of cyclic AMP and with the activation of phosphorylase (*167*). It is adequately explained by the well-known stimulation and dissociation of protein kinase by the nucleotide (see Section IV,D). Accordingly, after treatment with the hormone a larger proportion of this enzyme is in a form that does not require cyclic AMP for activity (*169*).

The inactivation of muscle glycogen synthetase by injection of epinephrine is impaired in adrenalectomized animals (*172*).

4. *The Events during Muscle Contraction*

During muscle contraction and subsequent recovery an interconversion between the *a* and *b* forms of glycogen synthetase occurs, together with important changes in the level of some metabolites; both effects seem to act together to modulate the activity of the enzyme in the working muscle. Upon electrical stimulation of resting muscle the amount of syn-

168. C. Villar-Palasi and J. I. Wenger, *Fed. Proc., Fed. Amer. Soc. Exp. Biol.* **26**, 563 (1967).
169. L. C. Shen, C. Villar-Palasi, and J. Larner, *Physiol. Chem. Phys.* **2**, 536 (1970).
170. E. Walaas and O. Walaas, *Diabetologia* **7**, 396 (1971).
171. B. J. Williams and S. E. Mayer, *Mol. Pharmacol.* **2**, 454 (1966).
172. C. Vilchez, M. M. Piras, and R. Piras, *Mol. Pharmacol.* **8**, 780 (1972).

TABLE III
CHANGES IN ENZYMIC ACTIVITY DURING MUSCLE CONTRACTION
AND RECOVERY[a]

Condition[b]	Synthetase a (% of total)	Phosphorylase a (% of total)	Glycogen (mg/g)
Rest	32 ± 4	22 ± 4	8.7 ± 0.3
S	30 ± 4	53 ± 6	5.8 ± 0.4
S + R	52 ± 5	5 ± 1	7.2 ± 0.3
S + R + S	22 ± 4	35 ± 6	5.4 ± 0.4

[a] From Staneloni and Piras (174).
[b] S denotes a 10-sec tetanic stimulation and R a 4-min recovery period. Results are expressed as mean ± S.E.M.

thetase a remains unchanged (159, 173). As shown in Table III, a transient activation of the enzyme occurs during the ensuing minutes of rest and a rapid inactivation is observed when recovering muscle, with higher levels of a enzyme, is stimulated; as a rule the amount of a form never drops below 20% of the total. Inverse changes occur in the activity of phosphorylase, and similar concerted interconversions can be produced during several cycles of stimulation and rest (174).

The actual rate of glycogen metabolism during muscle work, however, not only depends on changes in the level of synthetase a but also is further determined by changes in the concentration of some metabolites (173). During a 10-sec stimulation period, the level of glucose 6-phosphate increases up to tenfold, and the amount of creatine phosphate is halved; within 10 min of the subsequent recovery period, the concentration of these metabolites returns to the resting values. The levels of adenine or uridine nucleotides do not vary significantly throughout these experiments. When synthetases a and b, glucose 6-phosphate, creatine phosphate, P_i, and nucleotides are included in the assay system at concentrations similar to those prevailing in the tissue, the synthetase activity parallels the rate of glycogen synthesis *in vivo* (173). It has also been reported that electrical stimulation results in dissociation of the synthetase from particulate glycogen (175).

The rapid activation of phosphorylase during tetanic stimulation is adequately explained by the increase in free Ca^{2+}, which allows the non-activated phosphorylase kinase to become active at a physiological pH

173. R. Piras and R. Staneloni, *Biochemistry* **8**, 2153 (1969).
174. R. Staneloni and R. Piras, *BBRC* **36**, 1032 (1969).
175. R. Piras and R. Staneloni, *Fed. Proc., Fed. Amer. Soc. Exp. Biol.* **29**, 676 (1970).

(176–178). This mechanism does not account, however, for the simultaneous inactivation of the synthetase since muscle synthetase kinase is not known to be stimulated by Ca^{2+}.

VI. Glycogen Synthetase of Mammalian Heart

A. Properties

Glycogen synthetase b has been purified about 200-fold from rat heart *(47)* and synthetase a about 3000-fold from bovine heart *(178a)*. The kinetic properties of these enzymes are closely similar to those of skeletal muscle synthetases.

B. Control of Synthetase Activity in Heart

1. *Control by Glycogen*

The inverse relationship between glycogen content and the level of synthetase a, initially described for skeletal muscle (see Fig. 5), has also been found in perfused heart in which the glycogen content was either diminished by a low oxygen pressure *(130)* or by the absence of glucose *(144)*, or increased by sympathectomy *(179)*. Treatment of animals with glucocorticoids shifts the curve to higher glycogen values *(180)*, similar in this respect to the action of insulin on the isolated diaphragm (see Fig. 5).

2. *Control by Insulin*

Activation of cardiac synthetase by insulin can be demonstrated in the open-chested rat as early as 1 min after starting the infusion *(130, 171)*. However, the isolated perfused heart is unresponsive to the hormone *(130)*. It seems that simple perfusion of the heart produces an "insulinized state," evidenced by the presence of 49% of the synthetase kinase in the cyclic AMP-independent form as compared to 91% in nonperfused heart *(181)*. The latter value is very high with respect to skeletal muscle

---

176. G. I. Drummond, J. P. Harwood, and C. A. Powell, *JBC* **244**, 4235 (1969).
177. L. M. G. Heilmeyer, Jr., F. Meyer, R. H. Haschke, and E. H. Fischer, *JBC* **245**, 6649 (1970).
178. C. O. Brostrom, F. L. Hunkeler, and E. G. Krebs, *JBC* **246**, 1961 (1971).
178a. J. A. Thomas and J. Larner, *BBA* **293**, 62 (1973).
179. J. C. Daw and R. M. Berne, *Amer. J. Physiol.* **213**, 1480 (1967).
180. J. C. Daw, A. M. Lefer, and R. M. Berne, *Circ. Res.* **22**, 639 (1968).
181. F. Q. Nuttall and J. Larner, *BBA* **230**, 560 (1971).

(57) or diaphragm (169). More recently a clear-cut effect of insulin was observed in the isolated working heart (181a).

3. *Epinephrine and Glucagon*

Although epinephrine and glucagon are known to increase the level of cyclic AMP in heart (182), their effect on glycogen synthetase is far from clear. As a rule, epinephrine seems to be a less potent glycogenolytic agent in heart than in skeletal muscle (171); higher doses of epinephrine are required to produce a transient activation of phosphorylase. A paradoxical activation of glycogen synthetase within the first one or two minutes after epinephrine administration has been found by some authors (171, 183, 184) but not by others (185). This transient activation is followed by a return to the base line value during the next 5 min (171). Insufficient data are available concerning the level of cyclic AMP in heart more than 5 min after epinephrine administration. In one experiment with glucagon (171) the level of synthetase a was unchanged after 3 min whereas phosphorylase was activated. At that time, the glycogen content had been markedly reduced, and therefore the usual inverse relationship between glycogen and synthetase a had been disturbed. The inactivation of glycogen synthetase by glucagon was readily demonstrated when the level of synthetase a had been previously increased by the administration of insulin (185a).

VII. Glycogen Synthetase of Mammalian Liver

A. Properties of the Two Forms

1. *The Effect of Glucose 6-Phosphate*

Tables IV and V give kinetic data obtained with preparations of liver glycogen synthetase that can be reasonably assumed to be entirely or mostly in the a or in the b form. Inspection of these tables reveals very important variations, which indicate that, besides species differences,

---

181a. S. Adolfsson, O. Isaksson, and Å. Hjalmarson, *BBA* **279**, 146 (1972).
182. G. A. Robison, R. W. Butcher, and E. W. Sutherland, "Cyclic AMP." Academic Press, New York, 1971.
183. J. R. Williamson, *Pharmacol. Rev.* **18**, 205 (1966).
184. J. Belford and M. A. Cunningham, *J. Pharmacol. Exp. Ther.* **162**, 134 (1968).
185. G. A. Robison, R. W. Butcher, I. Øye, H. E. Morgan, and E. W. Sutherland, *Mol. Pharmacol.* **1**, 168 (1965).
185a. W. J. Bergstrom and F. Q. Nuttall, *BBA* **286**, 146 (1972).

TABLE IV
EFFECT OF GLUCOSE 6-PHOSPHATE ON KINETIC CONSTANTS OF LIVER SYNTHETASE b AND a^a

Species	Addition	Temp. (°C)	pH	Synthetase b			Synthetase a				Ref.
				K_m UDPG − G6P (mM)	K_m UDPG + G6P (mM)	$\dfrac{V_{max} + \text{G6P}}{V_{max} - \text{G6P}}$	K_m UDPG − G6P (mM)	K_m UDPG + G6P (mM)	$\dfrac{V_{max} + \text{G6P}}{V_{max} - \text{G6P}}$	$\dfrac{V_{max}\ a\ \text{form}}{V_{max}\ b\ \text{form} - \text{G6P} + \text{G6P}}$	
Rat[b]		30	8.9		0.9		1.1[c]	0.62	1	1	26
Rat	Na$_2$SO$_3$	38	8		8.3			0.85–2[d]		1[d]	186
Rat		37	7.4	16–32[c]	0.9		1.8[c]	0.2	1	1	100
Rat		37	7.4		0.3			0.17			187
Rat[b]		37	7.4		0.56						33
Rat[b]	Maleate	30	7.4	5	8.0	13	2.5[c]	0.4	1	13	188
			7.8		2.6						
			8.6		1.2						
Dog	Na$_2$SO$_3$	30	8.9		3.6			0.21[d]		1[d]	189
Mouse		20	7.4		0.35			0.06			102
		20						0.06[d]			
		37			0.35			0.07			
Mouse[b]	MgCl$_2$	25	7.5	2.9	1.6	4.2					142
Mouse C 57		37	8				2.8	0.90			190
Mouse I							8.4	0.74			
Rabbit		37	8	21[c]	0.7	1				1[d]	191

[a] Glucose-6-P, when added, was 2–10 mM.
[b] Partially purified preparations were used. The other results were obtained with crude liver extracts.
[c] Sigmoidal saturation curves.
[d] Synthetase a was obtained $in\ vivo$ by injection of hydrocortisone (186), glucose plus insuline (189), or prednisolone plus glucose (102), or by stimulation of the vagus nerve (191). The other results were obtained with synthetase a formed by activation $in\ vitro$.

many factors can play a role in the kinetics of the reaction. One of these factors is the ionic composition of the incubation medium. Sulfite, which is considered as a protector of the a form (26), counteracts the stimulation of the b enzyme by glucose 6-phosphate (102); its presence in the assay might explain the high K_m of the b enzyme for UDPG observed by some investigators (186, 189). Other seemingly innocuous substances like EDTA (192) and maleate (188) or glycerophosphate (193) buffers have a similar action. It has also been shown that synthetase a needs a significant ionic strength for full expression of its activity. Removal of salt results in a 50% reduction of V_{max} without change in affinity for UDPG (67).

The effect of glucose 6-phosphate on the two forms of liver glycogen synthetase can be summarized as follows:

1. It increases the affinity of synthetase b for UDPG; in the absence of the ligand, this affinity is extremely low and barely measurable. Whether or not a change in V_{max} occurs simultaneously is a controversial matter.

2. It also increases the affinity of synthetase a for UDPG without change in V_{max}.

3. Cooperative kinetics are usually observed with both enzyme forms in the absence of glucose 6-phosphate. Part of the effect of the ligand may be on the stability of the enzyme (see below).

4. At saturating concentration of glucose 6-phosphate the a and b forms can be differentiated by a significantly higher affinity of the former for UDPG. Their V_{max} is about the same. When measured in presumably saturating conditions, the change of enzymic activity during interconversion *in vitro* ranges from 0 to 20–30% (26, 28, 110, 111, 188, 194–196). Larger increases in "total" glycogen synthetase activity during activation *in vitro* were, however, noted by others (142, 143; see Section IV,F).

An important factor with regard to K_a value for glucose 6-phosphate

186. K. R. Hornbrook, H. B. Burch, and O. H. Lowry, *Mol. Pharmacol.* **2**, 106 (1966).
187. A. H. Gold, *Biochemistry* **9**, 946 (1970).
188. K. Sato, N. Abe, and S. Tsuiki, *BBA* **268**, 638 (1972).
189. J. S. Bishop and J. Larner, *JBC* **242**, 1355 (1967).
190. K. R. Hornbrook and J. B. Lyon, Jr., *BBA* **215**, 29 (1970).
191. T. Shimazu, *BBA* **252**, 28 (1971).
192. A. H. Gold, *BBRC* **31**, 361 (1968).
193. W. H. Glinsmann, E. P. Hern, L. G. Linarelli, and R. V. Farese, *Endocrinology* **85**, 711 (1969).
194. J. S. Bishop, *BBA* **208**, 208 (1970).
195. A. H. Gold, *JBC* **245**, 903 (1970).
196. K. Gruhner and H. L. Segal, *BBA* **222**, 508 (1970).

TABLE V
AFFINITY OF LIVER GLYCOGEN SYNTHETASES b AND a FOR GLUCOSE 6-PHOSPHATE

Species	pH	Temp. (°C)	UDPG (mM)	K_a glucose-6-P (mM) Synthetase b	Synthetase a	Ref.
Rat[a]	8.9	30	4.5	0.9		26
Rat	7.4	37	0.25	2[b]	0.06	100
Rat	7.4	37	0.25	1–2[b]		187, 192
Rat[a]	7.4	37	4	0.2		33
Rat[a]	7.4	30	1–5	0.5	0.1	188
Mouse	7.4	37	0.25	5–10[b]	0.06–0.3	34, 102
		20		0.5	0–0.02	
		20			0–0.1[c]	
Mouse[a]	7.5	25	1	0.5		142
Rabbit	8	37	0.5–5	0.7	0.2[c]	191

[a] Partially purified preparations were used. The other results were obtained with crude liver extracts.
[b] Sigmoidal saturation curves.
[c] Synthetase a was obtained in vivo by injection of prednisolone plus glucose (34, 102) or by stimulation of the vagus nerve (191). The other results were obtained with synthetase a formed by activation in vitro.

(Table V) is the temperature at which the assay was performed. Synthetase a was stimulated by glucose 6-phosphate at 37°, but nearly insensitive to it at 20° (102). The affinity of the b form for the ligand is also highly temperature-dependent. A similar influence of temperature was also observed with a mixed preparation (80). This finding may be related to the protection afforded by glucose 6-phosphate against thermal inactivation (4, 24, 80, 141, 197) and presumably results from a change in enzyme conformation. Saturation kinetics of synthetase a with glucose 6-phosphate are hyperbolic, whereas cooperative binding is often observed with the b enzyme at 37° and at low UDPG concentration (100, 102, 188, 192). Within certain limits, the affinity of the b enzyme seems independent of the concentration of UDPG (188, 191). Glucosamine 6-phosphate and galactose 6-phosphate are efficient stimulators (4), as well as 1,5-anhydroglucitol 6-phosphate (46). Glucose 6-sulfate and the 6-phosphate esters of mannose, 2-deoxyglucose, and sorbitol are inactive (46).

2. The Effect of Inorganic Phosphate

Mersmann and Segal (100) have shown that physiological concentrations of P_i stimulate synthetase a to nearly the same extent as does glu-

197. P. R. Weldon and D. Rubinstein, Can. J. Biochem. **44**, 591 (1966).

cose 6-phosphate, whereas the b enzyme remains inactive. Furthermore, the stimulation of the b form by glucose 6-phosphate is inhibited by inorganic phosphate (102), an observation which is presumably related to the fact that, at pH 7.8, the anion also inhibits muscle synthetase b competitively with glucose 6-phosphate (158). The effects of phosphate are shared by sulfate and by sulfite (102); they allow measurement of liver synthetase a without interference of b by performing the assay in the presence of 5–10 mM phosphate or sulfate (102, 198) or with a concentrate liver homogenate (199, 200).

3. The Effect of Nucleotides

In contrast to the muscle enzyme, cooperative binding of UDPG has been observed with both forms of the liver enzyme in the absence of glucose 6-phosphate (see Table IV) or in the presence of glucose 6-phosphate and ATP (187). Inhibition by UDP at pH 7.4 is similar to that observed with the muscle enzyme (46).

Adenosine nucleotides inhibit the a and b forms of glycogen synthetase, but inhibition of synthetase a is much less pronounced (26) and more easily reversed by glucose 6-phosphate and by Mg^{2+} (102, 187). Kinetic analysis of the inhibition (187) yields a pattern similar to that found with muscle synthetase (68): In the presence of glucose 6-phosphate, ATP inhibits both enzyme forms competitively with respect to UDPG; the inhibition is also substantially reversed by glucose 6-phosphate, and competition between ATP and glucose 6-phosphate is evident with the b form. Cooperative binding of glucose 6-phosphate is only observed with the b enzyme (see Table V), whereas cooperative effects with ADP are only evident with the a form (187).

4. The Effect of Magnesium and of Other Ligands

Magnesium stimulates the a enzyme to nearly the same extent as does glucose 6-phosphate, whereas its effect on the b form is less complete (102, 192). No cooperativity for UDPG was noted in the presence of this ion (142).

Magnesium increases the affinity of synthetase b for glucose 6-phosphate as well as for UDPG in the presence of a limiting amount of glucose 6-phosphate (187, 192). A kinetic analysis of the effect of Mg^{2+} on the liver enzyme in the absence of hexose phosphate has not been made. The effects of calcium and manganese ions are similar to those of Mg^{2+} (187). Some inhibitors have been reported (4, 187).

198. H. De Wulf and H. G. Hers, *Eur. J. Biochem.* **6,** 558 (1968).
199. H. De Wulf and H. G. Hers, *Eur. J. Biochem.* **2,** 50 (1967).
200. H. De Wulf and H. G. Hers, *Eur. J. Biochem.* **2,** 57 (1967).

5. The Effect of pH

The activity of synthetase a is influenced little by variations of pH between 6.5 and 9 (*26, 188*). Synthetase b has its optimal activity around pH 8.5 and retains only 20–25% of this activity at pH 7.5, at least when measured in the presence of glucose 6-phosphate plus sulfite (*26*) or maleate (*188*). Large variations in activity with differences in buffer ions and in other ionic conditions have been reported (*46*).

6. Activity of the Two Forms in Physiological Conditions

Mersmann and Segal (*100*) found that in the presence of physiological concentrations of UDPG, P_i, and glucose 6-phosphate, the a form is largely active and the b form virtually inactive, irrespective of small variations in the concentration of the ligands. They postulated accordingly that interconversion between b and a forms has the property of

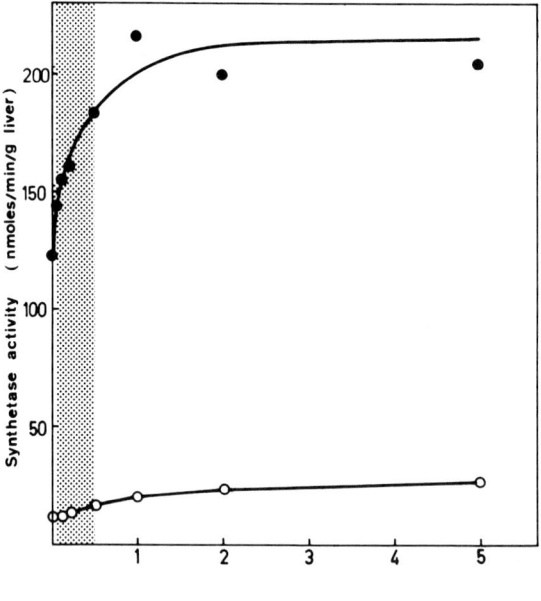

FIG. 6. The effect of glucose 6-phosphate concentration on the activity of glycogen synthetase a and b in a liver extract in the presence of a physiological mixture containing 5 mM P_i, 3 mM ATP, and 3 mM Mg acetate. [UDPG] was 0.25 mM and pH 7.4. The shaded area indicates the *in vivo* range of liver glucose 6-phosphate concentration, the highest values being observed after the administration of glucagon. After De Wulf et al. (*102*).

switching on and off the synthesis of glycogen. Essentially the same conclusion was reached from estimations of the activity of synthetase a and b in the presence of a more complete physiological mixture of substrates, stimulators, and inhibitors (102). As shown in Fig. 6, synthetase a has a high activity in the absence of glucose 6-phosphate, whereas synthetase b remains largely inactive even in the presence of glucose 6-phosphate concentrations that are severalfold higher than the physiological level. One can therefore conclude that, in contrast to the situation in muscle (Sections V,A,6 and V,B,4) and in yeast (Section IX,F,1), the concentration of glucose 6-phosphate plays no role in the control of the activity of glycogen synthetase in the liver.

B. Control of Synthetase Activity in Liver

Liver glycogen is mainly used as a reserve for the homeostasis of the blood glucose level, at the benefit of nonhepatic tissues. Its synthesis and degradation are under the control of extrahepatic factors such as the level of the glycemia and of various hormones, including glucagon, insulin, and glucocorticoids. For reasons described above (Section VII,A,6) the activity of glycogen synthetase b in the liver cell is not modified by glucose 6-phosphate.

1. *The Control by Glucose*

It has been known for a long time that the increase in blood glucose concentration resulting from food intake causes deposition of glycogen in the liver (201, 202). The mechanism of this glucose effect has been investigated by following the biochemical changes that occur in the liver of mice after an intravenous load of glucose (199). Within a few minutes, the rate of glucose to glycogen conversion and the amount of glycogen synthetase a in the liver increased to 40-fold and in a parallel manner. This increase was preceded by a latency of 1–2 min and reached its maximum after 5–10 min. At that time, the concentration of glucose 6-phosphate and that of UDPG were reduced to about 60% of their initial value, at least in fed mice. These data indicate that the large increase in glycogen synthesis is not the result of a push given by glucose on the metabolic pathway leading to glycogen but is entirely explained by the activation of glycogen synthetase.

Since a glucose load induces a rapid secretion of insulin by the pan-

201. S. Kuriyama, *JBC* **33**, 193 (1918).
202. C. F. Cori, *JBC* **70**, 577 (1926).

creas, the possible participation of this hormone in the glucose effect had to be considered. It is known, however, that a rise in blood glucose is followed by glycogen synthesis in fasted depancreatized (*203*), fasted alloxan-diabetic (*204, 205*), or anti-insulin treated animals (*206*). Furthermore, in the isolated perfused liver the level of synthetase a varies according to the concentration of circulating glucose (*207, 208*). Therefore, it seems clear that the glucose effect is not mediated by insulin. A glucose load slightly lowers the level of cyclic AMP in the liver of intact mice (*209*), although not in the perfused rat liver (*207, 208*).

Interestingly, the elevation of the blood glucose level also causes an immediate, partial, or complete inactivation of phosphorylase in the liver of the intact animal (*139, 198, 210*) and in the isolated perfused liver (*207, 208*). As shown in Fig. 7, the first change after the administration of glucose is the inactivation of phosphorylase, and the latency that precedes the activation of glycogen synthetase (Fig. 7A) corresponds to the time required to inactivate phosphorylase. As explained in Section IV,E,2, glucose favors the activation of glycogen synthetase *in vitro*, and this effect is mediated by the inactivation of phosphorylase a, which is an inhibitor of synthetase phosphatase. It seems reasonable that, *in vivo* like *in vitro*, phosphorylase a is the glucose receptor and that its conversion to phosphorylase b is the mechanism that allows the activation of glycogen synthetase. It is shown in Fig. 7B that a glucose load can induce an important decrease in the amount of phosphorylase a with no change in synthetase activity. This was observed each time that the level of phosphorylase a was not decreased below a threshold value that can be estimated to 10% of the total enzyme.

2. *The Effect of Insulin*

When insulin alone is given to normal mice or rats, glycogen synthetase remains in the b form within the following minutes (*34*) but is par-

203. R. W. Longley, R. J. Bortnick, and J. H. Roe, *Proc. Soc. Exp. Biol. Med.* **94**, 108 (1957).

204. B. Friedmann, E. H. Goodman, Jr., and S. Weinhouse, *JBC* **238**, 2899 (1963); *Endocrinology* **81**, 486 (1967).

205. K. R. Hornbrook, *Diabetes* **19**, 916 (1970).

206. H. G. Hers, H. De Wulf, W. Stalmans, and G. Van den Berghe, *Advan. Enzyme Regul.* **8**, 171 (1970).

207. H. Buschiazzo, J. H. Exton, and C. R. Park, *Proc. Nat. Acad. Sci. U. S.* **65**, 383 (1970).

208. W. Glinsmann, G. Pauk, and E. Hern, *BBRC* **39**, 774 (1970).

209. G. Van den Berghe, H. De Wulf, and H. G. Hers, *Eur. J. Biochem.* **16**, 358 (1970).

210. J. S. Bishop, N. D. Goldberg, and J. Larner, *Amer. J. Physiol.* **220**, 499 (1971).

8. GLYCOGEN SYNTHESIS FROM UDPG

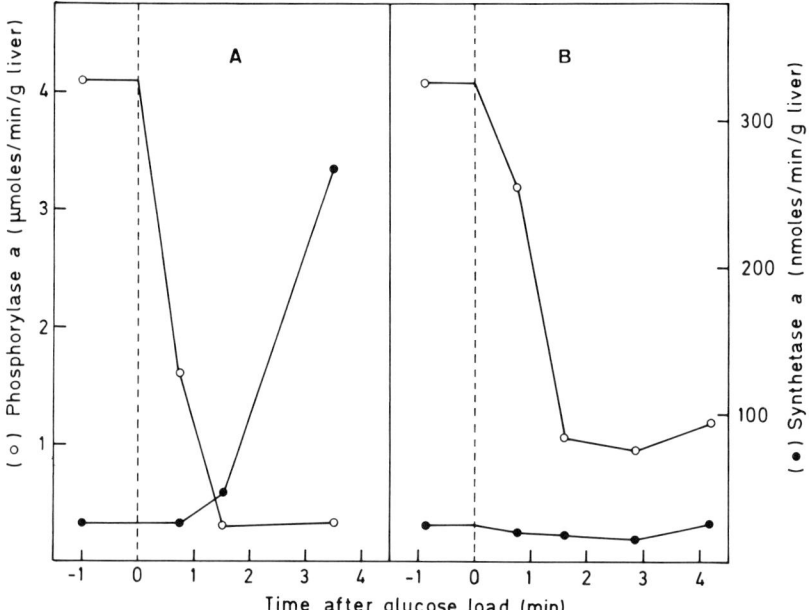

FIG. 7. Sequential changes in the level of phosphorylase a and of synthetase a in rat liver after the intravenous administration of glucose. In each experiment, several samples were taken at various time intervals from the liver of an anesthetized rat; they were quickly frozen and the two enzymic activities were measured as described elsewhere (139). Glucose was given at the dose of 1.5 mg/g body weight in A and 1 mg/g body weight in B (34).

tially activated after 2 or 3 hr (34, 143). Since many secondary regulatory changes may occur during this period, the interpretation of this delayed effect is difficult. When given together with glucose, insulin causes a rapid activation of glycogen synthetase (189), which can, however, be attributed in great part to glucose itself (see preceding paragraph). In the isolated perfused liver insulin antagonizes the effect of a small dose of epinephrine on the level of synthetase a (211). The hormone activates glycogen synthetase in cultured fetal liver (212).

In diabetic animals, liver glycogen synthetase is mostly in the inactive form despite a high level of glycemia. A rapid activation of glycogen synthetase occurs in recently diabetic rats (213), mice (34), or dogs

211. A. T. Hostmark, *Diabetologia* **7**, 396 (1971).
212. H. Eisen, J. Waters, and W. Glinsmann, *Fed. Proc., Fed. Amer. Soc. Exp. Biol.* **31**, 244 (1972).
213. C. Villar-Palasi, N. D. Goldberg, J. S. Bishop, F. Q. Nuttall, and J. Larner, in "Metabolic Regulation and Enzyme Action" (A. Sols and S. Grisolía, eds.), p. 149. Academic Press, New York, 1970.

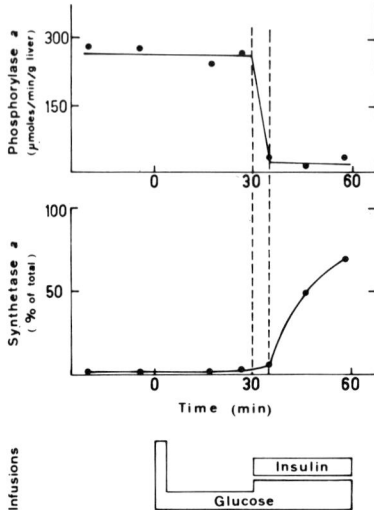

FIG. 8. Effect of an intravenous infusion of glucose and of glucose plus insulin on the level of phosphorylase a and of synthetase a in the liver of a pancreatectomized dog, maintained with daily insulin injections. The infusion pattern is shown below the graph. The vertical dotted lines emphasize the sequential character of the changes in enzymic activities. After Bishop et al. (210).

(210) following the administration of insulin. In the dog, the sequence of events was carefully analyzed (Fig. 8); it is particularly striking that the activation of glycogen synthetase occurs only after phosphorylase has been inactivated. This observation suggests that the effect of insulin on the activation of glycogen synthetase, like that of glucose, may be secondary to the inactivation of phosphorylase. The administration of insulin to chronically diabetic animals elicits an activation of liver glycogen synthetase only after about 1 hr (210, 214, 215). It is not clear whether the effects of diabetes and of insulin on liver glycogen metabolism can be explained (216, 217) or not (34, 210, 213, 218, 219) by changes in the level or in the production of cyclic AMP.

Synthetase phosphatase has been reported to be less active in the

214. D. F. Steiner, V. Rauda, and R. H. Williams, *JBC* **236**, 299 (1961).
215. D. F. Steiner and J. King, *JBC* **239**, 1292 (1964).
216. L. S. Jefferson, J. H. Exton, R. W. Butcher, E. W. Sutherland, and C. R. Park, *JBC* **243**, 1031 (1968).
217. J. H. Exton, S. B. Lewis, R. J. Ho, G. A. Robison, and C. R. Park, *Ann. N. Y. Acad. Sci.* **185**, 85 (1971).
218. N. D. Goldberg, S. B. Dietz, and A. G. O'Toole, *JBC* **244**, 4458 (1969).
219. J. E. Liljenquist, J. D. Bomboy, B. C. Sinclair-Smith, S. B. Lewis, P. W. Felts, W. W. Lacy, O. B. Crofford, and G. W. Liddle, *J. Clin. Invest.* **51**, 58a (1972).

liver of rats 2 days after alloxan treatment (*195*). A normal synthetase phosphatase was observed in mice treated in a similar way, except after high doses of alloxan, which killed all animals in 3 days (*34*). As seen above, the administration of insulin to these animals activates glycogen synthetase within a few minutes but restores synthetase phosphatase only after 1 hr (*34, 195*).

3. *The Effect of Glucagon and Epinephrine*

The administration of glucagon, epinephrine, or cyclic AMP to dog or mouse causes a rapid inactivation of liver glycogen synthetase (*189, 198*). The same effect of cyclic AMP is observed in the perfused liver (*134*). Injection of very small amount of glucagon into mice gives an immediate but transient rise in the concentration of cyclic AMP in the liver which precedes the inactivation of glycogen synthetase by 1–2 min (Fig. 9). There is therefore no doubt that the effect of glucagon on the inactivation of glycogen synthetase is mediated by the formation of cyclic AMP. Indeed, the activation of protein kinase by the cyclic nucleotide (see Section IV,D) causes the conversion of synthetase *a* into *b* and simultaneously the activation of phosphorylase kinase and the formation of phosphorylase *a*. Since the latter enzyme is an inhibitor of synthetase phosphatase, glucagon not only causes the inactivation of glycogen synthetase but also prevents its activation by the phos-

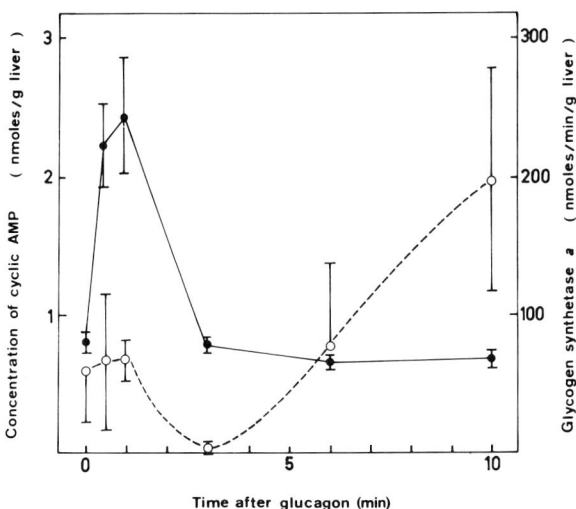

Fig. 9. Sequential changes in the level of (●) cyclic AMP and of (○) synthetase *a* in the liver of mice after the intravenous injection of glucagon (0.5 ng/g body weight). From Van den Berghe *et al.* (*209*).

phatase. The inactivation of liver synthetase phosphatase of diabetic dog by glucagon (194) can presumably be explained on a similar basis.

4. The Effect of Glucocorticoids

The administration of glucocorticoids is known to induce an important deposition of glycogen in the liver (220). It can be observed in fed and in fasted animals and is therefore not secondary to a stimulation of gluconeogenesis. The effect has also been obtained in diabetic animals (221, 222) and appears therefore not to be insulin-dependent.

The glucocorticoids require about 3 hr to stimulate liver glycogen synthesis. At that time, the concentration of glucose 6-phosphate and of UDPG are markedly reduced in the liver and the amount of synthetase a is greatly increased (186, 200, 223), whereas the activity of phosphorylase is decreased, although to a smaller extent (198, 205). These changes can be adequately explained by an increased activity of phosphorylase phosphatase as a result of the hormone (see Section IV,E,2). The effect of the steroid would therefore be the same as that of a protracted hyperglycemia, explaining that in the treated animal glycogen synthesis can occur when the level of glycemia is low.

The activity of synthetase phosphatase is normal in the liver of adrenalectomized animals except after prolonged fasting (28). Glycogen synthetase is not activated in response to a glucose load in the perfused liver of fasted adrenalectomized rats (208).

5. Other Effectors

Stimulation of the vagus nerve for 5–10 min results in an increased rate of glycogen synthesis, associated with a conversion of synthetase b into a, whereas splanchnic nerve stimulation has the reverse effect. The presence of the pancreas is not required for these effects of vagus nerve stimulation (191, 224). It is of interest to note that carbamylcholine increases the level of synthetase a in the isolated perfused liver (225).

The intravenous administration of prostaglandin E_1 to rats lowers the

---

220. C. N. Long, B. Katzin, and E. G. Fry, *Endocrinology* **26**, 309 (1940).
221. A. M. Miller, *Proc. Soc. Exp. Biol. Med.* **72**, 635 (1949).
222. W. Tarnowski, M. Kittler, and H. Hilz, *Biochem. Z.* **341**, 45 (1964).
223. W. Kreutner and N. D. Goldberg, *Proc. Nat. Acad. Sci. U. S.* **58**, 1515 (1967).
224. T. Shimazu and T. Fujimoto, *BBA* **252**, 18 (1971).
225. C. Ottolenghi, A. Caniato, and O. Barnabei, *Nature (London)* **229**, 420 (1971).

level of liver synthetase a within a few minutes (226). Prostaglandin E_1 does not increase the level of cyclic AMP in perfused liver (227).

VIII. Glycogen Synthetase of Other Mammalian Tissues

A. ADRENALS

The kinetic properties of synthetase b, which prevails in the fresh tissue, and of the a enzyme, obtained by incubation of an extract, have been studied at pH 7.3 (15). The saturation curves with UDPG and with glucose 6-phosphate are hyperbolic. The b enzyme has a relatively small V_{max} and, paradoxically, a small K_m for UDPG (0.07 mM). The two parameters are increased by the addition of glucose 6-phosphate or by conversion to the a form; the kinetic constants of synthetase a are not significantly influenced by the hexose phosphate. Both forms are inhibited by ATP, and this inhibition is much more efficiently reversed by glucose 6-phosphate in the case of the a form than of the b form. The kinetics of the ATP inhibition depend on the concentration of UDPG. Inorganic phosphate inhibits the b form but not the a enzyme.

The b into a conversion *in vitro* occurs without latency, is not sensitive to glucose, but is completely inhibited by 5 mg/ml glycogen; it is stimulated by glucose 6-phosphate at physiological concentration and by AMP (117). The glycogen inhibition is released by a combination of glucose 6-phosphate and Ca^{2+} (7.5 mM). Synthetase a is converted into b in the presence of ATP-Mg, and this conversion is stimulated by cyclic AMP.

B. BRAIN

Brain glycogen synthetase was reported to have similarities with the muscle enzyme (14, 228). The existence of interconvertible forms of the enzyme has been recognized by Goldberg and O'Toole (114). Fresh brain contains a glycogen synthetase that is poorly active in the absence of glucose 6-phosphate although it has a low K_m for UDPG. The activity that can be measured in the absence of glucose 6-phosphate increases upon incubation of a tissue extract; the reverse change occurs upon

226. R. T. Curnow and F. Q. Nuttall, *JBC* **247**, 1892 (1972).
227. J. H. Exton, G. A. Robison, E. W. Sutherland, and C. R. Park, *JBC* **246**, 6166 (1971).
228. D. K. Basu and B. K. Bachhawat, *BBA* **50**, 123 (1961).

the addition of ATP-Mg and is hastened by cyclic AMP. In conditions that are characterized by a low glycogen content, such as anoxia or hypoglycemia, the glucose 6-phosphate–independent activity increases severalfold. The finding that this enzyme form has a relatively higher K_m for UDPG was interpreted as indicating a lower physiological activity. Since the effects of nucleotides, inorganic phosphate, and magnesium have not been studied, it is, however, difficult to speculate on the physiological role of the two enzyme forms.

C. Spleen

The presence of two interconvertible forms of glycogen synthetase in bovine spleen has been demonstrated (115). Both forms are stimulated by glucose 6-phosphate. The b enzyme is characterized by a low affinity for the ligand and is inactive in its absence. The b to a conversion is stimulated by glucose 6-phosphate, particularly in the presence of magnesium.

D. Adipose Tissue

Exposure of rat epididymal adipose tissue to insulin reduces the adenyl cyclase activity and increases about twofold the activity of glycogen synthetase that can be measured in the absence of glucose 6-phosphate without change in total enzymic activity (229). Nutritional effects on the level of glycogen synthetase in adipose tissue have been described (230, 231).

E. Blood

1. *Leukocytes*

Glycogen synthetase is present in lymphocytes as a b form, and its kinetic properties have been described. Interconversion between b and a forms occurs in broken cells without major changes in total enzymic activity (118).

Glycogen synthetase is also present as a b form in freshly prepared homogenates of polymorphonuclear leukocytes of various mammals

229. R. L. Jungas, *Proc. Nat. Acad. Sci. U. S.* **56**, 757 (1966).
230. A. M. Chandler and R. O. Moore, *ABB* **108**, 183 (1964).
231. A. Gutman and E. Shafrir, *Amer. J. Physiol.* **207**, 1215 (1964).

(*232, 233*). The kinetic properties of this enzyme have been studied (*120, 234*). The human enzyme has been extensively purified (*17*). Interconversion between the *b* and *a* forms occurs in extracts of leukocytes from both normal and diabetic rats without changes in total activity (*120*). The failure to activate the enzyme in extracts of human leukocytes, except from insulin-controlled diabetics (*119*), remains unexplained since the activation takes place in normal intact leukocytes (*235*). The stimulation of this activation by a small amount of glucose (0.1 mg/ml) and by slightly larger amounts of other sugars is badly understood.

2. *Erythrocytes*

Glycogen synthetase of erythrocytes is entirely glucose 6-phosphate–dependent for activity (*34, 236, 237*). No synthetase phosphatase could be detected in these cells (*34, 237*).

3. *Platelets*

There is good indication that two interconvertible forms of glycogen synthetase exist in platelets (*238, 239*). Some kinetic properties of the native enzyme have been reported (*238, 240, 241*).

F. TISSUES SENSITIVE TO SEX HORMONES

The glycogen content of the levator ani muscle increases severalfold during the first day after the administration of testosterone to immature or castrated rats. A transient twofold rise in the level of synthetase *a* occurs without change in total enzymic activity (*242, 243*).

232. M. Rosell-Perez and V. Esmann, *Acta Chem. Scand.* **19**, 679 (1965).
233. C. Vanderwende, J. C. Johnson, and S. A. Thunberg, *Bull. N. J. Acad. Sci.* **13**, 35 (1968).
234. T. P. Stossel, F. Murad, R. J. Mason, and M. Vaughan, *JBC* **245**, 6228 (1970).
235. P. Wang, L. Plesner, and V. Esmann, *Eur. J. Biochem.* **27**, 297 (1972).
236. M. Cornblath, D. F. Steiner, P. Bryan, and J. King, *Clin. Chim. Acta* **12**, 27 (1965).
237. S. W. Moses, N. Bashan, and A. Gutman, *Eur. J. Biochem.* **30**, 205 (1972).
238. H. Vainer, P. Besson, C. Jeanneau, and J. Caen, *Nouv. Rev. Fr. Hematol.* **9**, 514 (1969).
239. H. Vainer, P. Besson, and J. Caen, *Nouv. Rev. Fr. Hematol.* **11**, 769 (1971).
240. H. Vainer and R. Wattiaux, *Nature (London)* **217**, 951 (1968).
241. S. Karpatkin, A. Charmatz, and R. M. Langer, *J. Clin. Invest.* **49**, 140 (1970).
242. S. Adolfsson and K. Ahren, *Acta Physiol. Scand.* **74**, 30A (1968).
243. E. Bergamini, G. Bombara, and C. Pellegrino, *BBA* **177**, 220 (1969).

The large increase in glycogen content of the human endometrium after midcycle is concomitant with a threefold increase in total glycogen synthetase activity but with no clear change in the percent of glucose 6-phosphate–independent activity (244, 245). The kinetic properties of the two enzyme forms have, however, not been studied. Much smaller changes in glycogen content and in glycogen synthetase activity occur during the estrous cycle in bovine endometrium (246) and in rat myometrium (244). The effect of a treatment with estrogen and with progesterone on myometrial glycogen synthetase has also been studied in the ovariectomized rat (247–249).

G. Tumors

In freshly harvested Novikoff hepatoma cells, glycogen synthetase is mostly in the b form and is converted into a within about 30 min upon incubation of the intact cells with glucose at 20° (34). Synthetase phosphatase and synthetase kinase can be demonstrated in broken-cell preparations from Novikoff (34) and Yoshida hepatomas (121). Synthetase kinase from the former cells is sensitive to cyclic AMP.

An inverse relationship between the level of synthetase a and the glycogen content has been noted in HeLa cells (250). The inhibition of synthetase phosphatase by glycogen seems of major importance in the control of glycogen synthesis in Yoshida hepatoma cells (16, 251). In a glycogen-rich strain glycogen synthetase remains largely in the a form, even when the cells contain ten times more glycogen than cells of a glycogen-deficient strain are able to accumulate (252). Synthetase phosphatases from the two strains display indeed a widely different sensitivity to inhibition by glycogen. Other factors, however, may add to the different capacity of the two strains for glycogen storage since glycogenolysis in the glycogen-rich cells occurs at a much slower rate than in the glycogen-deficient strain (252). Glycogen synthetases of

244. A. Rubulis, R. D. Jacobs, and E. C. Hughes, *BBA* **99**, 584 (1965).
245. E. C. Hughes, L. M. Demers, T. Csermely, and D. B. Jones, *Amer. J. Obstet. Gynecol.* **105**, 707 (1969).
246. L. L. Larson, G. B. Marion, and H. T. Gier, *Amer. J. Vet. Res.* **31**, 1929 (1970).
247. H. E. Williams and H. T. Provine, *Endocrinology* **78**, 786 (1966).
248. W. J. Bo, L. E. Maraspin, and M. S. Smith, *J. Endocrinol.* **38**, 33 (1967).
249. W. J. Bo and M. J. Ashburn, *Steroids* **12**, 457 (1968).
250. J. B. Alpers, *JBC* **241**, 217 (1966).
251. R. Saheki and S. Tsuiki, *BBRC* **31**, 32 (1968).
252. K. Sato and S. Tsuiki, *Cancer Res.* **32**, 1451 (1972).

both Yoshida strains resemble the muscle enzyme rather than the liver synthetase (121).

A tumor composed of immature granulocytes contains two types of synthetase b with different affinity for glucose 6-phosphate; both enzymes are converted into synthetase a upon incubation of cell extracts (253).

IX. Glycogen Synthetase of Nonmammalian Organisms

A. Frog Muscle

The native enzyme is dependent on glucose 6-phosphate. Some of its properties have been described (122, 254). Upon incubation the D activity rises without significant appearance of I activity; the reverse change occurs in an ATP-Mg-dependent, cyclic AMP-sensitive reaction (122, 123). These reactions could be understood as an interconversion between an a form that is glucose 6-phosphate-dependent and a b form that is inactive even in the presence of the ligand.

B. Tadpole Liver

Two forms of glycogen synthetase have been purified from the liver of premetamorphic *Rana catesbeiana* (48, 255). One, present in untreated animals, is supposed to be inactive *in vivo* and can thus be considered as a b form; the other appears after treatment of the animal with insulin and would be a physiologically active a form. The properties of these two enzyme forms have been reported in a series of papers by Kim and his co-workers (48, 255–257). Both enzymes have an absolute requirement for glucose 6-phosphate (K_a = 2.5–5 mM). In the presence of 20 mM glucose 6-phosphate they can be differentiated by a widely different affinity for UDPG (K_m of a form ≃0.1 mM; b form ≃2.5 mM). Citrate and other carboxylic acids (257) as well as inorganic phosphate (255) increase severalfold the affinity of synthetase a for glucose 6-phosphate. These compounds are less efficient in the case of the b enzyme.

Saturation curves of both enzyme forms with glucose 6-phosphate

253. S. A. Assaf and A. A. Yunis, *FEBS Lett.* **19**, 22 (1971).
254. M. Rosell-Perez and V. Villar-Palasi, *Rev. Espan. Fisiol.* **22**, 185 (1966).
255. K. H. Kim and L. M. Blatt, *Biochemistry* **8**, 3997 (1969).
256. J. S. Sevall and K. H. Kim, *JBC* **246**, 2959 (1971).
257. J. S. Sevall and K. H. Kim, *JBC* **246**, 7250 (1971).

or with UDPG (in the presence of glucose-6-P) are hyperbolic (*255*). As observed with other glycogen synthetases, the inhibition caused by ATP is more easily reversed by glucose 6-phosphate in the case of the *a* enzyme than with the *b* form. Cooperative effects are then observed with glucose-6-P and with ATP, but they are restricted to the *b* enzyme. Photooxidation of synthetase *b* abolishes the cooperative character of the inhibition by ATP, which is now purely competitive with both UDPG and glucose 6-phosphate (*256*). The photooxidized *b* enzyme retains, however, the low affinity for UDPG which characterizes the native *b* form.

Kim and Blatt (*255*) were unable to obtain activation of the synthetase *in vitro*. The activation by insulin requires about 3 hr *in vivo* with animals maintained at 24° (*255*). It has also been obtained with a minced liver preparation and is then unaffected by inhibitors of protein synthesis (*258*). It is only observed if the cellular integrity is preserved and does not require the entry of the hormone into the cell since it occurs also with Sepharose-bound insulin (*259*). A series of other agents, like glucose, thyroxine, hydrocortisone, glucagon, dibutyryl cyclic AMP, theophylline, and puromycin also induce the conversion of synthetase *b* into *a* when administered to tadpoles, but they do not have this effect on the minced liver preparation (*69*, *258*, *260*). Their inability to activate the synthetase in diabetic animals was taken as evidence for an action through insulin secretion.

C. Fish

The enzymes from toadfish muscle (*122*) and from trout liver (*20*) are partially dependent on glucose 6-phosphate. There is no clear evidence as yet for the existence of two forms.

D. Arthropoda

In insects, glycogen synthetase has been detected in muscle and fat body (*72*). The apparent inhibition of the enzyme in crude extracts by glucose 6-phosphate (*66*, *72*) probably results from utilization of the substrate by trehalose phosphate synthetase since purified glycogen synthetase is many times stimulated by the ligand (*21*). The unusual stimulation of crude glycogen synthetase by glucose 1-phosphate (*66*) was

258. L. M. Blatt, J. S. Sevall, and K. H. Kim, *JBC* **246**, 873 (1971).
259. L. M. Blatt and K. H. Kim, *JBC* **246**, 4895 (1971).
260. L. M. Blatt and K. H. Kim, *BBA* **192**, 286 (1969).

not observed with purified enzyme (*21*). The synthetase from bee larvae has been studied with special reference to its requirement for glycogen primer (*22*). There is no evidence as yet for the existence of two enzyme forms in insect tissues.

Glycogen synthetase activity in crustacean muscle is regulated by a hormone from the eyestalk. Removal of eyestalks results in an increased glucose-6-P-independent activity, and this change is reversed by the injection of an eyestalk extract (*261, 262*).

E. PROTOZOA

During starvation of *Amoeba proteus* the glycogen synthetase I activity almost disappears, whereas the total enzyme activity does not change appreciably; this pattern is consistent with the presence of two enzyme forms (*263*). In different batches of *Acanthamoeba castellanii* the enzyme is stimulated to a variable degree by glucose 6-phosphate (*264*). The fall in glycogen content during the first hours of encystment is paralleled by a paradoxical rise in glycogen synthetase activity measured with glucose 6-phosphate. Glycogen synthetase is the rate-limiting enzyme for glycogen synthesis in *Tetrahymena pyriformis* (*265*). The enzyme is usually only slightly stimulated by glucose 6-phosphate (*265, 266*). The activity of the enzyme rises severalfold when the cells are grown in the presence of glucose; this effect is potentiated by theophylline and reversed by triiodothyronine (*266*).

F. MOLDS AND YEAST

1. *Yeast*

Glycogen synthetase has been extracted from a haploid yeast strain in logarithmic growth as a *b* enzyme, whereas the active form was obtained during the early stationary phase (*23, 125*). Both enzyme preparations seem to contain a third type of enzyme, characterized by a very low affinity for UDPG ($K_m \simeq 5$ mM) in the presence of low concentrations of glucose 6-phosphate (*125*). The enzyme previously

261. R. Keller (1966), cited by Ramamurthi *et al.* (*262*).
262. R. Ramamurthi, M. W. Mumbach, and B. T. Scheer, *Comp. Biochem. Physiol.* **26**, 311 (1968).
263. B. Larner, *C. R. Trav. Lab. Carlsberg* **36**, 225 (1967).
264. R. A. Weisman, R. S. Spiegel, and J. G. McCauley, *BBA* **201**, 45 (1970).
265. D. E. Cook, N. I. Rangaraj, N. Best, and D. R. Wilken, *ABB* **127**, 72 (1968).
266. J. J. Blum, *ABB* **137**, 65 (1970).

purified from baker's yeast (*49*), and extensively studied (*267, 268*), has the characteristics of the *a* enzyme (*125*). The kinetic properties of the *a* and *b* forms are somewhat similar to those of the muscle enzymes. The effect of glucose 6-phosphate is to increase V_{max} of both forms, without change in K_m for UDPG, which is close to 0.5 mM for either enzyme form (*125*). Synthetase *a* has a 30-fold higher affinity for glucose 6-phosphate than the *b* enzyme. In each case, saturation with glucose 6-phosphate follows Michaelis–Menten kinetics, but cooperative binding becomes evident in the presence of ATP. Much smaller amounts of glucose 6-phosphate are required to reverse the inhibition by ATP in the case of the *a* form than with the *b* enzyme. Cooperative binding of ATP has been demonstrated, at least with synthetase *a* (*268*). The inhibition by ATP is more pronounced at pH 5.9 (considered as the physiological value for yeast) than at pH 7.5.

While the inhibition by both ATP and UDP is of the competitive type with respect to UDPG (*125, 267*), the *a* enzyme can be desensitized to ATP, but not to UDP, by dinitrophenylation (*268*). The finding that inhibition by UDP is not reversed by glucose 6-phosphate further substantiates the conclusion that UDP is a true competitive inhibitor, whereas ATP is an allosteric one. The inhibition caused by high concentrations (0.2 M) of several anions seems also of allosteric nature (*267*). At lower concentration, phosphate and sulfate stimulate the *a* enzyme.

Yeast also contains enzymes that catalyze the conversion of synthetase *b* into *a* in the presence of Mg^{2+} and of synthetase *a* into *b* in the presence of ATP-Mg. The interconversion occurs without important changes in total activity (*23, 125*).

The synthesis of glycogen in the yeast cell seems to depend on the level of both synthetase *a* and glucose 6-phosphate (*23*). The rapid synthesis of glycogen that occurs in the presence of glucose during late logarithmic and early stationary phases is adequately explained by the formation of a larger amount of total synthetase of which a larger percentage is in the *a* form (*23*). The mechanism that causes this activation of the enzyme is unknown. The amount of converter enzymes is the same during logarithmic growth and in stationary phase. In a mutant unable to accumulate glycogen no synthetase *a* was formed in stationary phase, but the analysis of the converter enzymes did not reveal a clear deficit (*23, 125*). The accumulation of glycogen in logarithmic cells early after the addition of glucose to the medium is attributed to a

267. L. B. Rothman and E. Cabib, *Biochemistry* **6**, 2098 (1967).
268. L. B. Rothman and E. Cabib, *Biochemistry* **6**, 2107 (1967).

large increase in the concentration of glucose 6-phosphate. Values as high as 4–8 mM can be reached (*269*), allowing even the *b* form to overcome inhibition by ATP (*125*). The secondary increase in the rate of glycogen synthesis, some 20 min after the addition of glucose, is associated with a *b* into *a* conversion (*23*).

2. *Molds*

Neurospora crassa grown on a sorbose medium contains little glycogen, and its glycogen synthetase is in the *b* form. On a sucrose medium more glycogen accumulates and the synthetase is present as an *a* enzyme. The enzyme forms can be interconverted in a broken-cell preparation. Conversion of synthetase *b* into *a* or addition of glucose 6-phosphate increases V_{max} without appreciable change in K_m for UDPG (*124*).

Evidence has been obtained for the presence of two forms of glycogen synthetase in *Dictyostelium discoideum* (*270*). The *b* enzyme has been purified (*271*). It acts not only on soluble glycogen but also on insoluble cell wall glycogen, being then less dependent on glucose 6-phosphate (*272*). Cooperative binding of UDPG has been observed with a glucose 6-phosphate–independent enzyme in the presence of ATP (*273*).

The stimulation of glycogen synthetase of *Blastocladiella emersonii* by glucose 6-phosphate may vary from 2.5-fold during growth phase to 97-fold for resting zoopores (*274*). A detailed kinetic study of the enzyme present in growing cells has been performed, indicating the presence of three binding sites, one for UDPG and UDP, the others for ATP and glucose 6-phosphate, respectively.

269. L. B. Rothman and E. Cabib, *Biochemistry* **8**, 3332 (1969).
270. P. A. Rosness, G. Gustafson, and B. E. Wright, *J. Bacteriol.* **108**, 1329 (1971).
271. B. E. Wright and D. Dahlberg, *Biochemistry* **6**, 2074 (1967).
272. B. E. Wright, D. Dahlberg, and C. Ward, *ABB* **124**, 380 (1968).
273. G. Weeks and J. M. Ashworth, *BJ* **126**, 617 (1972).
274. E. P. Camargo, R. Meuser, and D. Sonneborn, *JBC* **244**, 5910 (1969).

9

Lactose Synthetase

KURT E. EBNER

I. Introduction 363
    A. Historical 364
II. Molecular Properties 365
    A. Requirement for Two Proteins: Galactosyltransferase
       and α-Lactalbumin 365
    B. Galactosyltransferase 367
    C. Catalytic Properties 369
    D. Biological Significance 377

I. Introduction

Lactose (4-O-β-D-galactopyranosyl-α-D-glucopyranose) is the principal carbohydrate in the milk of most mammalian species and such milks usually contain from 2 to 7% lactose (1, 2). The biosynthesis of lactose has had a rather long and complex history even though it would appear at first glance that the biosynthesis of this disaccharide would be rather straightforward. The elucidation of the mechanism of the biosynthesis of lactose has revealed a new type of biological control system where an enzyme, a general galactosyltransferase involved principally in glycoprotein biosynthesis, is modified kinetically by the common milk whey protein, α-lactalbumin, so that lactose is formed at meaningful rates

1. E. R. Ling, S. K. Kon, and J. Porter, *in* "Milk: The Mammary Gland and Its Secretion" (S. K. Kon and A. T. Cowie, eds.), Vol. 1, p. 195. Academic Press, New York, 1961.
2. R. Jenness, E. A. Regehr, and R. S. Sloan, *Comp. Biochem. Physiol.* **13**, 339 (1964).

in the mammary gland as well as maintaining its ability to synthesize glycoproteins.

Lactose synthetase (UDPgalactose:D-glucose 1-galactosyltransferase, EC 2.4.1.22) is the name given to the enzyme system which catalyzes the formation of lactose from UDPgalactose and glucose. It is now clear that the name, though a misnomer, is useful for describing the biosynthesis of lactose even though two proteins are involved in the reaction and the main function of the galactosyltransferase in most tissues is concerned with glycoprotein biosynthesis rather than lactose biosynthesis. The principal reactions catalyzed by the galactosyltransferase in the absence or presence of α-lactalbumin are as follows:

$$\text{UDPgalactose} + \text{Glc} \xrightarrow[\text{Mn}^{2+}]{\alpha\text{-lactalbumin}} \text{lactose} + \text{UDP} \qquad (1)$$

$$\text{UDPgalactose} + \text{GlcNAc} \xrightarrow[\text{Mn}^{2+}]{} N\text{-acetyllactosamine} + \text{UDP} \qquad (2)$$

$$\text{UDPgalactose} + \text{ovalbumin} \xrightarrow[\text{Mn}^{2+}]{} \text{Gal-ovalbumin} + \text{UDP} \qquad (3)$$

A. HISTORICAL

Studies pertinent to the earlier work relating to the biosynthesis of lactose are available in reviews (3–5). Earlier studies by Gander et al. (6) indicated that a crude enzymic preparation from bovine mammary tissue could synthesize lactose 1-phosphate from UDPgalactose and glucose 1-phosphate and the lactose 1-phosphate from this reaction was subsequently hydrolyzed by a phosphatase to form lactose. These studies have not been confirmed, and it is now well established that the galactosyl acceptor from UDPgalactose is glucose as established by Hassid and his co-workers (7–9).

In 1962, Watkins and Hassid (7) isolated a particulate fraction from lactating bovine and guinea pig mammary glands which catalyzed the synthesis of lactose from UDPgalactose and glucose. In 1964, Babad and Hassid (8) discovered that the enzyme system occurred in a soluble form in bovine milk and were able to purify the enzyme some 70-fold

3. S. J. Folley, *Dairy Sci. Abstr.* **23**, 511 (1961).
4. L. F. Leloir and C. E. Cardini, in "Milk: The Mammary Gland and Its Secretion" (S. K. Kon and A. T. Cowie, eds.), Vol. 1, p. 421. Academic Press, New York, 1961.
5. E. A. Jones, *J. Dairy Res.* **36**, 145 (1969).
6. J. E. Gander, W. E. Petersen, and P. D. Boyer, *ABB* **60**, 259 (1956); **69**, 85 (1957).
7. W. M. Watkins and W. Z. Hassid, *JBC* **237**, 1432 (1962).
8. H. Babad and W. Z. Hassid, *JBC* **239**, PC946 (1964).
9. H. Babad and W. Z. Hassid, *JBC* **241**, 2672 (1966).

(9) before losing activity. This loss in activity resulted from the separation of the galactosyltransferase and α-lactalbumin, both of which are required for significant enzymic activity. The enzyme did not transfer galactose to glucose 1-phosphate but did transfer galactose to N-acetylglucosamine to form the product, N-acetyllactosamine. N-Acetylglucosamine was about 25% effective as glucose as the galactosyl acceptor. The reaction also required Mn^{2+} although this apparently could be partially replaced by Mg^{2+}. Karimoto and Reithel (10) were able to obtain from a bovine mammary powder a soluble extract which would catalyze the formation of lactose from UDPgalactose and glucose.

II. Molecular Properties

A. Requirement for Two Proteins: Galactosyltransferase and α-Lactalbumin

1. *Resolution of the Soluble Enzyme into Two Protein Fractions*

Lactose synthetase was very difficult to solubilize from mammary gland particles since there was a large loss of activity upon solubilization. Accordingly, attempts were made to purify the enzyme from skim milk where it appeared to act as a soluble enzyme although extensive purification also resulted in loss of activity. Brodbeck and Ebner (11) resolved these difficulties when they were able to separate a partially purified soluble lactose synthetase from bovine milk by chromatography on Bio Gel P-30 into two protein components, both of which were required for lactose synthetase activity. These proteins were called the A protein (of higher molecular weight) and the B protein (of lower molecular weight). Neither protein exhibited lactose synthetase activity when assayed under the conditions used, and activity was present only when both proteins were in the assay. These results showed that two proteins were required for the enzymic synthesis of lactose. Further work showed that lactose synthetase solubilized from bovine mammary tissue (12) and from the milk of the sheep, goat, and human (13) was resolved into two protein fractions by chromatography on Bio Gel P-30. Subcellular distribution studies showed that the A protein was mainly associated with a crude

10. R. S. Karimoto and F. J. Reithel, *Life Sci.* **4**, 919 (1965).
11. U. Brodbeck and K. E. Ebner, *JBC* **241**, 762 (1966).
12. U. Brodbeck and K. E. Ebner, *JBC* **241**, 5526 (1966).
13. U. Brodbeck, W. L. Denton, N. Tanahashi, and K. E. Ebner, *JBC* **242**, 1391 (1967).

microsomal fraction whereas the B protein was found in both the microsomal and soluble fraction (*12*), and it was also shown that the B protein was easily dissociated from the particulate fraction. The latter observation provided an explanation for the difficulty in obtaining an active enzymic preparation from tissue extracts since the B protein was readily dissociable and hence was rate limiting in the assay.

2. Identification of the B Protein as α-Lactalbumin

The lower molecular weight protein, or B protein, was crystallized, and it was shown that bovine α-lactalbumin substituted for the B protein in the lactose synthetase assay (*14*). Further work based on chemical, physical, and immunological studies showed that the B protein was α-lactalbumin, the common whey protein found in milk (*13*). Impure preparations of the A and B proteins from the cow, goat, sheep, and human are qualitatively interchangeable in the enzymic rate assay (*13*) and to date α-lactalbumin isolated from the milk of a variety of species all react enzymically with bovine galactosyltransferase (*15*).

3. Relationship between α-Lactalbumin and Lysozyme

Brew and Campbell (*16*) noted structural features which were similar between hen's egg-white lysozyme and guinea pig α-lactalbumin even though guinea pig α-lactalbumin did not have any lysozyme activity. These observations prompted work leading to the complete amino acid sequence and location of the disulfide bonds of bovine α-lactalbumin (*17*). Comparison of the linear amino acid sequence of bovine α-lactalbumin and hen's egg-white lysozyme showed that 49 amino acids are in the identical or corresponding positions and an additional 23 are conservative replacements of the 123 amino acids in the total sequence (*18*). These results indicate a high degree of structural similarity between lysozyme and α-lactalbumin, and indeed Browne et al. (*19*) were able to construct a three-dimensional model of α-lactalbumin based on the coordinates of hen's egg-white lysozyme. The structural relatedness be-

14. K. E. Ebner, W. L. Denton, and U. Brodbeck, *BBRC* **24**, 232 (1966).
15. N. Tanahashi, U. Brodbeck, and K. E. Ebner, *BBA* **154**, 247 (1968).
16. K. Brew and P. N. Campbell, *BJ* **102**, 258 (1967).
17. K. Brew, F. J. Castellino, T. C. Vanaman, and R. L. Hill, *JBC* **245**, 4570 (1970); K. Brew and R. L. Hill, *ibid.* p. 4559; T. C. Vanaman, K. Brew, and R. L. Hill, *ibid.* p. 4583.
18. R. L. Hill, K. Brew, T. C. Vanaman, I. P. Trayer, and P. Mattock, *Brookhaven Symp. Biol.* **21**, 139 (1968).
19. W. J. Browne, A. C. T. North, D. C. Phillips, K. Brew, T. C. Vanaman, and R. L. Hill, *JMB* **42**, 65 (1969).

tween α-lactalbumin and lysozyme suggest that these proteins are related in the evolutionary sense, and the proposal was made that both were derived from a common ancestral protein which by the process of gene duplication gave rise to the present proteins (18). Both lysozyme and α-lactalbumin participate in similar types of reactions in that lysozyme catalyzes the hydrolysis of a β-1,4-glycosidic bond whereas α-lactalbumin is apparently involved in the synthesis of a β-1,4-glycosidic bond. The latter statement is not strictly correct since α-lactalbumin does not appear to have any catalytic function but rather acts as a protein modifier of an existing reaction catalyzed by the galactosyltransferase. Neither lysozyme nor α-lactalbumin participates in nor interferes with each other's reaction (19), and to date there appears to be no evidence which suggests a functional relationship between lysozyme and α-lactalbumin even though their gross structures are very similar (20).

B. Galactosyltransferase

1. Purification and Properties

The galactosyltransferase involved in the lactose synthetase reaction occurs as a particulate enzyme in mammary (12) and other tissues (21). The enzyme is in a soluble form in milk and has been purified from bovine (22, 23) and human milk (24, 25). Fitzgerald et al. (22) purified the enzyme from bovine skim milk by conventional methods, and the enzyme had a specific activity between 3 and 5 μmoles lactose/min/mg at 25° and an apparent molecular weight of 70,000–75,000 as determined on Sephadex G-100. Trayer and Hill (23) also purified the galactosyltransferase from bovine skim milk by affinity chromatography where α-lactalbumin was covalently bound to Sepharose 4B. The specific activity of the final product was 14.1 at 37° (5.8 at 25°) and appeared to be homogeneous under denaturing conditions; for example, electrophoresis on polyacrylamide gels in 6.25 M urea, pH 3.2, gave a single band as did gel electrophoresis in sodium dodecyl sulfate. Under denaturing conditions of sodium dodecyl sulfate gel electrophoresis, gel filtration, and sedimentation equilibrium in 6 M guanidine hydrochloride, molecu-

20. R. Arnon and E. Maron, JMB **60**, 225 (1971).
21. D. K. Fitzgerald, L. McKenzie, and K. E. Ebner, BBA **235**, 425 (1971).
22. D. K. Fitzgerald, U. Brodbeck, I. Kiyosawa, R. Mawal, B. Colvin, and K. E. Ebner, JBC **245**, 2103 (1970).
23. I. P. Trayer and R. L. Hill, JBC **246**, 6666 (1971).
24. P. Andrews, FEBS Lett. **9**, 297 (1970).
25. T. Nagasawa, I. Kiyosawa, and N. Tanahashi, J. Dairy Sci. **54**, 835 (1971).

lar weights between 40,000 and 44,000 were obtained. Sucrose density gradient experiments gave a molecular weight of 42,000, and the results were consistent with a single polypeptide chain of about 42,000 molecular weight. The previous estimates of a molecular weight of 75,000 on Sephadex G-100 (*22*) appears to be too high and may result from the fact that the galactosyltransferase is a glycoprotein. A molecular weight of about 44,000 is obtained on Bio Gel P-200 columns and by sodium dodecyl sulfate gel electrophoresis (*26*). Trayer and Hill (*23*) also presented evidence to indicate that the galactosyltransferase appeared to exist as a variety of associated species (40,000–160,000) in dilute salt solutions as evidenced by sedimentation equilibrium studies.

The amino acid composition data (*23*) show that the protein has a high amide and proline content but rather low amounts of the polar amino acids. The protein contains 2.0% sialic acids, 1.1% glucosamine, 1.1% galactosamine, and 8.1% neutral sugars which consists of galactose, mannose, and glucose as determined qualitatively as their alditol acetates by gas–liquid chromatography. Recent work (*26*) showed that there are two principal forms of the bovine galactosyltransferase, one has a molecular weight of about 45,000 and the other of about 58,000. Both forms are enzymically active, contain carbohydrate, cannot be separated by affinity chromatography on Sepharose-α-lactalbumin columns, and appear to have similar catalytic sites.

The enzyme purified from human milk is less well characterized but appears to have properties similar to the bovine milk enzyme. Andrews (*24*) purified the enzyme from 4 to 7 day's postpartum human milk on a final Sepharose-α-lactalbumin column and obtained an enzyme preparation in 36% yield with a specific activity of 5.7 at 25°. The molecular weight was 40,000–42,000 as obtained from chromatography on Bio Gel P-200. Nagasawa *et al*. (*25*) purified the galactosyltransferase from Japanese human milk and obtained a preparation with a specific activity of 0.75, and it appeared as a single band on paper electrophoresis at pH 9.5. The amino acid composition showed a high percentage of proline and acidic amino acids. The protein is a glycoprotein and contained 1.10% sialic acid, 5.03% galactosamine, 2.47% glucosamine, 2.36% galactose, 0.87% mannose, and 0.87% fucose. The molecular weight as determined by Sephadex G-100 chromatography was 75,000 and 72,000 by electrophoresis at pH 9.5 in 0.2% sodium dodecyl sulfate. These molecular weights do not agree with those reported by Andrews (*24*)

26. S. C. Magee, R. Mawal, and K. E. Ebner, *Fed. Proc., Fed. Amer. Soc. Exp. Biol.* **31**, 499 (1972).

and may reflect difficulties in determining molecular weights of glycoproteins.

Palmiter (*27*) has reported enzymic activities associated with varying molecular weights of a crude ammonium sulfate fraction of mouse mammary galactosyltransferase as determined on Sephadex G-100. One form was at the exclusion volume, the major form (65%) was between 100,000 and 130,000, and another form was at 29,000. These results should be interpreted with some caution since they were obtained on Sephadex and may reflect in part anomalous migration of a glycoprotein on Sephadex as well as difficulty in truly solubilizing the tissue enzyme.

C. Catalytic Properties

1. Reactions Catalyzed

The general reaction catalyzed by the galactosyltransferase is as follows:

$$\text{UDPgalactose} + \text{acceptor} \xrightarrow{\text{Mn}^{2+}} \text{Gal-acceptor} + \text{UDP} \quad (4)$$

The major galactosyl acceptors are GlcNAc and terminal β-(1,4)-glycosides of GlcNAc including the carbohydrate side chain of glycoproteins. Glucose is a good acceptor only in the presence of α-lactalbumin and under these conditions lactose

$$\text{UDPgalactose} + \text{Glc} \xrightarrow[\alpha\text{-lactalbumin}]{\text{Mn}^{2+}} \text{lactose} + \text{UDP} \quad (5)$$

is formed at meaningful rates. This latter reaction has been termed *lactose synthetase,* and this term is useful for describing this reaction. The principal function of the galactosyltransferase is its involvement in glycoprotein biosynthesis except in the lactating mammary gland where the enzyme is also involved with the biosynthesis of lactose.

2. Substrate Specificity

a. Galactosyl Donor. Babad and Hassid (*9*) examined the galactosyl donor specificity of the bovine milk galactosyltransferase and found that dUDP-D-galactose was 80% as effective as UDP-D-galactose. No formation of lactose was observed when ADP-D-galactose, TDP-D-galactose, or GDP-D-galactose was used as a substrate.

b. Galactosyl Acceptor. Babad and Hassid (*9*) reported that neither

27. R. D. Palmiter, *BBA* **178**, 35 (1969).

α-D-glucose 1-phosphate, α-D-galactose 1-phosphate, L-glucose, D-xylose, maltose, nor α-methyl-D-glucoside was a galactosyl acceptor. D-Glucosamine was 0.3% as effective as D-glucose and cellobiose was 7.5% as effective as glucose, but N-acetyl-D-glucosamine was 25% as effective as glucose and the product of the latter reaction was identified as N-acetyllactosamine. The previous studies by Babad and Hassid (9) were carried out with the galactosyltransferase in the presence of α-lactalbumin.

Brew et al. (28) showed that the galactosyltransferase could transfer galactose to N-acetylglucosamine in the absence of α-lactalbumin and that α-lactalbumin inhibited this reaction. This inhibition was examined in more detail by Schanbacher and Ebner (29) who showed that it was hyperbolic (intercept) with respect to N-acetylglucosamine. Hill et al. (18) also showed that orosomucoid (galactose removed) was a galactosyl acceptor and this reaction was partially inhibited by α-lactalbumin.

The galactosyltransferase acceptor specificity of the enzyme from bovine milk has been examined in detail (29). In the absence of α-lactalbumin, the best substrates were GlcNAc and its β-(1,4)-glucosides GlcNAc, (GlcNAc)$_2$, (GlcNAc)$_3$, (GlcNAc)$_4$, N-acetylmuramic acid and ovalbumin are excellent substrates with apparent K_m values in the low millimolar range. Other β-glycosides are poorer acceptors and these include cellobiose, cellobiulose, β-methylglucose, β-indoxylglucose, glycosylmannose, and gentiobiose. α-Glycosides such as α-methylglucose and maltose were not active in the absence of, but were slightly active in the presence of, α-lactalbumin. Glucosamine and N-acetylmannosamine are poor acceptors, N-acetylmuramic acid is fair, and 2-deoxy-D-glucose was an acceptor only in the presence of α-lactalbumin. Mannose, fucose, melibiose, and UDP-GlcNAc are not acceptors even in the presence of α-lactalbumin. α-Lactalbumin inhibited markedly the transfer of galactose to GlcNAc (15 μg/ml gave 50% inhibition), whereas it was a much less effective inhibitor of β-(1,4)-glycosides such as (GlcNAc)$_3$ (1 mg/ml inhibited about 30%). These studies (29) showed that a variety of β-(1,4)-glycosides including ovalbumin are good galactosyl acceptors and that α-lactalbumin did not appreciably inhibit this transfer but inhibited markedly the transfer of galactose to GlcNAc.

c. *Metal Requirement.* Babad and Hassid (9) reported that in the crude bovine enzymic preparations Mg^{2+} was 25% as effective as Mn^{2+} which has an optimum concentration of 13.3 mM at pH 7.5; Ca^{2+} could

28. K. Brew, T. C. Vanaman, and R. L. Hill, *Proc. Nat. Acad. Sci. U. S.* **59**, 491 (1968).

29. F. L. Schanbacher and K. E. Ebner, *JBC* **245**, 5057 (1970).

replace Mg^{2+} but Co^{2+}, Na^+, K^+, and NH_4^+ had little stimulating activity. However, studies with the highly purified enzyme (30) showed that no activity was observed in the absence of Mn^{2+} or the presence of Mg^{2+} or Ca^{2+}, and concentrations above 4 mM were inhibitory. It was possible that Mg^{2+} or Ca^{2+} could displace Mn^{2+} from other proteins in the crude enzymic preparation (9) and therefore would appear to activate the enzyme.

3. Reaction Kinetics

a. Early Studies. The galactosyltransferase from bovine milk transfers galactose from UDPgalactose to GlcNAc, and this reaction is inhibited by α-lactalbumin; in the presence of α-lactalbumin the enzyme will transfer galactose to glucose to form lactose (28). On the basis of these observations α-lactalbumin was called a specifier protein in that it dictated the galactosyl acceptor specificity. However, it was subsequently shown that the galactosyltransferase can catalyze the transfer of galactose to glucose to form lactose in the absence of α-lactalbumin provided the glucose concentration was high ($K_m = 1.4\,M$) and that the maximum velocity of the reaction was the same in the absence or presence of α-lactalbumin (22). These results showed that the active site of the enzyme in the galactosyltransferase and the function of α-lactalbumin was to lower the apparent K_m of glucose to the low millimolar region. Similar results were obtained by Klee and Klee (31) with the bovine milk enzyme and by Andrews (32) with the human milk enzyme. In the assay for lactose synthetase, an optimum concentration of α-lactalbumin is required, and higher concentrations are inhibitory (22). Efforts to detect an α-lactalbumin–galactosyltransferase complex under conditions of maximum product formation have been unsuccessful (23, 33), which is supportive evidence for the absence of a long-lived kinetic complex.

b. Steady State Kinetics. The steady state kinetic analysis of the galactosyltransferase reaction was undertaken to determine the functional role of α-lactalbumin in the lactose synthetase reaction (30, 34, 35). The general approach was to study (a) the galactosyltransferase reaction with N-acetylglucosamine as the galactosyl acceptor in the

30. J. F. Morrison and K. E. Ebner, *JBC* **246**, 3977 (1971).
31. W. A. Klee and C. B. Klee, *BBRC* **39**, 833 (1970).
32. P. Andrews, *BJ* **111**, 14P (1969).
33. F. L. Schanbacher and K. E. Ebner, *BBA* **229**, 226 (1971).
34. J. F. Morrison and K. E. Ebner, *JBC* **246**, 3985 (1971).
35. J. F. Morrison and K. E. Ebner, *JBC* **246**, 3992 (1971).

absence of α-lactalbumin (*30*), (b) the galactosyltransferase reaction with glucose as the galactosyl acceptor at a fixed concentration of α-lactalbumin (*34*), and (c) the effects of α-lactalbumin on the reaction with either glucose or N-acetylglucosamine as the galactosyl acceptor (*35*).

The equilibrium of the reaction was essentially in the direction of product formation, and no reversibility of the reaction could be demonstrated. These results ruled out the use of product inhibition analysis for assisting in determining the order of addition of substrates and release of products. UDPglucose was an inhibitory analog of UDPgalactose, but no suitable inhibitory analog of N-acetylglucosamine was found from the compounds tested which were D-sorbitol, glycerol, sucrose, inositol, glucose, N-acetylgalactosamine, N-acetylmannosamine, galactosamine, mannosamine, glucuronic acid, and glucose 6-phosphate. It was also observed that N-acetylglucosamine caused substrate inhibition at high concentrations and that the inhibition was noncompetitive with respect to Mn^{2+} and UDPgalactose. It was postulated that the substrate, at high concentrations, formed inactive dead-end complexes on both the substrate addition side and product release side of the reaction mechanism. The steady state kinetic results based on initial velocity, dead-end inhibition, and substrate inhibition studies were consistent with an equilibrium ordered type of mechanism. Under conditions of thermodynamic equilibrium, Mn^{2+} adds first to the free enzyme and does not dissociate at each turn of the catalytic cycle. This is followed by an ordered addition of UDPgalactose and GlcNAc, followed by the ordered release of products N-acetyllactosamine and UDP. The weighted mean value with the standard error of the true kinetic constants were K_{im} (Mn^{2+}), 1.35 ± 0.14 mM; K_a (UDPgalactose), 0.060 ± 0.007 mM; K_{ia} (UDPgalactose), 0.065 ± 0.01 mM; and K_b (N-acetylglucosamine), 5.8 ± 0.6 mM; the weighted mean value of the true inhibition constant of UDPglucose was 0.089 ±0.006 mM. The initial rate equation for the proposed mechanism can be expressed as

$$v = \frac{VMAB}{K_{im}K_{ia}K_b + K_{ia}K_bM + K_{im}K_aB + K_aMB + K_bMA + MAB}$$

where V is the maximum velocity, and M, A, and B represent Mn^{2+}, UDPgalactose, and N-acetylglucosamine, respectively; K_{im} represents the dissociation constant of M with free enzyme, and K_{ia} is the dissociation constant of A with EM; K_a is the Michaelis constant for A, and K_b is the Michaelis constant for B (*30*).

Similar kinetic studies (*34*) were carried out with glucose as the galactosyl acceptor and with the concentration of α-lactalbumin fixed at 4 μM which is at a nonsaturating level. Substrate inhibition was observed with glucose. UDPglucose was a good inhibitor of the reaction, and it was observed that a number of pentoses at relatively high concentrations (0.4 M) inhibited the reaction: D-arabinose (13%), L-xylose (16%), D-xylose (36%), L-sorbose (45%), L-arabinose (54%), D-ribose (59%), and 2-deoxyglucose (68%). The inhibition by L-arabinose was competitive with respect to glucose and was used for dead-end inhibition analysis. The data from initial velocity and dead-end inhibition analysis provided support for the view that in the presence of glucose and α-lactalbumin the galactosyltransferase had an ordered addition of substrates and release of products, and it appeared that this sequence was not influenced by the galactosyl acceptor or by the presence of α-lactalbumin.

The weighted mean values of the conditional kinetic constants in the presence of 4 μM α-lactalbumin were K_{im} (Mn^{2+}), 1.24 ± 0.20 mM; K_a (UDPgalactose), 0.024 ± 0.004 mM; K_{ia} (UDPgalactose), 0.047 ± 0.009 mM; K_b (glucose), 5.3 ± 0.5 mM; and K_i (UDPglucose) 0.061 ± 0.002 mM. The true conditional K_i for L-arabinose was 0.27–0.34 M.

The previous studies (30, 34) showed that α-lactalbumin or the nature of the galactosyl acceptor did not affect the order of addition of substrate or release of products. The kinetic effects of α-lactalbumin on the galactosyltransferase reaction when glucose or N-acetylglucosamine were the galactosyl acceptors were investigated to provide an explanation for the role of α-lactalbumin in the lactose synthetase reaction (35). Previous studies had shown that N-acetylglucosamine (30) or glucose (34) at high

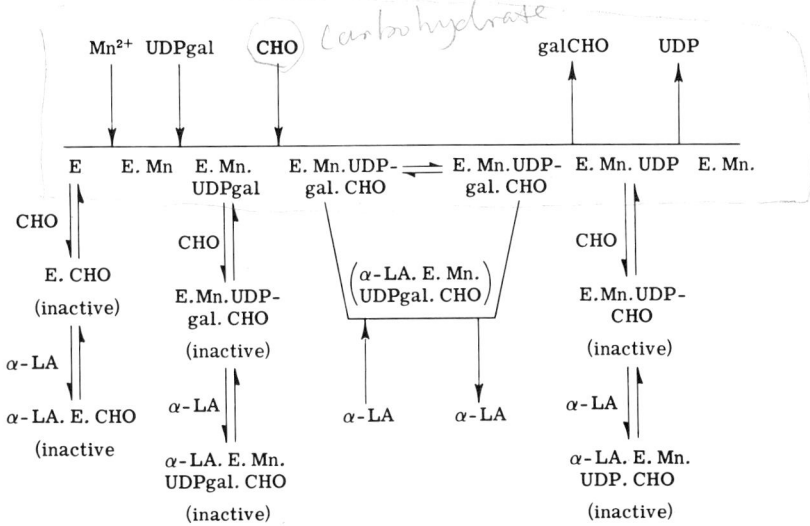

FIG. 1. A kinetic mechanism for galactosyltransferase.

concentrations would cause substrate inhibition. It was further observed that α-lactalbumin increased the extent of substrate inhibition observed either with glucose or N-acetylglucosamine, and it was concluded that α-lactalbumin in the presence of high concentrations of substrates formed additional dead-end complexes involving α-lactalbumin and the substrate dead-end complex (35). However, α-lactalbumin does not appear to increase inhibition observed with L-arabinose when glucose is the substrate. The proposal that α-lactalbumin forms additional dead-end complexes with the substrate provides an explanation for the observation that high levels of α-lactalbumin inhibit the enzymic assay for the galactosyltransferase.

A kinetic mechanism based on the steady state kinetic analysis for the galactosyltransferase and the role of α-lactalbumin in this reaction are illustrated in Fig. 1 where UDPgal represents UDPgalactose, α-LA is α-lactalbumin, and CHO represents a carbohydrate reactant which is either a substrate or an inhibitory substrate analog (35). The main features of the mechanism are as follows:

1. Mn^{2+}, UDPgalactose, and galactosyl acceptor (at nonsubstrate inhibiting concentrations) add to the enzyme in an ordered manner and the resulting complex may then combine with α-lactalbumin.

2. Mn^{2+} does not leave the enzyme at each turn of the catalytic cycle whereas α-lactalbumin does leave at each turn of the cycle. The products dissociate in the order: galactosyl acceptor and UDP.

3. At higher concentrations the galactosyl acceptor can add randomly to free enzyme, enzyme–manganese (not shown), and enzyme–manganese–UDPgalactose, but an active complex is produced only if there has been prior combination of Mn^{2+} and UDPgalactose. Otherwise, inactive dead-end complexes are formed which inhibit the overall reaction.

4. The effect of α-lactalbumin on inhibiting the reaction when N-acetylglucosamine is the galactosyl acceptor, activating the reaction when glucose is the galactosyl acceptor, increasing substrate inhibition by N-acetylglucosamine and glucose, and enhancing the dead-end inhibition by a substrate analog is essentially the result of its ability to combine with a complex containing carbohydrate and cause displacement of already established equilibria.

An example of the above is that when N-acetylglucosamine is the substrate the reaction proceeds well in the absence of α-lactalbumin (30). When the concentration of N-acetylglucosamine is nonsaturating, the addition of α-lactalbumin forces more of the reaction flux via the branched pathway; if this pathway is slower than the linear pathway which occurs in the absence of α-lactalbumin, then inhibition of the reaction is predicted which is consistent with the experimental findings. Similar arguments may be presented to explain how α-lactalbumin may act as an activator at concentrations of N-acetylglucosamine well below the K_m value. α-Lactalbumin can effectively increase the concentration of N-acetylglucosamine since it can reduce the apparent K_m of the substrate and cause part of the reaction flux to occur along the branched pathway. When the glucose concentration is 1–4 M, reaction occurs along the linear pathway in the absence of α-lactalbumin. The addition of α-lactalbumin reduces the K_m for glucose to the millimolar region, and virtually all the reaction proceeds along the branched pathway (35).

Additional dead-end complexes containing α-lactalbumin were postulated to explain the enhancement of substrate inhibition by glucose and N-acetylglucosamine by α-lactalbumin. The addition of α-lactalbumin would increase the steady state concentration of the total dead-end complexes, thereby reducing the overall reaction rate by reducing the amount of enzyme available for product formation.

From this kinetic study (30, 34, 35) it would appear that α-lactalbumin is best classified as a modifier protein rather than as a specifier protein as previously proposed (28) since the galactosyltransferase can by itself catalyze the formation of lactose in the absence of α-lactalbumin (high glucose) and that α-lactalbumin lowers the K_m for glucose so that glucose becomes an effective substrate. The fundamental role of α-lactalbumin appears to differ from the action of small organic molecules which can affect the binding of substrate and/or the maximum velocity of an enzyme by binding at a site distinct from the catalytic site. Under conditions of catalysis, α-lactalbumin requires the presence of Mn^{2+} and the two substrates on the enzyme prior to binding with the enzyme.

c. Detection of Reactant Complexes by Affinity Chromatography. The mechanism described in Fig. 1 provides an explanation for the failure to detect an enzyme–managanese–UDPgalactose–glucose–α-lactalbumin complex under conditions of maximum product formation since it is postulated that α-lactalbumin dissociates rapidly from the enzyme at each turn of the catalytic cycle. However, it does predict the formation of complexes under conditions of dead-end inhibition, e.g., high substrate and/or high α-lactalbumin concentrations and, indeed, evidence for the presence of such complexes has been reported (23, 24, 36).

Andrews (24) has purified the galactosyltransferase from human milk

36. R. Mawal, J. F. Morrison, and K. E. Ebner, *JBC* **246**, 7106 (1971).

using the affinity chromatography technique whereby α-lactalbumin was chemically coupled to Sepharose 6B which had been previously activated with CNBr. He observed that the galactosyltransferase was retained on such a column in the presence of 3 mM N-acetylglucosamine and released from the column when N-acetylglucosamine was removed from the eluting buffer. Trayer and Hill (23) have used a similar procedure of affinity chromatography to purify the galactosyltransferase from bovine milk. The buffer contained Mg^{2+} and glucose, and upon removal of glucose the galactosyltransferase was eluted from the column. The reason that the technique worked is that the Sepharose–α-lactalbumin columns contained a high concentration of α-lactalbumin, and in the presence of carbohydrate substrates dead-end complexes were formed by displacing established equilibria. In both the above cases no product can form because of the lack of Mn^{2+} and UDPgalactose in the reaction mixture. Upon removal of the substrate from the buffer, the complex dissociates and the galactosyltransferase is eluted in the buffer.

It became apparent that the presence of reactant complexes predicted for the mechanism outlined in Fig. 1 could be detected by affinity chromatography experiments using Sepharose-α-lactalbumin columns. The results of such studies (36) provided direct evidence for the formation of dead-end complexes involving α-lactalbumin which were predicted from the data obtained from the steady state kinetic investigations (30, 34, 35).

No complexes were detected with free enzyme or enzyme–manganese but complexes were formed with free enzyme and/or enzyme–manganese with the substrates, glucose, N-acetylglucosamine, and UDPgalactose. The addition of Mn^{2+} to substrates or substrate analogs causes tighter retardation of the galactosyltransferase but is not essential for complex formation. All of these complexes are not kinetically significant, and they arise as a consequence of the high concentration of α-lactalbumin bound to the Sepharose. They do appear to be important in explaining the possible point of inhibition of lactose formation by high concentrations of α-lactalbumin (22). Under identical conditions, glucose fails to retard the galactosyltransferase on a Sepharose–α-lactalbumin column at 20–25 mM, N-acetylglucosamine fails at 10–15 mM, whereas UDPgalactose fails at 2–3 μM. These results suggest that the inhibition of lactose formation by α-lactalbumin may principally result from the formation of enzyme–manganese–UDPgalactose–α-lactalbumin and enzyme–UDPgalactose–α-lactalbumin complexes. The extent of retardation is dependent upon the total and relative concentrations of Sepharose–α-lactalbumin and substrate or substrate analog. The product, lactose, does not cause retardation of the galactosyltransferase which is consistent with the lack of inhibition exhibited by lactose on the reaction.

D. Biological Significance

The principal physiological role of the galactosyltransferase involved in the biosynthesis of lactose appears to be to transfer galactose to a terminal N-acetylglucosaminyl residue of the carbohydrate side chain of a glycoprotein to form a product such as

$$\text{Gal} \xrightarrow{\beta\text{-}(1,4)} \text{GlcNAc} \xrightarrow{\beta} \text{glycoprotein}$$

Galactosyltransferase activity catalyzing this reaction is widely distributed (13, 18, 21, 37, 38) but is principally located in the Golgi apparatus (39) and has been used as an enzymic marker for this cellular organelle (40). The mammary gland is unique in that it has the ability to form α-lactalbumin whose synthesis is presumably under strict hormonal control. In the presence of α-lactalbumin, the galactosyltransferase can carry out at the same time both the synthesis of lactose and the transfer of galactose in incompleted glycoprotein carbohydrate side chains. Other tissue extracts than those from mammary glands can synthesize lactose provided α-lactalbumin is added to the reaction mixture (21). The enzyme is also present in porcine serum in a soluble form (41) and probably is identical to a soluble UDPgalactose:glycoprotein galactosyltransferase found in rat serum (42). Only ovalbumin was used as the galactosyl acceptor (42), and previous studies (29) have shown that ovalbumin is an excellent substrate for the galactosyltransferase isolated from milk.

Acknowledgments

This research was supported in part by grants from the National Institutes of Health, AM 10764; National Science Foundation, GB 7975; and a Career Development Award 1 KO4 GB 42396 from the National Institutes of Health.

---

37. E. J. McGuire, G. W. Jourdian, D. M. Carlson, and S. Roseman, JBC 240, PC4112 (1965).

38. T. Helting, JBC 246, 815 (1971).

39. R. G. Coffey and F. J. Reithel, BJ 109, 169 and 177 (1968); T. W. Keenan, D. J. Morré, and R. D. Cheetham, Nature (London) 228, 1105 (1970).

40. B. Fleischer, S. Fleischer, and H. Ozawa, J. Cell Biol. 43, 59 (1969); D. E. Leelavathi, L. W. Estes, D. S. Feingold, and B. Lombardi, BBA 211, 124 (1970).

41. R. L. Hudgin and H. Schachter, Can. J. Biochem. 49, 838 (1971).

42. R. R. Wagner and M. A. Cynkin, BBRC 45, 57 (1971).

10
Amino Group Transfer

ALEXANDER E. BRAUNSTEIN

I. Introduction	379
A. Recent Developments in the Study of Transamination	381
B. Formally Similar Processes of NH_2 Transfer	384
II. Basic Chemical Features of the Transamination Reaction	387
A. Congruent Nonenzymic Models	387
B. General Characteristics of Intermediate Steps	391
III. Aspartate:2-Oxoglutarate Aminotransferases	393
A. Isoenzymes and Multiple Subforms	393
B. Aspartate Transaminase$_{cyt}$ from Pig Heart and Related Forms from Animal Tissues	398
IV. Aminotransferases Acting on Other Substrates	462
A. Requiring Glutamate–Oxoglutarate (or Aspartate–Oxalacetate) as One Donor–Acceptor Pair	463
B. Acting on Two α-Amino–α-Oxo Monocarboxylic Substrate Pairs	473
C. Acting on ω-Amino and ω-Oxo Acids	473
D. Acting on Amino and Oxo Substrates Containing Noncarboxylic Acid Groups (or None)	475
E. Transaminations Occurring as Side Reactions in the Active Site of Pyridoxal-P-Linked Enzymes of Other Types	476
V. Concluding Remarks	480

I. Introduction

Single or coupled enzymic reactions resulting in intermolecular transfer of amino groups constitute essential links in intermediary nitrogen metabolism (1, 2). Their control by internal and environmental factors

1. A. E. Braunstein, in "Biochimie Comparée des Acides Aminés Basiques," Colloq. Int. Cent. Nat. Rech. Sci., Concarneau, 1959, p. 79. Paris, 1960.
2. A. Meister, "Biochemistry of the Amino Acids," 2nd ed., Vols. 1 and 2. Academic Press, New York, 1965.

plays an important role in the regulation of metabolic processes and physiological functions. Studies on molecular aspects of the control mechanisms are in rapid expansion and hold a prominent place in reviews concerning metabolic regulation (*2–6*).

Reversible enzymic transfer of the NH_2 group, together with a proton and an electron pair, from L-glutamic acid to pyruvic or other oxo acids [Eq. (1)], termed "transamination," was first demonstrated in 1937 in

$$HOOC\cdot CH_2\cdot CH_2\cdot CHNH_2\cdot COOH + HOOC\cdot CO\cdot R \rightleftharpoons$$
$$\rightleftharpoons HOOC\cdot CH_2\cdot CH_2\cdot CO\cdot COOH + HOOC\cdot CHNH_2\cdot R \quad (1)$$

muscle and other animal tissues (*7*); for early reviews, see Refs. *3, 8–10*. Thereafter, and especially following discovery (1945; Snell *et al.*, Gunsalus and co-workers) of the role of pyridoxal phosphate as the cofactor (prosthetic group) of enzymes catalyzing transamination reactions— the aminotransferases or transaminases (EC 2.6.1 . . .) (*11*)—an extensive literature accumulated concerning various chemical, biological, and medical aspects of pyridoxal-P-dependent transamination, as well as more or less similar metabolic reactions of NH_2 transfer.

This chapter reviews progress achieved during the last 12 years in investigations of pyridoxal-catalyzed transamination, including studies on model systems, the structure, physicochemical, and catalytic characteristics of aminotransferases, and current concepts about the molecular mechanism of enzymic transamination. Several formally analogous reactions of NH_2 transfer are listed in Section I,B.

Important details concerning earlier work on aminotransferases, as well as general or specific aspects of biological pyridoxal catalysis, are sur-

3. A. E. Braunstein, *Advan. Enzymol.* **19**, 355 (1957).
4. N. Katunuma, M. Okada, T. Katsunuma, A. Fujino, and T. Matsuzawa, in "Pyridoxal Catalysis: Enzymes and Model Systems" (E. E. Snell *et al.*, eds.), p. 255. Wiley (Interscience), New York, 1968.
5. G. Weber, ed., "Advances in Enzyme Regulation," Vols. 1–10. Pergamon, Oxford, 1962–1972. In particular, contributions by F. T. Kenney (Vol. 1, p. 137), C. A. Nichol and associates (Vol. 1, p. 341; Vol. 2, p. 115), W. E. Knox *et al.* (Vol. 2, p. 311; Vol. 3, p. 247), H. Pitot and associates, (Vol. 4, p. 199; Vol. 5, p. 303), M. Suda (Vol. 5, p. 181), H. Inoue and H. C. Pitot (Vol. 8, p. 289), B. Sheid and Y. S. Roth (Vol. 3, p. 355), N. Katunuma *et al.* (Vol. 4, p. 317), and B. L. A. Carter *et al.* (Vol. 9, p. 253), J. S. Nisselbaum *et al.* (Vol. 10, p. 275), and N. Katunuma *et al.* (Vol. 10, p. 289).
6. M. A. Kelsall, *Ann. N. Y. Acad. Sci.* **166**, Art. 1, 1–366 (1969).
7. A. E. Braunstein and M. G. Kritzmann, *Enzymologia* **2**, 129 (1937).
8. A. E. Braunstein, *Enzymologia* **7**, 25 (1939).
9. A. E. Braunstein, *Advan. Protein Chem.* **3**, 1 (1947).
10. A. Meister, *Advan. Enzymol.* **16**, 185 (1955).
11. "Enzyme Nomenclature" (Recommendations 1964 of I.U.B.). Elsevier, Amsterdam, 1965.

veyed in previous volumes of this treatise (*12–16*), in proceedings of special symposia (*17–20*), and other serial publications and monographs (*2, 5, 6, 21, 22*).

A. RECENT DEVELOPMENTS IN THE STUDY OF TRANSAMINATION

Lists and descriptions of many transamination reactions and of transaminases differing in substrate specificities or biological origin have been published (*2, 10, 11, 23–25*). From 20 aminotransferases included in 1964 as individual entries (subsubclass EC 2.6.1 . . .) in the EC List of Enzymes (*11*) their number has risen to more than 50 in 1972 (*25*), but the catalog is by no means definitive.

Reports continue to appear on the detection of novel transamination reactions and purification of the enzymes catalyzing them. On the other hand, classification of these enzymes presents difficulties caused by the relative rather than absolute substrate specificity of many aminotransferases. Depending on its biological source, a recognized group-specific aminotransferase acting on structurally related (e.g., branched-chain or aromatic) substrates may vary considerably with regard to the range and

12. P. D. Boyer, H. A. Lardy, and K. Myrbäck, eds., "The Enzymes," 2nd ed., Vol. 1, Chapter 6; Vol. 2, Chapter 6; and Vol. 6, Chapters 14 and 15. Academic Press, New York, 1959, 1960, and 1962, resp.

13. E. E. Snell and S. J. di Mari, "The Enzymes," 3rd ed., Vol. 2, p. 335, 1970.

14. E. A. Boeker and E. E. Snell, "The Enzymes," 3rd ed., Vol. 6, Chapter 7, p. 217, 1972.

15. E. Adams, "The Enzymes," 3rd ed., Vol. 6, Chapter 13, p. 479, 1972.

16. L. Davis and D. E. Metzler, "The Enzymes," 3rd ed., Vol. 7, Chapter 2, p. 33, 1973.

17. E. E. Snell, P. Fasella, A. E. Braunstein, and A. Rossi Fanelli, eds., "Chemical and Biological Aspects of Pyridoxal Catalysis." Pergamon, Oxford, 1963.

18. E. E. Snell, A. E. Braunstein, E. S. Severin, and Yu. M. Torchinsky, eds., "Pyridoxal Catalysis: Enzymes and Model Systems." Wiley (Interscience), New York, 1968.

19. "International Symposium on Vitamin B_6" (in honor of Prof. P. György; New York, 1964), *Vitam. Horm. (New York)* **22**, 361–885 (1964).

20. K. Yamada, N. Katunuma, and H. Wada, eds., "Symposium on Pyridoxal Enzymes." Maruzen, Tokyo, 1968.

21. P. Fasella, *Annu. Rev. Biochem.* **36**, 185 (1967).

22. P. Fasella and C. Turano, *Vitam. Horm. (New York)* **28**, 157 (1970).

23. H. Tabor and C. W. Tabor, eds., "Methods in Enzymology," Vols. 17A and 17B. Academic Press, New York, 1970 and 1971.

24. B. M. Guirard and E. E. Snell, *Compr. Biochem.* **15**, 138 (1964).

25. "Enzyme Nomenclature" (Revised Recommendations 1972 of I.U.B.). Elsevier, Amsterdam, 1973.

relative rates of reactions catalyzed. More than once, activities formerly attributed to a single transaminase were, on further preparative fractionation, assigned to different enzymes with narrower specificities. Conversely, an NH_2 transfer reaction ascribed to a special enzyme entity may eventually be identified as one of the functions of a previously known transaminase. Thus, tyrosine and phenylalanine aminotransferase activities of highly purified enzyme fractions from the mitochondria of mammalian liver and kidney have recently been recognized as subsidiary reactions catalyzed by mitochondrial aspartate:oxoglutarate aminotransferase (26, 27); in the cytosol, transaminations of aromatic amino acids and of aspartate are catalyzed by distinct enzymes (cf. Sections III,B,6a and IV,A).

Around 1959, procedures were published (28, 29) for extensive purification of L-aspartate:2-oxoglutarate aminotransferase (Asp-transaminase, EC 2.6.1.1) from pig heart, the first pyridoxal-P-dependent enzyme obtained in virtually pure state. These and newer, improved techniques (30, 31) made high purity preparations of the enzyme readily available.

Over the last decade, a variety of pyridoxal-linked enzymes from multi- and unicellular organisms, including many specific aminotransferases, were obtained in practically homogeneous condition; cf. Ref. 23. Some were shown to exist as distinct isoenzymes (32–34) and these, in turn, have eventually been fractionated into multiple subforms (35–38).

26. J. E. Miller and G. Litwack, JBC **246**, 3234 (1971).
27. R. Scandurra and C. Canella, Eur. J. Biochem. **26**, 196 (1972), cf. ibid. **3**, 219 (1967).
28. W. T. Jenkins, D. A. Yphantis, and I. W. Sizer, JBC **234**, 51 (1958).
29. H. Lis, BBA **28**, 191 (1958); H. Lis and P. Fasella, ibid. **33**, 567 (1959).
30. O. L. Polyanovsky and M. Telegdi, Biokhimiya **30**, 174 (1965).
31. B. E. C. Banks, S. Doonan, A. J. Lawrence, and C. A. Vernon, Eur. J. Biochem. **5**, 528 (1968).
32. Y. Morino and H. Wada, in "Chemical and Biological Aspects of Pyridoxal Catalysis" (E. E. Snell et al., eds.), p. 175. Pergamon, Oxford, 1963.
33. H. Wada and Y. Morino, Vitam. Horm. (New York) **22**, 44 (1964).
34. H. A. Lardy, V. Paetkau, and P. Walter, Proc. Nat. Acad. Sci. U. S. **53**, 410 (1965).
35. M. Martinez-Carrion, C. Turano, F. Riva, and P. Fasella, BBRC **20**, 206 (1965).
36. H. Wada, H. Kagamiyama, and T. Watanabe, in "Pyridoxal Catalysis: Enzymes and Model Systems" (E. E. Snell et al., eds.), p. 111. Wiley (Interscience) New York, 1968; Y. Morino, H. Kagamiyama, and H. Wada, JBC **239**, 943 (1964).
37. M. Martinez-Carrion and D. C. Tiemeier, Biochemistry **6**, 1715 (1967).
38. M. Martinez-Carrion, D. C. Tiemeier, and D. L. Peterson, Biochemistry **9**, 2574 (1970).

From the phenomenological description and formal kinetics of transamination reactions, the focus of attention has gradually shifted to studies of the molecular parameters, structure, and catalytic mechanism of the individual enzymes. Many aminotransferases, as well as other vitamin B_6-linked enzymes (*23, 33*) and some of their apoenzymes have been crystallized, but no X-ray diffraction data have thus far been reported on the three-dimensional molecular structure of any enzyme of this class.

In a few instances, short peptide sequences have been determined (*13*). The first complete sequence established (Fig. 7) is, apparently, that of Asp-transaminase (supernatant) from pig heart (Section III,B,3). This transaminase is the vitamin B_6-linked enzyme thus far investigated in greatest detail with the aid of a broad array of modern experimental approaches pertaining to enzymology, molecular physics, and bioorganic and biophysical chemistry. That is why our discussion of essential general features—structure and chemical topography of the active center and the stereochemistry and mechanism of enzymic transamination—will be based chiefly on data relating to this enzyme and some isodynamic Asp-transaminases from other sources (Section III), in conjunction with information derived from recent work on congruent chemical models (Section II,A).

A variety of modified structural analogs and derivatives of the B_6 vitamins and their phosphate esters have been synthesized (*39–42*, etc.). The physical and catalytic (coenzyme-like and inhibitory) properties of such compounds—in solution and in complexes with specific B_6-linked enzyme proteins and appropriate ligands—have been studied in detail (see reviews *11, 41–44*). The results of these studies shed light on structural (steric and electronic) requirements for binding of the cofactor and for its participation in sequential steps of the enzymic reactions.

Many types of substrate analogs (quasi-substrates and active-site-directed inhibitors) were designed and used to explore complementary

39. J. M. Osbond, *Vitam. Horm.* (*New York*) **22**, 367 (1964).
40. W. Korytnik and B. Paul, in "Pyridoxal Catalysis: Enzymes and Model Systems" (E. E. Snell *et al.*, eds.), p. 615. Wiley (Interscience), New York, 1968.
41. D. B. McCormick and L. D. Wright, eds., "Methods in Enzymology," Vol. 18A. Academic Press, New York, 1970. Contributions by W. Korytnik *et al.* (pp. 475, 483, 489, 500, and 524), V. L. Florentiev, V. I. Ivanov, and M. Ya. Karpeisky (p. 567), S. Fukui *et al.* (pp. 598 and 603), and J. Wursch *et al.* (p. 606).
42. E. E. Snell, *Vitam. Horm.* (*New York*) **28**, 265 (1970).
43. A. E. Braunstein, in "Enzymes and Isoenzymes: Structure, Properties and Functions" (D. Shugar, ed.), p. 101. Academic Press, New York, 1970.
44. V. I. Ivanov and M. Ya. Karpeisky, *Advan. Enzymol.* **32**, 21 (1969).

aspects of the same problem, viz., the mechanism and topography of interactions of substrate molecules in the catalytic center with pyridoxal-P and the enzyme protein (stereochemistry of enzymic transamination) (see Sections III,B,6–8 and Refs. *13, 21, 42, 44–46*). Valuable information on the location and functions of essential groups of the apo- and coenzyme was obtained with the aid of chemical probes [selective modification and chromophoric or isotopic tagging of functional groups, trapping of transient intermediates, etc. (Sections III,B,3 and III,B,6)].

The macromolecular geometry (conformation and quaternary structure) of aminotransferases and, on the other hand, dynamic aspects of the molecular mechanism of enzymic transamination have been explored by means of modern physical and physicochemical approaches involving the use of a variety of experimental tools. These include diverse electro- and hydrodynamic methods; optical studies (spectrophotometry of absorption, luminescence, ORD, CD, in the visible and ultraviolet range); magnetic resonance spectroscopy (ESR and NMR); physical and chemical probes of changes in conformation and conformational stability; and investigation of catalytic mechanism by the study of steady state, equilibrium, and fast-reaction kinetics of enzyme action (Sections III,A and III,B,1,2,4).

B. FORMALLY SIMILAR PROCESSES OF NH_2 TRANSFER

As mentioned before, there exist besides transamination proper (i.e., the transfer of amino nitrogen to a carbonyl group, associated with oxidoreduction and catalyzed by vitamin B_6-linked enzymes) several more or less analogous processes of metabolic NH_2 transfer by single or two-step (sequential) enzymic reactions. Such processes should be clearly distinguished from true transamination with which they are sometimes confused, for example, an enzyme classified mistakenly in the EC List (*11*) as hexosephosphate aminotransferase, EC 2.6.1.16.

The following comments serve to emphasize the distinction between typical transamination and such NH_2 transfer mechanisms. They may fall into two categories, differing in overall reaction balance or in the nature of sequential steps and intermediates.

Reactions in the first category include amido N transfer reactions, re-

---

45. A. E. Braunstein, *Vitam. Horm.* (*New York*) **22**, 451 (1964).
46. E. S. Severin, G. K. Kovaleva, and L. P. Sashchenko, *Biokhimiya* **37**, 469 (1972).

viewed by Meister (47). Amido NH_2 of glutamine is utilized, often as a preferred substitute for ammonia, in the synthesis of amino sugars (enzymes EC 2.6.1.16 and EC 2.4.2.14), or in the synthesis—coupled with ATP or GTP cleavage and catalyzed by amido-ligases (EC 6.3.5 . . .)— of other amides, amidines, or cyclic guanidines, e.g., NAD (from deamidoNAD), asparagine (48), N-formylglycine-amidine ribonucleotide, and GMP.

Reactions in the first category also include metabolically important two-step processes wherein L-aspartic acid acts as the source of NH_2 for the formation of carboxamide, amidino, or guanidino groups in the synthesis of AICAR (aminoimidazole-carboxamide ribonucleotide), AMP, and arginine, respectively [see review by Ratner (49)].

The transfers proceed in two reversible sequential steps [Eq. (2)]. The nitrogen of aspartic acid is first condensed with a C atom of the acceptor molecule by a specific GTP or ATP splitting C–N ligase (EC 6.3.4 . . .) to an N-substituted aminosuccinic acid. In the second step, this intermediate is cleaved by a C–N lyase (EC 4.3.2 . . .) detaching the C_4 chain (derived from Asp) as fumarate and releasing the end product of N transfer:

$$\begin{array}{c}\text{COOH}\\|\\\text{CHNH}_2\\|\\\text{CH}_2\\|\\\text{COOH}\end{array} + \text{HO}-\text{C}\!\!<\quad\underset{\substack{\text{(AMP;GDP)}\\\text{amido-ligase}\\\text{(EC 6.3.4...)}}}{\overset{\text{(ATP;GTP)}}{\xrightarrow{\hspace{2cm}}}}\quad\begin{array}{c}\text{COOH}\\|\\\text{CH}-\text{NH}-\text{C}\!\!<\\|\\\text{CH}_2\\|\\\text{COOH}\end{array}\quad\xrightarrow[\text{(EC 4.3.2...)}]{\text{C–N lyase}}\quad\begin{array}{c}\text{COOH}\\|\\\text{CH}\\||\\\text{CH}\\|\\\text{COOH}\end{array} + \text{H}_2\text{N}-\text{C}\!\!< \quad (2)$$

Recent additions to the list of such reactions are the synthesis of thiamine in *Escherichia coli* from aspartate and 4-oxythiamine via 4-succinothiamine (50), and a cycle of hydrolytic deamination of NAD (to NHD) followed by NAD resynthesis through NH_2 transfer from aspartate via succinoNAD. On the basis of indirect experimental evidence, the latter reaction is considered by Buniatian (51) as the key link in a pathway for ammonia formation from L-amino acids in mammalian tissues (see also Ref. 1).

47. A. Meister, "The Enzymes," 2nd ed., Vol. 6, p. 247, 1962.
48. M. K. Patterson, Jr. and G. R. Orr, *JBC* 243, 376 (1968).
49. S. Ratner, "The Enzymes," 2nd ed., Vol. 6, p. 495, 1962.
50. S. Fukui, N. Ohishi, S. Kishimoto, A. Tomizawa, and Y. Hamazima, *JBC* 240, 1315 (1963).
51. H. C. Buniatian, *in* "Handbook of Neurochemistry" (A. Lajtha, ed.), Vol. 3, Chapter 12, p. 339. Plenum, New York, 1970.

The second category includes reactions with overall equations identical to transamination but with different mechanisms.

Meister and co-workers (see Ref. *52*, p. 212), demonstrated reactions of coupled oxidative deamination and reductive amination [according to Eqs. (3)-(5)], catalyzed by L- or D-amino acid oxidases (EC 1.4.3.2-3), e.g.:

Phenylalanine + FAD-enzyme + H_2O ⇌ $FADH_2$-enzyme + phenylpyruvate + NH_3 (3)

$FADH_2$-enzyme + α-oxoisocaproate + NH_3 ⇌ FAD-enzyme + leucine + H_2O (4)

Sum: Phenylalanine + α-oxoisocaproate ⇌ phenylpyruvate + leucine (5)

The overall reaction [Eq. (5)] is equivalent to transamination, but it is *not an amino transfer* since intermediary release of ammonia is involved. The requirements for this mechanism in experimental models (*52*)—high concentration of NH_4^+ ions and strictly anaerobic conditions —make its physiological occurrence unlikely.

The process most closely analogous to true transamination is the unique, biologically important NH_2 transfer between L-lysine (ε-NH_2) and L-glutamate known as the "saccharopine pathway," occurring in yeasts and mammalian tissues (*53–56*). The process consists of two reversible steps [Eqs. (6) and (7)] (*54, 55*) catalyzed by different NAD- or NADP-requiring dehydrogenases (EC 1.5.1.7-9), see Ref. *25*.

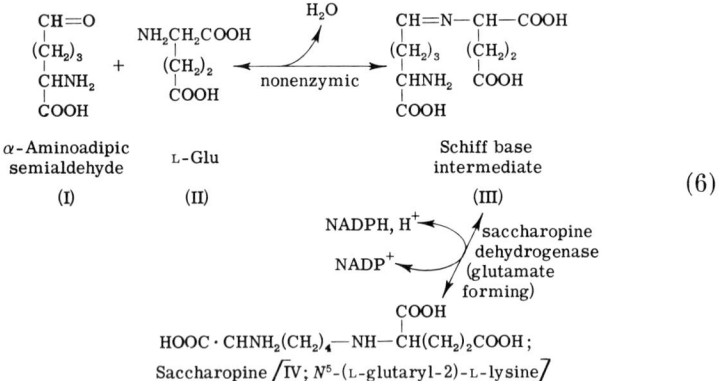

(6)

52. A. Meister, "The Enzymes," 2nd ed., Vol. 6, p. 193, 1962.
53. H. P. Broquist, "Methods in Enzymology," Vol. 17B, p. 124, 1971.
54. P. P. Saunders and H. P. Broquist, *JBC* **241**, 3435 (1966).
55. J. Hutzler and J. Dancis, *BBA* **158**, 62 (1968).
56. K. Higashino, K. Tsukada, and I. Lieberman, *BBRC* **20**, 285 (1965).

10. AMINO GROUP TRANSFER

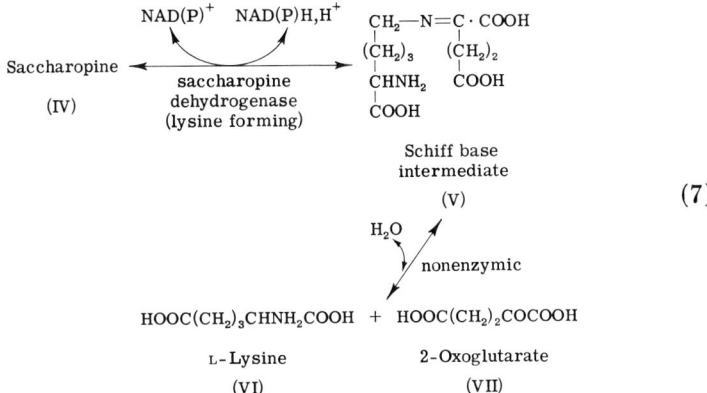

(7)

A classic pyridoxal-P-dependent transamination equivalent to the sum of reactions (6) plus (7) is actually catalyzed in bacteria (*Achromobacter* and *Flavobacterium* species) by an L-lysine:2-oxoglutarate 6-aminotransferase, EC 2.6.1.37 (*25*), see Ref. *57* and Section IV,C.

II. Basic Chemical Features of the Transamination Reaction

A. Congruent Nonenzymic Models

The well-known general theory of pyridoxal-dependent transformations of amino acids developed in 1952–54 by Braunstein, Snell, and their associates (*58–61*) need not be retold. Its basic principle, illustrated by the familiar schemes of the structure of aldimines (I) and tautomeric ketimines (II) formed from pyridoxal and amino acids (*13, 62*), is the weakening of bonds between the C^α atom of the amino acid and its substituents in these imines as a result of electron withdrawal from the α-carbon into the electron sink of the protonated pyridine ring:

57. K. Soda and H. Misono, *Biochemistry* **7**, 4110 (1968).
58. A. E. Braunstein and M. M. Shemyakin, *Biokhimiya* **18**, 393 (1953).
59. D. E. Metzler, M. Ikawa, and E. E. Snell, *JACS* **76**, 648 (1954).
60. E. E. Snell, *Vitam. Horm.* (*New York*) **16**, 77 (1958).
61. A. E. Braunstein, "The Enzymes," 2nd ed., Vol. 2, Part A, p. 113, 1960.
62. W. P. Jencks, "Catalysis in Chemistry and Enzymology," pp. 133–146. McGraw-Hill, New York, 1969.

$$\text{(I)} \rightleftharpoons \text{(II)} \tag{8}$$

Subsequent research lent ample support to the significance of these basic features for the functioning of B_6-linked enzymes and brought into light important additional aspects; for details see Refs. *13–16, 22, 43–45, 62–64*, etc.

The theory of Snell and Braunstein, as originally formulated, was actually a theory of catalytic potentialities of the coenzyme. In fact, it was largely based on studies concerning chemical reactions of amino acids in enzyme-free systems with various carbonyl compounds, e.g., 2-oxo acids, and especially pyridoxal, pyridoxal-P, or their analogs. Such model reactions were investigated in detail by many authors and are discussed in Refs. *12, 13, 17–21, 23, 43–45, 59–62, 64–69*. Their similarity to cognate enzymic reactions was striking. Yet considerable differences obviously existed between catalytic properties of the model systems [requirement for divalent metals as cofactors *(59, 60)*, simultaneous occurrence of reactions of several types, and broad tolerance for variation in structure of the carbonyl-containing catalysts] and those of pyridoxal-P-enzymes (no participation of metals, exceedingly fast reaction rates, high substrate and reaction specificities). It was obvious that the protein moieties played a decisive role in determining high catalytic efficiency and specificity of the enzymes, and the importance of studying the conformation of individual apoenzymes for elucidation of differences in the mode of binding and transformations of coenzyme and substrates was stressed years ago *(58, 60, 61)*.

When high purity preparations of aminotransferases and other B_6-

63. H. C. Dunathan, *Proc. Nat. Acad. Sci. U. S.* **55**, 712 (1966).
64. T. C. Bruice and S. J. Benkovic, "Bioorganic Mechanisms," Vol. 2, pp. 226–300. Benjamin, New York, 1966.
65. T. C. Bruice and R. M. Topping, *JACS* **85**, 1480 and 1488 (1963).
66. T. C. French, D. S. Auld, and T. C. Bruice, *Biochemistry* **4**, 77 (1965).
67. J. W. Thanassi, A. B. Butler, and T. C. Bruice, *Biochemistry* **4**, 1465 (1965).
68. D. S. Auld and T. C. Bruice, *JACS* **89**, 2083, 2090, and 2098 (1967).
69. P. Fasella, *in* "Pyridoxal Catalysis: Enzymes and Model Systems" (E. E. Snell *et al.*, eds.), p. 1. Wiley (Interscience), New York, 1968.

linked enzymes became available, new information rapidly accumulated about their functioning and structure. On this basis, more closely congruent models were devised, i.e., reaction schemes and nonenzymic systems realizing a given reaction through intermediates as similar as possible to those of the enzymic reaction.

Penetrating kinetic studies of the mechanism of model reactions between amino acids and pyridoxal or its analog meeting the minimum requirements for catalytic activity, 3-hydroxypyridine-4-aldehyde (*59, 60*), were made by Bruice and co-workers (*65–68*); cf. Refs. *13, 17, 18, 62, 64*. They showed (*64, 65*) that the rates of transamination reactions in such model systems are limited by the rate-determining prototropic rearrangement [aldimine (I) ⇌ ketimine (II), Eq. (8)] and can be partly enhanced by suitable buffers acting as general acid-base catalysts, e.g., imidazole-imidazolium or acetate buffers. In the presence of general bases the reaction can, under appropriate conditions, be fairly rapid in aqueous solution at 30°, requires no metal ions, and is directed quantitatively to NH_2 transfer, with suppression of pyridoxal-linked side reactions of other types. In studies relating to the initial reaction step—condensation of an amino acid (with NH_2 group nonprotonated) and the "minimal" coenzyme analog—individual rate constants for aldimine formation were about four orders higher with the cationic than with the anionic forms of 3-hydroxypyridine-4-aldehyde ($k_{cation} = 2.3 \times 10^6\ M^{-1}$ min^{-1}) (*66*). The estimated rate constant of this step in the case of Asp-transaminase is $\sim 2 \times 10^9\ M^{-1}$ min^{-1}. According to Bruice and Benkovic (*64, 70*), in nucleophilic attack on a carbonyl group a rate enhancement of 10^3–10^4 may be expected if this occurs in an intramolecular reaction in which the interacting groups are already constrained in a reactive orientation. If it is postulated that an adequate arrangement is provided in the active center and that the enzyme is able to sustain the reactants in their most reactive ionic forms, then values up to 10^9–$10^{10}\ M^{-1}$ min^{-1} seem reasonable for the rate constant of aldimine formation with the bound coenzyme in the cationic form (*44*).

Moreover, the coenzyme is known to exist in aminotransferases (and in pyridoxal enzymes generally) in the form of a Schiff base with the ϵ-NH_2 group of a lysine residue of the active site. This circumstance favors the very rapid aldimine bond formation between coenzyme and amino substrates in the catalytic centers (*13, 62*). As demonstrated by Jencks et al. (*71, 72*) in model experiments, an amino acid (with non-

70. T. C. Bruice and S. J. Benkovic, *JACS* **86**, 418 (1964).
71. E. H. Cordes and W. P. Jencks, *Biochemistry* **1**, 773 (1962).
72. W. P. Jencks and E. H. Cordes, in "Chemical and Biological Aspects of Pyridoxal Catalysis" (E. E. Snell et al., eds.), p. 57. Pergamon, Oxford, 1963.

protonated NH_2 group) will form a pyridoxylidene Schiff base much more rapidly by a reaction of carbonyl transfer from an imine of the coenzyme (transaldimination) than by condensation with the free aldehyde. Nucleophilic attack of the "donor" imine by the CO acceptor (substrate) is facilitated owing to the ease of protonation of Schiff bases ($pK_a \sim 7$) to highly reactive cationic imine intermediates. In the model systems of Bruice (68) and Jencks (72) and, presumably, in the active site of pyridoxal-P enzymes (13, 21, 44), reactivity of the intermediate Schiff bases is enhanced by hydrogen bonding between the 3-OH group of the coenzyme and the imino nitrogen, endowing the latter with partially cationic properties (71, 72). Both sequential stages of the transaldimination step formation and decomposition of a transient tetrahedral adduct (see Fig. 18, stages 3→4→5) are considered to be facilitated in the H-bonded Schiff base intermediates (62, 72), and especially in the fully protonated cationic imine (44).

A further essential consequence of hydrogen bonding (or metal chelation) of the imino N atom is promotion of the next catalytic step—the prototropic aldimine-ketimine interconversion—in transamination (and in other pyridoxal-linked reactions). One function of the hydrogen (or metal) bridge between the phenolic hydroxyl and imino N atom is to stabilize the required coplanar geometry of intermediate Schiff bases. The second, no less important, effect is strong polarization of the bonds between this nitrogen and adjacent C atoms (C^a of the substrate and $C^{4'}$ of coenzyme) in the cationic imine species. This contributes to bridging of the gap existing in the conjugated π-electron system of the pyridoxal–amino acid imine (73, 74), and thus enhances the electron-withdrawing effect of the pyridine cycle.

Dunathan (63, 75) has pointed out a most important factor controlling the direction of pyridoxal-linked reactions, namely, the orientation of C^a substituents with respect to the conjugated π system of the coenzyme–amino acid aldimine. Release of one of the substituents from the amino acid α-carbon with extension of the conjugated π system of the anionic intermediate is an early step common to all pyridoxal-dependent enzyme reactions; it is accompanied by an important increase in delocalization energy of the π-electron system. The gain in resonance energy will facilitate bond breaking if the geometry of the transition state is such that the σ bond to be activated is situated in a plane perpendicular to that of the coenzyme-imine system.

73. A. M. Pérault, B. Pullman, and C. Valdemoro, *BBA* **46**, 555 (1961).
74. B. Pullman, *in* "Chemical and Biological Aspects of Pyridoxal Catalysis" (E. E. Snell *et al.*, eds.), p. 103. Pergamon, Oxford, 1963.
75. H. C. Dunathan, *Vitam. Horm. (New York)* **28**, 399 (1970).

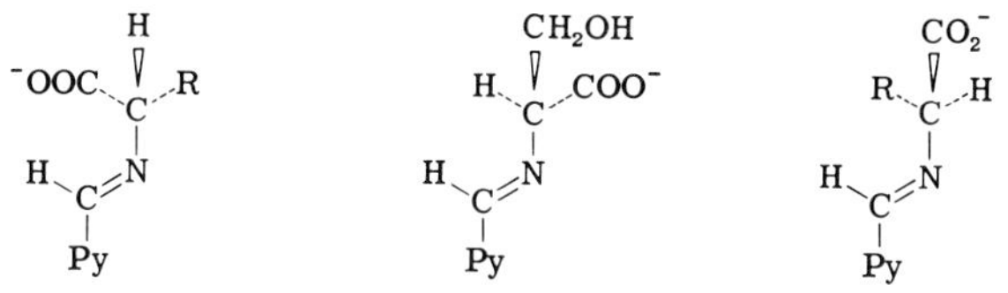

Fig. 1. Optimum conformations of PLP–amino acid aldimines for breaking of (a) the C^α–H bond (transamination, etc.); (b), the C^α–C^β bond (serine transhydroxymethylase), and (c) the C^α–COOH bond (α-decarboxylation). According to Dunathan (63).

Depending on the conformation around the C^α–N bond, the substrate bond subject to activation may be C^α–H, C^α–C^β, or C^α–CO_2H (Fig. 1a–c). The conformation in Fig. 1(c) is appropriate for α-decarboxylation, that in Fig. 1(b) is for release of the substrate's side chain, and that in Fig. 1(a) for transamination and other types of pyridoxal-linked reactions initiated by dissociation of the α-hydrogen.

The selective reaction specificity of pyridoxal enzymes is presumably determined largely by (a) the geometry of substrate binding, i.e., the position of anchoring sites for the α-carboxyl and other groups of the amino acid (63); (b) appropriate location of ionizable groups of the protein which act as acid-base catalysts, e.g., a proton-accepting group near the α-H atom in the case of transamination, and, of course, by (c) the particular steric and electronic structure of adequate substrate molecules.

The studies briefly considered in this section illustrate experimental and theoretical approaches which led to the design of increasingly congruent chemical models for enzymic transamination or, at the least, for sequential stages of this multistep reaction which require alternation of optimum conditions (44).

For further discussion and reaction schemes of mechanisms involved in chemical pyridoxal-linked transformations of amino acids and in cognate enzymic reactions see Refs. 13, 14–16, 21, 43, 62, 76 and other cited sources.

B. GENERAL CHARACTERISTICS OF INTERMEDIATE STEPS

In brief outline, the minimal series of sequential events in an enzymic transamination reaction (see Ref. 13, scheme 14, or Ref. 45, Fig. 1) includes the following:

76. H. C. Dunathan, *Advan. Enzymol.* **35**, 79 (1971).

1. Binding of amino substrate at its anchoring site (formation of the Michaelis complex).

2a. Condensation of bound amino substrate (with NH_2 group nonprotonated) and the internal pyridoxal–lysine imine (with protonated imino N) to a transient tetrahedral adduct (substrate aldimine).

2b. Release of lysine residue (nonprotonated) and enzyme-substrate aldimine (with protonated cationic imino N) from the tetrahedral adduct.

(Steps 2a and 2b constitute the transaldimination stage.)

3a. Removal of α-H atom from substrate aldimine (I) [see Eq. (8)] by a proton-accepting group (general base, B), yielding an intermediate with delocalized negative charge. The resonance states include the α-C^- carbanion of the aldimine (Ia), a negatively charged quinonoid structure with $\lambda_{max} \sim 490$ nm [(Ib), stage 6 in Fig. 17], and the 4'-C^- carbanion (IIa) of the tautomeric ketimine (II) [Eq. (9), cf. Ref. *62*]:

$$(\text{I}) \rightleftharpoons \begin{bmatrix} (\text{Ia}) & \leftrightarrow & (\text{Ib}) & \leftrightarrow & (\text{IIa}) \end{bmatrix} \rightleftharpoons (\text{II}) \quad (9)$$

[Occurrence of quinonoid intermediates of type (Ib) has been demonstrated in nonenzymic model systems (*68*), in enzymic transamination (in particular with slow reacting quasi-substrates, e.g., *erythro*-β-hydroxyaspartate, see Sections III,B,2 and 4) and in pyridoxal-linked reactions of other types (*14–16, 21, 44, 62*).]

3b. Protonation of carbanion (IIa) to ketimine (II) by the general acid, BH.

(Steps 3a and 3b constitute the prototropic aldimine–ketimine rearrangement—the rate-limiting key stage of transamination.)

4. Hydrolysis of (II), via a transient tetrahedral adduct with water (carbinolamine), yielding oxo acid and the pyridoxamine-P-protein (amino form of the enzyme).

5. Release of oxo acid from the anchoring site.

In the second half-reaction of transamination the same steps $\overrightarrow{(1-5)}$ proceed in reverse $\overleftarrow{(5-1)}$ with a different ketosubstrate molecule, oxo acid[2].

More detailed plausible versions of the reaction mechanism are discussed in Sections III,B,7 and 8.

III. Aspartate: 2-Oxoglutarate Aminotransferases

The presence of very active L-aspartate:2-oxoglutarate aminotransferases in organisms other than warm-blooded animals, e.g., in plants (77) and various microorganisms (78), has been demonstrated many years ago. Studies relating to the purification and properties of the enzyme from these sources are scanty and mostly outdated, with the exception of two somewhat better-characterized Asp-transaminases from cauliflower (79) and wheat germ (80). Transamination of D-aspartic acid in several species of *Bacilli*, investigated by Thorne and associates, is probably an activity of a D-stereospecific aminotransferase with relatively low amino donor selectivity (EC 2.6.1.21), obtained in high purity from *B. subtilis* (Section IV,A and Table XII).

This section is based entirely on studies of aspartate aminotransferases from the cytosol and mitochondria of mammalian and avian tissues.

A. Isoenzymes and Multiple Subforms

Aspartate:2-oxoglutarate aminotransferase (EC 2.6.1.1), like the alanine-oxoglutarate enzyme (7), is particularly active in heart muscle, and in most procedures for purification of the enzyme pig heart is used as a readily available source material. It was recognized early that crude extracts from minced myocardium, liver, or other animal tissues contain two major protein fractions having Asp-transaminase activity. They differ in electrophoretic mobility, in other physical, chemical, and functional characteristics, and in cellular localization (*32, 33, 36, 37, 69, 81–85*), (see Tables III and IV). The so-called anionic or supernatant Asp-transaminase is contained in the cytosol of animal cells, while the

77. D. G. Wilson, K. W. King, and R. H. Burris, *JBC* **208**, 863 (1954).
78. D. Rudman and A. Meister, *JBC* **200**, 591 (1953); B. N. Ames and B. L. Horecker, *ibid.* **220**, 113 (1956); cf. A. Meister (*2, 10*).
79. R. J. E. Ellis and D. D. Davies, *BJ* **78**, 615 and 623 (1961).
80. Z. H. M. Verjee, *Enzymologia* **37**, 110 (1969).
81. F. C. Decker and E. M. Rau, *Proc. Soc. Exp. Biol. Med.* **112**, 144 (1963).
82. J. W. Boyd, *BBA* **113**, 302 (1966).
82a. Y. Morino, H. Itoh, and H. Wada, *BBRC* **13**, 348 (1963).
83. J. S. Nisselbaum and O. Bodansky, *JBC* **239**, 4332 (1964); *ibid.* **241**, 2661 (1966).
84. M. Martinez-Carrion, C. Turano, E. Chiancone, F. Bossa, A. Giartosio, F. Riva, and P. Fasella, *JBC* **242**, 2397 (1967).
85. L. H. Bertland and N. O. Kaplan, *Biochemistry* **7**, 134 (1968).

cationic or mitochondrial enzyme, localized in the mitochondria, is released in soluble form upon disintegration of these organelles by freezing and thawing or by other means.

The isoelectric points of the two enzymes were estimated as ~ 5.0 (supernatant or cytosol enzyme) and ~ 6.8 (mitochondrial enzyme), respectively (32, 33, 86). Marked dissimilarities in primary structure, kinetic parameters, and immunochemical specificities (32, 33, 36, 37, 84) leave no doubt that Asp-transaminase$_{cyt}$ and Asp-transaminase$_{mit}$ are genetically distinct, true isoenzymes (25). The mitochondrial and cytosol isoenzymes have distinct metabolic functions and are, accordingly, controlled by different regulatory mechanisms (4, 5).

Aspartate transaminase prepared by the usual methods (28-31) contains only the cytosol enzyme, its mitochondrial counterpart being eliminated at the first purification steps. Extensively purified and crystallized Asp-transaminase$_{mit}$ has been isolated from heart and other organs of several animal species (27, 28, 32, 33, 37, 87); its physicochemical and functional characteristics are discussed below in comparison with those of Asp-transaminase$_{cyt}$ (86).

Preparations of both the cytosol and mitochondrial isoenzymes usually contain several molecular subforms differing in electrophoretic mobilities, catalytic activity, optical characteristics, and some other properties (38, 69, 84, 85, 87). From the cytosol Asp-transaminase of pig heart, Martinez-Carrion et al. (84) isolated no less than three main discrete protein subfractions, designated as alpha, beta, and gamma forms in the order of increasing anodic mobility. Specific catalytic activity is highest in the α, lower in β, and lowest in the γ (and δ) form. Marino et al. (88) achieved satisfactory separation of five fractions by ampholine electrofocusing: fraction A (corresponding to subform α) had its isoelectric point (pI) at 5.68 and specific activity = 1500 ± 100, and fraction B (subform β) had pI ~ 5.53 and specific activity 900-1370; fractions D and E (incompletely separated γ and δ forms) had pI ~ 5.38 and specific activity from 490 to 740.

The subforms do not differ significantly in molecular size, primary structure, and immunoprecipitation behavior. The main visible difference

---

86. In rat brain, the major difference in composition of mitochondrial and cytosol isoenzyme consists in the much higher amide-N content of the former (86a), which accounts, at least in part, for its cationic properties (see Section III,B,3 and Table V).

86a. S. C. Magee and A. T. Phillips, *Biochemistry* 10, 3397 (1971).

87. L. H. Bertland and N. O. Kaplan, *Biochemistry* 9, 2653 (1970).

88. G. Marino, M. de Rosa, V. Buonocore, and V. Scardi, *FEBS Lett.* 5, 347 (1969).

between them is in the mode of binding of coenzyme to the enzyme protein (84). In the α and β forms pyridoxal-P is bound mostly in the catalytically active mode, with absorption peaks at 362 or 430 nm (depending on pH) which shift to 330 nm on transamination with amino substrate (conversion to pyridoxamine-P). This fraction of the coenzyme can be reversibly resolved from the protein and is stably attached to it on reduction of the internal Schiff base (see Sections I,B and III,B,3) with sodium borohydride. Some of the coenzyme in the β form and most of it in the γ form is bound in a nonactive mode, characterized by λ_{max} 340 nm and incapacity to react with amino substrate or with sodium borohydride. This part of the coenzyme cannot be resolved from the apoenzyme by conventional procedures, but it is released as pyridoxal-P on denaturation of the protein, e.g., in alkaline solution.

The inert part of coenzyme (with λ_{max} 340 nm) is presumably bound as a substituted aldamine (as in phosphorylase) rather than in the normal unsubstituted lysine–aldimine form (84), see Sections III,B,3 and 8; in contrast to the latter, the 340-nm chromophore is optically inactive (cf. Section III,B,2). Decay in catalytic activity of the α and β subforms on prolonged aging or on brief exposure to concentrated urea is accompanied by partial transition of the coenzyme to the nonactive mode of binding. This transition, however, occurs without alteration of electrophoretic individuality of the subforms, which is preserved also after resolution of the α and β forms and after brief treatment of the apoenzymes with 8 M urea.

Prolonged (several days') exposure of the subforms or of the mixed apoenzymes to concentrated urea increases and equalizes their anodic mobilities, all subforms being gradually transformed into a species showing some resemblance to the γ form (89). Freshly resolved subforms α and β differ in the ease and mode of recombination with coenzyme. These facts and other features (spectra of absorption, fluorescence, and rotatory dispersion, etc.) suggest that the difference between the artificially produced forms may be based on conformational changes of the specific protein; differences in conformation possibly also exist between the natural subforms (see, however, Refs. 87 and 92).

Conventional high purity preparations of Asp-transaminase$_{cyt}$ usually display (in the presence of 2-oxoglutarate) low $OD_{340}:OD_{430}$ ratios, indicating predominance of the α form; this is the result of elimination of less active (and more acidic) subforms in fractionation steps involving ion exchange or electrophoresis (see below). Such preparations are satis-

89. B. E. C. Banks, S. Doonan, and C. A. Vernon, in "Pyridoxal Catalysis: Enzymes and Model Systems" (E. E. Snell et al., eds.), p. 305. Wiley (Interscience), New York, 1968.

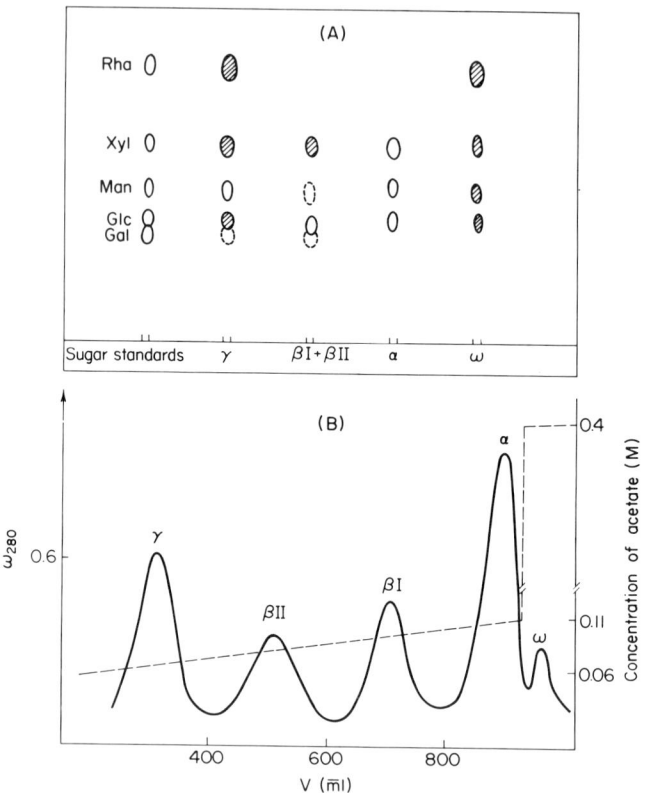

Fig. 2. Carbohydrate components in aspartate transaminase. After Polyanovsky and Denisova (*92*). (A) Scheme of paper chromatographic partition of neutral sugars in acid hydrolysates (3 N HCl; 3 hr; 100°) of subfractions of transaminase eluted from CM-Sephadex column. Separation in solvent mixture pyridine–isopentanol–water (1:1:0.8); sprayed with ammoniacal silver nitrate. (B) Fractionation of Asp-transaminase$_{cyt}$ on a column of CM-Sephadex A-50. Elution with a concentration gradient of sodium acetate (pH 5.4).

factory for most chemical or spectroscopic investigations; homogeneous preparations of the α subform are currently being used for precise quantitative measurements of kinetic and optical parameters of the enzyme.

Essentially similar data have been reported in recent papers devoted to the separation and properties of multiple subforms from the Asp-transaminase isoenzymes of chicken heart (*87*) and the mitochondrial isoenzyme of pig heart (*38, 90*).

More recently, Fasella et al. (*91*) detected additional subforms in pig

90. C. Michuda and M. Martinez-Carrion, *Biochemistry* **8**, 1095 (1969).
91. F. Bossa, R. A. John, D. Barra, and P. Fasella, *FEBS Lett.* **2**, 115 (1969).

heart Asp-transaminase fractionated by cation exchange chromatography, and characterized them as glycoproteins with unequal carbohydrate content (up to 10%) and variable catalytic activity (as high as 0.7 of the activity of α form, in a fraction strongly retained on the columns).

Polyanovsky and Denisova (92) studied the composition and relative amounts of carbohydrate constituents in subfractions obtained by slow gradient elution on a column of CM-Sephadex A-50 from a conventional high purity preparation of Asp-transaminase$_{cyt}$. The most acidic and least active subforms (δ and γ) of Martinez-Carrion et al. (84) were evidently eliminated in the preceding purification steps since the most anionic ("γ") fraction obtained in this laboratory is still nearly half as active as fraction "α" and has about one-half its absorbancy at 430 nm; thus, the fractions in Fig. 2 and Table I from Ref. 92 do not quite correspond to those designated by the same Greek letters in Refs. 84, 91. As seen in Fig. 2B, fraction ω is eluted from the cation exchange col-

TABLE I
Characterization of Asp-Transaminase$_{cyt}$ Subfractions[a]

Subfractions (see Fig. 2)	"γ"	"$\beta^I + \beta^{II}$"		α	ω
Specific activity[b]	18,000	I: 33,000	II: 26,000	40,000	10,000
E_{340}/mg	0.093	I: 0.040	II: 0.080	0.040	0.080
E_{430}/mg	0.080	I: 0.150	II: 0.120	0.170	0.100
Glucose	2.1 ± 0.7	4.9 ± 0.7		4.5 ± 1.0	7.1 ± 0.8
Mannose + Xylose + Rhamnose + Galactose[c]	3.2 ± 0.45	0.8 ± 0.45		0.4 ± 0.4	2.7 ± 1.0
Amino sugars[d]	0.5	—		—	1.0
Sialic acids[e]	2.6	0.25		0.17	0.1
E_{580}/mg in carbazole reaction[f]	1.2	1.8		1.0	8.5

[a] According to Polyanovsky and Denisova (92). Averaged from three fractionations.

[b] Oxalacetate formation, OD$_{280}$ milliunits/min/mg.

[c] Neutral sugars determined by the orcinol method [J. Brückner, BJ **60**, 200 (1955)] after acid hydrolysis of protein fraction (3 N HCl; 3 hr; 100°) and removal of peptides and amino acids on Dowex 50 × 2 (H$^+$ form); sugar content calculated as number of residues per subunit (45,000).

[d] Amino sugars (residues per subunit) determined according to C. Cessi and F. Piliego [Biochem. J. **77**, 508 (1960)].

[e] Sialic acids estimated by the resorcinol method of M. M. Svennerholm [(BBA **24**, 604 (1957)]; number of estimated residues per subunit.

[f] Relative optical densities at λ_{max} = 580 nm (α fraction = 1.0) in carbazole reaction [Z. Dische and C. Rothschild, Anal. Biochem. **21**, 125 (1967)].

92. O. L. Polyanovsky and G. F. Denisova, FEBS Lett. (1973) (in press).

umn only at elevated concentration of the eluant buffer. The diagram in Fig. 2A shows schematically the differences in paper chromatographic distribution of neutral sugars detected in acid hydrolysates of the transaminase subfractions. Approximate quantitative determinations of the sugars and estimations of sialic acids and amino sugars were done by means of conventional colorimetric procedures. As shown in Table I, the acidic properties of the fractions tested correlate remarkably well with their sialic acid contents, while the total sugar content increases in the sequence $\alpha < \beta < \gamma < \omega$, i.e., in the order of decreasing enzymic activity. Under quite mild conditions—in weakly acid (pH < 5) or weakly alkaline (pH > 8.5) solution—the carbohydrate components are gradually released from the protein fractions, indicating that the carbohydrate (or most of it) is attached by labile covalent bonds, presumably O-glycosidic bonds with hydroxyamino acid residues.

These results suggest (92) that the multiplicity of subforms present in Asp-transaminase preparations is determined by their qualitatively and quantitatively differing content of carbohydrate components and, in particular, of sugars containing ionogenic groups. Since the α subform has the lowest carbohydrate content, it is also evident that the artifactual "conformers" arising on its aging (storage) cannot be identical with the sugar-rich subforms (β, γ, and ω) obtainable on fractionation of fresh Asp-transaminase.

B. Aspartate Transaminase$_{cyt}$ from Pig Heart and Related Forms from Animal Tissues

1. *Physical Parameters and Macromolecular Structure*

Estimates of the molecular weight of aspartate transaminase from pig heart cytosol have varied widely, ranging from 116,000–120,000 (28, 33) to 75,000–76,000 (31, 89, 93). More recently, concordant values centering around 90,000 (88,000–93,000) have been derived independently from data relating to hydrodynamic characteristics and low-angle X-ray scattering (94, 95), content of bound coenzyme (35, 69), primary structure (96), gel filtration (97), osmometry (98), and dichrographic titration of

93. B. E. C. Banks, S. Doonan, J. Gauldie, A. J. Lawrence, and C. A. Vernon, *Eur. J. Biochem.* **6**, 507 (1968).
94. C. Michuda and M. Martinez-Carrion, *JBC* **244**, 5920 (1969).
95. O. L. Polyanovsky, Yu. M. Zagyansky, and L. A. Tumerman, *Mol. Biol. (USSR)* **4**, 458 (1970).
96. Yu. A. Ovchinnikov, A. A. Kiryushin, C. A. Egorov, N. G. Abdulaev, A. P. Kiselev, N. N. Modyanov, E. V. Grishin, A. P. Sukhikh, E. I. Vinogradova, M. Yu.

the apoenzyme with pyridoxal-P or its analogs (*99*). The correct MW of the dimeric holoenzyme, calculated from the complete peptide sequence recently established by Ovchinnikov, Braunstein and their associates (*99a*), is 93,250; cf. Section III,B,3, Table V, and Fig. 7.

From hydrodynamic and low-angle X-ray scattering data, Esipova *et al.* (*100*) calculated the parameters given in Table II, which also contains data from other sources for comparison.

One molecule of pure Asp-transaminase contains two molecules of coenzyme stably bound in specific aldimine linkage to lysine residues of the apoenzyme (*2*). The amino (or pyridoxamine-P) form of Asp-transaminase is resolved much more readily, and recombination of the apoenzyme is slower with pyridoxamine-P than with pyridoxal-P (*101, 102*). Phosphate ions facilitate resolution and inhibit reconstitution of the holoenzyme, evidently by competition with the coenzyme for a cationic binding site in the active center (*45, 102, 103*). Accordingly, most procedures for preparation of Asp-apotransaminase are based on conversion of the holoenzyme either to the amino form by treatment with amino substrate (L-aspartate or L-cysteinesulfinate) or to some other derivative with no covalent bond between coenzyme and protein (e.g., an oxime or hydrazone of the holoenzyme), followed by dialysis against phosphate buffer or by gel filtration (*38, 103–106*).

As indicated by coenzyme content (*28*) and primary structure data

---

Feigina, N. A. Aldanova, V. M. Lipkin, A. E. Braunstein, O. L. Polyanovsky, and V. V. Nosikov, *FEBS Lett.* **17**, 133 (1971). Cf. Ref. *96a*.

96a. Analogous data on partial resolution of the primary structure of Asp-transaminase$_{cyt}$ were reported by S. Doonan, M. J. Doonan, F. Riva, C. A. Vernon, J. M. Walker, F. Bossa, M. Carloni, and P. Fasella, *BJ* **130**, 443, 1972.

97. J. G. Farrelly and J. E. Churchich, *BBA* **167**, 280 (1968); J. E. Churchich and J. G. Farrelly, *JBC* **244**, 72 (1969).

98. N. Feliss and M. Martinez-Carrion, *BBRC* **40**, 932 (1970).

99. O. K. Mamaeva, E. S. Dementieva, V. I. Ivanov, M. Ya. Karpeisky, and V. L. Florentiev, *Mol. Biol. (USSR)* **4**, 762 (1970).

99a. Yu. A. Ovchinnikov, A. E. Braunstein, C. A. Egorov, O. L. Polyanovsky, N. A. Aldanova, M. Yu. Feigina, V. M. Lipkin, N. G. Abdulaev, E. V. Grishin, A. P. Kiselev, N. N. Modyanov, and V. V. Nosikov, *Dokl. Acad. Nauk SSSR* **207**, 728 (1972); also Yu. A. Ovchinnikov *et al., FEBS Lett.* **29**, 31 (1973).

100. N. G. Esipova, A. T. Dembo, V. G. Tumanyan, and O. L. Polyanovsky, *Mol. Biol. (USSR)* **2**, 527 (1968).

101. A. Meister, H. A. Sober, and E. A. Peterson, *JBC* **206**, 89 (1954).

102. Yu. M. Torchinsky, *Biokhimiya* **28**, 731 (1963).

103. V. Scardi, P. Scotto, M. Iaccarino, and E. Scarano, *BJ* **88**, 172 (1963).

104. H. Wada and E. E. Snell, *JBC* **237**, 127 and 133 (1962).

105. W. T. Jenkins and L. D'Ari, *BBRC* **22**, 376 (1966).

106. M. Arrio-Dupont, *BBRC* **36**, 306 (1969).

TABLE II

Physicochemical Properties and Molecular Parameters of
Asp-Transaminase from Pig Heart Cytosol
(6 mg/ml; pH 5.5; 20°)

Constant or parameter	Esipova et al. (100)	Other data	Ref.
Sedimentation constant, $s^0_{20,w}$	5.7 S	5.69 S	89
		5.60 S	69
		2.5 (pH < 3; 20°)	95
		2.8 (pH ⩾ 11; 4°)	95
Diffusion constant, $d^0_{20,w}$	6×10^{-7}	6.75×10^{-7}	89
(cm² sec⁻¹)		4.9×10^{-7} (at pH < 3)	95
Length of axes of approximating ellipsoid	$a = 50$ Å $b = 125$ Å		
Axial ratio a/b	1:2.5	1:6 (pH 11.5; 4°)	95
Hydration of molecule	0.30 g/1 g protein		
Dissymmetry coefficient:			
Of hydrated molecule f/f_0	1.230	1.5 (pH 11.5; 4°)	95
Of nonhydrated molecule f_e/f_0	1.122		
Molecular weight:			
From $s^0_{20,w}$ and $d^0_{20,w}$	87,000	77,000	89
From $s^0_{20,w}$ and f_e/f_0	90,000	94,000	69
		93,000 ± 2,000 (from osmometric data)	98
Volume of molecule (spherical approximation):			
From $s^0_{20,w}$ and $d^0_{20,w}$	106,500 Å³	107,000 Å³ (from fluorescence polarization data)	110
From low-angle scattering	168,000 Å³		
Surface-to-volume ratio	0.18 Å⁻¹		

[N- and C-terminal residues (*32, 33, 107*), peptide maps of tryptic digests (*37, 108*)], supernatant Asp-transaminase (*108*) and the mitochondrial isoenzyme (*37*) are both dimers, each consisting of two identical protein subunits (*37, 109, 110*).

Aspartate transaminase$_{cyt}$ is irreversibly disaggregated to denatured monomers on treatment with sodium dodecyl sulfate (*111*) or at extreme

107. Yu. M. Torchinsky, L. V. Abaturov, Ya. M. Varshavsky, and R. S. Nezlin, *Mol. Biol. (USSR)* **1**, 603 (1967).
108. O. L. Polyanovsky and N. E. Vorotnitskaya, *Biokhimiya* **30**, 619, 1965.
109. O. L. Polyanovsky, *BBRC* **19**, 364 (1965).
110. O. L. Polyanovsky, *in* "Pyridoxal Catalysis: Enzymes and Model Systems" (E. E. Snell *et al.*, eds.), p. 155. Wiley (Interscience), New York, 1968.
111. Yu. M. Torchinsky and V. O. Shpikiter, *Dokl. Akad. Nauk SSSR* **152**, 751 (1963).

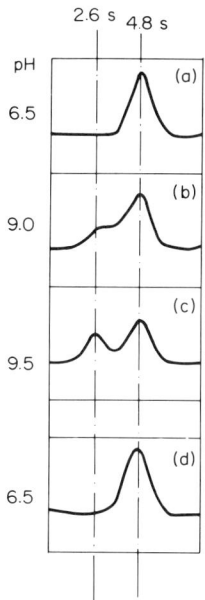

Fig. 3. Effect of pH on sedimentation pattern of succinylated Asp-transaminase (5 mg/ml; Spinco, model E, ultracentrifuge). (a)-(c) pH 6.5 → 9.4; (d) readjusted to pH 6.5. According to Polyanovsky (110).

pH values [>10 or <3] (112). Upon acylation of the enzyme with succinic anhydride, reversible dissociation of the succinylated transaminase (specific activity ~60% of original value) was achieved (109, 110). Owing to the increased repulsion between negatively charged groups, significant dissociation of the succinylated enzyme (protein concentration, 0.6–0.7%) can be observed ultracentrifugally at pH 8–9 (Fig. 3). Dissociation near neutrality will occur in 2 M urea and is suppressed at high ionic strength (in 2 M NaCl); this indicates the important role of hydrophobic interactions in stabilization of the dimeric structure. With the succinylated enzyme, the ratio of dimer (s_{obs} = 4.85 S) to monomer (s_{obs} = 2.68 S) decreases in the following sequence (110), which is also

PLP-enzyme > PMP-enzyme > oxime of PLP-enzyme > apoenzyme

valid for decrease in strength of coenzyme-protein linkage, lowering of conformational stability, and loosening of tertiary structure in the non-modified enzyme (107, 113).

112. O. L. Polyanovsky and V. O. Shpikiter, Dokl. Akad. Nauk SSSR 163, 1011 (1965).
113. L. V. Abaturov, O. L. Polyanovsky, Yu. M. Torchinsky, and Ya. M. Varshavsky, in "Pyridoxal Catalysis: Enzymes and Model Systems" (E. E. Snell et al., eds.), p. 171. Wiley (Interscience), New York, 1968.

Dissociation of intact Asp-transaminase at high dilutions (2.5–250 µg/ml, pH 8, $T = 20°$) was detected by measurements of fluorescence polarization of both the natural coenzyme chromophore and a covalently bound fluorescent dye marker (*95, 114*). The estimated dissociation constant, K_{eq}, was 0.6×10^7 at pH 8–9 (*110*). With a very sensitive sedimentation technique, Bertland and Kaplan (*85*) observed slight dissociation of the chicken heart enzyme (30 µg/ml) at pH 7.4 and substantial lowering of average molecular weight at pH 8 or 9. A moderate increase in specific activity of Asp-transaminase in very dilute solutions (*109, 115*) was attributed to 4–5-fold lowering of K_m values for amino and oxo substrate in the monomeric form of the enzyme (*110, 114*). Several authors (*31, 93, 98, 116*) have contested the data concerning dimer–monomer equilibrium of transaminase. Differences in experimental conditions constitute one source of the discrepancies; another resides in unforeseen complications rendering the interpretation of the fluorescence polarization data (*114*) unreliable, namely, the possible occurrence of independently rotating "nuclei" or "cores" in the subunit (*95*), and eventual fluorescence labeling of glycopeptide components of the enzyme (see above; cf. Ref. *92*).

Polyanovsky (*95*) has recently reconstituted native Asp-transaminase from subunits obtained at pH 11.5 in the cold, by way of neutralization with 8 M urea solution of pH 5 and dialysis in the presence of PLP.

Intact Asp-transaminase displays remarkable thermostability (*28, 117, 118*) and resistance to denaturing agents (e.g., 5 M urea) or to attack by proteinases. Studies of the kinetics of hydrogen–deuterium exchange (*107, 113*) revealed a high content (about 45% at pH 7.5 and 20°) of very slowly exchangeable peptide H atoms (considered to belong to both α helical and β structures); removal of the coenzyme lowers their content to 35–37% in the apoenzyme, and the destabilization of peptide H atoms is even larger in partially succinylated apotransaminase (Fig. 4). All criteria point to a compact, tight conformation of the enzyme molecule and to increasing destabilization of macromolecular (secondary, tertiary, and quaternary) structure along the sequence of forms shown above (*107, 110, 117, 118*). The enzyme contains no disulfide groups, and compact conformation of the protein subunit is mainly the result of hydrophobic interactions and hydrogen bonding.

114. O. L. Polyanovsky and V. I. Ivanov, *Biokhimiya* **29**, 728 (1964).
115. W. T. Jenkins and R. T. Taylor, *JBC* **240**, 2907 (1965).
116. J. E. Churchich, *BBA* **147**, 511 (1967).
117. O. L. Polyanovsky, Doctorate Thesis, Moscow, 1967.
118. O. L. Polyanovsky and V. Ya. Pikhelgas, *Dokl. Akad. Nauk SSSR* **171**, 1221 (1966).

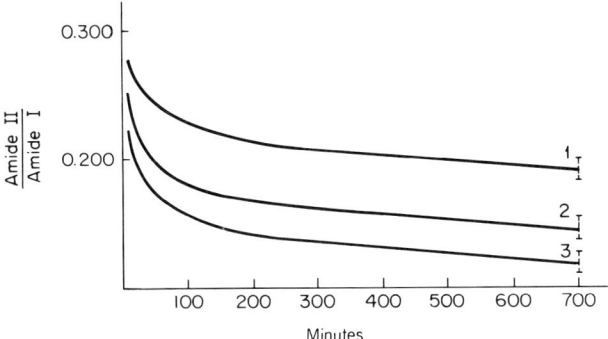

FIG. 4. Kinetics of hydrogen–deuterium exchange in aspartate transaminase protein (0.1 M phosphate buffer; 20°) (from Ref. *113*). Curves: 1, holoenzyme, pH 7.5; 2, apoenzyme, pH 7.5; 3, succinylated enzyme, pH 7.0.

Investigations of optical activity (ORD and CD parameters) have afforded some insight into conformational aspects of the structure of different molecular species and complexes of Asp-transaminase (*45, 119–127*).

As demonstrated first by Torchinsky and Koreneva (*119, 120*), apotransaminase and complexes of the holoenzyme with some carbonyl reagents (NH_2OH; NH_2-NH_2; HCN) have plain ORD curves in the 550–320-nm range, fitting one-term Drude equations. Almost all other forms have anomalies of ORD (induced Cotton effects) associated with absorption bands of the bound coenzyme. Early attempts (*121, 122*) to estimate α-helix content of Asp-transaminase from ORD parameters, calculated according to Yang and Doty or Moffit and Yang for species with plain ORD curves, gave unreliable results, although approximate parameters for the apoenzyme ($\lambda_c = 240$ nm; $a_0 = 180°$; $b_0 = -130°$) pointed to a relatively high content of ordered ($\alpha + \beta$) structure. For the oxime

119. Yu. M. Torchinsky and L. G. Koreneva, *Biokhimiya* **28**, 1087 (1963).

120. Yu. M. Torchinsky and L. G. Koreneva, *Biokhimiya* **29**, 780 (1964); *BBA* **79**, 426 (1964).

121. P. Fasella and G. G. Hammes, *Biochemistry* **3**, 530 (1964).

122. P. Fasella and G. G. Hammes, *Biochemistry* **4**, 801 (1965).

123. Yu. N. Breusov, V. I. Ivanov, M. Ya. Karpeisky, and Yu. V. Morosov, *BBA* **92**, 388 (1964).

124. V. I. Ivanov, Yu. N. Breusov, M. Ya. Karpeisky, and O. L. Polyanovsky, *Mol. Biol.* (*USSR*) **1**, 588 (1967).

125. Yu. M. Torchinsky, E. A. Malakhova, N. B. Livanova, and V. Ya. Pikhelgas, in "Pyridoxal Catalysis: Enzymes and Model Systems" (E. E. Snell *et al.*, eds.), p. 269. Wiley (Interscience), New York, 1968.

126. Yu. M. Torchinsky and G. A. Kogan, *Mol. Biol.* (*USSR*) **4**, 860 (1970).

127. C. Michuda and M. Martinez-Carrion, *JBC* **245**, 262 (1970).

TABLE III
Visible and Ultraviolet CD Data for Multiple Forms and Enzyme-Quasi-Substrate Complexes of Aspartate Transaminase Isoenzymes[a]

Enzyme form or complex	Substrate concn. (M)	pH	λ_{max} (nm)	$\Delta\epsilon/\epsilon \times 10^{-4}$ [b]	$[\Theta]$ [c]	$[\Theta']$ [d]
Cytosol Isoenzyme[e]						
Enz$_{PLP}$, α subform	None	8.5	365	19.2	122	
Enz$_{PLP}$, α subform	—	5.1	430	33.7	165	
Enz$_{PLP}$, α subform	—	5.1	221	—	−12,800	−9,760
Enz$_{PLP}$, α subform	—	5.1	207	—	−14,800	
Enz$_{PLP}$, α subform	—	5.1	193	—	23,150	
Enz$_{PLP}$, β subform	None	8.5	365	17.0	113	
Enz$_{PLP}$, β subform	—	5.1	430	32.0	100	
Enz$_{PLP}$, β subform	—	5.1	221	—	−13,400	−10,200
Enz$_{PLP}$, β subform	—	5.1	207	—	−15,000	
Enz$_{PLP}$, β subform	—	5.1	193	—	23,150	
Enz$_{PLP}$, γ subform	None	8.5	365	19.8	51.8	
Enz$_{PLP}$, γ subform	—	5.1	365	12.1	74	
Enz$_{PLP}$, γ subform	—	5.1	340	10.8	51.8	
Enz$_{PLP}$, γ subform	—	5.1	221	—	−13,400	−10,200
Enz$_{PLP}$, γ subform	—	5.1	207	—	−15,500	
Enz$_{PLP}$, γ subform	—	5.1	193	—	23,150	
apo-Enz, γ subform	None	5.1	340		14.5	
Mitochondrial Isoenzyme						
Enz$_{PLP}$, α subform	None	8.5	355	17.0		
Enz$_{PLP}$, α subform	—	5.1	436	32.7		
Enz$_{PLP}$, α subform	—	5.1	219	—	−15,700	−12,100
Enz$_{PLP}$, α subform	—	5.1	207	—	−16,700	
Enz$_{PLP}$, α subform	—	5.1	193	—	23,300	
Cytosol Isoenzyme, Nonfractionated						
Enz$_{PLP}$ + L-Asp	10^{-1}	8.9	(430)			
Enz$_{PLP}$ + L-Asp	10^{-1}	8.9	(365)			
Enz$_{PLP}$ + L-Asp	10^{-1}	8.9	336	5.9		
Enz$_{PLP}$ + L-Asp	10^{-1}	8.9	221	—	—	−9,760
Enz$_{PLP}$ + α-MeAsp(D,L)	10^{-1}	9.0	430	0	0	
Enz$_{PLP}$ + α-MeAsp(D,L)	10^{-1}	9.0	365	13.9		
Enz$_{PLP}$ + α-MeAsp(D,L)	10^{-1}	9.0	221	—	—	−9,760
Enz$_{PLP}$ + erythro-β-HO-Asp	7×10^{-3}	8.9	492	−5.8		
Enz$_{PLP}$ + erythro-β-HO-Asp	7×10^{-3}	8.9	336	7.15		

TABLE III (Continued)

Enzyme form or complex	Substrate concn. (M)	pH	λ_{max} (nm)	$\Delta\epsilon/\epsilon$ $\times 10^{-4}$ [b]	$[\Theta]$ [c]	$[\Theta']$ [d]
Enz$_{PLP}$ + erythro-β-HO-Asp	$7 + 10^{-3}$	8.9	221	—	—	−9,760
Enz$_{PMP}$	None	8.9	332	12.5	—	
Enz$_{PLP}$	None	8.9	221	—	—	−9,760
Mitochondrial Isoenzyme, Nonfractionated						
Enz$_{PLP}$ + α-MeAsp	10^{-1}	9.0	430	0	—	
Enz$_{PLP}$ + α-MeAsp	10^{-1}	9.0	368	8.4		
Enz$_{PLP}$ + α-MeAsp	10^{-1}	9.0	219	—	—	−12,000
Enz$_{PLP}$ + erythro-β-OH-Asp	4×10^{-2}	8.9	494	−4.84		
Enz$_{PLP}$ + erythro-β-OH-Asp	4×10^{-2}	8.9	335	10.5		
Enz$_{PLP}$ + erythro-β-OH-Asp	4×10^{-2}	8.9	219	—	—	−12,000
Enz$_{PMP}$	None	8.9	334	10.5	—	
Enz$_{PLP}$	None	8.9	219	—	—	−12,000

[a] Compiled from Martinez-Carrion et al. (38).
[b] $\Delta\epsilon/\epsilon \times 10^4$, Anisotropy factor; ϵ = absorbance at the given concentration; $\Delta\epsilon$ calculated from ellipticity, Θ, readings of Cary M-60 spectropolarimeter with CD attachment, using $\Delta\epsilon = \Theta/330$.
[c] $[\Theta]$, Molecular ellipticity values expressed in terms of mean residue weight = 114.
[d] $[\Theta']$, Molecular ellipticity values including refractive index correction.
[e] Subforms of Asp-transaminase$_{cyt}$ fractionated according to Ref. 84; fractionation of Asp-transaminase$_{mit}$, according to Refs. 37, 90.

and hydrazone of Asp-transaminase somewhat higher λ_c and b_0 values were estimated, indicating moderate changes in protein conformation on removal of the coenzyme (107, 126).

More recently, Torchinsky and Kogan (126) carefully recorded the ORD spectra of various forms of Asp-transaminase from 600 to 190 nm range, and Martinez-Carrion et al. (38) made a detailed comparative study of CD spectra of multiple subforms of the supernatant and mitochondrial isoenzymes in the visible and ultraviolet range.

Special attention was given to the "peptide" ORD and CD features in the far ultraviolet, characterizing electronic transitions in peptide groups, viz., the negative induced Cotton effects centering at 220 and 207 nm and the positive Cotton effects at 193–198 nm (38, 126, 128) (Table III). The negative ORD trough at 231 nm does not differ for the holoenzyme at pH 4.7 and 8.05 (no interference from pH-dependent long-wave

128. G. Holzwarth and P. Doty, JACS 87, 218 (1965).

Cotton effects in the coenzyme chromophore), the apoenzyme, the oxime of Asp-transaminase, or complexes with the quasi-substrates, α-methylaspartate and *erythro*-β-hydroxyaspartate (*126*). In all these species, as well as in subforms α, β, and γ of the PLP- and PMP-enzyme, Martinez-Carrion and co-workers observed no differences in parameters of the corresponding negative CD extremum at 221 nm (*38*). With the mitochondrial isoenzyme, the absolute CD parameters in the 500–219-nm range were slightly different, but equally invariant, in the far ultraviolet, for the different forms and complexes listed above (*38, 90*). These data testify to the absence, between the forms under study, of gross conformational transitions involving ordered (secondary) structures.

From the λ_c values in Drude equations fitting the plain ORD curves, Torchinsky (*107*) calculated (according to Yang and Doty) approximate α-helix contents amounting to 30% in the apoenzyme and 39% in the oxime and hydrazone of Asp-transaminase$_{cyt}$. By a procedure developed for the interpretation of contributions from α helix, β conformation, and random coil in CD spectra of polypeptides (*129*), Martinez-Carrion *et al.* (*38*) estimated as much as 37% α helix in Asp-transaminase$_{cyt}$ and 40% in mitochondrial enzyme, but little β conformation in either species. They added the judicious comment: "These values are, nevertheless, as absolute as the assumptions implicit in Greenfield and Fasman's work (*129*) prove to be."

In ORD spectra of Asp-transaminase (Fig. 5) showing identical troughs at 231 nm, the size of the positive peak at 198 nm is significantly higher for the holoenzyme (curve 1) and still higher for the oxime (curve 3) than for apoenzyme (curve 2); the Cotton effect at 198 nm could not be studied for complexes with quasi-substrates, showing strong absorption at this wavelength (*126*). The results indicate slight conformational changes in the enzyme protein, depending on its bonding with the coenzyme (*126*). The positive Cotton effect at 198 nm (reflecting π-π* transition in the amide group) is apparently more sensitive to such conformational change than is the negative Cotton effect at 231 nm (corresponding to $n \sim \pi^*$ transition). Torchinsky and Sinitsina (*130*) noted no change of the ORD curves of Asp-transaminase in the far ultraviolet region on alkylation of two nonessential –SH groups, and an increase in the positive Cotton effect at 198 nm upon mercaptidation of a third, functionally important thiol group; by other criteria, this latter modification was shown to cause moderate conformational alteration affecting the enzyme's binding site (see Section III,B,6).

129. N. Greenfield and G. D. Fasman, *Biochemistry* **8**, 4108 (1969).
130. Yu. M. Torchinsky and N. I. Sinitsina, *Mol. Biol.* (*USSR*) **4**, 256 (1970).

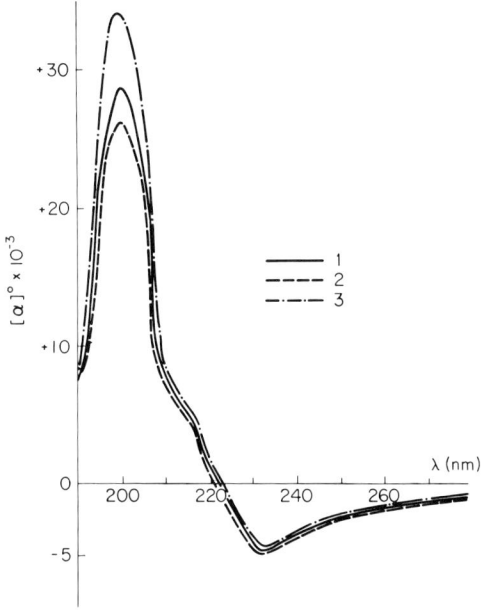

FIG. 5. ORD curves of Asp-transaminase in the 190–270-nm range (from Ref. 126). Curves: 1, solution of holoenzyme at pH 4.7 or 8.0; 2, solution of apoenzyme at pH 7.9; 3, same as 1, after addition of 0.02 M NH$_2$OH (pH 7.5).

2. Optical Properties of Aspartate Transaminase and of Its Complexes with Specific Ligands

Characteristic spectral features of the enzyme in the visible and near ultraviolet range result from the coenzyme chromophore. Individual forms of the transaminase, its substrate and inhibitor complexes differ in spectra of absorption, fluorescence emission, ORD and CD, depending on the chemical state and mode of binding of the coenzyme. Descriptions and tracings of the absorption spectra of purified Asp-transaminase first observed by Lis (29) and Jenkins (28) have been reported and discussed in many papers, e.g., Refs. 17, 18, 22, 37, 44, 84, 87, 94, 131, etc., see Table IV. Since the shape and exact position of spectral maxima is significantly affected by experimental conditions and minor contaminants, the importance of small deviations in published spectrophotometric curves for a given form of the enzyme should not be overrated.

Induced Cotton effects in coenzyme-linked absorption bands of Asp-transaminase and its specific-ligand complexes have been detected and studied in some detail by means of spectropolarimetry (119, 120) and—

131. P. Fasella, A. Giartosio, and G. G. Hammes, *Biochemistry* **5**, 197 (1966).

with higher sensitivity and better resolution of the peaks—by recording of CD spectra (*123–125*). Work done in the writer's laboratory (*99, 109, 123–125, 130*) and elsewhere (*38, 94, 121, 122*) has shown the value of differences in Cotton effects as a feature allowing distinction of derivatives or forms of Asp-transaminase (and other pyridoxal-P enzymes) with identical absorption spectra but dissimilar chemical structure, cf. Refs. *38, 43–45, 125*.

The pyridoxylidene form of Asp-transaminase, with the coenzyme bound to a lysine residue as an N^6-aldimine, has pH-indicator properties (*28*). The enzymically active form prevalent on the alkaline side of neutrality (at pH 7.5), with the characteristic absorption peak at 362 nm is converted on lowering of pH into an inactive yellow form with an absorption maximum at 430 nm, reaching full height near pH 5. The pH-dependent spectral shift is the result of binding of one proton by the coenzyme; the pK_a of this transition is 6.3 (under usual experimental conditions, cf. Sections III,B,4 and 6).

The absorption bands at 362 and 430 nm are optically active (*119, 125*), both are associated with positive Cotton effects induced by asymmetric interaction of pyridoxal-P with the enzyme protein (Fig. 6). On excitation at its absorption maximum (λ_{max}^A = 360 nm), the active unprotonated enzyme emits fluorescence with λ_{max}^F = 430 nm and low quantum yield, Q_{abs} = 2 × 10^{-3}; fluorescence of the inactive protonated form (λ_{max}^A = 430 nm) has its emission maximum at 520 ± 7 nm and much lower quantum yield, $Q_{abs} \sim$ 15 × 10^{-5} (*132*) (according to Ref. *133*, λ_{max}^F of this form is at 505 nm).

The amino or pyridoxamine-P form of Asp-transaminase produced on interaction of the enzyme with an excess of amino substrate has a pH-independent absorption maximum at 332 nm. Its relatively bright fluorescence, with λ_{max}^F = 395 nm and Q_{abs} = 4 × 10^{-4}, is markedly quenched in comparison to the fluorescence of free pyridoxamine-P—a feature attributed to hydrogen bonding between pyridine N of the coenzyme and a proton-donating group of the apoenzyme (*132, 134*). On the basis of comparisons with the absorption and fluorescence spectra of coenzyme derivatives (in model systems) with more or less well-defined ionic structures, Arrio-Dupont (*132*) tentatively assigned the following ionization states to the chromophore in the enzyme forms under study:

132. M. Arrio-Dupont, *Photochem. Photobiol.* **12**, 297 (1970).
133. Yu. N. Breusov, V. I. Ivanov, and M. Ya. Karpeisky, *Mol. Biol. (USSR)* **3**, 745 (1969).
133a. J. E. Churchich, *Biochemistry* **4**, 1405 (1965).
134. P. Fasella, C. Turano, A. Giartosio, and I. Hammadi, *G. Biochim.* **10**, 175 (1961).

Protonated, nonactive aldimine form (λ_{max}^A 430 nm): imino N protonated; phenolic hydroxyl ionized (hydrogen bonded?); pyridine N protonated.

Active aldimine form (λ_{max}^A 362 nm): imino N forming hydrogen bond with nonionized 3-hydroxyl; pyridine N either nonprotonated or cationic (protonated).

Amino form (λ_{max}^A 332 nm): phenolic group linked with imino N by chelate H bond; pyridine N protonated.

The validity of these assignments is, to some extent, debatable (cf. Section III,B,8).

In the amino form of the enzyme the absorption peak at 332 nm is associated with a positive Cotton effect, but its CD parameters are much lower than in the aldimine form at 362 and 430 nm, see Fig. 6 and Table IV.

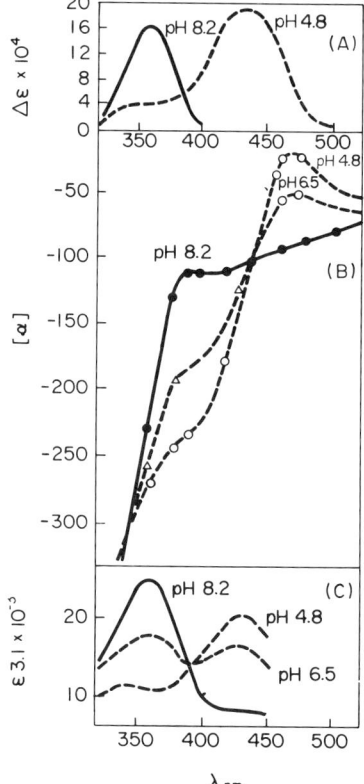

FIG. 6. (A) Circular dichroism; (B) optical rotatory dispersion, and (C) absorption spectra of the PLP form of Asp-transaminase at various values of pH (from Ref. 125).

TABLE IV
Parameters of Optically Active Absorption Bands of Aspartate Transaminase$_{\text{cyt}}$ and Its Complexes[a]

Enzyme form or complex	Ligand concn. (M)	pH	Absorption peaks		Optical activity parameters				Fluorescence parameters		
			$\lambda_{\max}$	ϵ	Anisotropy factor[b] $\Delta\epsilon/\epsilon \times 10^4$	Molar ellipticity[c] $[\Theta°]$	Reduced rotational strength[d] $[R]$	Ref.	Excitation λ^A (nm)	Emission λ^F (nm)	Ref.
Enz$_{\text{PLP}}$[e]	—	4.8	430	6700	27	59000	70.0	125	430	520 ± 7; 505 (very weak)	132, 133
Enz$_{\text{PLP}}$[e]	—	4.8	340[f]	2300	1.6	11900	11.3	125			
Enz$_{\text{PLP}}$[e]	—	8.2	362	8400	19	51000	58.5	125	360	430 (weak)	132, 133
Enz$_{\text{PMP}}$	—	4.8	333	8100	16	43900	33.4	125			
Enz$_{\text{PLP}}$	—	8.2	333	8250	17	45900	36.8	125			
Enz$_{\text{PLP}}$, NaBH$_4$-reduced	—	8.2	335	8200	16	44600	39.5	125			
Enz$_{\text{PLP}}$[e]	—	8.0	362	2400	28	—	—	123			
Enz$_{\text{PMP}}$	—	8.0	330	8500	14	—	—	123			
L-Cycloserine-inhibited Enz$_{\text{PLP}}$	10^{-3}	8.0	330	7600	7.2	—	—	123			
Enz$_{\text{PLP}}$[g]	—	4.6	430	—	33	—	89	141			
Enz$_{\text{PLP}}$	—	8.3	362	—	21.5	—	61	141			
Enz$_{\text{PLP}}$ + glutarate	2×10^{-1}	8.3	435	—	26	—	—	141			
Enz$_{\text{PLP}}$ + glutarate	2×10^{-1}	8.3	362	—	7	—	—	141			
Enz$_{\text{PLP}}$ + maleate	10^{-1}	8.3	435	—	26.5	—	—	141			

	Ionic strength	pH	λ_{max}			$[\Theta^\circ]$	Δ°	ϵ	Ref
Enz$_{PLP}$ + maleate	10^{-1}	8.3	362	—	—	12	—	—	141
Enz$_{PLP}$ + α-MeAsp	2×10^{-1}	8.3	430	392^h	330	0	—	—	141, 133a, 134
Enz$_{PLP}$ + α-MeAsp	2×10^{-1}	8.3	362	—	—	10	—	—	141
Enz$_{PMP}$	—	8.1	333	—	—	18	—	37	141
Enz$_{PMP}$ + glutarate	10^{-1}	8.1	333	—	—	16	—	34	141
Enz$_{PMP}$ + maleate	5×10^{-2}	8.1	333	—	—	9	—	16	141
Enz$_{PMP}$ + L-Asp	2×10^{-1}	8.1	333	—	—	15.5	—	33	141
Enz$_{PMP}$ + erythro-β-OH-Asp (D,L)	2×10^{-2}	8.1	333	—	—	13	—	27	141
Enz$_{PLP}$, NaBH$_4$-reduced	—	8.1	335	392^h	330	20	—	34	141, 133a, 134
Enz$_{PLP}$ + glutarate	2×10^{-1}	8.1	340	—	—	18	—	30	141
Enz$_{PLP}$ + maleate	2×10^{-1}	8.1	340	—	—	19	—	32	141

[a] Compiled from Refs. 123–125, 132–134, 142.
[b] ϵ = molar extinction coefficient at λ_{max}; $\Delta\epsilon$ (ellipticity) = $\epsilon_L - \epsilon_R$; optical anisotropy factor = $\Delta\epsilon/\epsilon \times 10^{-4}$.
[c] Molar ellipticity, $[\Theta^\circ] = 3300\Delta\epsilon$.
[d] Reduced rotational strength, $[R] = 0.75 \times 10^{-2} \pi [\Theta^\circ](\Delta^\circ/\lambda^\circ)$, where Δ° is the half-width of the CD band and λ° = the wavelength at the CD extremum.
[e] Purified Asp-transaminase$_{cyt}$ (30), mixture of subforms.
[f] The 340-nm peak disappears on removal of nonactive enzyme fractions (γ,δ) by chromatography on CM-Sephadex column, according to Ref. 35.
[g] α Subform of Asp-transaminase, purified according to Ref. 35.
[h] Binding to the apoenzyme strongly quenches the fluorescence of pyridoxamine-P (133a, 134), the quantum yield of fluorescence at 392 nm is reduced about 18-fold in the PMP-enzyme or the borohydride-reduced enzyme [Enz(H$_2$)]: $Q_{PMP} = 0.55$; $Q_{Enz(PMP)} = Q_{Enz(H_2)} = 0.03$ (133a). (According to Ref. 132, $Q_{Enz(PMP)}$ is 0.08.) The relative quantum yields of enzyme-bound pyridoxal-P in the active nonprotonated ($\lambda^A = 362$) and inactive protonated ($\lambda^A = 430$) forms are even lower.

pH-Independent maxima in the 330–340-nm range, with weak optical activity or none, are also observed in other derivatives of the enzyme having no double bond in conjugation to the coenzyme's pyridine ring, e.g., the PLP form after reduction with borohydride (Section III,B,3), the cyanide adduct (cyanohydrin), or other substituted aldamine species. Such a practically nondichroic peak near 340 nm (a shoulder in the amino form) is usually present in spectrophotometric tracings of purified Asp-aminotransferases; it belongs to the subform (γ) with coenzyme bound in the nonactive mode [Martinez-Carrion et al. (37, 38), see Section III,A].

The main absorption peaks of enzyme-bound Schiff bases of PLP with slow-reacting amino substrate analogs (external aldimines) have the same positions as those of the internal lysine-aldimine in the pyridoxylidene enzyme, viz., 362 and 430 nm. In contrast to the latter, the spectral peaks of the external enzyme–quasi-substrate aldimines are either nondichroic (at 430 nm) or with slight optical activity (at 362 nm).

Therefore, dichrography (or spectropolarimetry) is a convenient means for observation of the conversion of internal to external aldimines (the transaldimination step), e.g., in kinetic studies of interactions of Asp-transaminase with the inhibitory substrate analog, α-methylaspartate (107, 124), or competitive di- and monoanions. The oxo substrate, α-ketoglutarate, as well as maleate, glutarate, and some other nontransaminatable di- and monocarboxylic acids act as competitive inhibitors (28, 135). They enhance the proton affinity of the enzyme, shifting its pK_a from 6.3 to 7.8–8.0 or higher. The inhibited enzyme retains strong yellow color (increased absorption near 430 nm) at high pH values. The competitive anions are bound to those basic groups of the active center which bind the adequate substrates but do not affect the internal aldimine linkage.

Accordingly, the wavelengths (362 and 430 nm) of the absorption bands of Asp-transaminase and their ORD and CD parameters are not significantly altered in complexes of the enzyme with inhibitory anions (apart from the shift in pK_a) [(45, 125, 135–141), see Sections III,B,4 and 6].

When the competitive dicarboxylic amino acid, α-methylaspartate,

135. I. W. Sizer and W. T. Jenkins, in "Chemical and Biological Aspects of Pyridoxal Catalysis" (E. E. Snell et al., eds.), p. 123. Pergamon, Oxford, 1963.
136. W. T. Jenkins, JBC 239, 1742 (1964).
137. G. G. Hammes and P. Fasella, JACS 84, 4644 (1962).
138. G. G. Hammes and P. Fasella, JACS 85, 3929 (1963).
139. S. F. Velick and J. Vavra, JBC 237, 2109 (1962).
140. S. F. Velick and J. Vavra, "The Enzymes," 2nd ed., Vol. 6, p. 219, 1962.
141. Yu. M. Torchinsky, Biokhimiya 29, 534 (1964).

reacts with Asp-transaminase at pH 8.0, pK_a of the enzyme is similarly increased with lowering of the peak at 362 nm and appearance of the 430-nm maximum (the protonated species). However, the new 430 nm maximum of the transaminase-α-methylaspartate complex is optically inactive (see above). When added to the enzyme at pH 4.8, α-methylaspartate does not change the optical activity of the 430-nm peak (*119*), although the protonated enzyme is known to have increased affinity for dicarboxylic substrates and inhibitory anions. These findings showed that the binding of α-methylaspartate results in transaldimination to the "external" aldimine at pH 8.0 but not at pH 4.8 (*45, 119–125, 141, 142*), see Section III,B,6.

With certain quasi-substrates, in particular with *erythro*-β-hydroxy-L-aspartate (*136*), Asp-transaminase forms complexes showing an additional absorption peak near 490 nm associated with a strong negative Cotton effect (*44, 94, 120–125, 137, 138*); this peak is attributed to a quinonoid anionic intermediate of type (Ib) [Eq. (9)]. Spectrophotometric fast kinetic studies with transaminase and the substrate pair aspartate–oxalacetate at high concentrations indicate that normal transamination may proceed via short-lived quinonoid intermediates (see Sections III,B,6,*a* and 8).

According to Torchinsky and Koreneva (*119, 120, 125*) the characteristic absorption maxima of oximes and hydrazones of Asp-transaminase in the visible range are either optically inactive (in the case of small carbonyl reagents such as NH_2OH or NH_2–NH_2) or associated with negative Cotton effects [with CO reagents having bulky substituents, e.g., aminooxyacetate, aromatic hydrazines or hydrazides (*120, 125*)].

Karpeisky *et al.* (*44, 123, 124, 143*) detected a negative CD extremum in the near ultraviolet (at 295–300 nm). Its anisotropy parameters are largest in the protonated aldimine form of the enzyme, but it occurs also in the nonprotonated species and the amino form. Since free bipolar PLP-aldimines and pyridoxamine-P have no short-wave absorption peaks in this region, the CD band at 295–300 nm presumably belongs to the protein. It is absent in CD spectra of the apoenzyme and hence apparently induced in a chromophoric amino acid residue upon its association with the coenzyme. This CD band coincides in position and shape with the absorption maximum of tyrosine with ionized phenolic hydroxyl. In some enzyme derivatives, e.g., in complexes with certain coenzyme analogs, this negative Cotton effect is decreased with concomitant increase

142. Yu. M. Torchinsky and E. A. Malakhova, *Dokl. Akad. Nauk SSSR* **128**, 722 (1968).
143. A. L. Bocharov, V. I. Ivanov, M. Ya. Karpeisky, O. K. Mamaeva, and V. L. Florent'ev, *BBRC* **30**, 459 (1968).

TABLE V
AMINO ACID COMPOSITION OF ASPARTATE TRANSAMINASES OF DIFFERENT ORIGIN AND TERMINAL RESIDUES[a]

Amino acid	Pig heart (Refs. 37, 84, 90)		Pig heart (Ref. 99a)[c] Cytosol (unfractionated) MW 46,390	Beef heart (Ref. 148) Cytosol (unfractionated) MW 48,000	Rat brain[b] (Ref. 86a)		Pigeon breast muscle (Ref. 146) Cytosol (unfractionated) MW 45,000	Chicken heart (Ref. 87)	
	Mitochondria (unfractionated) MW 50,000	Cytosol (α subform) MW 47,000			Cytosol MW 40,000	Mitochondria (active subform II) MW 40,000		Cytosol (unfractionated) MW 50,000	Mitochondria MW 50,000
Ala	37	33	32	40	27	31	39	43	43
Arg	21	26	26	24	19	~18	23	24	22
Asx (or Asn + Asp)	40	42	Asn: 15 } 42 Asp: 27	44	39	33	45	56	50
Cys	5	5	5	5	3	6	4	[d]	[d]
Glx (or Gln + Glu)	43	45	Gln: 14 } 41 Glu: 27	50	33	37	33	48	46
Gly	36	28	28	32	26	33	31	41	39
His	12	8	8	8	6	8	8	8	13
Ile[e]	28	18	19	18	13	17	18	24	30
Leu[e]	30	40	38	36	31	29	32	41	40
Lys	31	20	19	21	18	25	21	22	31

10. AMINO GROUP TRANSFER

Met	12	6	6	6	9	11	12	15
Phe	20	23	23	22	18	22	26	24
Pro	19	24	24	32	16	20	23	20
Ser[f]	22	26	26	33	22	29	31	29
Thr[f]	16	24	25	25	17	28	30	27
Trp	10	10	9	9	6	10	9	7
Tyr	12	12	12	11	12	11	13	17
Val[e]	28	29	29	18	26	27	32	26
(NH$_3$)	d	d	d	(32)	(66)	(42)	d	d
Total[b]	422	420	412	434	387	417	483 + x[h]	479 + z[h]
N-Terminal residue	Ser (Refs. 36, 37)	Ala (Refs. 36, 85)	Ala	Ala (Ref. 148)	—	—	—	—
C-Terminal residue	Lys (Refs. 36, 37)	Gln (Ref. 99a)	Gln	Leu (Ref. 148)	—	—	—	—

[a] Number of residues calculated to nearest integer per subunit, see Ref. 148.
[b] Authors (147) calculated the number of residues per molecular weight 80,000 (dimer).
[c] Based on complete peptide sequence (99a).
[d] Not determined.
[e] In 72 hr hydrolysate.
[f] Extrapolated to 0 hr hydrolysis.
[g] See footnote 147.
[h] Here, "+x" and "+z" denote unknown numbers of Cys residues.

of a small positive CD peak near 275 nm, close to the $\lambda_{\max}^{A}$ of nonionized tyrosine. From these data and other evidence Karpeisky et al. (44) drew the inference that a tyrosine residue located in the active center may function as a proton-donating group, hydrogen bonded with the coenzyme in free Asp-transaminase and at some steps of the transamination reaction, and released in nonionized form at other stages of the catalytic cycle (cf. Sections III,B,3 and 8).

Circular dichroism curves in the 263–300-nm range of various forms and complexes of the Asp-transaminase isoenzymes have been recorded with higher resolution and described by Martinez-Carrion (127). Optically active bands in the far ultraviolet part of the spectra of various forms and complexes of the enzyme have been discussed above (Section III,B,1) in relation to three-dimensional structure.

3. Studies of the Primary Structure and Functionally Important Groups

Complete amino acid analyses of high purity preparations of pig heart Asp-transaminase$_{\text{cyt}}$ have been reported from several laboratories (86a, 144–147). More recently, Martinez-Carrion et al. (37, 84) have published comparative data on the amino acid composition of subforms of the supernatant and mitochondrial isoenzymes. The analyses revealed considerable differences in composition between the two isoenzymes but virtually none between all subforms of each. Therefore, only the data for nonfractionated Asp-transaminase$_{\text{mit}}$ and the α subform of the cytosol enzyme are presented in Table V, which also includes data on the composition of several Asp-transaminases from other biological sources (86a, 87, 144–148).

In chicken heart (87), the mitochondrial and cytosol transaminases differ much less in composition than in pig heart (37, 84); the percentage composition of the two avian isoenzymes is intermediate between those of the pig heart transaminases (147).

Identification of the N- and C-terminal residues (Table V) and com-

---

144. C. Turano, A. Giartosio, F. Riva, and P. Vecchini, in "Chemical and Biological Aspects of Pyridoxal Catalysis" (E. E. Snell et al., eds.), p. 149. Pergamon, Oxford, 1963.

145. Y. Morino and T. Watanabe, Biochemistry 8, 3397 (1971).

146. N. E. Vorotnitskaya. Science Candidate's Dissertation, Moscow, 1968.

147. Some inconsistencies can be noted in Table V between the total numbers of residues in each column and the respective molecular weights to which the analytical data were recalculated; there is better agreement in the relative (percentage) amino acid contents of the proteins.

148. G. Marino, V. Scardi, and R. Zito, BJ 99, 595 (1966).

parison of the peptide maps of tryptic digests confirmed marked dissimilarity in primary structure between the mitochondrial and cytosol coenzymes of pig heart (*37*). Differences between preparations of transaminase from different animal species are relatively less; thus, the pig and beef heart enzymes differed in the position of only 6 tryptic peptide spots among a total of 41. Only 50% of the spots in peptide maps of pig heart and pigeon muscle transaminases$_{cyt}$ had similar positions (*146*). Recent amino acid analyses and terminal residue determinations support earlier data (*144*), indicating that Asp-transaminase contains no disulfide bonds and has one single peptide chain in each of its two identical (*149, 150*) subunits.

Reduction of the "internal" Schiff base with sodium borohydride was used by Fischer *et al.* to label and identify the coenzyme-linked lysine residue (see below and Section III,B,6,c); from digests of the reduced enzyme they (*151*) and other authors (*146, 149, 152*) isolated peptide fragments of different length (4–16 residues) containing N^6-pyridoxyl-lysine, and reported tentative partial sequences (*149, 151, 152*) which all proved more or less incorrect. More recently, Morino and Watanabe (*145*) isolated from the Asp-transaminase isoenzymes two different tetrapeptide fragments, each including the coenzyme-linked lysine. The peptides had the following structures:

Mitochondrial enzyme: Ala–Lys(Pxy)–Asn–Met
Cytosol enzyme: Ser–Lys(Pxy)–Asn–Phe (cf. Ref. *152*)

The latter sequence recurs in the peptide sequence of Asp-transaminase resolved by Ovchinnikov *et al.* (*99a*) (Fig. 7, residues 257–260).

Several functional groups of the enzyme protein are considered to be important for catalytic activity or, at least, situated strategically in the active center or close to it (*22, 43, 69*). The experimental basis for identification of such groups is considered below, and further details concerning their position in the peptide chain and probable functions are presented in Sections III,B,6 and 8.

a. Lysine (ϵ-Amino Groups). As already mentioned, reduction with $NaBH_4$ of the postulated internal Schiff base in Asp-transaminases (and

149. N. E. Vorotnitskaya, V. A. Spivak, and O. L. Polyanovsky, *Biokhimiya* **33**, 375 (1968).
150. N. E. Vorotnitskaya, G. F. Lutovinova, and O. L. Polyanovsky, *in* "Pyridoxal Catalysis: Enzymes and Model Systems" (E. E. Snell *et al.*, eds.), p. 131. Wiley (Interscience), New York, 1968.
151. R. C. Hughes, W. T. Jenkins, and E. H. Fischer, *Proc. Nat. Acad. Sci. U. S.* **48**, 1615 (1962).
152. O. L. Polyanovsky and B. Keil, *Biokhimiya* **28**, 379 (1963).

```
                              10                                    20
Ala-Pro-Pro-Ser-Val-Phe-Ala-Glu-Val-Pro-Gln-Ala-Gln-Pro-Val-Leu-Val-Phe-Lys-Leu-Ile-Ala-Asp-Phe-Arg-
                  30                                    40                                    50
Glu-Asp-Pro-Asp-Pro-Arg-Lys-Val-Asn-Leu-Gly-Val-Gly-Ala-Tyr-Arg-Thr-Asp-Asp-Cys-Glu-Pro-Trp-Val-Leu-
                                    60                                    70
Pro-Val-Val-Arg-Lys-Val-Glu-Gln-Arg-Ile-Ala-Asn-Asp-Ser-Ser-Leu-Asn-His-Glu-Tyr-Leu-Pro-Ile-Leu-Gly-
                  80                                    90                                   100
Leu-Ala-Glu-Phe-Arg-Thr-Cys-Ala-Ser-Arg-Leu-Ala-Leu-Gly-Asp-Asp-Ser-Pro-Ala-Leu-Gln-Glu-Lys-Arg-Val-
                              110                                   120
Gly-Gly-Val-Gln-Ser-Leu-Gly-Gly-Thr-Gly-Ala-Leu-Arg-Ile-Gly-Ala-Glu-Phe-Leu-Ala-Arg-Trp-Tyr-Asn-Gly-
                 130                                   140                                   150
Thr-Asn-Asn-Lys-Asp-Thr-Pro-Val-Tyr-Val-Ser-Ser-Pro-Thr-Trp-Glu-Asn-His-Asp-Gly-Val-Phe-Thr-Thr-Ala-
                                   160                                   170
Gly-Phe-Lys-Asp-Ile-Arg-Ser-Tyr-Arg-Tyr-Trp-Asp-Thr-Glu-Lys-Arg-Gly-Leu-Asp-Leu-Gln-Gly-Phe-Leu-Ser-
                       180                                   190                                  200
Asp-Leu-Glu-Asx-Ala-Pro-Glx-Phe-Ser-Ile-Phe-Val-Ile-His-Ala-Cys-Ala-His-Asn-Pro-Thr-Gly-Thr-Asp-Pro-
                                  210                                  220
Thr-Pro-Glx-Glu-Trp-Lys-Gln-Ile-Ala-Ser-Val-Met-Lys-Arg-Arg-Phe-Leu-Phe-Pro-Phe-Phe-Asp-Ser-Ala-Tyr-
                      230                                  240                                  250
Gln-Gly-Phe-Ala-Ser-Gly-Asn-Leu-Glu-Lys-Asp-Ala-Trp-Ala-Ile-Arg-Tyr-Phe-Val-Ser-Glu-Gly-Phe-Glu-Leu-
                                   *260                                 270
Phe-Cys-Ala-Gln-Ser-Phe-Ser-Lys-Asn-Phe-Gly-Leu-Tyr-Asn-Glu-Arg-Val-Gly-Asn-Leu-Thr-Val-Val-Ala-Lys-
                        280                                  290                                 300
Glu-Pro-Asp-Ser-Ile-Leu-Arg-Val-Leu-Ser-Gln-Met-Glu-Lys-Ile-Val-Arg-Val-Thr-Trp-Ser-Asn-Pro-Pro-Ala-
                             310                                   320
Gln-Gly-Ala-Arg-Ile-Val-Ala-Arg-Thr-Leu-Ser-Asp-Pro-Glu-Leu-Phe-His-Glu-Trp-Thr-Gly-Asn-Val-Lys-Thr-
                        330                                  340                                 350
Met-Ala-Asp-Arg-Ile-Leu-Ser-Met-Arg-Ser-Glu-Leu-Arg-Ala-Arg-Leu-Glu-Ala-Leu-Lys-Thr-Pro-Gly-Thr-Trp-
                            360                                  370
Asn-His-Ile-Thr-Asp-Glu-Ile-Gly-Met-Phe-Ser-Phe-Thr-Gly-Leu-Asn-Pro-Lys-Gln-Val-Glu-Tyr-Leu-Ile-Asn-
                       380                                  390                                 400
Glu-Lys-His-Ile-Tyr-Leu-Leu-Pro-Ser-Gly-Arg-Ile-Asn-Met-Cys-Gly-Leu-Thr-Thr-Lys-Asn-Leu-Asp-Tyr-Val-
                           410
Ala-Thr-Ser-Ile-His-Glu-Ala-Val-Thr-Lys-Ile-Gln
```

FIG. 7. The primary structure of pig heart Asp-transaminase (*99a*) and location of functionally important amino acid residues.

The following are some major features of the peptide. Chain length, 412 residues; MW, 46,390 (apoenzyme subunit); N-terminal: Ala; C-terminal: Gln.

Lys-258: essential; forms aldimine bond with pyridoxal-P (see pp. 389, 417). Tyr-40: important for catalytically active conformation (*44, 161–164*); simultaneous modification with Cys-390 distorts active site (*166, 170a*) (see pp. 420, 422). Cys-390: semi-buried, structurally important (see above); blocking by thiol reagents lowers substrate affinity of the enzyme and causes considerable inactivation (see pp. 421–423 and references *161–170a*). Cys-45: exposed, functionally nonimportant; *N*-oxyl spin-label on its SH group responds to substrate-induced conformational changes in the active site (O. Polyanovsky *et al.*, unpublished).

other vitamin B_6-linked enzymes), followed by denaturation, digestion with proteinases, and isolation of fragments containing N^6-pyridoxyllysine [Lys(Pxy)] (*151, 152*) constituted the proof for linkage of the coenzyme by an aldimine bond to ϵ-NH_2 of a lysine residue of the active site (Lys-258, see also Sections III,B,2 and 6,*c*). Indirect evidence (*144*) suggested that the same ϵ-NH_2 group, upon its release in the transaldimination step, may function as proton acceptor (acid-base catalyst) in the aldimine–ketimine transition (*44, 75, 76, 153*), cf. Sections III,B,7, and 8. It is likely that other lysine (or arginine) residues of the enzyme provide cationic groups to bind the phosphate group of coenzyme and ω-carboxy groups of substrates.

b. Histidine (Imidazole Groups). Turano *et al.* (*22, 154, 155*) using selective alkylation by a bifunctional reagent (Section III,B,6,*c*), as well as Polyanovsky (*150*) and Martinez-Carrion (*156, 157*) with the aid of dye-sensitized photooxidation, independently demonstrated that blocking or destruction of 1 or 2 histidine residues (per subunit) leads to inactivation of Asp-transaminase. Kinetic analysis indicates that one imidazole group is essential (*156*). According to Martinez-Carrion (*157*), its function is to accept the α-hydrogen atom of amino substrates in the aldimine–ketimine interconversion. The essential lysine ϵ-NH_2 group is another likely candidate for this function, see above and Section III,B,8.) Martinez-Carrion (*158*) suggested that the same or a second imidazole group is the basic protein group which binds inhibitory anions competing at high pH values with a carboxylate group of substrates (the ω-carboxyl?—see Ref. *159* and Sections III,B,4 and 6,*b*). From the studies of Severin *et al.* (see Section III,B,6,*c*) on the interaction of transaminases with cycloserine derivatives, it seems more likely that an imidazole group participates in binding of the α-carboxyl of substrates.

153. E. E. Snell, *in* "Chemical and Biological Aspects of Pyridoxal Catalysis" (E. E. Snell *et al.*, eds.), p. 1. Pergamon, Oxford, 1963.

154. C. Turano, A. Giartosio, F. Riva, D. Barra, and F. Bossa, *in* "Pyridoxal Catalysis: Enzymes and Model Systems" (E. E. Snell *et al.*, eds.), p. 27. Wiley (Interscience), New York, 1968.

155. C. Turano, C. Borri, A. Orlacchio, and F. Rossa, *in* "Enzymes and Isoenzymes: Structure Properties and Functions" (D. Shugar, ed.), p. 123. Academic Press, New York, 1970.

156. M. Martinez-Carrion, C. Turano, F. Riva, and P. Fasella, *JBC* **242**, 1426 (1967).

157. D. L. Peterson and M. Martinez-Carrion, *JBC* **245**, 806 (1970).

158. S. Cheng, C. Michuda-Kozak, and M. Martinez-Carrion, *JBC* **246**, 3623 (1971).

159. C. Turano, D. Barra, F. Bossa, A. Ferraro, and A. Giartosio, *Eur. J. Biochem.* **23**, 349 (1972).

c. *Tyrosine (Phenolic Hydroxyl).* As stated earlier (Section III,B,2), evidence derived from the study of changes in CD spectra of Asp-transaminase complexes with some quasi-substrates and coenzyme analogs (Section III,B,5) has suggested the presence of an important tyrosine residue at the coenzyme-binding site (*44, 124*). Ivanov and Karpeisky (*44*) believed that the hydroxyl of this tyrosine residue is hydrogen bonded with the coenzyme's pyridine N atom at certain stages of the catalytic cycle (Section III,B,8).

A few years ago, Turano (*154, 155*) observed marked inactivation of Asp-apotransaminase (but not of holoenzyme) on specific blocking of one tyrosine residue (per subunit) with an active-site-directed bifunctional reagent, 1,5-difluoro-2,4-dinitrobenzene, or upon nitration of one tyrosine in the apoenzyme with tetranitromethane under special mild conditions (*159*); the latter reagent also oxidizes some SH groups (*160, 161, 164*), see below.

Christen and Riordan (*162, 162a, 163, 163a*) reported remarkable findings on a phenomenon of "syncatalytic" modification of 1.0–1.4 functional tyrosine residues (per dimer) in Asp-transaminase by $C(NO_2)_4$ in the presence of a substrate pair, with 90–95% inactivation of the enzyme. After this treatment the bound coenzyme is present entirely in the form of pyridoxamine-P which cannot be reconverted into pyridoxal-P. Upon resolution the modified protein is unable to recombine with coenzyme (*163, 164*), as is also the case with apoenzyme subjected to mild nitration with $C(NO_2)_4$ (*159*). The transient susceptibility of an essential tyrosine residue in the holoenzyme to nitration synchronous with the catalytic process indicates the occurrence of local transconformation in the functioning active center and seems compatible with the above-mentioned hypothesis of Ivanov and Karpeisky (*44*). The syncatalytic modification phenomenon has been reinvestigated in some detail in this Institute (*161, 164, 165*). While in essential agreement with Christen's experimental findings, the results point to a more complicated mechanism of the phenome-

160. W. Birchmeier and P. Christen, *FEBS Lett.* **18**, 209 (1971).
161. O. L. Polyanovsky, T. V. Demidkina, and C. A. Egorov, *FEBS Lett.* **23**, 262 (1972).
162. P. Christen, *Proc. Int. Congr. Biochem., 8th, 1969* Abstracts, p. 112 (1970).
162a. S. V. Shliapnikov, A. L. Bocharov, D. A. Gegelava, and M. Ya. Karpeisky, *Mol. Biol. (USSR)* **3**, 709 (1969).
163. J. F. Riordan and P. Christen, *Biochemistry* **9**, 3025 (1970).
163a. W. Birchmeier, K. J. Wilson, and P. Christen, *Abstr. Commun. 8th FEBS Meet. 1972* **405**; *FEBS Lett.* **26**, 113 (1972).
164. T. V. Demidkina, A. L. Bocharov, M. Ya. Karpeisky, and O. L. Polyanovsky, *Mol. Biol. (USSR)* **7**, 461, 620 (1973); cf. *BBRC* **50**, 377 (1973).
165. Yu. M. Torchinsky, R. A. Zufarova, M. B. Agalarova, and E. S. Severin, *FEBS Lett.* **28**, 302 (1972).

non. Under optimal mild conditions, syncatalytic modification with $C(NO_2)_4$ resulted in nitration of one tyrosine residue per subunit—the same residue which reacts on nitration of apoenzyme by the procedure of Turano *(159)*. From tryptic digests of the modified enzyme two $Tyr(NO_2)$-containing fragments were isolated; their cognate sequences [. . . Arg–]Lys–Val–Asn–Gly–Val–Gly–Ala–Tyr(NO_2)–Arg and Val–Asn–Gly–Val–Gly–Ala–Tyr(NO_2)–Arg *(161)* identify the syncatalytically susceptible residue as Tyr-40 of the enzyme's peptide chain *(99a)*, see Fig. 7.

Under the same experimental conditions, tetranitromethane irreversibly oxidizes one thiol group of the enzyme *(159, 161, 163a, 164)*. The same –SH group was identified as a functionally important one by selective modification under specific conditions (see the following paragraph). Modification of this SH group alone suffices to cause ~90% decrease in activity of the enzyme but, in contrast to syncatalytic modification by tetranitromethane, does not lead to irreversible amination of the coenzyme *(161, 163a, 164, 170)*. The latter effect, which precludes completion of the enzymic reaction cycle, is evidently associated with nitration of the functional tyrosine residue. Independent modification of the functional thiol group (alkylation, oxidation) followed by treatment with tetranitromethane in the absence of substrates results in the same chemical changes of Asp-transaminase as observed on "syncatalytic" nitration, viz., inactivation, irreversible amination of coenzyme, nitration of Tyr-40 *(164, 163a)*; the results cast some doubt on essentiality of Tyr-40 for the catalytic process (see below). The functions of this residue and of the important imidazole(s) in the enzymic reaction are in need of further investigation.

d. Cysteine (SH Groups). Native Asp-transaminase$_{cyt}$ contains 5 cysteine residues per subunit *(35, 96)* and no disulfide bonds. The reactivity and presumable role of the SH groups have been extensively studied *(22, 30, 130, 144, 150, 164–169,* etc.); the results can be interpreted as follows *(30, 130)*.

In the native enzyme three SH groups react randomly and readily (in weakly acid solution) with mercaptide-forming agents (Ag^+, *p*-mercuribenzoate). Two of these SH groups are exposed and react readily with

166. O. L. Polyanovsky, V. V. Nosikov, S. M. Deyev, A. E. Braunstein, E. V. Grishin, and Yu. A. Ovchinnikov, *FEBS Lett.* (1973) (in press); R. A. Zufarova, M. M. Dedyukina, L. V. Mamelova, and Yu. M. Torchinsky, *Khim. Prir. Soedin.* (1973) (in press).
167. O. L. Polyanovsky, *Biokhimiya* **27**, 734 (1962).
168. C. Turano, A. Giartosio, and F. Riva, *Enzymologia* **25**, 196 (1963).
169. C. Turano, A. Giartosio, and P. Fasella, *ABB* **104**, 524 (1964).

alkylating agents or oxidants; their modification does not impair transaminase activity, except the SH–SS exchange with Ellman's reagent (DTNB), which causes 30–40% inactivation (*130*). Following alkylation or oxidation of the two exposed SH groups, treatment of the modified enzyme with *p*-mercuribenzoate at pH 4.6 blocks a third SH group, with 95% inactivation; activity can be restored by addition of a thiol, e.g., glutathione or mercaptoethanol. The same SH group is blocked irreversibly by prolonged exposure to ICH_2COOH in concentrated urea. Kinetics of pseudosubstrate binding to transaminase preparations with the "functional" SH group thus modified reveal substantially impaired affinities (*130, 170*). When resolved from coenzyme, such preparations readily recombine with pyridoxal-P, but not with pyridoxamine-P (*155*). In Asp-transaminase$_{cyt}$ denatured with dodecyl sulfate or in concentrated urea, four SH groups are accessible to oxidation or alkylation, with irreversible change in protein conformation (*125, 130*). The fifth group reacts sluggishly with alkylating or mercaptidating agents under drastic conditions of denaturation.

Recently Torchinsky *et al.* (*165*) treated Asp-transaminase$_{cyt}$ with "cold" iodoacetate to block the two exposed SH groups, and then subjected it to syncatalytic alkylation with *N*-ethylmaleimide, according to Christen (*163a*), of the functionally important cysteine residue. The modified enzyme, with approximately 1% residual activity, can still slowly effect the full catalytic cycle, i.e., complete conversion to pyridoxamine-P enzyme by amino substrate (L-cysteinesulfinate) and quantitative restitution to the pyridoxal-P form by an excess of oxoglutarate. Using ^{14}C-labeled *N*-ethylmaleimide in such experiments, the authors (*165*) detected in proteolytic digests of the modified enzyme one radioactive peptide (residues 387–392), containing the modified, syncatalytically susceptible residue Cys-390 (see also Ref. *170*). Alternatively, two SH groups that become reactive in 8 M urea could be carboxymethylated with [^{14}C]iodoacetate (*166*); from digests of the labeled product two radioactive fragments were isolated—the one containing functional Cys-390 (see above) and peptide 249–256, comprising Cys-252 (Fig. 7). After drastic treatment in concentrated urea with highly radioactive iodoacetate, labeled S-alkylated cysteine (residue 191) was revealed also in a large tryptic fragment, the "core" peptide 168–214 (Fig. 7) (*166*). In similar experiments, preliminary modification of Cys-390 with cold ethylmaleimide prevented its labeling with[^{14}C]iodoacetate in 8 M urea (*165*).

Concordant definitive allocation of the five SH groups in Asp-trans-

---

170. W. Birchmeier. K. Y. Wilson, and P. Christen, *Abstr. Commun. 8th FEBS Meeting 1972* **405**; *JBC* **248**, 1751 (1973).

aminase$_{cyt}$ was completed independently by two teams in this institute, using different experimental approaches (*166*); the results are as follows:

Exposed: 1. Cys-45, selectively S-(1,2-dicarboxy)ethylated in maleate buffer; when modified with nitroxyl spin label, the ESR signal is sensitive to slight conformational changes resulting from ligand binding in the active site (see Section III,B,6); 2. Cys-82. Residues 82 and 45 are susceptible to alkylation, oxidation, etc., in the native holoenzyme without impairment of activity.

Semiburied, functionally important: 3. Cys-390, located near the active site, can be mercaptidated in weakly acid media, reacts with oxidants and alkylating agents in 8 M urea (or under syncatalytic conditions) with 95% inactivation of the enzyme, see above and p. 449.

Buried, apparently not involved in catalytic action: 4. Cys-252, susceptible to alkylation in concentrated urea or other denatured media; 5. Cys-191, difficultly accessible to modification under drastic conditions in denaturing media.

Martinez-Carrion (*171*) has reported on a comparative study of the –SH groups in mitochondrial and cytosol Asp-transaminases. The results relating to the latter isoenzyme were essentially confirmatory. The mitochondrial transaminase contained five SH groups per subunit; they could all be titrated by p-mercuribenzoate or slowly alkylated by iodoacetamide in 8 M urea with more extensive conformational change than in similarly treated Asp-transaminase$_{cyt}$. This points to less rigid macromolecular structure of the mitochondrial species.

Aspartate transaminase of mitochondria retained full catalytic activity and substrate affinity on carboxamidomethylation of five thiol groups and dialysis (*171*). This is in contrast with the nearly total inhibition of catalysis and impairment of substrate binding observed in the supernatant isoenzyme after modification of three SH groups.

Evidently, in the mitochondrial enzyme none of the five exposed SH groups is essential for catalysis. The same may apply to the cytosol enzyme, but in this case the "functional" SH group may be of critical significance for activity as a consequence of its steric proximity to the active site.

e. Threonine or Serine? (3-HO Group). Under certain conditions, e.g., upon reduction with NaBH$_4$ and in transaminases inhibited with cycloserine or its derivatives (Section III,B,6,*c*), the phosphate ester bond of the coenzyme tends to undergo cleavage in the active site [see E. Fischer *et al.* in Ref. *17*, p. 543; Braunstein (*45*)]. Part of the phosphate is re-

171. M. J. Stankewicz, S. Cheng, and M. Martinez-Carrion, *Biochemistry* **10**, 2877 (1977).

covered as P_i, and part appears to form stable covalent bonds with the enzyme protein (Yu. M. Torchinsky and E. A. Malakhova, personal communication, 1968).

Aspartate transaminase reacts with one equivalent (per subunit) of N-(5'-P-pyridoxyl)-L-glutamic acid—the borohydride-reduced PLP-glutamate aldimine—forming an irreversibly inactivated complex (172, 173). Khomutov and co-workers (172) reported preliminary data on the detection of free and peptide-bound phosphothreonine, together with P_i, in pepsin-pronase digests of the denatured EI complex. Further experiments with (^{35}P)-labeled inhibitor (174) apparently support the suggestion that a hydroxyamino acid residue situated in the active site in proximity to the coenzyme's phosphate ester bond may, under special conditions leading to labilization of this bond, accept the phosphate residue, see Section III,B,6,c. A cyclized analog of the inhibitor, N-(5'-P-pyridoxyl)-pyrrolidone carboxylic acid, slowly forms a (difficultly reversible) complex with apotransaminase without destabilization of the phosphate group (173).

4. Studies of the Steady State, Equilibrium, and Fast-Reaction Kinetics of Aspartate Transaminase

The kinetics of high purity Asp-transaminase have been investigated extensively with a view to elucidating the mechanism of enzymic transamination. A detailed experimental analysis of rate equations and steady state kinetic parameters, of competitive substrate and product inhibition and of pH dependence was reported in 1962 by Velick and Vavra (139, 140). Their papers, those of Banks et al. (89, 93, 175, 176), and the survey by Fasella (69) contain discussions of work on transaminase kinetics published till approximately 1966 by Jenkins et al. (115, 135, 136, 177, 178–182), Henson and Cleland (183), Nisselbaum and Bodansky

172. R. M. Khomutov, E. S. Severin, E. M. Khurs, and N. N. Gulyaev, *BBA* **171**, 201 (1969).
173. R. M. Khomutov, H. B. F. Dixon, L. V. Vdovina, M. P. Kirpichnikov, Yu. M. Morozov, E. S. Severin, and E. M. Khurs, *BJ* **124**, 99 (1971).
174. E. M. Khurs, E. S. Severin, H. B. F. Dixon, N. N. Gulyaev, and R. M. Khomutov, *Mol. Biol.* (*USSR*) **7** (1973) (in press).
175. B. E. C. Banks, A. J. Lawrence, C. A. Vernon, and J. F. Wooton, *in* "Chemical and Biological Aspects of Pyridoxal Catalysis" (E. E. Snell et al., eds.), p. 197. Pergamon, Oxford, 1963.
176. B. E. C. Banks, M. P. Bell, A. J. Lawrence, and C. A. Vernon, *in* "Pyridoxal Catalysis: Enzymes and Model Systems" (E. E. Snell et al., eds.), p. 191. Wiley (Interscience), New York, 1968.
177. W. T. Jenkins and I. W. Sizer, *JBC* **234**, 1179 (1959).
178. W. T. Jenkins and I. W. Sizer, *JBC* **235**, 620 (1960).
178a. W. T. Jenkins and L. D'Ari, *JBC* **241**, 2845, 5667 (1966).

(*83*), Nisonoff et al. (*184*), Fasella, Hammes, and associates (*69, 131, 137, 138, 185*), and others.

The data on steady state kinetic parameters (*139, 140*), rates of NH_2 exchange between a ^{14}C-labeled amino acid and its oxo analog (*177, 184*), spectrophotometric equilibrium studies of interactions between stoichiometric amounts of a single substrate and transaminase (*94, 115, 177–180*), and the fast-reaction studies of Hammes et al. (see below) all support the binary or "shuttle" mechanism of transamination, i.e., interaction of the enzyme in the forward and reverse half-reaction with alternatingly one molecule of amino and of oxo substrate at a time, as required by the Snell-Braunstein theory (cf. Sections II,B and III,B,8).

This kinetic mechanism is designated by the term "Ping-Pong Bi Bi" in Cleland's classification (*183*); it can be described by the condensed scheme:

$$\begin{array}{ccccc} AA^1 & & OA^1 & OA^2 & & AA^2 \\ \downarrow & & \uparrow & \downarrow & & \uparrow \\ \hline E_L & (AA^1 \cdot E_L \to OA^1 \cdot E_M) & E_M & (OA^2 \cdot E_M \to AA^2 \cdot E_L) & E_L \end{array}$$

where AA^1 and OA^1 are amino substrate and oxo product of forward half-reaction; OA^2 and AA^2 are oxo substrate and amino product of reverse half-reaction; and E_L and E_M are aldimine and amino form, respectively, of the enzyme.

In the studies of Velick and Vavra (*139, 140*), and others, experimentally measured steady state parameters (V_f, V_r, and K_m of the four substrates) and spectrophotometrically observed equilibria between individual substrates and the enzyme were in fair agreement with one another and with values calculated from kinetic equations. The relative values for individual kinetic parameters reported by different authors for cytosol Asp-transaminase and also for the mitochondrial enzyme (*36, 83, 94, 158*) are in qualitative correspondence, although the absolute numeric values vary considerably (see Table VI).

In the half-reactions, $E_L + AA \rightleftharpoons E_M + OA$, the equilibria favor the aldimine form of the enzyme (E_L), owing to high stability of the internal

---

179. W. T. Jenkins, in "Chemical and Biological Aspects of Pyridoxal Catalysis" (E. E. Snell et al., eds.), p. 139. Pergamon, Oxford, 1963.

180. W. T. Jenkins and R. T. Taylor, *JBC* **241**, 439 (1966).

181. W. T. Jenkins and L. D'Ari, *Biochemistry* **5**, 2900 (1966).

182. W. T. Jenkins and L. D'Ari, in "Pyridoxal Catalysis: Enzymes and Model Systems" (E. E. Snell, et al., eds.), p. 325. Wiley (Interscience), New York, 1968.

183. C. P. Henson and W. W. Cleland, *Biochemistry* **3**, 338 (1964).

184. A. Nisonoff, F. W. Barnes, Jr., T. Enns, and S. von Schuching, *Bull. Johns Hopkins Hosp.* **94**, 117 (1954).

185. P. Fasella and G. G. Hammes, *Biochemistry* **6**, 1798 (1967).

Schiff base. Therefore, the K_m values for amino substrates in the overall reaction are higher than for the cognate oxo substrates. The mitochondrial isoenzyme, in contrast to the cytoplasmic one, has preferential affinity for the C_4 substrates (*33, 94, 139, 140, 177*). Because oxalacetate is less stable chemically (and thermodynamically) than oxoglutarate, equilibrium of the overall transamination reaction favors formation of aspartate plus oxoglutarate.

Variations in the reported values of kinetic parameters (Table VI) result, in part, from differences in experimental conditions such as enzyme and substrate concentrations, assay techniques, and eventually, inadequate separation of isoenzyme (compare, for example, Refs. *183* and *83*). But the major source of divergence is the marked influence of the nature and concentrations of anions in the incubation medium upon the physical and catalytic properties of Asp-transaminases—spectral features, pK_a, pH optima, K_m values, and rate constants. It was noted early that the ionic composition of buffer solutions (in particular, the anion) affects the reaction rates in activity assays of Asp-transaminase (*21, 69, 175, 186*) as well as the K_m values of the supernatant and mitochondrial transaminase species from human liver (*187*), rat liver (*186*), and pig heart (*158*).

Complete deionization of the enzyme, e.g., on a mixed-bed ion exchange column (*21, 69, 188, 189*) lowers the pK_a of the coenzyme chromophore to ~5.25 and drastically reduces catalytic activity. On the other hand, binding of inhibitory dicarboxylic acids (maleate, 2-oxoglutarate, glutarate, etc.) is known to shift the enzyme's protolytic dissociation constant, pK_a, into the alkaline region (*28, 135, 139, 140, 178a, 182*). Kinetic measurements (*181, 182*) demonstrated that in weakly acid media the protonated (inactive) transaminase strongly binds two monoanions at the same basic sites which bind competitive dicarboxylic acids (and presumably the carboxyl groups of substrates). One of these sites was tentatively identified as the protonated internal aldimine nitrogen, the other is situated in proximity on the protein surface; at high pH values only one anionic group is bound at the latter cationic site (*158, 181*), see Section III,B,6.

Inorganic and organic monoanions vary considerably in their affinities to Asp-transaminases (to aldimine and amino forms of the supernatant and mitochondrial species): At saturating concentrations they raise the pK_a of Asp-transaminase to near 8, like the inhibitory dicarboxylates,

186. J. S. Nisselbaum, *Anal. Biochem.* **23**, 173 (1968).
187. T. C. Boyde, *BJ* **106**, 581 (1968).
188. G. Marino, A. M. Greco, V. Scardi, and R. Zito, *BJ* **99**, 589 (1966).
189. M. Bergami, G. Marino, and V. Scardi, *BJ* **110**, 471 (1968).

TABLE VI

Steady State Kinetic Parameters and Equilibria of Mammalian Mitochondrial and Cytosol Aspartate Transaminase Isoenzymes[a]

Enzyme source and type	Reaction	Buffer (M)	T (°C)	pH	K_m(app) values (mM)	V_{max}, V_r, V_f (Ref. 139)	K_{eq} (app)	Ref.
Pig heart (cyt)	Asp + KG ⇌ Glu + OA[b]	Arsenate (0.04)	20	7.4	K_{Asp} = 0.9; K_{KG} = 0.1; K_{Glu} = 4.0; K_{OA} = 0.04	V_f = 300; V_r = 1000	7 ± 1[c]	139
							6.2[d]	
Pig heart (mit)	Forward	Arsenate (0.06)	20	7.4	K_{Asp} = 0.08; K_{KG} = 0.8	V_f ~ 340		33
Pig heart (cyt)	Forward[e]	Pyrophosphate (0.1)	20	8.0	K_{Asp} = 4.0; K_{Glu} = 14.0			94
Pig heart (cyt)	Reverse[e]	Pyrophosphate (0.1)	20	8.0	K_{KG} = 0.4; K_{OA} = 0.02			94
Pig heart (mit)	Forward[e]	Pyrophosphate (0.1)	25	8.0	K_{Asp} = 2.8; K_{Glu} = 12.4			94
Pig heart (mit)	Reverse[e]	Pyrophosphate (0.1)	25	8.0	K_{KG} = 0.7; K_{OA} = 0.014			94
Pig heart (cyt)	Asp + KG ⇌ Glu + OA	Imidazole/HCl (0.1)	25	6.9	K_{Asp} = 2.8; K_{KG} = 0.13; K_{Glu} = 4.0; K_{OA} = 0.013			175
Pig heart (cyt)	Asp + KG ⇌ Glu + OA	Arsenate (0.1)	37	7.1	K_{Asp} = 3.9; K_{KG} = 0.43; K_{Glu} = 8.9; K_{OA} = 0.09	V_f/V_r = 0.51		183
Rat liver (cyt)	Reverse[f]	Barbital/HCl (0.04)	37	7.4	K_{Glu} = 7.8; K_{OA} = 0.037			186
Rat liver (mit)	Reverse[f]	Barbital/HCl (0.04)	37	7.4	K_{Glu} = 8.6; K_{OA} = 0.006			

[a] See also Ref. 94 for direct spectrophotometric titration of dissociation equilibria in half-reactions with stoichiometric amounts of the isoenzymes, and Ref. 158 for the effects of various anions on substrate affinities of the aldimine and amino forms of the cytosol and mitochondrial species.
[b] Abbreviations: KG, 2-oxoglutarate; OA, oxalacetate.
[c] Determined experimentally.
[d] Calculated from steady state kinetic parameters.
[e] Direct spectrophotometric titration of half-reactions.
[f] Assay with 2-oxoglutarate dehydrogenase.

and greatly increase the apparent K_m values of substrates. True K_s (and K_i) values for Asp-transaminase can be estimated by extrapolating the (linear) plots for anion-dependent apparent dissociation constants to zero anion concentration. Martinez-Carrion et al. (*158*) observed that sensitized photooxidation of 2 histidine residues abolishes the competitive effect of anions on the affinity of Asp-transaminase in the nonprotonated form (i.e., in the alkaline range) for glutarate or substrates. They suggested that an imidazole group might be the cationic site involved in binding of the substrate's carboxylate group (ω ?), see Section III,B,3. This suggestion is hard to reconcile with the pK_a of that cationic site, $\sim$11 (*178a, 182*), and with the fact that substrate affinity is increased, or at least unimpaired, in the photooxidized enzyme (*157, 158*).

The inhibitory power at pH $\sim$8 of the anions studied (*158, 181*), i.e., their affinity to the above-mentioned cationic site, diminishes along the following sequence:

Salicylate > sulfate > benzoate > butyrate > chloride > acetate > phosphate > cacodylate

Anion affinity is, in general, higher for the mitochondrial than for the supernatant isoenzyme, and in both species the anion-binding affinities of the amino form are higher than those of the aldimine form (*158*).

A noteworthy feature of enzymic transamination is the broad plateau of V_{max} plots versus pH, ranging at least from 6.3 to 9.5 (*139, 140*). This indicates that there is no overall release or uptake of protons between the medium and intermediate complexes at any stage of the reaction (*44*) (see Section III,B,8).

Studies of transamination in model systems (Section II,A) and measurements of the rates of enzyme-catalyzed equilibration of a [^{14}C]amino acid with its oxo analog (*176, 177*) indicated that tautomeric interconversion of the substrate-coenzyme Schiff bases was the slowest, rate-limiting step in each half-reaction. This was confirmed by Hammes and Fasella (*69, 137*) who studied the kinetics of fast intermediary steps in enzymic transamination by the temperature-jump technique. Using the α form of Asp-transaminase and improved apparatus, Fasella and Hammes (*185*) observed three distinct relaxation processes per half-reaction. Intermediates with λ_{max} at 430, 360, 490, and 330 nm are involved in each half-reaction, the rate-limiting step with rate constants k_l, k_{-l} occurring between the 490 and 330 nm intermediates. Lower bounds for the second-order rate constants of substrate binding were estimated as 10^7–10^8 M^{-1} sec^{-1}. Table VII presents the values of K'_s, k'_l and k'_{-l}, and the equilibrium constant $k'_l/k'_{-l} = R$, calculated from the relaxation times of Asp-transaminase interacting with Asp + OA or Glu + KG

TABLE VII

SUMMARY OF KINETIC PARAMETERS FOR Asp-TRANSAMINASE BASED ON TEMPERATURE-JUMP STUDIES (185)[a]

Constants[b]	Values for C_4 substrates	Values for C_5 substrates
$R = k'_l/k'_{-l}$	0.4	2.5
k'_l (sec^{-1})	530	2940
k'_{-l} (sec^{-1})	1330	1170
K'_{AA} (M)	4.2×10^{-3}	3.5×10^{-2}
K'_{OA} (M)	1.05×10^{-3}	3.6×10^{-4}

[a] Data from Fasella and Hammes (185).
[b] $R = k'_l/k'_{-l}$, equilibrium constant in rate-limiting step; k'_l and k'_{-l}, rate constants (lower bounds) for forward and reverse rate-limiting step, respectively; K'_{AA} and K'_{OA}, substrate constants (K'_s) for amino and oxo acids, respectively, in the rate-limiting step.

(185). In temperature-jump studies with the slow-reacting quasi-substrate, erythro-β-hydroxyaspartate, which forms a relatively long-lived quinonoid intermediate (λ_{max} ~490 nm), at least eight relaxation steps could be discerned in the overall reaction (190, 191).

5. Interactions with Coenzyme Analogs

In Asp-transaminase, as in other vitamin B_6–dependent enzymes, the coenzyme is in multiple linkage with functional groups of the active center. In addition to the reactive covalent aldimine bond, the linkage involves ionic binding at the phosphate group, hydrogen bonding at the pyridine N atom and phenolic oxygen, and hydrophobic interactions.

Binding at an active site, i.e., holoenzyme formation, enhances the weak catalytic properties of the free coenzyme by many powers of 10 and in a highly selective way. In the PLP-protein complexes the physical and chemical properties of both components undergo considerable alteration, depending on the structure of the specific protein and reflecting variations in the mode of attachment and specific interactions of the coenzyme. The nature and extent of such alterations is indicated by differences in rates and equilibria of protein-coenzyme binding, changes in optical and conformational parameters, in chemical reactivities of functional groups, etc.

Numerous variously modified analogs of pyridoxal and pyridoxal-P have been synthesized, and used as probes to elucidate the significance of structural features of the coenzyme molecule (a) for attachment to different apoenzymes, and (b) for catalytic activity (41–45); such studies

190. G. H. Czerlinski and J. Malkewics, Biochemistry 4, 1127 (1965).
191. G. G. Hammes and J. L. Haslam, Biochemistry 8, 1591 (1969).

TABLE

PROPERTIES OF ASPARTATE TRANSAMINASES

No.	Abbreviated name of analog	Modifications of PLP at positions						$\lambda_{max}{}^A$, nm	
		N-1	C-2	C-3	C-4	C-5	C-6	at pH $\sim$5.2	at pH $\geq$8.2
I	PLP	—	—	—	—	—	—	430	362
Ia	Pyridoxal (PL)	—	—	—	—	-CH$_2$OH	—	—	—
II	2-NorPLP	—	-H	—	—	—	—	425	360
III	2'-MePLP	—	-CH$_2$CH$_3$	—	—	—	—	435	365
IV	2'-PropPLP	—	-(CH$_2$)$_2$CH$_3$	—	—	—	—	440	370
V	2',2'-Me$_2$PLP	—	-CH(CH$_3$)$_2$	—	—	—	—	440	370
VI	6-MePLP	—	—	—	—	—	-CH$_3$	455	370
VII	6-Me-2-Nor-PLP	—	-H	—	—	—	-CH$_3$	455	370
VIII	2'-PhePLP	—	-CH$_2$C$_6$H$_5$	—	—	—	—	430 (CD+); 360 (CD+); 305 (CD negative)	360 (CD+)
IX	2'-OH-PLP	—	-CH$_2$OH	—	—	—	—	429	361
X	PLP N-oxide	N$\rightarrow$O	—	—	—	—	—	410–420^g (CD positive)	317^g
XI	N^+-MePLP	N^+-CH$_3$	—	—	—	—	—	—	—
XII	3-DeoxyPLP	—	—	-H	—	—	—	315 (CD positive)	315 (CD positive)^i
XIII	3-O-MePLP	—	—	-OCH$_3$	—	—	—	315 (CD positive)	315 (CD positive)^i
XIV	6-BrPLP	—	—	—	—	—	Br	390 (CD negative)^j	(CD abolished by amino substrate)
XV	6-ClPLP	—	—	—	—	—	Cl	390 (CD positive)^j	
XVI	4-Deformyl-4-vinylPLP	—	—	—	-CH=CH$_2$	—	—		358 (no CD)
XVII	5'-MePLP	—	—	—	—	-CHOPO$_3{}^{2-}$ | CH$_3$	—		
XVIII	5-DemethoxyPL-5-Me-C-phosphonate	—	—	—	—	-CH$_2$-PO$_3{}^{2-}$	—		
XIX	5-DemethoxyPL-5-Et-O-phosphonate	—	—	—	—	-(CH$_2$)$_2$-OPO$_3{}^{2-}$	—		
XX	PL-5'-O-Me-phosphonate	—	—	—	—	-CH$_2$-OPO$_3{}^{2-}$ | CH$_3$	—		
XXI	5-Demethoxy-5-carboxyethylPL	—	—	—	—	-(CH$_2$)$_2$CO$_2{}^+$	—		
XXII	5-DemethoxyPL-5-Prop-O-phosphonate	—	—	—	—	-(CH$_2$)$_3$-OPO$_3{}^{2-}$	—	400 (CD negative); 360 (CD positive); 300 (CD negative)	400 (CD negative); 360 (CD positive); 300 (CD negative)

^{a} Adapted from references 42, 44, 99, 104, 143, 192–201.
^{b} From Refs. 99, 143.
^{c} From Refs. 42, 104.
^{d} From Ref. 42.
^{e} From Ref. 143.
^{f} From Ref. 42.
^{g} Binding very slow with cytoplasmic, more rapid with mitochondrial apoenzyme.
^{h} Activity may result from reduction of analog to PLP on enzyme.

VIII
Reconstituted with Coenzyme Analogs[a]

pK_a	ΔD/D 420–450 nm	ΔD/D 360 nm	k_2 (apo→holo) M^{-1} min^{-1}	K_{Co} (app) $M \times 10^{-6}$	$K_{m(Asp)}$ (app) $M \times 10^{-3}$	$K_{m(KG)}$ $M \times 10^{-3}$	V_{max} (relative)	Ref.
6.25	28	27	1500[b]	0.15[c]	2.0[b]; 3.0[c]	0.1[b]; 0.16[c]	1	99, 143
—	—	—	—	>2,400	—	0.034	0.001	42, 104
5.8	29	21	450	0.07[d]	5.0[d]; 1.8[e]	0.5[d]; 0.24[e]	1.8[d]; 1.2[e]	99, 143
6.5	27	—	750	1.5[d]	0.5[f]	0.05[f]	0.32	99, 143
6.4	25	15	150	—	0.6	0.08	0.5	99, 143
7.1	25	8	100	—	10.0	1.0	0.9	99, 143
6.3	25	15	10,000	—	1.0	0.14	0.56	99, 143
6.4	25	10	600	—	>10	>10	0.46	99
	81						0.05	193
							1.0(in tris), 0.6(in phosphate); 0.25[g]; variable[h]	201, 199
								195, 195a
							0	195, 196
							0	196

	k_2 (apo→holo), M^{-1} min^{-1}		K_{Co} (app), $M \times 10^{-6}$				
	at pH 8.3 (0.01 M tris/HCl)	at pH 4.6 (0.1 M pyridine/HCl)	at pH ~8.2	at pH 4.6			
	8.9 × 10³	9 × 10⁴	0.9	0.2	0.01		193
	Binds slowly				0		202
6.1	1.6 × 10⁴	5.4 × 10⁴	4.1	2.6	0.03[h]; 0.4[k]		195
5.9	1.1 × 10³	2.7 × 10⁴	8.8	6.6	<0.002		195, 195a
	7 × 10⁴	5.8 × 10⁵			0.013		195, 195a
<4.0	30	2,300	270	60	<0.002		195, 195a
<4.0	280	5,900[l]	46	32	<0.002		195
	Binding atypical[m]		m				m

Slow binding at pH 5.2; no direct binding at pH 8.2; no resolution on dialysis. PLP can slowly displace XII or XIII.
Binds outside the active site. With large excess of XIV or XV a CD positive peak appears at 340 nm, abolished by amino substrate; indicates binding in active center.
One equivalent of racemic XVII is bound at a 2:1 ratio, i.e., only one enantiomer reacts.
Slow binding; with L-Glu slow half-reaction producing amino analog which binds very poorly.
Binds in 1:1 ratio; no resolution on dialysis. PLP interacts with the protein-XXII complex (Mamaeva et al., unpublished).
Binds in 1:1 ratio; no resolution on dialysis.

contribute to better understanding of the mechanism and stereochemistry of pyridoxal-linked enzymic reactions.

Analogs with high affinities to certain apoenzymes may produce complexes having specifically altered conformation but exhibiting weak or zero catalytic activity. When added simultaneously, pyridoxal-P competes with the analogs for the binding site, but once the apoenzyme-analog complex is formed, pyridoxal-P does not always easily displace the analog.

Data characterizing the interactions of Asp-apotransaminase with various coenzyme analogs are presented in Table VIII.

It seems appropriate to recollect some features of the bonding of pyridoxal-P, the natural coenzyme. The active holotransaminase$_{cyt}$ is practically nondissociated. Circular dichroism titrations of apoenzyme with pyridoxal-P or its alkyl-substituted analogs (99, 143, 192) and other data indicate stoichiometric binding; reported apparent dissociation constants, e.g., $K_{PLP} < 0.2$ μM (42), are of questionable validity. For such cases, second-order rate constants of complex formation are probably a better criterion for the ease of binding (143).

The pyridoxamine-P form of the enzyme (lacking the covalent aldimine bond) is formed distinctly slower and dissociates much more readily. It has a finite apparent dissociation constant, $K_{PMP} = 4.4 \times 10^{-6} M$ (101).

Apotransaminases from heart muscle (also from rat liver or *E. coli*) slowly catalyze reversible transamination half-reactions between free pyridoxamine and oxalacetate or 2-oxoglutarate. The ratio of half-saturating concentrations of PMP as the coenzyme (see above) and of pyridoxamine as substrate in the transamination with oxaloacetate [$K_{m(PM)} = 2.3 \times 10^{-3}$] (104) is

$$K_{PMP}/K_{m(PM)} = 4.4 \times 10^{-6} M/\sim 2.3 \times 10^{-3} M = \sim 2 \times 10^{-3} \qquad (10)$$

This ratio provides a rough estimate of the phosphate group's contribution to stability of the coenzyme–apoenzyme linkage. When saturated with pyridoxamine, apotransaminase catalyzes the complete NH_2 transfer reaction between aspartate and 2-oxoglutarate at 1/1000 of the turnover rate of intact holoenzyme. This demonstrates that the phosphate group is needed not only for firm binding of the coenzyme but also for effective catalysis, presumably by ensuring optimal positioning of essential groups of the coenzyme for interaction with groups of the protein and substrate (42, 44).

This is probably the reason why analogs with modified anionic sub-

192. V. L. Florentiev, V. I. Ivanov, and M. Ya. Karpeisky, "Methods in Enzymology," Vol. 18A, p. 567, 1970.

stituents at C-5', approximately but not exactly similar in orientation to the phosphate group of pyridoxal-P, mostly have low or zero coenzyme activity in spite of satisfactory binding, for example, compounds XVIII–XXII (Table VIII) and some others (*195–198*). Of the two enantiomers in racemic 5'-MePLP (compound XVII) only one binds with apotransaminase, giving a complex with fairly high enzymic activity (*195, 196*).

The minimal structure required for nonenzymic transamination, 3-hydroxy-4-pyridine aldehyde (cf. Section II,A), is sufficient for activity as substrate in the NH_2 transfer half-reaction catalyzed by pyridoxamine transaminase (*42*) but not as coenzyme for Asp-transaminase. Coenzyme analogs with absent or modified 3-OH group (Table VIII, compounds XII and XIII) bind at the enzyme's active site but produce catalytically inert complexes (*195, 196*).

Slight transaminase activity observed with the 3-O-Me analog (compound XIII) may result from release of, or contamination with, small amounts of PLP. Similarly, moderate activity obtained with PLP N-oxide (compound X), especially on binding to the mitochondrial apoenzyme (*199*), is possibly a result of decomposition of the analog in the assay system (*42*). N^+-MePLP (compound XI) slowly combines with apotransaminase to a practically inactive complex with a dichroic absorption maximum (*195*). Thus, a nonmodified heterocyclic nitrogen seems to be an essential requirement for activity.

Analogs with absent or modified alkyl substituents at position 2 (Table VIII; compounds II–V, VIII, and IX), as well as 6-MePLP (compound VI) and 2-Nor-6-MePLP (compound VIII), bind stoichiometrically at the active site, producing complexes with high to moderate transaminase activity (*42, 99, 104, 143, 192*, etc.). The binding rates decrease, and V_{max} values tend to diminish with increasing bulk of the alkyl substituents (*42, 143*).

193. S. Mora, Z. V. Nikolova, A. L. Bocharov, V. I. Ivanov, O. K. Mamaeva, N. Stambolieva, V. L. Florentiev, M. Ya. Karpeisky, *Mol. Biol. (USSR)* **6**, 559 (1972).
194. W. Korytnik, B. Lachmann, and N. Angelino, *Biochemistry* **11**, 722 (1972).
195. M. L. Fonda and R. J. Johnson, *JBC* **245**, 2709 (1970); E. S. Furbish, M. L. Fonda, and D. E. Metzler, *Biochemistry* **8**, 5169 (1969).
195a. M. L. Fonda, *JBC* **246**, 2230 (1971).
196. S. Mora, A. L. Bocharov, V. I. Ivanov, M. Ya. Karpeisky, O. K. Mamaeva, and N. Stambolieva, *Mol. Biol. (USSR)* **6**, 119 (1972).
197. O. K. Mamaeva, N. Sh. Padyukova, V. L. Florentiev, and M. Ya. Karpeisky, *Mol. Biol. (USSR)* **7**, No. 6 (1973) (in press).
198. N. S. Padyukova, Cand. Chem. Sci. Dissertation, Moscow, 1972.
199. S. Fukui, N. Ohishi, and S. Shimizu, "Methods in Enzymology," Vol. 18A, p. 598, 1970; cf. *ABB* **130**, 584 (1969).

2-NorPLP binds 3-fold slower than PLP, but has distinctly better affinity and somewhat higher co-transaminase activity than the latter. In many B_6 enzymes, among others in mammalian γ-aminobutyrate (β-alanine):2-oxoglutarate aminotransferase (*200*), 2-NorPLP is an inferior substitute for the natural coenzyme.

Introduction of a CH_3 group in position 6 increases the second-order rate constants for binding (cf. compound VI with I, and compound VII with II in Table VIII).

While optical properties, pK_a values, and kinetic parameters of the complexes vary with modification of the alkyl substituents in positions 2 and 6, hydrophobic interactions at these loci with adjacent areas of the coenzyme-anchoring site are clearly of importance for stabilization of the appropriate geometry of coenzyme binding. In particular, the hydrophobic site of the apoenzyme complementary to the 2-Me group and capable of accommodating the next higher homologous alkyls (the hydrophobic HC^2 corner in 2-NorPLP) is regarded by Ivanov and Karpeisky (*44*) as a kind of socket fixing the 2-5 axis of the coenzyme for pendulum-like changes in orientation associated with enzymic transamination (see Section III,B,8).

Spatial tolerance of Asp-apotransaminase for 2 and 6 substituents in PLP is relatively high (in comparison to that of other B_6-dependent enzymes tested). According to Fukui (*201*), 2'-OH-PLP (Table VIII, compound IX), in spite of slow binding has high affinity and full coenzyme activity (in tris buffer; phosphate buffer is inhibitory). In our laboratory, 6-Br-PLP was found to bind atypically, giving an inactive complex (*193*); the analog is apparently attached outside the active center (only with a 4:1 excess of compound XIV, some of it binds at the coenzyme-anchoring site). Changes in absorption and CD spectra in the presence of aspartate testify to the formation of aldimine intermediates. Since compound XIV is isosteric with the efficient analog, 6-Me-PLP (compound VI), changes in electronic properties of the molecule, caused by the halogen atom in position 6, are apparently responsible for the inadequacy of compound XIV (*193*). Modifications of the 4-CHO group produce derivatives inactive as coenzyme substitutes and usually with poor affinity, such as 4'-MePLP (*197*) or 4-deformyl-4-vinyl PLP (Table VIII, compound XVI) (*202*). Some N-substituted pyridoxamine-P derivatives (reduced coenzyme substrate, or quasi-substrate imines), which can be considered as stable analogs of the tetrahedral transaldimination inter-

200. V. Yu. Vasiliev and Z. K. Nikolaeva, *Biokhimiya* 38 (1973), in press.
201. S. Fukui, Y. Nakai, and F. Masugi, "Methods in Enzymology," Vol. 18A, p. 603, 1970.
202. I. Y. Yang, M. Fonda, and R. M. Khomutov, *Fed. Proc.* 30, 1087 Abs. (1971).

mediates, act as potent binding-site-directed inhibitors of transaminases, forming reversibly blocked or permanently inactivated complexes (154, 155, 173, 174; cf. Ref. 43), see also Section III,B,6,c.

6. *Interactions with Substrates, Quasi-Substrates, and Inhibitors*

a. *Substrate Specificity.* L-Aspartate:2-oxoglutarate aminotransferases act at widely differing rates on many analogs of the two main pairs of amino and oxo substrates (2, 10, 17, 18, 52). No accurate measurements of K_m and V_{max} values are available for most of the subsidiary substrates. Some of them react fairly rapidly at elevated concentrations (cf. Table II in Ref. 203). This points to low enzyme affinity as the cause of poor transamination.

L-Cysteinesulfinate is transaminated more rapidly than L-aspartate by pure Asp-transaminase$_{cyt}$; as a result of rapid breakdown of the oxo product, 3-sulfinylpyruvate, to SO_2 and pyruvate, the reaction is virtually irreversible (105, 204). γ-Methylglutamate (but not the β-methyl isomer), β-methylaspartate (205), *threo*-β-hydroxyaspartate (206), and the cognate α-oxo acids are fairly efficient as substitutes for the normal NH_2 donors and acceptors. With other dicarboxylic analogs (e.g., aminomalonate, *threo*-γ-hydroxyglutamate, cysteate, and homocysteate) and several monocarboxylic amino acids, reaction rates are considerably reduced, sometimes to a thousandth or less of the rate of normal transamination (203, 207); see Table IX.

Aspartate transaminase$_{mit}$ from mammalian tissues has relatively high activity on tyrosine or phenylalanine (26, 27), see Section I,A.

Preincubation of the enzyme at pH ~5 with cysteinesulfinate (106), aspartate (203), or another NH_2 donor produces apoenzyme since the PMP-enzyme is readily resolved in weakly acid solution.

b. *Quasi-Substrates and Other Competitive Inhibitors* (cf. Refs. 2, 9, 10, 17, 18, 43, 45, 52, 61, 136, 206, etc.). Compounds imitating the structure of a substrate molecule or its parts (amino or carboxyl groups, side chain fragments) compete with substrate for the active center of aminotransferases (and PLP-enzymes in general) in proportion to their affinities for complementary areas of the binding sites. Inhibitory capacity is enhanced where long-lived (stabilized) or atypical enzyme–inhibitor

203. A. Novogrodsky and A. Meister, *BBA* **87**, 605 (1964).
204. E. B. Kearney and T. P. Singer, *BBA* **11**, 276 (1953).
205. G. A. Galegov, *Biokhimiya* **26**, 635 (1961).
206. W. T. Jenkins, *JBC* **236**, 1121 (1961); *in* "Chemical and Biological Aspects of Pyridoxal Catalysis" (E. E. Snell *et al.*, eds.), p. 145. Pergamon, Oxford, 1963.
207. W. T. Jenkins, *JBC* **236**, 474 (1961).

TABLE IX

RELATIVE AMOUNTS OF LABELED GLUTAMATE FORMED BY ASPARTATE TRANSAMINASE FROM (^{14}C)-2-OXOGLUTARATE AND VARIOUS L-AMINO ACIDS (2×10^{-2} M)[a]

NH_2 donor	(^{14}C)-Glutamate formed
Aspartate	100.0
Glutamate	100.0
Methionine	0.13
Serine	0.12
Arginine	0.05
Phenylalanine	0.05
Leucine	0.024
Lysine	0.015
Alanine	0.015
D-Methionine	0
Glycine	0

[a] From Novogrodsky and Meister (*203*).

complexes are formed. Time-dependent irreversible inhibition (inactivation) may ensue if intermediates undergo conversion to stable secondary products with sterically obstructed active centers or covalently modified essential groups (see Section III,B,6,*c*,*d*).

1. *Substituted dicarboxylic amino and oxo acids.* α-Methylaspartate (*28, 94, 120, 126, 136, 141, 142, 177, 178, 208–210*), which is slowly bound to Asp-transaminase (see Table X) and transaldiminated, produces a stable enzyme–quasi-substrate aldimine. Lack of an α-H atom precludes its transamination. The pK_a of this complex is shifted to ~ 8; its absorption spectrum has a band at 360 nm with slightly positive Cotton effect, and an optically inactive band at 430 nm (see Table IV).

erythro-β-Hydroxyaspartate (*94, 120, 190, 191, 206*) reacts with the enzyme to form a mixture of intermediates with an intense absorption peak at 490 nm (*206*) in addition to (much smaller) usual maxima at 430 and 362 nm; very slow transamination follows, producing the PMP-enzyme (λ_{max} = 332 nm). The peak at 490 nm, associated with a strong

208. R. T. Taylor and W. T. Jenkins, *JBC* **241**, 4396 and 4406 (1966).
208a. W. T. Jenkins and R. T. Taylor, "Methods in Enzymology," Vol. 17A, p. 802, 1970.
208b. K. Aki, A. Yokojima, and A. Ichihara, *J. Biochem.* (*Tokyo*) **65**, 539 (1969); "Methods in Enzymology," Vol. 17A, p. 811, 1970.
208c. K. Aki, K. Ogawa, A. Shirai, and A. Ichihara, *J. Biochem.* (*Tokyo*) **62**, 610 (1967); K. Aki and A. Ichihara, "Methods in Enzymology," Vol. 17A, p. 807, 1970.
209. P. Fasella, A. Giartosio, and G. G. Hammes, *Biochemistry* **5**, 197 (1967).
210. G. G. Hammes and J. F. Tancredi, *BBA* **146**, 312 (1967); G. G. Hammes and J. L. Haslam, *Biochemistry* **7**, 1519 (1968).

TABLE X

BIMOLECULAR RATE CONSTANTS FOR ASSOCIATION OF ASPARTATE TRANSAMINASE WITH AMINO SUBSTRATES AND QUASI-SUBSTRATES

Amino acid	k_1 (app) M^{-1} sec^{-1}	Ref.
L-Glu	10^7	185
L-Asp	10^7	185
erythro-β-HO-Asp	3×10^6	191
α-MeAsp	1.2×10^4	210
β-ClGlu	1.44×10^3	239

negative CD extremum (Table IV), is assigned to a quinonoid anionic imine species (p. 392, Eq. 9, structure Ib), occurring as a transient intermediate also in normal transamination reactions (21, 44), see Section III,B,8. No intermediate with long-wave absorption maximum near 500 nm was noticeable in complexes of Asp-transaminase with threo-β-hydroxyaspartate (120, 206), which is transaminated at a moderate rate. Formation of such an intermediate in small amounts was observed with erythro-β-hydroxyglutamate and in significant amounts with threo-β-hydroxyglutamate (69); these two quasi-substrates undergo slow transamination. Interaction of Asp-transaminase with some other β-substituted analogs (211, 212) results in aberrant side reactions (Section III,B,6,d).

2. *Amines and related compounds.* This type of competitive inhibitors includes, besides the inhibitory amino substrates considered above, an extensive group of O- and N-amino compounds acting as carbonyl-binding reagents—hydroxylamine, hydrazine and their derivatives with unsubstituted NH$_2$ groups. With the enzyme at substrate level concentrations, the kinetics of interaction with carbonyl reagents can easily be followed by recording of changes in absorption spectra (213) and ORD or CD characteristics. In the spectra of carbonyl-blocked complexes of Asp-transaminase, optical dissymmetry of the coenzyme chromophore is mostly drastically diminished or inversed (45, 125). Carbonyl reagents with hydrophobic side chains (isoniazid, aromatic hydrazides and hydrazines) and those structurally related to substrate molecules (aminooxyacetate and other α-aminooxy acids, α-hydrazino acids) have increased enzyme affinity (low K_i values) (213, 214).

211. J. M. Manning, R. M. Khomutov, and P. Fasella, *Eur. J. Biochem.* **5**, 199 (1968).
212. E. Kun, D. W. Fanshier, and D. R. Grassetti, *JBC* **235**, 416 (1960).
213. W. T. Jenkins, S. B. Orlowski, and I. W. Sizer, *JBC* **234**, 2657 (1959).
214. P. Oehme, in "Pyridoxal Catalysis: Enzymes and Model Systems" (E. E. Snell et al., eds.), p. 677. Wiley (Interscience), New York, 1968.

The relatively weak binding of many carbonyl reagents reflects high stability of the internal PLP-lysine Schiff base, determined by constrained steric orientation of the groups (4'-CHO and ϵ-NH_2) constituting the aldimine link. A corollary of this feature is the fact that the isonicotinoyl hydrazone of PLP and several other stable carbonyl-blocked coenzyme derivatives react with apotransaminase and other PLP-dependent enzyme proteins not as inhibitory antagonists but as fully active substitutes for the coenzyme; see Ref. *17*, pp. 291 and 307; Gonnard and Nguyen-Phillipon (*215*). The PLP-hydrazones were shown (*45, 216, 216a*) to bind at the coenzyme-anchoring site and, by way of transaldimination with the reactive lysine residue, to restitute the specific internal aldimine. In a similar way, apotransaminase is rapidly converted to native holoenzyme by interaction with the fairly stable thiazolidine derivative obtained by condensation of PLP with D- or L-penicillamine (*217*).

Inhibition of specific aminotransferases and other PLP-enzymes *in vivo* (in microorganisms, and in the brain and other organs of animals) plays an important part in curative or toxic effects of many natural and synthetic products which are either carbonyl reagents per se, e.g., isoniazid, other hydrazine derivatives, penicillamine, L-canaline (*19, 61, 214, 217, 217a*), or eventual precursors of such agents—cycloserine and its analogs, for example, tricholomic acid (*218, 219*), linatine (*219a*), penicillin and L-canavanine; see Refs. *4, 5*.

3. *Competitive di- and monoanions* (*28, 135, 136, 139, 140, 178a–182*, etc.). Pronounced pH-dependent competitive inhibition of semipurified kynurenine transaminase (EC 2.6.1.7) and Asp-transaminase by certain mono- and dicarboxylic acids was observed by Mason (*220*), who pointed

215. P. Gonnard and C. Nguyen-Phillipon, *C. R. Acad. Sci.* **258**, 1911 (1964); cf. *BBA* **81**, 548 (1964).
216. Yu. M. Torchinsky, *BBRC* **10**, 401 (1963).
216a. M. Fujioka and E. E. Snell, *BBA* **81**, 548 (1964).
217. P. Siegmund, G. Hasenbank, and F. Koerber, *Hoppe-Seyler's Z. Physiol. Chem.* **344**, 1062 (1968).
217a. N. Katunuma, N. Okada, T. Matsuzawa, and N. Otsuka, *J. Biochem. (Tokyo)* **57**, 445 (1965); E. L. Rakiala, M. Kekomäki, J. Jänne, A. Raina, and N. C. R. Raina, *BBA* **227**, 337 (1971).
218. T. Takemoto and T. Nakajima, *Yakugaki Zasshi (Japan)* **84**, 1183 and 1230 (1964).
219. The L-isomer of γ-cycloglutamate was detected as a natural product named "tricholomic acid," the insecticidal agent from a Japanese mushroom (*218*), soon after a method for synthesis of the racemic compound had been developed by R. M. Khomutov and co-workers.
219a. H. J. Klosterman, G. L. Lamoureux, and J. L. Parsons, *Biochemistry* **6**, 170 (1967).
220. M. Mason, *JBC* **234**, 2770 (1959).

out that acids analogous in structure to the best oxo substrates were the strongest inhibitors. He suggested that inhibitory dicarboxylates were bound to the enzyme by both anionic groups at low pH values and only by one at high pH values. Subsequent kinetic and spectral studies with high purity Asp-transaminase (*28, 135, 139, 140*) confirmed that inhibitory dicarboxylates bind much more tightly at low pH, and that they cause a shift in protolytic pK of the coenzyme chromophore from 6.3 to 8 or above. Near neutrality the inhibitory power (enzyme affinity) and extent of shift toward the spectrum of protonated enzyme are parallel features, decreasing in the following sequence (*139, 140*).

Inhibitor: Maleate > glutarate > adipate > succinate > malate > fumarate
K_i (mM):   2.3   3   12   18   —   (very large)

2-Oxoglutarate at relatively high concentrations has similar inhibitory properties (*28, 135*).

The rigidly fixed anionic groups of maleate are apparently at the optimum intercarboxylate distance (cf. Section III,B,7), and enzyme affinities of other dicarboxylic acids correlate with the degree of conformational flexibility allowing their carboxyl groups to assume a spatially comparable position (*139, 140*). This postulation is at the basis of conjectures regarding the conformation of active-site-bound substrate molecules, see Sections III,B,7 and 8.

Mason's hypothesis on the mode of interaction between transaminase and anions was substantiated by the kinetic studies of Jenkins (*178a, 181, 182*) and Martinez-Carrion (*158*) and their associates (cf. Section III,B,4). It was shown that the cationic protein site that binds a carboxylate group both at high and low pH values is normally masked by a buffer anion (cf. Ref. *158*). The latter can be displaced by an inhibitory monoanion with higher affinity or by one anionic group of a dicarboxylic inhibitor or substrate (presumably ω-COO$^-$) with concomitant uptake of a proton by the complex (*178a, 181*). At low pH values, a further monoanion or the second anionic group of dicarboxylic ligands is attached to transaminase at a site tentatively identified as the protonated imino N atom of the internal PLP-lysine Schiff base (*181*). It seems quite likely, however, that the imidazole ring of an appropriately located essential histidine residue may act as a charge relay shifting protons between the substrate's α-carboxyl, the imino N, and 3-phenolic hydroxyl of coenzyme (cf. Sections III,B,3 and 6,*c*).

Spectrophotometric fast-reaction studies of the binding of glutarate with Asp-transaminase led Jenkins and D'Ari (*181, 182*) to the conclusion that attachment of the anionic groups of dicarboxylic ligands to the enzyme occurs in a compulsory order. Both in the case of the protonated complex (at low pH) with two carboxyl groups bound—at the cationic protein group (site "α") and at the aldimine lysine N atom (site "β")—and likewise at high pH values, where there is only one carboxyl group attached at "α," the bond at site "β" (with the imino N) must be formed first. In the latter case, the yellow acidic glutarate complex ("$\alpha\beta$") is a transient intermediate in the rapid conversion from initial "β" to final basic glutarate complex "α." Other kinetic aspects of the transaminase-anion interactions have been considered in Section III,B,4.

c. *Inhibition by Affinity Labeling. Modification of Essential Groups by Bifunctional Reagents.* Several types of active-site-directed inhibitors are included under this collective and not quite exact heading. Some of them react by stepwise or intermediate mechanisms, partially overlapping those discussed in Sections III,B,3, III,B,5, and III,B,6,*b* and *d*.

1. *Cycloserine and related compounds.* The antibiotic D-cycloserine (oxamycin, 4-aminoisoxazolidone-3) and analogous compounds have been used as tools for conformational inhibitor analysis of PLP-linked enzymes by Khomutov, Karpeisky, Severin, and co-workers (*17*, pp. 313 and 323; *18*, p. 631; *43, 44, 46, 221–227*).

D- and L-Cycloserine and their derivatives substituted in position 5 (Fig. 8) are rigid cyclic analogs of D- and L-alanine (*43, 228*) and of the

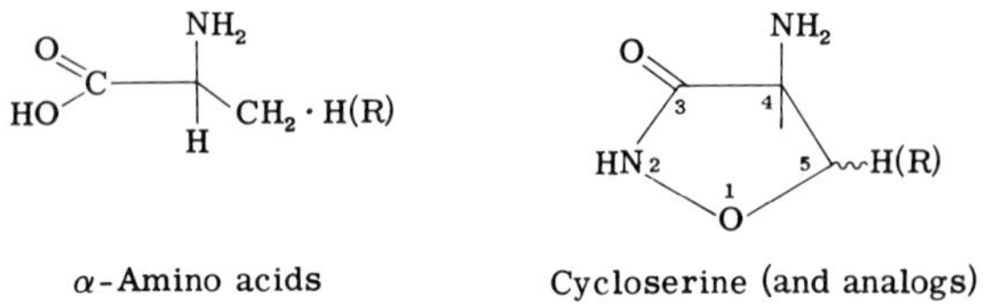

α-Amino acids   Cycloserine (and analogs)

FIG. 8. Structural similarity between α-amino acids and cycloserine or its derivatives.

221. R. M. Khomutov, M. Ya. Karpeisky, and E. S. Severin, *Biokhimiya* **26**, 772 (1961).
222. M. Ya. Karpeisky and Yu. N. Breusov, *Biokhimiya* **30**, 153 (1963).
223. R. M. Khomutov, G. K. Kovaleva, E. S. Severin, and L. Vdovina, *Biokhimiya* **32**, 908 (1967).
224. L. P. Sashchenko, E. S. Severin, and R. M. Khomutov, *Biokhimiya* **33**, 142 (1968).
225. G. K. Kovaleva and E. S. Severin, *Biokhimiya* **37**, 478 (1972).
226. G. K. Kovaleva and E. S. Severin, *Biokhimiya* **37**, 1282 (1972).
227. L. P. Sashchenko, E. S. Severin, G. K. Kovaleva, and R. M. Khomutov, *Biokhimiya* **33**, 1210 (1968); L. P. Sashchenko, E. S. Severin, D. E. Metzler, and R. M. Khomutov, *Biochemistry* **10**, 4888 (1971).
228. A. E. Braunstein, R. M. Azarkh, and T.-S. Hsü, *Biokhimiya* **26**, 882 (1961); see also *ibid.* **25**, 954 (1960).

higher α-amino acids, respectively (*18*, p. 631; *223*); the pseudo-acid cyclic hydroxamate group mimics the properties of a carboxyl group. Individual members of the series are time-dependent irreversible inhibitors with selective affinity for certain PLP-enzymes acting on the sterically cognate amino acids; see below, and footnotes *229*, *230*. In the initial, reversible stage these inhibitors bind at the active site as quasi-substrates and form PLP-aldimines (via the transaldimination step); hence, these compounds fail to react with the PMP form of transaminases. The PLP-inhibitor aldimine next undergoes conversion to a PMP-ketimine, accompanied by cleavage of the labile carboxamate bond. The resulting reactive acyl binds (covalently) to a nucleophilic group of the active site, thus completing the irreversible inactivation stage. From the inactive complex of (^{14}C)-cycloserine with Asp-transaminase ($EI_{340\ nm}$), Karpeisky and Breusov (*222*) isolated an imine consisting of pyridoxamine-P and degraded labeled inhibitor. These and other chemical and optical studies with several substrate-like cycloserine analogs (*46, 222, 225–227, 232*) substantiate the indicated inactivation mechanism (Fig. 9) which includes abortive transamination from inhibitor to coenzyme. Reverse transamination of the PMP-ketimine intermediate is prevented by acylation of an essential group, originally believed to be an acid-base catalyst functioning in this prototropic shift (*18*, p. 631; *223*). However, more recent studies (*46, 222*) point to acylation of a protein group normally involved in binding of the substrate α-carboxyl (see below and Fig. 11).

The mechanism just outlined is supported indirectly by evidence indicating that PLP-enzymes presumably forming (normally) no intermediate PMP-ketimine, e.g., glutamate α-decarboxylase (*224*) or replace-

229. The labile carboxamate bond is easily decyclized by hydrolysis or nucleophilic attack, in aqueous solution or at protein surfaces, to O-substituted hydroxylamines (3-aminooxylalanine and related products). These act as competitive inhibitors with low structural and steric selectivity, combining with PLP-enzymes to nonspecific carbonyl-blocked complexes (enzyme oximes); this occurs also with PLP-linked enzymes insensitive to the intact cyclic inhibitor. To avoid errors resulting from contamination by decyclization products, it is essential to use in inhibition studies only fresh solutions of recrystallized cycloserine derivatives.

230. Correct evaluation of the kinetics of this irreversible inactivation can be achieved (*46*) by using the procedure of K. Kitz and J. B. Wilson (*231*) for graphic estimation of K_i (equilibrium constant for formation of the reversible EI complex) and k_2 (rate constant for irreversible modification of the enzyme), see Section III,B,6,c.

231. K. Kitz and J. B. Wilson, *JBC* **237**, 3245 (1962).

232. V. Yu. Vasiliev, Z. K. Nikolayeva, L. P. Sashchenko, and E. S. Severin, *Biokhimiya* **38** (1973) (in press).

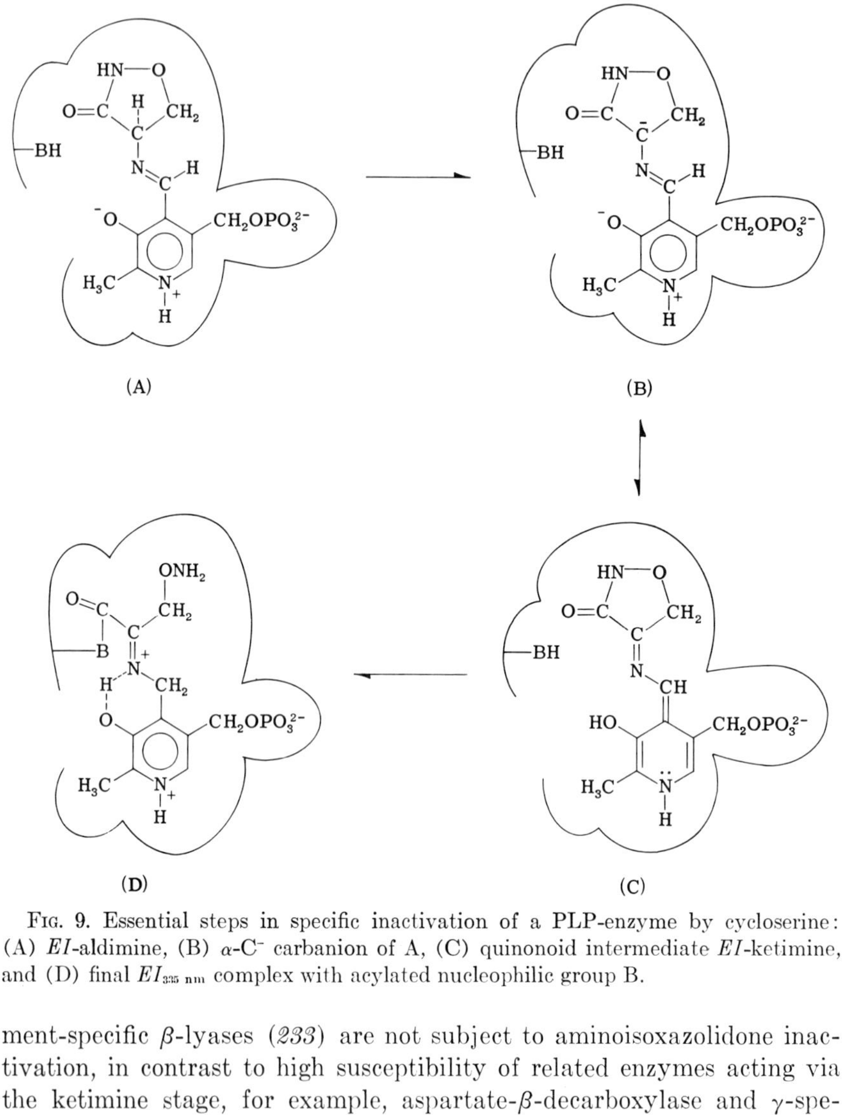

Fig. 9. Essential steps in specific inactivation of a PLP-enzyme by cycloserine: (A) EI-aldimine, (B) α-C^- carbanion of A, (C) quinonoid intermediate EI-ketimine, and (D) final $EI_{325\,nm}$ complex with acylated nucleophilic group B.

ment-specific β-lyases (233) are not subject to aminoisoxazolidone inactivation, in contrast to high susceptibility of related enzymes acting via the ketimine stage, for example, aspartate-β-decarboxylase and γ-specific lyases (see Ref. 234).

5-Substituted cycloserine analogs exist in diastereoisomeric forms with rigid configurations similar to rotational isomers of an amino acid. A

233. A. E. Braunstein, E. V. Goryachenkova, E. A. Tolosa, I. H. Willhardt, and L. L. Yefremova, *BBA* **242**, 247 (1971).
234. A. E. Braunstein, *Struct. Funct. Enzymes, Fed. Eur. Biochem. Soc. Symp. 8th* **29**, 135 (1972).

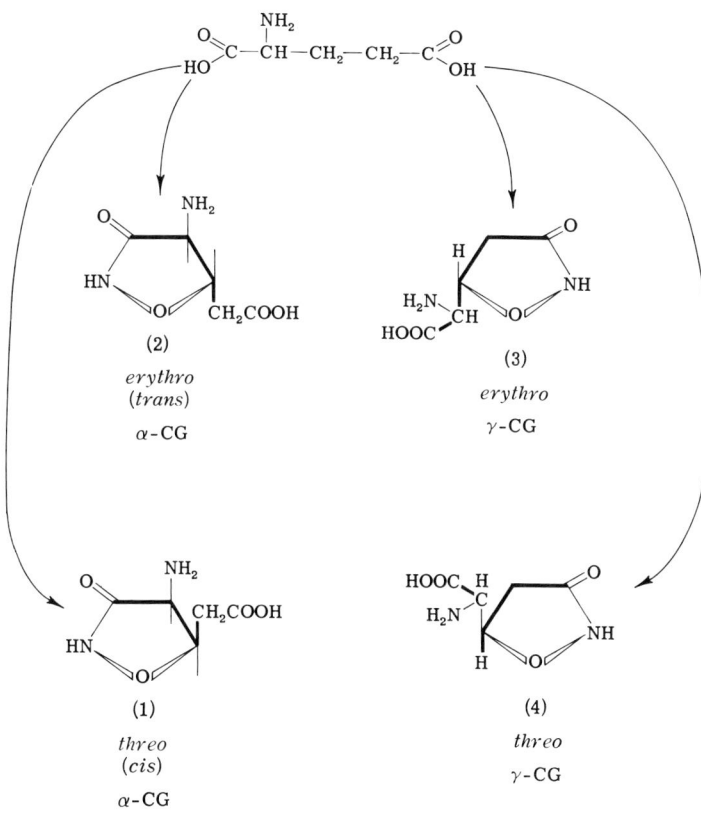

Fig. 10. Projection formulas of the isomeric cycloglutamic acids. [After Khomutov et al. in "Pyridoxal Catalysis: Enzymes and Model Systems" (E. E. Snell et al., eds.), p. 641. Wiley (Interscience), New York, 1968.] (1) and (2), *threo(cis)*- and *erythro(trans)*-α-cycloglutamates; (3) and (4), *erythro*- and *threo*-γ-cycloglutamates.

PLP-enzyme specific toward that amino acid usually has markedly preferential affinity for one of the cyclic diastereoisomers. Khomutov et al. (*18*, p. 631; *46, 223*) employed the active isoxazolidone derivatives as rigid templates of the complementary substrate-binding sites, i.e., as probes defining the conformation of the amino acid rotamer actually fixed. These authors synthesized the four possible isomeric isoxazolidones based on the skeleton of glutamic acid (Fig. 10). In compounds (1) and (2), conventionally named *threo*(or *cis*)- and *erythro*(or *trans*)-α-cycloglutamates (*46, 225, 235*), the α-carboxyl of glutamic acid forms

235. G. K. Kovaleva, E. S. Severin, P. Fasella, and R. M. Khomutov, *Biokhimiya* **38**, 370 (1973).

part of the cyclic carboxamate group, whereas the cycle embodies the γ-carboxyl in compounds (3) and (4), designated as *erythro-* and *threo-* γ-cycloglutamates, respectively. At least one of these structures would be expected to have high affinity for the substrate site of a glutamate-specific PLP-enzyme. Actually, compound (1) is a highly selective, powerful inhibitor of Asp-transaminase, alanine-glutamate (EC 2.6.1.2) and γ-aminobutyrate-glutamate transaminase (EC 2.6.1.19) *(193, 232, 234)*. The other isomers are much weaker inhibitors and produce qualitatively different *EI* complexes *(43, 223)*; see footnote *229*.

Important details of the kinetics and stereochemistry of interactions between Asp-transaminase and the α- and γ-cycloglutamates are presented and discussed in recent papers by Severin et al. *(46, 225, 226, 235)*; their main results and conclusions are summarized below and in Section III,B,7. The kinetic parameters estimated by the procedure of Kits and Wilson *(230, 231)* are shown in Table XI *(46)*. The K_i value for *threo*-α-cycloglutamate (CG) (1) was about one power lower than K_m for glutamate $(1 \times 10^{-2} M)$—evidently a consequence of fixed optimal conformation of this analog, which also had the highest rate constant, k_2, and shortest half-conversion time, $t_{1/2}$, for irreversible enzyme modification. *threo*-γ-Cycloglutamate (4) ranked second in affinity and had the slowest modification parameters; *erythro*-γ-cycloglutamate (3) and the *erythro*-α-isomer (2) had distinctly lower enzyme affinity and intermediate modification rates *(46)*. Studies of the mechanism and geometry of inhibition showed that each of the isomeric analogs initially produces (at pH 8) a PLP-quasi-substrate aldimine, as indicated by disappearance of CD in the 362-nm absorption peak of the enzyme *(225, 226, 235)*. *threo*-α-Cycloglutamate (1) is unique in producing as exclusive end product the typical irreversibly inactivated PMP-ketimine-containing complex (EI_{335}). The properties of EI_{335} and of products derived from it by NaBH₄ reduction, denaturation, and degradation, in conjunction with Dunathan's conclusions *(63, 76)* concerning geometry of substrate relative to coenzyme in transaminases, indicate (as mentioned above) that in EI_{335} the α-carboxyl of transaminated and decyclized compound (1) is in acyl linkage with a cationic site normally involved in attachment of the substrate's α-carboxyl *(235)*. This site was tentatively identified as an imidazole group because of marked instability of the acyl bond in denatured EI_{335} and other chemical and physical evidence. With the other cycloglutamates, the initial enzyme-quasi-substrate aldimines are further converted to mixtures of intermediates (in proportions depending on pH). In the case of the γ-cycloglutamates (3) and (4), the predominant end product, EI_{380}, has spectral and chemical properties of an oxime. The EI_{380} complex produced from isomer (3) (the one more readily available and having shorter $t_{1/2}$) is an oxime of transaminase with β-aminooxyglutamic acid *(226)*, evidently formed by decyclization of

TABLE XI
Kinetic Parameters of Inhibition of Asp-Transaminase by the Isomeric Cycloglutamates[a]

Inhibitor	K_i (M)	k_2 (min^{-1})	$t_{1/2}$ (min) at saturation
(1) threo-α-CG	1.3×10^{-3}	0.5	1.4
(2) erythro-α-CG	2.1×10^{-2}	0.125	5.5
(3) erythro-γ-CG	1×10^{-2}	0.125	5.5
(4) threo-γ-CG	8.2×10^{-3}	0.08	8.6

[a] Data from Snell et al. (18), p. 631.

the isoxazolidone ring at the PLP-aldimine stage, without any acylation of functional groups.

Either reduction with borohydride or exposure in weakly acidic media of EI_{335} complexes of Asp-transaminase with cycloserine (222) or threo-α-cycloglutamate (225) results in cleavage of the coenzyme's phosphate ester bond (see Sections III,B,3 and 4). No splitting at the phosphate bond occurred on similar treatment of any EI intermediates formed with erythro-γ-cycloglutamate (226). Space-filling three-dimensional models of the EI complexes showed that in intermediates formed by threo-α-cycloglutamate proximity between its free γ-carboxyl and the oxygen atom of the ester bond makes possible (at pH < 3) intramolecular acid catalysis, resulting in hydrolytic release (or transfer to a neighboring nucleophilic group) of the phosphate residue (*225*). In the *EI* intermediates with γ-cycloglutamates, absence of free γ-carboxyl groups precludes this mechanism of phosphate release, but the geometry of the pri-

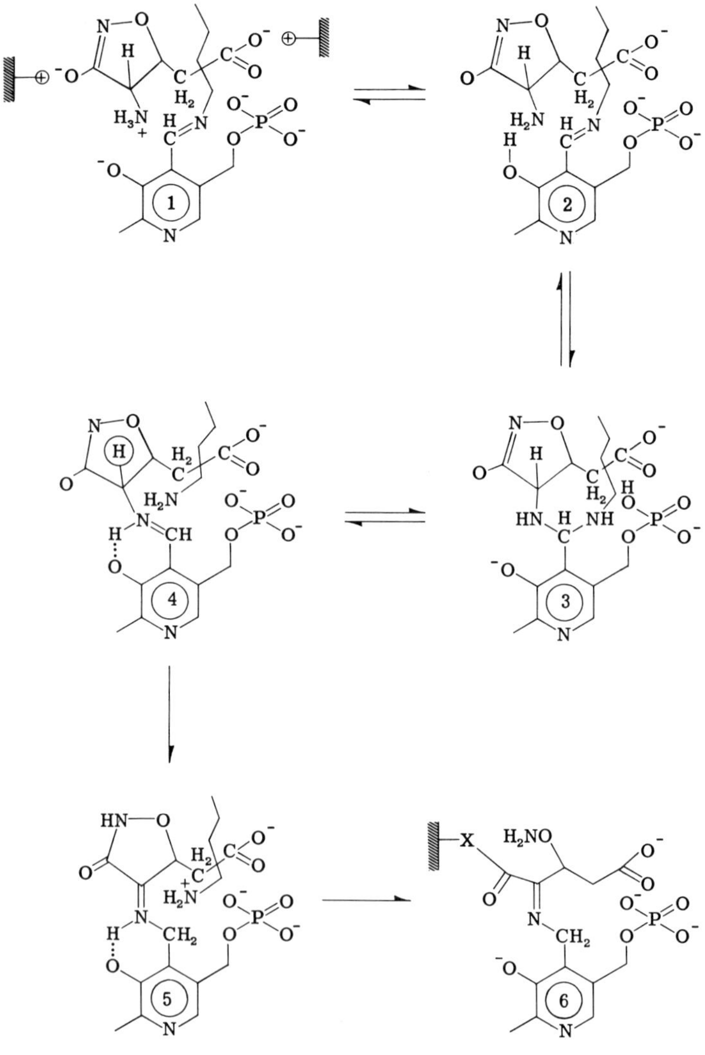

FIG. 11. Scheme of the mechanism of inactivation of aspartate transaminase by *threo*-α-cycloglutamate (*225, 235*).

mary enzyme quasi-substrate aldimine might allow intramolecular catalysis by the phosphate group of hydrolytic decyclization at the labile pseudo-acidic γ-carboxamate bond (46, 226). A similar mechanism of intramolecular acid catalysis by free γ-carboxyl might account for the previously mentioned (Section III,B,3) occurrence of de- and transphosphorylation (at pH < 2) in *EI* complexes of apotransaminase with *N*-(5'-P-pyridoxyl)-glutamic but not with *N*-(5'-P-pyridoxyl)-pyrrolidonecarboxylic acid (173, 174).

Severin *et al.* (Ref. 46, 235) suggested plausible schemes for the mechanisms of permanent inhibition of Asp-transaminase by *threo*-α-cycloglutamate [Fig. 11, closely similar to the path of inactivation by L-cycloserine, Fig. 9] and by the γ-cyclic analogs (Fig. 12), respectively. Their three-dimensional space-filling models of the *EI* complexes contributed significantly to clarification of the geometry of enzyme–substrate complexes in normal transamination, see Section III,B,7.

A series of open-chain *O*-alkyl derivatives of amino-carbhydroxamic acids, $R^3HN-CR^1H-CONH-OR^2$, were synthesized by this group of authors (43, 236). They detected a peculiar mechanism of enzyme inhibition by some of these linear *O*-alkyl carbhydroxamates, corresponding formally

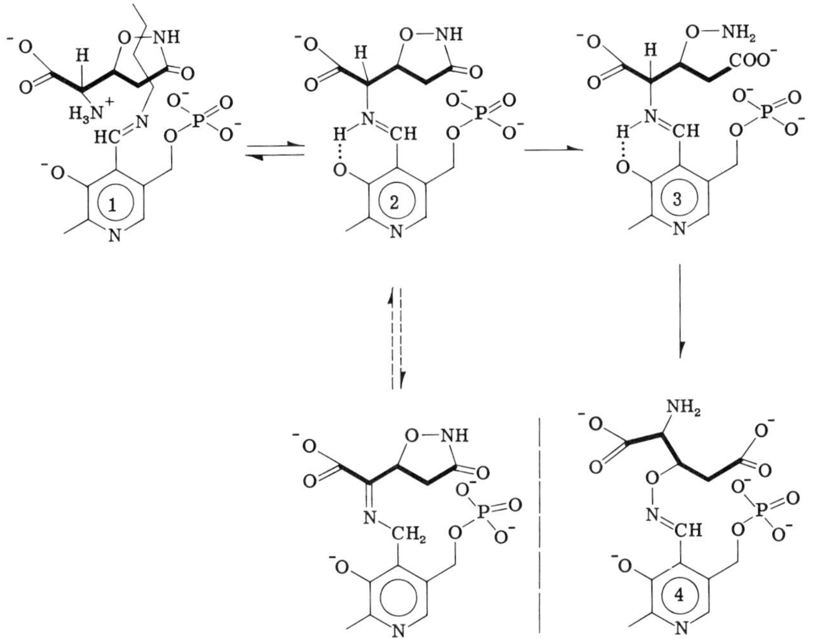

FIG. 12. Scheme of the presumable mechanism of inhibition of aspartate transaminase by γ-cycloglutamates (226).

to molecules of cycloserine analogs nicked at the 4–5 bond (equivalent to the C^α–C^β bond of an amino acid). Only those members of the open-chain series composed of nonmodified fragments of an amino substrate. For example, the O-alkyl glycine-hydroxamates (1)–(3) act as strictly selective inhibitors upon transaminases specific toward that very amino acid.

(1) $\quad NH_2-\overset{\alpha}{C}H_2-CO-NH-O-\overset{\beta}{C}H_3$

(2) $\quad NH_2-\overset{\alpha}{C}H_2-CO-NH-O-\overset{\beta}{C}H_2CH_2COOH$

(3) $\quad NH_2-\overset{\alpha}{C}H_2-CO-NH-O-\overset{\beta}{C}H_2C_6H_5$

Thus, all glutamate-specific transaminases are subject to time-dependent reversible inhibition by millimolar concentrations of (2), the open-chain analog of α-cycloglutamate; in addition, Ala-transaminase is sensitive to similar inhibition by compound (1) (dissected cycloserine) and Phe-transaminase by compound (3). In aqueous solution such compounds exist predominantly in the linear, extended conformation. On interaction with pyridoxal or PLP they exhibit no acylating properties, do not form oximes, and are fairly inert substances in other respects. The end products of transaminase inhibition by specifically active compounds of this series have the spectra of substituted PLP-oximes (λ_{max} = 380 nm) and are slowly reactivated by added oxo substrate. It was suggested (*236*) that the active carboxamate ethers with aminoacyl and alkyl groups complementary to parts of the substrate-anchoring locus are presumably bound in the enzyme's active site in a thermodynamically unfavorable pseudocyclic conformation with C^α and C^β brought into proximity, and form a PLP-aldimine. This structure and its transformations should closely resemble those of an aminoisoxazolidone *EI* complex. The inhibition scheme presented in Fig. 13 accounts for formation of protein-bound PLP-oxime as the final product.

In this noteworthy case of "compulsory fit" an appropriate enzyme forms an inactive complex by interaction with a constrained rotamer of a relatively simple molecule whose structure and properties per se would not suggest any specific inhibitory capacity. Agents of this type were designated as "conformational enzyme inhibitors" (*236*).

2. *Miscellaneous irreversible active-site-blocking agents.* In a broad sense this category includes agents inactivating Asp-transaminase by modification of an essential or strategically located group which is either selectively reactive or exposed under particular experimental conditions,

236. E. S. Severin, N. V. Gnuchev, G. K. Kovaleva, N. N. Gulyaev, and R. M. Khomutov, *in* "Pyridoxal Catalysis: Enzymes and Model Systems" (E. E. Snell *et al.*, eds.), p. 615. Wiley (Interscience), New York, 1968.

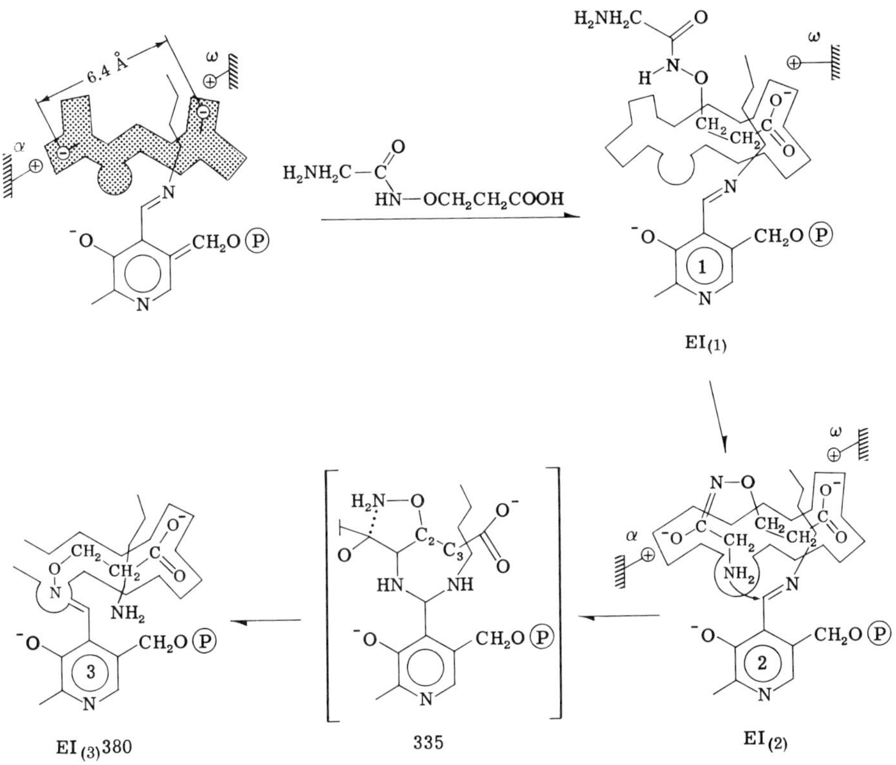

Fig. 13. Mechanism of inhibition of aspartate transaminase by glycyl-β-aminooxy-propionic acid (compound (3), pseudo-CG). After Severin et al. Ref. 236, p. 661. Shaded area: binding site for glutamate (see mechanism of action of cycloglutamates, Section III,B,7). Stages: 1, fixation of pseudo-CG in *trans* conformation at substrate-binding site; 2, S-*trans-cis* transconformation of complex $EI_{(1)}$ to $EI_{(2)}$; 3, formation of enzyme oxime, $EI_{(3)}$ (380 nm) via transient tetrahedral intermediate.

irrespective of any specific affinity of the reagent for the active site. Pertinent examples, considered in Section III,B,3, are reactions used for identification of functionally important groups of the enzyme, e.g., reduction of the PLP-lysine aldimine with NaBH₄, "syncatalytic" modification of tyrosine and cysteine residues, and the like.

In a few instances a known functional group of the active site (e.g., the reactive lysine) has been specifically cross-linked by a bifunctional reagent to a second group, thus demonstrating location of the latter in proximity.

Turano et al. (*22, 154, 155, 237*) treated apoenzyme from pre-

237. C. Turano, A. Giartosio, F. Riva, F. Bossa, and V. Baroncelli, *ABB* **117**, 678 (1966); in "Pyridoxal Catalysis: Enzymes and Model Systems" (E. E. Snell et al., eds.), p. 143. Wiley (Interscience), New York, 1968.

viously acetylated Asp-transaminase with 3-bromopropionylchloride and achieved acylation of the active-site lysine (ϵ-NH$_2$) with concomitant alkylation of one other amino acid residue. The latter was identified as histidine by isolation from the hydrolyzed product of N^1- and N^3-carboxyethyl histidines in a total yield of approximately 0.7 equivalent. No cross-linkage was obtained with bromoacetylchloride or 4-bromobutyryl chloride, evidently owing to unsuitable geometry of the C–halogen bonds. Using 1,5-difluoro-2,4-dinitrobenzene (FFDNB) as the cross-linking reagent in similar experiments, Turano (*22, 155*) observed disappearance of about one residue each of lysine and tyrosine from the modified protein; 2,4-dinitrophenylene-1-*N*-lysine-5-*O*-tyrosine and 1-fluoro-2,4-dinitrophenyl-5-*O*-tyrosine were the products recovered. The latter compound was obtained in small amount when the cross-linking reaction was prevented by preliminary acetylation of the active-site lysine. In mitochondrial Asp-transaminase the essential lysine residue undergoes N^6-carboxyethylation when the enzyme (in the PLP-form) is treated with the active-site-directed reagent, *p*-mercaptopropionate (*237a*).

In a few cases clearly in need of further study, inactivation of Asp-transaminase resulted from modification of strategic groups by affinity-labeling quasi-substrates. Thus, John and Fasella (*238*) observed that this enzyme slowly catalyzes normal transamination between L-serine-*O*-sulfate (an analog of dicarboxylic amino substrates) and 2-oxoglutarate. Concomitantly, the enzyme is gradually inactivated by a stoichiometric side reaction, involving release of inorganic sulfate from the quasi-substrate and stable blocking (alkylation?) by its organic moiety of an unknown site. Inactivation caused by transfer to a certain group of the enzyme of phosphate residues from the coenzyme (bound in a derivatized form or activated within the catalytic center by specific interactions, see Sections III,B,3 and 4) can also be viewed as a peculiar instance of affinity labeling.

d. Aberrant Side Reactions Catalyzed by Asp-transaminase. Asp-transaminase is known to catalyze atypical transformations of certain quasi-substrates with strongly electronegative β substituents. Interaction of the enzyme with the *threo* and *erythro* isomers of β-chloroglutamic acid was studied in detail by Manning *et al.* (*211*). The β-chloroglutamates were not transaminated with oxo substrates; the *threo* isomer caused competitive interference with NH$_2$ transfer from glutamate to oxalacetate. Pure Asp-transaminase in the aldimine form (not the PMP form, apoenzyme or free PLP) decomposed both isomers to equivalent

237a. Y. Morino and S. Okamoto, *Biochemistry* 11, 3196 (1972).
238. R. A. John and P. Fasella, *Biochemistry* 8, 4477 (1969).

amounts of NH_3, Cl^-, and 2-oxoglutarate, i.e., the transaminase catalyzed in this case exclusively an α,β elimination, similar in mechanism to the typical elimination reaction of amino substrates with electronegative β substitutents catalyzed by specific PLP-linked β-lyases (*16, 58–61*). It should be noted that in α,β elimination, just as in normal transamination, an early reaction step is removal of the α-H atom of the substrate (*63, 76*). The aberrant α,β elimination is much more active with *threo*-β-chloroglutamate ($K_s = 1.6 \times 10^{-2} M$; $V_{max} = 4$ µmoles/min/mg) than with the *erythro* isomer ($K_s > 1.5 \times 10^{-1} M$; $V_{max} > 0.5$ µmole/min/mg) (*211, 239*). What is apparently a similar β elimination reaction was observed much earlier by Kun *et al.* (*212*) in studies with crude extracts of mitochondrial Asp-transaminase. They found that β-fluoro-oxalacetate per se was stable on incubation with the enzyme; it acted as a competitive inhibitor ($K_i = 10^{-4} M$) of transamination between the normal substrates. When the enzyme was incubated with an amino donor (Asp, Glu, or cysteinesulfinate) and fluoro-oxalacetate, the latter was converted to oxalacetic acid with formation of ammonia in equivalent amount. The results suggest transamination of the quasi-oxo-substrate to β-fluoroaspartate, followed by defluoridation and deamination of the latter via α,β elimination.

7. *Stereochemistry of Enzymic Transamination and Chemical Topography of the Active Site*

Multipoint attachment of substrate and coenzyme at the active site of each individual PLP-linked enzyme protein imposes substrate, stereo, and reaction specificity on the system, in addition to greatly enhancing the reaction rate.

Significant progress in theoretical and experimental clarification of stereochemical aspects of biological pyridoxal-P catalysis is the result of work by Dunathan (*63, 75, 76, 240–241*). He has defined, in particular, the geometric variables determining the stereochemistry of enzymic transamination and established the actual parameters for the key step (Schiff-base interconversion).

The parameters to be considered (*63, 75, 76*) (see Fig. 14) are as follows:

239. E. Antonini, M. Brunori, P. Fasella, R. M. Khomutov, J. M. Manning, and E. S. Severin, *Biochemistry* **9**, 1211 (1970).
240. H. C. Dunathan, L. Davis, P. G. Kury, and M. Kaplan, *Biochemistry* **7**, 4532 (1968); see, also, *in* "Pyridoxal Catalysis: Enzymes and Model Systems" (E. E. Snell *et al.*, eds.), p. 325. Wiley (Interscience), New York, 1968.
241. J. E. Ayling, H. C. Dunathan, and E. E. Snell, *Biochemistry* **7**, 4537 (1968).

1. *Configuration at C^a* is defined by the absolute stereospecificity of practically all aminotransferases for L-amino acids [S symmetry for H^a].
2. *Conformation about the single bond C^a–N in aldimine* (I).
3. *Conformation about the bond $C^{4'}$–N in ketimine* (II). The arguments discussed earlier (Section II,A) require that in both cases the conformation be such as to place the C–H bond in a plane perpendicular to that of the conjugated π system. Two such conformations, 180° apart, are possible for each of the C–H bonds.
4. *Geometry of the prototropic interconversion* (I $\rightleftharpoons$ II). This will be *cis* if proton removal and proton addition occurs on the same side of the coenzyme's plane and *trans* if bond breaking and bond making occur on opposite sides.
5. *Configuration at $C^{4'}$ in ketimine* (II).

Dunathan used deuterium labeling to study the stereochemistry of reversible transamination between pyridoxamine and oxo substrate catalyzed by Asp-apotransaminase (*240*), see Section III,B,5, or by bacterial pyridoxamine-pyruvate transaminase (*241*). In both instances the reaction was strictly stereospecific: only one of the two (labeled) H atoms was lost from $C^{4'}$ of pyridoxamine in the forward and incorporated in the same position in the reverse reaction, leaving the original H–$C^{4'}$ bond of pyridoxal intact. Using (S)-($C^{4'}$-d_1)-pyridoxamine prepared by asymmetric synthesis, Dunathan could demonstrate that it is the pro-S proton of pyridoxamine which is enzyme-labile. Besmer and Arigoni (*243*) have verified this assignment of absolute symmetry: pyridoxamine with tritium in the position labile to apotransaminase was chemically degraded to glycine and further to glycolate; demonstration of S symmetry of the glycolate confirmed the S symmetry of the original ($C^{4'}$-t_1)-pyridoxamine.

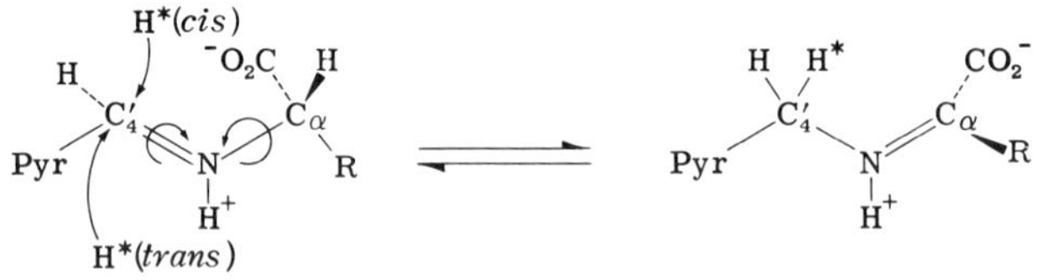

FIG. 14. Stereochemical variables in transamination (from reference *76*).

The actual geometry of prototropic shift (I $\rightleftharpoons$ II) was established as follows. In Dunathan's experiments (*240, 241*) retention of a significant amount of label in the hydrogen transferred indicated (a) that the proton was transferred by a single general-base group and its protonated acid form, and (b) that geometry of the transfer was *cis* (*242*). Arigoni and co-workers (*244*) have shown that in the active site of Asp-transaminase both the aldimine (I) and ketimine (II) coenzyme–substrate Schiff base are reduced by NaB^3H_4 with the same stereochemistry as in the proton transfer step, i.e., C^a is triated to the amino acid derivative of L(S)-symmetry while $C^{4'}$ is reduced from the *si* face to give $C^{4'}-t_1$ of S symmetry. Thus, proton transfer and borohydride reduction proceed from the same exposed face (*si*) of the Schiff base complex; this requires *cis* geometry for the proton transfer in Asp-transaminase. It is of interest that the same stereochemistry of protonation (enzymic formation of PMP tritiated in S symmetry) was observed in our laboratory (*245*) for the abortive transamination occurring as a side reaction in the active site of glutamate decarboxylase (see Section IV,E) and by Bailey *et al.* (*246*) for decarboxylation-dependent transamination catalyzed by "dialkyl amino acid transaminase" (cf. also Section IV,E).

Thus, four of the five variables of Fig. 14 are determined; namely, the C^a configuration is L(S), the symmetry of the labile proton in (II) was found to be pro-S, the conformation about the $C^{4'}=N$ bond was assumed *trans* because *cis* would result in steric hindrance to planarity, and the geometry of proton transfer was shown experimentally to be *cis* (see above). Hence, the fifth variable, conformation about C^a-N in (I), is defined as shown in Fig. 15, with the α-carboxyl of the substrate lying to the same side as the coenzyme's phenolic group. Figure 15 represents the stereochemical path of (I $\rightleftharpoons$ II) transformation (via an intermediate ion pair) in enzymic transamination (*76*). Independent evidence, arguing in favor of the same C^a-N conformation in the coenzyme-substrate aldimines, is to be found in papers by other authors (*44, 46*).

The inhibition studies of Khomutov, Severin, and associates with sterically rigid glutamate analogs (isoxazolidone derivatives) and other quasi-substrates (Section III,B,6,*c*) shed light on the presumable conformation of the whole substrate molecule in the Michaelis complex (*ES*)

---

242. Dunathan's suggestion that the ϵ-NH_2 group of the active-site lysine might function as the catalytic general base accounts for the observed extent of loss of label transferred (since the hydrogen accepted randomizes with two other H atoms in the conjugate acid form of this group), but partial dilution of the mobile proton would be possible during *cis* transfer by any other single acid-base group.

243. P. Besmer and D. Arigoni, *Chimia* 23, 190 (1969).

244. P. Besmer, E. Bertola, and D. Arigoni, cited from Dunathan (*76*); see, also, paper presented by D. Arigoni in *Int. Congr. Biochem., 8th, 1969*, Symposium 3 (1970).

245. B. S. Sukhareva, H. Dunathan, and A. E. Braunstein, *FEBS Lett.* 15, 241 (1971).

246. G. B. Bailey, T. Kusamrarn, and K. Vuttivej, cited in Dunathan (*75, 76*).

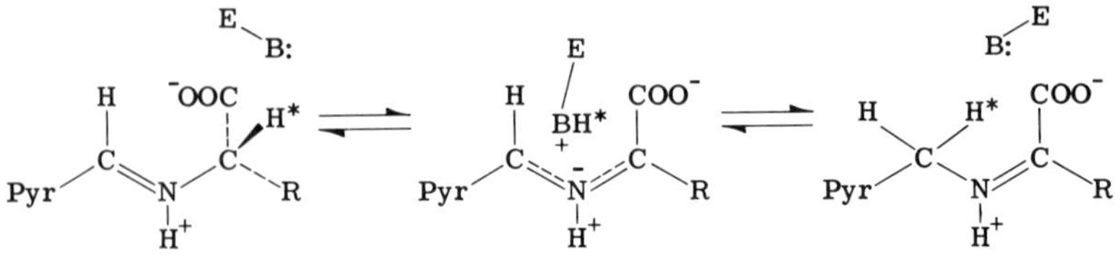

FIG. 15. Conformation about the C^α–N bond and stereochemical path of PLP-aldimine (I) ⇌ PMP-ketimine (II) transformation in enzymic transamination. From Ref. 76.

with Asp-transaminase (46). The two cycloglutamates with highest affinity—the *threo-α* and *threo-γ* isomers [Fig. 10, compounds (1) and (4)]—are evidently those whose rigid cyclic molecules fit best into the substrate-binding site. Inspection of space-filling models shows that the carbon chain of compound (4) can readily assume, by adjustment of its flexible part, the fixed geometry of atoms C^1–C^4 in compound (1); that of atom C^5 is fixed in compound (4), and in isomer (1) free rotation about the C^3–C^4 bond allows bringing atom C^5 into the only position that makes the skeletons of (1) and (4) completely superimposable. This conformation (projection formulas A and B in Fig. 16) is probably very close to that of glutamate in the *ES* complex (formula C).

Assuming a distance of approximately 6.4 Å between centers of the two carboxyl-binding cationic groups of the enzyme, we find that these groups will be within effective reach for ion pair formation with carboxyl groups α and γ of glutamate in the extended conformation shown in Fig. 16D, (C-1)O_2 and (C-5)O_2, on the one hand, and with carboxyls α and β of aspartate (or inhibitory C^4-dicarboxylates) in the *cis* (maleinoid) conformation, (C'-1)O_2' and (C'-5)O_2' on the other (46).

It should be noted that while the geometric factors optimal for substrate binding may not be exactly the same which favor rapid reaction (75, 76), they are probably quite similar. If we assume that the conformation of protein-bound C_4- and C_5-amino substrates does not undergo substantial change on formation of their PLP-aldimines, then the unique geometry of glutamate and aspartate allowing anchoring of their carboxylate groups at the same two cationic sites (superimposed formulas on Fig. 16 D) also accounts for similarities in peculiar properties of the imines formed by Asp-transaminase with *erythro*-β-hydroxy-L-aspartate (formula F), on the one hand (Section III,B,6,*b*), and *threo*-β-hydroxy (formula E), resp. *threo*-β-chloro, derivatives of glutamic acid, on the other; the stereochemistry for the β-substituent relative to H–C^α bond is similar (*trans*) in the compounds mentioned.

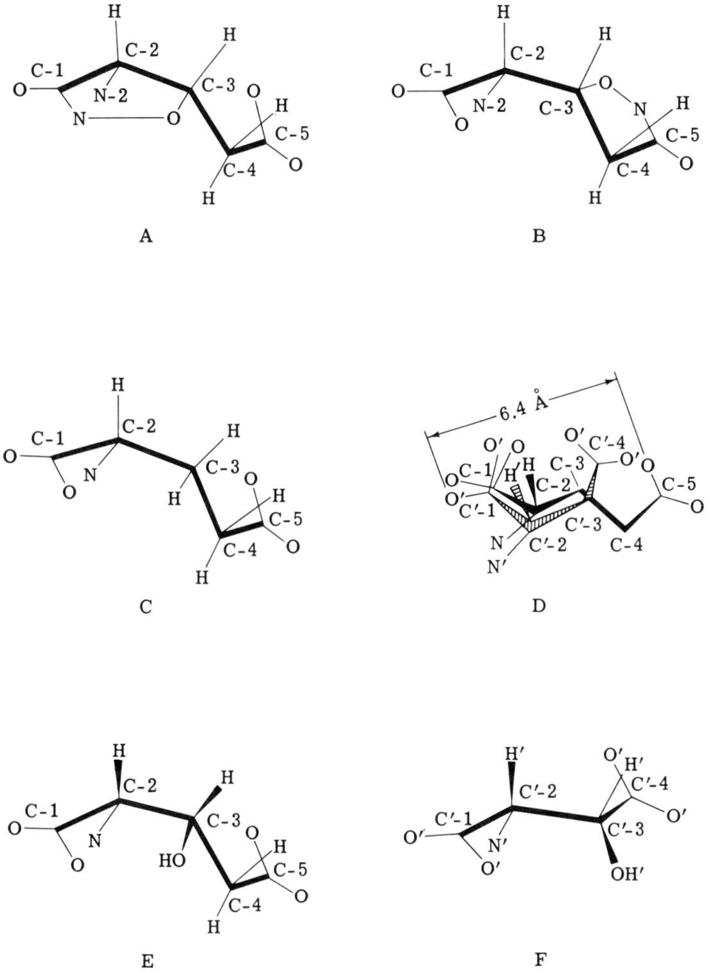

Fig. 16. Projection formulas of the cyclic glutamate analogs and presumable conformations of amino substrates and quasi-substrates at the active site in Michaelis complexes of aspartate transaminase. After Severin et al. (46). A, threo-α-cycloglutamate; B, threo-γ-cycloglutamate; C, glutamate (in superimposable extended conformation); D, binding of α and ω carboxyls of aspartate (symbols of all atoms primed) and of glutamate at the same cationic sites (6.4 Å apart); E, threo-β-hydroxyglutamate; and F, erythro-β-hydroxyaspartate.

8. *Dynamic Spatial Aspects of Enzymic Transamination*

Like most enzyme-catalyzed reactions, transamination proceeds through a series of consecutive steps, each requiring a particular set of optimal conditions—alternating or even conflicting—with regard to ioni-

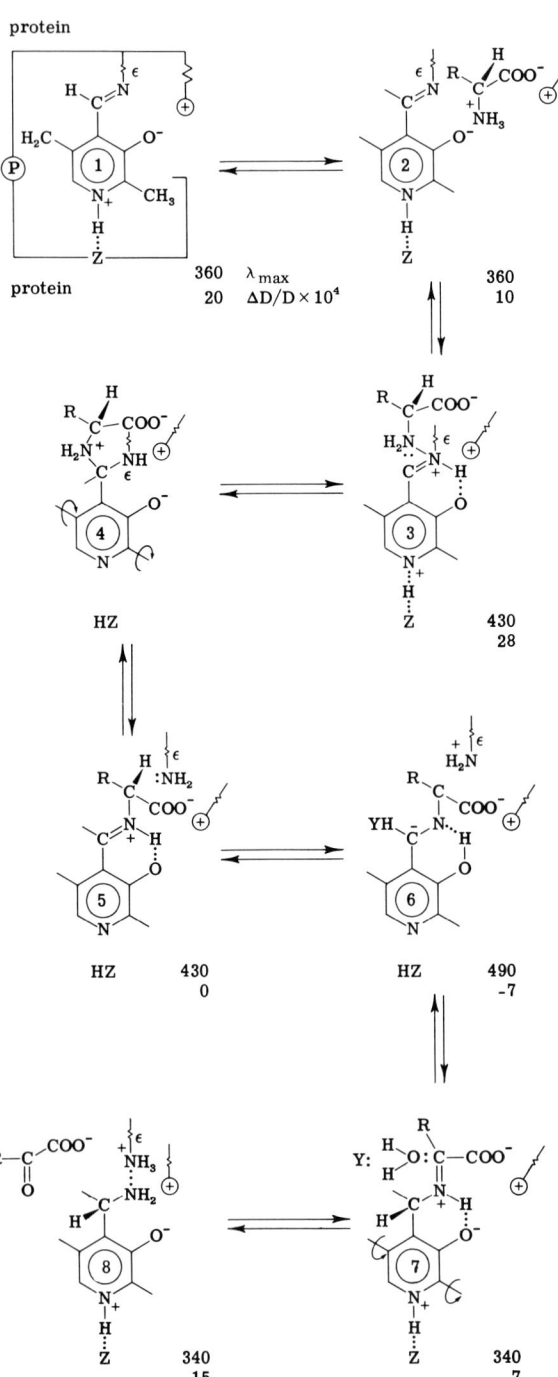

zation states, orientation and interactions of functional groups of the protein, cofactor, and substrates. The catalytic mechanism can be interpreted adequately only in terms of the sequential transitions in electronic and geometrical patterns of enzyme–substrate intermediates (43).

A dynamic molecular model for enzymic transamination, embodying essential physical-organic and stereochemical inferences from recent experimental data with general features of reaction schemes proposed earlier, was developed by Karpeisky and associates (43, 44, 247, 248). Their model is presented below in concise outline with comments and partial revision of certain features in the light of newer evidence considered earlier in this chapter, e.g., concerning the stereochemistry of substrate binding and tautomeric proton shift, and functions of essential imidazole groups.

A crucial feature of the dynamic model is the assumption that cooperative proton shifts involved in the transaldimination and subsequent reaction steps are linked with reversible translocation (reorientation) of the coenzyme. The postulated sequential steps (44) are illustrated in the (condensed) scheme of Fig. 17 (43). Assignment of absorption maxima (λ_{max}) and factors of anisotropy ($\Delta D/D$) to definite structures shown in the scheme rests on observed optical properties (a) of stable forms of the enzyme; (b) of transient normal reaction intermediates, studied by fast-reaction techniques (69, 134, 183, 209); and (c) of stabilized complexes with similar structures between Asp-transaminase and inhibitory quasi-substrates, e.g., α-methylaspartate (119–122, 210) cycloserine or threo-α-cycloglutamate (222, 235), and erythro-β-hydroxyaspartate (69, 121, 122, 185, 191, etc.; cf. Sections III,B,2,4, and 7).

Let us consider the reaction steps shown in Fig. 17.

1. In the active nonprotonated aldimine form of Asp-transaminase (stage 1) the phenolic group of PLP is ionized. Its pK_a is 6.2 in contrast to pK_a of ~11.0 for free PLP-aldimines. This large difference in pK values is partially accounted for by the postulated hydrogen bonding of the pyridine N atom with a proton-donating group (HZ) of the protein (135, 136, 177), but a positive charge on this N atom would lower the pK_a only to ~8.0. Further lowering of pK_a to 6.2 is attributed (44) to

247. M. Ya. Karpeisky and V. I. Ivanov, *Nature* (*London*) **210**, 494 (1966); A. E. Braunstein, V. I. Ivanov, and M. Ya. Karpeisky, in "Pyridoxal Catalysis: Enzymes and Model Systems" (E. E. Snell et al., eds.), p. 291. Wiley (Interscience), New York, 1968.

248. V. I. Ivanov and M. Ya. Karpeisky, *Mol. Biol.* (*USSR*) **1**, 288 and 588 (1967).

FIG. 17. Dynamic model of enzymic transamination. According to Ivanov and Karpeisky (44) (see Ref. 43).

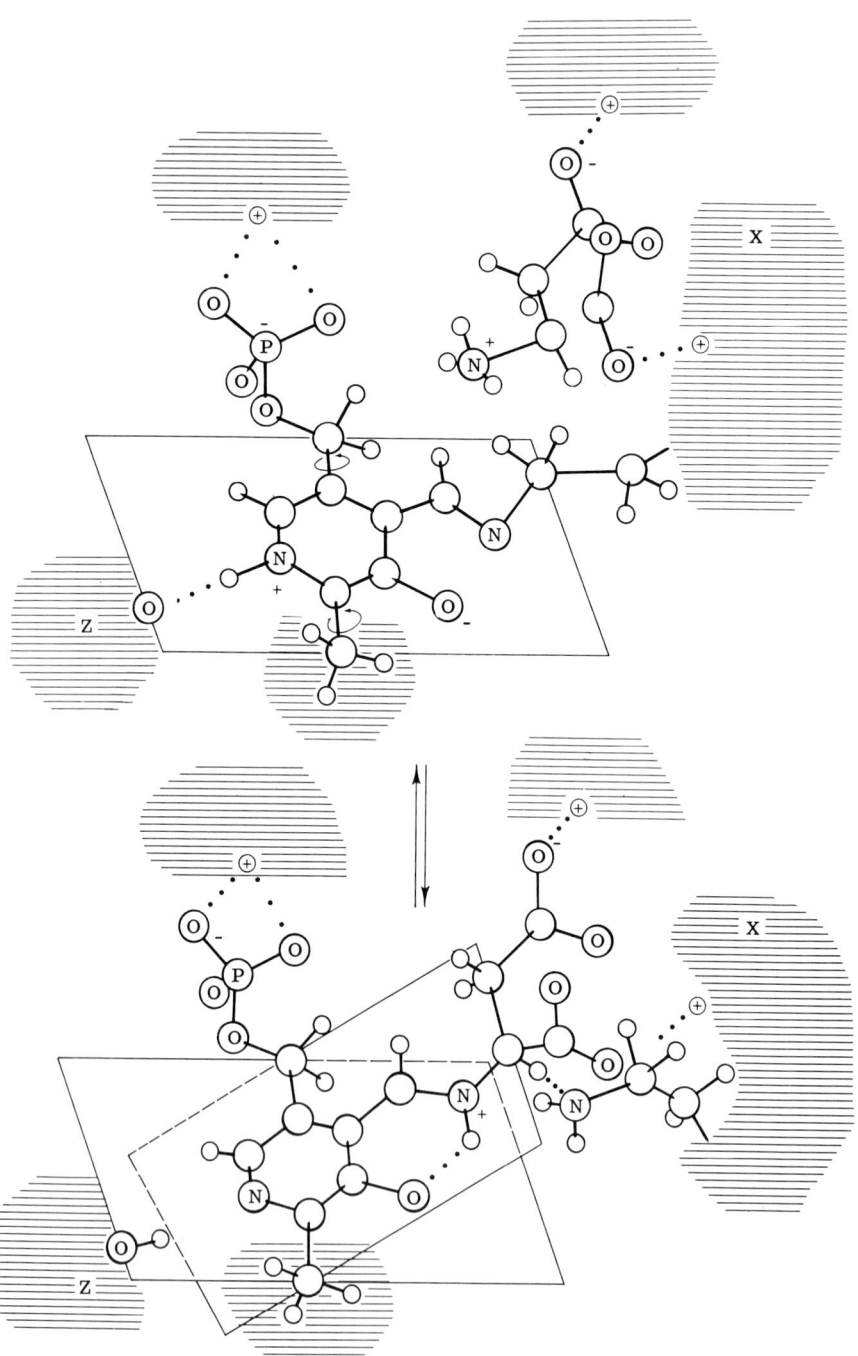

the presence of a cationic group ⊕ in close proximity to the phenolic group.

2. In the optimum range of pH for enzymic transamination all dissociable groups of the amino substrate are ionized and, as indicated by specific affinity of the enzyme for dicarboxylic substrates, the two negatively charged carboxylate groups bind to suitably located cationic groups of the enzyme. It is assumed that the α-carboxylate group binds to the same cationic site ⊕ (stage 2) whose interaction with the phenolic group was considered above as one of the factors lowering pK_a of the internal aldimine. The effect of this factor is eliminated, and the pK_a value rises to ~8.0 when the cationic group ⊕ is neutralized by the α-COO⁻ group of amino substrate. At the same time, neutralization of the ionized α-carboxyl lowers the basicity of the substrate's amino group, shifting its pK from 9.6 to 7.4. This makes possible the transfer of a proton from the substrate NH_3^+ group to the coenzyme (stage 3). Now the un-ionized, nucleophilic amino group is in proximity to the polarized, highly reactive C=N bond of the protonated Schiff base.

As noted earlier, Jenkins (160) tentatively suggested the imino N of this Schiff base as the cationic site ⊕ binding the α-carboxyl, but the data of Martinez-Carrion (158, 159a) and evidence obtained by Severin et al. (46, 226) point rather to identity of this site with an essential imidazole group. The latter assignment is attractive because a suitably located imidazole ring might easily provide, at this and later reaction steps, for the required shunting of protons between α-carboxyl, phenolic hydroxyl, the N atoms of internal and external Schiff bases and amino substrates, with concomitant poising of pK_a values of the structures involved.

The geometry of substrate binding has been discussed in Sections III,B,6 and 7.

3. For the next step (transaldimination), this geometry admits only one orientation, allowed by the stereochemistry of nucleophilic addition (25); namely, the orbital of the lone electron pair of the substrate NH_2 group must be directed toward the $C^{4'}$ atom along the reaction coordinate in a plane perpendicular to that in which the imino group is located. The initial distance between the centers of amino N and $C^{4'}$ is no less than 3.5 Å (sum of van der Waals radii). To produce the intermediate tetrahedral adduct (stage 4) and, further, the protonated substrate aldimine (stage 5), this distance must be reduced to 1.5 Å (length of a C–N single bond), see Fig. 18 (44). This requires a change in disposition either of

---

FIG. 18. Scheme of the spatial structure of ES complex, A (stage 3 of Fig. 17) and of the enzyme-substrate aldimine, B, (stage 5 of Fig. 17). From Ref. (44), Figs. 7 and 8.

substrate or of coenzyme (or of both) relative to the protein. Inspection of three-dimensional molecular models shows that the necessary reorientation can be achieved easily by swinging of the coenzyme ring through an angle of $\sim 40°$ around an axis passing through $C^{2'}$ and $C^{5'}$ of the coenzyme. One end of the axis is fixed by the 5'-phosphate-protein bond; its other end faces the hydrophobic site normally holding the 2-methyl group. To provide freedom of rotatory displacement, the hydrogen bond between the pyridine N and group Z has to be broken. This requirement is met in the preceding step (stage 3) when the ionized phenolic group accepts a proton from the NH_3^+ group; the associated lowering of pK of the ring nitrogen to ~ 6.5 results in proton transfer to group Z with breaking of the hydrogen bond (Figs. 17 and 18).

With the pyridine ring rotated by $\sim 40°$, covalent binding of atom $C^{4'}$ to the substrate N atom produces the tetrahedral intermediate 4; localization of the cationic positive charge on the lysine ϵ-N atom facilitates its elimination and completion of the transaldimination step, i.e., formation of the substrate aldimine (stage 5). A minor additional rotation of the formyl carbon about the $C^{4'}-C^4$ bond may occur in the transaldimination step (*44*). The change in mutual orientation of coenzyme and protein associated with this step is indicated by disappearance of CD in the main absorption peak of enzyme-bond substrate aldimine. Further confirmation comes from recent studies by Arigoni (*249*). On reduction of pure Asp-transaminase with sodium borotritide he demonstrated that carbon atom $C^{4'}$ had added ^{3}H in R configuration, i.e., that in the internal Schiff base the *re* face of the imine bond was exposed, in contrast to the substrate aldimine, in which the opposite, *si* face was shown to be exposed in similar earlier studies (*243*), see above.

That the conformational change associated with transaldimination is not confined to reorientation of the formyl carbon atom ($C^{4'}$) alone is indicated by the phenomena of "syncatalytic" unmasking of a tyrosyl and a cysteinyl residue (see Section III,B,3). It will be remembered that Ivanov and Karpeisky (*44*) had tentatively identified HZ as the hydroxyphenyl group of a tyrosine residue hydrogen bonded to the pyridine N, mainly on the basis of changes in the CD spectra in the near-ultraviolet range in stable quasi-substrate aldimines (similar to stage 5) and in complexes of apotransaminase with certain PLP analogs (see Section III,B,3). These spectra revealed expected reciprocating changes in the negative CD extremum at 295 nm (λ_{max}^A of the tyrosine anion) and a positive CD peak at 280 nm (λ_{max}^A of nonionized tyrosine) (Fig. 19). The assignment is in need of additional substantiation.

249. D. Arigoni, *Analys. Simul. Biochemical Systems, Fed. Eur. Biochem. Soc. Symp., 8th,* **25** (1972).

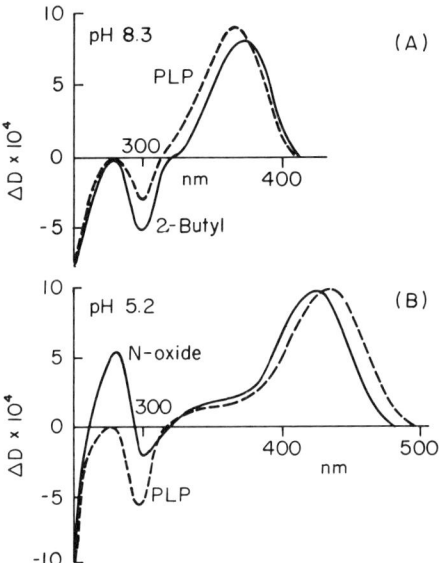

FIG. 19. Circular dichroism spectra of complexes of apotransaminase with PLP and some coenzyme analogs. According to Ivanov and Karpeisky (see Ref. *43*). (A) At pH 8.3: 1, native holoenzyme; 2, complex with 2-*n*-butyl analog of PLP. (B) At pH 5.2: 3, complex with PLP *N*-oxide; 4, native holoenzyme.

4. The mechanism and stereochemistry of prototropic aldimine–ketimine rearrangement (stages 5–7) was depicted in the dynamic scheme (Fig. 17) as involving acid-base catalysis by two functional groups—one accepting H⁺ from C^α (assumed tentatively to be ϵ-NH₂ of the active site lysine), and another which protonates $C^{4'}$ at stage 7. This assumption is no longer tenable; the correct mechanism and geometry of this step, elucidated by Dunathan (*76*)—*cis* proton transfer by a single acid base group (ϵ-NH₂ of reactive lysine residue or an imidazole group?)—has been discussed in Section III,B,6. This stereochemistry is entirely compatible with the dynamic model and accounts for transient formation of an anionic (quinonoid) intermediate (stage 6).

The $C^{4'}$ atom acquires tetrahedral configuration on protonation (stage 7). This allows the coenzyme ring to rotate back into the original plane with concomitant reprotonation of the pyridine N by group HZ and increase in acidity of the phenolic hydroxyl.

5. Ionization of the phenolic group and binding of its proton to the imino N favors nucleophilic addition of a molecule of water at the C^α atom. Via the transient hydrated ketimine, the α-oxo acid and protein-bound pyridoxamine-P are formed (stage 8), completing the first half-reaction of transamination.

6. For interaction with the CO group of an oxo substrate molecule in

the second half-reaction of transamination, the amino group of PMP must be in the un-ionized form. In the enzyme (stage 8) this group is in the vicinity of two cationic groups—the lysine ε-amino group and the ⊕ site that binds substrate α-carboxyl groups. Neutralization of one of the cationic groups by the ionized α-carboxyl of oxo substrate will raise the nucleophilic state of the PMP amino nitrogen.

The article of Ivanov and Karpeisky (*44*) discusses further details of the dynamic model and important inferences relating to peculiar features of enzymic transamination and to general aspects of enzyme catalysis.

The molecular mechanism outlined above, manifestly conjectural in many points (including the concrete geometry of alternating orientation of the coenzyme), attempts to integrate currently available information derived from indirect evidence. The scheme is subject to improvement as new information accumulates. Accurate description of the reaction mechanism will have to await precise knowledge of the enzyme's three-dimensional chemical topography.

IV. Aminotransferases Acting on Other Substrates

An extensive literature has accumulated about a broad variety of aminotransferases differing in substrate specificities and biological origin. Competent surveys are available, e.g., Refs. *2, 5, 6, 23, 25, 52*.

In the text of this section and in the cumulative Table XII these aminotransferases are arbitrarily divided into subgroups based on substrate specificity, as follows: (A) Requiring Glu–oxoglutarate (or Asp–oxalacetate) as one donor–acceptor pair; (B) acting on two α-amino–α-oxo monocarboxylic substrate pairs; (C) acting on ω-amino and ω-oxo acids; and (D) acting on amino and oxo substrates containing non-carboxylic acid groups (or none). A special section (E) deals with transaminations occurring as side reactions in pyridoxal-P-linked enzymes of other types.

Most of the published papers are concerned with the occurrence and general description of the enzymes and with biological and medical aspects that cannot be dealt with adequately within the scope of this chapter, such as metabolic functions in biosynthetic and biodegradative pathways, biochemical genetics and evolution, and regulation of the action and synthesis of individual aminotransferases, their pathogenetic and diagnostic significance.

In recent years, a number of aminotransferases from animal tissues and microorganisms have been extensively or partially purified, and an excellent compendium of the relevant experimental techniques has been published (23). Thus far, these researches have not contributed substantially to our knowledge of molecular aspects of enzyme-catalyzed transamination.

Accordingly, the present section is restricted, in the main, to a cumulative list of properties of the better known aminotransferases (number in Enzyme List from Revised IUB Recommendations (1972) on Enzyme Nomenclature; see Table XII). Additional comments on their noteworthy features are given below, together with mention of some interesting enzymes not yet thoroughly studied.

A. REQUIRING GLU–OXOGLUTARATE (OR ASP–OXALACETATE) AS ONE DONOR–ACCEPTOR PAIR

In addition to Asp-aminotransferase (EC 2.6.1.1), considered at length in Section III of this chapter, a few other enzymes requiring two dicarboxylic NH_2-donor–acceptor pairs were claimed to exist as independent entities. For the enzyme transferring NH_2 from glutamate to 2-oxoadipate in the biosynthesis of lysine in yeasts (250) and other fungi, individual existence seems to be proven (see also Ref. 251 on partial purification of a similar enzyme from rat liver mitochondria). Transamination from γ-hydroxyglutamate to oxoglutarate or oxalacetate in the course of 3-hydroxyproline catabolism (252) is the result of group-specific activity of EC 2.6.1.1 rather than of a special transaminase, EC 2.6.1.23. Phosphoserine-glutamate transaminase (EC 2.6.1.52, Table XII, group A) has a dianionic monocarboxylic donor–acceptor couple as the second substrate pair (253, 254).

A micrococcal glycine:oxalacetate aminotransferase (EC 2.6.1.35, Table XII, group A), functioning in glycine biosynthesis, is thus far unique in requiring aspartate as the preferred dicarboxylic NH_2 donor (255, 256).

Among the most active and best-characterized high purity aminotransferases from mammalian tissues are the alanine transaminases (EC

250. M. Matsuda and M. Ogur, JBC **244**, 3352 (1969).
251. Y. Nakatani, M. Fujioka, and K. Higashino, BBA **198**, 219 (1970).
252. A. Goldstone and E. Adams, JBC **237**, 3476 (1962); cf. BBA **78**, 629 (1963); H. Wada and Y. Morino (33).
253. H. Hirsch and D. M. Greenberg, JBC **242**, 2283 (1967); "Methods in Enzymology," Vol. 17B, p. 331, 1971.
254. D. E. Walsh and H. J. Sallach, JBC **241**, 4068 (1966).
255. R. G. Gibbs and J. G. Morris, BBA **85**, 501 (1964); "Methods in Enzymology," Vol. 17A, p. 981, 1970.
256. H. L. Kornberg and J. G. Morris, BJ **95**, 577 (1965); R. G. Gibbs and J. G. Morris, ibid. **99**, 27P (1966).

TABLE
PROPERTIES OF THE BETTER KNOWN

No. in Enzyme List	Source	"Standard" (EC) NH₂ donors and acceptors, and their K_m (mM)	Other substrates (relative reaction rates; rate with "standard" donor = 100)	K_{eq} (app) (arrow indicates forward or reverse direction of reaction)	Degree of purification	Specific activity (μmoles/min/mg)
						A. Require Glu-Oxoglutarate (or
2.6.1.2	Pig heart (cyt)	Ala(38) + 2-oxo-glutarate(0.4) ⇌ Glu + pyruvate	α-Abu(1.4); α-Aad(0.4)	← 1.6	380×	340 (turnover no. = 1100 sec⁻¹)
2.6.1.2	Rat liver	Ala(34) + 2-oxo-glutarate(1.1) ⇌ Glu + pyruvate	None found	← 1.6	550× (crystalline; homogeneous)	276
2.6.1.35	M. denitrificans	Gly + oxalacetate ⇌ Asp(1.9) + glyoxylate (0.43)	Ser(16); Asn(7); threo-β-HO-Asp (2)	← 61 (pH 7.1; 25°)	41× (partially purified)	12.3 (V_{max} = 43)
2.6.1.52	Sheep brain	Ser(P) + 2-oxo-glutarate ⇌ Glu(0.25) + 3-P-hydroxy-pyruvate(0.7)	Glu(100); Ala(10)	→ 0.128	>400× (ultracentr. homogeneous)	1.29
2.6.1.43	Pig heart (cyt)	Leu(5) + 2-oxo-glutarate(5) ⇌ Glu + ketoLeu	Ile(100); Val(90); Leu(66); Abu (22); Met(11); SMeCys(8), etc.	→ 1.75	5000× (~90% pure)	51
2.6.1.42	Pig heart (mit)	Leu(0.4) + 2-oxo-glutarate(2.2) ⇌ Glu + ketoLeu	Ile(110); Leu(100); Val(50); Nva, Nleu (weak)		250× (partially purified)	5.1
2.6.1.42	S. typhimurium	Leu + 2-oxoglutarate ⇌ Glu + ketoLeu	Ile(110); Val(60)		150× (pure, crystalline)	77
2.6.1.6	Rat liver	Leu(25) + 2-oxo-glutarate(0.07) ⇌ Glu + ketoLeu	None found		500×	1.03
2.6.1.5	Rat liver (cyt)ᶠ	Tyr(1.4) + 2-oxo-glutarate(0.7) ⇌ Glu + ketoTyr	Phe(8); Trp(1.6); Tyr(I₁)(100); Tyr(3HO)(16)		2500× (>95% pure)	500
2.6.1.5 (2.6.1.27)	A. euridyceᵃ	Phe(Tyr) + 2-oxo-glutarate (oxalacetate, phenylpyruvate) ⇌ α-amino acid + ketoPhe (ketoTyr)	Broad specificity: Tyr,Phe,Trp (104 → 34); Asp (22); kynurenine(4) — activity ratios constant on purification		210× (partially purified)	~49 (Phe → Glu)

* Symbols for some less common substrates according to CBN Recommendations: α-Abu, α-aminobutyrate; α-Aad, α-aminoadipate Nva, norvaline; Nleu, Norleucine; Tyr(3HO), 3′,4′-dihydroxyphenylalanine (dopa); β-AisoBu, β-aminoisobutyrate; α,γ-Dbu, α,γ-di aminobutyrate; γ-Abu, γ-aminobutyrate; δ-Aval, δ-aminovalerate; ε-Acap, ε-aminocaproate; Thia-Lys, S-(2-aminoethyl)-L-cysteine.

XII
PURIFIED AMINOTRANSFERASES*

Optimal pH	MW and subunit structure	K_{PLP} (μM)	Absorption peaks (nm) at low pH	Absorption peaks (nm) at high pH	Ref.	Footnotes and comments
Asp-Oxalacetate) as One Donor–Acceptor Pair						
8.0	115,000 (sedimentation),[a] 80,000 (dimer?) (PLP content)	a	425, 325	330, 430 ($pK_a = 7.4$)	259	[a] Reversible resolution not achieved. Sensitive to thiol reagents, L-cycloserine, carbonyl reagents, cysteine. Mitochondrial isoenzyme see Ref. 260.
7.0–9.0	114,000 (dimer)	b	430	335; 400–450 ($pK_a = 7.2$–7.4)	257, 258	[b] Reversibly resolved after p-chloromercuribenzoate treatment. Inhibitors: aminooxyacetate ($K_i < 10^{-7}\,M$); thiol reagents; NH_2OH (10^{-4}–$10^{-3}\,M$).
~7.0		Requires PLP, dissociable			255, 256	Forms part of the β-hydroxyaspartate system of glycine metabolism (256). Inhibitors: β-hydroxy-Asp, maleate, acetonitrile, NH_4^+ ions.
8.15	96,000	Slow resolution	415	—	253	Also found in liver and other animal tissues (254) and in bacteria.
8.3–8.5	~75,000 (sedimentation) (monomer?)	High PLP affinity[c]	414 (spectrum pH-independent)	—	208, 208c	[c] Easily resolved in PMP form. Inhibitors: NH_2OH ($10^{-7}\,M$) and other CO reagents; carboxylic substrate analogs; thiol reagents (reversed by thiols; they also activate the pure enzyme). Aggregation in absence of thiols. A similar soluble enzyme purified 300× from hog brain, Ref. 208b.
8.6	~75,000	14 μM			208c	Requires presence of thiols for activity.
7.8		d	425, 320	320 (425 bleached) ($pK_a \sim 8.2$)	261	[d] PMP form resolved on dialysis in presence of mercaptoethanol. Functions in biosynthesis.
8.7		4 μM[e]			262	[e] Resolved in the presence of NH_2OH. Not activated by thiols, but inactivated completely by $10^{-3}\,M$ p-chloromercuribenzoate.
7.5–7.6	~115,000 (tetramer)	0.017 (4 eq. per mole)	415	415	263, 263a	[f] The mitochondrial aromatic transaminase is identical with Asp-transaminase$_{mit}$ (26, 27). The enzyme from pig brain was purified 900-fold and studied in some detail (263b). Inhibitors: P_i($10^{-3}\,M$); aromatic 2-hydroxy acids; thiol reagents. Inducible by glucocorticoids. Amino acid analysis in Ref. 263.
7.5–8.0		Requires PLP			263c	[g] Very broad donor and acceptor specificity. Preferred acceptors are: oxoglutarate > oxalacetate > phenylpyruvate. Inducible by Phe in nutrient medium.

TABLE XI

No. in Enzyme List	Source	"Standard" (EC) NH₂ donors and acceptors, and their K_m (mM)	Other substrates (relative reaction rates; rate with "standard" donor = 100)	K_{eq} (app) (arrow indicates forward or reverse direction of reaction)	Degree of purification	Specific activity (μmoles/min/mg)
2.6.1.24-2.6.1.26	Rat kidney (mit)	Halogenated Tyr + 2-oxoglutarate ⇌ Glu + halogenated ketoTyr	Thyroxine(100); I₃-Thyronine (66); Tyr(I₂) (70); Tyr(Br₂) = Tyr(Cl₂) (140); Tyr(I₁) (100). OxoGlu (100); oxaloacetate(58); pyruvate(5)		32× (partially purified)	1.6 [Tyr(I₂) → Glu]
2.6.1.27	C. sporogenes	Trp(2.7) + 2-oxoglutarate(0.16) ⇌ Glu + ketoTrp	Tyr; Phe		~200× (partially purified)	—
2.6.1.4	Liver: [Rat (Ref. 280)] [Human (Ref. 281)]	Gly + 2-oxoglutarate ⇌ Glu (4.16) + Glyoxalate(8.3)	Ala(16); Glu(39); Met(17); Arg(12)	Practically irreversible	~290× (partially purified)	—
2.6.1.21	B. subtilis	D-Ala(3.2) + 2-oxoglutarate (1.2) ⇌ D-Glu + pyruvate	D-Isomers of Glu (39), Abu(100), Asp(41), Asn (62); Orn, Nle, Ser, Met (all <15)	⇌ 1.6	1,100× (nearly homogeneous)	90 μmoles/10′/mg

B. Acting on Two α-Amino/α-O

| 2.6.1.44 | Human liver | Ala(1.0) + glyoxylate → Gly + pyruvate | Donors (weak): Ser,Arg,Trp Acceptors: 3-hydroxypyruvate | Reverse reaction not demonstrated | 800× | 10.8 |
| 2.6.1.15 | Rat liver (cyt) | Gln(6) + an α-oxo acid ⇌ an α-amino acid + 2-oxoglutaramate | Donors (weak): most monocarboxylic α-amino acids. Acceptors: most monocarboxylic α-oxo acid | Reaction reversible | 700× (virtually homogeneous) | 243 |

C. Acting on ω-Ami

| 2.6.1.18 | P. fluorescens | β-Ala(62) + pyruvate(14) ⇌ α-Ala + malonic semialdehyde | Donors: γ-Abu (116), δ-Aval (118), ε-Acap (134), β-Aisobut(160). Acceptors: glyoxylate(17); α-oxobut(9) | → ~0.2 (pH 8; 35°) | 84× (partially purified) | 56.6 |
| 2.6.1.19 | P. fluorescens | γ-Abu + 2-oxoglutarate ⇌ Glu + succinic semialdehyde | None found | → ~0.1 | 57× (partially purified) | 3.4 |

(Continued)

Optimal pH	MW and subunit structure	K_{PLP} (μM)	Absorption peaks (nm) at low pH	at high pH	Ref.	Footnotes and comments
^{h}	80,000 (filtration through Sephadex G-100)	4.1^h			265, 266	^{h} Purified enzyme is completely resolved; pH optimum varies with substrate from 6.0 with Tyr(I$_2$) to 7.0–7.5 with Tyr(I$_1$). K_m for Tyr(I$_2$) = 1.3 mM, for Tyr(I$_1$) = 6.5 mM. Hormone transaminase in thyroid gland (EC 2.6.1.26) probably identical with haloaromatic transaminase (EC 2.6.1.24) (264).
8.4	~97,000	2.18				
7.3					280, 281	
8.2–8.8	53,000 ± 4,000 (monomer)		<u>415</u> (CD negative) (spectrum pH-independent)	<u>415</u> (CD negative)	276	Inhibitors: D-cycloserine (10^{-7} M); NH$_2$OH ($<10^{-4}$ M). Insensitive to thiol reagents. Probably identical with D-Asp-transaminase (EC 2.6.1.10) and possibly with the D-Met:pyruvate enzyme (EC 2.6.1.41).

Monocarboxylic Substrate Pairs

Optimal pH	MW and subunit structure	K_{PLP} (μM)	at low pH	at high pH	Ref.	Footnotes and comments
8.4		~1.0			279	Native enzyme is easily resolvable, can be activated and stabilized with PLP. Inhibition: 85% by 10^{-3} M NH$_2$OH.
8.5		^{i}	<u>420</u>	<u>420</u>	283, 284	^{i} Partially resolved; can be stabilized by PLP, thiols, and oxo acids. Strong inhibition by SH reagents, by phenylpyruvate (90% at 10^{-1} M). The similar L-asparagine:α-oxoacid aminotransferase (EC 2.6.1.14) has not yet been adequately purified (285). L-Cycloserine is selectively inhibitory ($I_{58} = 10^{-5}$ M) for EC 2.6.1.14 and EC 2.6.1.27 but not for EC 2.6.1.15 (228).

nd ω-Oxo Acids

Optimal pH	MW and subunit structure	K_{PLP} (μM)	at low pH	at high pH	Ref.	Footnotes and comments
~9.2		~20^j			286	^{j} Resolved as the oxime; full reactivation by 5×10^{-5} M PLP. Inactive with dicarboxylic donors or acceptors.
>8.8		^{k}			287	^{k} Not resolved; no activation by PLP. Inhibitors: KCN ($I_{50} = 10^{-5}$ M); NH$_2$OH ($I_{50} = 3 \times 10^{-6}$ M).

TABLE XII

No. in Enzyme List	Source	"Standard" (EC) NH_2 donors and acceptors, and their K_m (mM)	Other substrates (relative reaction rates; rate with "standard" donor = 100)	K_{eq} (app) (arrow indicates forward or reverse direction of reaction)	Degree of purification	Specific activity (μmoles/min/mg)
2.6.1.19	Rat brain	γ-Abu(4.5) + 2-oxoglutarate $\rightleftharpoons$ Glu + succinic semialdehyde	β-Ala(130), γ-Abu(100), D,Lβ-$Aiso$Bu (37), α,γ-Dbu(8.5)	$\rightarrow$ $\sim$0.2	300–400$\times$	$\sim$5.0
2.6.1.43	R. spheroides	L-Ala(8) + 1,5-dioxovalerate (0.4) $\rightleftharpoons$ 5-aminolevulinate + pyruvate	Donors: γ-Abu(74), δ-Aval(59), ϵ-Acap(73)	Reversible	26$\times$ (partially purified)	36.0
2.6.1.13	Pig kidney	Orn + a 2-oxoacid $\rightarrow$ an α-amino acid + L-glutamate semialdehyde[a]	Donors: (very slow): α,δ-Dbu, δ-Aval, Val, $allo$Ile. Acceptors: λ-OxoGlu(100), glyoxylate(16), 2-oxoadipate(9), pyruvate(7), oxalacetate(3)	Irreversible (cyclization of Glu semialdehyde)	10,000$\times$ (pure, crystalline)	52.0
2.6.1.13	Rat liver (mit)	Orn(2.8) + 2-oxoglutarate(0.28) $\rightarrow$ Glu + L-glutamate semialdehyde	Acceptors: glyoxylate, pyruvate Donors: none found	$\leftarrow$ 70 (virtually irreversible)	1000$\times$ (pure, crystalline)	Units spectrophotometric, equivalent to approx. 400 μmoles/20 min/mg
2.6.1.11	E. coli	N^2-AcOrn(0.3) + 2-oxoglutarate (1.0) $\rightleftharpoons$ Glu + N-Ac-glutamate semialdehyde	None found	Reaction reversible (functions in Arg biosynthesis)	100$\times$ (partially purified)	$\sim$100 I.U.
2.6.1.36	A. liquidum	Lys(2.8) + 2-oxoglutarate(0.5) $\rightarrow$ Glu + α-aminoadipate semialdehyde	Donors: Orn(55), Thia-Lys(16), N^6-AcLys(5.7), N^5-AcOrn(4) Acceptors: 2-oxocaproate (8.7); oxalacetate (5.3), oxovalerate(5.4), 2-oxobutyrate(5.0)	Virtually irreversible owing to cyclization of aminoaldehyde product	180$\times$ (homogeneous, crystalline)	18.6 (turnover number 2,200 moles/min)

(Continued)

Optimal pH	MW and subunit structure	K_{PLP} (μM)	Absorption peaks (nm) at low pH	Absorption peaks (nm) at high pH	Ref.	Footnotes and comments
~8.4		1.5^l (K_{PMP} 20 mM)	<u>416</u>, 335	416, <u>335</u>	288	^{l} Resolved in PMP form by filtration through Sephadex G-50. Inhibitors: butyrate($K_i = 2.9\,\mu M$) and other fatty acids; p-chloromercuribenzoate (10^{-6} M); aminooxyacetate (10^{-7} M). Similarly purified enzyme from pig kidney (289) has other K_m (true) values: γ-Abu (0.36 mM), 2-oxoGlu (0.31), Glu (5.0) and succinic semialdehyde (0.01).
~7.0 (broad)		100^m			290	^{m} Partial resolution by means of Cys. Inhibition (percent at 10^{-3} M): L-penicillamine (72%), homocysteine (30), and other aminothiols; aminooxyacetate (100%) and other CO reagents; iodoacetate (30), p-chloromercuribenzoate (78% inactivation at 1.4×10^{-4} M); dioxo acids. Enzyme found in most uni- and multicellular organisms.
8.0	~248,000 (tetramer) (4 eq. PLP)		<u>415</u>, 335	415, <u>335</u>	291	^{n} K_m values not reliable owing to substrate activation and inhibition effects. Inhibitors: Same as for preceding enzyme (EC 2.6.1.43); sensitivity to carbonyl reagents higher; e.g., NH$_2$OH—40% inhibition at 5×10^{-7} M.
~8.0	132,000 (tetramer)	o	<u>416</u>, 330	416, <u>330</u>	292, 293	^{o} High coenzyme affinity; added PLP stabilizes the enzyme. Isoelectric point, pI = 5.38. Has 4 SH groups per subunit (1 exposed, 3 unreactive). Amino acid analysis and electron microscopy of crystals in Ref. 292. Inhibitors: L-canavanine, δ-Aval (100% at 2.5×10^{-3} M) and other competitive substrate analogs; sensitive to thiol and carbonyl reagents. Occurs also in other mammalian organs. The liver enzyme is nutritionally and hormonally inducible. Similar enzyme from C. reinhardtii, see Ref. 294.
8.0	[$S_{20,w} = 9.2$ S]	1.7^p			296	^{p} Purified enzyme completely resolved; PLP, EDTA, and thiols stabilize. Sensitive to inactivation by heavy metals and reagents alkylating SH groups. Repressible by arginine. Purification of similar enzyme from C. reinhardtii, see Ref. 297.
Varies with NH$_2$ Donors: 8.4(Lys); 7.5(Orn); 9.0(Thialysine)	116,000 ± 3,000 (dimer)	0.36^q; K_{PMP} = 7.3 μM	<u>415</u>, 340 (Spectrum pH-independent)	<u>415</u>, 340	295	^{q} One of the two PLP's bound in nonactive unresolvable mode (λ_{max} 340 nm). Enzyme also occurs in Flavobacterium sp.; it is inducible by lysine-containing media. Inhibitors: competitive substrate analogs, e.g., δ-aminovalerate. δ-hydroxylysine ($K_i = 7.7 \times 10^{-5}$ M and 1.4×10^{-2} M, respectively).

TABLE XII

No. in Enzyme List	Source	"Standard" (EC) NH₂ donors and acceptors, and their K_m (mM)	Other substrates (relative reaction rates; rate with "standard" donor = 100)	K_{eq} (app) (arrow indicates forward or reverse direction of reaction)	Degree of purification	Specific activity (μmoles/min/mg)
					D. Acting on Amino and Oxo Substrates	
2.6.1.9	*S. typhimurium* (derepressed mutant)	L-Histidinol-1-P (0.25) + 2-oxoglutarate(0.29) ⇌ Glu + imidazolylacetol-1-P(1.2)	His (K_m = 7.5 mM)		~170× (essentially pure)	~22.2
2.6.1.30	*Pseudomonas* sp. (pyridoxine-induced)	Pyridoxamine (0.04) + pyruvate(0.1) ⇌ L-Ala + pyridoxal	Donors: several analogs of pyridoxamine, (Ref. *42*) Acceptors: none found	Reaction reversible	60× (homogeneous, crystalline)	7.03
2.6.1.29	*E. coli* B (putrescine-induced selected mutant)	Putrescine + 2-oxoglutarate (0.9) ⇌ Glu + 4-aminobutyraldehyde	Donors: other diamines (C₄→C₇) (100→30%). Acceptor: pyruvate	Reverse reaction hindered by cyclization of aminoaldehyde	70× (partially purified)	1.63

2.6.1.2; Table XII, group A) from liver (*257, 258*) and myocardium (*259, 260*). They are similar in most regards, except the strict substrate specificity of the hepatic enzyme. Isodynamic L-alanine transaminases from *Bacilli* and other microorganisms have not been investigated systematically. An extensively studied family of group A aminotransferases of diverse biological origin are those using monocarboxylic substrates with hydrophobic side chains as one donor–acceptor pair: the group-specific branched-chain amino acid transaminases (EC 2.6.1.42, Table XII, group

257. T. S. Matsuzawa and H. L. Segal, *JBC* **243**, 5929 (1968); "Methods in Enzymology," Vol. 17A, p. 153, 1970.
258. P. N. Gatehouse, S. Hopper, L. Schatz, and H. L. Segal, *JBC* **242**, 2319 (1967); cf. *in* "Pyridoxal Catalysis: Enzymes and Model Systems" (E. E. Snell et al., eds.), p. 215. Wiley (Interscience), New York, 1968.
259. M. H. Saier and W. T. Jenkins, *JBC* **242**, 91 (1967); "Methods in Enzymology," Vol. 17A, p. 159, 1970.
260. N. Katunuma, S. Matsuda, and M. Igumi, see *Chem. Abstr.* **59**, 6660 (1965).

(*Continued*)

Optimal pH	MW and subunit structure	K_{PLP} (μM)	Absorption peaks (nm) at low pH	Absorption peaks (nm) at high pH	Ref.	Footnotes and comments
Containing Noncarboxylic Acid Groups (or None)						
8.6	59,000 ± 3,000[r,s] (1 eq. PLP); tends to polymerize in the absence of thiols	—		425–430; 340 (pH = 7.5)	298, 299, 300	[r] By equilibrium sedimentation; has two nonidentical subunits of approx. equal size with only 1 PLP-containing catalytic site, 1 N-terminal (Ser) and 1 C-terminal (Val) residue. Amino acid analysis (299). Contains 13 Cys residues, no disulfide. See Ref. 300 for kinetic parameters, specificity, etc.
					301	[s] Nonidentity of subunits questioned and a single species of peptide chain (MW 30,000) assumed, on genetic evidence from mutant strains of *S. typhimurium*.
8.5	148,000(?) (tetramer)		410[t] (pH 7)		104, 307, 308	[t] The purified enzyme contains loosely bound pyridoxal, readily removable by dialysis or conversion to pyridoxamine (addition of L-Ala); apoenzyme also binds PLP, but the complex is inactive. No inhibition of reaction by P_i or PLP.
9–10					304, 305	Sensitive to carbonyl reagents (100% inhibition with NH_2OH, HCN, semicarbazide, at $5 \times 10^{-4} M$); activated by PLP after ultraviolet irradiation or treatment with phenylhydrazine and Norit. Diamine (and monoamine) aminotransferases have wide distribution in bacteria and plants (306).

A) (*208, 261*) and monospecific leucine-glutamate transaminase (EC 2.6.1.6, see Table XII) (*262*). A related series of aromatic amino acid transaminases with relatively broad substrate specificities (EC 2.6.1.5 and 2.6.1.27) (*263–264*) includes also one or two specialized thyroid-hormone (or halogenated tyrosine) transaminases (EC 2.6.1.24–26, see

261. M. S. Coleman and F. B. Armstrong, *BBA* **227**, 56 (1971).
262. K. Aki, K. Ogawa, and A. Ichihara, *BBA* **159**, 276 (1968); cf. "Methods in Enzymology," Vol. 17A, p. 814, 1970.
263. F. Valeriote, F. Auricchio, D. Riley, and G. Tomkins, *JBC* **244**, 3618 (1969); D. K. Granner and G. M. Tomkins, "Methods in Enzymology," Vol. 17A, p. 633, 1970.
263a. G. A. Jacoby and B. N. LaDu, *JBC* **239**, 419 (1964).
263b. H. George and S. Gabay, *BBA* **167**, 555 (1968).
263c. M. Fujioka, Y. Morino, and H. D. Wada, "Methods in Enzymology," Vol. 17A, p. 585, 1970.
264. S. R. O'Neill and R. D. DeMoss, *ABB* **127**, 361 (1968).

Table XII) (*265, 266*). Similar, insufficiently characterized enzymes, formally belonging to group B, transfer NH_2 groups between hydrophobic monocarboxylic substrates, e.g., a valine-isoleucine transaminase from green plants (EC 2.6.1.32) (*267*) and a similar transaminase shifting NH_2 among several aromatic and branched-chain substrates (EC 2.6.1.28) (*268*).

While not yet prepared in a state of high purity, several enzymes in group A, important in the catabolism of amino acids in uni- and multicellular organisms, have been studied phenomenologically in some detail, for example, histidine-glutamate transaminases (EC 2.6.1.38) (*269–271*), kynurenine transaminases (EC 2.6.1.7) responsible for the formation of kynurenic and xanthurenic acids (*272–274*), and cysteine transaminase (EC 2.6.1.3) (*275*).

D-Specific aminotransferases of restricted occurrence—in *Bacilli*, and possibly in other bacteria incorporating D-amino acids into cell-wall mucopeptides—are of considerable theoretical interest. The only well-defined member of the group, D-Ala-D-Glu transaminase (EC 2.6.1.21, Table XII, group A), isolated virtually pure from *B. subtilis* (*276*), has fairly broad specificity for a number of mono- and dicarboxylic D-amino acids. Thus, the existence of separate transaminases for D-Asp (EC 2.6.1.10) (*277*), is open to doubt [cf., however, a recent communication about a D-Met-pyruvate transaminase, EC 2.6.1.41, from *B. anthracis* (*278*)]. Noteworthy features of the *B. subtilis* D-Ala-transaminase (*276*) are its remarkable, unusual resistance to SH-blocking reagents, and

265. M. Nakano, *JBC* **242**, 73 (1967); "Methods in Enzymology," Vol. 17A, p. 660, 1970.
266. M. Nakano and T. S. Danovski, *BBA* **85**, 18 (1964).
267. Z. S. Kagan, A. S. Dronov, and V. L. Kretovich, *Dokl. Akad. Nauk SSSR* **175**, 1173 (1967); **179**, 1236 (1968).
268. N. K. Sukanya and C. S. Vaidyanathan, *BJ* **92**, 594 (1964).
269. J. G. Coote and H. Hassal, *BJ* **91**, 82 (1964).
270. R. Wakramasinghe, J. Hedegaard, and J. Roche, *C. R. Soc. Biol.* **161**, 1891 (1967).
271. W. L. Albritton and A. P. Levin, *BJ* **114**, 662 (1969).
272. O. Wiss, *Hoppe-Seyler's Z. Physiol. Chem.* **293**, 106 (1953).
273. M. Mason, *JBC* **227**, 61 (1957); Y. Ueno, K. Hayashi, and R. Shukuya, *J. Biochem. (Tokyo)* **54**, 75 (1963).
274. W. B. Jakoby and D. M. Bonner, *JBC* **221**, 689 (1956).
275. F. Chatagner and G. Sauret-Ignazi, *Bull. Soc. Chim. Biol.* **38**, 415 (1956).
276. M. Martinez-Carrion and W. T. Jenkins, *JBC* **240**, 3538 and 3547 (1965); "Methods in Enzymology," Vol. 17A, p. 167, 1970.
277. C. B. Thorne, C. B. Gomez, and R. D. Housewright, *J. Bacteriol.* **69**, 357 (1955); C. B. Thorne and D. M. Molnar, *ibid.* **70**, 420 (1955).
278. I. W. Mapson, J. E. March, and D. A. Wardale, *BJ* **115**, 653 (1969).

the inverse (negative) sign of CD extremum in the absorption band at 415 nm, pointing to a peculiar chemical topography of the active site, cf. Ref. 76.

B. ACTING ON TWO α-AMINO–α-OXO MONOCARBOXYLIC SUBSTRATE PAIRS

Two transaminases of this group, acting on substrate pairs with hydrophobic side chains, were mentioned above. Another is the liver Ala-glyoxalate transaminase (EC 2.6.1.44, Table XII, group B) (279). In this reaction, and in all similar ones where glyoxalate accepts NH_2 groups from the aminodicarboxylates (EC 2.6.1.35 and 2.6.1.4; Table XII, group A) (255, 256, 280, 281) or from serine (EC 2.6.1.45) (282), the thermodynamic equilibrium strongly favors glycine formation, which probably is the biological function of these transaminases in intermediate steps of the biosynthesis of purines, pteridines, creatine, and other nitrogen-containing bases.

Transaminases of widespread occurrence catalyze the transfer of $α-NH_2$ from glutamine (EC 2.6.1.15, Table XII, group B) (283, 284) and asparagine (EC 2.6.1.14) (285) to a variety of monocarboxylic oxo acids; only EC 2.6.1.15 has been obtained in nearly homogeneous form (283). Glutamine and asparagine are the most common storage and transport forms of nondifferentiated metabolic nitrogen; in mobilizing both N atoms (amino and amido) of these compounds, the transaminases just mentioned play an important part (4).

C. ACTING ON ω-AMINO AND ω-OXO ACIDS

Table XII, group C, comprises a number of transaminases, highly or partially purified from microorganisms or animal tissues, which catalyze

279. J. S. T. Thompson and K. E. Richardson, JBC 242, 3614 (1967); "Methods in Enzymology," Vol. 17A, p. 163, 1970.
280. H. Nakada, JBC 239, 468 (1964).
281. J. S. T. Thompson and K. E. Richardson, ABB 117, 599 (1966).
282. J. King and E. R. Waygood, Can. J. Biochem. 46, 771 (1968).
283. A. J. L. Cooper and A. Meister, "Methods in Enzymology," Vol. 17A, p. 951, 1970; Biochemistry 11, 661 (1972).
284. A. E. Braunstein and T.-S. Hsü, Biokhimiya 25, 758 (1960); cf. ibid. p. 1113; H. Z. Kupchik and W. E. Knox, "Methods in Enzymology," Vol. 17A, p. 951, 1970.
285. A. Meister and P. E. Frazer, JBC 210, 37 (1954).

the reversible transfer of NH_2 groups from and to carbon atoms in terminal position—β, γ, δ, ω—of amino–oxo acid pairs; the second substrate pair is sometimes also ω-substituted, more usually it is a mono- or dicarboxylic α-amino–α-oxo pair. As a rule, the substrate specificity is relative rather than absolute. Thus, the so-called β-Ala–α-Ala transaminase (EC 2.6.1.18) (286) and several γ-aminobutyrate transaminases (EC 2.6.1.19) (287–289), δ-aminolevulinate transaminases (EC 2.6.1.43), e.g., Ref. 290, pure crystalline ornithine δ-aminotransferases (EC 2.6.1.13) (291–294), and the unique bacterial L-lysine:2-oxoglutarate ϵ-aminotransferase (EC 2.6.1.36) (57, 295) are all capable of reacting with ω-amino donors and acceptors of varying carbon-chain length, at rates sometimes exceeding the reaction rate with the "nominal" best substrate.

Transfer of NH_2 from the ω position of diamino acids tends to be unidirectional owing to ready cyclization of the α-amino–ω-oxo acid reaction product. Ornithine δ-transaminases function in the conversion of ornithine to proline and in its catabolism via glutamate. The reverse process of ornithine (and arginine) biosynthesis from glutamate or proline is ensured by another enzyme, N^2-acetylornithine transaminase (EC 2.6.1.11, Table XII, group C) (296, 297), the ω-NH_2 acceptor, N-acetylglutamate semialdehyde, being protected from cyclization by acylation of the α-amino group.

286. O. Hayaishi, Y. Nishizuka, M. Tatibana, M. Takoshita, and S. Kuno, *JBC* **236**, 787 (1961). See R. A. Stinson and M. S. Spencer [*BBRC* **34**, 120 (1969)] about a similar enzyme.
287. E. M. Scott and W. B. Jakoby, *JBC* **234**, 932 (1959).
288. I. A. Sytinsky and V. Yu. Vasiliev, *Enzymologia* **39**, 1 (1970); V. Yu. Vasiliev, I. A. Sytinsky, and Z. K. Nikolaeva, *Biokhimiya* **35**, 556 (1970).
289. V. Yu. Vasiliev, V. P. Yeremin, E. S. Severin, and I. A. Sytinsky, *Biokhimiya* **38**, 355 (1973).
290. J. M. Turner and A. Neuberger, "Methods in Enzymology," Vol. 17A, p. 188 (1970); see, also, *BBA* **67**, 345 (1963).
291. W. T. Jenkins and H. Tsai, "Methods in Enzymology," Vol. 17A, p. 281, 1970.
292. C. Peraino, L. G. Bunville, and T. N. Tahmisian, *JBC* **244**, 2241 (1969); T. Matsuzawa, T. Katsunuma, and N. Katunuma, *BBRC* **32**, 161 (1968).
293. H. J. Strecker, *JBC* **240**, 1225 (1965).
294. J. Südi and G. Dénes, *Acta Biochim. Biophys.* **2**, 291 (1967).
295. K. Soda and H. Misono, *Biochemistry* **7**, 4102 (1968) (see, also, p. 4110); "Methods in Enzymology," Vol. 17B, 222, 1971.
296. A. M. Albrecht and H. J. Vogel, *JBC* **239**, 1872 (1964); H. J. Vogel and E. E. Jones, "Methods in Enzymology," Vol. 17B, p. 260, 1971.
297. J. Südi and G. Dénes, *Acta Biochim. Biophys.* **2**, 279 (1967); "Methods in Enzymology" Vol. 17B, p. 277, 1971.

D. ACTING ON AMINO AND OXO SUBSTRATES CONTAINING
NONCARBOXYLIC ACID GROUPS (OR NONE)

A few enzymes have been described, and certainly others exist, which catalyze NH_2 transfer to 2-oxoglutarate (resp., other acceptors) from primary amino compounds either containing an acidic group other than carboxyl, e.g., an O- or C-phosphonate group, or having no anionic group at all.

Among the PLP-dependent enzymes in this subgroup, histidinol phosphate aminotransferase, EC 2.6.1.9 (Table XII, group D) has been extensively purified from mutant strains of *Salmonella typhimurium* (*298, 299*). Ames and co-workers (*299*) and other authors (*300, 301*) have studied in detail biochemical and genetic aspects of the function and control of this enzyme in microbial histidine biosynthesis.

C-Phosphonoacetaldehyde was detected as an intermediate in the degradation of 2-aminoethylphosphonate by *B. cereus;* its formation, dependent on PLP and pyruvate, is attributed to the action of a specific aminotransferase, EC 2.6.1.37 (*302*), and there is some evidence for broader occurrence of aminoalkyl-C-phosphonate transaminases in a variety of organisms (*303*).

Aliphatic diamines are the NH_2 donors, and 2-oxoglutarate or pyruvate are the acceptors, for a diamine transaminase, EC 2.1.6.29 (Table XII, group D) purified 80-fold from *E. coli* and detected in many other bacteria (*304, 305*), fungi, and higher plants (*306*). Investigation of these enzymes does not appear to have progressed, although it has been made evident that the tissues of various plants contain PLP-linked amine

298. R. G. Martin and R. F. Goldberger, *JBC* **242**, 1168 (1967); R. G. Martin, M. J. Voll, and E. Apella, *ibid*. p. 1175.
299. R. G. Martin, *ABB* **138**, 239 (1970); cf. "Methods in Enzymology," Vol. 17B, p. 33, 1971.
300. W. L. Albritton and A. P. Levin, *BJ* **144**, 662 (1969); *JBC* **245**, 2525 (1970).
301. M. M. Rechler and C. B. Bruni, *JBC* **246**, 1806 (1971); T. Kohno and J. Yourno, *ibid*. p. 2203.
302. J. M. La Nauze and H. Rosenberg, *BBA* **165**, 438 (1968).
303. E. Roberts, D. G. Simonsen, M. Horiguchi, and J. S. Kittredge, *Science* **159**, 886 (1968).
304. K.-H. Kim, *J. Bacteriol.* **86**, 320 (1963); *JBC* **239**, 783 (1964); K. Michaels and K.-H. Kim, *BBA* **115**, 58 (1966).
305. K.-H. Kim and T. T. Tchen, "Methods in Enzymology," Vol. 17B, p. 812, 1971.
306. K. Hasse and G. Schmid, *Biochem. Z.* **337**, 69, 438 (1968); see *in* "Pyridoxal Catalysis: Enzymes and Model Systems" (E. E. Snell *et al.*, eds.), p. 229. Wiley (Interscience), New York, 1968.

transaminases catalyzing reversible NH_2 transfer between diamines or alkyl (resp., aralkyl) monoamines and oxoglutarate or pyruvate (306). Such enzymes can obviously be of importance in biosynthesis of amines via aldehydes in plant species known to lack decarboxylases for the cognate α-amino acids.

An important specific amine transaminase mentioned earlier in this chapter (Section III,B,5) is the pyridoxamine aminotransferase, EC 2.6.1.30 (Table XII, group D), obtained in pure crystalline form from *Pseudomonas* sp. MA and studied by Snell and associates (104, 307, 308). It does not require PLP as cofactor, but acts on pyridoxamine, and on some of its analogs (42), as a substrate in reversible NH_2 transfer to pyruvate.

Two peculiar aminotransferases, detected in bacterial cells, function in the synthesis of specific nucleoside diphosphate 4-amino sugars by way of NH_2 transfer from L-glutamate. These are EC 2.6.1.33, TDP-4-amino-4,6-dideoxy-D-glucose transaminase, partially purified from *E. coli* (309), and EC 2.6.1.34, UDP-4-amino-2-acetamido-2,4,6-trideoxyglucose transaminase (310); they may be the first representatives of a larger family of related enzymes.

A remarkable feature in streptomycin biosynthesis is the participation of aminotransferases in two different steps of the formation of streptidine (in *Streptomyces bikiniensis*) from *myo*-inositol (311). The NH_2 acceptors are keto derivatives of cyclitols in both reactions. The first is reversible transamination of *scyllo*-inosose to *scyllo*-inosamine by a PLP-enzyme utilizing L-glutamine (α-NH_2) and, less efficiently, a variety of L-α-amino acids as NH_2 donors (312). At a later stage, another transaminase is believed to produce *N*-amidino-streptamine by NH_2 transfer from L-alanine or L-glutamate (but not from L-glutamine) to *N*-amidino-3-keto-inosamine (311). As yet, neither of the two enzymes has been purified or studied in detail.

E. Transaminations Occurring as Side Reactions in the Active Site of Pyridoxal-P-Linked Enzymes of Other Types

Interconversion of PLP-aldimines and tautomeric PMP-ketimines (I ⇌ II, Eqs. 8 and 9) is a common key step in transamination and in

307. J. E. Ayling and E. E. Snell, *Biochemistry* **7**, 1616 and 1626 (1968).
308. H. Kolb, R. D. Cole, and E. E. Snell, *Biochemistry* **7**, 2946 (1968).
309. M. Matsuhashi and J. L. Strominger, *JBC* **241**, 4738 (1966).
310. J. Distler, B. Kaufman, and S. Roseman, *ABB* **116**, 466 (1966).
311. J. B. Walker and M. S. Walker, *Biochemistry* **6**, 3821 (1967).
312. J. B. Walker and M. S. Walker, *BBA* **170**, 219 (1968).

reactions catalyzed by certain other PLP-enzymes, e.g., the α-amino acid racemases, aspartate β-decarboxylase, and γ-eliminating lyases (*13, 58, 61*). Rapid removal of a proton from carbon atoms $C^α$ in enzyme-bound PLP-substrate aldimines (I) or $C^{4'}$ in the PMP-ketimines (II) requires catalysis by proton acceptors (general base groups) suitably located in the active sites. Protonation of the Schiff base carbanions (Ia) and (IIa, Eq. 9) need not necessarily be catalyzed; it can occur spontaneously by direct uptake of H^+ ions from the solvent medium. Eventual hydrolysis of ketimine (II) completes a transamination half-reaction, i.e., formation of a pyridoxamine-P protein. Only in the transaminases are such PMP-forms normal intermediates; in enzymes of other types they are catalytically inert and mostly dissociate readily to apoenzyme and PMP.

In the case of kynureninase (*313, 314*) and of aspartate β-decarboxylase (*315*), interaction with substrate or other amino acids was early found to result in gradual enzyme inactivation, which could be counteracted by addition of pyruvate (or another α-oxo acid) or of pyridoxal-P; the coenzyme also reversed inactivation gone to completion. The correct interpretation of these observations was first given for Asp-β-decarboxylase by Meister and associates (*316*). They demonstrated gradual transamination of the enzyme with aspartate or other amino acids, resulting in conversion of the decarboxylase to the PMP form; this can either revert to the active aldimine form through slow transamination with a 2-oxo acid, or undergo resolution and recombine with added pyridoxal-P.

The long-suspected occurrence of a similar phenomenon of inactivating transamination in the case of kynureninase was substantiated recently by Soda and co-workers (*317*). They showed that incubation of pure crystalline kynureninase (from *Pseudomonas marginalis*) with L-alanine (a reaction product) resulted in production of an inactive enzyme species with $λ_{max} = 332$ nm (the PMP-protein) which was resolved to apoenzyme

313. E. V. Goryachenkova, *Dokl. Akad. Nauk SSSR* **80**, 643 (1965); W. E. Knox, *BJ* **53**, 379 (1953).
314. W. B. Jakoby and D. M. Bonner, *JBC* **205**, 709 (1953).
315. J. Cattaneo-Lacombe, J. C. Senez, and P. Beaumont, *BBA* **30**, 458 (1958); J. S. Nishimura, J. M. Manning, and A. Meister, *Biochemistry* **1**, 442 (1962); also *in* "Chemical and Biological Aspects of Pyridoxal Catalysis" (E. E. Snell *et al.*, eds.), p. 229. Pergamon, Oxford, 1963.
316. A. Novogrodsky and A. Meister, *JBC* **239**, 879 (1964); E. W. Miles, A. Novogrodsky, and A. Meister, *in* "Pyridoxal Catalysis: Enzymes and Model Systems" (E. E. Snell *et al.*, eds.), p. 425. Wiley (Interscience), New York, 1968.
317. M. Moriguchi, T. Yamamoto, and K. Soda, *BBRC* **44**, 1416 (1971).

by dialysis. Soda and his group (*318*) reported also evidence for abortive transamination in a PLP-dependent racemase; it was observed that crystalline arginine racemase prepared from *Pseudomonas graveolens* slowly racemized ornithine, with concomitant inactivation of the enzyme as a result of transfer of NH_2 groups from quasi-substrate to coenzyme. The resulting fairly stable PMP-enzyme can be reconverted to the active aldimine species by displacement of PMP with added PLP, as well as by incubation with pyruvate or other 2-oxo acids; ornithine is transaminated with pyruvate by the enzyme (optimal pH = 11.0) at 4.1×10^{-5} of the rate of ornithine racemization.

The so-called 2,2-dialkyl amino acid:pyruvate aminotransferase (decarboxylating) (see Ref. *319*) should correctly be classified as a carboxylase (EC 4.1.1.64) since the first enzyme-catalyzed reaction step is decarboxylation (*25*). The occurrence of an acid-base-catalyzed reversible aldimine–ketimine transformation step is made evident by ability of this enzyme to catalyze equilibrium exchange of NH_2 groups between labeled alanine and pyruvate (*319*). The reaction mechanism of decarboxylation-dependent transamination is analogous to that of a similar nonenzymic reaction studied by Kalyankar and Snell (*320*).

In principle, neither base-catalyzed primary removal of a proton from the C^α atom nor occurrence of an enzyme-bound PMP-Schiff base as a normal reaction intermediate is prerequisite for abortive transamination in the active site of PLP-dependent enzymes of amino acid metabolism. In the action of all these enzymes the first chemical step is the formation of a PLP-aldimine anion with negative charge on the C^α atom by removal of any one of its substituents: –H, –COOH or –R (side chain). In every case, one of the resonance states of the Schiff-base anion is the protein-bound PMP-imine carbanion capable of reprotonation at $C^{4'}$.

Accordingly, NH_2 transfer to the coenzyme should be possible as a side reaction for any PLP-linked enzyme. As stated above, the protonation step does not require catalysis by a proton donor group, but base catalysis is needed to remove a proton from $C^{4'}$ for the reverse tautomeric shift. Pyridoxal enzymes normally forming no ketimine intermediate presumably have no acid-base group in appropriate position for catalysis of the prototropic shift; abnormal transamination leading to PMP formation in the active site of such an enzyme would be expected to result in permanent inactivation not reversible by added oxo acids, whereas re-

318. T. Yorifuji, H. Misono, and K. Soda, *JBC* **246**, 5093 (1971).
319. G. B. Bailey and W. B. Dempsey, *Biochemistry* **6**, 1526 (1967); "Methods in Enzymology," Vol. 17A, p. 829, 1970.
320. G. D. Kalyankar and E. E. Snell, *Biochemistry* **1**, 594 (1962).

placement of the weakly bound PMP by added pyridoxal-P might regenerate active holoenzyme. For an α-decarboxylase this situation was first demonstrated by Huntley and Metzler (*321*) in the case of inactivation of glutamate decarboxylase (*E. coli*) on interaction with α-methylglutamate. Slow abortive transamination of PLP to PMP in the active site of this decarboxylase was also shown to be the cause of its inactivation on incubation with other quasi-substrates (e.g., L-aspartate) and in the course of decarboxylation of the adequate substrate, L-glutamate (*322*): the stereochemistry of proton addition at $C^{4'}$ of PMP proved to be the same in this aberrant side reaction (*245*) as in normal transamination reactions (*76*), see Section III,B,7.

Abnormal transamination has further been observed in a PLP-linked enzyme effecting cleavage of C^β–C^α (or, more exactly, R–C^α) bonds, namely, in serine transhydroxymethylase, EC 2.1.2.1 (*323*). The enzyme catalyzes reversible conversion of L-serine and homologous α-hydroxyamino acids into glycine, and the formation of D-alanine from α-methyl-L-serine (*324*). It was demonstrated that in these reactions replacement of the side chain (R–) of an L-amino acid by a proton of the same symmetry, i.e., in the position of α-H in a D-amino acid, proceeds via a PLP-aldimine anion with negative charge at the C^α atom. In the spectra of intermediate complexes in the forward and reverse reactions there occur the typical absorbance maxima (in the 495–505-nm region) of PLP-imine anions (*323*), and the enzyme catalyzes exchange of labeled H atoms in glycine in the pro-*S* position (*76*).

Schirch and Jenkins (*323*) found that serine transhydroxymethylase and its apoenzyme catalyze a transamination half-reaction between D-alanine and pyridoxal-P [Eq. (12).]

$$\begin{array}{c} \text{D-Ala} + \text{PLP-enzyme} \longrightarrow \text{enzyme-PMP} + \text{pyruvate} \\ \diagup \quad \diagdown \quad \diagup \quad \diagdown \\ \text{PLP} \quad \text{apoenzyme} \quad \text{PMP} \end{array} \quad (12)$$

The reaction proceeds at a rate three decimal orders slower than the rate of serine–glycine interconversion. The NH_2 transfer is practically unidirectional as a consequence of strong association of the PLP-protein and very weak association of the PMP-protein.

This abnormal side reaction is entirely concordant with the other examples, considered above, of transaminations in active sites, resulting from

321. T. E. Huntley and D. E. Metzler, *in* "Symposium on Pyridoxal Enzymes" (K. Yamada, N. Katunuma, and H. Wada, eds.), p. 81. Maruzen, Tokyo, 1968.
322. B. S. Sukhareva and A. E. Braunstein, *Mol. Biol. (USSR)* **5**, 302 (1971); see also B. S. Sukhareva *et al.* (*245*).
323. L. Schirch and W. T. Jenkins, *JBC* **239**, 3797 and 3801 (1964).
324. L. Schirch and M. Mason, *JBC* **237**, 2578 (1962); **238**, 1032 (1963).

eventual protonation of anionic Schiff base intermediates in the resonance structure of a PMP-imine carbanion (protonation at $C^{4'}$ atom) and hydrolysis of this "wrong" tautomeric Schiff base.

V. Concluding Remarks

Aspartate transaminase is prominent, if not exceptional, in the extent of insight into structural and functional features of the enzyme derived from indirect chemical and physical approaches, in advance of elucidation of its primary and three-dimensional structure. There is a considerable lag in the study of other, less readily available aminotransferases; comparable information about these will help to clarify structural and evolutionary aspects of substrate specificity, efficiency, control of catalytic activity, etc.

Delay in X-ray structural investigation of aspartate transaminases is apparently caused by technical rather than principal difficulties, mainly, in the preparation of sufficiently large and resistant enzyme monocrystals. One may expect that the difficulties will be surmounted in the near future, and accurate knowledge of the enzyme's spatial atomic topography will provide a reliable basis for elaboration of the correct reaction mechanism.

To survey at any length the important literature that has accrued about biological aspects of enzymic amino group transfer is beyond the scope of this chapter. Some sources of reference to work in this area are quoted in Section I, e.g., Refs. *1–6, 19, 20*. The problem of general and specific functions of aminotransferases is a very broad one. Brief mention of some major trends of research will serve to emphasize their diversity.

A field still in active development concerns the different roles of transaminases and their isoenzymes in the paths of biosynthesis and degradation of amino acids in uni- and multicellular organisms, and in the control of these pathways by familiar regulatory mechanisms—both "metabolic," i.e., enzyme-directed (competitive and feedback inhibition, etc.), and at the level of gene expression (induction and repression). The area of this field of research is expanding, owing to the detection of aminotransferases catalyzing transformations of a broader range of nitrogen compounds such as alkylamines, amino sugars, and other bioactive nitrogen bases (Section IV,D).

Numerous studies relating to nutritional and hormonal induction of the synthesis, in mammalian organs, of certain aminotransferases—in particular those acting on alanine, aspartate, or aromatic amino acids—have made evident the role of these induction phenomena in the stimulation of gluco- and ketoneogenesis, and, concomitantly, of general nitrogen catabolism and ureogenesis, under various conditions of stress.

More specialized correlations, discovered in recent years and not yet conclusively explained, appear to exist in procaryotes and eucaryotes between enhanced biosynthesis of ornithine transaminases (and ornithine decarboxylase), i.e., of enzymes involved in the biosynthesis of aliphatic diamines, and high rates of cell growth and, *eo ipso*, of the diamine-dependent synthesis of RNA. Another special aspect currently under active investigation (*6, 19*) concerns the physiological interplay, and eventually, regulation by chemical agents, of γ-aminobutyrate transaminases and L-glutamate decarboxylase in brain and synapses, i.e., of PLP-linked enzymes controlling the concentrations and metabolism in the nervous system of γ-aminobutyrate and glutamate, two of the most important neuroactive nitrogen metabolites.

A topic under continuing experimental and theoretical study is the presumable significance of mitochondrial and cytosol aspartate aminotransferases in integrated regulation of the concentrations of ions, individual respiratory intermediates, and co-dehydrogenases in the oxidized and reduced form within the mitochondria and outside.

Promising fields for biomedical research include exploration of the role of aminotransferase defects in inborn metabolic errors and the biochemical pharmacology and toxicology of transamination-dependent physiological activities in health and disease.

In the latter area, a solid basis for rational experimentation is provided by the many available detailed studies relating to selective action of coenzyme and substrate analogs and other synthetic (and natural) specific effectors on isolated aminotransferase systems (cf. Section III).

ACKNOWLEDGMENTS

The author wishes to thank his colleagues, in particular, Drs. V. Ivanov, M. Karpeisky, O. Polyanovsky, E. Severin, and Yu. Torchinsky, for stimulating discussion and critical comment. He also wishes to thank Mrs. V. Bukanova for assistance in preparation of the manuscript.

11

Coenzyme A Transferases

W. P. JENCKS

I. Introduction 483
II. Properties 485
III. Catalytic Properties 486
    A. Specificity 486
    B. Assay 487
    C. Thermodynamics 487
    D. Mechanism and Kinetics 488

I. Introduction

The coenzyme A transferases are members of an unusual class of enzymes that catalyze the transfer of the *leaving group* (X) of one activated acyl or phosphoryl group to another [Eq. (1)].

$$Ac_1-X + Ac_2-O^- \rightleftharpoons Ac_1-O^- + Ac_2-X \qquad (1)$$

Reactions that result in the transfer of an *acyl* or *phosphoryl group* are well known and may be easily explained by the displacement of a leaving group (X) by a nucleophilic reagent (Y^-), which may proceed through the intermediate formation of a tetrahedral addition intermediate [Eq. (2)]. Leaving group transfer is a far more complicated reaction

$$^-Y \overset{O}{\underset{}{\overset{\|}{C}}}-X \rightleftharpoons \left[Y-\overset{O^-}{\underset{|}{C}}-X \right] \rightleftharpoons Y-\overset{O}{\underset{}{\overset{\|}{C}}} + X^- \qquad (2)$$

that cannot be explained by a simple displacement or addition-elimination mechanism. It is more properly regarded as a form of acyl or phosphoryl activation. In the reactions catalyzed by coenzyme A transferases, for example, an initially low energy, unreactive carboxylate ion is converted to an "energy-rich" thiolester at the expense of the energy-rich bond of the original thiolester substrate [Eq. (3)] and, in fact, the physiological role of these enzymes is just this—the activation of acyl compounds. In this sense these reactions are comparable to many other

$$R_1-\overset{O}{\underset{\|}{C}}\sim SCoA + R_2-\overset{O}{\underset{\|}{C}}-O^- \rightleftharpoons R_1-\overset{O}{\underset{\|}{C}}-O^- + R_2-\overset{O}{\underset{\|}{C}}\sim SCoA \quad (3)$$

complex acyl activation reactions, but they differ from the others in that the activating compound (the thiolester) that acts as a dehydrating agent not only extracts oxygen from the carboxylate ion but also donates a group (coenzyme A) that replaces this oxygen atom and becomes a part of the activated acyl product [Eq. (3)]. The proteolytic enzyme pepsin catalyzes a somewhat similar transfer of the leaving amine group of a peptide to the carboxylic acid group of a second substrate molecule [Eq. (4)] (1, 2), but this reaction probably does not represent a physi-

$$\underset{H}{ZNH\overset{O}{\underset{\|}{C}}C}-\underset{H}{NH\overset{R}{C}COO^-} + \underset{H}{ZNH\overset{R^*}{C}COO^-} \longrightarrow \underset{H}{ZNH\overset{R}{C}COO^-} + \underset{H}{ZNH\overset{\overset{*O}{\|}}{C}}-\underset{H}{NH\overset{R}{C}COO^-} \quad (4)$$

ologically significant acyl group activation and may not even represent an essential part of the normal mechanism of action of the enzyme (3).

The occurrence of this unusual type of enzyme-catalyzed acyl group activation with coenzyme A transfer was reported from several laboratories in 1953 (4–7). The best known enzyme is succinyl-CoA:3-ketoacid (or 3-oxoacid) CoA-transferase that serves to activate acetoacetate, an important energy source for several organs in the mammal, to acetoacetyl-CoA at the expense of the energy-rich bond of a normal intermediate of the citric acid cycle, succinyl-CoA [Eq. (5)] (5, 6, 8, 9). The oxida-

1. H. Neumann, Y. Levin, A. Berger, and E. Katchalski, *BJ* **73**, 33 (1959).
2. J. S. Fruton, S. Fujii, and M. H. Knappenberger, *Proc. Nat. Acad. Sci. U. S.* **47**, 759 (1961).
3. M. S. Silver and M. Stoddard, *Biochemistry* **11**, 191 (1972).
4. E. R. Stadtman, *JBC* **203**, 501 (1953).
5. J. R. Stern, M. J. Coon, and A. del Campillo, *Nature (London)* **171**, 28 (1953).
6. D. E. Green, D. S. Goldman, S. Mii, and H. Beinert, *JBC* **202**, 137 (1953).
7. H. R. Whiteley, *Proc. Nat. Acad. Sci. U. S.* **39**, 779 (1953).
8. J. R. Stern, M. J. Coon, and A. del Campillo, *JBC* **221**, 1 (1956).
9. J. R. Stern, M. J. Coon, A. del Campillo, and M. C. Schneider, *JBC* **221**, 15 (1956).

$$^-\text{OCCH}_2\text{CH}_2\overset{\text{O}}{\overset{\|}{\text{C}}}\sim\text{SCoA} + \text{CH}_3\overset{\text{O}}{\overset{\|}{\text{C}}}\text{CH}_2\overset{\text{O}}{\overset{\|}{\text{C}}}\text{O}^- \rightleftharpoons {}^-\text{OCCH}_2\text{CH}_2\overset{\text{O}}{\overset{\|}{\text{C}}}\text{O}^- + \text{CH}_3\overset{\text{O}}{\overset{\|}{\text{C}}}\text{CH}_2\overset{\text{O}}{\overset{\|}{\text{C}}}\sim\text{SCoA} \quad (5)$$

tion of acetoacetate, after activation to its coenzyme A derivative, provides a particularly important source of energy for heart muscle and it may be significant that the K_m for acetoacetate of the heart muscle enzyme appears to be significantly lower than that of the skeletal muscle enzyme (10, 11). There is little or no activity of this enzyme in liver, which serves as a source rather than as a site of utilization of acetoacetate; there is activity in kidney (2).

A coenzyme A transferase that activates short chain fatty acids and related compounds, including succinate, at the expense of acetyl- or propionyl-CoA is found in *Clostridium kluyveri* and other microorganisms (4, 7, 12–14). These bacterial enzymes play an important role in the activation of carboxylic acids in a number of fermentation pathways such as the propionic acid fermentation. Other enzymes of this class have been reported to bring about the interconversion of the coenzyme A derivatives of β-ketoadipate and succinate (15, 16), adipate and succinate (16), γ-hydroxybutyrate and acetate (17), β-hydroxy-β-methylglutarate and succinate (18), citramalate and succinate (19), and oxalate and succinate (20, 21).

These enzymes have been denoted by a number of elegant but confusing names, including *transphorase* and *thiophorase*. The choice of the general name *coenzyme A transferase* by a recent Committee on Biochemical Nomenclature represents a positive contribution by that body in supplying an understandable, descriptive nomenclature.

II. Properties

Most of the coenzyme A transferases have not been extensively purified. Succinyl-CoA:3-ketoacid CoA-transferase has been purified from

10. L. B. Hersh and W. P. Jencks, *JBC* **242**, 3468 (1967).
11. J. B. Blair, *JBC* **244**, 951 (1969).
12. E. A. Delwiche, E. F. Phares, and S. F. Carson, *J. Bacteriol.* **71**, 598 (1956).
13. G. K. K. Menon and J. R. Stern, *JBC* **235**, 3393 (1960).
14. S. H. G. Allen, R. W. Kellermeyer, R. L. Stjernholm, and H. G. Wood, *J. Bacteriol.* **87**, 171 (1964).
15. M. Katagiri and O. Hayaishi, *JBC* **226**, 439 (1957).
16. P. P. Hoet and R. Y. Stanier, *Eur. J. Biochem.* **13**, 71 (1970).
17. J. K. Hardman and T. C. Stadtman, *JBC* **238**, 2088 (1963).
18. R. E. Burch, H. Rudney, and J. J. Irias, *JBC* **239**, 4115 (1964).
19. R. A. Cooper and H. L. Kornberg, *BBA* **62**, 438 (1962).
20. W. B. Jakoby, E. Ohmura, and O. Hayaishi, *JBC* **222**, 435 (1956).
21. J. R. Quayle, D. B. Keech, and G. A. Taylor, *BJ* **78**, 225 (1961).

pig heart to an almost pure state with a specific activity of 1400 Stern units (3.5 IU/mg) by the addition to the standard purification procedure of a step based upon sedimentation of an enzyme-Blue Dextran complex by preparative ultracentrifugation (9, 10, 22–24). The skeletal muscle enzyme has been purified to a smaller extent (11). Sedimentation coefficients at pH 7.6 ($s_{20,w}$) of 5.5 (7 mg/ml) and 5.1 (12 mg/ml) have been reported (9, 25). The molecular weight in 0.05 M phosphate buffer, pH 7.4, is 92,000 by sedimentation equilibrium (23); a value on the order of 78,000 was obtained previously by molecular exclusion chromatography (25). In 6 M guanidine hydrochloride the molecular weight by sedimentation equilibrium is 48,000 in the presence or absence of mercaptoethanol, suggesting that the native enzyme consists of two subunits joined by noncovalent bonds. The enzyme contains 1–3% carbohydrate. Isoelectric focusing reveals one weak and three strong bands with isoelectric points evenly spaced in the range 5.7–6.5 (23). Frozen solutions of the enzyme are stable for months or years (9, 10). The propionate-acetate-succinate enzyme from *Propionibacterium* has been purified to an estimated 56% purity (14). The β-ketoadipate-succinate enzyme from *Pseudomonas fluorescens* has been purified 30-fold (15).

III. Catalytic Properties

A. SPECIFICITY

Succinyl-CoA:3-ketoacid CoA-transferase exhibits a high degree of specificity toward its substrates. In the β-ketoacid series it has maximal activity toward acetoacetate and shows significant activity toward β-methyl acetoacetate, β-ketovalerate, β-ketocaproate, β-ketoisocaproate, and malonic semialdehyde but not toward β-ketoadipate, β-ketooctanoate, or benzoylacetate (9, 26). The activity with malonate is 2% of that with succinate, but no activity was observed with methylmalonate, methylsuccinate, 2,2-dimethylmalonate, α- or β-methylglutarate, α,α- or

22. L. B. Hersh and W. P. Jencks, "Methods in Enzymology," Vol. 13, p. 75, 1969.
23. H. D. White, Ph.D. Thesis, Brandeis University, Waltham, Massachusetts, 1973; H. White, F. Solomon, and W. P. Jencks, *Fed. Proc., Fed. Amer. Soc. Exp. Biol.* **32**, 606, Abs. (1973).
24. H. D. White and W. P. Jencks, *Abstr. 160th ACS Meet., 1970.*
25. L. B. Hersh and W. P. Jencks, *JBC* **242**, 3481 (1967).
26. G. K. K. Menon, J. R. Stern, F. P. Kupiecki, and M. J. Coon, *BBA* **44**, 602 (1960).

β,β-dimethylglutarate, crotonate, β-methylcrotonate, glutarate, malate, β-hydroxybutyrate or saturated fatty acids (*9, 13*). However, at high enzyme concentrations, most carboxylic acids without substituents at the α position react at rates approximately 0.01–1.0% of that with succinate (*23*). There is also a strict specificity with respect to the coenzyme A moiety of the substrate (*9, 27*), which will be commented on later.

The bacterial enzyme, which is utilized for fatty acid activation, displays a much broader specificity, exhibiting activity toward fatty acids with chain length up to octanoate, including formate, and toward succinate, lactate, and α-vinylacetate (*4, 7, 14, 28, 29*). The enzyme from *Propionibacterium* is inactive toward acetoacetate, methylmalonate, β-hydroxy-β-methylglutaryl-CoA, and fluoroacetate (*14*).

B. ASSAY

Succinyl CoA:3-ketoacid CoA-transferase is usually assayed by measuring the initial rate of formation or disappearance of the ultraviolet absorption of the enolate form of acetoacetyl-CoA (*22, 30*). This absorption is increased in the presence of metal ions that stabilize the enolate species (*30*). Since the formation of acetoacetyl-CoA is thermodynamically unfavorable and gives rise to product inhibition, it is convenient to couple the reaction to β-hydroxyacyl-CoA dehydrogenase and follow the disappearance of NADH absorption spectrophotometrically (*10*).

C. THERMODYNAMICS

Values of the equilibrium constant for acetoacetyl-CoA formation have been reported ranging from 0.004 to 0.067 (*10, 11, 30–32*). It is experimentally difficult to approach equilibrium from both directions because of inhibition by acetoacetyl-CoA, and some of these crude estimates have been made from the kinetic constants by application of the Haldane relationships (*10, 11*). A more accurate value of 0.026 has been obtained

27. F. Solomon, Ph.D. Thesis, Brandeis University, Waltham, Massachusetts, 1970.
28. I. Lieberman, *ABB* **51**, 350 (1954).
29. W. S. Sly and E. R. Stadtman, *JBC* **238**, 2632 (1963).
30. J. R. Stern, *JBC* **221**, 33 (1956).
31. J. R. Stern, M. J. Coon, and A. del Campillo, *JACS* **75**, 1517 (1953).
32. F. Lynen and S. Ochoa, *BBA* **12**, 305 (1953).

from the equilibrium constants for the formation of enzyme–CoA from acetoacetyl-CoA and from succinyl-CoA in the two half-reactions (*23*).

D. Mechanism and Kinetics

The obvious part of the reaction catalyzed by the CoA-transferases is the transfer of the thiol group from a thiolester to an "acceptor" carboxylate ion [Eq. (6)], but the important part with respect to driving

$$R_1\overset{O}{\underset{}{\overset{\|}{C}}}-SCoA + R_2\overset{O}{\underset{}{\overset{\|}{C}}}-\overset{*}{O}^- \rightleftharpoons R_1\overset{O}{\underset{}{\overset{\|}{C}}}-\overset{*}{O}^- + R_2\overset{O}{\underset{}{\overset{\|}{C}}}-SCoA \qquad (6)$$

force and mechanism is the extraction of oxygen from the acceptor carboxylate ion R_2COO^- and its transfer to the product carboxylate ion. This transfer has been demonstrated directly, by the use of isotopically labeled oxygen, for the succinate-acetoacetate enzyme (*31*). The mechanistic problem in these reactions is the activation of a carboxylate ion, which is the least activated form of an acyl compound, to the energy-rich form of the thiolester product; thus, it is available for metabolic utilization. As in all such activations, this is done by removing a carboxylate oxygen atom with a dehydrating agent and replacing it with a good leaving group, in this case the thiol group of the CoA moiety of the original thiolester. The unusual feature of these reactions is that the dehydrating agent, instead of ATP or some other phosphate anhydride, is the thiolester substrate itself. In principal, however, these reactions may be regarded as members of the same class of activation reactions which, in at least some cases, proceed by a two-step mechanism. The first step involves the transfer of phosphate or AMP to the carboxylate ion. This is followed by the displacement of this phosphate group and the bridging oxygen atom that was derived from the carboxylate ion substrate by an acceptor (Y^-) such as the thiol group of coenzyme A or the hydroxyl group of transfer RNA in the second step [Eq. (7)]. In the

$$R_1\overset{O}{\underset{}{\overset{\|}{C}}}-\overset{*}{O}^- + R_2O\overset{O}{\underset{O^-}{\overset{\|}{P}}}\sim X \underset{}{\overset{\pm X^-}{\rightleftharpoons}} \left[R_1\overset{O}{\underset{}{\overset{\|}{C}}}-\overset{*}{O}-\overset{O}{\underset{O^-}{\overset{\|}{P}}}-OR_2 \right] \underset{}{\overset{\pm Y^-}{\rightleftharpoons}} \begin{array}{c} R_1\overset{O}{\underset{}{\overset{\|}{C}}}-Y \\ + \\ ^-\overset{*}{O}-\overset{O}{\underset{O^-}{\overset{\|}{P}}}-OR_2 \end{array} \qquad (7)$$

case of CoA-transferase, we do not know for certain the nature of the activating and dehydrating agent that is initially added to the carboxylate ion, but we may surmise that it is an activated acyl group of a thiolester.

The simplest mechanism for the extraction of oxygen from the carboxylate ion is through an anhydride intermediate [Eq. (8)] (*33*). The

$$\underset{\underset{O}{\overset{\|}{\underset{-O^{*}-CR_{2}}{\overset{O}{\|}}}}}{\overset{O}{\overset{\|}{R_{1}C-SCoA}}} \rightleftarrows \left[\underset{\underset{O}{\overset{\|}{\underset{O^{*}-CR_{2}}{\overset{}{|}}}}}{\overset{O}{\overset{\|}{R_{1}C\overset{-}{}SCoA}}} \right] \rightleftarrows \underset{\underset{O}{\overset{\|}{\underset{-O^{*}C-R_{2}}{\overset{}{|}}}}}{\overset{O}{\overset{\|}{R_{1}CSCoA}}} \quad (8)$$

formation of such an anhydride intermediate is thermodynamically unfavorable for a bimolecular reaction in solution but would not be unreasonable at the active site of the enzyme where the reactants are bound with the loss of most of their translational and rotational entropy (*34, 35*). The driving force for the reaction is clear-cut—the carboxylate ion expels the thiol anion, which then attacks the newly activated acyl group to expel the product carboxylate ion. This mechanism could be identified if evidence could be obtained for the formation of either the anhydride intermediate or the free thiol group of CoA (which probably would not dissociate from the enzyme), but attempts to trap such intermediates have so far been unsuccessful (*27*). The alternative mechanism involves a concerted four-center reaction [Eq. (9)], in which the driving forces for the reaction cannot be so clearly identified (*36*). An inter-

$$\underset{\underset{O}{\overset{\|}{\underset{-O^{*}-CR_{2}}{\overset{}{|}}}}}{\overset{O}{\overset{\|}{R_{1}C-SCoA}}} \rightleftarrows \underset{\underset{O}{\overset{\|}{\underset{-O^{*}CR_{2}}{\overset{}{|}}}}}{\overset{O}{\overset{\|}{R_{1}CSCoA}}} \quad (9)$$

mediate and somewhat more attractive mechanism involves addition of the carboxylate ion to the carbonyl group of the thiolester, followed by thiol anion expulsion and attack on the carbonyl group of the original carboxylate ion [Eq. (10)] (*37*). The difference between these mechanisms, as in the case of acyl activations by ATP, is largely a function

33. A. B. Falcone and P. D. Boyer, *ABB* **83**, 337 (1959).
34. S. Milstien and L. A. Cohen, *Proc. Nat. Acad. Sci. U. S.* **67**, 1143 (1970).
35. M. I. Page and W. P. Jencks, *Proc. Nat. Acad. Sci. U. S.* **68**, 1678 (1971).
36. W. P. Jencks, "The Enzymes," 2nd ed., Vol. 6, p. 384, 1962.
37. L. Jaenicke and F. Lynen, "The Enzymes," 2nd ed., Vol. 3, Part B, p. 47, 1960.

$$R_1\overset{O}{\underset{\underset{O}{\overset{|}{-O^*-CR_2}}}{\overset{||}{C}-SCoA}} \rightleftharpoons \left[R_1-\overset{O^-}{\underset{\underset{O}{\overset{||}{O^*-CR_2}}}{\overset{|}{C}-SCoA}}\right] \rightleftharpoons R_1-\overset{O}{\underset{\underset{O}{\overset{||}{-O^*}}}{\overset{||}{C}}}\quad \underset{\underset{O}{\overset{||}{CR_2}}}{SCoA} \qquad (10)$$

of the lifetime of the various possible intermediates and will be difficult to distinguish experimentally. However, the finding that the reactivity of substituted acetates with the enzyme-CoA intermediate increases directly with increasing basicity indicates that most of the charge on the attacking oxygen atom is lost in the transition state; this is consistent with a stepwise mechanism that proceeds through an anhydride intermediate [Eq. (8)] and is inconsistent with a fully concerted cyclic mechanism [Eq. (9)], which should be relatively insensitive to substituent effects (*23*).

As might be expected for such a complex reaction, there are few chemical models for this type of acyl activation and those that do exist are not well characterized with respect to mechanism and kinetics. The closest analogy is the formation of thioacetic acid from thiobenzoic acid in acetic acid at 88° over a period of hours (*38*). The requirement of small amounts of water for this reaction may involve a facilitation of the ionization of acetic acid to the carboxylate ion to attack thiobenzoic acid and form an anhydride intermediate. There is some evidence in favor of anhydride or anhydride-like intermediates in other reactions of this class (*39–44*). Precedent for acyl transfer through a four-center mechanism in aqueous solution [Eq. (9)] exists in the intramolecular transfer of a benzoyl group from sulfur to nitrogen in an isothiourea derivative [Eq. (11)] (*45*).

$$\underset{\underset{CH_3}{|}}{\overset{\overset{O}{||}}{Ar\overset{|}{C}}\text{-S}}_{N=C-NCH_3}^{} \longrightarrow \underset{\underset{CH_3}{|}}{\overset{\overset{O}{||}}{Ar\overset{|}{C}}}\quad \overset{\overset{}{S}}{\underset{}{\underset{H}{\overset{||}{C}-NCH_3}}} \qquad (11)$$

Although the exchange of labeled succinate and acetoacetate into their

38. I. A. M. Ford and S. A. M. Thompson, *Nature (London)* **185**, 96 (1960).
39. D. Davidson and M. Karten, *JACS* **78**, 1066 (1956).
40. R. N. Ring, J. G. Sharefkin, and D. Davidson, *J. Org. Chem.* **27**, 2428 (1962).
41. G. W. Anderson, J. Blodinger, R. W. Young, and A. Welcher, *JACS* **74**, 5304 (1952).
42. G. W. Anderson, J. Blodinger, and A. D. Welcher, *JACS* **74**, 5309 (1952).
43. R. W. Young, K. H. Wood, R. J. Joyce, and G. W. Anderson, *JACS* **78**, 2126 (1956).
44. C. Kutzbach and L. Jaenicke, *Justus Liebigs Ann. Chem.* **692**, 26 (1966).
45. R. F. Pratt and T. C. Bruice, *Biochemistry* **10**, 3178 (1971).

respective CoA thiolester derivatives suggested the possibility that the enzymic reaction might proceed through an enzyme–CoA intermediate (46, 47), there appeared to be no significant advantage, on chemical grounds, for the formation of such an intermediate. Since an anhydride or four-center mechanism almost certainly must be involved at some point in the reaction, it seemed more likely that an acyl–enzyme intermediate would be formed and detected. It was a surprise, therefore, to find the following almost incontrovertable evidence that an enzyme–CoA compound is formed from the acyl-CoA substrate and is a kinetically significant intermediate in the reaction [Eqs. (12) and (13)] (10, 11, 25).

$$\text{AcAcSCoA} + \text{E} \rightleftharpoons \text{AcAcSCoA} \cdot \text{E} \rightleftharpoons \text{AcAc} + \text{E-SCoA} \quad (12)$$

$$\text{E-SCoA} + \text{Succ} \rightleftharpoons \text{SuccSCoA} \cdot \text{E} \rightleftharpoons \text{SuccSCoA} + \text{E} \quad (13)$$

(a) The reaction follows ping-pong kinetics with parallel lines on double reciprocal plots of velocity against substrate concentration. This is characteristic of enzymic reactions that proceed through two half-reactions in which the group that is transferred (CoA in this case) is bound to the enzyme in an intermediate (enzyme–CoA) which does not contain the rest of the first substrate (the acyl group of the original acyl-CoA).

(b) Inhibition by each of the carboxylate products is competitive with respect to the other carboxylate substrate and is noncompetitive with respect to the acyl-CoA substrate, as expected for such a mechanism.

(c) The expected Haldane relationships that relate the kinetic constants to the overall equilibrium constant are obeyed.

(d) At a given concentration of acetoacetyl-CoA the maximum rate of acetoacetate exchange into acetoacetyl-CoA is equal to the maximum rate of formation of succinyl-CoA in the presence of a saturating concentration of succinate. In other words, when the formation of the enzyme–CoA intermediate is rate determining (r.d.s.) it can react either with succinate to give product or with acetoacetate to regenerate acetoacetyl-CoA at the same rate [Eq. (14)].

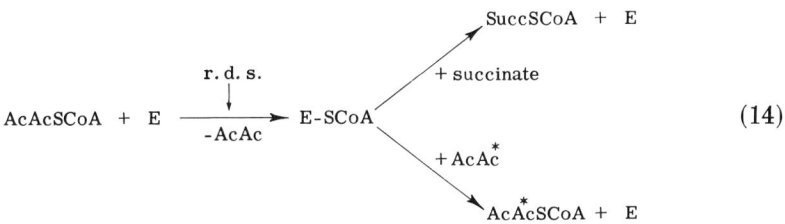

(e) Under the same conditions, if both succinate and acetoacetate are present the amount of inhibition by acetoacetate of the formation of succinyl-CoA is equal to the amount of incorporation of acetoacetate into acetoacetyl-CoA; thus, the total rate of the reaction remains constant [Eq. (14)]. In other words, the decrease in succinyl-CoA formation by acetoacetate is caused by the reaction of the enzyme–CoA intermediate with acetoacetate to give acetoacetyl-CoA instead of succinyl-CoA; thus, the amount of back reaction to regenerate starting material must be equal to the amount of inhibition.

(f) At low concentrations of succinate, the incorporation of acetoacetate into acetoacetyl-CoA is greater than the amount of inhibition of succinyl-CoA formation. When the succinate concentration is reduced, the reaction of the enzyme–CoA intermediate with succinate becomes rate determining and its formation becomes relatively fast. Under these conditions, enzyme–CoA is formed fast enough so that it can react with acetoacetate to give exchange into acetoacetyl-CoA with relatively little inhibition of succinyl-CoA formation; i.e., the amount of acetoacetate exchange becomes larger than the amount of inhibition of succinyl-CoA formation.

(g) An enzyme–CoA intermediate is formed from acetoacetyl-CoA or succinyl-CoA that can be isolated by molecular exclusion chromatography. It reacts with succinate to give succinyl-CoA. It undergoes hydrolysis through pH-independent and base-catalyzed pathways at a rate that is much faster ($t_{1/2}$ ca. 30 min at pH 6.5) than expected for a thiolester, indicating that its reactivity is increased by groups at the active site of the enzyme. It is reduced by borohydride to give an inactive enzyme and tritium is incorporated into the protein when the reduction is carried out with tritium-labeled borohydride.

(h) Labeled oxygen is incorporated into the enzyme upon incubation with oxygen-labeled succinate and either succinyl-CoA or acetoacetyl-CoA, as required by the previously observed transfer of labeled oxygen between carboxylate substrates and by the enzyme–CoA mechanism (48). The overall transfer of labeled oxygen from carboxylate substrate to carboxylate product occurs through two turnovers of the enzyme—in the first turnover the labeled enzyme is formed and in the second this labeled oxygen atom is transferred to the carboxylate product. This is shown for an enzyme with a carboxylate group at the active site in Eqs. (15) and (16).

---

46. Experiments of C. Gilvarg, quoted in reference 30.
47. Experiments of J. R. Stern and A. del Campillo, quoted in reference 13.
48. R. W. Benson and P. D. Boyer, JBC **244**, 2366 (1969).

$$R_1\overset{O}{\overset{\|}{C}}-SCoA + R_2C\overset{*}{O}O^- + E\text{-}COO^- \longrightarrow R_1COO^- + R_2\overset{O}{\overset{\|}{C}}-SCoA + E\text{-}C\overset{*}{O}O^- \quad (15)$$

$$R_1\overset{O}{\overset{\|}{C}}-SCoA + R_2C\overset{*}{O}O^- + E\text{-}C\overset{*}{O}O^- \longrightarrow R_1C\overset{*}{O}O^- + R_2\overset{O}{\overset{\|}{C}}-SCoA + E\text{-}C\overset{*}{O}O^- \quad (16)$$

Of these criteria, the quantitative succinate and acetoacetate exchange and exchange-inhibition experiments (d–f) provide the strongest support for the enzyme–CoA mechanism, because they provide quantitative evidence that such an intermediate is formed at a rate that is adequate to account for the normal catalysis by the enzyme. The isolation of enzyme–substrate "intermediates" can always be interpreted as the result of the formation of inactive, nonphysiological complexes in the absence of a quantitative evaluation of their reactivity.

The oxygen-containing site of CoA binding in the enzyme–CoA intermediate has been identified as the γ-carboxyl group of glutamate by reduction of the intermediate with tritium-labeled borohydride [Eq. (17)]

$$E\text{-}COO^- + R\overset{O}{\overset{\|}{C}}SCoA \rightleftharpoons E\text{-}\overset{O}{\overset{\|}{C}}SCoA + RCOO^- \quad (17)$$
$$\downarrow \overset{*}{B}H_4^-$$
$$E\text{-}\overset{*}{C}H_2OH + HSCoA$$

and identification of the product as α-amino-δ-hydroxyvaleric acid after hydrolysis (49). Thus, the intermediate, like the substrate and product, is a thiolester.

The identification of this thiolester intermediate tells us about the reaction pathway followed by the enzyme in some detail but does not help us to understand the chemistry of the catalytic mechanism. Exactly the same problems are involved in forming a thiolester intermediate with a carboxyl group covalently linked to the enzyme [Eq. (17)] as in forming a thiolester product directly from a carboxylate group bound to the enzyme in a Michaelis complex. The preceding discussion regarding anhydride and four-center mechanisms, therefore, applies equally to the formation of the enzyme–CoA intermediate from the thiolester substrate and the carboxylate group in the active site. Perhaps it is easier for the enzyme if one carboxyl group is fixed as a part of the active site, with its translational and rotational entropy largely lost as a consequence of its

49. F. Solomon and W. P. Jencks, *JBC* **244**, 1079 (1969).

covalent bonding to the enzyme, than to fix both substrates at the same time in looser Michaelis complexes. It still is not known whether there are one or two sites for the two carboxylate substrates. It is probable, even if there are two sites corresponding to the specificity-determining structures of the substrates, that they share a common site for the reacting carboxylate group so that the enzyme does not need to have two sets of catalyzing groups at the active site.

1. *Other Properties of the Active Site*

The enzyme shows a decrease in activity with decreasing pH that suggests a requirement for a group in the basic form at the active site for activity (*9, 10, 27*). The pK of this group, from the pH dependence of the maximal velocity of the acetoacetyl-CoA–succinate reaction, is approximately 7.7. On chemical grounds the obvious candidate for this group is the carboxylate group of the enzyme with its pK perturbed upward by a poor ion-solvating environment. Such a perturbation would be expected to increase its nucleophilic reactivity and to stabilize an anhydride intermediate. The same perturbation would be expected for the substrate carboxylate group that attacks the enzyme–CoA intermediate in a very similar environment. The pK of a sensitive thiol group is over 8.3; thus, this group is not a candidate for the kinetic pK unless it is perturbed downward by the presence of substrate. No effect of pH on the K_m for acetoacetyl-CoA has been observed between pH 7.3 and 8.8 (*25*). The bacterial acetate-propionate-succinate enzyme shows a rather sharp pH optimum between pH 6.5 and 7.8 (*14*).

Several types of indirect evidence suggest the presence in the active site of a cationic group, which may be a protonated amine that provides a binding site for a substrate carboxylate group:

(a) The enzyme is strongly inhibited by monovalent anions. The order of effectiveness follows the Hofmeister series, as expected for binding to a cationic site, and inhibition by chloride ion is at least partially competitive with respect to succinate (*10*).

(b) The enzyme is irreversibly but incompletely inhibited by incubation with acetoacetyl-CoA. This inhibition is stimulated by metals that are expected to increase the reactivity of this thiolester as an acylating agent, and it is decreased by chelating agents. A similar inhibition is observed with reactive chemical acylating agents such as *p*-nitrophenyl acetate and acetylimidazole. This inhibition is decreased in the presence of substrate and exhibits the dependence on pH that is expected for the acylation of a group that is in the unreactive, protonated form at neutral pH (*10*).

(c) At low concentrations of succinate, at which the half-reaction of enzyme-bound succinate with the enzyme–CoA intermediate is rate determining, the pH-activity curve is bell-shaped. The pK of the upper limb is approximately 8.4, consistent with the ionization of a protonated amine (27).

The enzyme contains a sulfhydryl group that reacts with p-chloromercuribenzoate, 5,5'dithiobis(2-nitrobenzoic acid) (DTNB, Ellman's reagent), and N-ethylmaleimide with loss of enzymic activity (27). With concentrated enzyme solutions the loss of activity is initially reversible with dithiothreitol but rapidly becomes irreversible. Acetoacetate and succinate (0.15–0.2 M) have no effect on the rate of inactivation by DTNB, but the inactivation rate is dramatically increased by succinyl-CoA or acetoacetyl-CoA (27). At pH 8.1 the same maximal increase in inactivation rate of 20-fold is obtained with concentrations of acetoacetyl-CoA or succinyl-CoA that are far below the K_m or K_I values for these substrates. This suggests that the increase in rate results from an enhanced rate of inactivation of the product of a chemical reaction of the enzyme with these substrates, i.e., the formation of the enzyme–CoA intermediate [Eqs. (12) and (13)], rather than from binding of acyl-CoA to the enzyme. This increase in rate is prevented by low concentrations of acetoacetate or succinate, which drive the reaction backward and decrease the equilibrium concentration of enzyme–CoA. By using the rate of inactivation as an indicator of the fraction of enzyme in the enzyme–CoA form in the presence of different concentrations of acyl-CoA and the corresponding acid substrate, the equilibrium constants for the two half-reactions [Eqs. (12) and (13)] were found to be 40 ± 10 and 1.0 ± 0.1 for the formation of enzyme–CoA from acetoacetyl-CoA and succinyl-CoA, respectively, in the presence of 4×10^{-4} M DTNB (23). The fact that the latter equilibrium constant is close to unity means that the free energy of hydrolysis of the enzyme–CoA intermediate is the same as that of succinyl-CoA; thus, there is little or no extra stabilization of the enzyme–CoA from noncovalent binding interactions of the CoA moiety with the CoA binding site of the enzyme.

The enzyme exhibits a remarkable degree of specificity toward its acyl-CoA substrates. Succinyl pantetheine and acetoacetyl pantetheine are inactive at high concentrations (9). S-Acetoacetyl-N-acetylcysteamine is at least 10^4-fold less reactive than acetoacetyl-CoA at a concentration of 0.013 M in turnover experiments and appears to be at least 10^6 less reactive in single turnover experiments, based on the sensitivity to inactivation by borohydride of the enzyme–thiolester intermediate (27). One might expect, therefore, that this high degree of specificity would be reflected in a tight binding of the CoA moiety of the specific substrates. Just the opposite is observed—CoA and acetyl-CoA at a concentration of 10^{-3} M have little effect on the rate of the pig heart enzyme and give only about 50% inhibition of a muscle enzyme (*11*, *27*). In marked contrast, 1.4 mM oxyCoA (*50*), which differs from CoA only by the substitution of a hydroxyl for a sulfhydryl group, gives a 93% inhibition of the pig heart enzyme (*27*). A similar situation obtains for sulfhydryl inactivation. Although the rate of this inactivation is increased markedly by acyl-CoA substrates, 10^{-3} M acetyl-CoA has no effect on the initial rate and 7×10^{-4} M oxyCoA causes a 4-fold decrease in inactivation rate (*27*). It has been noted above that the equilibrium constant for enzyme–CoA formation gives no evidence for a net favorable binding interaction with CoA. The favorable free energy of binding of acyl-CoA substrates must reflect mainly the binding energy of the acetoacetyl or succinyl groups. All of this suggests that this enzyme represents an extreme example of the utilization of the binding energy of specific substrates to bring about some thermodynamically unfavorable change in the system that increases the rate of the catalyzed reaction (lowers the free energy of activation). Apparently this utilization of binding energy decreases the observed binding energy so drastically in this case that little or no binding is observed even for CoA itself, which must contain many of the specific binding sites that are reflected in the high degree of specificity of the enzyme toward acyl-CoA substrates. This free energy that is obtained from the binding interactions could be utilized to force structural changes in either the enzyme or the substrate, or both, that are advantageous for the catalytic process.

50. T. L. Miller, G. L. Rowley, and C. J. Stewart, *JACS* **88**, 2299 (1966). We are very grateful to Dr. Stewart for a generous gift of oxyCoA.

12

Amidinotransferases

JAMES B. WALKER

I. Introduction	497
II. Glycine Amidinotransferase	498
A. Biological Distribution	498
B. Catalytic Properties	499
C. Regulation	503
III. Inosamine-P Amidinotransferase	505
A. Biological Distribution	505
B. Catalytic Properties	506
C. Regulation	508

I. Introduction

Amidinotransferases catalyze the reversible transfer of an amidino group ($-C(=NH_2^+)NH_2$) from the guanidino moiety of the physiological donor, L-arginine, to the amino moiety of certain acceptor compounds (1). Only two amidinotransferases have been extensively studied. These are L-arginine:glycine amidinotransferase (EC 2.1.4.1), which catalyzes the first of the two reactions involved in creatine biosynthesis by vertebrates (2–4), and L-arginine:inosamine-P amidinotransferase, which catalyzes two nonconsecutive reactions in streptomycin biosynthesis by Streptomyces soil bacteria (5, 6). Amidinotransferases have also been

1. S. Ratner, "The Enzymes," 2nd ed., Vol. 6, p. 267, 1962.
2. K. Bloch and R. Schoenheimer, JBC 138, 167 (1941).
3. H. Borsook and J. W. Dubnoff, JBC 138, 389 (1941).
4. J. B. Walker, Advan. Enzyme Regul. 1, 151 (1963).
5. J. B. Walker, JBC 231, 1 (1958).
6. J. B. Walker, Lloydia 34, 363 (1971).

implicated in the biosynthesis of taurocyamine (7), arcaine (8), lombricine (9), viomycin (10), galegin (11), and certain other guanidino compounds in higher plants (12). In exploratory work where the physiological acceptors are not known, amidinotransferases can be most simply detected by assaying for either arginine produced by canavanine:ornithine transamidination (13), or hydroxyguanidine produced by arginine:NH_2OH transamidination (5, 14). The first assay is particularly useful for extracts containing arginase, the hydrolytic counterpart of amidinotransferase (13).

II. Glycine Amidinotransferase

$$\underset{\text{L-Arginine}}{\overset{\overset{+}{N}H_2}{\underset{H_3\overset{+}{N}-CH-COO^-}{\overset{|}{\underset{|}{\overset{NH-C-NH_2}{\underset{(CH_2)_3}{|}}}}}}} + \underset{\text{Glycine}}{H_2N-CH_2-COO^-} \rightleftharpoons \underset{\text{L-Ornithine}}{\overset{NH_2}{\underset{H_3\overset{+}{N}-CH-COO^-}{\overset{|}{\underset{|}{(CH_2)_3}}}}} + \underset{\text{Guanidinoacetate}}{\overset{NH_2}{\underset{HN-CH_2-COO^-}{\overset{|}{C=\overset{+}{N}H_2}}}} \quad (1)$$

A. Biological Distribution

Glycine amidinotransferase, which participates in creatine biosynthesis, was first discovered by Borsook and Dubnoff in kidneys of mammals (3), and then reported by Nakatsu to occur in liver of cow, hog, chicken, frog, and lizard (15). The highest concentrations occur in pancreas of man (13) and other mammals (16), and in rat decidua and deciduoma (17, 18). It is possible that pancreatic amidinotransferase also functions in control of insulin release, which is markedly stimulated by guanidino-

7. N. V. Thoai, S. Zappacosta, and Y. Robin, *Comp. Biochem. Physiol.* **10**, 209 (1963).
8. Y. Robin, C. Audit, and M. Landon, *Comp. Biochem. Phy iol.* **22**, 787 (1967).
9. R. J. Rossiter, T. J. Gaffney, H. Rosenberg, and A. I . Ennor, *BJ* **76**, 603 (1960).
10. K. Raczynska-Bojanowska, M. Tyc, L. Pass, and Z. Kotula, *Bull. Acad. Pol. Sci.* **17**, 203 (1969).
11. G. Reuter, *Flora (Jena)* **154**, 136 (1964).
12. J. Miersch and H. Reinbothe, *Phytochemistry* **6**, 485 (1967).
13. J. B. Walker, *BBA* **73**, 241 (1963).
14. J. B. Walker, *JBC* **235**, 2357 (1960).
15. S. Nakatsu, *Mem. Fac. Sci., Kyushu Univ., Ser. C* **3**, 1 (1958).
16. J. B. Walker, *Proc. Soc. Exp. Biol. Med.* **98**, 7 (1958).
17. J. B. Walker and W. T. Gipson, *BBA* **67**, 156 (1963).
18. J. B. Walker, *Ciba Found. Study Group [Pap.]* **19**, 43 (1965).

acetate (*19*). It would be interesting to determine whether glycine potentiates the stimulation of insulin release by arginine. The developing chick embryo synthesizes its own creatine *de novo* utilizing this enzyme (*20*). Van Pilsum has recently published an extensive survey of amidinotransferase activities in animal tissues (*21*). The important observation was made that creatine present in certain invertebrates probably results from absorption from the environment and not *de novo* biosynthesis (*21*). However, negative results obtained with carnivores, including sharks (*21*), as well as absolute enzymic activities obtained for all species, must be carefully evaluated, taking into account the previous dietary history, including prior repression of the enzyme by dietary creatine from ingested muscle tissues (*14*), and lowering of enzyme levels by subnormal diets (*22–24*). These two effects are synergistic (*14, 18*): Any carnivore feeding on vertebrates but with inadequate food intake would probably have extremely low levels of amidinotransferase.

B. Catalytic Properties

1. *Substrate Specificity*

The physiological donor of the amidino group is L-arginine, as shown by experiments with intact animals (*2, 25, 26*) as well as studies *in vitro* (*3*). Guanidinoacetate can donate its amidino group *in vitro* (*27*) and *in vivo* (*26*), although its major fate is irreversible methylation to form creatine (*28–30*). L-Canavanine, a guanidinoxy analog of arginine occurring in jack bean seeds, is an excellent amidino donor *in vitro* (*31*) and *in vivo* (*18*). 4-Guanidinobutyrate and 3-guanidinopropionate are weak

19. R. N. Alsever, R. H. Georg, and K. C. Sussman, *Endocrinology* **86**, 332 (1970).
20. M. S. Walker and J. B. Walker, *JBC* **237**, 473 (1962).
21. J. F. Van Pilsum, G. C. Stephens, and D. Taylor, *BJ* **126**, 325 (1972).
22. Y. Bauerova and F. Sorm, *Biokhimiya* **21**, 397 (1956).
23. J. F. Van Pilsum, D. A. Berman, and E. A. Wolin, *Proc. Soc. Exp. Biol. Med.* **95**, 96 (1957).
24. J. B. Walker, *JBC* **236**, 493 (1961).
25. D. Stetten, Jr. and B. Bloom, *JBC* **220**, 723 (1956).
26. W. H. Horner, *JBC* **220**, 861 (1956).
27. M. Fuld, *Fed. Proc., Fed. Amer. Soc. Exp. Biol.* **13**, 215 (1954).
28. G. L. Cantoni and P. J. Vignos, Jr., *JBC* **209**, 647 (1954).
29. H. J. Almquist, E. Mecchi, and F. H. Kratzer, *JBC* **141**, 365 (1941).
30. J. B. Walker and S. H. Wang, *BBA* **81**, 435 (1964).
31. J. B. Walker, *JBC* **218**, 549 (1956).

donors (*13*); additional donors can be detected by the more sensitive radiochemical assays.

The physiological acceptors are glycine, for creatine biosynthesis (*3*), and, we suggest, L-ornithine, for metabolic control purposes. Additional acceptors include L-canaline, the aminoxy analog of ornithine (*31*), 4-aminobutyrate (*32*), glycylglycine (*18*), 3-aminopropionate (*32*), hydroxylamine (*5, 33*), and water (*34, 35*). With water as acceptor and arginine as donor, the reaction resembles that catalyzed by arginase, but apoarginase does not have amidinotransferase activity. Kinetic studies indicate that the acceptor amino group must be in the unprotonated form (*36*). It is therefore of interest that the terminal amino group of ornithine is more basic than that of glycine, and the optimal pH of transamidination with ornithine as acceptor is higher than that with glycine as acceptor (*37*). Among its substrates and products, amidinotransferase has its highest affinity for L-ornithine (*37*); consequently, ornithine exerts strong product inhibition, and, as predicted by the Haldane relationship, the reverse reaction proceeds more slowly than the forward reaction (*27*). L-Norvaline is a good competitive inhibitor (*13, 37*). Since any of the above donors can transfer an amidino group to any of the above acceptors, a number of assays for enzymic activity can be employed.

2. Reaction Mechanism

a. Indirect Evidence for an Enzyme–Amidine Intermediate. The occurrence of partial or half-reactions, in addition to chemical trapping of the amidino group in the absence of added acceptor substrates, strongly suggested a double-displacement reaction mechanism (*33*). The first evidence was provided by detection of canavanine:ornithine and canavanine:glycine transamidinations in hog kidney extract (*31*). Since canavanine is an analog of arginine, it was suggested that canavanine:ornithine transamidination occurred as the sum of reactions (2) and (3):

$$\text{Canavanine} + \text{enzyme} \rightleftharpoons \text{canaline} + \text{enzyme—C}(= NH_2^+)NH_2 \quad (2)$$

$$\text{Enzyme—C}(= NH_2^+)NH_2 + \text{ornithine} \rightleftharpoons \text{enzyme} + \text{arginine} \quad (3)$$

$$\text{Enzyme—C}(= NH_2^+)NH_2 + HONH_2 \rightarrow \text{enzyme} + HO—NH—C(= NH_2^+)NH_2 \quad (4)$$

32. J. J. Pisano, C. Mitoma, and S. Udenfriend, *Nature* (*London*) **180**, 1125 (1957).
33. J. B. Walker, *JBC* **224**, 57 (1957).
34. S. Ratner and O. Rochovansky, *ABB* **63**, 277 (1956).
35. F. Conconi and E. Grazi, *JBC* **240**, 2461 (1965).
36. G. Ronca, V. Vigi, and E. Grazi, *JBC* **241**, 2589 (1966).
37. S. Ratner and O. Rochovansky, *ABB* **63**, 296 (1956).
38. J. B. Walker, *JBC* **221**, 771 (1956).

Conclusive evidence was soon obtained for an arginine:ornithine exchange transamidination (*38*). Minimal kinetic criteria were satisfied since the reaction rates of these partial reactions were equal to or greater than rates of the overall transamidination reactions. Reaction of the postulated intermediate with hydroxylamine, as the sum of reactions (3) and (4), provided further evidence for a double-displacement mechanism (*33*). Guanidinoacetate:NH_2OH (*13, 24*) and guanidinoacetate:glycine (*33, 37*) transamidinations were expected and detected. The observation that susceptibility of pancreatic amidinotransferase to inhibition by a small labile sulfhydryl reagent, formamidine disulfide, was enhanced by preincubation with ornithine or glycine, and decreased by arginine, was compatible with the existence of the equilibrium shown in reaction (3), particularly if the intermediate were enzyme—S—$C(=NH_2^+)NH_2$ (*39*). In the amidino form the enzyme would presumably be unable to form a mixed disulfide; acceptors would remove the protecting amidino group. The situation is evidently more complex when a more bulky disulfide, 5,5′-dithiobis-2-nitrobenzoate, is employed as inhibitor (*40*).

Detailed kinetic studies by Grazi's laboratory are also consistent with a double-displacement mechanism (*36*).

b. Direct Evidence for an Enzyme–Amidine Intermediate. Grazi and co-workers have purified glycine amidinotransferase 900-fold from hog kidney and obtained an apparently homogeneous preparation with an approximate molecular weight of 100,000 (*35, 41*). Treatment of this preparation with DL-([^{14}C]*guanidino*) arginine or L-(U-^{14}C)arginine gave labeled enzyme which was separated from labeled substrate on a Sephadex G-25 column at 4°. Heating the labeled enzyme at 100° decomposed the complex and gave labeled urea but no arginine or ornithine. Upon incubation of the labeled enzyme with amidino acceptors at 20°, label was transferred to glycine or ornithine, forming guanidinoacetate or arginine, respectively (*42*). This evidence suggested that the labeled enzyme was the previously postulated (*33*) enzyme–amidine intermediate. Unfortunately, slow hydrolysis of the enzyme–amidine complex at neutral pH hindered kinetic studies of complex formation and determination of its chemical characteristics.

Subsequently, however, it was found that the enzyme–amidine complex is stable in 0.1 N HCl, but catalytically inactive both as a trans-

39. J. B. Walker and M. S. Walker, *ABB* **86**, 80 (1960).
40. E. Grazi, V. Vigi, and N. Rossi, *Eur. J. Biochem.* **1**, 182 (1967).
41. E. Grazi and F. Conconi, "Methods in Enzymology," Vol. 17A, p. 1007, 1970.
42. E. Grazi, F. Conconi, and V. Vigi, *JBC* **240**, 2465 (1965).

ferase and hydrolase (*43*). Amidinotransferase could be labeled with L-([^{14}C]*guanidino*) arginine, the reaction stopped at the desired time with 0.1 N HCl, and the labeled enzyme–amidine complex isolated by separation on a Sephadex G-25 column. By this means it was estimated that one amidino group was combined per mole of enzyme, and the equilibrium constant for enzyme–amidine formation from arginine was 0.31 (*43*).

Chemical characterization of the intermediate was next undertaken. Hydrolysis of the radioactive enzyme–amidine intermediate in 6 N HCl gave a labeled compound which was identified as 2-amino thiazoline 4-carboxylic acid (I) (*43*), a compound expected to be formed under acidic conditions on cyclization of S-amidinocysteine with elimination of ammonia (*44*), as shown in reaction (5). Label was located in position

$$\text{Enzyme-S-}^*\text{C}\begin{smallmatrix}\text{NH}_2^+\\ \text{NH}_2\end{smallmatrix} \xrightarrow{\underset{110°}{6\,N\,\text{HCl}}} \text{Amino acids} + \begin{smallmatrix}\text{H}_2\text{C-S}\\ |\quad\quad\;\;^*\\ \text{HC-N}\\ \text{HOOC}\end{smallmatrix}\text{C-NH}_2 + \text{NH}_4 \quad (5)$$

(I)

2, as determined by oxidation with bromine, followed by alkaline hydrolysis (*43*). It was therefore concluded, as suggested earlier (*33, 39*), that transamidination occurs as follows, with cysteine at the catalytic site:

L-Arginine + enzyme—SH $\rightleftharpoons$ L-ornithine + enzyme—S—C(=NH$_2^+$)NH$_2$ (6)

Enzyme—S—C(=NH$_2^+$)NH$_2$ + glycine $\rightleftharpoons$ enzyme—SH + guanidinoacetate (7)

Sum: L-Arginine + glycine $\rightleftharpoons$ L-ornithine + guanidinoacetate (8)

It should be noted, however, that the reactive cysteine is only one of several thiol groups present in the enzyme (*40*). It is not known whether the terminal amino groups of both ornithine and glycine occupy the same site on the enzyme (*13*). Grazi's laboratory had earlier noted that reactions of amidinotransferase with dinitrofluorobenzene (*45*) and 5,5'-dithiobis-2-nitrobenzoate (*40*) resulted in complex kinetic behavior with respect to transferase and hydrolase activities. In future studies, by use of guanidino-labeled guanidinoacetate, canavanine, and arginine as amidino donors, suboptimal incubation temperatures, and stabilization of the enzyme–amidine intermediate in 0.1 N HCl it should be possible to clarify those complex interactions.

43. E. Grazi and N. Rossi, *JBC* **243**, 538 (1968).
44. J. X. Khym, R. Shapira, and D. G. Doherty, *JACS* **79**, 5663 (1957).
45. V. Vigi, G. Ronca, and E. Grazi, *BBRC* **20**, 757 (1965).

The very interesting inactivation of glycine amidinotransferase with CO_2 in the presence of amidino donor (46) was interpreted by Grazi's group on the basis of the model system of Khym et al. (44, 47). It will be important to determine if an amino group near the thiol at the catalytic site, which Grazi suggested might participate in a relatively irreversible CO_2-mediated cyclization reaction (46, 47), also functions in reversible amidino group interchange during normal catalysis. If such is the case, experimental conditions might be devised which result in isolated enzyme containing in its polypeptide chain ornithine, guanidino-labeled arginine, or homoarginine. It is evident that although much progress has been made on reaction mechanism since the last review (1), additional work remains to be done. Comparisons of the hog kidney enzyme with amidinotransferases from mammalian pancreas and chick liver might also prove of interest.

C. Regulation

Creatine is required by all vertebrates for muscle and nerve tissues; for example, an average adult man contains 120 g of creatine. Biosynthesis of creatine in excess of need would result in wasteful depletion of three of the twenty amino acids required for protein synthesis: arginine, glycine, and methionine. This is particularly serious for those vertebrates, such as birds and reptiles, which cannot synthesize arginine. A requirement for methionine in the diet is, of course, general among higher animals.

Glycine amidinotransferase, which catalyzes the first committed step specific for creatine biosynthesis, was the first enzyme whose tissue levels were found to be regulated by end product repression in intact higher animals (48) and during embryonic development (20). Both dietary creatine (14), at concentrations comparable to that of ingested vertebrate skeletal muscle, and endogenously synthesized creatine (30) lower the level of chick liver amidinotransferase. The degree of enzyme repression in a tissue is proportional to concentration of creatine in that tissue (18, 30). The physiological importance of control of this enzyme can be readily demonstrated by feeding or injecting (30, 49) guanidinoacetate,

46. E. Grazi, G. Ronca, and V. Vigi, JBC **240**, 4269 (1965).
47. D. G. Doherty, R. Shapira, and W. T. Burnett, Jr., JACS **79**, 5667 (1957).
48. J. B. Walker, BBA **36**, 574 (1959).
49. J. B. Walker, Proc. Soc. Exp. Biol. Med. **112**, 245 (1963).

a product of the regulated enzyme. Endogenously synthesized creatine increases tremendously (*29, 30*) with concurrent repression of amidinotransferase (*14, 30*), and fatty liver and other toxic symptoms appear. The toxic symptoms presumably result, at least in part, from depletion of methyl groups in the animal as a result of the irreversible methylation of guanidinoacetate to form creatine (*4*). Feeding or injecting similar quantities of creatine has no adverse effect, and might even be beneficial under some circumstances. The detailed mechanism by which creatine lowers the steady state (*14, 18*) level of amidinotransferase has not been determined. It is believed to inhibit synthesis of active enzyme, in which case the rate of decrease in amidinotransferase activity during creatine feeding would be a measure of the half-life of this enzyme (*4, 48*). The rate of restoration of amidinotransferase activity follows similar kinetics after removal of creatine from the diet and is inhibited by inhibitors of protein synthesis (*24*). Amidinotransferase levels also decline on fasting (*23, 24*), or an inadequate or toxic diet, at about the same maximal rate as creatine repression (*24*); but these two effects are not additive at their respective maximal rates, again indicating that the same process, e.g., inhibition of synthesis of active enzyme, is inhibited in both cases. The lowest amidinotransferase levels occur on a suboptimal, creatine-containing diet, conditions often encountered in nature by carnivores which feed on vertebrates. On such a dietary regimen, minimal endogenous synthesis of creatine is required. Herbivores and insectivores do not normally encounter creatine in the diet. During rapid growth of the musculature the creatine requirement is particularly high, but because of the nonenzymic, first-order formation of creatinine from creatine phosphate and creatine throughout life to the extent of 2% of the body pool per day (*2*), creatine must be continually synthesized by adults or provided in their diets.

Most enzymes operate *in situ* at substrate concentrations below their respective K_m values (*50*). In the closed system of the developing chick embryo, it has been established that arginine is the rate-limiting substrate in creatine biosynthesis (*30, 49*). When additional citrulline or arginine is provided, glycine then becomes rate-limiting (*30, 49*). Although ornithine is a potent inhibitor *in vitro*, ornithine injected into eggs does not inhibit synthesis of creatine by the developing chick embryo (*4*). It is possible that compartmentalization or rapid metabolism of ornithine prevents inhibition of amidinotransferase in this system.

Methfessel has recently written a comprehensive review of glycine

50. D. E. Atkinson, *Curr. Top. Cell. Regul.* **1**, 29 (1969).

amidinotransferase (51); regulation of this enzyme has been most recently reviewed by the author (52).

III. Inosamine-P Amidinotransferase

$$\underset{\text{L-Arginine}}{\underset{\overset{+}{H_3N}-CH-COO^-}{\overset{\overset{+}{NH_2}}{\underset{(CH_2)_3}{HN-\overset{\|}{C}-NH_2}}}} + \underset{(II)}{\text{(II)}} \rightleftharpoons \underset{\text{L-Ornithine}}{\underset{\overset{+}{NH_3}}{\underset{CH-COO^-}{\overset{NH_2}{(CH_2)_3}}}} + \underset{(III)}{\text{(III)}} \qquad (9)$$

R = (a) —OH; (b) —NH—C($=NH_2^+$)NH_2; (c) —NH_2

A. BIOLOGICAL DISTRIBUTION

Inosamine-P amidinotransferase participates in biosynthesis of the streptomycin family of antibiotics (53) and therefore occurs in numerous strains of *Streptomyces*, filamentous bacteria widely distributed in soils (54). Strains with the highest amidinotransferase activities include *S. griseus* ATCC 12475, *S. griseus* ATCC 11984, *S. griseocarneus* ATCC 12628, *S. galbus* ATCC 14077, *S. bluensis* var. *bluensis*, *S. bikiniensis* ATCC 11062, *S. humidus* ATCC 12760, and *S. hygroscopicus forma glebosus* ATCC 14607. In these strains amidinotransferase activity cannot be readily detected during the early phase of rapid growth on complex media; activity rapidly appears as the growth rate decreases, provided that conditions such as media composition, pH, and temperature are favorable (55). Activity often reaches a peak and then declines later in the growth cycle (55). Because of the important influence of growth conditions on enzymic activity, it is not known how many other strains of *Streptomyces* have the gene coding for this amidinotransferase. The growth conditions required by one strain for development of amidinotransferase activity are often not favorable for a different strain. The possibility exists that streptidine-P and streptomycin are synthesized in small amounts by many strains of *Streptomyces*, with correspondingly low levels of amidinotransferase, and only in certain strains is the biosyn-

51. J. Methfessel, *Z. Inn. Med.* **25**, 80 (1970).
52. J. B. Walker, *Symp. Fundam. Cancer Res.* **19**, 317 (1965).
53. A. L. Demain and E. Inamine, *Bacteriol. Rev.* **34**, 1 (1970).
54. S. A. Waksman, "The Actinomycetes." Ronald Press, New York, 1967.
55. J. B. Walker and V. S. Hnilica, *BBA* **89**, 473 (1964).

thetic pathway derepressed and streptidine-P and streptomycin secreted into the medium (56).

B. Catalytic Properties

1. Substrate Specificity

There are several specificity problems. A fundamental problem concerns whether one amidinotransferase, or two amidinotransferases with different substrate specificities, catalyze the two nonconsecutive transamidinations involved in biosynthesis of the streptidine moiety of streptomycin (57). Plá has reported column separation of two amidinotransferase activities with different heat stabilities, in markedly different yields (58); however, the respective fractions were not tested with specific acceptors. The possibility should be considered that the minor component might be derived from the major component during the prolonged extraction procedure. In favor of two different enzymes, besides the data of Plá, is the analogy with the two dehydrogenases, two aminotransferases (59), and two phosphatases (60) also involved in streptidine-P biosynthesis (6). In favor of one enzyme, or two closely related isoenzymes, are the facts that (a) purification by our methods (61, 62) has not yet separated two activities of different substrate specificities; (b) dialyzed extracts catalyze transamidination from the 2-deoxy derivative of compound IIIb to compound IIa; and (c) S. bluensis, which synthesizes bluensomycin, a monoguanidinated analog of dihydrostreptomycin (63), has amidinotransferase activity toward both acceptors IIa and IIb (57), yet does not appear to synthesize streptidine-P or other diguanidinated derivatives in vivo. In this discussion it will be assumed that there is a single amidinotransferase or two structurally related enzymes with similar substrate specificities.

The physiological amidino donor in streptomycin biosynthesis is L-arginine, as suggested by experiments with intact mycelia (64–66) and

56. M. S. Walker and J. B. Walker, JBC 241, 1262 (1966).
57. M. S. Walker and J. B. Walker, BBA 136, 272 (1967).
58. L. C. Plá, BBA 242, 541 (1971).
59. J. B. Walker and M. S. Walker, Biochemistry 8, 763 (1969).
60. M. S. Walker and J. B. Walker, JBC 246, 7034 (1971).
61. J. B. Walker and M. S. Walker, "Methods in Enzymology," Vol. 17A, p. 1012, 1970.
62. A. L. Miller and J. B. Walker, J. Bacteriol. 99, 401 (1969).
63. B. Bannister and A. D. Argoudelis, JACS 85, 119 (1963).
64. G. D. Hunter, M. Herbert, and D. J. D. Hockenhull, BJ 58, 249 (1954).
65. W. H. Horner, JBC 239, 578 (1964).

cell-free preparations (*56*). Enzymic studies showed that, indeed, both guanidino groups of streptomycin were synthesized by transamidination reactions, and these transamidinations occurred prior to formation of the glycosidic linkages (*67*). L-Canavanine (*68*), streptidine-P (IIIb) (*56*), and 2-deoxystreptidine-P can serve as amidino donors *in vitro*. 1-Guanidino-1-deoxy-*scyllo*-inositol 4-phosphate (IIIa) is a surprisingly weak amidino donor relative to streptidine-P (IIIb); this difference might be related to different charge-induced conformations of the two compounds. The physiological amidino acceptors are compounds IIa and IIb (*57*) and possibly L-ornithine for catabolic or regulatory purposes. Compound IIc and its 2-deoxy derivative are excellent acceptors (*56*), as is 2-amino-2-deoxy-*neo*-inositol 5-phosphate, which differs from compound IIa in the configurations of all four hydroxyl groups. Other good acceptors include L-canaline, glycylglycine (*18, 69*), 1,4-diaminobutyl-1-phosphonate, and hydroxylamine (*5, 6*). The structural requirements for an amidino donor appear to be more strict than for an amidino acceptor.

One of the most difficult problems concerned the location of the phosphate group relative to the amino group to be transamidinated. Another problem was whether the R groups in compounds IIb and IIc were at position 3 or 6. A comparison of the behavior of compounds IIb, IIIa, and IIIc toward periodate oxidation helped solve both problems (*60*). Other evidence too diverse to discuss here is also compatible with these assignments. During streptomycin biosynthesis the phosphate groups of acceptors IIa and IIb are introduced at the indicated positions by a specific kinase (*67, 70*). For studies *in vitro*, nonspecific chemical phosphorylation of inosamines, streptamine, and streptidine can also be employed to synthesize amidino acceptors or donors; amidinotransferase selects the correct positional isomer from the complex reaction mixture (*56*). The specificity of amidinotransferase is such that compound IIIc cannot be transamidinated to form IIIb, nor can compound IIc be converted to IIIb, even at high arginine concentrations (*56, 57*).

2. *Reaction Mechanism*

Inosamine-P amidinotransferase has been partially purified from derepressed *S. bikiniensis* to high specific activity (*61*), but detailed studies

---

66. I. I. Tovarova, E. Y. Kornitskaya, S. A. Pliner, V. A. Puchkov, N. S. Wulfson, and A. S. Khokhlov, *Izv. Akad. Nauk SSSR, Ser. Biol.* **31**, 911 (1966).
67. J. B. Walker and M. S. Walker, *Biochemistry* **6**, 3821 (1967).
68. J. B. Walker, *Fed. Proc., Fed. Amer. Soc. Exp. Biol.* **18**, 346 (1959).
69. Y. Li, *Sheng Wu Hua Hsueh Yu Sheng Wu Wu Li Hsueh Pao* **5**, 94 (1965); *CA* **63**, 6050 (1965).
70. J. B. Walker and M. S. Walker, *BBRC* **26**, 279 (1967).

analogous to those of Grazi's group on glycine amidinotransferase (43) have not yet been performed. The bacterial enzyme is also a sulfhydryl enzyme and is more susceptible than the animal enzyme to inhibition by cystamine (55). The two enzymes might also differ in the availability of built-in amidino acceptors or donors near the active sulfhydryl group; susceptibility of the bacterial enzyme to inactivation by CO_2 in the presence of arginine (46) has not been tested. Inosamine-P amidinotransferase catalyzes both arginine:ornithine exchange transamidination and reaction of arginine with hydroxylamine to form hydroxyguanidine (5, 6), which suggests that the bacterial enzyme also forms an isothiouronium intermediate during catalysis, but conclusive proof has not yet been obtained. Inasmuch as the amino group of the amidino acceptor of the animal enzyme must be in the unprotonated form for reaction (36), it is interesting to note that the pK_a values of amino groups of inosamines are approximately 7.6 (63), significantly lower than that of glycine or ornithine.

C. Regulation

Streptomycin, like most other antibiotics, is secreted into the medium following a phase of rapid mycelial growth. Inosamine-P amidinotransferase activity is low during the early phase of rapid growth on complex media and increases dramatically later in the growth phase (55), along with other enzymes in the biosynthetic pathway from *myo*-inositol to streptidine-P (6). If mycelia of *S. griseus* ATCC 12475 are harvested early, when amidinotransferase activity is low, and then resuspended and shaken overnight in 1% NaCl solution, the specific activity increases markedly (55). The increase in specific activity occurring during starvation is inhibited by cyanide and inhibitors of protein synthesis (55). Differentiation to the antibiotic-synthesizing state evidently can be accomplished utilizing endogenous resources exclusively. Many strains have lower amidinotransferase activity when grown at temperatures above 30° or in high concentrations of glucose or inorganic phosphate. It is not known whether genes coding for any or all of the streptidine-synthesizing enzymes (6) are organized in one or more operons or occur in episomes. The mechanism controlling the increase in enzymic activity remains unknown. Other complex differentiative processes in bacteria, such as spore formation and germination, are known to be controlled by many genes, loss of any one of which can stop the process, and this might also be true for streptomycin biosynthesis. Mutations which prevent streptomycin biosynthesis might occur either in a gene coding for a biosynthetic enzyme or in a gene whose function is required for the differentiative process.

In this connection the A factor studied by Khokhlov and other Russian workers (71, 72) and the C factor of Szabo and co-workers (73) appear to fill important functions in differentiation to streptomycin production. Amidinotransferase activity has been used by the Russian groups as a convenient marker because of the ease of assay, but it will not be known whether A factor acts on formation of this enzyme alone until activities of other enzymes in the pathway have been assayed.

Amidinotransferase acceptors accumulate in mycelia under certain conditions (56, 57), and feeding of *myo*-inositol increases their concentrations. These observations suggest that arginine is often the rate-limiting substrate *in vivo*, as was the case of glycine amidinotransferase and the developing chick embryo (49). Possible functions of streptomycin in *Streptomyces* have recently been discussed (6).

71. I. I. Tovarova, E. Y. Kornitskaya, S. A. Pliner, L. A. Shevchenko, L. N. Anisova, and A. S. Khokhlov, *Izv. Akad. Nauk SSSR, Ser. Biol.* **8,** 427 (1970).
72. G. A. Penzikova, M. M. Levitov, L. N. Anisova, N. S. Ivkina, and I. A. Rapoport, *Antibiotiki (Moscow)* **16,** 27 (1971).
73. G. Szabo, I. Bekesi, and S. Vitalis, *BBA* **145,** 159 (1967).

13

N-Acetylglutamate-5-Phosphotransferase

GÉZA DÉNES

I. Introduction 511
II. Molecular Properties 513
   A. Enzyme Purification 513
   B. Physical Properties 513
III. Catalytic Properties 513
   A. The Catalytic Reaction 513
   B. Allosteric Inhibition 516

I. Introduction

Udaka and Kinoshita, studying the enzymic steps of arginine biosynthesis, first reported that in crude extract of *Micrococcus glutaminicus* the reduction of N-acetyl-L-glutamate to N-acetyl-glutamate-5-semialdehyde requires both NADPH and ATP and that the reaction takes place through the formation of a carboxyl-activated intermediate (*1*). Indirect evidence suggested that the carboxyl-activated derivative of N-acetyl-L-glutamate was N-acetyl-L-glutamate-5-phosphate and that the formation of the intermediate was inhibited by arginine, the end product of the pathway (*2*). A similar N-acetyl-L-glutamate reducing system was detected in crude extracts of *Saccharomyces cerevisiae* (*3*). The N-acetyl-L-glutamate reducing system of *Escherichia coli* was separated

1. S. Udaka and S. Kinoshita, *J. Gen. Appl. Microbiol.* **4**, 272 (1958).
2. S. Udaka and S. Kinoshita, *J. Gen. Appl. Microbiol.* **4**, 283 (1958).
3. R. H. De Deken, *BBRC* **8**, 462 (1962).

into two enzymes, but the activity of neither of these enzymes was inhibited by L-arginine (*4*).

The first enzyme of the N-acetyl-L-glutamate reducing system is the N-acetyl-L-glutamate-5-phosphotransferase (proposed name, ATP:N-acetyl-L-glutamate-5-phosphotransferase). This enzyme was purified from both *Chlamydomonas reinhardti* (*5*) and from *E. coli* (*6*). The phosphotransferase purified from *C. reinhardti* is an allosteric enzyme, and its activity is inhibited by arginine. The purified phosphotransferase from *E. coli* has no allosteric properties, and its activity is influenced neither by arginine nor by any of the intermediates of the arginine biosynthetic pathway.

From the viewpoint of allosteric or nonallosteric properties of phosphotransferase and the feedback control of arginine biosynthesis, two different types of arginine biosynthetic pathway were recognized. In *E. coli* and some other gram-negative bacteria the only way for the synthesis of N-acetyl-L-glutamate is the acetylation of L-glutamate by acetyl coenzyme A, and the activity of this acetylating enzyme is inhibited by arginine (*7*). This appears to have two consequences: (a) in the *coli*-type arginine pathway the phosphotransferase does not need to have allosteric properties and (b) the acetyl group of α-N-acetyl-L-ornithine is hydrolyzed to acetate by acetylornithine (*8*). In *C. reinhardti*, yeast, plant-type organisms, and some bacteria, the acetyl groups of α-N-acetyl-L-ornithine is not hydrolyzed to acetate during the arginine biosynthetic pathway but transferred to glutamate by the ornithine acetyltransferase (*1, 9*). Because of the transfer reaction the acetyl group is used in a cyclic manner by the pathway, and in the plant-type arginine pathway the allosteric enzyme of the pathway is the next one in sequence to the enzyme responsible for the synthesis of N-acetyl-L-glutamate which is the phosphotransferase. Since the phosphotransferase has an important role in the biosynthesis of arginine, it is safe to assume that the enzyme is present in all living organisms having an arginine biosynthetic pathway.

4. A. Baich and H. J. Vogel, *BBRC* **7**, 491 (1962).
5. A. Farago and G. Dénes, *BBA* **136**, 6 (1967); G. Dénes, "Methods in Enzymology," Vol. 17A, p. 269, 1970.
6. H. J. Vogel and W. L. McLellan, "Methods in Enzymology," Vol. 17A, p. 251, 1970.
7. W. K. Maas, G. D. Novelli and F. Lipmann, *Proc. Nat. Acad. Sci. U. S.* **39**, 1004 (1953).
8. H. J. Vogel and D. M. Bonner, *JBC* **218**, 97 (1956); H. J. Vogel and W. L. McLellan, "Methods in Enzymology," Vol. 17A, p. 265, 1970.
9. M. Staub and G. Dénes, *BBA* **128**, 82 (1966); G. Dénes, "Methods in Enzymology," Vol. 17A, p. 273, 1970.

II. Molecular Properties

A. ENZYME PURIFICATION

1. *Allosteric Phosphotransferase from C. reinhardti*

The procedure for purification of the enzyme from extract of *C. reinhardti* takes advantage of the unusual heat stability in the presence of arginine. The enzyme was purified 30-fold with repeated heat treatment in the presence of arginine followed by ammonium sulfate fractionation (*5*). The estimated molecular weight of the enzyme based on Sephadex G-200 chromatography is 400,000 (*10*).

2. *Nonallosteric Phosphotransferase from E. coli*

The enzyme was purified from extracts of *E. coli* (ATCC 25542) 67-fold by ammonium sulfate precipitation and by chromatography on hydroxyapatite, DEAE-cellulose, and Sephadex G-100 (*6*). The estimated sedimentation constant of the enzyme is 4.2 S.

B. PHYSICAL PROPERTIES

The partially purified enzymes from *C. reinhardti* were not pure enough for the study of their physical properties.

III. Catalytic Properties

A. THE CATALYTIC REACTION

1. *Stoichiometry*

The only metabolic reaction catalyzed by the phosphotransferase is the transfer of the terminal phosphate group of ATP to the γ-carboxyl group of N-acetyl-L-glutamate, the synthesis of N-acetyl-L-glutamate-5-phosphate [Eq. (1)]. In the presence of neutralized hydroxylamine in

N-Acetyl-L-glutamate + ATP → N-acetyl-L-glutamate-5-phosphate + ADP (1)

N-Acetyl-L-glutamate-5-phosphate + ATP + NH_2OH →
$\qquad N$-acetyl-L-glutamate-5-hydroxamate + ADP + P_i (2)

10. A. Farago and G. Dénes, *BBA* **178**, 400 (1969).

the reaction mixture, N-acetyl-L-glutamate-5-hydroxamate is formed from the carboxyl-activated derivative of N-acetyl-L-glutamate [Eq. (2)]. Studies on the stoichiometry of the transfer reaction have shown that in the presence of hydroxylamine for every mole of N-acetyl-L-glutamate-5-hydroxamate formed, one mole of P_i and one mole of ADP appeared (5). These findings support Eq. (1) described previously (1). In the absence of hydroxylamine according to Eq. (1), the formation of ADP and of the acylphosphate of N-acetyl-L-glutamate occurs. It was found that N-acetyl-L-glutamate-5-phosphate is as unstable as β-aspartylphosphate in acidic solutions and is completely hydrolyzed under conditions as used for the determination of inorganic phosphate according to Fiske and SubbaRow (11).

The equilibrium of the reaction may not favor the formation of N-acetyl-L-glutamate-5-phosphate and ADP because the forward reaction is linear only for a short period of time (5). For this reason the use of hydroxylamine as a trapping agent is required for kinetic analysis of the reaction catalyzed by the enzyme.

2. pH Optima and Activating Ions

The pH optimum of the phosphotransferase purified from *C. reinhardti* is 5.5, but the enzyme is active over a fairly broad pH range between 5.5 and 8.5. The enzyme requires magnesium or cobalt(II) ions for activity (5).

The phosphotransferase purified from *E. coli* has a pH optimum between 6.8 and 7.8, and it is activated by magnesium ions (6).

3. Substrate Specificity

The substrate specificity of the enzyme purified from either *C. reinhardti* or *E. coli* is the same. ATP is the only phosphorylating agent that has been reported to phosphorylate N-acetyl-L-glutamate. GTP has been found inactive when substituted for ATP (6). Similarly, only N-acetyl-L-glutamate has been found to act as an acceptor of the terminal phosphate group of ATP (5, 6). L-Glutamate, D-glutamate, and N-benzoyl-L-glutamate were inactive when substituted for N-acetyl-L-glutamate (6).

4. Assays

Two products, ADP and N-acetyl-L-glutamate-5-phosphate, are formed in the reaction catalyzed by the phosphotransferase, and two dif-

11. C. H. Fiske and Y. SubbaRow, *JBC* **66**, 375 (1925).

ferent methods have been used for the determination of the activity of enzyme.

The amount of ADP formed in the reaction mixture was measured with a coupled assay system using phosphoenolpyruvate and pyruvate kinase. In the presence of ADP and pyruvate kinase, stoichiometric amount of pyruvate is formed from phosphoenolpyruvate and the pyruvate produced was determined colorimetrically as its dinitrophenylhydrazone (6).

The activity of the phosphotransferase was usually estimated by measurement of N-acetyl-L-glutamate-5-hydroxamate formation from N-acetyl-L-glutamate-5-phosphate in the presence of neutralized hydroxylamine in the reaction mixture; the amount of hydroxamate formed was determined colorimetrically (1, 4, 5).

5. Kinetics

N-Acetyl-L-glutamate-5-phosphotransferase is the key enzyme in the regulation of arginine biosynthesis in the freshwater alga, *C. reinhardti*. The catalytic activity of the enzyme is controlled allosterically by arginine. The enzyme performs its catalytic function in a straightforward manner, in the sense that the steady state kinetics of the reaction catalyzed in the absence of allosteric inhibitor do not suggest the involvement of catalytic site interactions of the sort that would be manifested in changes of K_m as a function of substrate concentration (5). The apparent Michaelis constant of the enzyme at pH 5.5 is 15 mM for N-acetyl-L-glutamate and 1.6 mM for ATP.

The analysis of initial velocity was carried out according to Cleland (12). Hydroxylamine was present in the reaction mixture; therefore, the concentration of one of the products, N-acetyl-L-glutamate-5-phosphate, was always zero. The enzyme has an absolute requirement for magnesium or cobalt(II) ion, but in the experiments described 10 mM magnesium chloride was used. The magnesium ion concentration used in the experiments had no inhibitory effect on the reaction even in the presence of the smallest ATP concentration used, but it was enough to ensure that the nucleotide is complexed at all times. Magnesium chloride above 20 mM, however, caused inhibition regardless of the concentration of ATP. Although the activity of phosphotransferase is about 50% higher at pH 5.5 than at pH 8.0 (5), the initial velocity analysis was carried out at pH 8.0 because the dissociation constants of magnesium nucleotides were known at this pH.

The double reciprocal plots obtained with various acetylglutamate concentrations at constant ATP levels (and with ATP as the varied sub-

12. W. W. Cleland, *BBA* **67**, 104 (1963).

strate at constant acetylglutamate levels, respectively) versus initial velocity yielded in both cases linear intersecting curves (13). This kinetic pattern is consistent only with a sequential reaction mechanism (12, 14) according to Eq. (3) indicating that both substrates must be

$$v = \frac{VAB}{K_{ia}K_b + K_aB + K_bA + AB} \quad (3)$$

added to the enzyme before either of the products is released; A and B are the concentrations of the substrates, K_{ia} is the dissociation constant for EA complex, and K_a and K_b are the Michaelis constants for A and B, respectively. The concentration of one of the substrates does not influence the Michaelis constant of the other. The Michaelis constant of the enzyme at pH 8.0 and 37° for acetylglutamate is 4.0 mM, and for ATP it is 3 mM. The reaction is inhibited by ADP. The inhibition caused by ADP is competitive with respect to ATP and noncompetitive with acetylglutamate. The results are not sufficient to indicate whether the binding of the substrates occurs in ordered or random sequence (13).

The initial velocity pattern of the reaction catalyzed by the phosphotransferase from *E. coli* was not studied in detail. The apparent K_m value of the enzyme for N-acetyl-L-glutamate is 6.0 mM and for ATP it is 1.0 mM (6).

B. Allosteric Inhibition

The allosteric properties of the phosphotransferase purified from *C. reinhardti* have been studied in detail.

1. *Kinetics*

The catalytic activity of the enzyme and its inhibition by arginine vary independently of each other as a function of pH. The pH optimum of the inhibition of enzymic activity by arginine is 7.5 (5).

The hyperbolic shape of the substrate saturation curves does not change either for N-acetyl-L-glutamate or for ATP in the presence of different concentrations of L-arginine. The kinetics of L-arginine inhibition as a function of N-acetyl-L-glutamate concentration is competitive in the sense that V_{max} is unaffected by the inhibitor concentration. Noncompetitive inhibition kinetics were obtained for arginine as a function of varying ATP concentrations (5). The catalytic activity of the enzyme

13. A. Farago and G. Dénes, *Acta Biochim. Biophys. Acad. Sci. Hung.* **4**, 251 (1969).
14. W. W. Cleland, *Annu. Rev. Biochem.* **36**, 77 (1967).

is inhibited not only by arginine but also by some other arginine-like compounds such as L-canavanine and L-citrulline. However, these arginine analogs are inhibitory only at high concentrations as compared with arginine (5).

Allosteric enzymes have separate binding sites for their substrates and allosteric effectors (15). A distinction between substrate and allosteric binding sites has been made for some allosteric enzymes by desensitization with heavy metals or heat treatment. In the case of phosphotransferase all these efforts failed consistently to achieve desensitization of the arginine binding site of the enzyme, e.g., treatment with p-mercuribenzoate or heat. It was observed, however, that urea at low concentrations (0.1–1.5 M), while not inhibiting the activity of the enzyme, suppressed the inhibitory effect of arginine. Since the inhibitory effect of L-arginine in the presence of urea depends only on the ratio of the concentrations of the compounds, the kinetic relation between urea and arginine seems to be competitive (5). The reversible desensitization of the enzyme with urea against the inhibitory effect of arginine strongly suggests that the phosphotransferase has a separate, specific binding site for the allosteric inhibitor.

Studies of the influence of arginine concentration on the reaction rate in the presence of a nearly saturating concentration of N-acetyl-L-glutamate (67.5 mM) have shown a linear Lineweaver-Burk plot only when the reciprocal of the initial velocity data was plotted against the reciprocal of the square of the arginine concentration (5). According to classic reaction kinetics, this result suggests that two molecules of arginine are required for the inhibition, or at least two arginine binding sites per molecule of enzyme are involved in the inhibition process. The apparent inhibitor constant, K_i, of the enzyme for L-arginine in the presence of 67.5 mM of N-acetyl-L-glutamate at pH 7.5 is 0.35 mM. Properties of allosteric enzymes suggest that the observed anomalies avail from cooperative interaction between the subunits of the enzyme molecule binding the substrate or the allosteric effector (16, 17). Plots of the data to the Hill equation, as described by Monod et al. (15), gave a straight line with a slope of 2.0 (18). In other experiments at different pH values between 5.5 and 7.5 the slope was exactly the same. Similar analysis of the substrate binding yielded straight lines with a slope of 1.0, suggesting that the substrate binding sites of the enzyme molecule are independent from each other with no cooperative interactions between them.

The phosphotransferase fits into the group of the K-type allosteric en-

15. J. Monod, J. P. Changeux, and F. Jacob, *JMB* **6**, 306 (1963).
16. J. Monod, J. Wyman, and J. P. Changeux, *JMB* **12**, 88 (1965).
17. D. E. Koshland, Jr., G. Némethy, and D. Filmer, *Biochemistry* **5**, 365 (1966).

TABLE I
The Hill Coefficient and the Arginine Concentration Causing a 50% Inhibition, $I_{1/2}$, under Different Experimental Conditions

pH	Temp. (°C)	Acetylglutamate (mM)	n	$I_{1/2}$ (μM)
6	37	75	2	400
7	37	75	2	400
7	25	75	2	100
7	15	75	2	20
7.5	37	75	2	400
7.5	37	25	2	300
7.5	37	5	2	250
7.5	37	2.5	2	180
8	37	75	2.5	430
8.5	37	75	3.3	300
9	37	75	2.3	500

zymes as classified by Monod et al. (16). According to the symmetry model (16) in those systems in which an allosteric effector modifies the apparent affinity of the substrate, the substrate should also exhibit homotropic cooperative interactions. In the case of N-acetylglutamate-5-phosphotransferase, the substrate does not obey this prediction. Further, Hill plots of arginine inhibition at various competitive substrate, N-acetyl-L-glutamate concentrations gave parallel straight lines (18). The decrease in the substrate concentration was accompanied by a decrease in the arginine concentration causing 50% inhibition, $I_{1/2}$, but the slope of the line remained 2. There was no indication that the cooperative effect of arginine was weakening with a decreasing substrate concentration, although this would be expected on the basis of the symmetric model. The concentration of the noncompetitive substrate ATP does not influence the position or the slope of the straight line in the Hill plot.

The inhibition of the activity of N-acetylglutamate-5-phosphotransferase is exothermic; from 37° to 15° there is a 20-fold increase in the apparent affinity of the enzyme toward arginine ($I_{1/2}$ 0.4 mM and 20 mM, respectively) but without any alteration in the cooperativity as shown in Table I. While at and below pH 7.5 the interaction coefficient remains two, a shift in the pH from 7.5 to alkaline direction causes a change in the n value without great difference in $I_{1/2}$. The Hill coefficient n increases with increasing pH and reaches its maximum at pH 8.3 and decreases again. The measured maximal n value is about 3.5.

18. A. Farago and G. Dénes, BBA 185, 263 (1969).

2. Effect of Temperature

As mentioned previously, the inhibitory effect of arginine strongly depends on the temperature, and arginine (Table I) protects the enzyme against the inactivating effect of heat. The protective effect of arginine against the inactivating effect of elevated temperature was employed to see that Hill coefficient, $n = 2$, obtained from the Hill plots between pH 5.5 and 7.5 are related with the number of arginine molecules bound to the enzyme, or they are apparent values with no connection to the stoichiometry of the enzyme inhibitor complex formation caused by the presence of the substrates or some steps of catalyzed reaction (19). At 55°, the inactivation of the purified free enzyme follows first-order kinetics and the rate constant, k_1, is 0.086 min^{-1}. In the presence of arginine the heat inactivation of the enzyme follows apparent first-order kinetics, and the rate constants, k', in the presence of 0.03, 0.05, 0.07, 0.1, and 0.15 mM of arginine are 0.06, 0.046, 0.033, 0.02, and 0.012 min^{-1}, respectively. From Eq. (4) it is possible to determine the overall dissociation constant, K, of the E-I_n complex and the number of the inhibitor molecules bound to the enzyme, n.

$$\frac{k'}{k_1} = K\left(\frac{1 - (k'/k_1)}{I^n}\right) + \frac{k_2}{k_1} \quad (4)$$

where k_2 is the rate constant of the inactivation of E-I_n complex. The plot of k'/k_1 against $[1-(k'/k_1)]/I^n$ gives a straight line only if $n = 2$, and in this case K is equal to $2.2 \times 10^{-9}\ M$. Since the numerical value of n obtained from the heat inactivation experiment in the absence of substrates and from enzyme inhibition experiments in the presence of the substrates is the same, this result permits the conclusion that the value

TABLE II

THERMODYNAMIC CHARACTERISTICS OF THE REACTION $E + 2\ \text{ARG} = E - \text{ARG}_2$; $\Delta H° = -50{,}600$ cal/mole; $\Delta F° = RT \ln K_i$; $\Delta S° = (\Delta H° - \Delta F°)/T$

Temp. (°C)	K_i (apparent overall)	$\Delta F°$ (cal/mole)	$\Delta S°$ (cal/mole deg)
37	1.6×10^{-7}	$-9{,}700$	-132
30	4.0×10^{-8}	$-10{,}300$	-132
28	8.0×10^{-9}	$-11{,}100$	-130
20	1.6×10^{-9}	$-11{,}800$	-133
15	4.0×10^{-10}	$-12{,}450$	-132

19. A. Farago and G. Dénes, BBA 139, 521 (1967).

of n is connected with the stoichiometry of the phosphotransferase–arginine complex formation and that the complex is E-arginine$_2$. The apparent inhibitor constants, K_i for arginine, determined by the inhibition of the enzymic reaction at different temperatures are shown in Table II.

It is noteworthy that the apparent K_m for N-acetyl-L-glutamate is $5.0 \pm 1 \times 10^{-3}\ M$ and does not change in the temperature range studied. The standard enthalpy change, $\Delta H°$, for the formation of E-arginine$_2$ complex calculated according to the van't Hoff equation is $-50,600$ cal/mole. The arginine inhibition of the phosphotransferase is an exothermic exergonic reaction as shown in Table II. The large decrease in entropy suggests that the arginine-inhibited enzyme has a more ordered conformation than the active enzyme, the allosteric transition involving a conformational change of the enzyme structure.

Author Index

Numbers in parentheses are reference numbers and indicate that an author's work is referred to, although his name is not cited in the text.

A

Abaturov, L. V., 400, 401(107), 402(107, 113), 403(113), 405(107), 406(107), 412(107)
Abbott, J. C., 132, 153(54)
Abd-El-Al, A., 295
Abdulaev, N. G., 398, 399, 415(99a), 417(99a), 418(99a), 421(96)
Abdullah, M., 320(95), 321
Abe, N., 326, 332(121), 333(121), 335(121), 342(188), 343, 344(188), 346(188), 356(121), 357(121)
Abeles, R. H., 163
Abood, L. G., 34
Acs, G., 54, 181
Adair, L. B., 297, 298(125), 301
Adams, E., 381, 386(15), 389(15), 390(15), 463
Adams, J. M., 208
Adler, E. Z., 76
Adolfsson, S., 341, 355
Agalarova, M. B., 418(165), 420, 422(165)
Agarwal, K. C., 83, 84(193), 85(191, 193), 86(193)
Agarwal, R. P., 83, 84(193), 85(193), 86(193)
Agosin, M., 76, 77(165)
Agren, G., 23
Ahren, K., 355
Ainslie, G. R., 30
Aitken, D. M., 304
Akhtar, M., 219
Aki, K., 436, 465(208b, 208c, 262), 471
Albert, J. L., 326, 327(123), 357(123)
Albrecht, A. M., 214, 469(296), 474
Albritton, W. L., 471(300), 472, 475

Aldanova, N. A., 398(96), 399, 415(99a), 417(99a), 418(99a), 421(96, 99a)
Alder, E., 2
Alexander, N., 216
Algranati, I. D., 313, 360(49)
Allan, P. W., 54, 55
Allen, C. M., Jr., 108, 109(49), 110(49), 243
Allen, S. H. G., 485, 486(14), 487(14), 494(14)
Allison, D. P., 48
Almquist, H. J., 499, 504(29)
Alpers, J. B., 356
Alsever, R. N., 499
Amberg, R., 221
Amelunxen, R., 249
Ames, B., 184, 185
Ames, B. N., 104, 393
Anders, M., 170, 174(8, 9), 176(9), 187
Anderson, E. P., 56(64), 57, 58(75), 59(64, 75, 78, 79, 80, 81, 82), 60(75), 61(52, 75, 78), 87, 90(200), 91(200, 206)
Anderson, G. W., 490
Anderson, P. M., 101, 295
Anderson, R. L., 48
Anderson, W. M., Jr., 187, 188(90)
Andrews, P., 367, 368, 371, 375
Angelino, N., 430(194), 433
Anisova, L. N., 509
Anthony, R. S., 23, 115
Antonini, E., 437(239), 451
Aposhian, H. V., 210
Appaji Rao, N., 75
Appleman, M. M., 326, 327, 331(129)
Apps, D. K., 76, 77(157, 164), 78(157, 164), 79(157, 170), 80(164), 81(157, 169, 170, 172), 82(164, 178)

Arber, W., 191, 192
Archibald, R. M., 108
Argoudelis, A. D., 506, 508(63)
Arigoni, D., 453, 460
Arima, T., 70, 71(125), 73(125)
Armstrong, D. J., 176
Armstrong, F. B., 465(261), 471
Armstrong, R. L., 184
Arnon, R., 367
Arrio-Dupont, M., 399, 408, 435(106)
Ascione, R., 183
Asensio, C., 48
Ashburn, M. J., 356
Ashworth, J. M., 361
Assaf, S. A., 357
Atkins, I. C., 229
Atkinson, D. E., 504
Audit, C., 498
Auld, D. S., 387(66, 68), 388, 390(68)
Auricchio, F., 465(263), 471
Austrian, R., 282, 293(101), 295(101)
Ayling, J. E., 451, 452(241), 471(307), 476
Azarkh, R. M., 440, 467(228)

B

Babad, H., 364, 365(9), 369, 370, 371(9)
Babior, B., 142
Bachhawat, B. K., 353
Baer, A. A., 172
Baginsky, M. L., 221, 222(160)
Baglioni, C., 208
Baguley, B. C., 171, 172(16), 173, 174
Baich, A., 512, 515(4)
Bailey, G. B., 453, 478
Baily, S. K., 2
Baker, B. R., 214
Baliga, B. S., 184, 185(70)
Balis, M. E., 55
Banaschak, H., 55
Banks, B. E. C., 382, 392(31), 395, 398(31, 89), 400(89), 402(31, 93), 424, 426 (175), 427(175), 428(176)
Bannister, B., 506, 508(63)
Barber, A. A., 312
Bardos, T. J., 70
Barile, M. F., 116
Barker, H. A., 122(5), 123
Barnabee, O., 352
Barnard, E. A., 4, 5(20), 7, 9, 19, 26
Barner, H. D., 214

Barnes, F. W., Jr., 425
Baroncelli, V., 449
Barra, D., 396, 397(91), 419, 420(154, 159), 421(159), 435(154), 449(154)
Basford, R. E., 36, 37(136), 38, 41(143)
Bashan, N., 355
Basu, D. K., 353
Bauchop, T., 98
Bauerova, Y., 499
Bauer, Š., 318, 322(78)
Baum, H., 34
Baumann, C. A., 123
Beard, C. B., 257
Beaumont, P., 477
Beck, C. F., 56, 57(53), 58(53), 59(53), 61 (53), 70(53), 87)
Beckwith, J. R., 284, 293(101), 295
Beinert, H., 484
Bekesi, I., 509
Belford, J., 341
Bell, M. P., 424, 428(176)
Bello, L. J., 83, 84(183), 86(183), 94(183), 95(215, 218)
Belocopitow, E., 326, 331(129), 332(147), 335(147)
Belozersky, A. N., 193
Belser, W. L., 301
Beltz, R. E., 62
Bender, M. L., 248
Benisek, W. F., 232, 246(41), 263(41), 264 (41)
Benkovic, S. J., 51, 219, 387(64), 388, 389
Bennett, L. L., 51, 52(9), 53(9), 54(9), 55
Benson, R. W., 492
Berdahl, B. J., 190
Bergami, M., 426
Bergamini, E., 355
Berger, A., 484
Berger, L., 2, 5(7), 6
Bergstrom, W. J., 341
Berlin, C. M., 101, 104(23), 105(23), 109 (23), 112(23), 116(23), 117(23)
Berman, D. A., 499, 504(23)
Berman, P. H., 55
Berne, R. M., 340
Bernfield, M. R., 184
Bernhardt, S., 303
Bernhart, P. H., 102
Bernlohr, R. W., 118
Bernofsky, C., 82(181), 83

AUTHOR INDEX 523

Bertland, L. H., 393, 394, 395(87), 396 (87), 407(87), 415(85), 416(87)
Bertola, E., 453
Besmer, P., 453
Bessman, M. J., 83, 84(183, 194), 85(194), 86(183), 88, 92(197), 94(183), 95(215, 216, 217, 218), 96(219)
Besson, P., 355
Best, N., 359
Betheil, J. J., 76, 82(154), 200
Bethell, M. R., 229, 246, 255(58), 259(58), 269, 276(58), 297, 298(124), 301(124)
Bethge, P. H., 291
Bhargava, A. K., 19
Bhatti, A. R., 304
Biely, P., 318, 322(78)
Biempica, L., 179, 180(49)
Bieri, J. G., 163, 164(107)
Bilik, E., 179, 180(49)
Billen, D., 193, 194
Bindstein, E., 311, 353(15)
Birchmeier, W., 418(163a, 170), 420, 421 (163, 170), 422(163a), 459(160)
Birchall, T., 248
Birnbaumer, L., 327, 331, 332(131, 147), 335(147)
Bishop, J. S., 315, 325, 326(111), 327(111), 328(135), 338(57), 341(57), 342(189), 343(111), 348, 349(189), 350(213), 351 (189), 352(194)
Bishop, S. H., 102, 103, 104(41), 105(41), 107(41), 109(40, 45), 111(40), 112(41)
Bjork, G. R., 170
Blair, J. B., 485, 486(11), 487(11), 491 (11), 496(11)
Blakley, R. L., 122(7), 123, 134(7), 161(7), 198(8), 199, 200(8), 210(8), 211(8), 212(8), 214, 215(8), 216(8)
Blangy, D., 289(108), 290
Blatt, L. M., 317, 331, 349(143), 357, 358 (69)
Bloch, K., 497, 499(2), 504(2)
Blodinger, J., 490
Bloom, B., 499
Blout, E. R., 250, 253(73), 255(73), 267 (73), 275(99), 276, 277(73), 278(73)
Blum, J. J., 359
Bo, W. J., 356
Bocharov, A. L., 413, 418(162a, 164), 420, 421(164), 430(143, 193, 196), 431 (143, 193, 196), 432(143), 433(143), 434(193), 444(193)
Bodansky, O., 393, 424, 425(83), 426(83)
Bode, W., 186
Boeker, E. A., 381, 386(14), 389(14), 390 (14)
Boezi, J. A., 184
Bojarski, T. B., 92
Bombara, G., 355
Bomboy, J. D., 350
Bone, A. D., 31, 44(101)
Bonner, D. M., 472, 477, 512
Bonnet, R., 153
Borek, B. A., 206
Borek, E., 168, 170(1), 174(1), 176, 177, 179(40), 180(40, 50), 181(48, 51), 182, 183, 184(40), 185(40, 48, 50, 51, 53, 54, 70, 71, 73, 74, 79), 186(53), 190, 191, 192, 194
Borel, M., 14, 15(49, 50), 20, 21(68)
Born, J., 45
Borri, C., 419, 420(155), 422(155), 435 (155), 449(155), 450(155)
Borsook, H., 497, 498(3), 499(3), 500(3)
Bortrick, R. J., 348
Bos, C. J., 34
Bossa, F., 392(84), 393, 396, 397(84, 91), 399, 405(84), 407(84), 416(84), 419, 420(154, 159), 421(159), 435(154), 449 (154)
Botsford, J. R., 160
Bourget, P. A., 307
Bowman, B., 55
Bowser, H. R., 206
Boxer, G. E., 54
Boyd, J. W., 393
Boyde, T. C., 426
Boyer, P. D., 13, 14(41), 16(41), 17(41), 23, 364, 381, 386(12), 489, 492
Boyle, C., 82(180), 83
Bozhkov, V. M., 76, 77(156), 78(156)
Bratvold, G. E., 314
Braunstein, A. E., 219, 379, 380, 381, 383, 384, 385(1), 386(17, 18, 43, 45, 58, 61), 387(18), 389(43, 45), 391(7), 398(96), 399(45), 407(18), 408(43, 45), 412(45), 413(45), 415(99a), 417(43, 99a), 418 (99a, 166), 421(96, 99a), 422(116), 423, 427(116), 429(43, 45), 435(9, 17, 18, 45, 61), 437(45), 438(17, 45, 61),

447(43), 440(17, 18, 43), 441(18, 442), 443(18), 444(234), 445(18), 447(43), 449(18), 451(58, 61), 453, 457(43), 461 (43), 467(228, 284), 473, 477(58, 61), 479(245), 480(1, 3)
Breckenridge, B. M., 311, 317(14), 353(14)
Breitman, T. R., 62, 71
Brenner, M., 184
Brent, T. P., 64, 67(103), 68(103)
Bresnick, E., 70, 71(124), 72(124, 126), 74, 306
Breusov, Yu, N., 403, 408(123), 410(123, 133), 411(123, 133), 413, 440, 441 (222), 445(222), 457(222)
Brew, K., 366, 367(18, 19), 370(18), 371 (28), 375(28), 377(18)
Brewer, C. F., 214
Brierly, G. P., 34
Britton, H. G., 14, 15, 16, 17, 43
Brockman, R. W., 54, 56, 58(83), 59(81, 82), 61(52)
Brodbeck, U., 365, 366(12, 13), 367(12), 368(22), 371(22), 376(22), 377(13)
Brodie, J. D., 132, 134(52), 163(52), 165
Broquist, H. P., 386
Brostrom, C. O., 340
Brot, N., 125, 127, 128(28), 134(55), 152 (19), 159
Brown, B. I., 319
Brown, D. H., 313, 316, 318(37), 319, 320 (82), 321(37), 322(82), 335(37)
Brown, F., 209
Brown, G. M., 187, 189(91), 190
Brown, N. B., 315, 338(57), 341(57)
Brown, N. E., 313, 314(44, 45), 315(47), 319(45), 321, 326(47), 340(47)
Brown, P. R., 83
Brown, S. S., 163
Browne, W. J., 366, 367(19)
Bruice, T. C., 219, 387(64, 65, 66, 67, 68), 388, 389, 390(68), 490
Bruni, C. B., 471(301), 475
Brunori, M., 437(239), 451
Bruns, R., 56(67), 57, 59(67)
Bryan, P., 355
Brzozowski, T., 101
Buc, H., 289(108), 290
Buc, J., 48
Buccino, R., 70, 72(126)

Buccino, R. J., Jr., 83, 84(189), 85(189), 86(189)
Buch, L., 177
Buchanan, J. M., 122, 123, 125, 130(3, 8, 9), 131(8, 9), 132(9), 134(9), 135(9, 36), 144(9, 34), 152(9, 22), 158, 160 (3, 36), 162(3), 163(3, 36), 164(106), 201, 203, 204(38), 205
Buckman, T., 271, 274, 275
Bueding, E., 48, 312, 319
Buniatian, H. C., 385
Bunville, L. G., 469(292), 474
Buonocore, V., 394
Burch, H. B., 342(186), 343, 352(186)
Burch, R. E., 485
Burchall, J. J., 212
Burdon, R. H., 193
Burke, G. T., 165
Burleson, S. S., 70, 72(126), 74
Burnett, W. T., 503
Burns, R. O., 186
Burris, R. H., 393
Burrows, W. J., 176
Burt, M. E., 101
Burton, E. G., 160, 161(94, 95, 96), 162 (94, 95, 96)
Buschiazzo, H., 348
Bussel, J. B., 280
Butcher, R. W., 323, 341, 350
Butler, A. B., 387(67), 388
Butler, R. C., 75

C

Cabib, E., 312, 313, 317, 326(23), 327 (125), 332, 333(154), 334(68, 154), 335 (68, 154), 336(68), 359(23, 125), 360 (23, 49, 125), 361(23, 125)
Caen, J., 355
Caldwell, I. C., 54
Calin, M., 248
Calvin, M., 255
Camargo, E. P., 361
Campbell, J. W., 101
Campbell, P. N., 366
Canella, C., 382, 392(27), 435(27), 465(27)
Canellakis, E. S., 56
Caniato, A., 352
Cannata, J. J. B., 142, 152(63)
Cantoni, G. L., 499
Capecchi, M. R., 208

Capra, J. D., 184, 186(65)
Caputto, R., 50, 51(3), 52(3, 8), 56(3), 75(3)
Caravaca, J., 101, 102(22), 109(22), 111 (22, 40)
Cardeilhac, P., 271
Cardini, C. E., 310, 311(2), 319(1), 364
Carloni, M., 399
Carlson, D. M., 377
Carminatti, H., 32, 310, 316(3), 317(3), 319(3), 320(3), 332(3), 334(3), 335(3)
Caroline, D. F., 302
Carreras, J., 245
Carriuolo, J., 261
Carroll, W. R., 101, 104(23), 105(23), 109 (23), 112(23), 116(23), 117(23)
Carson, S. F., 485
Carter, B. L. A., 380, 381(5), 392(5), 438 (5), 462(5), 480(5)
Carter, C. E., 92, 93(209)
Carter, J., 119
Castellino, F. J., 366
Cathou, R. E., 123, 130(8), 131(8)
Catsimpoolas, N., 234
Cattaneo-Lacombe, J., 477
Cauthen, S. E., 160, 162(91), 163(102)
Chabas, A., 245
Chambaut, A. M., 337
Chamberlin, M. J., 236, 274(48)
Chandler, A. M., 354
Chaney, S., 177
Chang, P. K., 69, 70(121), 72(121)
Changeux, J.-P., 231, 232, 236, 246, 251, 269(31), 271(31), 273, 274, 275, 281, 286(98), 287(98), 288, 289, 517, 518 (16)
Charmatz, A., 355
Chatagner, F., 472
Cheetham, R. D., 377
Cheng, L., 19, 25(62), 26(62), 44(62)
Cheng, S., 419, 423, 425(158), 426(158), 428(158), 435(158), 459(158)
Cheng, Y.-C., 83
Chíancone, E., 392(84), 393, 397(84), 405 (84), 407(84), 416(84)
Chou, A. C., 37, 39(141), 41
Christen, P., 418(162, 163, 163a, 170), 420, 421(163a, 170), 422(163a), 459(160)
Christensen, J. J., 110
Chu, M. Y., 63, 64(98)

Chumley, S., 55
Chung, A. E., 76, 77(161, 163), 78(161, 163), 79(161, 166), 80(161, 163), 81 (161, 163)
Churchich, J. E., 398(97), 399, 402, 408, 411(133a)
Ciardi, J. E., 57, 58(71), 59(71), 60(69)
Čihák, A., 56(65, 66), 57, 59(65, 66, 73, 74), 60(74), 61(74)
Cirillo, V. J., 33
Clark, B. F. C., 208
Clark, J. B., 14, 15, 16, 17, 34, 43, 81
Clark-Turri, L., 32
Clauser, H., 337
Cleaver, J. E., 69, 71(116), 72(116), 73 (116), 74(116)
Cleland, W. W., 245, 321, 424, 425, 426 (183), 427(183), 457(183), 515, 516 (12)
Clow, J. E., 34
Coffey, R. G., 377
Cohen, L. A., 489
Cohen, P., 314, 315(56)
Cohen, P. P., 97, 101, 102, 103(43), 104 (43), 109(43), 110(43), 111(43), 112 (43), 113(43), 114(43), 229, 306
Cohen, S. S., 209, 214
Cohlberg, J. A., 231, 235, 237, 239, 240, 242, 274, 275
Cohn, M., 20
Cole, R. D., 471(308), 476
Coleman, M. S., 297, 298(127), 299, 465 (261), 471
Collat, J. W., 132, 153(54)
Collins, K. D., 232, 246(36), 247, 248(62), 251, 253, 254, 255(76), 256, 257(62), 258(62), 259(62), 262(62), 267(62), 269(36), 280(62), 281(36), 288, 289
Colman, P. D., 255, 271(75), 277
Colombo, B., 208
Colowick, S. P., 2, 3, 4(15), 5(7, 11, 13, 19), 6(7, 13, 16), 7(16, 22), 8(16, 22, 38, 63), 9(15, 22, 26), 10(15, 16, 22), 11(15, 19), 12(15, 38), 13(15, 22), 18 (13), 19, 20(15, 55, 63, 64), 21(63), 23 (54, 55), 24(54, 55), 25(55, 62), 26(15, 16, 24, 62), 27(22, 26, 63), 29, 30(92), 31(9, 92), 44(62), 46(91, 92)
Colvin, B., 367, 368(22), 371(22), 376(22)

Conconi, F., 500, 501(35)
Conrad, H. E., 317, 321
Contreras, G., 183
Cook, D. E., 359
Cook, R. A., 275
Coon, M. J., 484, 486(9), 487(9), 488(31), 494(9), 495(9)
Cooper, A. J. L., 473
Cooper, C., 14, 15(53), 17, 22(53)
Cooper, L. W., 313
Cooper, R. A., 314, 485
Coote, J. G., 472
Copley, M., 41
Corbin, J. D., 324
Cordes, E. H., 388(71, 72), 389
Cori, C. F., 2, 5(7), 6(7), 319, 321, 347
Cori, G. T., 323, 324(100)
Cornblath, M., 355
Cossins, E. A., 160, 161(97)
Craig, J. W., 332, 337, 338
Crane, R. K., 1, 31(1), 34, 35, 38(129), 44(100), 45(99, 100)
Craven, P. A., 36, 37(136), 41
Crawford, E. J., 311, 317(14), 353(14)
Crofford, O. B., 350
Crokaert, R., 229
Crusberg, T., 212
Csermely, T., 356
Culp, L. A., 187, 189(91), 190
Cunningham, M. A., 341
Curci, M. R., 306
Curnow, R. T., 353
Cynkin, M. A., 377
Cysyk, R., 74
Czerlinski, G. A., 429, 436(190)

D

Dahlberg, D., 361
Dahlberg, J. E., 189
Dale, B., 186
Dancis, J., 55, 386
Danenberg, P. V., 215
Danneberg, P. B., 214
Danforth, W. H., 332, 336, 337, 338(159), 339(159)
Daniel, L. J., 162
Danovski, T. S., 467(266), 472
D'Ari, L., 399, 424(181, 182), 425, 426 (181, 182), 428(181, 182), 435(105), 438(178a, 179–182), 439(178a, 181, 182)

Darrow, R. A., 3
Davidson, D., 490
Davies, D. D., 216, 393
Davies, G. E., 229, 232, 235, 252, 256, 262 (16)
Davies, J. E., 189
Davis, L., 381, 386(16), 389(16), 390(16), 451(16), 452(240)
Davis, R. H., 102, 302, 303(141, 142), 305 (141)
Daw, J. C., 340
Day, R. A., 203
De Backer, M., 184
de Barsy, T., 329, 348(139), 349(139)
Decker, F. C., 393
De Deken, R. H., 511
De Hauwer, G., 118
Deibel, R. H., 102, 103(42), 109(40), 111 (40), 115(42)
Dela Fuente, G., 18, 21(58), 22(59), 24 (61), 25(61), 27(59)
DeLange, R. J., 314
del Campillo, A., 484, 486(9), 487(9), 488 (31), 494(9), 495(9)
Del Giacco, R., 318
Delwiche, E. A., 485
Demain, A. L., 505
Dembo, A. T., 399
Dementieva, E. S., 399, 408(99), 430(99), 431(99), 432(99), 433(99)
Demerec, M., 293
Demers, L. M., 356
Demidkina, T. V., 418(161, 164), 420, 421 (161, 164)
DeMoss, R. D., 467(264), 471
Dempsey, W. B., 478
Dénes, G., 186, 469(294, 297), 474, 512, 513(5), 514(5), 515(5), 516(5), 517(5, 18), 518, 519
Deng, Q., 73
Denis-Duphil, M., 304
Denisova, G. F., 395(92), 396, 397, 398 (92), 402(92)
Denton, W. L., 365, 366(13), 377(13)
Deodar, S., 216
Derechin, M., 7, 9, 19, 26
de Rosa, M., 394
Deutsch, E., 153
DeWulf, H., 311, 317, 318(67), 323, 325, 326(110), 327(110), 328(9, 67), 329 (9), 330(9, 67, 110), 331, 342(102),

343(67, 102, 110), 344(102), 345(102), 346, 347(102, 199), 348(139, 198), 349 (139), 351(198, 209), 352(198, 200)
Deyev, S. M., 418(166), 421, 422(166), 427(166)
Dickerman, H., 123, 126, 130(14), 132(14), 135(51), 143(14), 154(14), 159, 163 (14), 164(107), 208, 209(60, 61, 62)
Diehl, H., 124
Dietrich, L. S., 76, 77(160), 78(160), 79 (160), 80(160), 81(160)
Dietz, S. B., 350, 354(218)
Diller, A., 218
Dilworth, M. J., 130, 132, 135(53), 144 (35), 160(53), 163(53)
di Mari, S. J., 381, 383(13), 384(13), 385 (13), 386(13), 387(13), 388(13), 389 (13), 477(13)
DiMauro, S., 312
Dipietro, D., 31
Dische, Z., 184, 185(72)
Distler, J., 476
Divekar, A. Y., 51, 52(12), 53(12), 54(12)
Dixon, H. B. F., 424, 435(173, 174), 447 (173, 174)
Dixon, M., 25
Dodd, W. A., 160, 161(97)
Doherty, D. G., 502, 503(44)
Doi, R. H., 208
Donachie, W. D., 302, 306
Donaldson, K. O., 123, 131(13), 146
Doonan, M. J., 399
Doonan, S., 382, 392(31), 395, 398(31, 89), 399, 400(89), 402(31, 93), 424 (89, 93)
Dore, E., 190
Doty, D. M., 109
Doty, P., 405
Downing, M., 297, 298(129), 301(129)
Drahouský, D., 193
Dratz, E. A., 255
Dreyfus, P. M., 163
Drochmans, P., 312
Dronov, A. S., 472
Drummond, G. I., 331, 340
Dryer, W. J., 114
Dubbs, D. R., 56, 61(60), 74, 183, 185(60)
Dubnoff, J. W., 497, 497(3), 499(3), 500(3)
Duckworth, D. H., 94, 95(216)
Duda, E., 186
Duffield, P. H., 101, 103, 104(45), 105 (45), 110(45), 111(45), 112(45), 113

Dunathan, H. C., 219, 388, 389, 390, 391, 419(75, 76), 444(63, 76), 451, 452(76), 453(76), 454(75, 76), 461, 473(76), 479(76, 245)
Duncan, L., 331
Dunlap, R. B., 212, 214
Dunn, D. B., 190
Duphil, M., 303
Durham, J. P., 62, 63(91, 94), 64(91, 94, 97), 65(91, 94, 97), 66(97), 69
Dussoix, D., 191
Dvořák, M., 57, 59(73)

E

Easterby, J. S., 9, 38, 39(145), 40, 41(145)
Ebisuzaki, K., 186
Ebner, K. E., 365, 366(12, 13), 367(12), 368(22), 370, 371(22), 372(30, 34, 35), 373(30, 34, 35), 374(30, 35), 375(30, 34, 35), 376(22, 30, 34, 35, 36), 377 (13, 21, 29)
Eboué-Bonis, D., 337
Eckfeldt, J., 287
Edsall, J. T., 110
Edwards, B. F. P., 239, 242(50), 263(50), 264(50), 283(50), 284(50), 291, 292 (108b)
Egorov, C. A., 398, 399, 415(99a), 417 (99a), 418(99a, 161), 420, 421 (96, 99a, 161)
Eigen, M., 247
Eisen, H., 349
Elford, H. L., 122, 130(3), 135(36), 160 (3, 36), 162(3), 163(3, 36), 164(106)
Ellis, R. J. E., 393
Elsden, S. R., 98
Emmelot, P., 34
England, S., 75
Engstrom, L., 23
Enns, T., 425
Ennor, A. H., 498
Enter, N., 83, 84(186), 85(186), 86(186), 87(186)
Epstein, S. M., 82(180), 83
Ertel, R., 125, 152(19)
Erwin, M. J., 205
Esipova, N. G., 399
Esmann, V., 311, 313(17), 326, 331(119), 355 (17, 119, 120)
Estes, L. W., 377
Euler, H., 2

Evans, D. R., 239, 242(50), 263(50), 264(50), 265(85), 283(50), 284(50), 291, 292(108b)
Evans, H. J., 161
Exton, J. H., 348, 350, 353

F

Fabiano, R., 13(45), 14, 17(45), 21(45)
Fahrney, D. E., 4
Falaschi, A., 194
Falcone, A. B., 489
Fanshier, D. W., 437, 451(212)
Farago, A., 512, 513(5), 514(5), 515(5), 516(5), 517(5, 18), 518, 519
Farese, R. V., 343
Farinas, B., 76, 77(160), 78(160), 79(160), 80(160), 81(160)
Farkaš, V., 318, 322(78)
Farrelly, J. G., 398(97), 399
Fasella, P., 25, 26(84), 381, 384(21), 386(17, 21, 22), 387(17), 388(21), 389(21), 390(21), 391(69), 392(84), 393, 396, 397(84, 91), 398(69), 399, 400(69), 405(84), 407(17, 22, 84), 408(121, 122), 411(134), 412, 413(121, 122, 123), 416(84), 417(22, 69, 138), 418(169), 419(22), 421(22), 423(17), 424, 425(138), 426(21, 69), 428, 429, 435(17), 436, 437(21, 69, 185, 230), 438(17), 440(17), 441(22), 442, 444(235), 446(235), 450(21, 211), 451(211), 457(69, 121, 122, 134, 185, 209, 235)
Fasman, G. D., 406
Faurholt, C., 99, 110
Fearon, W. R., 108
Feigina, M. Yu., 398(96), 399, 415(99a), 417(99a), 418(99a), 421(96)
Feingold, D. S., 377
Feldman, M., 170, 174(14), 176(14)
Feliss, N., 398(98), 399, 400(98), 402(98)
Felts, P. W., 350
Fernandez, M. C., 331, 332(147), 335(147)
Ferraro, A., 419, 420(159), 421(159)
Field, J. B., 76, 77(159), 78(159), 81(159), 82(180)
Filmer, D., 517
Fischer, E. H., 314, 315(55, 56), 320(95), 321, 328, 331, 340, 417, 419(151)
Fischer, G. A., 56(63), 57, 59(63), 62, 63(93), 64(93, 98), 65(93), 67(93)

Fishman, W. H., 312, 320
Fiske, C. H., 100, 109, 514
Fittler, F., 184
Fitzgerald, D. K., 367, 368(22), 371(22), 376(22), 377(21)
Flaks, J. G., 205, 209, 214
Flatt, N. C., 187, 189(92)
Fleischer, B., 377
Fleischer, S., 377
Fleissner, E., 168, 170(1), 174(1)
Fletterick, R. J., 28
Florentiev, V. L., 383, 399, 408(99), 413, 429(41), 430(99, 143, 192, 193, 197), 431(99, 143, 193), 432(99, 143), 433(99, 143, 192), 434(193, 197), 444(193)
Focesi, A., Jr., 142, 152(63)
Folley, S. J., 364
Fonda, M., 431(202), 434
Fonda, M. L., 430(195, 195a), 431(195, 195a), 433
Ford, I. A. M., 490
Fosset, M., 314, 315(56)
Foster, M. A., 122, 130, 144(35), 160(1a, 2), 162(1a, 91)
Fox, P. R., 177, 185(42)
Franks, W. A., 239, 242(50), 263(50), 264(50), 283(50), 284(50)
Frazer, P. E., 467(285), 473
Frearson, P. M., 74
Frederiksen, S., 54
Freisheim, J. H., 212
French, T. C., 387(66), 388
Fridland, A., 212
Fridlender, B., 177, 183, 185(41)
Frieden, C., 30
Friedkin, M., 189, 210(5), 211, 212, 215, 216(5)
Friedman, D. L., 310, 313(6), 325(6), 326, 332(6)
Friedman, S., 132, 135(53), 160(53), 163(53)
Friedmann, B., 200, 348
Friedmann, T., 55
Friedrich, W., 144
Friesen, J. D., 186
Fromm, H. J., 13(44, 45, 46, 47, 48), 14(40), 15(41), 16, 17(44, 45, 46, 47, 48), 21(45), 22(40), 23(48), 33, 41, 42(148), 43(153), 44(150, 151, 152, 154)
Fruton, J. S., 484, 485(2)
Fry, E. G., 352

Fujii, S., 70, 71(125), 73(125), 484, 485(2)
Fujimoto, D., 191
Fujimoto, T., 352
Fujino, A., 380, 392(4), 438(4), 473(4), 480(4)
Fujioka, M., 216, 217(138, 139), 218(138, 139), 438, 463, 465(263c), 471
Fukui, S., 152, 162, 163(104), 383, 385, 429(41), 430(199, 201), 431(199, 201), 433, 434
Fuld, M., 499, 500(27)
Furbish, E. S., 430(195), 431(195), 433

G

Gabay, S., 465(263b), 471
Gaffney, T. J., 498
Gahan, L. C., 321
Galegov, G. A., 435
Galivan, J., 130
Gallagher, R. E., 175, 183, 185(25)
Gallo, R. C., 175, 183(25), 184, 185(25, 75)
Galloway, E., 163
Gambetti, P., 372
Gamble, J. L., Jr., 24
Gancedo, C., 323, 324
Gander, J. E., 364
Gantt, R. R., 169
Garren, L. D., 73
Garvey, T. B., III, 58, 59(78), 61(78), 90, 91(206)
Gatehouse, P. N., 465(258), 470
Gauldie, J., 398, 402(93), 424(93)
Gazith, J., 3, 4(15), 9(15), 10(15), 11(15, 36), 12, 13, 20(15), 26(15)
Gefter, M., 182, 184, 185(56), 186(83), 190(83), 192(83), 194(56, 83)
Gegelava, D. A., 418(162a), 420
Gell, J., 83
Gennari, G., 267
Georg, R. H., 499
George, H., 465(263b), 471
Gerhart, J. C., 226, 227, 228, 230, 231, 232, 234(28), 235(28), 236(31), 237, 238 (28), 239, 243, 246(31), 250(47), 251 (31, 47), 255(47), 263(47), 266, 268 (10, 91), 269(31), 270, 271(31), 272, 273(31), 274(31), 275, 276(47), 277 (28), 278(10), 281(31), 288, 290, 291, 293, 295(11), 296(11, 28), 297, 298 (123), 299(123), 300(123)
Giartosio, A., 392(84), 393, 397(84), 405 (84), 407(84), 408, 411(134), 416(84), 417(144), 418(168, 169), 419(144), 420(154, 159), 421(144, 159), 425 (131), 436, 449, 457(134, 209)
Gibbs, R. G., 463, 465(255, 256), 473(255, 256)
Gibson, F., 123
Gier, H. T., 356
Gillespie, D., 233
Gillespie, R. J., 248
Gipson, W. T., 498
Giri, K. V., 75
Glansdorff, N., 294, 295, 302(120)
Glaser, L., 119
Glasky, A. J., 170, 177(15), 185(15)
Glasziou, K. T., 101, 110(20), 115
Glick, J. M., 58, 59(79)
Glinsmann, W. H., 327, 328, 343, 348, 349, 351(134), 352(208)
Gnuchev, N. V., 447(236), 448
Gödeken, O. G., 320
Gold, A. H., 319, 325(83), 328(83), 342 (187), 343, 344(187, 192), 345(187, 192), 351(195)
Gold, A. M., 4
Gold, E., 194
Gold, M., 170, 174(8, 9), 176(9), 182, 185 (56), 186(83), 187, 190(83), 191, 192 (83), 194(56, 83)
Goldberg, I. H., 318
Goldberg, N. D., 315, 326, 337, 338(57), 341(57), 348, 349, 350(210, 213), 352, 353, 354(218)
Goldberger, R. F., 471(298), 475
Goldblatt, P. J., 36
Goldenberg, S. H., 32, 310, 311, 312(4), 313(4), 316(3), 317(3, 4), 318, 319(3, 73), 320(3, 73), 322(73), 332(3), 334 (3), 335(3), 344(4), 345(4)
Goldman, D. S., 484
Goldstone, A., 463
Goldthwait, D. A., 203
Goldwasser, E., 56
Gomez, C. B., 472
Gonnard, P., 438
Gonzalez, C., 32, 83, 84(186), 85(186), 86 (186), 87(186)
Gooding, R. H., 3, 4(15), 9(15), 10(15), 11(15), 12(15), 13(15), 20(15), 26(15)
Goodman, E. H., Jr., 348
Goodrich, M. E., 271

Gordon, M., 181
Gorini, L., 101, 109(33), 111
Goryachenkova, E. V., 442, 477
Got, C., 14, 15(49, 50), 20, 21(68)
Gotto, A. M., 55
Grace, J. T., Jr., 184
Grado, C., 183
Granner, D. K., 465(263), 471
Grassetti, D. R., 437, 451(212)
Grassl, M., 99
Grav, H. J., 70, 92(123), 93(123)
Gray, C. W., 274
Gray, R. M., 29, 46(92a)
Grazi, E., 500, 501(35, 36), 502(40), 503, 508(36, 43, 46)
Greco, A. M., 426
Green, A. A., 323, 324(100)
Green, D. E., 34, 484
Greenbaum, A. L., 81
Greenberg, D. M., 201(27, 28), 202, 211, 212, 216, 463, 465(253)
Greenberg, E., 317
Greenberg, G. R., 186, 199, 201, 203, 210, 214
Greenberg, H., 187, 189(93)
Greene, R. C., 160
Greenfield, N., 406
Greengard, O., 181
Greenwell, P., 265, 267(86)
Gregory, D. S., 262, 267(83), 269(83), 278(83), 279(83)
Grenson, M., 56, 58(55), 59(55)
Griffin, J. H., 232, 250, 253, 255(73), 267(73), 269(38), 274(38), 275(99), 276, 277(73), 278(73)
Griffith, T. J., 83, 84(187), 85(187), 86(187), 87(187), 92(187)
Griffiths, J. M., 162
Griffiths, M. M., 82, 83
Grinson, M., 305
Grippo, P., 195
Grishin, E. V., 398, 399, 415(99a), 417(99a), 418(99a, 166), 421(96, 99a), 422(166), 427(166)
Grisolia, S., 76, 100, 101, 102(22), 103(38), 104(41), 105(41), 106(38), 107(38), 109(22, 40), 111(22, 40), 112(38, 41), 114(38), 245, 249
Gross, T., 215, 218(122)
Grossbard, L., 33, 37, 38(117), 39(117), 40(117), 42(117), 43(117)

Gruhner, K., 343
Guest, J. R., 122, 132, 135(53), 160(1a, 2, 53), 162(1a), 163(53)
Guirard, B. M., 381
Gulyaev, N. N., 424, 435(174), 447(174, 236), 448
Gundersen, L. E., 212
Gunsalus, I. C., 221
Gustafson, G., 361
Guthöhrlein, G., 108
Gutman, A., 354, 355

H

Hablanian, A., 312
Hacker, B., 179
Hager, S. E., 306
Hakala, M. T., 51, 52(12), 53(12), 54(12)
Halbreich, A., 208
Hall, L. M., 101
Hall, R. H., 184
Hall, T. C., 56(67), 57, 59(67), 63
Halleux, P., 294
Halpern, B. C., 177
Halpern, R. M., 177, 178
Hamada, K., 92, 93(207)
Hamazima, Y., 385
Hammadi, I., 408, 411(134), 457(134)
Hammes, G. G., 14, 15, 25, 26(84), 113, 226, 232, 236, 247, 256, 257, 269, 274, 275(97), 285, 287(97), 288(3), 403, 407, 408(121, 122), 412, 413(121, 122, 137), 417(138), 425(137, 138), 428, 429, 436(191), 437(185, 191, 210), 457(121, 122, 185, 191, 209, 210)
Hancock, R. L., 177, 185(42)
Handschumacher, R. E., 56(63), 57, 59(63)
Hanna, M. L., 130, 131(32), 133(32), 136(48), 137(48), 149(32), 151(48), 153
Hansen, K., 51, 52(10), 53(10), 54(10)
Hansen, R., 33, 40(116)
Hanshoff, G., 245
Hanson, T. L., 33, 42, 43(153), 44(154)
Harding, N. G. L., 212, 214
Hardman, J. K., 485
Harmon, J., 102, 103(38), 106(38), 107(38), 112(38), 114(38)
Harrison, L. W., 287
Harrison, W. H., 29, 46(92a)
Hartman, P. E., 185

Hartman, S. C., 201, 203, 205
Hartmann, G. R., 51, 52(13), 53(13)
Hartmann, K.-U., 212, 214(98)
Harwood, J. P., 340
Haschemeyer, R. H., 101
Haschke, R. H., 340
Hasenbank, G., 438
Hashimoto, T., 70, 71(125), 73(125)
Haslam, J. L., 429, 436(191), 473(191, 210), 457(191, 210)
Hass, L. F., 23
Hassal, H., 472
Hasse, K., 471(306), 475, 476(306)
Hassid, W. Z., 364, 365(9), 369, 370, 371(9)
Hatch, F. T., 123
Hatefi, Y., 200, 201(29), 202, 216
Hatfield, G. W., 186
Hattman, S., 194
Hauk, R., 316
Hausen, P., 56, 61(62)
Hausmann, R., 182, 185(56), 186(83), 190(83), 191(83), 194(56, 83)
Hawker, J. S., 321
Hay, J., 190
Hayaishi, O., 467(286), 474, 485, 486(15)
Hayashi, K., 472
Haynes, G. R., 73
Hayward, W. S., 301
Hedegaard, J., 472
Hedeskov, C. J., 326, 331(119), 355(119, 120)
Heidelberger, C., 212, 214(98), 215
Heilmeyer, L. M. G., Jr., 340
Heinrikson, R. L., 56
Helmreich, E., 311
Helser, T. L., 189
Helting, T., 377
Helleiner, C. W., 83, 84(187), 85(187), 86(187), 87(187), 92(187)
Henderson, J. F., 54
Henson, C. P., 424, 425, 426(183), 427(183), 457(183)
Heppel, L. A., 83, 86(185), 87(185)
Her, M. O., 70, 71(127), 72(127), 74(127)
Herbert, M., 506
Hern, E., 348, 352(208)
Hern, E. P., 327, 343, 351(134)
Hernandez, A., 35, 38(129)
Herriott, S. T., 95, 96(219)
Herrlich, P., 185, 186(82)
Herrman, J. L., 48

Herrmann, R. L., 101, 102(24), 105(24), 109(24), 117(24), 203
Hers, H. G., 311, 317, 318(67), 319, 323, 325, 326(110), 327(110), 328(9, 67), 329(9, 67), 330(9, 67, 110), 331, 342(102), 343(67, 102, 110), 344(102), 345(102), 346(102), 347(102, 199), 348(139, 198), 349(139), 351(198, 209), 352(198, 200)
Hersh, L. B., 485, 486(10), 487(10, 22), 491(10, 25), 494(10, 25)
Hervé, G. L., 231, 233(30), 284, 294, 296
Heyde, E., 51, 61(6)
Hiatt, H. H., 92
Hickenbottom, J. P., 311, 312(8), 313(8), 314(8), 326(8), 327(8), 332(8)
Hickenbottom, R. S., 330
Hidalgo, J. L., 311, 312(18), 326(18), 331(18)
Higashino, K., 386, 463
Hill, D. L., 51, 52(9), 53(9), 54(9)
Hill, J. A., 125, 153(21)
Hill, J. M., 303
Hill, R. L., 366, 367(18, 19), 368, 370, 371(23, 28), 375(23, 28), 376, 377(18)
Hilz, H., 354
Hinds, K. R., 163
Hiraga, K., 223
Hiraga, S., 70, 83, 84(184), 86(184), 87(184)
Hirsch, H., 463, 465(253)
Hitchings, G. H., 212
Hizukuri, S., 312, 313(26), 317(42), 325(26), 326, 330, 331(26), 342(26), 343(26), 344(26), 345(26), 346(26), 354(115)
Hjalmarson, Å., 341
Hnilica, V. S., 505, 508(55)
Ho, D. H. W., 54
Ho, R. J., 350
Hockenhull, D. J. D., 506
Hoet, P. P., 485
Hoffmeyer, J., 55
Hogenkamp, H. P. C., 125, 140, 142, 143(60, 65), 152
Hohnadel, D. C., 14, 15(53), 17, 22(53)
Hokin, L. E., 111
Hokin, M. R., 111
Holler, B. W., 187, 189(94)
Holloway, B. W., 301

Holloway, C. T., 160
Holmes, S., 152
Holoubek, H., 228, 242(11), 295(11), 296(11), 297, 298(123), 299(123), 300(123)
Holzer, H., 324
Holzwarth, G., 405
Hoogenraad, N. J., 306, 307
Hopper, S., 465(258), 470
Hoppe-Seyler, P., 168
Horecker, B. L., 393
Horiguchi, M., 475
Horikoshi, K., 208
Hornbrook, K. R., 330, 342(186, 190), 343, 348, 352(186, 205)
Horner, W. H., 499, 506
Hoskins, A. P. V., 55
Hostmark, A. T., 349
Housewright, R. D., 472
Housman, D., 233
Hsu, T.-S., 440, 467(228, 284), 473
Huang, H. T., 54
Hudgin, R. L., 377
Huennekens, F. M., 130, 131, 132, 134(52), 163(52), 189, 200(7), 201(29), 202, 203(7), 210(2), 211(6, 7), 212(6, 7), 214, 215(2), 216(2, 7), 221, 222(7, 160)
Hughes, E. C., 356
Hughes, R. C., 417, 419(151)
Huijing, F., 313, 315, 317(42), 326, 377(130), 330(130), 331, 337, 338(57), 340(130), 341(57)
Hultin, H. O., 34
Hummel, J. P., 114
Humphreys, G. K., 211
Humphreys, J. S., 101, 104(21), 107(21), 111(21), 112(21)
Hunkeler, F. L., 311, 312(8), 313(8), 314(8), 326(8), 327(8), 332(8), 340
Hunter, G. D., 506
Huntley, T. E., 479
Hurd, S. S., 328
Hurlbert, R. B., 168, 187, 189(92)
Hurwitz, J., 87, 92(201), 170, 174(8), 176, 182, 185(56), 186(83), 187, 190(83), 191, 192(83), 194(56, 83)
Hurwitz, R., 306
Huston, R. B., 331
Hutchison, D. J., 214
Hutson, T. Y., 297, 298(129), 301(129)

Hutzler, J., 386
Hwang, K. J., 28

I

Iaccarino, M., 195, 399
Ichihara, A., 436, 465(208b, 208c, 262), 471
Igarashi, K., 70
Igumi, M., 465(260), 470
Ihl, M., 183
Ikawa, M., 386(59), 387, 451(59)
Ilivicky, J., 76, 77(165)
Illingworth, B., 319, 320(82), 321, 322(82)
Inagaki, A., 306
Inagami, T., 19, 25(62), 26(62), 44(62)
Inamine, E., 505
Ingraham, J. L., 56, 57(53), 58(53), 59(53), 61(53), 70(53, 87), 87, 295
Ingraham, L. L., 261
Ingram, P., 311, 319(20), 358(20)
Ingram, V., 189
Inoue, H., 380, 381(5), 392(5), 438(5), 462(5), 480(5)
Irias, J. J., 485
Irwin, V., 190
Isaac, J. H., 301
Isaksson, O., 341
Issaly, A. S., 101, 102(25), 105(25), 109(25), 112(25), 117(25), 118(25), 302
Issaly, I. M., 101, 102(25), 105(25), 109(25), 112(25), 117(25), 118(25), 302
Ito, K., 307
Itoh, H., 393
Ivanov, V. I., 383, 384(44), 386(44), 387(44), 388(44), 389(44), 390(44), 399, 402, 403, 407(44), 408(44, 99, 123, 124), 410(123, 133), 411(123, 124, 133), 412(124), 413(123, 124), 416, 418, 419(144), 420(124), 428(44), 429(41, 44), 430(44, 99), 431(99, 143, 193, 196), 432(44, 99, 143), 433(99, 143, 192), 434(44, 193), 435(44), 437(44), 440(44), 444(44, 193), 457(44), 459(57), 460, 461
Ives, D. H., 62, 63(91, 94), 64(91, 94, 97), 65(91, 94, 97), 66(97), 69, 71, 73
Ivkino, N. S., 509
Iwatsuki, N., 71, 72(133)
Izatt, R. M., 110

J

Jackson, J. F., 229
Jacob, F., 284, 293(101), 295(101), 517
Jacobs, R. D., 356
Jacobson, G. R., 234, 263, 264(84), 265(84), 273(84), 281(84), 282(84), 284(84)
Jacobson, K. B., 186
Jacoby, G. A., 294, 465(263a), 471
Jaenicke, L., 127, 130(26), 131(26), 132, 141, 150, 151(39, 70, 71), 189, 199, 203, 489, 490
Jagganathan, V., 38, 39(146), 43(142)
Jakoby, W. B., 467(287), 472, 474, 477, 485
Jänne, J., 438
Jeanneax, C., 355
Jefferson, L. S., 350
Jemionek, J. F., 36
Jencks, W. P., 100, 211, 248, 258, 261, 386(62), 387, 388(62, 71, 72), 389(62), 390(62), 485, 486(10), 487(10, 22), 489, 491(10, 25), 493, 494(10, 25)
Jenkins, W. T., 216, 218(136), 219, 382, 392(28), 398(28), 399(28), 402(28), 407, 408(28), 412(28), 413(136), 417, 419(151), 424(115, 135, 136, 179, 180, 181, 182), 425(115, 177, 178), 426(28, 135, 177, 181, 182), 428(177, 178a, 181, 182), 435(105, 136), 436(28, 128, 177, 178, 206), 437(206), 438(28, 135, 136, 178a, 179–182), 439(28, 135, 178a, 181, 182), 457(135, 136, 177), 465(208, 259), 467(26), 469(291), 470, 471(208), 472, 474, 479
Jenkinson, P., 178
Jenness, R., 363
Jewett, S. L., 265, 267(86)
John, R. A., 396, 397(91), 450
Johnson, A. W., 144
Johnson, J. C., 355
Johnson, R. J., 430(195), 431(195), 433
Jones, D. B., 356
Jones, E. A., 364
Jones, J. G., 26
Jones, M. E., 98, 99(2), 100(2, 4), 101(2), 102(11), 105(4), 106(39), 107(39), 108(4), 109(2, 17, 49), 110(4, 49), 111(4, 39), 112(39), 115(4), 229, 243, 246, 248(52), 255(58), 259(58), 269(21), 276(58), 297, 298(124, 125, 126, 127, 128), 299(128), 301(124), 305, 306, 307(126, 162, 163)
Jones, O. W., 73
Jordan, P. M., 219
Jørgensen, A. Ø., 54
Jori, G., 267
Josse, J., 94, 95(214), 191
Jourdian, G. W., 377
Joyce, R. J., 490
Junga, I. G., 55
Jungas, R. L., 354

K

Kagamiyama, H., 382, 391(36), 392(36), 415(36), 416(36), 417(36), 425(36)
Kagan, Z. S., 472
Kahan, E., 210
Kahan, F., 210
Kahle, P., 168
Kaji, A., 3, 5(13), 6(12, 13), 18(13), 20(55), 23(55), 24(55), 25, 26
Kalckar, H. M., 2
Kaldor, G., 214
Kallen, R. G., 210, 211
Kalman, S. M., 101, 103, 104(45), 105(45), 109(45), 110(45), 111(45), 112(45), 113
Kalman, T. I., 70, 214
Kalousek, F., 193
Kalyankar, G. D., 478
Kamel, M. Y., 48
Kammen, H. O., 214
Kaplan, J. G., 301, 303, 304, 305
Kaplan, M., 451, 452(240)
Kaplan, N. O., 76, 77(155), 78(155), 80(155), 81(155), 393, 394, 395(87), 396(87), 407(87), 415(85), 416(87)
Kappler, J. W., 193
Kara, J., 57
Karimoto, R. S., 365
Karpatkin, S., 355
Karpeisky, M. Ya., 383, 384(44), 386(44), 387(44), 388(44), 389(44), 390(44), 399, 403, 407(44), 408(44, 99, 123, 124), 410(123, 133), 411(123, 124, 133), 412(124), 413(44, 124), 416(44), 418(44, 162a, 164), 419(44), 420(124), 421(164), 428(44), 429(41, 44), 430(44, 99, 143, 192, 193, 196, 197), 431(99, 143, 193, 196), 432(44, 99,

143), 433(99, 143, 192), 434(44, 193, 197), 435(44), 437(44), 440(44), 441(222), 443(222), 444(44), 457(44, 222), 459(44), 460, 461
Karten, M., 490
Katagiri, M., 485, 486(15)
Katchalski, E., 484
Katchman, B., 76, 82(154)
Kato, K., 327, 328(135)
Katsunuma, T., 380, 392(4), 438(4), 469 (292), 473(4), 474, 480(4)
Katunuma, N., 380, 381(5), 386(20), 392 (4, 5), 438(4, 5), 462(5), 465(260), 469 (292), 470, 473(4), 474, 480(4, 5, 20)
Katzen, H. M., 32, 33(112, 113), 34, 40 (113), 130, 135(36), 158, 160(36), 163 (36)
Katzin, B., 352
Kaufman, B., 476
Kaufman, B. T., 131
Kaufman, S., 24
Kawasaki, H., 222
Kay, L. D., 200, 201(29), 202, 216
Kaye, A. M., 174, 177, 185(41), 190
Kearney, E. B., 75, 435
Keech, D. B., 485
Keenan, T. W., 377
Keil, B., 417, 419(152)
Keir, H. M., 69, 71(118), 73(118)
Kekomäki, M., 438
Keller, R., 359
Kellermeyer, R. W., 485, 486(14), 487(14), 494(14)
Kelsall, M. A., 380, 381(6), 462(6), 480 (6), 481(6)
Kemp, R. G., 314
Kenkare, U. W., 3, 7, 9(26), 27(26), 38, 39 (144), 41, 43(144), 45(144)
Kenney, F. T., 380, 381(5), 392(5), 438 (5), 462(5), 480(5)
Kerbiriou, D., 284
Keresztesy, J. C., 123, 131(13), 146, 207
Keir, H. M., 190
Kerr, S. J., 175, 177, 178, 180(24), 181 (51), 183(24), 184, 185(24, 51, 72)
Kersten, H., 170
Kersten, W., 170
Kerwar, S. S., 132, 134(52), 163(52)
Kessel, D., 56(67), 57, 59(67), 63(101), 64 (96), 65(96)

Khokhlov, A. S., 506(66), 507, 509
Khomutov, R. M., 424, 431(202), 434, 435 (173, 174), 437(239), 438, 440, 441 (223, 224, 227), 442, 444(235), 446 (235), 447(173, 174, 236), 448, 450 (211), 451(211), 457(235)
Khurs, E. M., 424, 435(173, 174), 447 (173, 174)
Khym, J. X., 502, 503(44)
Kielley, R. K., 89, 92, 93(210), 94(210)
Kikuchi, G., 222, 223
Kim, K.-H., 313, 317, 331, 349(143), 357 (48), 358(69, 256), 471(304, 305), 475
Kindt, T. J., 317
King, J., 313, 315(46), 330(46), 331(46), 334(46), 344(46), 345(46), 346(46), 350, 355
King, K. W., 393
King, J., 473
Kingdon, H. S., 324
Kinoshita, S., 511, 512(1), 514(1), 515(1)
Kirpichnikov, M. P., 424, 435(173), 447 (173)
Kirschner, K., 226, 285
Kirschner, M. W., 250, 253, 255(71), 258 (71), 267, 278, 280(71, 88), 281(88), 289
Kiryanov, G. I., 193
Kiryushin, A. A., 398, 421(96)
Kiselev, A. P., 398, 399, 415(99a), 417 (99a), 418(99a), 421(96, 99a)
Kishimoto, S., 385
Kisliuk, R. L., 160, 162(90), 211, 212, 216
Kit, S., 56, 61(60), 74, 183, 185(60)
Kittler, M., 352
Kittredge, J. S., 475
Kitz, K., 441, 444
Kiyosawa, I., 367, 368(22, 25), 371(22), 376(22)
Kjellin-Straby, K., 170
Klagsbrun, M., 183
Klee, C. B., 371
Klee, W. A., 152, 154(74), 371
Klein, G., 59
Klein, S. M., 221, 222
Klemperer, H. G., 73
Klenow, H., 51, 52(10), 53(10), 54(10), 66
Kleppe, K., 259
Kliewer, M., 161
Klosterman, H. J., 438

Knappe, J., 108
Knappenberger, M. H., 484, 485(2)
Knivett, V. A., 99, 102(8)
Knox, W. E., 380, 381(5), 392(5), 438(5), 462(5), 467(284), 473, 477, 480(5)
Kobayashi, S., 62(95), 63(92), 64(92, 95), 65(92, 95), 89(106), 92, 93(207)
Kochav, D., 14, 15
Kochi, H., 222, 223
Koch, J., 161, 163(99)
Kochavi, D., 113
Koerber, F., 438
Kogan, G. A., 403, 405, 406(126), 407(126), 436(126)
Kolb, H., 471(308), 476
Kon, H., 142
Kon, S. K., 363
Kong, C. M., 83
Königk, E., 144
Korbecki, M., 56
Koreneva, L. G., 403, 407(119, 120), 408(119), 413, 436(120), 457(119, 120)
Koritz, S. B., 229
Kornberg, A., 51, 52(7), 69, 70(120), 71(120), 76, 77(152), 78(152), 82(152), 84, 88, 92(197), 94, 95(214), 191, 194, 212
Kornberg, H. L., 463, 465(256), 473(256), 485
Kornberg, S. R., 94, 95(214), 191
Kornfeld, R., 313, 316, 318(37), 319, 320(82), 321(37), 322(82), 335(37)
Kornitskaya, E. Y., 506(66), 507, 509
Korytnik, W., 383, 429(41), 430(194), 433
Korzenovsky, M., 99
Koshland, D. E., Jr., 29, 517
Koskimies, O., 306
Kosow, D. P., 13, 14(42), 16, 17, 29, 30, 31, 38, 42, 43(42, 147), 44(42), 45, 46(43)
Kotula, Z., 498
Kovaleva, G. K., 384, 440(46), 441(46, 223, 225, 226, 227), 443(46, 223, 225), 444(46, 223, 235), 445(225), 446(225, 235), 447(46, 226, 236, 248), 453(46), 454(46), 455(46), 457(235), 459(46, 226)
Kozai, Y., 62(95), 63(104), 64(95), 65(95), 89(106), 92, 93(207)
Krahl, M. E., 33, 40(116)
Kratzer, F. H., 499, 504(29)

Krebs, E. G., 311, 312(8), 313(8), 314(8), 315(55), 324, 326(8), 327(8), 331, 332(8), 340
Kretchmer, N., 306, 307
Kretovich, V. L., 472
Kreutner, W., 352
Krishnaswamy, P. R., 75
Krisman, C. R., 319, 321, 322(84)
Kritzmann, M. G., 380, 391(7)
Kroeger, H., 168
Kropp, E. S., 36
Krygier, V., 63(103), 64, 66, 67(103, 107, 108), 68(103, 107, 115)
Krystal, G., 57, 58(76), 59(76), 60(72, 76), 61(76)
Kuby, S. A., 113
Kuchino, Y., 171, 172, 173(17)
Kudryashova, I. B., 193
Kuhnlein, U., 192
Kun, E., 437, 451
Kundig, W., 48
Kung, H. F., 159
Kunitz, M., 2, 5(6), 7, 12
Kuno, S., 467(286), 474
Kupchik, H. Z., 467(284), 473
Kupiecki, F. P., 486
Kuriyama, S., 347
Kury, P. G., 451, 452(240)
Kusamrarn, T., 453
Kushner, D. J., 301
Kushner, V. P., 76, 77(156), 78(156)
Kutzbach, C., 163, 200, 490

L

Labitan, A., 63(103), 64, 67(103), 68(103)
Lachmann, B., 430(194), 433
Lacroute, F., 303, 304, 305
Lacy, W. W., 350
LaDu, B. N., 465(263a), 471
Lagunas, R., 18, 22(59), 27(59)
Lai, C. J., 189
Laishley, E. J., 118
Lal, B. M., 193
Laloux, M., 329, 348(139), 349(139)
Lamoureux, G. L., 438
Landon, M., 498
Landsberg, B., 176
Langenbach, R. J., 215
Langer, R. M., 355
Lardy, H. A., 18, 50, 76(2), 200, 381, 382, 386(12)

Lark, C., 193
Lark, K. G., 66
Larner, J., 310, 311, 312, 313(6, 26), 314 (44, 45), 315(47, 54), 316(59), 317 (42), 319(45), 321, 322(5, 39), 323, 324, 325(6, 26), 326(47, 58, 111), 327 (111, 136), 328(39, 41), 330(130, 237), 331(26), 332(6, 39, 41), 333(39, 40), 334(40, 41, 160), 335(39, 40, 158, 160), 336, 337(130), 338(57, 167), 340 (47, 136), 341(57), 342(189), 343(26, 111), 344(26), 345(26, 158), 346(26), 348, 349(189), 350(210, 213), 351 (189), 357(122), 358(122), 359
Larrabee, A. R., 123, 130(8), 131(8)
Larson, L. L., 356
Lascelles, J., 162, 163(102)
Lavalle, R., 118
Law, L. W., 58, 59(80)
Lawrence, A. J., 382, 392(31), 398, 402 (31, 93), 424(93), 426(175), 427(175), 428(176)
Lazarus, N. R., 4, 5(20), 7, 9, 26
Leary, R., 212
Leboy, P. S., 174, 175
Lecar, H., 142
Lederberg, S., 185
Lederer, B., 329
Leelavathi, D. E., 377
Lefer, A. M., 340
Legrain, C., 294
Lehman, I. R., 84, 88, 92(197)
Leibach, T. K., 51, 52(13), 53(13)
Leloir, L. F., 32, 310, 311(2), 312(4), 313 (4), 316(3), 317(3, 4), 319(13), 320 (3), 322(84), 332, 334(3), 335(3), 344 (4), 345(4), 364
Lengyel, P., 208
LePage, G. A., 55
Levenberg, B., 101, 203
Levin, A. P., 205, 471(300), 472, 475
Levin, Y., 484
Levine, L., 246, 255(58), 259(58), 276(58)
Levine, R. L., 306, 307
Levison, S. A., 212
Levitov, M. M., 509
Levitski, A., 29
Levy, H. L., 163
Lewis, S. B., 350
Li, C-C., 208
Li, Y., 507

Liacouras, A. S., 57, 58(75), 59(75, 78), 60(75), 61(75, 78)
Liau, M. C., 168, 187, 189(92)
Lichtenstein, J., 214
Liddle, G. W., 350
Lieberman, I., 386, 487
Liebl, V., 301
Liljenquist, J. E., 350
Linarelli, L. G., 343
Lindberg, B., 51, 52(10, 11), 53(10, 11), 54(10, 11)
Ling, E. R., 363
Linn, S., 192
Lipkin, V. M., 398(96), 399, 415(99a), 417 (99a), 418(99a), 421(96, 99a)
Lipmann, F., 98, 99(2), 100(2, 4), 101(2), 105(4), 106, 108(4), 109(2, 17), 110 (4), 111(4), 115(2, 4), 208, 311, 313, 317(38), 328(38), 335(38), 512
Lipscomb, W. N., 239, 242(50), 263(50), 264(50), 283(50), 284(50), 291, 292 (108b)
Lipshitz-Wiesner, R., 180, 181(51), 185 (51)
Lis, H., 382, 392(29), 407
Litvak, S., 76, 77(165)
Littauer, U. Z., 170, 174(14), 176(14)
Litwack, G., 382, 435(26), 465(26)
Livanova, N. B., 403, 408(125), 409(125), 410(125), 411(125), 412(125), 413 (125), 422(125), 437(125)
Lodish, H. F., 233
Loeb, J. R., 214
Loeb, L., 184, 185(73)
Lomax, M. I. S., 214
Lombardi, B., 377
Lombardini, J. B., 178
London, R. E., 270, 271(93), 272, 273
Long, C. N., 352
Longley, R. W., 348
Lopez, J. A., 100, 109
Lorenson, M. Y., 211, 212(93)
Lou, M. F., 101, 102(24), 105(24), 109(24), 117(24)
Loughlin, R. E., 122, 130(3), 160(3), 162 (3), 163(3), 164
Love, S. H., 187, 188(90)
Low, M., 190
Lowenstein, J. M., 306
Lowry, O. H., 34, 45, 100, 109, 119, 342 (186), 343, 352(186)

M

Lucas, Z. J., 56
Lucas-Lenard, J., 208
Luce, J. K., 54
Luck, D. J. L., 312, 344(24)
Lue, P. F., 304, 305
Lueck, J. D., 48
Lundahl, P., 170
Lutovinova, G. F., 417, 419(150), 421 (150)
Lutwak-Mann, C., 2
Lyman, K., 72
Lynen, F., 487, 489
Lyon, J. B., Jr., 342(190), 343

M

Maas, W. K., 512
McAuslan, B., 163
McCauley, J. G., 359
McClintock, D. K., 246, 255, 271(59), 276 (59), 280(60)
McCormick, D. B., 75, 383, 421(41)
McDonald, M., 2, 5(6), 7, 12
McDougall, B. M., 122, 130(3), 160(3), 162(3), 163(3), 211, 212
McFarlane, E. S., 175
McFarland, P., 177, 185(42)
McGarrahan, J. F., 318
McGuire, E. J., 377
McKenzie, L., 367, 377(21)
McKinley, S., 99
MacKinnon, J. A., 48
McLean, P., 81
McLellan, W. L., 512, 513(6), 514(6), 515 (6), 516(6)
McMurray, C. H., 239, 242(50), 263(50), 264(50), 265(55), 283(50), 284(50), 291
Maddaiah, V. T., 312
Maden, E. H., 189
Madsen, N. B., 312, 317
Mäenpää, P. H., 184
Magasanik, B., 55, 205, 206
Magee, S. C., 368, 394, 416(86a)
Magnusson, P.-H., 59
Mahoney, M. J., 163
Mahowald, T. A., 113
Mahy, B. W. J., 56
Mainigi, K. D., 70, 72(126)
Maitra, U., 194
Malakhova, E. A., 403, 408(125), 409 (125), 410(125), 411(125, 142), 412 (125), 413(125), 422(125), 436(142), 437(125)
Maley, F., 18, 62, 65(90), 66, 71(90), 87, 211, 212(93)
Maley, G. F., 62, 65(90), 66, 71(90), 211, 212(93)
Malkewics, J., 429, 436(190)
Mamaeva, O. K., 399, 408(99), 413, 430 (99, 143, 193, 196), 431(99, 143, 193, 196), 432(99, 143), 433(99, 143), 444 (193)
Mangum, J. H., 132, 134(52), 163(52), 164, 165(112)
Manjeshwar, R., 32
Mann, T., 2
Mann, K. M., 18
Manning, J. M., 437(239), 450, 451(211), 477
Mano, Y., 111
Mapson, I. W., 472
Maley, F., 318
Maraspin, L. E., 356
March, J. E., 472
Marcker, K. A., 208
Marco, R., 32, 35, 36(130)
Marino, G., 394, 426
Marion, G. B., 356
Markus, G., 246, 255, 271(59, 75), 276 (59), 280(60)
Marmur, J., 191, 192(112), 210
Maron, E., 367
Marsh, A., 79, 81(172)
Marshall, M., 101, 102, 103(43), 104(43), 109(43), 110(43), 111(43), 112(43), 113(43), 114(43)
Marshall, R. O., 100
Martin, B. T., 193
Martin, J. B., 109
Martin, R. G., 104, 471(298, 299), 475
Martinez-Carrion, M., 382, 391(37), 392 (37, 38), 393, 396(38), 397, 398(35, 98), 399(38), 400(37, 98), 402(198), 403, 405(37, 84, 90), 406(90), 407(37, 84, 94), 408(38, 94), 411(35), 412, 413 (94), 415(37), 416, 419, 421(35), 423, 425(94, 158), 426(94, 158), 427(94), 428(187), 435(158), 436(94), 439, 459, 467(276), 472
Marvin, S. V., 295
Mason, M., 216, 218(135), 438, 472, 479

Mason, R. J., 355
Massaro, E. J., 9
Masugi, F., 430(201), 431(201), 434
Masui, H., 73
Mathews, C. K., 131, 209
Matsuda, M., 463
Matsuda, S., 465(260), 470
Matsuhashi, M., 476
Matsumoto, S., 274, 275(97), 287(97)
Matsuzawa, T., 380, 392(4), 438(4), 469(292), 473(4), 474, 480(4)
Matsuzawa, T. S., 465(257), 470
Mattock, P., 366, 367(18), 370(18), 377(18)
Mawal, R., 367, 368(22), 371(22), 375, 376(22, 36)
Maxwell, E. S., 83, 86(185), 87(185)
Mayer, S. E., 338, 340(171), 341(171)
Mays, L. L., 179, 180(50), 181, 185(50)
Mazumder, R., 142, 152(63)
Mecchi, E., 499, 504(29)
Mechie, M. J., 190
Mecke, D., 324
Medoff, G., 194
Mehler, A. H., 76, 206
Meighen, E. A., 232, 242(33), 266(33), 279(33)
Meikle, A. W., 55
Meister, A., 101, 295, 379, 380(2), 381(2, 10), 385, 386, 393, 399(2), 432(101), 435(2, 10, 52), 436, 462(2, 52), 467(283, 285), 473, 477, 480(2)
Meloni, M. L., 54
Mell, G. P., 212
Mendicino, J., 311
Menon, G. K. K., 485, 486, 487(13), 491(47), 492(13)
Mergeay, M., 295, 302(120)
Mersmann, H. J., 312, 322, 323(100), 324(100), 330(28), 342(100), 343(28), 344, 346, 352(28)
Mervyn, L., 144
Messmer, I., 304
Methfessel, J., 505
Metzenberg, R. L., 101, 111, 113
Metzler, D. E., 381, 386(16, 59), 387, 389(16), 390(16), 430(195), 431(195), 433, 440, 441(227), 451(16, 59), 479
Meunier, J. C., 48
Meuser, R., 361

Meyer, F., 340
Meyer, W. L., 331
Meyerhof, O., 2
Meyskens, F. L., 52
Michaels, K., 471(304), 475
Michuda, C., 396, 398, 403, 405(90), 406(90), 407(94), 408(94), 413(94), 416, 425(94), 426(94), 427(94), 428, 435, 436(94), 459
Michuda-Kozak, C., 419, 425(158), 426(158), 428(158), 435(158), 459(158)
Middleton, B., 81, 82(178)
Miech, R. P., 83, 84(188), 85(188, 190), 86(188, 190), 90(188)
Miersch, J., 498
Migita, L., 208
Mii, S., 484
Mildvan, A. S., 20
Miles, E. W., 221, 477
Millar, F. K., 58, 59(78), 61(78), 90, 91(206)
Miller, A., 208
Miller, A. M., 352
Miller, J. E., 382, 435(26), 465(26)
Miller, T. L., 496
Miller, W. G., 111
Milner, L., 158, 159
Milstien, S., 489
Misono, H., 387, 469(295), 474(57), 478
Mitoma, C., 500
Mittelman, A., 184
Miyoshi, Y., 110
Modebe, M. O., 229, 245(15), 246(15), 247(15), 249(15), 252(15), 256(15), 258(15), 259(15), 260(15), 261(15), 262(15), 265(15), 267(15), 268(15), 270(15), 289(15)
Modyanov, N. N., 398, 399, 415(99a), 417(99a), 418(99a), 421(96, 99a)
Mohr, S. C., 287
Mokrasch, L. C., 101, 102(22), 109(22), 111(22, 40)
Molnar, D. M., 472
Momparler, R. L., 62, 63(93, 103), 64(93), 66, 67(93, 103, 107, 108), 68(103, 107, 115), 70, 71(127), 72(127), 73(127)
Moner, J. G., 56, 57(57), 58(57), 59(57), 60(57), 61(57)
Monod, J., 275, 286, 287, 289(108), 290, 517, 518

Montalvo, F., 48
Montes de Oca, F., 169
Montgomery, J. A., 54
Moore, B. G., 176
Moore, R. O., 354
Mora, S., 430(193, 196), 431(193, 196), 433, 434(193), 444(193)
Mordoh, J., 319, 322(84)
Morgan, H. E., 341
Mori, M., 54
Moriguchi, M., 477
Morino, Y., 382, 383(33), 391(32, 33, 36), 392(32, 33, 36), 393, 398(33), 400(32, 33), 415(36, 148), 416(36), 417(36), 425(36), 426(33), 427(33), 450, 463, 465(263c), 471
Morningstar, J. F., 160, 162(90)
Morosov, Yu. V., 403, 408(123), 410(123), 411(123), 413(123)
Morozov, Yu. M., 424, 435(173), 447(173)
Morré, D. J., 377
Morris, J. G., 463, 465(255, 256), 473(255, 256)
Morris, N. R., 193
Morrison, J. F., 51, 61(6), 371, 372(30, 34, 35), 373(30, 34, 35), 374(30, 35), 375 (30, 34, 35), 376(30, 34, 35, 36)
Morrow, G., 163
Morse, P. A., Jr., 71
Moses, S. W., 355
Mosse, H., 306
Motokawa, Y., 222, 223
Mudd, S. H., 152, 154(74), 163
Muenster, L. J., 248
Muir, L. W., 314, 315(56)
Mukherjee, K. L., 214
Müller, G., 139, 144(59), 151(59)
Müller, O., 139, 144(59), 151(59)
Mumbach, M. W., 359
Munch-Petersen, A., 66
Murad, F., 355
Murai, J. T., 178
Murer, E., 34
Murie, R., 124
Murphy, T. A., 312, 317(21), 358(21), 359 (21)
Murray, A. W., 52
Murray, B. K., 163
Myrbäck, K., 381, 386(12)

N

Nagabhushanan, A., 201(28), 202
Nagasawa, T., 367, 368
Najjar, V. A., 24
Nakada, H., 467(280), 473
Nakada, H. I., 200
Nakai, Y., 430(201), 431(201), 434, 463
Nakajima, K., 183, 185(60)
Nakajima, T., 438
Nakamura, H., 62, 63(92), 64(92), 65(92), 89(106)
Nakano, H., 80
Nakano, M., 467(265, 266), 472
Nakano, Y., 216, 217(38), 218(138)
Nakatsu, S., 498
Nakazawa, T., 163
Neal, G. E., 163
Neet, K. E., 30
Needham, D. M., 25
Nelbach, M. E., 231, 234, 235, 236, 238, 277(28, 49), 279, 296(28)
Nelson, D. J., 92, 93(210)
Nemchinskaya, V. L., 76, 77(156, 158), 78(156, 158), 80(158), 81(158), 82
Nemethy, G., 517
Neuberger, A., 469(290), 474
Neudecker, T. J., 51, 52(13), 53(13)
Neuhard, J., 55, 56, 57(53), 58(53), 59 (53), 61(53, 54), 69(54), 70(53, 54, 87), 87, 226, 292, 293(5), 295(5), 297(5), 298(5), 299(5)
Neumann, H., 484
Neumann, J., 297, 298(126), 307(126)
Nezlin, R. S., 400, 401(107), 402(107), 405 (107), 406(107), 412(107)
Nguyen-Phillipon, C., 438
Nichol, C. A., 380, 381(5), 392(5), 438(5), 462(5), 480(5)
Nielsen, J., 4, 5(19), 8(63), 11(19), 19, 20 (63), 21(63), 27(63)
Niemeyer, H., 32
Nigam, V. N., 318
Nikolaeva, Z. K., 430(207), 434, 441, 444 (232), 469(288, 289)
Nikolova, Z. V., 430(193), 431(193), 433, 434(193), 444(193)
Ning, J., 42, 44(150, 151)
Nishimura, J. S., 477
Nishimura, S., 171, 172, 173(17)

Nishizuka, Y., 467(286), 474
Nisonoff, A., 425
Nisselbaum, J. S., 380, 381(5), 392(5), 393, 424, 425(83), 426(83), 427(186), 438(5), 462(5), 480(5)
Nitowsky, H. M., 32, 33(112)
Noat, G., 14, 15(50), 20, 21(68)
Noda, L., 50, 83(4), 87(4)
Nolan, C., 314, 315(55)
Noltmann, E. A., 113
Norberg, P., 301
Nordlie, R., 50, 76(2)
Nordlie, R. C., 48
North, A. C. T., 366, 367(19)
North, J. A., 163, 164, 165(112, 114)
Nosikov, V. V., 398(96), 399, 415(99a), 417(99a), 418(99a, 166), 421(96, 99a), 422(166), 427(166)
Novelli, G. D., 184, 512
Novoa, W. B., 314, 315(55)
Novogrodsky, A., 435, 436(203), 477
Nuttall, F. Q., 326, 327(130), 330(130), 337(130), 340(130), 341, 349, 350(213), 353

O

Ochoa, S., 76, 87, 142, 152(63), 487
Oda, K., 191, 192(112)
O'Donovan, G. A., 56, 61(54), 69(54), 70(54), 226, 292, 293(5), 295(5), 296, 297(5), 298(5, 123), 299(5, 123), 300
Oehme, P., 437, 438(214)
Oeschger, M. P., 83, 84(194), 85(194)
Ogawa, K., 436, 465(208c, 262), 471
Oginsky, E. L., 99
Ogur, M., 463
Ohishi, N., 385, 430(199), 431(199), 433
Ohmori, H., 162, 163(104)
Ohmura, E., 485
Oka, H., 76, 77(159), 78(159), 81(159)
Okada, M., 380, 392(4), 438(4), 473(4), 480(4)
Okamoto, S., 450
Okazaki, R., 69, 70(120), 71(120, 122), 72(133)
Okubo, S., 210
Okuda, H., 70, 71(125), 73(125)
Olah, G. A., 248

Olavarría, J. M., 32, 310, 316(3), 317(3), 319(3), 320(3), 332(3), 334(3), 335(3)
Oliver, I. T., 306
O'Neill, S. R., 467(264), 471
Ong, B. L., 229, 307
Orengo, A., 56(68), 57(59), 58(59, 68), 59(59, 68), 61(59, 68), 62(68)
Orlacchio, A., 419, 420(155), 422(155), 435(155), 449(155), 450(155)
Orlowski, S. B., 437
O'Rourke, C. M., 168
Orr, G. R., 385
Orr, M. J. V. B., 95, 96(219)
Orrell, S. A., Jr., 312, 319
Orringer, B. P., 79
Osaki, S., 327
Osbond, J. M., 383
Osborn, M. J., 189, 200, 201(29), 202, 210(2), 211, 215(2), 216(2)
Oski, F. A., 46
O'Toole, A. G., 326, 350, 353, 354(218)
Otsuka, N., 438
Ottolenghi, C., 352
Ovchinnikov, Yu. A., 398, 399, 415(99a), 417, 418(99a, 166), 421(96, 99a), 422(166), 427(166)
Øye, I., 326, 341
Ozawa, H., 377
Ozbun, J. L., 321

P

Padyukova, N. Sh., 430(197, 198), 431(197), 433, 434(197)
Paetkau, V., 382
Page, M. I., 489
Paik, W. K., 179, 181(48), 185(48)
Pailes, W. H., 142, 143(65)
Palmiter, R. D., 369
Paranjpe, S. V., 38, 39(146)
Pardee, A. B., 227, 228, 229, 230, 242(9), 268(10), 269, 270, 271, 278(10), 284, 295(9)
Parisi, E., 195
Park, C. R., 348, 350, 353
Parks, C. C., 83
Parks, L. W., 160
Parks, R. E., Jr., 83, 84(188, 193), 85(188, 190, 191, 193), 86(188, 190, 193), 90(188)

AUTHOR INDEX

Parodi, A. J., 319, 322(84)
Parry, M. J., 37, 42, 43(155)
Parsons, J. L., 438
Pass, L., 498
Passoneau, J. V., 45
Pasternak, C. A., 56(63), 57, 59(63)
Pastore, E. J., 211
Paterson, A. R. P., 54
Patterson, M. K., Jr., 385
Pattison, J. R., 162, 163(102)
Pauk, G., 348, 352(208)
Paul, B., 383
Paul, J., 92
Payne, M. R., 55
Peabody, R. A., 203
Pearce, F. K., 214
Peel, J. L., 130
Pegg, A. E., 175, 176
Pellegrino, C., 355
Penman, S., 187, 189(93)
Penzikova, G. A., 509
Peraino, C., 469(292), 474
Pérault, A. M., 390
Perbal, B., 294, 296
Peschke, G. J., 51, 52(13), 53(13)
Peterkofsky, A., 123, 130(14), 132(14), 143 (14), 154(14), 163(14), 184, 186(65)
Peters, J. M., 201(27), 202
Petersen, W. E., 364
Peterson, D. L., 382, 392(38), 396(38), 399 (38), 405(38), 406(38), 408(38), 412 (38), 419, 428(157)
Peterson, E. A., 399, 432(101)
Phares, E. F., 485
Phear, E. A., 211
Phillips, A. T., 394, 416(86a)
Phillips, D. C., 366, 367(19)
Phillips, J. H., 170
Pierard, A., 295, 302(120, 121), 305
Pierre, K. J., 55
Pigiet, V. P., 231, 232, 234(28), 235(27, 28), 236(27, 28), 237(27), 238(28), 239(27), 240(27), 242(27, 33), 250, 253, 255(72), 266(33), 267(72), 274 (27), 275(27), 276, 277(28), 278(27), 279(33), 284, 296(28)
Pih, K. D., 193
Pikhelgas, V. Ya., 402, 403, 408(125), 409 (125), 410(125), 411(125), 412(125), 413(125), 422(125), 437(125)

Pilkis, S. J., 33, 37, 40(116, 140), 43(140)
Pillinger, D. J., 179, 181, 184, 185(48, 71)
Pinjani, M., 311
Piras, M. M., 311, 326, 327(117), 338, 353 (15, 117)
Piras, R., 311, 313, 314(50), 317, 326, 327 (117), 332, 333(154), 334(68, 154), 335, 336(173), 338, 339, 353(15, 117)
Pisano, J. J., 500
Pitot, H. C., 380, 381(5), 392(5), 438(5), 462(5), 480(5)
Pizer, L. I., 209
Plá, L. C., 506
Plagemann, P. G. W., 56
Plaut, G., 81
Plaut, G. W. E., 200
Plesner, L., 311, 313(17), 355(17)
Pliner, S. A., 506(66), 507, 509
Plotnick, A., 194
Plunkett, W., 56, 57(57), 58(57), 59(57), 60(57), 61(57)
Poels, C. L. M., 56
Polatnick, J., 183
Polyanovsky, O. L., 382, 392(30), 395(92), 396, 397, 398(92, 96), 399, 400(95), 401(109), 402(92, 109, 110, 113), 403 (113), 408(109, 124), 412(124), 413 (124), 415(99a), 417(99a), 418(99a, 161, 164, 166, 167), 419(152), 420 (124), 421(30, 96, 99a, 150, 161, 164), 422(166), 427(166)
Porter, J., 363
Porter, R. W., 229, 232, 236(37), 245, 246, 247, 249, 252(15), 256(63), 257(63), 258(15), 259(15), 260(15), 261(15), 262(15), 265(15), 267(15), 268, 270 (15), 281(37), 289(15)
Potter, V. R., 35, 71, 195
Powell, C. A., 340
Prager, M. D., 229
Pratt, J. M., 125, 153(21)
Pratt, R. F., 490
Preiss, J., 317, 321
Prescott, D. M., 73
Prescott, L. M., 108, 229, 297, 298(128), 299(128)
Pricer, W. E., Jr., 51, 52(7), 201, 202
Pringle, J. R., 9
Provine, H. T., 356

Prusoff, W. H., 66, 69, 70(121), 72(121), 74
Puchkov, V. A., 506(66), 507
Puchwein, G., 51, 52(13), 53(13)
Pullman, B., 390
Purich, D. L., 13(48), 14, 17(48), 23(48), 42, 44(151, 152)

Q

Quayle, J. R., 485
Qureshi, M. Y., 320(95), 321

R

Rabajille, E., 32
Rabinowitz, J. C., 189, 200(3), 201(3), 202(3, 21), 203(30), 206(3, 30), 208, 210(3), 215(3), 216(3), 217(141)
Rabinowitz, M., 318
Racker, E., 45
Raczynska-Boyanowska, K., 498
Radford, A., 303
Raijman, L., 102, 103(38), 106(38), 107(38), 112(38), 114(38), 249
Raina, A., 438
Raina, N. C. R., 438
Rakiala, E. L., 438
Rall, T. W., 323, 337, 338(167)
Ramamurthi, R., 359
Ramasastri, B. V., 211
Ramel, A. H., 4, 5(20), 7, 9, 10(32), 26
Rangaraj, N. I., 359
Rapoport, I. A., 509
Rasmussen, A. H., 54
Ratner, S., 385, 497, 500, 501(37), 503(1)
Rau, E. M., 393
Rau, J., 55
Rauda, V., 350
Ravel, J. M., 101, 104(21), 107(21), 111(21), 112(21)
Rebhun, L. I., 315, 316
Rechler, M. M., 471(301), 475
Redfield, B. G., 123, 126, 130(14), 132(14), 135(51), 143(14), 154(14), 159, 163(14), 164(107), 208, 209
Redkar, V. D., 38, 39(144), 41, 43(144), 45(144)
Reeves, R. E., 48
Regehr, E. A., 363
Reich, E., 54
Reichard, P., 56, 57, 59(70), 245

Reid, E., 81
Reimann, E. M., 311, 312(8), 313(8), 314(8), 324, 326(8), 327(8), 332(8)
Reinbothe, H., 498
Reissig, J. L., 101, 102(25), 105(25), 109(25), 112(25), 117(25), 118(25), 302, 305
Reithel, F. J., 365, 377
Rejal, T. H., 170, 177(15), 185(15)
Remer, A. K., 82(180), 83
Remy, C. N., 187, 188(90)
Reuter, G., 498
Révész, L., 59
Reyes, P., 212
Reynard, A. M., 23
Reynolds, M. S., 123
Ricard, J., 14, 15(49, 50), 20, 21(68), 48
Richards, K. E., 232, 239(39), 241, 291
Richardson, K. E., 467(279, 281), 473
Richardson, S. H., 34
Richert, D. A., 221
Riddick, D. H., 184, 185(75)
Riddle, B., 210
Riddle, M., 168, 169(4)
Riley, D., 465(263), 471
Riley, W. D., 314
Riman, J., 65
Ring, R. N., 490
Riordan, J. F., 418(163), 420
Riva, F., 382, 392(84), 393, 397(84), 398(35), 399, 405(84), 407(84), 411(35), 416(84), 417(144), 418(168, 169), 419(144), 420(154), 421(35, 144), 435(154), 449(154)
Robbins, P. W., 311, 317
Roberts, E., 475
Roberts, N. R., 34
Robin, Y., 498
Robison, G. A., 323, 341, 350, 353
Roche, J., 472
Rochovansky, O., 500, 501(37)
Rodeh, R., 170, 174, 176
Roe, J. H., 348
Ronca, G., 500, 501(36), 502, 503, 508(36, 46)
Ropp, M., 218
Roscoe, D. H., 210
Rose, I. A., 13, 14(42), 16, 17, 24, 29, 30, 31, 35(128), 36(128), 38, 39, 42, 43(42, 147), 44(42), 45(128), 46(43)
Rosell-Perez, M., 311, 312(18), 313(17),

317(42), 322(39), 326(18), 327(123), 328(39, 41), 331(18, 119), 332(39, 41), 333(39, 40, 41, 156), 334(40), 335(39, 40, 41, 154, 158), 340(144), 345(158), 355(17, 119, 120), 357(122, 123), 358 (122)
Roseman, S., 48, 377, 476
Rosemeyer, M. A., 9
Rosenberg, H., 498
Rosenberg, L. E., 163
Rosenbusch, J. P., 230, 231, 232, 234(24), 235, 236, 237, 238(26), 240, 250, 253 (73), 255(73), 267(73), 269(38), 274 (38), 275(99), 276, 277(73), 278(73), 296(26)
Rosenthal, S., 122, 123, 130(3, 8, 9), 131 (8), 132(9), 134(9), 135(9, 36), 144 (9, 34), 152(9), 160(3, 36), 162(3), 163(3, 36)
Rosness, P. A., 361
Ross, P. D., 152, 154(74)
Rossa, F., 419, 420(155), 422(155), 435 (155), 449(155), 450(155)
Rossi, N., 501, 502(40), 508(43)
Rossi Fanelli, A., 381, 386(17), 387(17), 407(17), 423(17), 435(17), 438(17), 440(17)
Rossini, L., 335
Rossiter, R. J., 498
Roth, J. R., 293
Roth, J. S., 69, 71(117), 72(117), 73(117), 74(117), 83, 84(189), 85(189), 86(189)
Roth, Y. S., 380, 381(5), 392(5), 438(5), 462(5), 480(5)
Rothman, J. B., 317, 334(68), 335(68), 336(68)
Rothman, L. B., 317, 332, 333(154), 334 (68, 154), 335(68, 154), 336(68), 360, 361
Rothman-Denes, L. B., 312, 326(23), 327 (125), 359(23, 125), 360(23, 125), 361 (23, 125)
Rowbury, R. J., 158
Rowland, L. P., 312
Rowley, G. L., 496
Roy, K. L., 176
Rubin, M. M., 287, 288, 289
Rubinstein, D., 344
Rubulis, A., 356
Rudman, D., 393
Rudney, H., 485

Rudolph, F. B., 13(47), 14(40), 17(40, 47), 21, 22(40)
Rüdiger, H., 130, 131, 142, 150, 151(39, 70, 71), 154(62)
Ruffner, B. W., Jr., 87, 90(200), 91(200)
Ruiz-Amil, M., 47
Russell, R. L., 184
Rustum, Y. M., 4, 5(20), 9
Ryan, A. M., 194
Ryman, B. E., 311, 331

S

Saari, J. C., 314, 315(56)
Sabater, B., 48
Sacristán, A., 331, 340(144)
Sagers, R. D., 221, 222
Saheki, R., 311, 353(16), 356
Saier, M. H., 465(259), 470
Sakami, W., 131, 160, 161(94, 96), 162 (94, 96), 163(41), 216
Sakai, T. T., 215
Salas, M., 31, 32(104), 37, 42(138), 43(138)
Salas, T., 37, 42(138), 43(138)
Salim, M., 189
Sallach, H. J., 463, 465(254)
Salmon, B., 34
Salomon, R., 177, 185(41)
Salsas-Leroy, E., 311, 313(17), 355(17)
Sampson, S. D., 54
Sanada, Y., 312, 313(33), 315(33), 342 (33), 344(33)
Sanadi, D. R., 76, 82(154)
Sanchez, R., 32
Sandruss, R., 320
Sanger, F., 314, 315(54)
Santi, D. V., 214, 215
Sapag-Hagar, M., 32
Sapico, V., 48
Sashchenko, L. P., 384, 440(46), 441(46, 224, 227), 443(46), 444(46, 232), 447 (46), 453(46), 454(46), 455(46), 459 (46)
Sasko, H., 315, 337, 338(57), 341(57)
Sato, K., 162, 163(104), 311, 326, 332 (121), 333(121), 335(121), 342(188), 343, 344(188), 346(188), 356(16, 121), 357(121)
Sato, T., 222
Sauberlich, H. E., 123
Sauerbier, W., 185, 186
Saunders, P. P., 386

Sauret-Ignazi, G., 472
Sawyer, B., 77
Scamahorn, J. O., 317, 358(69)
Scandurra, R., 382, 392(27), 435(27), 465 (27)
Scarano, E., 195, 399
Scarborough, G. A., 184, 185(74)
Scardi, V., 394, 399, 415(148), 416, 426
Schachman, H. K., 7, 9, 10(32), 155, 230, 231, 232, 234(28), 235(27, 28), 236 (27, 28, 31), 237(27), 238(28), 239(23, 27), 240(27), 242(27, 33), 243, 246 (31), 250(47), 251(31, 47), 253, 255 (47, 71), 258(71), 263(47), 266(33), 269(31), 273(31), 274(27, 31), 275 (27), 276(47), 277(28), 278(27), 279 (33), 280(91), 281(31), 296(28)
Schachter, H., 377
Schaffer, M. H., 234, 245
Schanbacher, F. L., 370, 377(29)
Schatz, L., 465(258), 470
Scheer, B. T., 359
Schein, A., 190
Schepartz, A. I., 76, 82(154)
Schimke, R. T., 32, 33(113), 37, 38(117), 39(117), 40(113, 117), 42(117), 43 (117), 101, 104(23), 105(23), 109(23), 112(23), 116(23), 117(25)
Schirch, L., 215, 216, 217(140), 218(122, 135, 136, 140), 219, 479
Schlender, K. K., 311, 317, 326(7), 327(7), 332, 334(160), 335(160)
Schmid, G., 471(306), 475, 476(306)
Schmid, R., 317
Schmidt, J. J., 4, 7(22), 8(38), 9(22), 10 (22, 29), 11, 12, 13, 27(22)
Schmidt, P. G., 234, 246, 247, 251(46, 61), 252, 257(61), 259(61), 262(46), 267 (46), 270, 271(93), 272, 273, 288(61)
Schnebli, H. P., 51, 52(9), 53(9), 54(9)
Schneider, M. C., 484, 486(9), 487(9), 494 (9), 495(9)
Schoenheimer, R., 497, 499(2), 504(2)
Scholar, E. M., 83, 84(193), 85(193), 86 (193)
Scholefield, P. G., 57, 58(76), 59(76), 60 (76), 61(76)
Scholtissek, C., 55, 58(36, 37)
Schram, E., 229
Schrauzer, G. N., 153, 154
Schray, K. J., 51

Schrecker, A. W., 63(102), 64(100), 68
Schulman, J. D., 163
Schulz, D. W., 34
Schulze, I. T., 3, 4(15), 6(16), 7(16), 8(16), 9(15), 10(15, 16), 11(15), 12(15), 13 (15), 20(15), 26(15, 16)
Schulze-Wethmar, F. H., 324
Schwartz, G. P., 38, 41(143)
Schweiger, M., 185, 186(82)
Scoffone, E., 267
Scott, E. M., 467(287), 474
Scott, R. B., 313
Scotto, P., 399
Scrimgeour, K. G., 132, 134(52), 163(52), 198, 211(6), 212(6)
Sebastian, J., 35, 36(130), 48
Seegmiller, J. E., 55
Segal, H. L., 312, 313(33), 315(33), 319, 322, 323(100), 324(100), 325(83), 328 (83), 330(28), 342(33, 100), 343(28), 344(33), 346, 352(28), 465(257, 258), 470
Seifert, J., 65
Selhub, J., 160, 161(94), 162(94)
Senez, J. C., 477
Seno, T., 172
Sevall, J. S., 313, 357(48), 358(256)
Severin, E. S., 381, 384, 386(18), 387(18), 407(18), 418(165, 166), 420, 421, 422 (165, 166), 424, 427(166), 435(18, 173, 174, 237, 239), 440(18, 46), 441 (18, 46, 223, 224, 225, 226, 227), 442, 443(18, 223, 225), 444(223, 232, 235), 445(18, 225), 446(225, 235), 447(46, 173, 174, 226, 236), 448, 449, 451, 453 (46), 454(46), 455, 457(235), 459(225), 474
Shafrir, E., 354
Shapira, R., 502, 503(44)
Shapiro, B. M., 324
Sharefkin, J. G., 490
Sharma, C., 31, 32
Sharma, O. K., 177, 179(40), 180(40), 181, 184(40), 185(40, 51, 73)
Shaw, N., 144
Shedden, W. I. H., 73
Sheid, B., 177, 178(49), 179, 192, 380, 381 (5), 392(5), 438(5), 462(5), 480(5)
Shemyakin, M. M., 386(58), 387, 451(58), 477(58)
Shen, L. C., 326, 338, 341(169)

AUTHOR INDEX 545

Shepherdson, M., 228, 242(9), 295(9)
Shershneva, L. P., 172
Shevchenko, L. A., 509
Shigeura, H. T., 54
Shill, J. P., 30
Shimada, K., 65, 70, 71(127a), 89(106), 92, 93(207)
Shimazono, N., 111
Shimazu, T., 342(191), 343, 344(191), 352 (191)
Shimizu, S., 152, 162, 163(104), 430(199), 431(199), 433
Shimono, H., 62, 63(92), 64(92), 65(92), 83, 84(192), 85(192), 86(192), 87 (192, 198), 88(198), 89(106, 198), 92 (198)
Shirai, A., 436, 465(208c)
Shive, W., 101, 104(21), 107(21), 111(21), 112(21)
Shliapnikov, S. V., 418(162a), 420
Shoaf, W. T., 106, 307
Shoup, G. D., 73
Shpikiter, V. O., 400, 401
Shugart, L., 184
Shukuya, R., 472
Sie, H. G., 312, 320
Siegmund, P., 438
Siekevitz, P., 35
Siersma, P. W., 185
Siess, E., 324
Sillero, A., 48
Silver, M. S., 484
Silverman, M., 199, 207, 208(49)
Silverstein, E., 14(41), 16(41), 17(41)
Simms, E. S., 84, 88, 92(197)
Simon, L. N., 170, 177(15), 179, 185(15)
Simon, M., 210
Simonsen, D. G., 475
Sinclair-Smith, B. C., 350
Singer, T. P., 435
Sinitsina, N. I., 406, 408(130), 421(130), 422(130)
Sipe, J. E., 187, 188(90)
Sizer, I. W., 382, 392(28), 398(28), 399 (28), 402(28), 407(28), 408(28), 412 (28), 424(135), 425(177, 178), 426 (28, 135, 177), 428(177), 436(28, 177, 178), 437, 438(28, 135), 439(28, 135), 457(135, 177)
Sköld, O., 56, 57, 59(69, 70), 60(69)

Skoog, F., 176
Slade, H. D., 99
Slater, T. F., 77
Slein, M. W., 2, 5(7), 6(7)
Slessor, K. N., 320(95), 321
Sloan, R. S., 363
Sloninski, P., 303
Sly, W. S., 487
Smellie, R. M. S., 70, 92(123), 93(123)
Smirnova, T. B., 76, 77(158), 78(158), 80 (158), 81(158)
Smith, B. C., 208, 209(60, 62)
Smith, C., 315, 316(59)
Smith, C. H., 313, 314
Smith, E. L., 144
Smith, I., 187
Smith, J. D., 190, 192
Smith, K. E., 229, 269(21)
Smith, L. C., 123, 130(9), 132(9), 134(9), 135(9, 36), 144(9), 152(9), 160(36), 163(36)
Smith, M., 212
Smith, M. A., 181
Smith, M. S., 356
Smith, P. F., 116, 117
Smith, R. A., 48, 177, 178
Smith, R. C., 176
Sneider, T. W., 195
Snell, E. E., 98, 219, 220(149), 381, 382 (17), 383(13), 384(13, 42), 385(13), 386(13, 14, 18, 59, 60), 387(13, 17, 18), 388(13), 389(13, 14), 390(14), 399, 407 (17, 18), 419, 423(17), 429(42), 430 (42, 104), 431(42, 104), 432(42, 104), 433(42, 104), 435(17, 18), 438(17), 440(17, 18), 441(18), 443(18), 445, 449 (18), 451(59, 60), 452(241), 471(307, 308), 476(104), 478
Sober, H. A., 399
Soda, K., 387, 469(295), 474(57), 477, 478
Soderling, T. R., 311, 312(8), 313(8), 314 (8), 326(8), 327(8), 332(8)
Soderman, D. D., 32, 33(112), 34
Söll, D., 176, 208
Solomon, F., 24, 487, 489(27), 493, 494 (27), 495(27), 496(27)
Solomon, L. R., 62
Sols, A., 18, 21(58), 22(59), 27(59), 31, 32 (104), 34(100), 37, 42(138), 43(138), 44(100), 45(99, 100), 47, 323, 324
Somerville, R., 186

Sonoda, S., 62(95), 63(92), 64(92, 95), 65 (92), 89(106)
Sonneborn, D., 361
Šorm, F., 56(65, 66), 57, 59(65, 66, 74), 60 (74), 61(74), 499
Southard, J. H., 34
Søvik, O., 326, 332, 333(155), 335(155), 337
Spears, C., 159, 163
Spector, L. B., 23, 24, 98, 99(2), 100(2), 101(2), 109(2, 17), 115(2)
Spencer, M. S., 467(286), 474
Spiegel, R. S., 359
Spielvogel, R. L., 193
Spiess, G. I., 51, 52(13), 53(13)
Spivak, V. A., 417
Srinivasan, P. R., 168, 175, 176, 182, 184, 185(53, 54, 70), 186(53), 191, 192
Stadtman, E. R., 324, 484, 485(4), 487(4)
Stadtman, T. C., 216
Staehelin, M., 171, 172(16), 173, 174(20)
Stafford, M. A., 73
Stalmans, W., 311, 312, 313(29), 317, 323, 328(9, 67), 329(9, 67), 330(9, 67), 331, 342(102), 343(67, 102), 344(102), 345 (102), 346(102), 348(139), 349(139)
Stalon, V., 294
Stambolieva, N., 430(193, 196), 431(193, 196), 433, 434(193), 444(193)
Staneloni, R. J., 313, 314(50), 336(173), 339
Stanier, R. Y., 485
Stankewicz, M. J., 423
Stark, G. R., 229, 231, 233(30), 234, 235, 243, 245(15), 246(15, 36), 247(15, 46), 248, 249(15), 251, 252(15, 16, 46), 253, 254, 255(36), 256(15, 16, 63), 257 (61, 62, 63), 258(15, 62), 259(15, 61, 62), 260(15), 261(15, 79), 262(15, 16, 46, 62), 263(35), 264(84), 265(15, 35, 84), 267(15, 46, 62, 86), 268(15), 219(36), 270(15), 273(84), 278(35), 280(62), 281(36, 84), 282(84), 284 (84), 288(61), 289(15)
Staub, M., 186, 512
Stavrianopoulos, J., 127, 130(26), 131(26), 132, 141
Steel, B. F., 123
Steers, E., 208, 209
Steers, E. J., Jr., 154, 155(79), 156(79), 157(79), 158(79)
Stein, H., 56, 61(62)
Steiner, D. F., 313, 315(46), 330(46), 331 (46), 334(46), 344(46, 141), 345(46), 346(46), 350, 355
Steitz, T. A., 27(90), 28
Stellwagen, E., 9, 10(32)
Stephens, G. C., 499
Stern, H., 73
Stern, J. R., 484, 485, 486(9), 487(9, 13), 488(31), 491(46, 47), 492(13, 30), 494 (9), 495(9)
Stetten, D., Jr., 499
Stewart, B. W., 163, 165(114)
Stewart, C. J., 496
Stewart, R., 248
Stinson, R. A., 467(286), 474
Stjernholm, R. L., 485, 486(14), 487(14), 494(14)
Stoddard, M., 484
Stodolsky, M., 210
Stokstad, E. L. R., 161, 163(99), 200
Stolzenbach, F. E., 80
Stossel, T. P., 355
Stout, M. G., 177
Strauss, B., 210
Strecker, H. J., 469(293), 474
Streeter, D. G., 177, 179
Stromberg, K. J., 169
Strominger, J. L., 83, 86(185), 87(185), 476
Stulberg, M. P., 184
Su, C. H., 160
Subak-Sharpe, J. H., 55
SubbaRow, Y., 100, 109, 514
Suda, M., 380, 381(5), 392(5), 438(5), 462 (5), 480(5)
Südi, J., 469(294, 297), 474
Sugino, Y., 62(95), 63(92, 104), 64(92, 95), 65(92, 95), 83, 84(184, 192), 85 (192), 86(184, 192), 87(184, 192), 88 (198), 89(106, 198), 92(198), 93(207), 110
Sukanya, N. K., 472
Sukhareva, B. S., 453, 479(245)
Sukhikh, A. P., 398, 421(96)
Summers, D. F., 189
Sund, R. F., 104
Sussman, K. C., 499
Sutherland, E. W., 323, 341, 350, 353

Suzuki, S., 80
Suzuki, K., 80
Svensson, I., 170
Sweeney, E. W., 101, 104(23), 105(23), 109(23), 112(23), 116(23), 117(23)
Swiatak, K. R., 179
Sykes, B. D., 246, 251, 257(61), 259, 264, 265(85), 288(61)
Sytinsky, I. A., 469(288), 474
Syvanen, M., 229, 293(18), 294(18), 296(18)
Szabo, G., 509

T

Tabor, C. W., 381, 382(23), 383(23), 386(23), 462(23), 463(23)
Tabor, H., 201, 202(21, 22), 206(22), 207(22), 381, 382(23), 383(23), 386(23), 462(23), 463(23)
Tackett, S. L., 132, 153(54)
Tahmisian, T. N., 469(292), 474
Takasugi, M., 163
Takeda, Y., 326, 354(115)
Takeishi, K., 184
Takemoto, T., 438
Takeyama, S., 125, 152(22)
Takoshita, M., 467(286), 474
Talbert, P. T., 211
Tallalay, P., 178
Tanahashi, N., 365, 366(13), 368(25), 377(13)
Tanaka, R., 34, 111
Tancredi, J. F., 436, 437(210), 457(210)
Tarnowski, W., 352
Tarr, H. L. A., 53, 55(17)
Tatibana, M., 306, 307, 467(286), 474
Tavitian, A., 54
Taylor, A. T., 73
Taylor, D., 499
Taylor, G. A., 485
Taylor, R. T., 122(4, 6), 123, 124(16), 125 (15, 17), 126(15, 23a), 127, 128(15, 27), 129(27), 130(15, 25), 131(15, 32), 132(15), 133(15, 31, 32), 134(15, 16, 27, 55), 135(15, 16, 17), 136(48), 137(48), 138, 139(31, 57), 140(27, 58), 141(17, 57, 61), 142(57, 58), 143(16, 17, 58), 144(27), 145(49, 61), 146, 147, 148(17, 58, 61), 149(31, 32, 49), 150, 151(17, 31, 48, 49, 58), 152(15, 19), 153(15, 27, 31, 32, 49), 154(58), 164(15, 16), 165(49), 198(9), 199, 402, 424(115, 180), 425(115), 436, 438 (180), 465(208), 471(208)
Tchen, T. T., 471(305), 475
Teipel, J., 29
Tejerina, G., 122, 160(2)
Telegdi, M., 382, 392(30), 421(30)
Teller, D., 328
Tellez-Iñon, M. T., 326, 361(124)
Teraoka, H., 86, 87(198), 88(198), 89(198), 92(198)
Terenzi, H., 326, 361(124)
Thanassi, J. W., 387(67), 388
Theisen, M. C., 48
Thiebe, R., 184
Thoai, N. V., 498
Thomas, J. A., 317, 332, 334(160), 335(160), 340
Thomassen, E., 56, 57(53), 58(53), 59(53), 61(53), 70(53)
Thompson, J. S. T., 467(279, 281), 473
Thompson, S. A. M., 490
Thompson, U. B., 70, 71(124), 72(124)
Thorne, C. B., 472
Thorne, K. J. I., 102, 106(39), 107(39), 111(39), 112(39)
Thunberg, S. A., 355
Tiedemann, H., 45
Tiemeier, D. C., 382, 391(37), 392(37, 38), 396(38), 399(38), 400(37), 405(37), 406(38), 407(37), 412(37, 38), 415(37), 416
Ting, R. C. Y., 175, 183(25), 185(25)
Tolosa, E. A., 442
Tomisek, A. J., 55
Tomkins, G., 465(263), 471
Tomkins, G. M., 465(263), 471
Toniolo, C., 267
Tønnesen, T., 54
Topping, R. M., 387(65), 388
Torchinsky, Yu. M., 381, 386(18), 387(18), 399, 400, 401(107), 402(107, 113), 403(113), 405(107), 406(107, 126), 407(18, 119, 120), 408(119, 125, 130), 409(125), 410(125, 141), 411(125, 141, 142), 412(107, 125), 413(141), 416(141), 418(165, 166), 421(130), 422(125, 130, 166), 427(166), 435(18), 436(120, 126, 142), 437(120,

125), 438, 440(18), 441(18), 443(18), 445(18), 449(18), 457(120)
Torres, H. N., 320, 326, 327, 331, 332(131, 147), 335(147), 361(124)
Touster, O., 55
Tovarova, I. I., 506(66), 507, 509
Tramell, P. R., 101
Traut, R. R., 311, 313, 317(38), 328(38), 335(38)
Trayer, I. P., 366, 367(18), 368, 370(18), 371(23), 375(23), 376, 377(18)
Trayser, K. A., 3, 5(11, 13), 6(13), 18(13), 23(54), 24(54), 26(24)
Tremblay, G. C., 307
Trivelloni, J. C., 317, 358(72)
Trojaborg, W., 312
Trotta, P. P., 101
Truman, J. T., 54
Tsai, H., 469(291), 474
Tsuiki, S., 311, 326, 332(121), 333(121), 335(121), 342(188), 343, 344(188), 346(188), 356(16, 121), 357(121)
Tsukada, K., 386
Tsukamoto, K., 62, 63(92), 64(92), 65(92), 89(106)
Tucker, R. G., 210
Tukachinsky, S. E., 76, 77(156), 78(156)
Tumanyan, V. G., 399
Tumerman, L. A., 398, 400(95), 402(95)
Turano, C., 381, 382, 386(22), 392(84), 393, 397(84), 398(35), 405(84), 407 (22, 84), 408, 411(35, 134), 416(84), 417(22, 144), 418(168, 169), 419(144), 420(154, 155, 159), 421(22, 35, 144, 159), 422(155), 435(154, 155), 441 (22), 449, 450, 457(134)
Turchenko, E. I., 76, 77(156), 78(156)
Turkington, R. W., 168, 169(4), 175, 179, 185(22), 193
Turner, J. M., 469(290), 474
Tuttle, L. C., 99
Tye, M., 498

U

Udaka, S., 511, 512(1), 514(1), 515(1)
Udenfriend, S., 500
Ueno, Y., 472
Uhlendorf, B. W., 163
Ukita, T., 184
Ukstins, I., 131, 163(41)
Ureta, T., 32

Uretsky, S. C., 54
Urshel, M. J., 63, 64(100)
Utter, M. F., 54, 82(181), 83
Uyeda, K., 45, 202, 203(30), 206, 207(30), 216, 217(141)

V

Vaidyanathan, C. S., 472
Vail, M. H., 54
Vainer, H., 355
Valdemoro, C., 390
Valeriote, F., 465(263), 471
Valladares, Y., 56, 61(60)
Vallejo, C. G., 35, 36
Vanaman, T. C., 229, 232, 233(35), 234, 252(16), 256(16), 262(16), 263(35), 264, 265(35), 278(35), 366, 367(18, 19), 370(18), 371(28), 375(28), 377 (18)
Van Bibber, M. J., 83, 84(183), 86(183), 94(183)
van Demark, P. J., 117
Van den Berghe, G., 348, 351
Vanderwende, C., 355
Vande Woude, G. F., 183
Van Pilsum, J. F., 499, 504(23)
Vanyushin, B. F., 193
Vardanis, A., 312, 313, 317, 319, 320(80, 87), 331, 342(142), 344(80, 142), 345 (142), 358(66), 359(22)
Varshavsky, Ya. M., 400, 401(107), 402 (107, 113), 403(113), 405(107), 406 (107), 412(107)
Vasil, I. K., 73
Vasiliev, V. Yu., 430(200), 434, 441, 444 (232), 469(288, 289), 474
Vaughan, M., 355
Vavra, J., 412, 424, 425, 426(139, 140), 427(139, 140), 428(139, 140), 438 (139, 140), 439(139, 140)
Vdovina, L. V., 424, 435(173), 440, 441 (223), 443(223), 444(223), 447(173)
Vecchini, P., 416, 417(144), 419(144), 421 (144)
Velick, S. F., 412, 424, 425, 426(139, 140), 427(139), 428(139, 140), 438(139, 140), 439(139, 140)
Venetianer, P., 186
Venkstern, T. V., 172
Verhue, W., 319
Verjee, Z. H. M., 393

Vernon, C. A., 382, 392(31), 395, 398(31, 89), 399, 400(89), 402(31, 93), 424 (89, 93), 426(175), 427(175), 428(176)
Vesely, J., 56(65, 66), 57, 58, 59(65, 66, 73, 74), 60(74), 61(74)
Vigi, V., 500, 501(36), 502(40), 503, 508 (36, 46)
Vignos, P. J., 499
Vilchez, C., 338
Villar-Palasi, C., 310, 311, 313, 315(47), 317(42), 322(5, 39), 323, 324, 326(7, 47), 327(7, 130), 328(39), 330(130, 136), 331, 332(39), 333(39), 334, 335 (39), 336, 337(130), 338(57), 340(47, 130), 341(57, 169), 349, 350(213)
Villar-Palasi, V., 334, 357
Vinogradova, E. I., 398, 399, 421(96)
Vinuela, E., 31, 32(104), 37, 42(138)
Vitalis, S., 509
Vogel, H. J., 469(296), 474, 512, 513(6), 514(6), 515(4), 515(6), 516(6)
Volcani, B. E., 98
Volfin, P., 337
von Euler, H., 76
von Fellenberg, R., 246, 255, 259(58), 276(58)
von Funcke, H. G., 324
von Schuching, S., 425
Vorotnitskaya, N. E., 400, 416, 417(146), 419(150), 421(150)
Voytek, P., 69, 70(121), 72(121)
Vuttivej, K., 453

W

Wachsman, J. T., 190
Wada, H., 216, 217(138), 218(138), 382, 383(33), 386(20), 391(32, 33, 36), 392 (32, 33, 36), 393, 398(33), 399, 400 (32, 33), 415(36), 416(36), 417(36), 420(20), 425(36), 426(33), 427(33), 430(104), 431(104), 432(104), 433 (104), 463, 465(263c), 471, 476(104)
Waelsch, H., 206, 208
Wagner, E., 189
Wagner, F., 125, 127, 128(23), 143(23), 153(23)
Wagner, R. R., 377
Wahba, A. J., 212
Wainfan, E., 176, 182, 185(53, 54), 186(53)
Wakramasinghe, R., 472
Waksman, S. A., 505
Walaas, E., 327, 338
Walaas, O., 327, 338
Walker, D. G., 31, 32, 37, 42, 43(155)
Walker, J. B., 476, 497, 498(5), 499(14, 18), 500(5, 13, 18, 31), 501(13, 24, 33, 38), 502(13, 33, 39), 503(14, 17, 20, 30), 504(4, 14, 18, 24, 30, 48, 49), 505, 506(6), 507(5, 6, 18, 56, 57, 60, 61), 508(5, 6, 55), 509(6, 49, 56, 57)
Walker, J. M., 399
Walker, M. S., 476, 499, 501, 502(39), 503 (20), 506, 507(56, 57, 60, 61), 509 (56, 57)
Waller, J., 208
Walsh, C. T., Jr., 23, 24
Walsh, D. A., 311, 312(8), 313(8), 314(8), 324, 326(8), 327(8), 332(8)
Walsh, D. E., 463, 465(254)
Walter, P., 382
Wang, F. K., 161, 163(99)
Wang, P., 311, 313(17), 355(17)
Wang, S. H., 499, 503(30), 504(30)
Wang, T. P., 76, 77(155), 78(155), 80 (155), 81(155)
Wanka, F., 56, 73
Wanson, J. C., 312
Ward, C., 359
Ward, G. A., 56
Ward, J. B., 119
Wardale, D. A., 472
Warms, J. V. B., 34, 35(128), 36(128), 39, 45(128), 46
Warner, R. D., 142, 152(63)
Warren, L., 203, 204(38), 205
Warren, S. G., 239, 242(50), 263(50), 264 (50), 283(50), 284(50), 291, 292
Watanabe, T., 382, 391(36), 392(36), 415 (36), 416(36), 417(36), 425(36)
Waters, J., 349
Watkins, W. M., 364
Watson, D. H., 73
Wattiaux, R., 355
Waygood, E. R., 473
Webb, E. C., 2
Webb, T. E., 57, 60(72)
Weber, G., 380, 381(5), 392(5), 438(5), 462(5), 480(5)
Weber, K., 230, 231, 232(29), 233(29), 234 (24), 235, 236, 237, 238(26), 239(32),

240, 242, 267(29), 274, 275(99), 276, 296(26)
Weeks, G., 361
Wehrli, W., 173, 174(20)
Wei, S. H., 311, 326(7), 327(7)
Weil-Malherbe, H., 31, 44(101)
Weinhouse, S., 31, 32, 200, 348
Weisblum, B., 189
Weisman, R. A., 359
Weissbach, H., 122(4, 6), 123, 124, 125 (15), 126(15), 127, 128(15, 27, 28), 129(27), 130(14, 15), 131(15), 132 (14, 15), 133(15, 31), 134(15, 27, 55), 135(15, 51), 137, 138, 139(31, 57), 140 (27, 58), 141(57, 61), 142(57, 58), 143 (14, 58), 144(27), 145(49, 61), 146, 147, 148(58, 61), 149(31, 49), 151 (31, 49, 58), 152(15, 19), 153(15, 27, 31, 32, 49), 154(14, 58), 155(79), 156 (79), 157(79), 158(79), 159(86), 162 (82), 163(14), 164(15, 107), 165(49), 198(9), 199, 208, 209(61)
Weissman, S. M., 92
Weitzman, P. D. J., 231, 278(25)
Weker, K. K., 250, 253(73), 255(73), 267 (73), 277(73), 278(73)
Welch, M. K., 4, 5(19), 11(19)
Welcher, A. D., 490
Weldon, P. R., 344
Wellner, D., 220
Wenger, J. I., 315, 338(57), 341(57)
Westheimer, F. H., 261
Whelan, W. J., 311, 320(95), 321, 331
White, H. D., 486, 487(23), 488(23), 490 (23), 495(23)
White, J. S., 229, 269(21)
Whitehead, E., 30, 226
Whiteley, H. R., 200, 484, 485(7), 487(7)
Whitfield, B. L., 194
Whitfield, C. D., 129, 154, 155(79), 156 (79), 157(79), 158(79), 159(86), 162 (82)
Wiame, J. M., 118, 295, 302(120), 305
Wiebelhaus, V. D., 18
Wieland, O., 324
Wiley, C. E., 34
Wiley, D. C., 239, 242, 263, 264(50), 283 (50), 284, 291, 292(108b)
Wilkin, D. R., 359
Wilkinson, A. P., 216
Willhardt, I. H., 442

Williams, B. J., 338, 340(171), 341(171)
Williams, D. C., 163
Williams, H. E., 52, 356
Williams, L. G., 102, 302, 303(142)
Williams, R. C., 232, 239(39), 241, 291
Williams, R. H., 350
Williams, R. J. P., 125, 153(21)
Williams, S. S., 70
Williamson, J. R., 341
Wilson, D. G., 393
Wilson, E. M., 219, 220(149)
Wilson, I. B., 231, 262, 267, 269(83), 278 (75, 83), 279(83)
Wilson, J. B., 441, 444
Wilson, J. E., 34, 36, 37, 39(141), 41
Wilson, K. J., 418(163a), 420, 421(163a), 422(163a)
Wilson, K. Y., 418(170), 421(170), 422
Wilson, M., 221
Wilson, S. M., 177
Windgassen, R. J., 153, 154
Winkler, A., 57
Winlund, C. C., 236, 274(48)
Winocur, E., 190
Winton, B., 324
Wiss, O., 472
Wodinsky, I., 63
Wolfenden, R., 260
Wolin, E. A., 499, 504(23)
Womack, F. C., 3, 4(15), 5(19), 8(63), 9 (15), 10(15), 11(15, 19), 12(15), 13 (15), 19, 20(15, 63, 64), 21(63), 26 (15), 27(63), 29, 30(92), 31(92), 46 (91, 92)
Wong, R. S. L., 184, 185(74)
Wood, H. G., 485, 486(14), 487(14), 494 (14)
Wood, J. L., 234
Wood, K. H., 490
Wood, W. B., 192
Woods, D. D., 122, 123, 130, 132, 135(53), 144, 158, 160(1a, 2, 53), 162(1a, 91), 163(53)
Woodward, V. W., 303
Wooton, J. F., 424, 426(175), 427(175)
Wosilait, W. D., 323
Wright, B. E., 216, 361
Wright, L. D., 383, 429(41)
Wu, C.-W., 226, 232, 236(37), 269, 274 (37), 285, 287, 288(3)
Wulff, K., 324

Wulfson, N. S., 506(66), 507
Wursch, J., 383, 429(41)
Wyatt, G. R., 190, 209, 312, 317(21), 358(21), 359(21)
Wykes, J. R., 73
Wyman, J., 110, 275, 286(98), 287(98), 517, 519(16)
Wyngarden, L., 201, 202(22), 206(22), 207(22)
Wynston, L. K., 201(28), 202

Y

Yamada, K., 381, 386(20), 480(20)
Yamada, R., 152
Yamamoto, T., 477
Yamamoto, Y., 76, 77(162), 78(162), 79(162), 80(162), 81(162)
Yan, Y., 293
Yang, I. Y., 431(202), 434
Yashphe, J., 101, 109(33), 111
Yates, R. A., 227, 229, 295
Yeh, Y. C., 210
Yeremin, 474
Yero, I. L., 76, 77(160), 78(160), 79(160), 80(160), 81(160)
Yip, A. T., 326
Yohn, D. S., 184
Yokojima, A., 436, 465(208b)
Yon, R. J., 307
Yorifuji, T., 478
York, R., 83, 85(190), 86(190)

Yoshiba, K., 163
Yoshida, M., 184
Yoshida, T., 222
Young, D. V., 175, 176
Young, J. E., 229
Young, R. W., 490
Younger, L., 313, 315(46), 330(46), 331(46), 334(46), 344(46), 345(46), 346(46)
Yphantis, D. A., 155, 382, 392(28), 398(28), 399(28), 402(28), 407(28), 408(28), 412(28), 426(28), 436(28), 438(28), 439(28)
Yu, C-T., 208
Yunis, A. A., 357
Yura, T., 70

Z

Zachau, H. G., 184
Zagyansky, Yu. M., 398, 400(95), 402(95)
Zappacosta, S., 498
Zemek, J., 318, 322(78)
Zewe, V., 13(44, 45), 14, 17(44, 45), 21(45), 41, 42(148)
Zieve, F. J., 328
Zimmerman, E. F., 187, 189(94)
Zimmerman, S. B., 94, 95(214), 191
Zito, R., 415(148), 416, 426
Zufarova, R. A., 418(165, 166), 420, 422(165, 166), 427(166)

Subject Index

A

Absorption bands, optically active, aspartate transaminase, 410–411
Acetaldehyde, serine hydroxymethyltransferase and, 218
Acetate
  aspartate transaminase and, 428
  aspartate transcarbamylase and, 251, 252
  carbamate kinase and, 107
  substituted, coenzyme A transferase and, 490
Acetate kinase
  carbamate and, 102, 106
  carbamate kinase and, 107
  phosphorylation of, 115
Acetoacetate
  activation of, 484–485, 486
  coenzyme A transferase inhibition and, 491–492
S-Acetoacetyl-N-acetylcysteamine, coenzyme A transferase and, 495
Acetoacetyl coenzyme A
  coenzyme A transferase and
    assay, 487
    inhibition, 494
Acetoacetyl pantetheine, coenzyme A transferase and, 495
Acetyl coenzyme A
  coenzyme A transferase inhibition by, 496
  fatty acid activation and, 485
N-Acetylglucosamine, lactose synthetase and, 364, 365, 369, 370, 371, 372, 374–375, 376
N-Acetylglutamate-5-hydroxamate, formation of, 514, 515
N-Acetylglutamate-5-phosphotransferase
  allosteric inhibition
    kinetics, 516–518
    temperature effects, 519–520

catalytic reaction
  assays, 514–515
  kinetics, 515–516
  pH optima and activating ions, 514
  stoichiometry, 513–514
  substrate specificity, 514
  purification, 513
N-Acetylglutamate semialdehyde
  aminotransferase and, 468, 474
  formation of, 511
Acetyl groups, arginine biosynthesis and, 512
Acetylimidazole, coenzyme A transferase and, 494
N^6-Acetyllysine, aminotransferase and, 468
N-Acetylmannosamine, galactosyltransferase and, 370
N-Acetylmuramic acid, galactosyltransferase and, 370
N^2-Acetylornithine, aminotransferase and, 468, 474
N^5-Acetylornithine, aminotransferase and, 468
Acetyl phosphate
  aspartate transcarbamylase and, 243–244, 249, 251, 252, 265
  carbamate kinase and, 103, 107
Acetylpyridine adenine dinucleotide, nicotinamide adenine dinucleotide kinase and, 79, 80, 81
Acridine orange, transfer ribonucleic acid methyltransferases and, 177
Actinomycin D
  deoxyribonucleic acid methyltransferase and, 191
  transfer ribonucleic acid methyltransferase levels and, 179
Active site
  aspartate transaminase, 417
    chemical topography, 451–455
  coenzyme A transferase, 492, 493–496

functional groups, aspartate transcarbamylase, 262–268
N-Acyl glucosamines, yeast hexokinases and, 18, 24, 47
Adenine
  carbamate kinase and, 114
  ribosomal ribonucleic acid methyltransferase and, 188
  transfer ribonucleic acid methyltransferases and, 176
Adenosine
  analogs, adenosine kinase and, 53–54
  aspartate transcarbamylase and, 271
  transfer ribonucleic acid methyltransferases and, 176
Adenosine deaminase, adenosine kinase assay and, 52
Adenosine deoxynucleotides, deoxyadenosine kinase and, 67–68
Adenosine diphosphate
  N-acetylglutamate-5-phosphotransferase and, 514–515, 516
  adenosine kinase assay and, 52
  arginine degradation and, 98
  assay, carbamate kinase assay and, 109
  binding, carbamate kinase and, 103, 113, 114
  citrulline formation and, 99–100
  deoxythymidine monophosphate kinase and, 93–94
  glycogen synthetase and, 345
  mammalian hexokinase and, 33, 42–43, 44, 46
  nicotinamide adenine dinucleotide kinase and, 81
  uridine kinase assay and, 58
  uridine monophosphate kinase and, 91
  yeast hexokinase and, 15–17, 21, 22, 23–24, 29, 30–31
Adenosine diphosphate glucose, glycogen synthetase and, 318, 321
Adenosine kinase
  assay, 51–52
  distribution and purification, 51
  kinetic and molecular properties, 52–53
  substrate specificity, 53–54
Adenosine monophosphate
  aspartate transcarbamylase and, 271, 272
  deoxycytidine monophosphate kinase and, 88
  glycogen synthetase and, 353
  glycogen synthetase phosphatase and, 329
  guanosine monophosphate kinase and, 86
  mammalian hexokinase and, 43
  nicotinamide adenine dinucleotide kinase and, 81
  synthesis of, 385
  yeast hexokinase and, 17, 22
Adenosine-N^1-oxide, adenosine kinase and, 54
Adenosine triphosphatases
  hexokinase and
    kinetic studies, 18–19, 21
    sulfhydryl groups and, 25
Adenosine triphosphate
  N-acetylglutamate-5-phosphotransferase and, 514, 515, 516, 518
  N-acetylglutamate-5-semialdehyde formation and, 511
  adenosine kinase and, 53
  amino group transfer and, 385
  analogs, hexokinase and, 15, 17, 22
  arginine degradation and, 98
  aspartate transcarbamylase and, 228, 265, 269, 270–273, 275, 276–277, 285, 287, 288, 289, 291, 295, 298, 300
  binding, hexokinase and, 15–17, 20–22, 25, 29
  carbamyl phosphate formation and, 100, 101, 111–113
  guanosine monophosphate kinase and, 84, 85, 86
  citrulline formation and, 98–99
  deoxyadenosine kinase and, 67, 68
  deoxycytidine kinase and, 65, 66
  deoxythymidine kinase and, 71, 72
  deoxythymidine monophosphate kinase and, 93
  formation
    carbamate kinase assay and, 109
    carbamyl phosphate and, 103, 106, 111, 115
  5-formyltetrahydrofolate cyclodehydrase and, 201, 202
  10-formyltetrahydrofolate synthetase and, 199

glycinamide ribonucleotide transformylase and, 203
glycogen synthetase and, 314, 315, 361
  inactivation by, 325–326, 331
  inhibition by, 334–335, 345, 353, 354, 358, 360, 361
mammalian hexokinase and, 33, 35, 41, 42–43, 44
mitochondrial hexokinase and, 35
nicotinamide-adenine dinucleotide kinase and, 78, 79, 80, 81
phage deoxyribonucleotide kinases and, 94–95, 96
pseudouridine kinase and, 62
riboflavin kinase and, 75
uridine-cytidine kinase and, 57, 58, 59, 61
uridine or cytidine monophosphate kinase and, 91
S-Adenosylethionine, transfer ribonucleic acid methyltransferases and, 176, 177
S-Adenosylhomocysteine
  B_{12} methyltransferase and, 132, 144
  deoxyribonucleic acid methyltransferase and, 191
  glycine methyltransferase and, 178
  transfer ribonucleic acid methyltransferases and, 176, 177, 178
S-Adenosylmethionine
  B_{12} methyltransferase and, 123, 128–129, 130, 131, 132, 133, 134, 137, 138, 143–151, 152–153, 164–165
  deoxyribonucleic acid methyltransferases and, 190–191, 192, 194
  enzyme repression and, 160
  ribosomal ribonucleic acid methyltransferases and, 188
  transfer ribonucleic acid methyltransferases and, 169, 172, 177, 178, 182, 185
Adipate
  activation of, 485
  aspartate transaminase and, 439
Adipose tissue, glycogen synthetase of, 354
Adrenalectomy, glycogen synthetase and, 312, 330, 352
Adrenal gland, glycogen synthetase of, 353

Adrenocorticotropic hormone, deoxythymidine kinases and, 73
Aerobacter aerogenes
  aspartate transcarbamylase of, 298
  glucokinase of, 48
Affinity chromatography
  lactose synthetase, detection of reactant complexes and, 375–377
Affinity labeling, aspartate transaminase, 440–451
Alanine
  aminotransferase and, 463, 464, 466, 468, 470, 473, 474, 476, 478
  aspartate transaminase and, 436
  kynureninase and, 477
D-Alanine
  aminotransferase and, 466, 472–473
  α-methylserine hydroxymethylase and, 221
  serine hydroxymethyltransferase and, 218, 219, 479
β-Alanine, aminotransferases and, 466, 468, 474
Alanine transaminase, inhibition of, 448
Albumin, glycogen synthesis and, 321
Aldamine, inactive aminotransferase and, 395
Aldimine
  aspartate transaminase form, stability of, 425–426
  conformation, transaminase active site and, 452, 453, 457, 459–460
  formation rate, 389
  ketimine transformation, 390
Allose 6-phosphate, glycogen synthetase and, 332
Allosteric effectors
  aspartate transcarbamylase
    binding site, 270–273
    conformational changes and, 276–277
    enzyme properties in presence, 269–270
Allothreonine, serine hydroxymethyltransferase and, 215, 218
Alloxan, glycogen synthetase phosphatase and, 351
Amethopterin
  resistance, thymidylate synthetase and, 213, 214

SUBJECT INDEX 555

Amides, synthesis of, 385
Amidines, synthesis of, 385
S-Amidinocysteine, glycine amidinotransferase and, 502
Amidino-enzyme
  evidence for
    direct, 501–503
    indirect, 500–501
N-Amidino-3-keto-inosamine, aminotransferase and, 476
N-Amidinostreptamine, aminotransferase and, 476
Amidinotransferases, reactions catalyzed, 497
Amino acid(s)
  aromatic, aspartate transcarbamylase and, 267–268
  aspartate transaminase composition, 414–415
  aspartate transcarbamylase composition
    catalytic subunit, 232–234
    regulatory subunit, 235
  carbamate kinase composition, 103
  galactosyltransferase composition, 368
  α-lactalbumin sequence, 366
  mammalian hexokinase composition, 41
  non-B_{12} methyltransferase composition, 155
  sequence, aspartate transcarbamylase regulatory subunit, 234, 235
  synthesis, transfer ribonucleic acids and, 186
  yeast hexokinases, composition, 10–11
Amino acid oxidases, amino group transfer and, 386
2-Aminoadenosine, adenosine kinase and, 53
α-Aminoadipate, aminotransferases and, 464
α-Aminoadipate semialdehyde, aminotransferases and, 468
4-Aminobutyraldehyde, aminotransferase and, 470
α-Aminobutyrate, aminotransferases and, 464, 466
D-α-Aminobutyrate, α-methylserine hydroxymethylase and, 221

γ-Aminobutyrate
  aminotransferases and, 466, 468, 474, 481
  glycine amidinotransferase and, 500
ϵ-Aminocaproate, aminotransferases and, 466, 468
Amino-carbhydroxamates
  O-alkyl, aspartate transaminase and, 447–448
2-Amino-2-deoxy-neo-inositol 5-phosphate, inosamine phosphate amidinotransferase and, 507
S-(2-Aminoethyl)cysteine, aminotransferase and, 468
2-Aminoethylphosphonate, aminotransferase and, 475
Amino groups
  aspartate transcarbamylase active site and, 265–267
  substrate, aspartate transcarbamylase and, 256–257, 261
Amino group transfer
  basic chemical features
    congruent nonenzymic models, 387–391
    general characteristics of intermediate steps, 391–392
    conformation around C^{α}-H bond, 391
  current studies, 480–481
  formally similar processes, 384–387
  historical background, 379–381
  recent developments in, 381–384
  as side reactions, 476–480
α-Amino-δ-hydroxyvalerate, coenzyme A transferase and, 493
4-Amino-5-imidazolecarboxamide ribonucleotide
  adenosine kinase and, 54
  synthesis of, 385
5-Amino-4-imidazole carboxamide ribonucleotide transformylase, reaction catalyzed and properties, 204–205
β-Aminoisobutyrate, aminotransferases and, 466, 468
5-Aminolevulinate, aminotransferase and, 468, 474
Aminomalonate
  aspartate transaminase and, 435
  aspartate transcarbamylase and, 262

6-Aminonicotinamide adenine dinucleotide, nicotinamide adenine dinucleotide kinase and, 80
Aminoxyacetate, aspartate transaminase and, 413, 437
3-Aminopropionate, glycine amidinotransferase and, 500
Aminopterin
 methionyl-transfer ribonucleic acid transformylase and, 209
 thymidylate synthetase and, 211, 214
Amino sugars
 aspartate transaminase and, 397
 synthesis of, 385
Aminotransferases
 acting on monocarboxylic substrate pairs, 466–467, 473
 ω-amino and ω-keto acid substrates, 466–469, 473–474
 noncarboxylic substrates, 470–471, 475–476
 requiring dicarboxylic amino or keto acids, 463–467, 470–473
 subgrouping of, 462
δ-Aminovalerate, aminotransferases and, 466, 468
Ammonia
 carbamyl phosphate formation and, 100, 101, 105, 106, 108
 glycine biosynthesis and, 222
 glycine decarboxylation and, 221, 222
Ammonium ions
 guanosine monophosphate kinase and, 86
 pseudouridine kinase and, 62
 ribosomal ribonucleic acid methyltransferases and, 188
 transfer ribonucleic acid methyltransferases and, 174–175
Ammonium sulfate, carbamate kinase stability and, 104
α-Amylase
 glycogen synthetase assay and, 317–318
 glycogen primer ability and, 320
 oligosaccharide formation and, 320
β-Amylase
 limit dextrin, glycogen synthetase and, 319, 322

1,5-Anhydroglucitol 6-phosphate
 glycogen synthetase and, 344
 mitochondrial hexokinase and, 35
Anions, transfer ribonucleic acid methyltransferases and, 176
Antipyrine, carbamyl aspartate assay and, 229
Antisera
 aspartate transcarbamylase and, 255, 276
 yeast hexokinases and, 11–12
Arabinose, lactose synthetase and, 373, 374
D-Araboflavin, riboflavin kinase and, 75
Arcaine, biosynthesis, 498
Arginase, glycine amidinotransferase and, 500
Arginine
 N-acetylglutamate-5-phosphotransferase and, 512, 515, 516–520
 amidinotransferases and, 497, 498, 499, 501, 502, 504, 506, 508
 aminotransferase and, 466, 474
 aspartate transaminase and, 436
 biosynthesis of, 385, 503, 511–512
 carbamyl phosphate synthetase and, 118
 dissimilation of, 98, 118
 growth requirement for, 115–116
Arginine dihydrolase, occurrence of, 116
Arginine racemase, ornithine and, 478
Arsenate, glycogen synthetase and, 334
Arthropoda, glycogen synthetase of, 358–359
Asparagine
 aminotransferases and, 464, 466, 473
 synthesis of, 385
Aspartate
 amino group, utilization of, 385
 aminotransferases and, 463, 464–467, 472
 analogs, binding to aspartate transcarbamylase and, 250
 binding to aspartate transcarbamylase, 228, 246–247, 252, 253, 256–257, 258–259
  cooperativity, 268–269, 270, 278, 279, 285, 288–289, 299
 exchange, aspartate transcarbamylase and, 245

SUBJECT INDEX 557

glutamate decarboxylase and, 479
radioactive, aspartate transcarbamylase assay and, 229
Aspartate-β-decarboxylase, inactivation of, 477
Aspartate:oxoglutarate aminotransferases
  isoenzymes and multiple subforms, 393–398
  pig heart and animal tissues
    coenzyme analogs and, 429–435
    dynamic spatial aspects, 455–462
    kinetics, 424–429
    optical properties, 407–416
    physical parameters and macromolecular structure, 398–406
    primary structure and functionally important groups, 416–424
    stereochemistry and active site topography, 451–455
    substrates, quasi-substrates and inhibitors, 435–451
    substrate specificity, 382
Aspartate transcarbamylase(s)
  biosynthesis and genetics
    control of, 295–297
    location of genes, 292–293
    single operon for both kinds of chain, 293–294
  carbamate kinase assay and, 108
  carbamyl phosphate synthetase complex, 303, 304, 307
  conformation
    carbamyl phosphate binding and, 250–251
    catalysis and, 257–259
    cooperative properties and, 275–277
    dicarboxylic acid binding, 253–256
  cooperative properties
    allosteric effectors, 269–273
    comparison with subunits, 277–278
    kinetics of ligand binding, 285–287
    ligand-induced conformational change, 275–277
    possible mechanisms, 287–292
    stoichiometry of ligand binding, 273–275
    structural modifications and, 278–285
    substrate binding, 268–269

  feedback control, 227
  isolation and characterization
    assay procedures, 228–230
    properties associated with regulation, 227–228
    purification, 228
    size and subunit composition, 230–231
  isolation and characterization of subunits
    catalytic, 231–232
    comparison with native enzyme, 277–278
    regulatory, 234–237
    primary structures, 232–234, 235
  mechanism of catalysis
    details, 243–262
    functional groups at active site, 262–268
  other than Escherichia coli
    bacterial, 297–302
    fungal, 302–306
    mammalian, 306–307
    plant, 307–308
  reaction catalyzed, 226
  reconstitution from subunits, 242–243
    metals and, 237–238
    methods, 237
    other metals than zinc, 238–239
  repression of biosynthesis, 227–228
Aspergillus oryzae, hexokinase, 47
Atractyloside, mitochondrial hexokinase and, 35, 36
Avian myeloblastosis virus
  virion, transfer ribonucleic acid methyltransferase in, 169
5-Azacytidine, uridine-cytidine kinase and, 58, 59
5-Aza-2'-deoxycytidine, tumor resistance to, 59
Azadeoxythymidine, deoxythymidine kinase and, 70
8-Azaguanosine monophosphate, guanosine monophosphate kinase and, 83, 85–86
6-Azauridine, uridine-cytidine kinase and, 57, 58–59
6-Azauridine triphosphate, uridine phosphorylation and, 61

Azotobacter vinelandii, aspartate transcarbamylase of, 298

B

Bacillus subtilis
  aspartate transcarbamylase of, 298, 299, 302
  carbamyl phosphate enzymes of, 118–119
Bacteria, aspartate transcarbamylases of, 297–302
Bacteriophage
  host restriction, deoxyribonucleic acid methyltransferase and, 191–192, 194
  induction or infection
    deoxyribonucleic acid methyltransferase and, 193–194
    transfer ribonucleic acid methyltransferase levels and, 182, 185–186
  T-even, mononucleotide kinases of, 94–95
  T-5, deoxyribonucleotide kinase of, 95–96
Base catalysis, aspartate transcarbamylase and, 261–262
Benzoate, aspartate transaminase and, 428
6-Benzylaminopurine riboside, transfer ribonucleic acid methyltransferases and, 176
Bicarbonate, carbamyl phosphate formation and, 100, 101, 105, 106, 109
Biotin, carbamate kinase and, 107
Blastocladiella emersonii, glycogen synthetase of, 361
Blue Dextran, coenzyme A transferase and, 486
B_{12} methyltransferase
  alkylation studies with radioactive 5-methyltetrahydrofolate, 137–143
  assay of, 122–123
  catalytic properties
    binding of radioactive folate substrate, 136–137
    inhibition by propyl iodide, 127–129
    methyl group transfer reactions catalyzed, 129–135
  mechanism of homocysteine transmethylation, 151–154
  occurrence of, 162–164
  physical properties
    absorption spectrum and B_{12} chromophore, 124–125
    resolution-reconstitution and molecular weight, 125–127
  purification, 123–124
  role of S-adenosyl-L-methionine, 143–151
Borohydride
  aspartate:oxoglutarate aminotransferase and, 395, 410, 412, 417, 423, 444, 445, 449
  aspartate transcarbamylase derivative and, 266
  enzyme-coenzyme A intermediate and, 492, 493
Borotritide, aspartate transaminase and, 453, 460
Brain
  aminotransferases of, 464–465, 468–469
  hexokinases of, 31, 32, 36
  B_{12} methyltransferase of, 165
  glycogen synthetase of, 353–354
3-(Bromoacetyl)-pyridine, nicotinamide-adenine dinucleotide kinase and, 77, 79–80, 82
Bromocytidine triphosphate, aspartate transcarbamylase and, 230, 236, 274, 276–277, 286–287
5-Bromodeoxycytidine triphosphate, deoxythymidine kinase and, 72
5-Bromodeoxyuridine, deoxythymidine kinase and, 70
5-Bromodeoxyuridine
  resistance, deoxythymidine kinase and, 69
5-Bromodeoxyuridine monophosphate, phage deoxyribonucleotide kinases and, 94, 96
8-Bromoguanosine monophosphate, guanosine monophosphate kinase and, 85–86
3-Bromopropionylchloride, aspartate transaminase and, 450
6-Bromopyridoxal phosphate, aspartate transaminase and, 434

SUBJECT INDEX 559

Bromosuccinate, aspartate transcarbamylase and, 267, 279–280
Buffers
  aspartate transaminase and, 426–428
  non-B_{12} methyltransferase and, 157
  nonenzymic transamination and, 389
  yeast hexokinase and, 29
Butyrate, aspartate transaminase and, 428

C

Cacodylate, aspartate transaminase and, 428
Cadmium ions, aspartate transcarbamylase and, 238, 273, 276, 277, 296
Caffeine, glycogen synthetase phosphatase and, 328, 329
Calcium ions
  deoxyadenosine kinase and, 67
  galactosyl transferase and, 370–371
  glycogen synthetase and, 331, 335
  mitochondrial hexokinase and, 35
  phosphorylase and, 339–340
  transfer ribonucleic acid methyltransferases and, 174
  uridine kinase and, 59
L-Canaline
  aspartate transcarbamylase and, 438
  glycine amidinotransferase and, 500
  inosamine phosphate amidinotransferase and, 507
Canavanine
  N-acetylglutamate-5-phosphotransferase and, 517
  amidinotransferase assay, 498
  aspartate transaminase and, 438
  glycine amidinotransferase and, 499, 500
  inosamine phosphate amidinotransferase and, 507
Carbamate, carbamate kinase and, 106, 108
Carbamate kinase
  assays, 107–108
    forward reaction, 108–109
    reverse reaction, 109–110
  distribution, 101–102
  function and relation to other enzymes, 115–119

  historical background, 97–100
  metabolite control of activity, 116–119
  molecular properties
    composition, size and subunit structure, 103–104
    purification, 102–103
    stability, 104–105
    sulfhydryl reagent effects, 105
  reactions catalyzed, 105–107
  specificity and cofactors, 107
  thermodynamics, kinetics and catalytic mechanism, 110–115
Carbamyl aspartate
  assay
    colorimetric, 228–229
    radioactive, 229–230
  formation of, 100
Carbamylcholine, glycogen synthetase and, 352
Carbamyl phosphate
  adenosine triphosphate formation and, 106
  availability to ornithine transcarbamylase, 303, 304, 305
  binding, 282
    catalytic subunit, 232
    conformational changes and, 250–251, 275, 276, 286–287
    cooperativity and, 269
    stoichiometry, 274
    succinate and, 290
  binding to carbamate kinase, 114
  bromocytidine triphosphate binding and, 236
  cytidine triphosphate binding and, 274–275
  labeled
    aspartate transcarbamylase assay and, 229–230
    isotope effects and, 259
  stability of, 108
  transfer to aspartate, mechanism, 243–251, 258
Carbamyl phosphate synthetase
  adenosine diphosphate and, 113
  control of, 295, 302
  regulation of, 302–303
  mammalian, 307
  substrates of, 101

Carbohydrate
  aspartate:oxoglutarate aminotransferase subforms and, 396–398
  coenzyme A transferase and, 486
  galactosyltransferase composition, 368
Carbon dioxide
  evolution, carbamate kinase assay and, 110
  exchange, glycine decarboxylation and, 221–222
  formation, 10-formyltetrahydrofolate deacylase and, 200
  glycine amidinotransferase and, 503
  glycine biosynthesis and, 222
Carbonic anhydrase, carbamate kinase and, 105–106
Carboxylation enzyme, glycine biosynthesis and, 222
Carboxypeptidase A
  aspartate transcarbamylase and, 232
  hexokinase and, 9, 10
Catecholamines
  hexokinase and
    mammalian, 46
    yeast, 29
Cathepsin
  lysosomal, hexokinase and, 35
Cellobiose, galactosyltransferase and, 370
Cellobiulose, galactosyltransferase and, 370
Chloride ions
  aspartate transaminase and, 428
  coenzyme A transferase and, 494
  transfer ribonucleic acid methyltransferases and, 176
β-Chloroglutamate, aspartate transaminase and, 450–451, 454
p-Chloromercuribenzoate
  coenzyme A transferase and, 495
  α-methylserine hydroxymethylase and, 221
1-(3-Chloromercuri-2-methoxypropyl)urea, aspartate transcarbamylase and, 231
2-Chloromercuri-4-nitrophenol, aspartate transcarbamylase and, 263, 264
Chymotrypsin
  formiminoglutamate formiminotransferase and, 207
  glycogen synthetase and, 315
  hexokinase and, 3

Cinchona alkaloids, glycogen synthetase and, 335
Circular dichroism
  aspartate transaminase, 404–406, 407–409, 412–413, 416, 437, 460
Citramalate, activation of, 485
Citrate
  glycogen synthetase and, 357
  hexokinase and
    mammalian, 46
    yeast, 29
Citrobacter freundii, aspartate transcarbamylase and, 298, 299–300
Citrulline
  N-acetylglutamate-5-phosphotransferase and, 517
  arsenolysis of, 98
  formation of, 98
  phosphorolysis of, 98, 115
Cobalamin(s), see B_{12}
Cobalt ions
  aspartate transcarbamylase reassociation and, 238, 239
  deoxythymidine monophosphate kinase and, 93
  nicotinamide adenine dinucleotide kinase and, 78
  uridine kinase and, 59
Coenzyme A
  coenzyme A transferase inhibition by, 496
  enzyme-bound intermediate, 491–494
Coenzyme A transferases, reactions catalyzed, 483–485
Concanavalin A, glycogen synthetase and, 313
Cooperativity, structural models for, 290–292
Cordycepin, adenosine kinase and, 53
Corrinoids, B_{12} methyltransferase reconstitution and, 125–126
Creatine
  biosynthesis, 497, 499, 500
  regulation, 503–505
  glycine amidinotransferase and, 503–504
Creatine kinase, adenosine diphosphate and, 113
Creatine phosphate, muscle contraction and, 339

Creatinine, rate of formation, 504
Cyanate
  aspartate transcarbamylase and, 243, 244–245, 279
  carbamyl phosphate formation and, 115
  formation from carbamyl phosphate, 109
Cyanide
  aspartate transaminase and, 403, 412
  aspartate transcarbamylase and, 234, 264, 265, 281
  B_{12} methyltransferase and, 125
  hexokinase and, 27
Cyano B_{12}
  B_{12} methyltransferase and, 129, 130, 131
  repression, 158, 159, 160
Cyclic adenosine monophosphate, glycogen synthetase and, 311, 326–327, 337, 338, 340, 348, 350, 351, 353, 354, 356, 357, 358
Cyclic nucleotides, glycogen synthetase kinase and, 327
α-Cycloglutamates, aspartate transaminase and, 443–447, 453–455, 457
γ-Cycloglutamate
  isomers, aspartate transaminase and, 443–447, 453–455
Cycloheximide, transfer ribonucleic acid methyltransferase levels and, 179
Cycloserine
  aspartate transaminase and, 410, 419, 423, 438, 440–442, 445, 448, 457
  α-methylserine hydroxymethylase and, 221
Cystamine, inosamine phosphate amidinotransferase and, 508
Cysteate, aspartate transaminase and, 435
Cysteine
  aminotransferase and, 472
  B_{12} methyltransferase and, 131
  carbamate kinase and, 105
  nicotinamide adenine dinucleotide kinase and, 77
  non-B_{12} methyltransferase and, 156, 157
Cysteine residues
  aspartate transaminase and, 421–423, 460

aspartate transcarbamylase, 232, 233–234
  metal binding site, 236
  glycine amidinotransferase and, 502
Cysteinesulfinate, aspartate transaminase and, 399, 422, 435
Cytidine
  aspartate transcarbamylase and, 270–271, 272
  deoxyadenosine kinase and, 67
  deoxycytidine kinase and, 64
  derivatives
    aspartate transcarbamylase and, 271
Cytidine deoxynucleotides, deoxyadenosine kinase and, 67
Cytidine diphosphate, uridine monophosphate kinase and, 90, 91
Cytidine diphosphate glucose, glycogen synthetase and, 318
Cytidine kinase, see Uridine-cytidine kinase
Cytidine monophosphate
  deoxycytidine monophosphate kinase and, 88
  uridine monophosphate kinase and, 90
Cytidine monophosphate kinase, see Uridine monophosphate kinase
Cytidine triphosphate
  aspartate transcarbamylase and, 226, 228, 230, 236–237, 247, 265, 287, 293, 298, 299, 301
    conformational changes and, 276–277
    cooperative effects and, 269–273, 284, 285, 288
    stoichiometry of binding, 274–275
    structural modifications, 278–280
  deoxyadenosine kinase and, 67
  nicotinamide adenine dinucleotide kinase and, 80
  riboflavin kinase and, 75
  uridine monophosphate kinase and, 90
  uridine phosphorylation and, 61, 62
Cytidine triphosphate synthetase
  mutants, aspartate transcarbamylase and, 295
Cytosine, aspartate transcarbamylase and, 270, 271, 272
Cytosine arabinoside, deoxycytidine kinase and, 63–64
Cytosine arabinoside triphosphate, deoxyadenosine kinase and, 68

Cytosol
  aspartate:oxoglutarate aminotransferase, 393–394
    amino acid composition, 414–415, 417

D

7-Deazaadenosine, transfer ribonucleic acid methyltransferases and, 176
Decarboxylation, conformation around C^{α}-CO_2H bond, 391
2'-Deoxyadenosine
  adenosine kinase and, 53
  deoxycytidine kinase and, 63, 64, 65
Deoxyadenosine-deoxythymidine-deoxyguanosine-deoxycytidine monophosphate kinase, occurrence and properties, 95–96
Deoxyadenosine diphosphate
  deoxythymidine kinase and, 71
  deoxythymidine monophosphate kinase and, 93–94
  uridine monophosphate kinase and, 91
Deoxyadenosine kinase
  distribution, purification and assay, 66
  kinetic and molecular properties; reaction mechanism, 67–68
  substrate specificity, 66–67
Deoxyadenosine monophosphate kinase, properties, 86–87
Deoxyadenosine triphosphate
  adenosine kinase and, 53
  aspartate transcarbamylase and, 269
  deoxyadenosine kinase and, 67
  deoxycytidine monophosphate kinase and, 88
  deoxythymidine kinase and, 71
  deoxythymidine monophosphate kinase and, 93
  guanosine monophosphate kinase and, 86
  phage deoxyribonucleotide kinases and, 94, 96
  uridine-cytidine kinase and, 58
  uridine or cytidine monophosphate kinase and, 91
  yeast hexokinase and, 21, 22
Deoxycytidine, deoxyadenosine kinase and, 67
Deoxycytidine diphosphate
  deoxycytidine kinase and, 65
  deoxythymidine kinase and, 71, 72, 74
  uridine monophosphate kinase and, 90
Deoxycytidine kinase
  distribution, purification and assay, 62–63
  kinetic, molecular and allosteric properties, 64–66
  substrate specificity, 63–64
Deoxycytidine monophosphate, deoxycytidine kinase and, 65
Deoxycytidine monophosphate-cytidine monophosphate-uridine monophosphate kinase, substrate specificities, 87–88
Deoxycytidine monophosphate kinase, purification and properties, 88–89
Deoxycytidine triphosphate
  deoxycytidine kinase and, 65, 66
  deoxythymidine kinase and, 71
  uridine-cytidine kinase and, 58
  uridine monophosphate kinase and, 90
Deoxycytidylate hydroxymethyltransferase, reaction catalyzed and properties, 209–210
2-Deoxy-D-glucose, galactosyltransferase and, 370, 373
2-Deoxyglucose 6-phosphate
  glycogen synthetase and, 334
  mammalian hexokinase and, 45
  mitochondrial hexokinase and, 35
Deoxyguanosine
  deoxyadenosine kinase and, 66–67
  deoxycytidine kinase and, 63, 64, 65
Deoxyguanosine kinase, deoxycytidine kinase and, 68
Deoxyguanosine monophosphate kinase, *see* Guanosine monophosphate kinase
Deoxyguanosine triphosphate
  adenosine kinase and, 53
  deoxyadenosine kinase and, 67
  deoxythymidine monophosphate kinase and, 93
  guanosine monophosphate kinase and, 86
  uridine-cytidine kinase and, 58, 61
Deoxyribonucleic acid
  species specificity of, 168
  synthesis
    deoxyadenosine and, 66

deoxythymidine kinase and, 74
transfer ribonucleic acid methyltransferases and, 176
Deoxyribonucleic acid methyltransferases
  biological significance, 194–195
  occurrence, 190
  properties
    bacteria, 190–192
    eucaryotes, 192–193
    regulation, 193–194
2-Deoxystreptidine phosphate, inosamine phosphate amidinotransferase and, 507
Deoxythymidine diphosphate, deoxycytidine kinase and, 65
Deoxythymidine kinase
  distribution, purification and assay, 69–70
  kinetic, molecular and allosteric properties, 71–74
  reaction mechanism and active site, 74
  substrate specificity, 70
Deoxythymidine monophosphate kinase
  assay and stability, 92
  distribution and purification, 91–92
  kinetic and molecular properties, 93–94
  substrate specificity, 92–93
Deoxythymidine triphosphate
  deoxyadenosine kinase and, 67
  deoxycytidine kinase and, 65, 66
  deoxythymidine kinase and, 71, 72, 73, 74
  deoxythymidine monophosphate kinase and, 93
  uridine-cytidine kinase and, 58
Deoxyuridine, deoxythymidine kinase and, 70
Deoxyuridine diphosphate galactose, lactose synthetase and, 369
Deoxyuridine monophosphate
  deoxycytidine monophosphate kinase and, 88
  deoxythymidine monophosphate kinase and, 92, 93
  phage deoxyribonucleotide kinase and, 94–95
  thymidylate synthetase and, 211, 212, 213, 214

Deoxyuridine triphosphate
  deoxycytidine kinase and, 65
  uridine-cytidine kinase and, 58
Deoxyuridylate hydroxymethyltransferase, reaction catalyzed and properties, 210
Deoxyxanthosine monophosphate, guanosine monophosphate kinase and, 86
Detergent, hexokinase isolation and, 38
Deuterium, aspartate transaminase and, 452
Deuterium oxide, aspartate transcarbamylase and, 261
Dextrins, glycogen synthetase and, 319, 322
Diabetes, hexokinase and, 32, 33
Diacetylmonoxine, carbamyl aspartate assay and, 229
2,2-Dialkyl amino acid:pyruvate aminotransferase, reactions catalyzed, 478
Diamines, aminotransferases and, 475–476
1,4-Diaminobutyl-1-phosphonate, inosamine phosphate amidinotransferase and, 507
α,γ-Diaminobutyrate, aminotransferase and, 468
α,δ-Diaminobutyrate, aminotransferase and, 468
Diaphorases, B_{12} methyltransferase and, 130
Dibromotyrosine, aminotransferase and, 466
Dicarboxylic acids
  binding, aspartate transcarbamylase and, 251–253, 262
Dichloroflavin, riboflavin kinase and, 75
2,6-Dichlorophenolindophenol, nicotinamide adenine dinucleotide kinase assay and, 78
Dichlorotyrosine, aminotransferase and, 466
Dictyostelium discoideum, glycogen synthetase of, 361
Diethylaminoethylcellulose
  mitochondrial hexokinase and, 37
  yeast hexokinase and, 3, 4, 6, 12

Diethylstilbestrol
  deoxythymidine monophosphate kinase and, 92
  transfer ribonucleic acid methyltransferase levels and, 179, 181
Differentiation
  deoxyribonucleic acid methyltransferases and, 195
  transfer ribonucleic acid methyltransferases and, 185
1,5-Difluoro-2,4-dinitrobenzene, aspartate transcarbamylase and, 420, 450
Dihydrofolate, thymidylate synthetase and, 211, 212
Dihydrofolate reductase, thymidylate synthetase and, 211
Dihydrolipoamide dehydrogenase, "hydrogen carrier protein" and, 222, 223
Dihydroorotase, aspartate transcarbamylase complex and, 307
Dihydroorotate dehydrogenase, regulation of, 302
Dihydroxyphenylalanine, aminotransferase and, 464
Diiodothyronine, aminotransferase and, 466
Diiodotyrosine, aminotransferases and, 464, 466
N,N-Dimethyl carbamyl phosphate, aspartate transcarbamylase and, 249, 252
1,5-Dimethyl-2-phenyl-3-pyrazolone, carbamyl aspartate assay and, 229
Dimethyl suberimidate, aspartate transcarbamylase and, 232
Dimethylsulfoxide, carbamyl phosphate synthetase complex and, 307
Dinitrofluorobenzene, glycine amidinotransferase and, 502
1,5-Dioxovalerate, aminotransferase and, 468
5,5′-Dithiobis(2-nitrobenzoate)
  aspartate transaminase and, 422
  aspartate transcarbamylase and, 234, 263, 264, 277–278, 281–284
  coenzyme A transferase and, 495
  glycine amidinotransferase and, 501, 502
  glycine decarboxylation and, 222

Dithioerythritol, carbamate kinase and, 105
Dithiols
  vicinal, glycine decarboxylation and, 221
Dithiothreitol
  adenosine kinase and, 52, 53
  B_{12} methyltransferase and, 125, 126, 128, 130, 131, 132, 137, 138, 143, 151, 153
  carbamate kinase and, 103
  coenzyme A transferase and, 495
  deoxythymidine monophosphate kinase and, 92
  non-B_{12} methyltransferase and, 156, 157

E

Elastase, mitochondrial hexokinase and, 38
Electron microscopy
  aspartate transcarbamylase, 232, 239–241, 290
  glycogen synthetase, 315, 316
Embryogenesis
  deoxyribonucleic acid methyltransferase and, 195
  transfer ribonucleic acid methyltransferase levels and, 185
Endometrium, glycogen synthetase of, 356
Entamoeba histolytica, hexokinase of, 48
Epinephrine, glycogen synthetase and, 336, 338, 341, 349, 351
Equilibrium, aspartate transaminase, 425–426
Equilibrium constant
  carbamate kinase, 110–111
  coenzyme A transferase, 487–488, 495
Erythrocytes, glycogen synthetase of, 355
Erythromycin
  resistance, ribosomal ribonucleic acid methyltransferase and, 189
Escherichia coli
  aspartate transcarbamylase
    biosynthesis and genetics, 292–297
    cooperative properties, 268–292
    detailed subunit structure, 239–243

isolation and characterization, 227–231
mechanism of catalysis, 243–268
reconstitution, 237–239
subunit isolation and characterization, 231–237
B_{12} methyltransferase
assay and purification, 122–124
catalytic properties, 127–137
mechanism, 151–154
physical properties, 124–127
radioactive 5-methyltetrahydrofolate and, 137–143
role of S-adenosylmethionine, 143–151
carbamate kinases of, 101
mannokinase of, 48
non-B_{12} methyltransferase
assay and purification, 154–155
catalytic properties, 156–158
physical properties, 155–156
repression of, 158–160
Estradiol, transfer ribonucleic acid methyltransferase levels and, 179–180, 181, 185
Estrogen, glycogen synthetase and, 356
Ethanolamine deaminase, light stability, 142
Ethidium bromide, transfer ribonucleic acid methyltransferases and, 177
Ethylenediaminetetraacetate
carbamate kinase and, 105
glycogen synthetase and, 315, 343
hexokinase reactivation and, 27, 30
Ethyleneimine, aspartate transcarbamylase and, 263
N-Ethylmaleimide
aspartate transaminase and, 422
aspartate transcarbamylase and, 263
coenzyme A transferase and, 495
5-formyltetrahydrofolate cyclodehydrase and, 202
hexokinase and, 26
hydrogen carrier protein and, 223
α-Ethylserine, α-methylserine hydroxymethylase and, 221
Ethylthioadenosine, transfer ribonucleic acid methyltransferases and, 176
Eucaryotes, deoxyribonucleic acid methyltransferases of, 192–193

F

Fat pad, hexokinases of, 32, 33
Fatty acids, activation of, 485, 487
Ferricyanide, aspartate transcarbamylase and, 263
Ferrous ions, uridine kinase and, 59
Fetal tissues, transfer ribonucleic acid methyltransferases in, 177–178, 185
Fish, glycogen synthetase of, 358
Flavin adenine dinucleotide
glycine decarboxylation and, 221
reduced, B_{12} methyltransferase and, 130, 151, 152
Flavin mononucleotide
reduced, B_{12} methyltransferase and, 128, 129, 130, 131, 132, 137, 138, 143, 145, 147–148, 149–150, 152, 153, 164, 165
Fluorescence, aspartate transaminase, 408
Fluorescence polarization, aspartate transaminase, 402
Fluoride
carbamyl phosphate formation and, 100
glycogen synthetase and, 330
glycogen synthetase phosphate and, 328, 330
2-Fluoroadenosine, adenosine kinase and, 53
5-Fluorocytidine, uridine-cytidine kinase and, 58
5-Fluorodeoxyuridine, deoxythymidine kinase and, 70
5-Fluorodeoxyuridylate, thymidylate synthetase and, 214, 215
β-Fluoro-oxalacetate, aspartate transaminase and, 451
5-Fluorouracil, insensitivity to, 293
5-Fluorouridine
insensitivity to, 293
uridine-cytidine kinase and, 57, 58, 59
5-Fluorouridine triphosphate, uridine phosphorylation and, 61
Foot-and-mouth disease virus, transfer ribonucleic acid methyltransferases and, 183
Formaldehyde (and congeners)
transfer

deoxycytidylate hydroxymethyltransferase, 209–210
deoxyuridylate hydroxymethyltransferase, 210
glycine decarboxylation, 221–223
serine hydroxymethyltransferase,
215–221
thymidylate synthetase, 210–215
Formamide disulfide, glycine amidinotransferase and, 501
Formate (and congeners)
transfer
5-amino-4-imidazole carboxamide
ribonucleotide transformylase,
204–205
formiminoglutamate formiminotransferase, 206–207
formiminoglycine formiminotransferase, 206
5-formiminotetrahydrofolate cyclodeaminase, 202–203
N-formylglutamate transformylase,
207–208
5-formyltetrahydrofolate cyclodehydrase, 201–202
10-formyltetrahydrofolate deacylase
and, 200
10-formyltetrahydrofolate synthetase, 198–200
glycinamide ribonucleotide transformylase, 203–204
5,10-methenyltetrahydrofolate cyclohydrolase, 201
methionyl transfer ribonucleic acid
transformylase, 208–209
Formiminoalanine, formiminoglycine
formiminotransferase and, 206
Formiminoglutamate formiminotransferase
N-formylglutamate transformylase
and, 208
reaction catalyzed and properties, 206–
207
Formiminoglycine formiminotransferase,
reaction catalyzed and properties,
206
5-Formiminotetrahydrofolate cyclodeaminase, reaction catalyzed and
properties, 202–203

N-Formylglutamate transformylase, reaction catalyzed and properties, 207–
208
N-Formylglycine-amide ribonucleotide,
synthesis of, 385
Formyl methionine residues, aspartate
transcarbamylase, 233
5-Formyltetrahydrofolate, N-formyl-
glutamate transformylase and, 207
5-Formyltetrahydrofolate cyclodehydrase, reactions catalyzed, 201–202
10-Formyltetrahydrofolate
5-amino-4-imidazole carboxamide
ribonucleotide transformylase and,
205
methionyl-transfer ribonucleic acid
transformylase and, 208
10-Formyltetrahydrofolate deacylase,
cofactor requirement, 200
10-Formyltetrahydrofolate synthetase,
reaction catalyzed, 198–199
Frog
muscle, glycogen synthetase of, 357
Fructose
mammalian hexokinase and, 31, 32, 42
yeast hexokinases and, 5, 6, 15
Fructose 6-phosphate, tumor hexokinase
and, 45
Fumarate
aspartate transaminase and, 439
aspartate transcarbamylase and, 252

G

Galactose, aspartate transaminase and,
397
Galactose 6-phosphate, glycogen synthetase and, 332, 344
Galactosyltransferase, see also Lactose
synthetase
kinetic studies
early, 371
reactant complex detection, 375–376
steady state, 371–375
lactose synthesis and, 363
purification and properties, 367–369
reactions catalyzed, 369
substrate specificity
galactosyl acceptor, 369–370

galactosyl donor, 369
  metal requirement, 370–371
Galegin, biosynthesis, 498
Gene(s)
  aspartate transcarbamylase, location, 292–293
Gentiobiose, galactosyltransferase and, 370
Glucagon, glycogen synthetase and, 312, 341, 351–352, 358
Glucocorticoids
  glycogen synthetase and, 352
  glycogen synthetase phosphate and, 328–329
Glucokinase
  inactivation of, 40
  purification of, 37
Glucosamine, galactosyltransferase and, 370
Glucosamine 6-phosphate, glycogen synthetase and, 332, 344
Glucose
  adenosine kinase assay and, 52
  carbamate kinase assay and, 109
  glycogen synthetase control and, 347–348, 353, 358, 359, 360–361
  glycogen synthetase phosphate and, 328, 329
  lactose synthetase and, 364, 369, 371, 372–373, 374, 375, 376
  mammalian hexokinase and, 31, 32, 39, 40, 41–42, 43, 44
  yeast hexokinases and, 5, 6, 8, 13–14, 15–17, 22, 24
    conformation and, 27
    saturating concentrations, 19–20
    sulfhydryl groups and, 25–26
Glucose 1-phosphate, lactose synthetase and, 364, 365, 370
Glucose 6-phosphate
  brain hexokinase and, 31
  glucocorticoids and, 352
  glycogen synthetase and, 310, 334, 347, 358, 359, 361
    aggregation, 314, 315
    cooperative binding, 334, 345
    "inactive" synthetase, 330, 331
    kinetics, 332–334, 341–344, 353, 360, 361
    magnesium ions and, 335
  stimulation by, 322–323, 324–325, 335, 344–345, 353, 357–358, 359, 361
  synthetase kinase and, 327
  synthetase phosphatase and, 328
  hexokinase kinetics and, 14, 15–17, 22, 24
  mammalian hexokinase and, 32, 33, 41, 42, 44–46
  mitochondrial hexokinase and, 34–35, 36, 37
  muscle contraction and, 339
  nicotinamide adenine dinucleotide kinase assay and, 78
Glucose-6-phosphate dehydrogenase
  adenosine kinase assay and, 52
  carbamate kinase assay and, 109–110
  nicotinamide adenine dinucleotide kinase assay and, 78
Glucose residues, aspartate transaminase and, 397
Glutamate
  aminotransferases and, 464–470, 472, 476
  aspartate transaminase and, 436, 437, 444
  aspartate transcarbamylase and, 262
  saccharopine pathway and, 386
Glutamate decarboxylase, 481
  inactivation of, 479
  stereochemistry of protonation, 453
Glutamate residues, coenzyme A transferase, 493
Glutamic semialdehyde, aminotransferase and, 468
Glutamine
  amido group, synthetic reactions of, 385
  aminotransferases and, 466, 473, 476
  carbamyl phosphate synthetase and, 101, 117, 118, 303, 304, 306
Glutamine synthetase, nomenclature and, 323–324
Glutarate
  aspartate transaminase and, 410, 411, 412, 426, 428, 439–440
  aspartate transcarbamylase and, 262
Glutathione
  B_{12} methyltransferase and, 131
  carbamate kinase and, 105

Glycerol
  adenosine kinase purification and, 52
  carbamate kinase and, 105, 117
Glycerol kinase, mitochondrial hexokinase and, 35
Glycerophosphate, glycogen synthetase and, 343
Glycinamide ribonucleotide transformylase, reaction catalyzed and properties, 203–204
Glycine
  aminotransferases and, 463, 466, 473
  aspartate transaminase and, 436
  biosynthesis of, 222–223
  methylating enzyme, transfer ribonucleic acid methyltransferases and, 178–179
  oxidative decarboxylation, proteins involved, 221–222
  serine hydroxymethyltransferase and, 215, 217, 218, 219–220
Glycine amidinotransferase
  biological distribution, 498–499
  catalytic properties
    reaction mechanism, 500–503
    regulation, 503–505
    substrate specificity, 499–500
Glycogen
  glycogen synthetase and
    association with, 311–312, 330
    control, 336, 340, 353
  glycogen synthetase phosphate and, 328, 330, 356
  synthesis de novo, 320–321
Glycogen synthetase
  adrenal, 353
  dinitrophenylation of, 360
  general catalytic properties
    assay methods, 317–318
    donor specificity, 318
    glucosyl acceptor, 319–321
    mechanism, 321–322
    reaction catalyzed, 316–317
  heart
    control, 340–341
    properties, 340
  historical background, 310–311
  liver
    control, 347–353
    properties of the two forms, 341–347
  mammalian muscle
    control of, 336–340
    properties of the two forms, 332–336
  molecular properties
    association with glycogen, 311–312
    physicochemical properties, 313–316
    purification, 312–313
  nonmammalian organisms, 357–361
  other mammalian tissues, 354–357
  solubilization of, 312
  two forms of, 310–311, 313, 360
    activity in physiological conditions, 335–336, 346–347
    basic system of interconversion, 325–326
    general properties, 324–325
    "inactive" form, 330–331
    muscle contraction and, 339
    nomenclature, 322–324
    proteolytic inactivation, 331
    synthetase kinase, 326–327
    synthetase phosphatase, 327–330
Glycogen synthetase kinase
  insulin and, 337–338
  properties of, 326–327
  tumors and, 356
Glycogen synthetase phosphatase
  liver, properties of, 328–330
  muscle, properties of, 327–328, 336
  tumors and, 356
Glycoproteins, galactosyltransferase and, 369, 377
Glycosylmannose, galactosyltransferase and, 370
Glycylglycine
  glycine amidinotransferase and, 500
  inosamine phosphate amidinotransferase and, 507
Glyoxylate, aminotransferases and, 464, 466, 468, 473
Golgi apparatus, lactose synthetase and, 377
Guanidines
  cyclic, synthesis of, 385
Guanidinium chloride
  aspartate transcarbamylase regulatory subunit and, 234, 237
  coenzyme A transferase and, 486
  galactosyltransferase and, 367–368

hexokinase and, 9
non-B_{12} methyltransferase and, 155
Guanidinoacetate
  glycine amidinotransferase and, 499, 501, 503–504
  insulin release and, 498
4-Guanidinobutyrate, glycine amidinotransferase and, 499–500
1-Guanidino-1-deoxy-*scyllo*-inositol 4-phosphate, inosamine phosphate amidinotransferase and, 507
3-Guanidinopropionate, glycine amidinotransferase and, 499–500
Guanosine
  adenosine kinase and, 53
  guanosine monophosphate kinase and, 86
Guanosine deoxynucleotides, deoxyadenosine kinase and, 67–68
Guanosine diphosphate
  guanosine monophosphate kinase and, 86
  hexokinase and, 29, 30, 31
Guanosine diphosphate glucose, glycogen synthetase and, 318
Guanosine kinase, *see* Inosine-guanosine kinase
Guanosine monophosphate, synthesis of, 385
Guanosine monophosphate-deoxyguanosine monophosphate kinase
  assay, 84
  distribution and purification, 82–84
  kinetic and molecular properties, 84–85
  substrate specificity and reaction mechanism, 85–86
Guanosine triphosphate
  adenosine kinase and, 53
  amino group transfer and, 385
  aspartate transcarbamylase and, 271, 273, 298
  carbamate kinase and, 107
  deoxyadenosine kinase and, 67
  deoxythymidine monophosphate kinase and, 93
  guanosine monophosphate kinase and, 86
  hexokinase and, 19
  nicotinamide adenine dinucleotide kinase and, 80

riboflavin kinase and, 75
uridine-cytidine kinase and, 57, 58, 61

H

Halobacterium cutirubrum, aspartate transcarbamylase of, 301
Heart
  aminotransferases of, 464–465
  aspartate transaminases
    chicken, 414–415, 416
    pig, 414–415, 416
  glycogen synthetase of, 340–341
  hexokinases, 32
  purification of, 38
Heat, aspartate transcarbamylase and, 228, 231, 277, 304
α-Helix, aspartate transcarbamylase, 403, 406
Hepatoma, hexokinase and, 34
Hexokinase
  adenosine diphosphate and, 113
  adenosine kinase assay and, 52
  assay of, 4–5
  carbamate kinase assay and, 109
  comparative aspects, 46–48
  historical, 1–2
  mammalian
    aggregation phenomenon, 40–41
    amino acid composition and essential groups, 41
    enzyme-substrate interactions, 43–44
    isoelectric point, 41
    kinetic studies, 41–43
    mechanism, 41–44
    molecular weight and subunit structure, 39–40
    occurrence of multiple forms, 31–33
    purification procedures, 37–38
    regulation of activity, 44–46
    relation of soluble to insoluble forms, 33–37
  yeast
    chemical nature of proteolytic modification, 12–13
    comparison of native isozymes, 10–12
    conformational changes and enzymic activity, 27–28
    equilibrium measurement of enzyme-substrate interaction, 19–22

kinetic studies of adenosine triphosphatase reaction, 18–19
mode of action
conformational change and enzymic activity, 27–28
equilibrium measurement of enzyme-substrate interaction, 19–22
kinetic studies of adenosine triphosphatase reaction, 18–19
kinetic studies of hexokinase reaction, 13–17
modification by added proteases, 6–7
molecular weight in nondenaturing solvents, 7–8
native isoenzymes and modification by endogenous proteases, 2–6
question of a phosphoenzyme intermediate, 23–25
reaction kinetics, 13–17
regulation of activity, 29–31
role of sulfhydryl groups, 25–27
ultimate subunit size in denaturing solvents, 8–10
Hexose phosphate aminotransferase, classification of, 384
Histidine, aminotransferases and, 470, 472
Histidine residues
aspartate transaminase, 419, 428, 439, 444, 450, 461
aspartate transcarbamylase, 267
B_{12} methyltransferase and, 143
Histidinol-1-phosphate, aminotransferase and, 470, 475
Histone phosphatase, glycogen synthetase phosphatase and, 327
Homocysteate, aspartate transaminase and, 435
Homocysteine
methionine biosynthesis and, 121–122, 123, 129, 131, 132, 134, 139, 140, 143, 145, 147, 149, 151–154
non-B_{12} methyltransferase and, 154, 156, 157
Hormones, transfer ribonucleic acid methyltransferases and, 179–182
Hydrazine(s), aspartate transaminase and, 403, 413, 437, 438
2-Hydrazinoadenosine, adenosine kinase and, 53

Hydrocortisone
glycogen synthetase and, 358
transfer ribonucleic acid methyltransferase levels and, 179
Hydrogen
exchange, aspartate transaminase, 402
transfer, thymidylate synthetase and, 211–212, 214–215
Hydrogen bonding
amino group transfer and, 390
aspartate transaminase, 402, 429, 457, 460
Hydrogen carrier protein, glycine biosynthesis and, 222–223
Hydrophobic interactions, aspartate transaminase and, 401, 402, 429, 434
β-Hydroxyacyl coenzyme A dehydrogenase, coenzyme A transferase assay and, 487
β-Hydroxy-L-aspartate
isomers, aspartate transcarbamylase and, 252
erythro-β-Hydroxyaspartate
amino group transfer and, 392
aspartate transaminase and, 404, 405, 406, 411, 413, 429, 436–437, 454, 457
threo-β-Hydroxyaspartate
aminotransferase and, 464
aspartate transaminase and, 435, 437, 454
γ-Hydroxybutyrate, activation of, 485
γ-Hydroxyglutamate, aminotransferase and, 463
threo-γ-Hydroxyglutamate, aspartate transaminase and, 435, 437
Hydroxylamine
N-acetylglutamate-6-phosphotransferase and, 513–514, 515
aspartate transaminase and, 403, 413, 437
carbamyl phosphate and, 99
glycine amidinotransferase and, 500, 501
inosamine phosphate amidinotransferase and, 507, 508
p-Hydroxymercuribenzoate, carbamate kinase and, 105
5-Hydroxymethyl deoxycytidine-deoxyguanosine-deoxythymidine mono-

phosphate kinase, occurrence and properties, 94–95
β-Hydroxy-β-methylglutarate, activation of, 485
α-Hydroxymethylserine
  α-methylserine hydroxymethylase and, 220, 221
  serine hydroxymethyltransferase and, 219
3-Hydroxyproline, catabolism, 463
3-Hydroxy-4-pyridine aldehyde
  enzymic transamination and, 433
  nonenzymic transamination and, 389
2′-Hydroxypyridoxal phosphate, aspartate transaminase and, 434
3-Hydroxypyruvate, aminotransferase and, 466

I

Imidazole, methylcobinamide photolysis and, 142–143
Imidazoylacetol-1-phosphate, aminotransferase and, 470
2-Iminoriboflavin, riboflavin kinase and, 75
β-Indoxylglucose, galactosyltransferase and, 370
Inosamine phosphate amidinotransferase
  biological distribution, 505–506
  catalytic properties
    reaction mechanism, 507–508
    regulation, 508–509
    substrate specificity, 506–507
Inosine
  adenosine kinase and, 53
  aspartate transcarbamylase and, 271
Inosine diphosphate glucose, glycogen synthetase and, 318
Inosine-guanosine kinase
  existence, distribution and properties, 54–56
Inosine monophosphate, guanosine monophosphate kinase and, 86
Inosine triphosphate
  adenosine kinase and, 53
  aspartate transcarbamylase and, 271, 273
  mammalian hexokinase and, 33
  nicotinamide adenine dinucleotide kinase and, 80

uridine-cytidine kinase and, 58
yeast hexokinase and, 22
Inosinicase, 5-amino-4-imidazole carboxamide ribonucleotide transformylase and, 204, 205
myo-Inositol, amidinotransferase acceptors and, 509
scyllo-Inosose, aminotransferase and, 476
Insulin
  glycogen synthetase and, 310, 332, 336–338, 340–341, 347–351, 354, 357, 358
  hexokinase and, 32, 33
  release, guanidinoacetate and, 498
  transfer ribonucleic acid methyltransferase levels and, 179
Intestine, hexokinases of, 32
Iodoacetamide
  aspartate transaminase and, 423
  aspartate transcarbamylase and, 264
  carbamate kinase and, 105
Iodoacetate
  aspartate transaminase and, 422
  aspartate transcarbamylase and, 263
  hexokinase and, 9, 26
5-Iododeoxycytidine triphosphate, deoxythymidine kinase and, 72
5-Iododeoxyuridine
  deoxythymidine kinase and, 70, 74
  resistance, deoxythymidine kinase and, 69
5-Iododeoxyuridine monophosphate, deoxythymidine monophosphate kinase and, 93
Isocitrate dehydrogenase, nicotinamide adenine dinucleotide kinase assay and, 77, 78, 81
Isoleucine, aminotransferase and, 464, 472
allo-Isoleucine, aminotransferase and, 468
Isoniazid, aspartate transaminase and, 437, 438
Isonicotinoyl hydrazone, pyridoxal phosphate, aspartate transaminase and, 438
N^6-(Δ^2-Isopentenyl) adenosine, transfer ribonucleic acid methyltransferases and, 176

Isothiourea
  derivative, benzoyl group transfer and, 490

K

Kasugamycin
  resistance, ribosomal ribonucleic acid methyltransferase and, 189–190
Ketimine
  conformation, transaminase active site and, 452, 453, 461–462
  formation of, 390
β-Ketoadipate, activation of, 485, 486
β-Ketocaproate, coenzyme A transferase and, 486
Ketoglutarate, see Oxoglutarate
β-Ketoisocaproate, coenzyme A transferase and, 486
β-Ketovalerate, coenzyme A transferase and, 486
Kidney
  aminotransferase of, 468
  B_{12} methyltransferase of, 164–165
  hexokinases of, 32
Kinetin riboside, transfer ribonucleic acid methyltransferases and, 176
Kynureninase, inactivation of, 477–478
Kynurenine, aminotransferase and, 464, 472
Kynurenine transaminase, inhibition of, 438

L

α-Lactalbumin
  coupling to Sepharose 6B, 376
  formation of, 377
  galactosyltransferase inhibition by, 370, 371, 374–375
  lactose synthesis and, 363, 364, 365, 366, 371, 373
Lactate, coenzyme A transferase and, 487
Lactate dehydrogenase
  adenosine kinase assay and, 52
  carbamate kinase assay and, 109
  deoxythymidine monophosphate kinase assay and, 92
  guanosine monophosphate kinase assay and, 84
  uridine kinase assay and, 58
  uridine monophosphate-cytidine monophosphate kinase assay and, 90
Lactobacillus leichmannii, aspartate transcarbamylase of, 298, 301
Lactose, galactosyltransferase and, 376
Lactose synthetase, see also Galactosyltransferase
  biological significance, 377
  historical background, 364–365
  reaction catalyzed, 364
  requirement for two proteins
    identification of one as α-lactalbumin, 366
    relationship to lysozyme, 366–367
    resolution, 365–366
Leucine
  aminotransferases and, 464, 471
  aspartate transaminase and, 436
Leuconostoc mesenteroides, hexokinase of, 48
Leukocytes, glycogen synthetase of, 354–355
Levator ani muscle, glycogen synthetase of, 355
Light
  methyl-B_{12} methyltransferase and, 140–141
  propyl-B_{12} methyltransferase and, 128, 140
Linatine, aspartate transaminase and, 438
Lipase, nomenclature and, 323, 324
Lipoamide dehydrogenases, B_{12} methyltransferase and, 130
Lipopolysaccharide, deoxyribonucleic acid methyltransferase and, 194
Lithium ions, transfer ribonucleic acid methyltransferases and, 175
Liver
  aminotransferases of, 464–465, 466–467, 468–469
  B_{12} methyltransferase of, 164
  glycogen synthetase
    control of, 347–353
    properties of the two forms, 341–347
    tadpole, 357–358
  glycogen synthetase phosphatase of, 328–330
  hexokinases of, 31–33
Lombricine, biosynthesis, 498

Lymphocytes
  phytohemagglutinin induction, transfer ribonucleic acid methyltransferases and, 185
Lysine
  ε-amino group transfer, 386–387
  aminotransferase and, 463, 468, 474
  aspartate transaminase and, 436
Lysine residues
  amino transferases, 389
  aspartate transaminase, 399, 417, 419, 438, 450, 453, 460, 461, 462
  aspartate transcarbamylase, 266–267
Lysozyme, relationship to α-lactalbumin, 366–367
Lyxose, hexokinase and, 18, 21, 22, 24, 30–31

M

Magnesium ions
  N-acetylglutamate-5-phosphotransferase and, 514, 515
  adenosine kinase and, 52
  carbamate kinase and, 107, 111, 113
  carbamyl phosphate synthetase and, 304
  deoxyadenosine kinase and, 67
  deoxycytidine kinase and, 64, 65
  deoxycytidine monophosphate kinase and, 89
  deoxythymidine kinase and, 71
  deoxythymidine monophosphate kinase and, 93
  galactosyltransferase and, 370–371
  glycogen synthetase and, 315, 325, 330, 331, 334, 335, 342, 345
  guanosine monophosphate kinase and, 85
  mammalian hexokinase and, 43
  methionyl-transfer ribonucleic acid transformylase and, 209
  mitochondrial hexokinase and, 35
  nicotinamide adenine dinucleotide kinase and, 78, 79, 80, 81
  non-B_{12} methyltransferase and, 156, 162
  phage deoxyribonucleotide kinases and, 95, 96
  ribosomal ribonucleic acid methyltransferases and, 188
  thymidylate synthetase and, 213, 214
  transfer ribonucleic acid methyltransferases and, 170, 174, 175
  uridine kinase and, 59
  yeast hexokinase and, 16
Malate
  aspartate transaminase and, 439
  binding, aspartate transcarbamylase and, 251, 254, 256, 257, 259, 261, 269, 286
Maleate
  aspartate transaminase and, 410–411, 412, 426, 439
  aspartate transcarbamylase and, 252, 269, 280, 289
  glycogen synthetase and, 342, 343
  5,10-methenyltetrahydrofolate cyclohydrolase and, 201, 204, 205
Malonate
  aspartate transcarbamylase and, 262, 269
  coenzyme A transferase and, 486
Malonic semialdehyde
  aminotransferase and, 466
  coenzyme A transferase and, 486
Maltose
  galactosyltransferase and, 370
  glycogen synthetase and, 320
Maltotriose, glycogen synthetase and, 320
Mammals
  aspartate transcarbamylases of, 306–307
  B_{12} methyltransferases of, 162–163
Manganese ions
  aspartate transcarbamylase reassociation and, 238–239
  carbamate kinase and, 107
  deoxyadenosine kinase and, 67
  deoxycytidine monophosphate kinase and, 89
  deoxythymidine kinase and, 71
  deoxythymidine monophosphate kinase and, 93
  guanosine monophosphate kinase and, 85
  glycogen synthetase phosphatase and, 328
  lactose synthetase and, 364, 365, 369, 370–371, 372, 373, 374, 376
  mitochondrial hexokinase and, 35

nicotinamide adenine dinucleotide kinase and, 78, 80
non-B_{12} methyltransferase and, 156
phage deoxyribonucleotide kinases and, 95, 96
uridine kinase and, 59
yeast hexokinase and, 20
Mannose
  aspartate transaminase and, 397
  mammalian hexokinase and, 31, 45
  yeast hexokinases and, 5, 15, 27
Mannose 6-phosphate, mitochondrial hexokinase and, 35
6-Mercaptoadenosine triphosphate, aspartate transcarbamylase and, 287
4-Mercaptobutyrate, B_{12} methyltransferase and, 131
5-Mercaptodeoxyuridine, deoxythymidine kinase and, 70
Mercaptoethanol
  aspartate transcarbamylase reassociation and, 237
  B_{12} methyltransferase and, 123, 129–130, 131, 134
  carbamyl aspartate assay and, 229
  carbamate kinase and, 105
  deoxycytidine monophosphate kinase and, 89
  deoxythymidine monophosphate kinase and, 92
  hexokinase and, 26–27
  nicotinamide adenine dinucleotide kinase and, 77
  non-B_{12} methyltransferase and, 156, 157
2-Mercaptoethylamine, B_{12} methyltransferase and, 131
Mercaptopropionate, aspartate transaminase and, 450
6-Mercaptopurine riboside, adenosine kinase and, 54
Mercurials, aspartate transcarbamylase and, 228
p-Mercuribenzoate
  N-acetylglutamate-5-phosphotransferase and, 517
  adenosine kinase and, 52–53
  aspartate transaminase and, 421, 422, 423
  aspartate transcarbamylase and, 230, 231, 235, 236, 250, 255, 263, 276, 299
  5-formyltetrahydrofolate cyclodehydrase and, 202
  uridine-cytidine kinase and, 60, 61
  yeast hexokinase and, 25–26
Mercuric ions, deoxyadenosine kinase and, 67
Mercury, aspartate transcarbamylase and, 234–235, 236, 238, 242, 275
Mestranol, transfer ribonucleic acid methyltransferase levels and, 180–181
Metal ions
  aspartate transcarbamylase reassociation and, 237–238
  deoxycytidine kinase and, 64
  divalent, nonenzymic transamination and, 388, 389, 390
  formiminoglutamate formiminotransferase and, 206
  formiminoglycine formiminotransferase and, 206
  α-methylserine hydroxymethylase and, 221
Metamorphosis
  insect, transfer ribonucleic acid methyltransferases and, 185
5,10-Methenyltetrahydrofolate
  formation and measurement of, 123
  glycinamide ribonucleotide transformylase and, 203–204
5,10-Methenyltetrahydrofolate cyclohydrolase
  anions and, 201
  formiminoglutamate formiminotransferase and, 207
  5-formiminotetrahydrofolate cyclodeaminase and, 203
Methionine
  aminotransferases and, 464, 466, 472
  aspartate transaminase and, 436
  assay of, 123
  biosynthesis, enzymes involved, 121–122
  methyltransferase repression and, 158, 159–160
Methionyl-transfer ribonucleic acid transformylase, reaction catalyzed and properties, 208–209

SUBJECT INDEX 575

β-Methylacetoacetate, coenzyme A transferase and, 486
N^6-Methyladenine
  occurrence of, 190
  ribosomal ribonucleic acid methyltransferase and, 188
N^6-Methyladenosine, adenosine kinase and, 54
α-Methylaspartate, aspartate transaminase and, 404, 405, 406, 411, 412-413, 436, 437, 457
β-Methylaspartate, aspartate transaminase and, 435
Methyl B_{12}
  B_{12} methyltransferase reconstitution and, 125-126
  methionine biosynthesis and, 132, 134-135, 140
Methyl-B_{12} methyltransferase
  absorption spectrum, 139, 141
  formation, S-adenosylmethionine and, 145, 152, 165
  photolysis of, 140-142
N-Methyl carbamyl phosphate, aspartate transcarbamylase and, 249, 252
Methylcobinamide
  B_{12} methyltransferase and, 135
  photolysis of, 142-143
S-Methylcysteine, aminotransferase and, 464
3-Methylcytidine, aspartate transcarbamylase and, 271
5-Methylcytosine
  deamination of, 195
  occurrence of, 190
5-Methyldeoxycytidine monophosphate, phage deoxyribonucleotide kinase and, 94
5,10-Methylenetetrahydrofolate
  deoxycytidylate hydroxymethyltransferase and, 209
  deoxyuridylate hydroxymethyltransferase and, 210
  glycine decarboxylation and, 221, 222
  α-methylserine hydroxymethylase and, 221
  serine hydroxymethyltransferase and, 215, 216, 217, 218
  thymidylate synthetase and, 211, 212, 213, 214-215

Methylenetetrahydrofolate dehydrogenase
  glycine biosynthesis and, 222
  repression of, 158
α-Methylglucose, galactosyltransferase and, 370
β-Methylglucose, galactosyltransferase and, 370
α-Methylglutamate, glutamate decarboxylase and, 479
γ-Methylglutamate, aspartate transaminase and, 435
7-Methylguanine methyltransferase, organ specific variation of, 168
7-Methyl inosine, aspartate transcarbamylase and, 271
Methyl iodide
  aspartate transcarbamylase and, 264, 265, 281
  B_{12} methyltransferase methylation and, 147-148, 152
  B_{12} methyltransferase propylation and, 144
Methylmalonyl mutase, light stability, 142
Methylmercuric iodide, hexokinase and, 9
Methylmercury, yeast hexokinase and, 26
N-Methylphosphonacetamide, aspartate transcarbamylase and, 247, 248, 249, 252
Methyl phosphonate
  aspartate transcarbamylase and, 248, 249, 252, 259, 267
6-Methylpurine riboside, adenosine kinase and, 54
5′-Methylpyridoxal phosphate, aspartate transaminase and, 433
6-Methylpyridoxal phosphate, aspartate transaminase and, 433, 434
N^+-Methylpyridoxal phosphate, aspartate transaminase and, 433
3-O-Methylpyridoxal phosphate, aspartate transaminase and, 433
α-Methylserine, serine hydroxymethyltransferase and, 218, 219
α-Methylserine hydroxymethylase, occurrence and properties, 220-221

α-Methyl succinate
  binding, aspartate transcarbamylase
    and, 256
5-Methyltetrahydrofolates
  methionine biosynthesis and, 121–122,
    123, 130–131, 133, 134, 143, 145,
    146–147, 149, 151–154, 164, 165
  non-B_{12} methyltransferase and, 154,
    156, 157–158, 161
  radioactive
    binding by B_{12} methyltransferase,
      136–137
    synthesis of, 123
6-Methylthiopurine riboside, adenosine
  kinase and, 54
Microorganisms
  aminotransferases of, 464–465, 466–467,
    468–469, 470, 472, 475
  B_{12} methyltransferases in, 162, 163
  carbamate kinase in, 101
  deoxyribonucleic acid methyltransferases of, 190–192
  non-B_{12} methyltransferases in, 160
  uridine-cytidine kinase in, 56, 58
Microsomes
  lactose synthetase and, 366
  membranes, hexokinase of, 34
Milk, galactosyltransferase in, 367
Mitochondria
  aminotransferase of, 382, 393–394
  aspartate transaminase
    amino acid composition, 414–415, 417
    sulfhydryl groups, 423
  B_{12} methyltransferase in, 163
  membranes, hexokinase of, 34–36
Monoiodotyrosine, aminotransferase
  and, 466
Muscle
  contraction, events during, 338–340
  glycogen synthetase
    control of, 336–340
    frog, 357
    physicochemical properties, 313–315
    properties of the two forms, 332–336
  glycogen synthetase phosphatase of,
    327–328
  hexokinases of, 32, 33
Mutations, aspartate transcarbamylase,
  293

Mycoplasma, arginine metabolism in,
  116–117
Myokinase
  adenosine kinase assay and, 52

N

Neohydrin, aspartate transcarbamylase
  and, 231
Nerves
  stimulation, glycogen synthetase and,
    352
Neurospora, carbamate kinase of, 101–
  102 117
Neurospora crassa
  aspartate transcarbamylase of, 302–305
  glycogen synthetase of, 361
Nickel ions, uridine kinase and, 59
Nicotinamide, transfer ribonucleic acid
  methyltransferases and, 177, 178–179
Nicotinamide adenine dinucleotide
  adenosine kinase assay and, 52
  carbamate kinase assay and, 109
  coenzyme A transferase assay and, 487
  deoxythymidine monophosphate kinase
    assay and, 92
  glycine decarboxylation and, 221, 222
  guanosine monophosphate kinase assay and, 84
  α-isomer, phosphorylation of, 80
  reduced, nicotinamide adenine dinucleotide kinase and, 81, 82
  synthesis, 385
  thymidylate synthetase and, 211
  uridine kinase assay and, 58
Nicotinamide adenine dinucleotide
  kinase
    assay, 77–78
    distribution, purification and stability,
      76–77
    kinetic and molecular properties, 78–79
    reaction mechanism, 79–80
    substrate specificity, 80–82
Nicotinamide adenine dinucleotide
  phosphate
    adenosine kinase assay and, 52
    carbamate kinase assay and, 109
    dihydrofolate reductase and, 212
    10-formyltetrahydrofolate deacylase
      and, 200

reduced, nicotinamide adenine dinucleotide kinase and, 81
serine hydroxymethyltransferase and, 216
Nicotinamide 2'-deoxyadenine dinucleotide, nicotinamide adenine dinucleotide kinase and, 81
p-Nitrophenylacetate, coenzyme A transferase and, 494
Non-B_{12} methyltransferase
  assay and purification, 154–155
  catalytic properties and folate binding, 156–158
  occurrence, 160–161
  physical properties, 155–156
  repression of synthesis, 158–160
Norleucine, aminotransferases and, 464, 466
2-Nor-6-methylpyridoxal phosphate, aspartate transaminase and, 433
2-Norpyridoxal phosphate, aspartate transaminase and, 434
Norvaline
  aminotransferase and, 464
  glycine amidinotransferase and, 500
Nuclear magnetic resonance spectroscopy, aspartate transcarbamylase, 247–249, 251, 257, 259, 262
Nuclei, transfer ribonucleic acid methyltransferases in, 168
Nucleoli
  ribosomal ribonucleic acid methyltransferases in, 187, 189
  transfer ribonucleic acid methyltransferases in, 168–169
Nucleoside phosphorylases, inosine-guanosine kinase and, 55
Nucleoside triphosphates, deoxythymidine kinase and, 70

O

Oligosaccharides, glycogen synthetase and, 320
Operon(s), aspartate transcarbamylase, 293–294
Optical rotatory dispersion, aspartate transaminase, 403, 405–407, 409, 412, 437

Ornithine
  amidinotransferase assay and, 498
  aminotransferase and, 466, 468, 474, 481
  arginine racemase and, 478
  carbamyl phosphate synthetase and, 302
  glycine amidinotransferase and, 500, 504
  inosamine phosphate amidinotransferase and, 507, 508
Ornithine transcarbamylase
  carbamyl kinase assay and, 108
  carbamyl phosphate availability to, 303, 304–305
  relation to aspartate transcarbamylase, 294
Orotidylate decarboxylase
  deficiency, aspartate transcarbamylase and, 228
Ovalbumin, lactose synthetase and, 364, 370, 377
Oxalacetate
  aminotransferases and, 463, 464–468
  aspartate transaminase and, 413, 426, 432
Oxalate, activation of, 485
2-Oxoadipate, aminotransferase and, 463, 468
α-Oxobutyrate, aminotransferase and, 466, 468
2-Oxocaproate, aminotransferase and, 468
2-Oxoglutaramate, aminotransferase and, 466
α-Oxoglutarate
  aminotransferases and, 463, 464–470, 474, 475, 476
  aspartate transaminase and, 412, 426, 432, 436, 439
Oxovalerate, aminotransferase and, 468
Oxycoenzyme A, coenzyme A transferase and, 496
Oxygen
  methyl-B_{12} methyltransferase and, 149, 153
  transfer, coenzyme A transferase and, 488–489, 492
4-Oxythiamine, amino group transfer to, 385

P

Penicillamine, aspartate transaminase and, 438
Penicillin, aspartate transaminase and, 438
Pepsin-pronase, aspartate transaminase and, 424
Peptides, aspartate transaminase active site, 417
Periodate, inosamine phosphate amidinotransferase and, 507
Permanganate, aspartate transcarbamylase and, 263, 264, 268
pH
  N-acetylglutamate-5-phosphotransferase and, 514, 515, 516, 517–518
  adenosine kinase and, 52
  aminotransferases and, 465, 467, 469, 471
  aspartate transaminase and, 401, 402, 408, 412, 426, 428, 438–440, 459
  aspartate transcarbamylase and, 248, 249, 251–252, 256, 257, 259, 262, 265, 267, 278
  B_{12} methyltransferase and, 135
  carbamate kinases and, 111, 112
  citrulline formation and, 99
  coenzyme A transferase and, 494
  deoxyadenosine kinase and, 67
  deoxycytidine kinase and, 65
  deoxycytidine monophosphate kinase and, 88
  deoxycytidylate hydroxymethyltransferase and, 209
  deoxyribonucleic acid methyltransferases and, 191, 192
  deoxythymidine kinase and, 71
  deoxythymidine monophosphate kinase and, 93
  formiminoglutamate formiminotransferase and, 206
  formiminoglycine formiminotransferase and, 206
  5-formiminotetrahydrofolate cyclodeaminase and, 203
  5-formyltetrahydrofolate cyclodehydrase and, 202
  glycinamide ribonucleotide transformylase and, 204
  glycine amidinotransferase and, 500
  glycogen synthetase and, 332, 333, 334, 335, 342, 346, 360
  guanosine monophosphate kinase and, 84–85
  hexokinase and, 8, 10, 27, 29, 33
  methionyl-transfer ribonucleic acid transformylase and, 209
  α-methylserine hydroxymethylase and, 221
  nicotinamide adenine dinucleotide kinase and, 78
  non-B_{12} methyltransferase and, 157, 161
  phage deoxyribonucleotide kinases and, 95, 96
  serine hydroxymethyltransferase and, 217, 218
  thymidylate synthetase and, 213, 214
  transfer ribonucleic acid methyltransferases and, 170, 174
  uridine kinase and, 59
  uridine monophosphate kinase and, 91
Phenazine methosulfate, nicotinamide adenine dinucleotide kinase assay and, 78
Phenylalanine
  aminotransferases and, 464, 466
  aspartate transaminase and, 435, 436
Phenylalanine aminotransferase, substrate specificity, 382
Phenylmethanesulfonyl fluoride, hexokinase preparation and, 4
Phenylpyruvate, aminotransferase and, 464
Phosphate
  assay, carbamate kinase and, 108–109
  glycogen synthetase content of, 314, 315
  glycogen synthetase phosphatase and, 328
Phosphate group, transamination catalysis and, 432–433
Phosphate ions
  aspartate transaminase and, 399, 424, 428
  aspartate transcarbamylase and, 251, 259, 261, 275
  exchange, aspartate transcarbamylase and, 245

SUBJECT INDEX 579

glycogen synthetase and, 324, 334, 344–345, 353, 357, 360
mammalian hexokinase and, 45, 46
5,10-methenyltetrahydrofolate cyclohydrolase and, 201
mitochondrial hexokinase and, 35
non-B_{12} methyltransferase and, 156, 161, 162
yeast hexokinase and, 8, 20, 28, 29
Phosphatidylethanolamine, mitochondrial hexokinase and, 36
Phosphoenolpyruvate, citrulline formation and, 99–100
6-Phosphogluconate dehydrogenase, nicotinamide adenine dinucleotide kinase assay and, 78
3-Phosphoglycerate, hexokinase and, 29
3-Phosphohydroxypyruvate, aminotransferase and, 464
Phospholipase C, mitochondrial hexokinase and, 36
Phosphomonoesterase
  deoxycytidine monophosphate kinase assay and, 88
  guanosine monophosphate kinase assay and, 84
C-Phosphonoacetaldehyde, aminotransferase and, 475
Phosphonacetamide, aspartate transcarbamylase and, 248, 249, 251, 252, 261, 262
Phosphonacetate, aspartate transcarbamylase and, 251, 252
N-(Phosphonacetyl)-L-aspartate
  aspartate transcarbamylase and, 232, 246, 254–255, 265, 266, 269, 273–274, 291
  modified enzyme and, 281–284
  as transition state analog, 260–261
Phosphorylase
  blood glucose levels and, 348
  glycogen synthetase assay on, 317–318
  limit dextrin, glycogen synthetase and, 319
  muscle contraction and, 339
  nomenclature and, 323, 324
Phosphorylase a
  glycogen synthetase and, 311
  glycogen synthetase phosphatase and, 329, 330, 348, 350, 351

Phosphorylase kinase, nomenclature and, 323
Phosphorylase kinase phosphatase, glycogen synthetase phosphatase and, 328
Phosphorylase phosphatase
  glucocorticoids and, 352
  glycogen synthetase phosphatase and, 330
Phosphoserine, aminotransferase and, 463, 464
Phosvitin, synthesis by rooster, 181, 185
Photoinactivation, aspartate transcarbamylase, 267
Photooxidation
  aspartate transaminase, 419, 428
  glycogen synthetase, 335, 358
pH stat, aspartate transcarbamylase assay and, 230
Plants
  aspartate transcarbamylases of, 307–308
  non-B_{12} methyltransferase in, 160–162
Plasma membranes, hexokinase and, 34
Platelets, glycogen synthetase of, 355
Poliovirus, transfer ribonucleic acid methyltransferase and, 183
Polyethylene glycol, aspartate transcarbamylase and, 301
Polyoma virus, transfer ribonucleic acid methyltransferases and, 183
Polyribonucleotides
  synthetic, transfer ribonucleic acid methyltransferases and, 176
Polysaccharide
  bull sperm, transfer ribonucleic acid methyltransferase and, 177
  glycogen synthetase glucosyl acceptor activity, 319–320
Polyvinylsulfate, ribosomal ribonucleic acid methyltransferase and, 188
Potassium chloride, nicotinamide adenine dinucleotide kinase and, 78
Potassium cyanate, carbamyl phosphate synthesis and, 100
Potassium dihydrogen phosphate, carbamyl phosphate synthesis and, 100
Potassium ions
  5-amino-4-imidazole carboxamide ribonucleotide transferase and, 205

formiminotetrahydrofolate cyclodeaminase and, 202
guanosine monophosphate kinase and, 86
pseudouridine kinase and, 62
transfer ribonucleic acid methyltransferases and, 175
Pregnancy, transfer ribonucleic acid methyltransferase levels and, 179
Proflavin
  deoxyribonucleic acid methyltransferase and, 191
  transfer ribonucleic acid methyltransferases and, 177
Progesterone, glycogen synthetase and, 356
Prolactin, transfer ribonucleic acid methyltransferase levels and, 179
Proline, metabolism, 474
Propionyl coenzyme A, fatty acid activation and, 485
Propyl iodide
  B_{12} methyltransferase inhibition by, 127–129, 134–135
  S-adenosylmethionine and, 144
Prostaglandin E_1, glycogen synthetase and, 352–353
Protease(s)
  aspartate transcarbamylase and, 255, 271, 276
  endogenous, hexokinase and, 3, 9–10
  hexokinases and, 38, 40
Protein(s)
  synthesis, control of, 184–185
  transfer ribonucleic acid methyltransferase inhibition and, 177–178, 180
Protein kinase
  glycogen synthetase and, 311, 314, 326, 351
  regulation of, 327
Proteus vulgaris, aspartate transcarbamylase of, 298, 301
Protons
  exchange, serine hydroxymethyltransferase and, 219–220
  shifts, transaminase active site and, 452–453, 457, 459, 461
  transfer, aspartate transcarbamylase and, 261
Protozoa, glycogen synthetase of, 359

Pseudomonas aeruginosa, aspartate transcarbamylase of, 298, 301
Pseudomonas fluorescens, aspartate transcarbamylase of, 298, 301
Pseudouridine diphosphate glucose, glycogen synthetase and, 318
Pseudouridine kinase, occurrence and properties, 62
Pseudouridylate, deoxycytidine monophosphate kinase and, 88
Purine riboside, adenosine kinase and, 54
Puromycin, glycogen synthetase and, 358
Putrescine
  aminotransferase and, 470
  transfer ribonucleic acid methyltransferases and, 175
Pyridoxal
  nonenzymic models of transamination and, 387–391
Pyridoxal phosphate
  amino group transfer and, 380
  analogs, aspartate transaminase and, 430–435
  aspartate:oxoglutarate aminotransferase
    bonding of, 429
    content of, 399
    optical properties and, 408, 412
    subforms and, 395
  aspartate transcarbamylase and, 265–266
  glycine biosynthesis and, 222, 223
  glycine decarboxylation and, 221, 222
  α-methylserine hydroxymethylase and, 221
  serine hydroxymethyltransferase and, 216, 218, 219
Pyridoxamine
  aminotransferase and, 470, 476
  aspartate transaminase and, 432, 452
Pyridoxamine phosphate
  abortive transamination and, 477–480
  aspartate transaminase and, 399, 462
  optical properties, 408
  serine hydroxymethyltransferase and, 218
N-(5'-P-Pyridoxyl)-L-glutamate, aspartate transaminase and, 424, 447
N-(5'-P-Pyridoxyl)-pyrrolidone car-

SUBJECT INDEX 581

boxylate, aspartate transaminase
  and, 424, 447
Pyrophosphate
  aspartate transcarbamylase and, 247
  glycogen synthetase and, 334
  glycogen synthetase phosphatase and,
    328
Pyruvate
  aminotransferases and, 464, 466, 468,
    470, 472, 475, 476, 478
  arginine racemase and, 478
  decarboxylase reactivation and, 477
  serine hydroxymethyltransferase and,
    218
Pyruvate dehydrogenase, nomenclature
  and, 323, 324
Pyruvate kinase
  N-acetylglutamate-5-phosphotrans-
    ferase assay and, 515
  adenosine kinase assay and, 52
  carbamate kinase assay and, 109
  deoxythymidine monophosphate ki-
    nase assay and, 92
  glycogen synthetase assay and, 317
  guanosine monophosphate kinase and,
    84
  uridine kinase assay and, 58
  uridine monophosphate-cytidine mono-
    phosphate kinase assay and, 90

R

Regulatory subunit
  aspartate transcarbamylase
    metal binding site, 236
    primary structure, 235
    regulation and, 236–237
    size and substructure, 234–235
Rhamnose, aspartate transaminase and,
  397
Rhodopseudomonas spheroides, aspartate
  transcarbamylase of, 298
Riboflavin kinase, distribution, purification and properties, 74–75
Ribonuclease, deoxythymidine kinase
  and, 73
Ribonucleic acid, see Ribosomal, Transfer
  messenger, aspartate transcarbamylase
  and, 296
Ribose, lactose synthetase and, 373

Ribosomal ribonucleic acid
  methyltransferases
    biological significance, 189–190
    isolation and properties, 187–189
    occurrence, 187
9-β-Ribosyl-2,6-diaminopurine, ribosomal
  ribonucleic acid methyltransferase
  and, 188
Rubidium ions, guanosine monophosphate kinase and, 86

S

Saccharomyces cerevisiae, aspartate
  transcarbamylase of, 303–305
Saccharopine pathway, occurrence of,
  386–387
Salicylate, aspartate transaminase and,
  428
Salmonella typhimurium, aspartate transcarbamylase of, 292–296, 298, 299,
  301–302
Salt
  concentration, glycogen synthesis and,
    321, 343
  hexokinase isolation and, 38
Schiff base
  aminotransferases and, 389–390, 417,
    438, 453, 459, 460
  aspartate transcarbamylase and, 266,
    267
  glycine decarboxylation and, 222
  serine hydroxymethyltransferase and,
    219
Schistosome mansoni, hexokinases of, 48
Sedoheptulose 7-phosphate, glycogen synthetase and, 332
Serine
  aminotransferases and, 466, 473
  aspartate transaminase and, 436
  glycine biosynthesis and, 222
  thymidylate synthetase and, 211
D-Serine, α-methylserine hydroxymethylase and, 221
Serine hydroxymethyltransferase
  abnormal transamination by, 479
  assay of, 215–216
  glycine biosynthesis and, 222
  mechanism, 219–220
  properties, 217

reaction catalyzed, 215
thymidylate synthetase and, 211
Serine phosphate
  glycogen synthetase, hexapeptide map, 314–315
Serine residues, aspartate transaminase, 423–424
L-Serine-O-sulfate, aspartate transaminase and, 450
Serratia marcescens, aspartate transcarbamylase of, 298, 301
Serratia marinorubra, pyrimidine synthesis in, 301
Sialic acids, aspartate transaminase and, 397, 398
Silver ions
  aspartate transcarbamylase, 421
  carbamate kinase and, 105
  yeast hexokinase and, 25
Slime mold
  colonizing, transfer ribonucleic acid methyltransferases in, 185
Sodium dodecyl sulfate
  adenosine kinase and, 53
  aspartate transaminase and, 400, 422
  aspartate transcarbamylase and, 232, 234, 235, 255, 271, 277, 299, 304
  galactosyltransferase and, 367–368
  glycogen synthetase and, 314
  mammalian hexokinase and, 39
  yeast hexokinase and, 9, 10
Sodium ions, transfer ribonucleic acid methyltransferases and, 175
1,5-Sorbitan 6-phosphate, glycogen synthetase and, 332
Sorbose
  glycogen accumulation and, 361
  lactose synthetase and, 373
Sorbose 1-phosphate, yeast hexokinase and, 18
Spermidine, transfer ribonucleic acid methyltransferases and, 175
Spermine, transfer ribonucleic acid methyltransferases and, 174, 175
Spleen, glycogen synthetase of, 354
Spores
  germination, transfer ribonucleic acid methyltransferases and, 185
Starch, glycogen synthetase and, 319
Starvation, hexokinase and, 32, 33

Streptidine, aminotransferase and, 476
Streptidine phosphate, inosamine phosphate amidinotransferase and, 507
Streptococcus faecalis, aspartate transcarbamylase of, 298, 299
Streptomyces, inosamine phosphate amidinotransferases in, 505–506
Streptomyces violaceoruber, hexokinase of, 48
Streptomycin
  biosynthesis, 497, 505–506, 507
  regulation of, 508–509
Succinate
  activation of, 485
  aspartate transaminase and, 439
  binding, aspartate transcarbamylase and, 231, 232, 242, 246, 251–252, 257, 258, 259, 264–265, 267, 280–281, 290
    conformational changes and, 253–255, 276, 277, 286
    cooperativity, 269, 289
    stoichiometry, 273
  carbamyl phosphate decomposition and, 244, 245
  coenzyme A transferase inhibition and, 491–492
Succinic anhydride
  aspartate transaminase and, 401
  aspartate transcarbamylase and, 266–267, 279
Succinic semialdehyde, aminotransferase and, 466
Succinyl coenzyme A, acetoacetate activation and, 484–485
Succinyl coenzyme A: 3-ketoacid coenzyme A transferase
  catalytic properties
    assay, 487
    mechanism and kinetics, 488–496
    specificity, 486–487
    thermodynamics, 487–488
  properties of, 485–486
Succinyl pantetheine, coenzyme A transferase and, 495
Sucrose, glycogen accumulation and, 361
Sulfate ions
  aspartate transaminase and, 428
  glycogen synthetase and, 324, 334, 345, 360

SUBJECT INDEX 583

Sulfhydryl groups
  aspartate transaminase, 406, 420, 421–423
  aspartate transcarbamylase, 236, 255, 263–265, 291
  coenzyme A transferase, 495, 496
  glycogen synthetase, 314
  inosamine phosphate amidinotransferase, 508
  mammalian hexokinase, 41
  nicotinamide adenine dinucleotide kinase and, 77
  thymidylate synthetase and, 214
  uridine-cytidine kinase, 60
  yeast hexokinase activity and, 25–27
3-Sulfinylpyruvate, aspartate transaminase and, 435
Sulfite
  glycogen synthetase phosphatase and, 328
  "inactive" glycogen synthetase and, 330, 342, 343, 345
Sulfito-B_{12}, B_{12} methyltransferase and, 125
SV 40 virus, transfer ribonucleic acid methyltransferase and, 183

T

Tadpole
  liver, glycogen synthetase of, 357–358
Taurocyamine, biosynthesis, 498
Temperature
  N-acetylglutamate-5-phosphotransferase and, 517, 518, 519–520
  glycogen synthetase and, 330–331, 344
Testosterone, glycogen synthetase and, 355
Tetrahydrofolate, S-adenosylmethionine and, 132, 133, 134
  B_{12} methyltransferase and, 130, 143, 146, 147
  deoxycytidylate hydroxymethyltransferase and, 210
  formaldehyde and, 209
  formation, measurement of, 122–123
  formiminoglutamate formiminotransferase and, 206
  formiminoglycine formiminotransferase and, 206
  glycine biosynthesis and, 222, 223
  glycine decarboxylation and, 221, 222
  α-methylserine hydroxymethylase and, 220
  proton exchange and, 219
  and related compounds, methionyl-transfer ribonucleic acid transformylase and, 209
  serine hydroxymethyltransferase and, 215–216, 217, 218
  thymidylate synthetase and, 211–212
Tetranitromethane
  aspartate transaminase and, 420–421
  aspartate transcarbamylase and, 267, 280–281
Tetrathionate, aspartate transcarbamylase and, 263
Theophylline, glycogen synthetase and, 358, 359
Thiamine, biosynthesis, 385
Thioacetic acid, formation from thiobenzoic acid, 490
Thioglycolate, B_{12} methyltransferase and, 131
6-Thioguanosine monophosphate, guanosine monophosphate kinase and, 85–86
Thiols
  deoxycytidine kinase and, 64, 65
  mammalian hexokinase and, 39
  thymidylate synthetase and, 213
  yeast hexokinase reactivation and, 27
Thionicotinamide adenine dinucleotide, nicotinamide adenine dinucleotide kinase and, 80, 81
Thioredoxin, deoxycytidine monophosphate kinase and, 89
2-Thiouracil, aspartate transcarbamylase and, 284–285
2-Thiouridine monophosphate, aspartate transcarbamylase and, 271, 273
Threonine, serine hydroxymethyltransferase and, 215, 218
Threonine residues, aspartate transaminase, 423–424
Thymidine, *see also* Deoxythymidine
Thymidine diphosphate-4-amino-4,6-dideoxyglucose, aminotransferase and, 476
Thymidine diphosphate glucose, glycogen synthetase and, 318

Thymidylate synthetase
  assay of, 212
  mechanism of action, 214–215
  occurrence and properties, 213–214
  reaction catalyzed, 211
Thymine
  synthesis, macromolecular level and, 195
Thyroxine
  aminotransferase and, 466
  glycogen synthetase and, 358
  transfer ribonucleic acid methyltransferase levels and, 181–182, 185
Toyocamycin, adenosine kinase and, 54
Transamination, see Amino group transfer
Transfer ribonucleic acid
  conformation, methylation sites and, 173
  methyl acceptor capacities, 171, 172
  structure and methylation sites, 173
Transfer ribonucleic acid methyltransferases
  biological significance, 184–186
  ionic stimulation, 174–175
  occurrence, 168–169
  purification, 169–172
  regulation
    bacteriophage induction or infection, 182
    hormones, 179–182
    inhibitors, 176–179, 180, 182, 183
    nononcogenic or oncogenic virus infection, 182–183
    tumor tissue, 183–184
  substrate specificity, 170, 171, 172–174
Trehalose phosphate, glycogen synthetase assay and, 317, 358
Tricholomate, aspartate transaminase and, 438
5-Trifluoromethyldeoxyuridine, deoxythymidine kinase and, 70
Triiodothyronine, glycogen synthetase and, 359
Trypsin
  aspartate transaminases and, 417
  aspartate transcarbamylase and, 280, 306
  glucokinase and, 40
  glycogen synthetase and, 314, 331
  hexokinase and, 3, 6–7, 8, 9, 10, 11–13
Tryptophan, aminotransferases and, 464, 466
Tryptophan pyrrolase
  inhibition, transfer ribonucleic acid and, 186
Tryptophan residues
  aspartate transcarbamylase, 267, 276, 278
  non-B_{12} methyltransferase and, 155–156
Tubercidin, adenosine kinase and, 54
Tumors
  glycogen synthetase of, 356–357
  hexokinase, purification of, 38
  transfer ribonucleic acid methyltransferases in, 183–184
  uridine-cytidine kinase in, 56–57, 58–59, 60
Tyrosine
  aminotransferases and, 464, 466
  aspartate transaminase and, 435
Tyrosine aminotransferase, substrate specificity, 382
Tyrosine residues
  aspartate transaminase, 413, 416, 420–421, 450, 460
  aspartate transcarbamylase, 267–268, 276, 278, 280–281, 288
  non-B_{12} methyltransferase and, 155

U

Uracil, aspartate transcarbamylase and, 228, 294, 295–296, 301, 303
Urea
  N-acetylglutamate-5-phosphotransferase and, 517
  adenosine kinase and, 52, 53
  aspartate:oxoglutarate aminotransferase and, 395, 401, 402, 422, 423
  aspartate transcarbamylase and, 228
  catalytic subunit, 231
  cooperative properties, 278
  B_{12} methyltransferase and, 125, 126, 137, 148
  formation, carbamyl phosphate assay and, 109
  galactosyltransferase and, 367
  glucokinase and, 40

glycine amidinotransferase and, 501
hexokinase and, 9
Uridine, aspartate transcarbamylase and, 302, 307
Uridine-cytidine kinase
  assay of, 57–58
  distribution and purification, 56–57
  kinetic and molecular properties, 59–60
  reaction mechanism, 60–61
  regulatory properties, 61–62
  substrate specificity, 58–59
Uridine diphosphate
  deoxycytidine kinase and, 65
  glycogen synthetase and, 314, 316, 361
    assay, 317
    inhibition by, 325, 334, 345, 360
  uridine monophosphate kinase and, 90–91
Uridine diphosphate-4-amino-2-acetamido-2,4,6-trideoxyglucose, aminotransferase and, 476
Uridine diphosphate deoxyglucose, glycogen synthetase and, 318, 322
Uridine diphosphate galactose, lactose synthetase and, 364, 369, 372, 373, 374, 376
Uridine diphosphate glucosamine, glycogen synthetase and, 318
Uridine diphosphate glucose
  galactosyltransferase and, 372, 373
  glucocorticoids and, 352
  glycogen synthetase and, 310, 316, 361
    aggregation, 314
    cooperative binding, 345, 361
    kinetics, 321, 332–333, 334–335, 342, 343, 353, 354, 357–358, 359–360, 361
    magnesium ions and, 335
    two enzyme forms, 324–325
  labeled, glycogen synthetase assay and, 317–318
Uridine monophosphate
  aspartate transcarbamylase and, 298
  carbamyl phosphate synthetase and, 295, 302
  cytidine monophosphate kinase and, 90
  deoxycytidine monophosphate kinase and, 88

Uridine monophosphate kinase
  mutants, aspartate transcarbamylase and, 295
Uridine monophosphate kinase-cytidine monophosphate kinase, purification and properties, 90–91
Uridine triphosphate
  adenosine kinase and, 53
  aspartate transcarbamylase and, 271, 277, 298, 303, 304, 305
  carbamyl phosphate synthetase and, 118
  carbamyl phosphate synthetase and, 303, 304, 305, 306
  deoxyadenosine kinase and, 67
  deoxycytidine kinase and, 65
  glycogen synthetase and, 331, 334
  nicotinamide adenine dinucleotide kinase and, 80
  uridine monophosphate kinase and, 90–91
  uridine phosphorylation and, 61

V

Vaccinia virus, transfer ribonucleic acid methyltransferases and, 183
Valine, aminotransferases and, 464, 468, 472
α-Vinylacetate, coenzyme A transferase and, 487
Viomycin, biosynthesis, 498
Viruses
  deoxyribonucleic acid, deoxythymidine kinase induction and, 69, 73
  nononcogenic and oncogenic infections, transfer ribonucleic acid methyltransferase levels and, 182–183, 185
Vitamin B_{12}, see B_{12}

W

Water, glycine amidinotransferase and, 500
Wheat germ, hexokinase, 48

X

X-ray crystallography
  aspartate transaminase, 398, 400
  aspartate transcarbamylase, 239, 242, 263, 290, 291
  yeast hexokinase, 27–28

Xylose
  aspartate transaminase and, 397
  lactose synthetase and, 373
  yeast hexokinase and, 19, 24, 25, 26

Y

Yeast
  glycogen synthetase of, 359–361
  hexokinases
    chemical studies on native and modified forms, 10–13
    mode of action, 13–28
    modification by added protease, 6–7
    modification by endogenous proteases, 2–6
    molecular weight and subunit structure, 7–10
    regulation of, 29–31
  non-B_{12} methyltransferase of, 161

Z

Zeatin riboside, transfer ribonucleic acid methyltransferases and, 176
Zinc ions
  aspartate transcarbamylase and, 234, 235, 236, 237, 277
  biosynthesis, 296
  effector binding and, 273, 275, 277
  reassociation and, 238–239
  guanosine monophosphate kinase and, 85
  nicotinamide adenine dinucleotide kinase and, 78, 80
  riboflavin kinase and, 75

QP
601
E523
v.9

DEC 20 1973